MW01639857

Wireless Mesh Networking

About the Author

George Aggélou, B.Sc., B.Eng., Ph.D., is currently a staff member of the Greek Ministry of Transport and Communications, Headquarters—General Directorate of Air Navigation, Electronics Division—Telecommunication Facilities Section. For 5 years, he was Assistant Professor of Electronic and Electrical Engineering at the Institute of Technology in Greece, and the Director of G-Alpha Telecomms, a company that builds next-generation wireless networks for enterprises and industry. Dr. Aggélou is also the cofounder of Mobile E-Commerce Technologies, Ltd., in London. He has been involved in the standardization of wireless networking for many years, first as a researcher at the IBM T. J. Watson Research Center in New York, and then as a team leader at Cisco Systems in London. He is also the author of the book *Mobile Ad Hoc Networks: From Wireless LANs to 4G Networks*, published by McGraw-Hill in 2004. Dr. Aggélou was the recipient of the London-based RACAL Prize for Research Excellence in 2000.

Wireless Mesh Networking

George Aggélou, B.Sc., B.Eng., Ph.D.

Ministry of Transport and Communications
Headquarters—General Directorate of Air Navigation
Electronics Division—Telecommunication Facilities Section
Athens, Greece

New York Chicago San Francisco
Lisbon London Madrid Mexico City
Milan New Delhi San Juan
Seoul Singapore Sydney Toronto

The McGraw-Hill Companies

Library of Congress Cataloging-in-Publication Data

Aggélou, George.
Wireless mesh networking / George Aggélou.
p. cm.
ISBN 978-0-07-148256-1 (alk. paper)
1. Wireless communication systems. I. Title.
TK5103.2.A415 2008
621.384—dc22 2008023925

1 2 3 4 5 6 7 8 9 0 DOC/DOC 0 1 4 3 2 1 0 9 8

ISBN 978-0-07-148256-1
MHID 0-07-148256-3

Sponsoring Editor
Stephen S. Chapman

Production Supervisor
Pamela A. Pelton

Editing Supervisor
Stephen M. Smith

Project Manager
Sandhya Joshi, Aptara®, Inc.

Copy Editor
Subathra Natarajan, Aptara®, Inc.

Proofreader
Lee A. Young

Indexer
Slavka Zlatkova

Art Director, Cover
Jeff Weeks

Composition
Aptara®, Inc.

Printed and bound by RR Donnelley.

This book is printed on acid-free paper.

To my wife Vaya and my daughters Artemis and Stella.

Contents

Preface

Over the last 10 years, the Internet, the web, broadband wired and wireless communications, and other early information and communication technology (ICT) innovations have gradually embraced the whole of business and society and are increasingly permeating every aspect of our life. Driven by miniaturization and the accelerating convergence between computing, communications, media, and knowledge technologies, a new generation of ICTs is emerging that will likely foster profound changes for at least one or two more decades. There are plenty of blossoming technological breakthroughs simmering beneath the surface as in labs all around the world researchers pursue the technology race for ever-smaller size, cheaper and higher-gigahertz computing power and terabit memory capacity, and femtosecond optical pulses and gigabits per second communication bandwidths. If information technology continues to develop as it has—and the available science suggests that it will—then a "second wave" of new products and capabilities will transform society again. The next transformation will, however, involve more than mere increased power and device sophistication. We will see a proliferation of devices based on novel physics and engineering—electronics based, for example, in plastic and other organic materials. We will see one of the oldest information technologies—paper—being replaced by electronic rivals that share its advantages (affordability and ease of use) but bring others as well (word processing, links to the World Wide Web, etc.). We will go beyond miniature electronics to micro-machinery, and to devices that assemble and possibly even design themselves. The next generation of devices will also exploit insights from molecular and systems biology, among other sciences, to become more "intelligent," and to bring perception, learning, and reasoning into their routine functions.

Most significantly, however, information technology is set to experience a massive and unprecedented increase in systems complexity, as developers strive to integrate a vast spectrum of diverse technologies and "intelligent devices by the billions" into connected networks. In tomorrow's world, the environment will brim with pervasive sensors and other devices. Communications traffic will increase enormously as these devices share information in order to carry out the "housekeeping" chores of an information-centric world. The Internet will be everywhere, and will be a vastly deeper and more powerful environment than we know today, with multiple layers and inhabited by a population of intelligent software agents aiming to support its health and efficient function. The information society of tomorrow will be, first and foremost, a networked society, with individuals and businesses always linked to a global web of technology, and an economy founded on a seamless environment of networked information resources. These networks will aim to provide socially beneficial functions—from monitoring individual health to supporting global enterprise—with efficiency and resilience. In this vein, the new challenges include (http://cordis.europa.eu/ist/fet.comms.htm) further exploring the new miniaturization and computing frontiers; harnessing the increasing complexity of networked computing and communication systems, comprising myriads of interconnected heterogeneous components with terabytes of flowing data, spanning from the nanoscale to the planetary scale; and designing and building ever more intelligent systems and personalized products and services.

The realization of such connectivity captures the promise of the next wireless frontier: wireless mesh networking. The signature of a mesh network is that there is no central orchestrating device. Instead, each node is outfitted with radio communications gear and acts as a relay point for other nodes. In this regard, wireless mesh networking is an economical and reliable way to extend next-generation communications and Wi-Fi coverage. Indeed, wireless mesh is a new, high-performance broadband solution providing unprecedented reliability to WiFi carriers and wireless Internet service providers. It allows engineers to build a broadband network beyond the range of a single access point, increasing reach and coverage through multiple hops, without compromising performance or reliability.

In the coming years, the truly mobile mesh network may take to the highways, as swarms of cars equipped with the technology serve as nodes. There are a number of recent examples that show the potential of mesh networking: Researchers at Microsoft are working to create wireless technologies that allow neighbors to connect their home networks together. There are many advantages to enabling such connectivity and forming a community mesh network. Community-based multihop wireless networks are disruptive to the current broadband Internet access paradigm, which relies on cable and digital subscriber lines (DSLs) being deployed in individual homes. This is important because mesh networking allows the free flow of information without any moderation or selective rate control. Compared to the large DSL and cable modem systems that are centrally managed, mesh networking is organic—everyone in the neighborhood contributes network resources and cooperates.

Moteran, a partnership between Mitsubishi and Deutsche Telekom, is outfitting cars in some German cities with high-bandwidth mesh networking equipment for entertainment and communications. MeshNetworks, a U.S.-based company, is working with U.S. car makers on an application to alert a driver when a car in front deploys its airbags, winning precious seconds to avert a crash. The technology does not stop there. Intel is planning to mesh home PCs, TV sets, and stereos via low-cost hardware add-ons. Radiant Networks has already installed roof-mounted, high-bandwidth wireless nodes with steerable antennas throughout the city of Salem, Virginia, so customers connect for online access via other customers rather than using wireless base stations.

Wireless Mesh Networking is a complete, self-contained book that introduces background theory and applications of this revolutionary new technology. It provides a thorough grounding on emerging technologies and concepts using a wide range of practices, on-going research and development efforts pursued in Europe, the United States, and Asia, and worldwide standardization activities to allow its use in high-level research endeavors. Offering a deep, rich treatment of critical topics on wireless mesh networking, this unique reference concentrates on "how" and "why" the technology works in addition to providing descriptions of technology.

This skills-building tool encompasses a number of recurring themes on integrating WiFi, WiMAX, and ZigBEE into seamless wireless networks, the latest methods for ensuring security across the wireless network, in-depth coverage of data fusion principles, and vital information on SmartMesh networking in SensorNets, and introduction of situated selfware communication smart artifacts, whether in the form of sensors, actuators, robotic devices, tags, or whatsoever, that boost the creation of new ambient environments that increasingly link the real and the virtual world.

Guidance for the Reader

The book chapters and their intended audiences are summarized in the following table:

	Chapter	Audience
0	Preface	All readers
1	Introduction	All readers
2	Wireless Mesh Communication Architectures and Protocols	Most readers
3	Energy-Aware WM^{2}Net Communications	Technical readers
4	Principles of Communications Coverage in WM^{2}Nets	Technical readers
5	Medium Access Control Principles in WM^{2}Nets	Technical readers
6	Capacity Principles In WM^{2}Nets	Technical readers
7	Security Issues in WM^{2}Nets	Technical readers
8	Autonomic Selfware WM^{2}Net Communications	Most readers

Glossary

AC	autonomic communications
ACCA	autonomic communication: coordination action
ACL	access control list
AMR	adaptive multi-copy routing
ANA	autonomic network architectures
AoA	angle-of-arrival
AOES	adaptive on-line energy saving
AP	access point
APS	ad hoc positioning system
BACnet	building automation and control networks
BAR	battery-aware routing
BD	bounded delay
BIB	broadcast incremental bandwidth
BIP	broadcast incremental power
BiSNET	biologically inspired architecture for sensor networks
BS	base station
BSAS	basic sequential algorithmic scheme
CAC	command-after-command
CAN	controller area network
CAPTRA	coordinated packet traceback protocol
CAQ	command-after-query
CBT	core-based tree
CGF	contention-based geographic forwarding
CH	cluster-head
CoNet	cooperative networking
COTS	commercial off-the-shelf
CPLD	complex programmable logic device
CRC	cyclic redundancy check
CRDP	cluster radius decision point
CRREP	client request reply
CSMA/CA	carrier sense multiple access with collision avoidance
CTF	confirm-to-forward
CTS	clear-to-send
DCC	dynamic cluster control
DC-CTO	dynamic clustering scheme for coverage-time optimization
DCF	distributed coordination function
DECIDUOUS	decentralized source identification system
DFG	data flow graph
DGA	distributed genetic algorithms
DIFS	distributed interframe space
DLP	discrete logarithm problem

DRA	degree radian area
DRP	dynamic routing protocol
DSDV	destination-sequenced distance-vector
DVMRP	distance vector multicast routing protocol
ECDLP	elliptic curve discrete logarithm problem
EEMU	energy efficient multi unicast
EICT	emerging information and communication technology
E-NEXT	network of excellence in emerging networks experiments and technologies
EP	evolutionary programing
ERP	effective radiated power
ES	evolution strategy
ESS	evolutionary stable strategy
EWMA	exponentially weighted moving average
FH/TH-PPM	frequency hopping/time hopping-pulse position modulated
FHR	first hop router
FMM	fully multi-rate multicast
GA	genetic algorithm
GEDIR	compass routing and geographic distance routing
GP	genetic programing
GPS	global positioning system
GW	gateway nodes
HOL	head-of-line
HSP	handshake silence period
HVAC	heating, ventilating and air conditioning
IBE	identity-based encryption
IGW	Internet gateway
ISI	inter symbol interference
KMT	Kiyon's multi-channel TDMA
LEMA	localized energy-efficient multicast algorithm
LHR	last hop router
LMT	locally parallelized, multi-radio WCDS tree
MAC	medium access control
MAI	multiple access interference
MAN	metropolitan area networks
MAP	mode assignment with probability
MBAA	multi-bean adaptive array
MBSAS	modified BSAS
MBSs	mesh base stations
MCA	maximum communication area
MCDS	minimum CDS
MCREQ	multicast client request
MDR	maximum data retrieval
MEMS	microelectromechanical systems
MFA	maximum forwarding area
MFR	most forward within radius
MH	mobile host
MIB	management information base

MIMO	multiple-input-multiple-output
MISO	multiple-input-single-output
MLB	minimum latency broadcasting
MLD	multicast listener discovery
MLME	MAC layer management entity
MQAM	multiple quadrature amplitude modulation
MR	mobile router
MR^2-MC	multicasting in multi-radio, multi-rate, multi-channel
MRTS	multicast RTS/CTS
MSH_CSCH	mesh centralized scheduling
MST	minimum spanning tree
MT	mobile terminal
MTCM	multiple trellis-coded modulation
MWT	multiple-radio weighted-connected-dominating-set tree
NAV	network allocation vector
NCAF	network coding with amplify-and-forward
NCDF	network coding with decode-and-forward
NCDNF	network coding with denoise-and-forward
NCJDF	network coding with joint decode-and-forward
NDP-Proxy	neighbor discovery protocol proxy
NEMS	nanoelectromechanical systems
NES-C	nested C
NFP	nearest with forward progress
NLP	nonlinear programing
NPDA	network disconnection prediction algorithm
OFDM	orthogonal frequency division multiplexing
OGF	on-demand geographic forwarding
OLSR	optimized link state routing
OOK	on-off keying
O-QPSK	orthogonal-quadrature phase shift keying
OWL	web ontology language
PAMT	parallelized, approximate-shortest, multi-radio WCDS tree
PCF	point coordination function
PDFG	probabilistic data-flow graph
PDGP	parallel distributed genetic programing
PIM-SM	protocol independent multicast-sparse mode
PLME	physical layer management entity
PMP	point-to-multipoint
PSR	partial source routing
PWM	pulse-width modulation
QAC	query-after-command
RMT	remote mesh terminal
RSS	received signal strength
RSSI	RSS indicator
RTA	Reuleaux triangle area
RTF	request-to-forward
RTT	round-trip-time
S&F	store-and-forward

SBM	single best-rate multicast
SCAN	sensing, computing, and networking
SCBF	space-code bloom filter
SENIT	Security Expert INITiative
SIFS	short interframe space
SIMO	single input multiple output
SINR	signal-interference-noise ratio
SKKE	symmetric-key key establishment
SME	station management entity
SOFDMA	scalable OFDMA
SPT	shortest path tree
STBC	space–time block codes
STBF	space–time bloom filter
STC	space–time coding
STTC	space–time trellis codes
TDMA	time division multiple access
ToA	time-of-arrival
TPA	transmission power allocation
TRC	time-reversal communication
UDDI	universal description, discovery, and integration
UWB	ultra wideband
WABA	window based adaptive backoff algorithm
WCDS	weighted connected dominating set
WFA	wait-for-all
WiMAX	world interoperability for microwave access
WiMeSAS	wireless mesh sequential algorithmic scheme
WLANs	wireless local area networks
WM^2Net	wireless mobile mesh network
WM^2SAnet	wireless mobile mesh sensor and actuator network
WM^2Snet	wireless mobile mesh sensor network
WMR	wireless mesh router
WMU	wireless mesh users
WPAN	wireless personal area networks
WSNet	wireless sensor network

Wireless Mesh Networking

CHAPTER 1

Introduction

1.1 The Emerging Information and Communication Technology Landscape

Over the past years, the information and communication technology (ICT) industry has focused principally on rolling out evermore sophisticated mobile and wireless technologies that increasingly permeate all aspects of everyday life. We are getting progressively dependent on computers and communications networks. As individual devices have become smaller in size and more powerful, they have increasingly been linked together within complex networks. Before the end of this decade (2005–2015), seamless broadband communication networks will be spanning from the personal area to the regional and global area. This will be made possible by *meshing* all sorts of different computing and communication networks, whether these are fixed wired and wireless networks; third or higher generation mobile networks; wireless personal area networks (PANs) and local area networks (LANs); satellites; or any other.

The distinct processes of communication and computation will be linked together within miniature artefacts and objects—"smart" devices—with sensing, computational, pervasive, autonomic, and self-organizing capabilities. These artefacts will be embedded or "hidden" in the environment and communicate with each other. Their linked communications will underlie the creation of distributed and self-regulating systems that are adapted to human needs. These semi-intelligent and highly adaptive networks will augment the physical environment with new properties, enhancing its interaction with people, while keeping the underlying system out of sight. The aim will be to hide the overall system complexity, preserve human attention by delivering us only information that is rich with meanings and contexts and provide stable functionality, with functions being revealed only "*on-demand, anywhere, and anytime.*" We expect—indeed, it seems virtually certain—that these new network resources will, especially in combination, stimulate a new generation of personalized applications and services. Such *situated and cooperating smart artefacts*, whether in the form of sensors, actuators, robotic devices, tags, or whatsoever will boost the creation of new ambient environments that are tailored to individual needs and will increasingly link the real and the virtual world. These massively distributed systems will all *mesh* together to form "dynamic ecosystems" that are immersed in computerized ambient environments, and growing and adapting themselves to the evolving needs of individual users and communities.

Breakthroughs are also expected by pursuing ICT research in combination with other disciplines, for example, those related to new materials, bio and life sciences, and from the knowledge base of the cognitive, biological, and social sciences. Convergence, in particular, is expected to change rapidly the ICT landscape. Over the next 5–10 years, virtually every home and enterprise will have access to broadband with speeds much greater than those available today. Interactive digital TV will become more widespread; 3G services will be amply established; and context aware, especially location-based services will be widely available. All of this will make possible a plethora of new, accessible, and affordable mobile and networked services and experiences for consumers that will further embed ICT within their daily lives.

On the other hand, divergence is inevitable due to diverse technologies and localized controls. Although current efforts on ICT focus on simplification, universality, and convergence, a more mature shift towards a rich and diversified interconnectedness now appears to be taking place that will span all geographical areas. Divergence will bring a multiplicity of regionalized, autonomous, local, and tailor-made solutions to store, process, and communicate at lower costs and much closer to real use.

These waves of convergence and divergence that we are witnessing are progressively contributing to an ICT infrastructure that starts permeating all sectors of human activity while reshaping the ICT landscape. The increasing higher density of communication systems requires more and more distributed and self-organizing structures; simple and dependable elements, exhibiting complex bio-inspired behaviors.

These areas can be seen as a long-term evolution of what is currently considered under the name of the emerging information and communication technology (EICT) landscape. The future EICT landscape that may / will become available or will be desirable in the next 15–20 years' time encompasses new telecommunications and computing paradigms, including bio-computing and data grids, pervasive computing, mesh and overlay networks, virtual reality and ambient environments. A cartography of EICT is illustrated in Fig. 1.1. This sequel highlights the specifics of the EICT paradigm, exposing its principles, intricacies as well as a synopsis of its internal mechanics. A full treatment of the theory and practices of EICT paradigms could itself be an elaborated topic and our treatment here is necessarily brief. Our objective here is to develop an intuitive feel for the EICT vision and illustrate the position of wireless mesh networking in this emerging communication landscape.

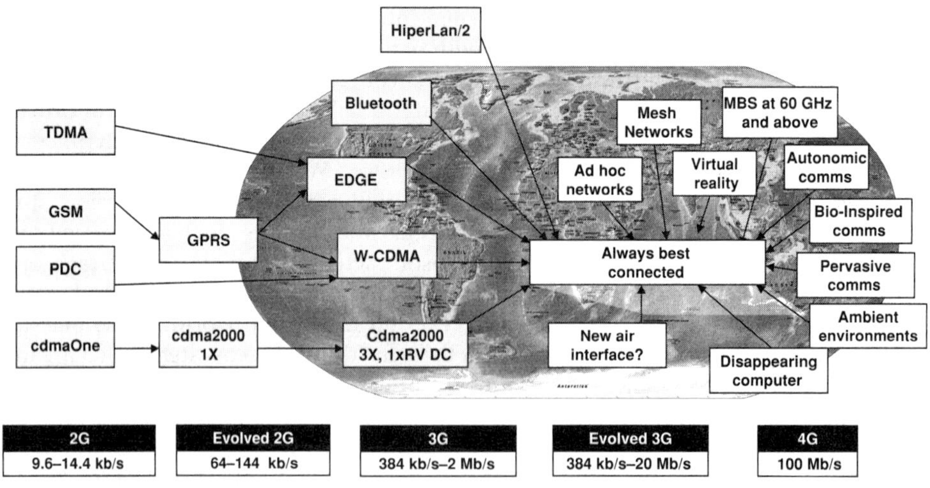

FIGURE 1.1 The landscape of emerging information and communication technology (EICT).

1.1.1 Autonomic Communications

In *autonomic communications* (AC), applications and services are not ported onto a pre-existing network, but the network itself grows out of the applications and the services that the end users desire.

AC is loosely defined through *a set of system features*, centered on networking **selfware**. Selfware itself is the common name for all the "self" system properties, that is, for a number of tightly coupled processes—sensing, data handling, decision making and communication—that are used to achieve system properties of self-awareness, self-healing, self-configuration, self-optimization, and that can be instantiated in a multitude of variations.

Selfware principles and technologies borrow largely from autonomous distributed systems research, nonconventional networking (ad hoc, sensor, mesh, peer-to-peer, group communication, active networks, etc.), formal methods, etc. They have a broad scope, addressing all facets of communication—human-to-human, human-to-cyber, business-to-business, cyber-to-cyber, etc.—by empowering network elements to best fit communication intentions, to observe and to react by self-organization to context changes without explicit user interaction.

1.1.2 Bio-Inspired Communication Systems

Advances in life sciences and biotechnology of the last 20 years have remarkably enhanced our understanding of biological systems in all their complexity. This vast body of knowledge can be exploited for designing and implementing new ICT systems. In contrast to technological systems, biological systems inspired from life forms show a high capability to grow, adapt, emerge, self-organize, self-assemble, replicate, heal, self-organize and evolve. In particular, *bio-inspired communication* architectures replicate the economic and social system supported by the communication network itself. In this respect, the network can be seen as the nervous system of a larger "parent" system, therefore its characteristics and functionalities should be studied and understood in relation to the behavior of the parent system that the network mediates. Rather than analyzing the structure of the communication network, the accent is more on the *cause-effect relationship* between the structural features of the communication network and the resulting behavior of the "parent" system it supports (be it economic, business, social, financial, etc.).

"Bio-inspired" materials or "artefacts" can thus benefit from the emerging technology of synthetic biology to create "life-like" communication and computing paradigms that have the abilities to grow, self-repair and adapt.

1.1.3 Pervasive Computing and Communications

Prospective advances in microprocessor, communication, and sensor technologies envision a whole new era of computing systems, referred to as *pervasive computing*. In pervasive computing, ICTs seamlessly and invisibly pervade into the "fabric of everyday life" to deliver services that are adapted to the person, the environment and the context of their use.

Economic, social, and technological trends are moving towards greater mobility. In banking, Internet and telephone use, people increasingly expect high-speed and multimedia capabilities. In the near future, an increasing number of objects—from automobiles to books, from houses to hairbrushes—are expected to carry both communications and computational technology. Devices will become far smaller and more powerful, and will fade invisibly into the environment. On the basis of wireless technologies, these devices will communicate with one another—your laptop sending information (maps, addresses etc.) to your car, devices inside the house communicating with your watch (reminding you of appointments). The information technology environment will become pervasive—everywhere and always on.

The challenges of pervasive computing and communications are dominated by the ubiquity of a vast manifold of heterogeneous, small, embedded, and mobile devices, all enabled to communicate in a seamless way with heterogeneous technologies; the self-organization and evolvability of their population and interoperation, the ability of perceiving and interpreting their situation locally or via distribute communications, their overcoming of traditional end-to-end paradigm for connections and their ability of taking advantage of communication opportunities, the autonomy of their goal-oriented behavior, the dynamicity and context adaptation of services they offer, the ad hoc interoperability of services, and the different modes of user interaction upon those services.

1.1.4 Artificial Intelligence and Natural Cognition

The vision of *artificial intelligence (AI) and cognition* is to build artificial cognitive systems inspired by biology, in particular neuroscience, under the following two assumptions: (1) cognition by systems interacting with the real world depends upon and is facilitated by their body and (2) the structure of this body, the environment, and the body-environment interaction are inseparable from one another.

Traditional cognitive science, cognitive psychology, and AI make no commitment to the form of a cognitive system's implementation. Today, especially in cognitive neuroscience and robotics, the infrastructure (i.e., embodiment) is considered much more crucial to the understanding of cognition. One obvious difference between IT systems and biological cognition is the extent to which biology has extendable processing able to make analogies and cope with novel percepts.

Natural cognition is aimed at taking a relatively radical step away from classical AI-based IT approaches to cognition towards research on self-organization and development as a natural framework for cognition. In this context, cognition is seen as more than just an inferential process. It is a property that results from the interaction of an organism with its environment. Embodiment, as the central notion in this vision, is characterized by a number of attributes, such as: Embodiment is intrinsically developmental and is structured by interaction with the environment. It enables affective interaction, the acquisition of meaning from percepts created through sensory-action integration and the grounding of "concepts" in the agent's sensory–motor and social interaction. It facilitates learning by formation of cross-modal associations through induction and generation of correlations. It includes continuous dynamics with discrete attractor states, provides the basis for grounding and maintains the distinction between the description of cognition by external observers, and its underlying mechanisms.

1.1.5 Virtual Reality

One of the most exciting possibilities for tomorrow's society is the opportunity to use ICT to bridge the real and virtual worlds. Reflecting the fact that ICT is no longer seen as a bolt-on, organizations are capitalizing on digital technologies to customize their products, access wider markets and exploit cross-synergies between physical and online activities. These include industrial design, manufacturing, logistics, information services, entertainment, and games.

In industrial design, for instance, clay models are being replaced by "digital factories" in which products are built physically only when all constituting elements and the full manufacturing process have been simulated and optimized with *virtual reality* techniques. In transport and logistics, the real-time monitoring of shipments leads to leaner supply chains, lower costs, and happier customers. Similar paybacks can be expected for mobility in general; once cars, trucks, and roads are linked to tracking, monitoring, and visualization software through sensors and cameras.

The entertainment industry increasingly makes use of the simulation and "mixed-reality" techniques that bridge the real world of scenes and actors with the flexible and limitless space of virtual reality where everything is possible. At the other end of the chain, the virtual world of films and games is extended to the real world of toys and various gadgets creating huge leverage. New virtual communities reduce the need for mobility and offer new opportunities for community building.

Clearly, we are only seeing the very first benefits in terms of enhanced creativity, lower costs, and diversity. Much more is ahead of us. In health, for example, there is a prospect of the *in silico* human and the associated vision of a much more efficient healthcare system seamlessly connecting patient, doctors, and hospitals with simulation, databases, health monitoring software, and embedded actuators to provide us with permanent monitoring, full accessibility of record and customization of treatment.

1.1.6 Ambient Environments

The communication systems currently implemented or under development, either wired or wireless, suffer from several limitations. For example, the need for a cable or the spectrum and bandwidth available may restrict the user mobility and limit the range of possible services. Security, privacy, and ease of use are also huge challenges. In perspective, and coherently with the scenario "the real world will become our interface," our interaction in future *ambient environments* will increasingly be enabled through all senses and using distributed devices (sensors, actuators, smart tags, etc.) embedded everywhere. Multimediality will then be extended beyond the transmission of voice, data and images to include, among others, position, haptics, and "feelings." Future communications will have to make this possible.

A future communication scenario may involve for example generalized parallel data transmission through a multitude of different channels, including radio, optical, or other novel transport and transmission mechanisms and physical media. Embedded distributed intelligent systems can be required to perceive, forward, reconstruct, and coherently present this information to the user(s), in the presence of a number of constraints dictated by the environment, the size, weight, and power consumption of the devices, etc.

1.1.7 Nanoscale Materials

In the coming decade, ambient intelligence applications focusing on health, comfort and leisure, communication, mobility, safety, and security will guide application-driven research in components. Opportunity-driven innovations will increasingly be based on novel nanodevice building blocks and related nanofabrication techniques, on new component design and architectures, and likely on bio-inspired approaches and concepts. With *nanoscale devices* reaching approximately 10 nanometers (nm) in 2015–2020, new opportunities will emerge to combine ultimate "top-down" semiconductor platforms with "bottom-up" developments in materials, physics, chemistry, and biology.

The ever-increasing computing performance and storage capacities achievable with existing technologies will eventually reach a plateau in another 10–15 years of time, with storage capacities of terabytes and peak performance of teraflops for a standard chip. On the other hand, the power consumption of high-performance chips is estimated to rise to intolerably high values. These predictions have highlighted the need to explore technological alternatives to extend IT capabilities beyond the limitations of current CMOS technology and underpin the current trend towards diversification and increased complexity of technologies hybridized onto CMOS platforms.

To differentiate future research directions, however, it is useful to draw a distinction between the *More of Moore, More Than Moore* and *Beyond Moore* technology drivers. The *More of Moore* approach is focused on delivering the ITRS roadmap for late CMOS and post-CMOS systems and in particular, how to continue Moore's law beyond the predicted 22 nm node in 2011. The *More Than Moore* approach is focused on delivering greater functionality through heterogeneous integration of nanosystems with electronic, optical, magnetic, chemical, biological, mechanical, and other functions. Research themes, which extend *Beyond Moore* in terms of their potential to deliver disruptive technologies, include new paradigms such as intra-molecular computing and engineered coherent solid-state quantum systems.

1.1.8 The Disappearing Computer

It seems like a paradox but it will soon become reality: the rate at which computers disappear will be matched by the rate at which information technology increasingly permeates our environment and determine our lives. The increasing ubiquity of computers and related devices (e.g., sensors) and their diffusion into our environment requires a rethinking of the complex interplay between technology and humans. The often-quoted observation by Mark Weiser, that "The most profound technologies are those that disappear. They weave themselves into the fabric of everyday life until they are indistinguishable from it" sets the course for ubiquitous computing and the stage for the vision of an unobtrusive, calm technology.

A vision of the future is one in which our world of everyday objects and places becomes infused and augmented with information processing and exchange. In this vision, the technology providing these capabilities is unobtrusively merged with real world objects and places, so that in a sense it disappears into the background, taking on a role more similar to electricity—an invisible pervasive medium. As a consequence, human-centered notions, such as real objects and everyday settings, can come into the foreground, rather than the computer-centric ones, which have determined the evolution of the computer-as-we-know-it. It offers the opportunity of seeing how objects can become augmented with new properties and qualities and how these can be designed to enrich everyday living in completely different ways. Artefacts will be able to adapt and change, not just in a random fashion but based on how people use and interact with them. Together, new functionalities and new forms of use will emerge that will enrich everyday life, resulting in an everyday world that is more "alive" and "deeply interconnected" that our current day understanding.

1.2 Meshing Large-Scale Wireless Network Elements: The Vision of Wireless Mobile Mesh Networking (WM^{2}Net)

One of the major trends in the communication and computing fields is related to the arising of pervasive communication/computing environments, characterized by an extremely large number of embedded devices (Weiser, 1999; Kahn et al., 1999). As stressed above, such devices will possess computing, sensing, identifying, and communicating capabilities, making it possible for user-situated services to interface directly with the surrounding environment, entailing thus the possibility of introducing radically novel services, with a major impact on the way people–technology interactions are conceived today.

Let us consider the following scenario: several thousands of tiny devices [e.g., *nanoelectromechanical systems* (NEMS)] being deployed on public transport means (metro, buses, taxis) capable of performing a wide spectrum of functionalities, including the detection of their location, road, and weather conditions, and so on. As vehicles pass each other, they exchange information summaries. These summaries eventually diffuse across different sections of the metropolis. Drivers can plan alternate routes, estimate trip times, send queries, locate and call the nearest available cab driver, and be warned of dangerous driving conditions. This mesh of wireless devices eventually forms a massively populated network, commonly named a *wireless mesh network*. Should the WMNet constitute mobile mesh terminals, it is then referred to as *wireless mobile mesh network* or, simply, *WM^{2}Net*. A WM^{2}Net is a peer-to-peer multihop wireless cooperative self-organized, self-configured, and self-managed communication network among a huge number of wireless transceivers that all have network routing capabilities. Using wireless multihopping techniques, nodes are enabled to route their own calls and also to re-route other's traffic to neighbor mobiles via mobile-to-mobile, or *multihop*, relaying. A mobile with a poor link to a central database [or a *base station* (BS), or *access point* (AP)], needs not increase its transmission power to compensate for link losses but instead hand over its call to another mobile terminal, which lies in a more advantageous position and can help as intermediate (relaying) node to forward the communication between the mobile terminal and the serving AP or another peer (mesh) node. In this context, continuous communications around broken or blocked paths is assured by "hopping" from node to node. An in-depth treatment of the theory and practices of wireless ad hoc networking is the topic in (Aggélou, 2004). As the author demonstrated in Chap. 5, as long as sufficient node density exists for establishing multihop paths, the multihop relaying concept becomes a promising approach to enhance and/or improve communications coverage and robustness against radio link failures yet with limited infrastructure costs.

A snapshot of WM^{2}Net (Fig. 1.2) would look like an archipelago, where a myriad of islands (each composed of a limited number of connected nodes) will exist. Through advanced mesh networking protocols, wireless mesh devices form a sea of connectivity. As water flows to fill every room of a submerged ship, the mesh networking connectivity will seek out and exploit any possible communication path by hopping data from node to node in search of its destination. While the capabilities of any single device are minimal,

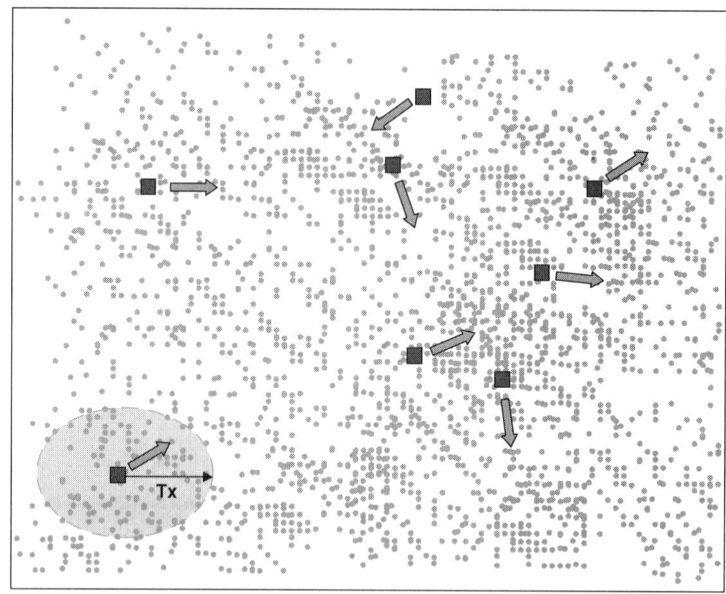

Figure 1.2 A snapshot of massively populated metropolitan WM^{2}Net.

the composition of thousands or millions of devices offers radical new technological possibilities.

Comparing to a communication system of today, the most touted benefits of a *WM^{2}Net* are first its ability to network massive populations of network elements, from a few thousands to a few millions, and, second, to route around obstructions, such as buildings and trees, and pervade within all scales of communications even in highly shadowed areas where communications could barely survive using the conventional networking paradigms of today.

Notably, a massively populated WM^{2}Net of nanosensor devices [denoted hereafter as *wireless mobile mesh sensor network* (WM^{2}Snet)] materializes the emerging *Smart Environment* paradigm. Smart environments represent the next evolutionary development step in building, utilities, industrial, home, shipboard, and transportation systems automation. Like any sentient organism, the smart environment relies first and foremost on sensory data from the real world. Sensory data comes from multiple sensors of different modalities in distributed locations.

Sensors are battery-powered small-sized *microelectromechanical systems* (MEMS)/ NEMS devices, equipped with on-chip integrated computation, communication, sensing, storage, and possible positioning capabilities. A *wireless sensor network* (WSNet) is a network of hundreds or even thousands of such sensor devices that sense the data of interest in a surrounding. The sensed data is communicated to a central node, often called the *sink* node, where the data is analyzed, further processed or fused. An enriched version of a WSNet is the *wireless sensor and actuator network* (WSANet). Whereas WSNets perform only sensory activities, wireless sensor and actuator (or actor) networks perform both sensing and operating tasks. Actors are resource-rich devices equipped with higher processing and communications capabilities than a common sensor device; an actor node could thus assume the responsibilities of the sink node too.[1]

Typical examples of battery-powered small sized sensor nodes are the *Crossbow-Berkeley Mica Mote* and the *Microstrain's X-Link Measurement Sensors*. For completeness sake, the information presented in the following paragraphs is intended to provide a synopsis of these technologies.

Crossbow Berkeley Motes may be the most versatile WSNet devices on the market for prototyping purposes. Crossbow (www.xbow.com) makes three Mote processor radio module families—MICA (first generation), MICA2, and MICA2-DOT (second generation) (de Silva, 1989; Frank, 2000). Nodes come with five sensors installed—temperature, light, acoustic (microphone), acceleration/seismic, and magnetic. These are especially suitable for surveillance networks for personnel and vehicles. Different sensors can be installed, if desired. Low power and small physical size (25 mm) enable placement virtually anywhere. Since all sensor nodes in a network can act as BSs, the network can self configure and has multihop routing capabilities. The operating frequency is ISM band, either 916 MHz or 433 MHz, with a data rate of 40 kbps and a range of 30–100 ft. Each node has a low power microcontroller processor with speed of 4 MHz, a flash memory with 128 kb, and SRAM and EEPROM of 4 kb each. The operating system is Tiny-OS, a tiny micro-threading distributed operating system developed by University of California, Berkeley with a *nested* C (NES-C) source code language (similar to C). Installation of these devices requires a great deal of programming.

Microstrain's X-link Measurement Sensor system (www.microstrain.com) may be the easiest system to get up and running and to program. The sensor nodes are multichannel, with a maximum of 8 sensors supported by a single wireless node. There are three types of sensor nodes—S-link (strain gauge), G-link (accelerometer), and V-link (supports any sensors generating voltage differences). The sensor nodes have a preprogrammed EPROM, so a great deal of programming by the user is not needed. Onboard data storage is 2 MB. Sensor nodes use a 3.6-volt lithium ion internal battery (9 V rechargeable external battery is supported). A single receiver (BS) addresses multiple nodes. Each node has a unique 16-bit address, so a maximum of 2^{16} nodes can be addressed. The RF link between BS and nodes is bi-directional and the sensor nodes have a programmable data logging sample rate. The RF link has a 30-meter range with a 19200-baud rate. The baud rate on the serial RS-232 link between the BS and a terminal PC is 38400.

1.2.1 WM²Net Application Areas

Today, WM²Nets are proposed mainly to provide broadband Internet services to *residential* clients in some very limited application scenarios, such as broadband home networking, community and enterprise networks, as well as metro scale public Internet access (Bruno et al., 2005). These are highlighted in the following paragraphs. Extended scenarios of usage accompany each application paradigm.

[1] In general terms, one should distinguish between an actor and an actuator. An actuator is a device that converts an electrical-control signal to a physical action. An actor is a device that is able to act on the environment by means of one or several actuators; it is also a single network entity that performs networking-related functionalities, that is, receive, transmit, and relay data. For instance, a robot may interact with the physical environment by means of several motors and servomechanisms (actuators). However, from a networking perspective, the robot constitutes a single entity (actor).

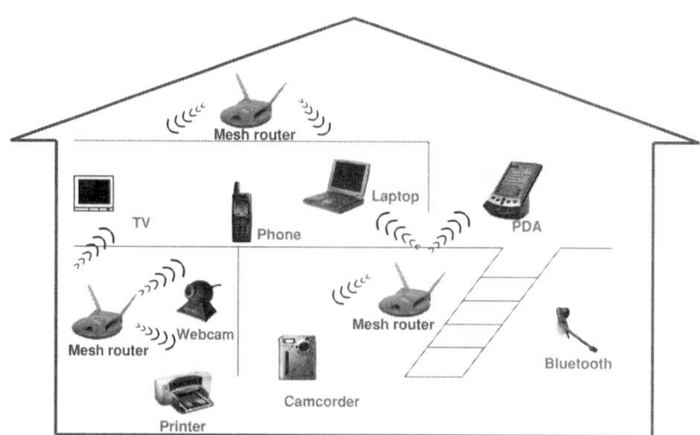

Figure 1.3 WM^{2}Nets for broadband home networking.

1.2.1.1 Broadband Home Networking

Currently, wireless broadband home networking is realized through IEEE 802.11 *wireless local area networks* (WLANs). To ensure adequate coverage of all areas within the home zone, a process called *siting*, or *radio planning*, calculates the right position of APs. Siting is a costly and rather time-consuming process. WM^{2}Net technology on the other hand can make it possible for service providers, such as cable or telecommunications companies, to provide next-generation broadband services to homeowners without the need to install new cabling or conduct a siting study. Using an embeddable multihop mesh routing software with wireless hardware for the home, a broadband wireless multihop mesh solution provides wireless delivery of high-definition video PTV, audio, VoIP, gaming, and any other IP traffic throughout the home.

In this context, the access points are replaced with wireless mesh routers, as shown in Fig. 1.3. Communication between these nodes is established in a multihop mesh fashion. Altering the locations of mesh routers or adjusting their power levels accordingly can easily eliminate areas with poor coverage.

A good example of broadband home networking is the Advanced Multi-Hop UWB Mesh for Home Entertainment Networking, which is based on Kiyon's Multi-channel TDMA (KMT) software (www.kiyon.com/products). KMT features dynamic routing protocol (DRP) embedded in any RF device, including wireless access points, routers, and switches to transform a wireless network into a high-performance mesh for quality-of-service (QoS) sensitive media traffic such as IPTV and HD video.

UWB mesh basically adjusts channel (time slot) allocations across the network based on changing traffic patterns, whereas also levelizes UWB bandwidth throughput over several hops.

1.2.1.2 Community and Neighborhood Networking

The common wireless access architecture for community networking comprises two core segments: the fixed segment, which connects the AP/GW via wired connections to the Internet (e.g., DSL), and the wireless segment, which connects the AP/GW with the mesh terminals. This type of network access infrastructure poses several drawbacks:

- Although the information is shared within the community or neighborhood, all traffic, however, must flow through the GW/AP and/or the Internet.

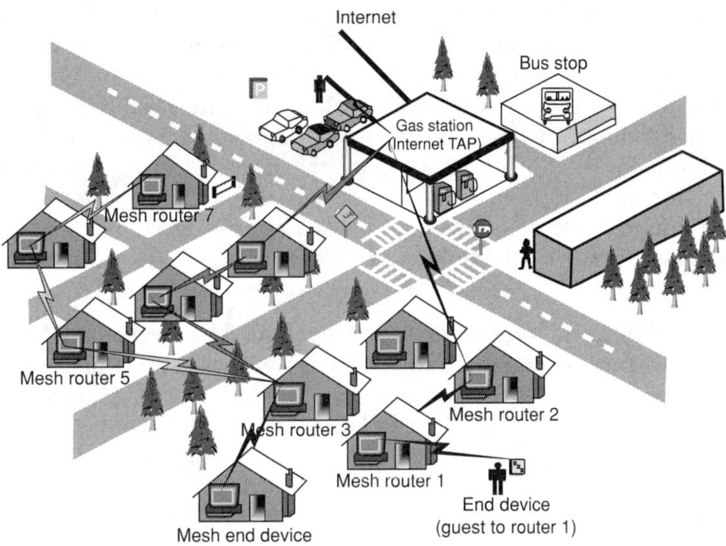

FIGURE 1.4 WM^{2}Nets for community networking.

- Large percentage of areas in between houses is not covered by wireless services.
- An expensive but high bandwidth GW between multiple homes or neighborhoods may not be shared and wireless services must be set up individually. Inherently, this drives up network service costs.

The WM^{2}Net paradigm addresses these concerns through flexible mesh network configurations among homes, as shown in Fig. 1.4. There are indeed many advantages to enabling such connectivity and forming a community mesh network. For example, when enough neighbors cooperate and forward each other's packets, they do not need to individually install an Internet "tap" (GW) but instead can share faster, cost-effective Internet access via GWs that are distributed in their neighborhood. Packets dynamically find a route, hopping from one neighbor's node to another to reach the Internet through one of these GWs. Another advantage is that neighbors can cooperatively deploy backup technology and never have to worry about losing information due to a catastrophic disk failure. A third advantage is that this technology allows locally created information to be used locally without having to go through a service provider and the Internet. Neighborhood community networks allow faster and easier dissemination of cached information that is relevant to the local community.

Community-based multihop wireless networking is disruptive to the current broadband Internet access paradigm, which relies on cable and DSL being deployed in individual homes. It is important because it allows free flow of information without any moderation or selective rate control. Compared to the large DSL and cable modem systems that are centrally managed, mesh networking is **organic**—everyone in the neighborhood contributes network resources and cooperates.

Locust World is a good example of shared ADSL services in community networks: Remote communities can use the Mesh to share a single expensive Internet link, like a satellite or leased line, among enough users to make the service affordable. A T1 or satellite connection is often out of reach of individual small businesses and personal users. However, if there is enough local interest then their purchasing power can tip the balance and help to provide excellent value within the local community.

Other community-based wireless mesh deployments include those in low-income neighborhoods of Houston, Texas [e.g., the Transit Access Points (TAPs) Project, Rice University: http://taps.rice.edu], in rural parts of India [e.g., the Digital Gangetic Plains (DGP) project: http://cse.iitk.ac.in/users/braman/dgp.html], and the MIT Roofnet project (www.roofnet.mit.edu).

1.2.1.3 Enterprise Networking in Metropolitan Area Networks (MANs)

Enterprise networking can be a small network within an office or a medium-size network in an entire building or even a large-scale network that expands to all offices in multiple buildings. Currently, IEEE 802.11 WLANs are widely used. However, these wireless networks are still isolated islands. Connections among them can be achieved through wired Ethernet connections, which is the key driver for the increased cost of enterprise networks. Adding more backhaul access modems will increase capacity only locally, but would not improve robustness to link failures, network congestion, and other problems of the entire enterprise network. If mesh routers replace APs, Ethernet wires can be eliminated. Multiple backhaul access modems can be shared by all nodes in the entire network and, thus, improve robustness and resource utilization of enterprise networks. WM^2Nets can grow easily as the size of enterprise expands. The service model of enterprise networking can be applied to a number of other public and commercial service networking scenarios, such as airports, hotels, shopping malls, convention centers, sport centers, etc., as well as to MANs that cover a potentially much larger area than home, enterprise, building, or community networks (Fig. 1.4).

1.2.1.4 Building Automation

In a building, various electrical devices including power, light, elevator, air conditioner, etc., need be controlled and monitored. Currently this task is accomplished through standard wired networks.

WiFi-based networks are the wireless alternative to reduce the deployment and maintenance overhead of such networks. However, the WiFi approach has not demonstrated the cost-versus-performance foreseen as the cost of wiring turns WiFi deployment for this application a rather expensive solution. If mesh routers replace building automation and control networks (BACnet) APs, the deployment cost is significantly reduced and the deployment process becomes also much simpler.

The KAN354B for BACnet WiFi Mesh (Fig. 1.5) (http://www.kiyon.com/products) is a good example of wireless building automation. The KAN354B BACnet WiFi Mesh enables new and retrofitted buildings to be outfitted with state-of-the-art building controls with minimal or no new wiring required. In addition to reducing the cost and labor associated with laying wire, the technology also makes it easier and less expensive for customers to make additions or changes to their systems, or accommodate tenant moves. Utilizing highly reliable, self-managing mesh networking technology, building systems such as heating, ventilating and air conditioning (HVAC), lighting, safety and security, and IT are easily installed and integrated into the centralized building management system, and field accessible using a standard wireless laptop computer.

FIGURE 1.5 KAN354B BACnet wireless mesh router.

1.3 Technical Challenges and Book Structure

To make the vision of wireless mesh networking a reality, a number of new and challenging issues need be efficiently addressed, including *cross-layer energy optimization techniques, energy-efficient geographic communications, connectivity, coverage and capacity principles, biologically-inspired communications, spatiotemporal correlation principles, distributed key management, privacy and security, autonomic operation, self-management and resilience, and WM^{2}Net testbeds and prototypes.*

Currently, these and other technical as well as operational challenges of WM^{2}Nets constitute the forefront research activities worldwide. A clear understanding of the relationship between the aforementioned technical challenges and factors such as network topology, traffic patterns, network node density, transmission power level, and node mobility provides a guideline for protocol development, architecture design, deployment and operation of the network. The theory and practices behind these constitute the core of this book. A high-level description of each chapter is outlined below.

Chapter 2: Wireless Mesh Communication Architectures and Protocols The malfunctioning of nodes and node mobility are the two frequent causes of the fragility of network infrastructures in radio networks characterized by intermittent service availability. As a consequence, some network segments may become completely disconnected from other network segments during operation. Operating in arbitrary, temporary, and unsupervised conditions, to maintain a certain service level WM^{2}Net devices need be capable of self-organizing themselves in areas with semi- or quasi-static network topologies. The limitations on power consumption imposed by portable wireless radios coupled with the fact that the communication infrastructure does not rely on the assistance of centralized stations, implies that WM^{2}Net nodes must communicate with each other either directly or indirectly using multihop routing techniques. As nodes may also be moving about, this results to a distributed multihop wireless mesh network with a time-varying topology. Adding to these, large-scale deployments, anywhere from hundreds to millions of mesh terminals, are orders of magnitude larger than that in traditional wireless networks. Given that WM^{2}Net nodes are devices with very limited memory, communication bandwidth, and processing resources, implementation of traditional networking protocols with no concern about energy consumption, scalability and self-awareness is not practical in these networks. To ensure efficiency, survivability, and adaptivity in WM^{2}Nets, it is imperative to adopt efficient and lightweight data delivery (routing) as well as resource allocation (medium access control (MAC)) techniques.

Chapter 3: Energy-Aware WM^{2}Net Communications A WM^{2}Net is expected to operate and be functional for days, weeks and even months after deployment. WM^{2}Net devices, however, are battery-powered devices with on-chip integrated capabilities, including communications, computing and processing. The battery capacity of mesh devices is a very important issue in the design of WM^{2}Nets as it restricts the operation time of a wireless radio. In some application scenarios, replenishment of power resources might be impossible. Consequently, carefully scheduling and budgeting battery power has become a critical issue in WM^{2}Net design.

Chapter 4: Principles of Communications Coverage in WM^{2}Nets Key factors for a best-connected WM^{2}Net are the development of optimal node placement strategies, antenna gain, antenna pattern, battery lifetime, transmitted power, receiver sensitivity, and the number of mesh devices per unit of surface (i.e., nodal density). Limited battery power restricts the communications coverage as well as the operation time of wireless devices. Below some critical threshold for remaining battery power, a multihop node will not be able to function as a router, thus immediately affecting the network connectivity, possibly isolating one or more segments of the network. Fewer routers almost always mean fewer routes and therefore increased likelihood of degraded performance in the network. In fact, communication becomes meaningless if a node is not able to communicate owing to low battery power. In the same line of thoughts, one tacit assumption on the WM^{2}Net nodal population is that mesh node density should be high enough to ensure network-wide connectivity at any time. Higher nodal populations enhance network connectivity and fault-tolerance, but drive-up MAC-level contention too. Fault-tolerance is defined as the existence of multiple internally vertex-disjoint (or edge-disjoint) paths between each pair of mesh nodes. The challenge is to calculate the minimum number of relay nodes deployed, which could still induce a best-connected communication graph.

Chapter 5: Medium Access Control Principles in WM^{2}Nets In wireless communications, channel transmissions are overheard from all nodes in close proximity to the transmitting node. A data packet collision occurs when more than two users are transmitting at the same time to the same node. The MAC protocol is responsible for controlling access to the physical medium as well as for accounting for the available resources. Efficient channel access during call setup is thus vital for minimizing the blocking/dropping rates. Contention Asymmetry, Traffic Asymmetry, and the Hidden and Exposed Terminal problems, though not unique to wireless networks, turn the design of a MAC protocol tailored for operation in a massively populated WM^{2}Net environment into a very challenging task.

Chapter 6: Capacity Principles in WM^{2}Nets In a WM^{2}Net context the issue of capacity is considered in the light of two intrinsically different types of traffic: the extra-mesh and the intra-mesh traffic. When extra-mesh becomes the overwhelming traffic in a WM^{2}Net, the network performance is very akin to that of a traditional cellular network where the available capacity of each cell is fixed and largely defined from the BS. When intra-mesh becomes the overwhelming traffic, it is argued that the underlying network capacity increases as nodal population increases. In pure mesh configurations with no central entity being in place to instrument the allocation of resources, a number of unwanted conditions may occur. The so-called "tragedy of the commons," for instance, is a reality when resources are shared among multiple users. Such a tragedy relates to the days when common land was used for the grazing of livestock with free access for all. The danger is that free access to a finite resource can result in that resource being fully consumed or compromised further such that it loses its usefulness to all. What then, if users could somehow add grazing capacity as they joined the common?

Chapter 7: Security Issues in WM^{2}Nets WM^{2}Nets are often deployed in open, unattended environments, which lack physical protection and are thus vulnerable to malicious attacks (Karlof et al., 2003). To ensure the desirable behavior from our communication infrastrucuture, preventive security management schemes that are resilient to external and impersonation attacks are needed. Security mechanisms are essential to ensure the authenticity, confidentiality, freshness, and integrity of the information exchanged and processed by such networks.

Chapter 8: Autonomic Selfware WM^{2}Net Communications In massively populated WM^{2}Net deployments, the manual configuration of individual nodes is certainly not an option. Predeployment configuration is also infeasible because some configuration parameters such as node location and network neighborhood are typically unknown prior to deployment whereas other parameters may change over time, necessitating self-tuning and self-configuration. Ideally, for wireless mesh deployments to be successful, the system shall automatically configure itself for any possible physical node placement. AC is an emerging paradigm where the network itself grows out of the applications and the services that end users want. AC is centered around networking selfware—a novel approach to perform network control, as well as management, middle box communication, service creation and composition of network functionalities, etc. based on universal and fine-grained multiplexing of numerous policies, rules and events that is done autonomously but facilitates desired behavior of groups of network elements.

CHAPTER 2

Wireless Mesh Communication Architectures and Protocols

2.1 Introduction

A wireless mobile mesh network (WM^2Net) is a network of wireless, possibly mobile, devices that can freely and dynamically self-organize in arbitrary and temporary network topologies. A WM^2Net may be left unattended after deployment for days, months or even years in areas without any preexisting communication infrastructure. Various sources of failure (e.g., malfunction of nodes and mobility) produce a time-varying communication infrastructure. Communication protocols tailored for operation in a WM^2Net need thus be designed to operate in such dynamic contexts and cope with this frequently changing multihop network topology.

The limitations on power consumption imposed by portable wireless radios, coupled with the fact that the communication infrastructure does not rely on the assistance of centralized stations, implies that terminals must communicate with each other either directly or indirectly using multihop routing techniques. Unlike traditional wireless networks, WM^2Nets do not rely on predeployed infrastructure; instead, each individual WM^2Net node becomes part of the overall infrastructure. In this context, not all WM^2Net nodes need communicate directly with a high-power control tower or base station or access point, but only with their local peers, instead. Peer-to-peer networking protocols provide a mesh-like interconnect to shuttle data between the thousands of tiny devices in a multihop fashion.

With these unique characteristics in mind coupled with the fact that WM^2Net nodes are small sized objects with built in nonreplaceable batteries, a networking mechanism to be applicable in a WM^2Net must be simple, energy efficient, scalable, and robust. Hence, implementation of traditional networking protocols such as routing and medium access control (MAC), with no concern about energy consumption, scalability, and fault tolerance, is not practical in these networks.

Examining the structure and operation of WM^2Net networks can offer insights into ways in which communications could be supported within massively populated wireless mesh configurations. Given that WM^2Net nodes are devices with very limited memory, communication bandwidth, and processing resources, in order to support large-scale network deployments, anywhere from hundreds to millions of mesh nodes, efficient and lightweight WM^{2}Net protocols are needed. The design of communication protocols tailored for operation in such dynamic contexts turns out to be a very challenging engineering goal.

2.2 WM^{2}Net Configurations

A WM^{2}Net configuration can be either *hierarchical, flat or hybrid*. In a *hierarchical* or *infrastructure* network, nodes are partitioned into groups, often called *clusters*. Generally, there are three kinds of nodes in a cluster, namely, the cluster-head node, the gateway (GW) node, and the cluster-member (internal) node. Cluster-head (CH) nodes basically emulate the functionalities of an AP. All nodes in a cluster can communicate with their CH and (possibly) with each other (of the same cluster) (Iwata et al., 1999).

Various different heuristics can be used for the CH election. These may include node addresses, node degrees (neighbor connectivity), transmission power and mobility, or more sophisticated node weights combining the above attributes (see Chatterjee et al., 2002; Singh and Raghavendra, 1998).

GW nodes are used to provide connectivity among clusters. To communicate within a cluster, a GW must select the frequency or code used by that cluster. GW nodes can communicate with multiple CHs and thus can be reached via multiple paths. Consequently, similar to a router in the wireline Internet, which is equipped with multiple subnet addresses, a GW may have multiple hierarchical addresses.

Depending on the number of hierarchies (levels), the depth of the network can vary from a single tier to multiple tiers. Figure 2.1 illustrates a two-tier example. At Level = 0, there appear 4 physical-level clusters C0-1, C0-2, C0-3, and C0-4. Level 1 and level 2 clusters are generated by recursively selecting CHs (Iwata et al., 1999).

For a node A in a cluster X (Fig. 2.2), to establish communication with some node B at some different cluster, say Y, its traffic must first be routed to its CH. From the CH, traffic is then routed to a GW node, to another CH, and so on until the CH of the destination node (cluster Y) is reached. Traffic is then delivered to the destination node.

Furthermore, infrastructured WM^{2}Net architectures enable the integration with existing wireless networks (see Fig. 2.1) through GW/bridge functionalities built in mesh routers.

In a flat, or client, WM^{2}Net architecture there is no grouping and all nodes have equal responsibilities. Connections are established between nodes that are in close enough proximity to allow sufficient radio propagation conditions to establish connectivity. In this form of architecture, client nodes constitute the actual network to perform routing and configuration functionalities as well as to provide customers with end-user applications. Clients thus function both as mesh router for routing and self-configuration, and as end user. A packet destined to a node in the network hops through multiple nodes to reach the destination. An example of a flat network is depicted in Fig. 2.3.

The *hybrid* WM^{2}Net architecture is the combination of infrastructure and client meshing as shown in Fig. 2.4. Mesh clients can access the network through mesh routers as well as through meshing with other mesh clients. While the infrastructure provides connectivity to other networks such as the Internet, Wi-Fi, WiMAX (see Section 5.4, for details), cellular, and sensor networks, the routing capabilities of clients provide improved connectivity and coverage inside the WM^{2}Net.

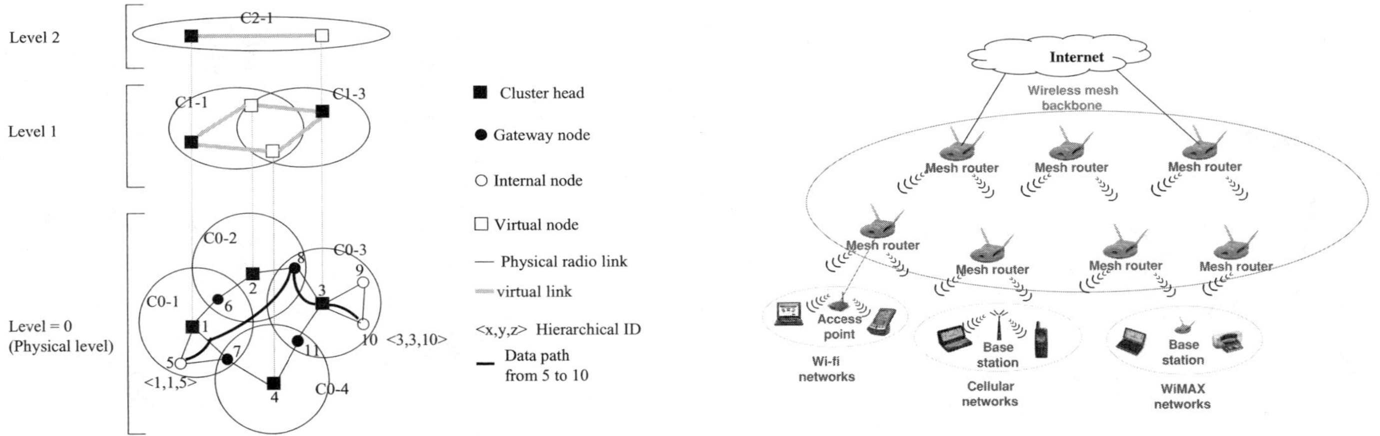

FIGURE 2.1 An example of physical/virtual clustering.

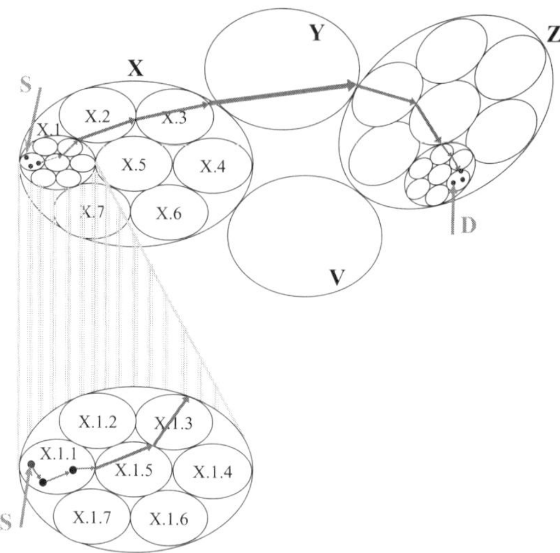

FIGURE 2.2 Hierarchical routing.

2.3 Hierarchical versus Flat WM²Net Architecture

The major advantage of the *hierarchical* architecture is the ease of the mobility and resource management process. Mobility of nodes within an infrastructure-less mobile wireless network raises organizational problems quite different and rather more challenging than those for wireless communication (cellular) networks (Mouly and Pautet, 1992). As there is no centralized administrative control, rapid response to nodal movement requires *adaptive, autonomous*, and *distributed* organization mechanisms that involve minimal manual intervention.

The aggregation of nodes into clusters provides a convenient framework for the development of important features, including channel reuse among clusters (in terms of frequency, time, or spreading code) (Gilhousen et al., 1991), channel access and channel and bandwidth allocation (Gerla and Tsai, 1995; Gilhousen et al., 1991). In cluster-based

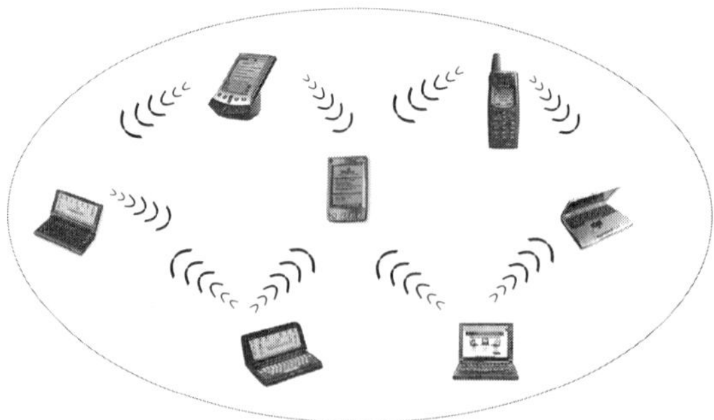

FIGURE 2.3 Flat or client WM²Net.

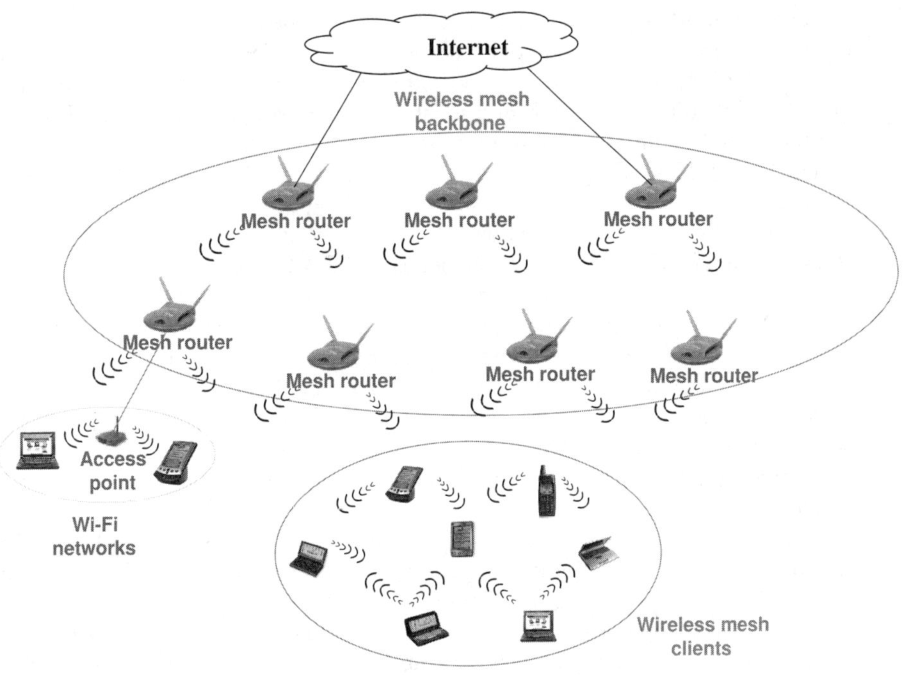

FIGURE 2.4 Hybrid WM^{2}Nets.

networks, a CH node acts as a local coordinator within its cluster: it keeps track of the member nodes of its cluster (network management), accounts for resources so that bandwidth reservations can be placed on them (in a deterministic or statistical sense), and assists in locating nodes outside its cluster during the course of a data transmission (Gerla and Tsai, 1995). This approach is in line to that followed in cellular communication networks, where such resource accountability is facilitated by the fact that all stations learn of each other's requirements through a control station (e.g., base station in cellular systems). In addition, to reduce the overhead inherent to routing and processing (Iwata et al., 1999), complete routing information is maintained only for intracluster routing (Lin and Gerla, 1997), whereas for intercluster routing the topology details are hidden through hierarchical aggregation techniques. For networks composed of a large number of mobile devices, a hierarchical network configuration would be a practical and scaleable solution (Klein et al., 1997; Xu et al., 1998).

Furthermore, clustering can provide a nice methodology for accomplishing quality-of-service (QoS) guarantees in a mobile network. Ensuring QoS communications in a wireless mesh network is apparently highly dependent upon routing and resource management control: conforming to QoS measures, such as delay bounds, depends on the quality of the chosen route whereas, in addition, complying with QoS guarantees imposes the use of a MAC method that guarantees the successful transmission of packets under high mobility and/or heavy load circumstances. This can hardly be made feasible without a central regulatory authority to carry out the significant functions of routing and channel (resource) management in WM^{2}Nets.

The major advantage of clustering stems from the fact that some information about the state of the network is kept local (Broch et al., 1998; McDonald and Znati, 1999). Even in very dynamic situations, not all changes need be propagated throughout the network. Specifically, by confining location update propagation to the lowest level (in the hierarchy) containing the moving endpoint, costs can be made proportional to the distance moved. As with most large organizational problems, specialized node roles and regional node addressing help hierarchical routing protocols to scale with network size, especially when there is a structure in the underlying network connectivity (e.g., group mobility) that can be exploited (McDonald and Znati, 2000).

Notably, mobility and traffic have a great impact on cluster size and design. In large-scale WM^{2}Nets, local conditions can vary significantly, so in some regions a larger cluster size might be advisable. However, there is not a single or default parameter, or a set of parameters, such as cluster size and cluster merge/split threshold, that can be fitted for all real networks. Besides, it is unlikely that a single default parameter would be acceptable for all clustered WM^{2}Nets. In these terms, a single initial value would be a good starting point and during operation, different heuristics are used to locally optimize cluster parameters.

There are also several features in cluster-based wireless networks, which are potentially complex to implement (Iwata et al., 1999). First, cluster IDs are dynamically assigned. This assignment must be unique—not an easy task in dynamic contexts, where the hierarchical topology is continuously changing. Second, each cluster can dynamically merge and split, based on the number of nodes in the cluster. This feature causes frequent changes of CH, thus degrading the network performance significantly. Since the diameter of a cluster is variable, it is also difficult to predict the time it takes to propagate clustering control messages among nodes. As a consequence, it is difficult to bound the convergence time of the clustering algorithm. Third, the paging and query/response approach used to locate mobile nodes may lead to a nonnegligible amount of control message overhead. Fourth, if the CH leaves its current cluster, this function migrates to another location manager. This requires a complex consistency management between original and new cluster.

Finally, the determination of the CHs shall be done in such a way that the reconfigurations of the network topology are minimized. This is an important issue, since an essential criterion in cluster-based algorithms is *cluster stability*. Frequent CH changes may adversely affect the performance of other functions such as scheduling and resource allocation, which rely on it.

Under the "extreme" scenario where mobility rates are high and mobility patterns random, such that all nodes in the field are moving very rapidly in different random directions, each cluster stays intact for only a very short amount of time and, under this scenario, it seems that clusters would need to be constantly created/modified, thus rendering a lot of cluster maintenance overhead. Therefore, the control signaling generated from the cluster maintenance becomes the bulk of overhead.

In *flat* architectures, each node maintains a routing table with entries for all nodes in the network (Iwata et al., 1999). Flat networks require only one "type" of equipment, as all nodes have to perform the same operation. That is, all nodes are treated as network members, similar to Cluster-Member nodes in hierarchical clustered architectures, with all having the same responsibilities. In addition, nodes in flat networks transmit at a significantly lower power than the transmission power of a CH, which is reasonably higher in order to cover its cluster territory. Operating a network at low power levels has several implications: first, the battery power of the nodes in WM^{2}Nets is preserved. Second, the wireless spectrum can be better reused, leading to more network capacity. Third, and possibly most importantly, a larger degree of low probability of interception and detection can be achieved, resulting in a more secure network operation.

On the other hand, a flat architecture is acceptable if the user population is small. As the number of mobile hosts increases, however, so does the overhead, thus creating scalability concerns when applied to large networks.

To conclude, there are surely circumstances under which a flat WM^{2}Net architecture is preferable to a hierarchical one, but one always has the size factor backwards. Hierarchal clustering tends to localize the impact of state changes, so it tends to be more useful in larger networks than in smaller ones. If each node is one or two hops away from every other node, clustering may not be the optimal solution in terms of signaling overhead generated. However, if the network diameter is 10 or 20 hops, a methodology is needed to reduce the size of the routing problem, and clustering seems to be an effective way to achieve this.

2.4 Routing in Mobile Wireless Networks

In WM^{2}Nets, where communication terminals are mobile and the transmission medium is wireless, routing is a major problem. The limitations on power consumption imposed by portable wireless radios, coupled with the fact that the communication infrastructure does not rely on the assistance of centralized stations, imply that terminals must communicate with each other either directly or indirectly using multihop routing techniques. As nodes move about, this results in a distributed multihop wireless network with a time-varying topology.

Before delving into the details of the properties underlying dynamic routing in wireless networks, our primary issue is to find out whether a conventional routing protocol, like link-state or distance-vector, could apply in a wireless multihop environment. To respond, we need first to list several outstanding structural differences that exist between wireline and wireless mobile networks and make routing very different in the two environments (Johnson, 1994; Perkins, 2000; Toh, 2002; Prakash, 1998):

- In a mobile wireless network, the **rate of topological changes** is relatively very high compared to that of wireline networks. As is the case in wireline networks, the procedures for route selection and traffic forwarding in wireless mobile networks require accurate information about the current state of the network (e.g., node interconnectivity and link quality) and the session (e.g., traffic rate, end-point locations), in order to direct traffic along paths that are consistent with the services requirements of the session and the service restrictions of the network.

  However, changes in network or traffic sessions are likely to occur more frequently in mobile wireless networks than in stationary wireline networks. The degree of dynamism in route selection depends on several factors, including the type and frequency of changes in network and session state; the limitations on response delay imposed in assembling, propagating, and acting upon this state information; the amount of network resources available for these functions; and the expected performance degradation resulting from a mismatch between selected routes and the actual network and session state.

The routing mechanism must be able to quickly detect and respond to such state changes in order to minimize service degradation of existing traffic sessions whereas, at the same time, the algorithm must do so using a minimal amount of network resources, in order to maximize the overall network performance (Ramanathan and Steenstrup, 1996).

On one hand, the effectiveness of a routing protocol increases as network topology information becomes more detailed and up to date. To maintain up-to-date routing tables, a conventional routing protocol should be forced to continuously send and receive topology updates. In WM^2Nets, however, the topology may change quite often, requiring *frequent exchanges of control information* (e.g., routes, route updates, or routing tables) among the network nodes. In an event-triggered Link-State protocol (Aggélou, 2004) any topological change would trigger a flooding, resulting in a flooding rate equal to the topological change rate. In this scenario, a blind route update mechanism could unnecessarily waste network resources since updates are sent even when no data transmission at all occurs in the network. In addition, as the number of network nodes can be large, the potential number of destinations is also large, thus requiring *a high volume of control information* exchanged among the network nodes. As a consequence, the amount of update traffic can be even higher, the distribution of which can eventually saturate the network.

Notably, radio spectrum is a scarce resource, which means that packet-radio networks typically have limited bandwidth available. Because the wireless devices must share access to the radio channel, the bandwidth available to any node is even more limited. Relatively low bandwidth combined with the potential for routing algorithms to generate large numbers of packets means that efficiency is paramount in designing packet-radio routing algorithms. This observation is, however, in contradiction with the fact that all updates in the wireless communication environment travel over the air and are then costly in resources. An even more disappointing fact is that as the network size increases and as the nodal mobility increases, smaller and smaller fraction of this total amount of control traffic will be even used. This is so, since the more mobile nodes become, the shorter the residual lifetime of a link turns out to be. Thus, the period in which the routing information remains valid decreases as well. Since the rate of link failure is directly related to node mobility, greater mobility increases both the volume of control traffic required to maintain routes and the congestion due to traffic backlogs. Thus, a crucial algorithm design objective in order to achieve routing responsiveness and efficiency is the minimization of reaction to mobility and of exchange of information.

On the other hand, when the rate of topological changes is extremely high, little can be done to ensure that routing algorithms converge fast enough to track topological changes (Iwata et al., 1999; Corson and Ephremides, 1995). In this situation, flooding-based routing algorithms may be the only viable routing option. In this regard, we should also note that under extreme conditions, where the changes in network topology occur too frequently, finding a loop-free path may become impossible too. We conclude then that the topology changes shall occur sufficiently slowly in order to allow successful propagation of topology updates.

- **Broadcast transmissions are unreliable.** Since broadcast packets are not receiver directed, there is no way to reserve the wireless medium at the receivers before transmitting a broadcast packet (e.g., with the use of an RTS/CTS exchange—see Section 5.3.1). Consequently, broadcast packets are inherently less reliable than unicast packets.

  This difference does not exist in wireline networks, and presents a fundamental limitation of wireless networks that must be accounted for in the design of WM^{2}Net routing protocols. Broch et al. (1998) demonstrate that over any single hop, 99.8% of *unicast data* packets are received successfully, while only 92.6% of *broadcast packets* are received. The difference between the two numbers is attributed to collisions.

- **Wireless links can be asymmetric and unidirectional** (Haas and Tabrizi, 1998; Chambers, 2002; Prakash, 1998). A link between two nodes i and j is called unidirectional when node i can properly receive traffic from a node j (in this sense, i can receive data from j above a certain BER threshold and thus properly decode it), but j cannot receive properly traffic from i. As a consequence, a transmission that requires a handshake between i and j fails.

  A link between two nodes i to j is called asymmetric, when the transmission quality of the link (e.g., data rate) from i to j is different from that of the link from j to i.

- Wireless mesh devices present **technological limitations** on the use of resources, namely, battery power, transmission bandwidth, and CPU time, compared to their wireline counterparts.

With these constraints in mind, routing protocols designed for wireline networks cannot directly apply in WM^{2}Nets. In fact, conventional routing protocols would perform very badly (Krishna et al., 1997; Barret et al., 2001; Chambers, 2002), both from a practical standpoint of building such a network, and from a theoretical standpoint in terms of what there seems to be promising routing algorithms, if used in a dynamic environment.

Link-state and distance-vector would probably work very well in a WM^{2}Net with low mobility, that is, a network where the topology is not changing very often. The main problem with link-state and distance-vector is that they are designed for a static topology, which means that they would have problems to converge to a steady state in a WM^{2}Net with a frequently changing topology. In addition, the problem that still remains that the link-state and distance-vector are highly dependent on periodic control messages. As the number of network nodes can be large, the potential number of destinations is also large. This requires large and frequent exchange of data among the network nodes. This is in contradiction with the fact that all updates in a wireless interconnected mesh network are transmitted over the air and thus are costly in resources such as bandwidth, battery power and CPU.

Based upon these considerations, the desirable qualitative properties of a wireless mesh networking protocol are: (1) to cope with frequently changing network infrastructures; (2) to ensure small convergence time, based on high collaboration among nodes; (3) to be robust given a common spectrum of WM^{2}Net conditions, such as high channel congestion and frequently changing topologies; (4) to scale well in large node populations, in terms of storage, computational and transmission overhead; and (5) to be simple.

Besides, other more evident desirable qualitative properties include:

- **Scalability:** In its general view, scalability of a networking protocol is its ability to support the continuous increase of the network parameters (as for example traffic rate, network size, etc.) without degrading network performance (Iwata et al., 1999; Santivanez et al., 2002). Notably, one shall distinguish, however, between network scalability and routing protocol scalability. In its general context, network scalability is what the network can support, whereas routing protocol scalability is what the routing protocol can handle provided that the network can. Simply speaking, if the network can support thousands of nodes for a given traffic load, then for a routing protocol to be considered scalable, it should not break when run over that network of thousands of nodes with that traffic load. So, basically, routing protocol scalability means matching (or improving) the network scalability properties.

  From (Iwata et al., 1999), it is argued that the routing protocol scalability is dependent on the scalability properties of the network the protocol runs over. That is, the network's own scalability properties provide the reference level as to what to expect from a routing protocol. Obviously, if the overhead induced by a routing protocol grows faster than the network rate but slower than the minimum traffic load, the routing protocol is not degrading network performance. The latter is, in fact determined by the minimum traffic load.

  Furthermore, scalability of a routing protocol does not solely depend on its performance (e.g., packet delivery ratio) versus network density, or traffic load, or control overhead, or some combination of performance measures. While we may assume that a protocol A that uses control signaling (control packets) more efficiently (number of packets delivered per control packet) than a protocol B, or, else, both protocols deliver the same number of packets, but one must work harder to do so, can we say that protocol A is more scalable than protocol B, given that the performance of both protocols is the same (at least from an external view)? Measuring the protocol's control overhead does not necessarily provide enough information to extrapolate the results to what will happen when network parameters (size, mobility, etc.) are increased. This is so, since there are other factors, as route suboptimality for instance, that may become more relevant as traffic and network size increase. To this end, a routing protocol that produces less control overhead may be forming longer paths, which may not be an issue at your current traffic rate, but as the traffic rate increases the extra hops may be comparable to (or greater than) the control overhead.
- **Distributed operation:** The protocol should not be dependent on a centralized controlling node.
- **Demand-based operation:** To minimize the control overhead in the network and thus not wasting network resources more than necessary, the protocol should be reactive. This means that the protocol should only react when needed and that the protocol should not periodically broadcast control information.
- **Multiple routes information:** To reduce the number of reactions to topological changes and congestion, multiple routes could be used. If one route has become invalid, it is possible that another stored route could still be valid and thus saving the routing protocol from searching for a new route.

- **QoS support:** The general goal of efficient routing is to get packets reliably from the source to the destination while maximizing the capacity of the network and minimizing the delivery delay. Optimal routing algorithms that maximize capacity or minimize delay typically need an estimate of the network flows, network topology, information about the residual capacity of links, and so on. Some sort of QoS support is then necessary to incorporate into the routing protocol.
- **Power conservation and Sleep period operation:** WM^{2}Net nodes can be laptops and thin clients, such as PDAs, as well as micro/nanodevices such as microelectromechanical systems (MEMS) and nanoelectromechanical systems (NEMS) with very limited battery power resources. It is therefore important for the network protocols to support sleep and temporarily inactive modes.
- **Network partition support:** Mobility of nodes together with wireless links of varying quality could lead to overly frequent topology changes. These could further cause to some network segments to become completely disconnected from other network segments.

2.5 Routing Protocol Categories

Routing has traditionally used the knowledge of the instantaneous connectivity of the network with emphasis on the state of network links. This is the so-called *topology-based* approach (Mauve et al., 2001). An alternative to the topology-based approach, called *location-based (or position-based)* routing, uses information related to the physical position of nodes to help the task of routing (Basagni et al. (1998, 1998a); Bose et al., 1999; Mauve et al., 2001; Capkun et al., 2001; Xue and Li, 2001; De Couto and Morris, 1998; Haas and Liang, 1999; Karp, 2001; Karp and Kung, 2000; Ko and Vaidya (1998, 1999); Li et al., 2000).

An alternative to these approaches, called *power/energy-aware routing*, uses information related to the remaining battery lifetime of mobiles with the goal to produce paths that comprise nodes with a high value of remaining lifetime as well as to help them adjust their transmission power so that to keep the energy required to complete the routing task at minimum levels. Power/energy-aware routing is covered in Chapter 3.

This sequel highlights the specifics of each category, exposing their advantages as well as potential limitations. To develop an intuitive feel of the concepts behind each category, we tabulate a few typical routing techniques per category.

2.5.1 Topology-Based Routing Protocols

In the topology-based approach, the associated routing protocols can be classified into the following three general categories, based on the timing when the routes are discovered and updated: *proactive* (also called *table-driven*), *reactive* (also called *on-demand*), and *hybrid*.

2.5.2 Proactive (Table-Driven) Routing

The proactive approach is similar to the connectionless approach of traditional datagram networks. In proactive schemes, nodes, based on a periodic update process (Royer and Toh, 1999; Bruce McDonald and Taieb Znati, 1999), attempt to compute a priori and maintain consistent, up-to-date routing information to all nodes in the network, regardless of whether the routes are being used for carrying packets.

A routing protocol is then proactive in the sense that nodes calculate all possible paths to all destinations independently of their effective use such that when a packet needs to be forwarded, the route is already known and can be immediately used.

As is the case for wireline networks, each node maintains a routing table. Routing tables typically contain a list of addresses for all possible destinations, next-hop nodes, and the number of hops to reach each destination node. The routing table is constructed using either link-state or distance-vector algorithms.

2.5.2.1 Properties

The main advantage of proactive routing protocols is that, when an application needs to initiate a data call, routing information is immediately available thus eliminating route acquisition delays. In fact, this can be useful in various cases, as in interactive applications.

Although this approach can ensure high quality routes in a static topology, as of a wireline network, it does not scale well to large, highly dynamic networks. In fact, proactive protocols require each node to maintain a large table to store routing information for the entire network. The constant propagation of routing incurs substantial signaling traffic and power consumption. Given that both bandwidth and battery power are scarce resources in mobile devices, pure proactive schemes may not be the appropriate solution for a mobile wireless environment with a large number of nodes (Royer and Toh, 1999; Pearlman and Haas, 1999). A more disappointing observation, however, is that the overhead expended to establish and/or maintain a route between a source-destination pair is wasted if the data source never requires a data path.

Also, since proactive schemes rely on periodic broadcasts, they need some time to converge before a route can be used. This convergence time is probably negligible in a static wireline network, where the topology is not changing so frequently. In a mobile wireless network, on the other hand, where the topology is expected to be very dynamic, this convergence time will probably mean a lot of dropped packets before a valid route is detected.

2.5.3 Reactive (On-Demand) Routing

The philosophy behind on-demand routing protocols is to evaluate the network on an as-needed basis and create routes only when there is a need for carrying data traffic. If no data traffic is generated, then the routing activity shall be totally absent. Based on the assumption that not all the routes are used at the same time, a need for route triggers thus a route search in a reactive routing strategy.

Reactive protocols are characterized by the elimination of the conventional routing tables at nodes, and consequently the need of routing table updates to track changes in the network topology. As a result, an on-demand process for discovering routes is a prerequisite; a path discovery is triggered asynchronously when there is a need for data packet and no path to the intended node is known.

The discovery procedure is often based upon a query-reply cycle: the data source node floods the network with a query packet to discover a route to the data destination. Assuming no network disconnections, the destination will be eventually reached by the query. Upon receiving the query, the destination sends a reply back to the source. Since multiple copies of the same query may arrive at the destination via alternate paths, the destination may send more than one reply back to the source, each producing a different route. The source selects then the optimum route, using its own route selection criteria.

The process is completed once a route is found or all possible route permutations have been examined. Once established, the data source uses the route to send its data packets to the destination.

During the data-forwarding phase, routing information is maintained by a maintenance procedure until either the destination becomes inaccessible along all paths from the source or until the route is no longer desired (Royer and Toh, 1999). Should at any time during the data exchange the route maintenance indicates that the route is broken a new route discovery cycle is triggered to set up a new route.

When the communication ends, the source does not attempt to further maintain the connectivity to the destination. Hence, network resources are not expended to maintain routes that are not actively used for data traffic.

2.5.3.1 Properties

On-demand routing strategies create and maintain routes between a pair of source-destination only when necessary. Therefore, in contrast to the proactive approach, in reactive protocols the control overhead as well as routing table storage is drastically reduced, when traffic is sparse (Aggélou, 2004 (Chapter 3); Broch et al., 1998; Das et al., 1998; Das et al., 2000; Das et al., 2000a; Jacquet and Viennot, 2000; Johansson et al., 1999; Maltz et al., 1999; Johnson and Maltz, 1996). On-demand routing does scale well thus to large populations, as it does not maintain a permanent routing entry to each destination.

However, similar to connection-oriented communications, a route is not available when needed but upon completion of the Route Discovery phase. This may introduce an initial route setup latency. This latency may in fact be detrimental to certain applications such as interactive applications (e.g., distributed database queries). Moreover, the quality of the data path (e.g., bandwidth, delay etc.) is not known prior to call setup. It can be discovered only while setting up the path, and must be monitored by all intermediate nodes during the session, thus paying the related latency and overhead penalty. Such a priori knowledge is, however, desirable in multimedia applications for call acceptance control decisions as well as bandwidth negotiations (Iwata et al., 1999).

Because of the long route setup delays as well as the lack of path quality information prior to call set-up, pure reactive routing protocols may not be applicable to real-time communications (Giordano, 2000). Table 2.1 lists some of the basic differences between the two classes of algorithms.

2.5.4 Hybrid Routing (Haas, 1997; Ramanathan and Steenstrup, 1998; Krishna et al., 1997; Lin and Gerla, 1997; Gerla and Tsai, 1995)

Mobility of nodes in infrastructure-less wireless networks raises organizational problems quite different and rather more challenging than those for infrastructured wireless networks. Mobile wireless networks differ in the frequency and degree at which the topology changes. A protocol that works well in one WM^{2}Net may not work well in another with a different density, size, etc. The diverse applications of WM^{2}Nets pose, however, a challenge for a single protocol that operates efficiently across a wide range of operational conditions and network configurations. Purely proactive or purely reactive protocols perform well in a limited region of this range. For example, reactive routing protocols are well suited for networks where the "call to mobility" ratio is relatively low. Proactive routing protocols, on the other hand, are well suited for networks where this

Parameters	On-Demand	Table-Driven
Availability of routing Information	Available when needed	Always available regardless of need
Routing philosophy	Flat	Mostly flat except for cluster switch gateway routing (CSGR)
Periodic route mobility	Not required	Yes
Coping with mobility	Using localized route discovery and in ABR and SSR	Inform other nodes to achieve consistent routing table
Signaling traffic generated	Grows with increasing mobility of active routes (as in ABR)	Greater than that of on-demand routing
QoS support	Few can support QoS	Mainly shortest path as QoS metric

Table 2.1 Comparisons of On-Demand Versus Table-Driven Routing Protocols

ratio is relatively high. The performance of either class of protocols degrades when the protocols are applied to regions of WM^{2}Net space between the two extremes.

Regardless of what type of routing protocol is preferred there will be a set of circumstances under which it will not perform well. Consequently, despite being designed for the same type of underlying network, it does not seem that a routing strategy based exclusively on proactive or reactive routing can achieve the objectives required for WM^{2}Net routing (Royer and Toh, 1999). A desirable design objective for an architectural framework capable of supporting routing in large WM^{2}Nets subject to high rates of node mobility shall balance the tradeoff between reactive and proactive routing while minimizing the shortcomings of each (Bruce McDonald and Taieb Znati, 1999). So, what is ideally needed is a single routing protocol that has the intelligence to adjust its behavior *dynamically* based on the rate of changes (mobility) and the activity (rate of data) as to match the specific mobility/activity ratio.

Researchers advocate that the issue of "efficient operation over a wide range of conditions" can be addressed by a *hybrid* routing approach, where the "proactive" and "reactive" behavior is mixed in the amounts that best match these operational conditions. Given multiple protocols, each suited for a different region of the WM^{2}Net design space, it does makes sense to capitalize on each protocol's strengths by combining them into a single framework (that is, hybridization). In the most basic hybrid framework, one of the protocols would be selected based on its suitability for the specific network's characteristics. Although not an elegant solution, such a framework would perform as well as the best-suited protocol for any scenario and outperform either protocol over the entire WM^{2}Net design space. However, by not using both protocols together, this approach fails to capitalize on the potential synergy that would make the framework perform as well or better than either protocol for any given scenario.

A more promising approach for protocol hybridization is to have the base protocols operate simultaneously, but with different "scopes." For the case of a two-protocol framework, protocol A could operate locally, while the operation of protocol B would be global. The key to this framework is that the local information acquired by protocol A is used by protocol B to operate in a more efficient manner. This framework can be tuned to network behavior simply by adjusting the size of the protocol A's scope. In one extreme configuration, the scope of protocol A is reduced to nothing, leaving protocol B to run by itself. As the scope of protocol A is increased, the information provided to protocol B increases as well, thereby decreasing protocol B's overhead. At the other extreme, protocol A is made global, eliminating the load of protocol B altogether. So, at either extreme, the framework defaults to the operation of an individual protocol. In the wide range of intermediate configurations, the framework performs better than either protocol on its own.

It is worth remarking that routing protocols exhibit, to some extent, some degree of multiscope behavior. Certain proactive routing protocols for instance monitor the status of neighbor connectivity through broadcast beacons, which occur at a faster rate than the global Link-State (or Distance-Vector) advertisements.

To highlight all these issues in practical terms, this sequel describes a new dynamic clustering scheme for coverage-time optimization (DC-CTO) in two-tier WM^{2}Nets. The coverage-time is defined as the time elapsed until the first CH runs out of power. In regulating cluster regions, DC-CTO scheme achieves balanced power consumption among CHs such that energy-rich CHs progressively increase their cluster region to enlarge their coverage area with higher populations of member nodes.

2.5.4.1 Coverage-Time Optimized Dynamic Clustering for Two-Tiered WM^{2}Nets[1]

Dynamic Clustering for DC-CTO Scheme DC-CTO aims to minimize the energy consumption in CHs, while the entire mesh network remains fully connected. To achieve energy efficiency, cluster ranges are adjustable. Energy-efficient radii are calculated based on the results of DC-CTO. DC-CTO consists of three phases: the initial phase, the dynamic cluster control (DCC) phase, and the transmission power allocation (TPA) phase. The following assumptions are made:

- The architecture of WM^{2}Net has a two-tiered hierarchical structure.
- The upper layer comprises the CHs and the lower layer comprises the SNs.
- A sink node is aware of the position of all CHs.

Initial Phase In the initial phase, CHs are deployed randomly to construct a triangle that determines "cluster radius decision points (CRDPs)" as shown in Fig. 2.5. The distance between CRDP and each CH is assumed to be equal to the radius of each cluster. To construct a triangle Delaunay triangulation (de Berg et al., 2000; Aurenhammer, 1991) is used, as it guarantees the construction of equilateral triangles. The construction of equilateral triangles leads to balanced energy consumption of each CH.

[1]Excerpt from the invited article *"Coverage-time optimized dynamic clustering for two-tiered wireless mesh networks," †Joongheon Kim, ‡Wonjun Lee, §Eunkyo Kim, and ¶Timothy K. Shih (*This work was jointly supported by grants from the Korea Science and Engineering Foundation (KOSEF) [R01-2005-000-10267-0] and SK Telecom [KU-R040572]; †Research engineer, Digital Media Research Lab., LG Electronics, Seoul, Korea; ‡Faculty member of the Department of Computer Science and Engineering, Korea University, Seoul, Korea, E-mail: wlee@korea.ac.kr); §Research engineer, LG Electronics Institute of Technology, LG Electronics, Seoul, Korea; ¶Faculty member of the Department of Computer Science and Information Engineering Tamkang University, Taiwan).

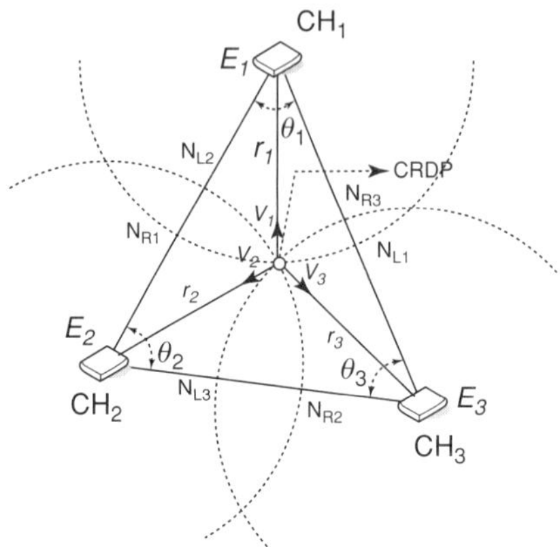

Figure 2.5 System model for a DC-CTO scheme.

Dynamic Cluster Control (DCC) Phase Upon completion of the initial phase, DC-CTO goes to the "dynamic cluster control (DCC) phase." The cluster radii of the three CHs can be dynamically controlled using a CRDP as a pivot. The CH and CRDP distance becomes the radius of each cluster. In this context, the synergy of DC-CTO with DCC can achieve balanced energy consumption among CHs.

In addition to balancing energy consumption among CHs, another goal of the DCC phase is to determine the positions of CRDPs that contribute to the minimum energy consumption in CHs. To achieve this goal, energy-efficient cluster radii are needed. As shown in Fig. 2.5, the triangle is composed of three CHs. If the size of a cluster increases, its CH will consume more energy attributed to the increased population. Therefore, if the overlapping areas of each sector are larger than the optimal, CHs consume more energy than required. It is important thus to find the fewest overlapping areas while full network connectivity is ensured. A CRDP is determined using an energy-constrained objective function that is based on nonlinear programming (NLP) methods with iteration policy. The objective function is illustrated in Eq. (2.1).

$$\min: f = \frac{3}{2}\sum_{i=1}^{3}\theta_i r_i^2 \frac{E_i}{\sum_{j=1}^{3} E_j} - S$$

$$d_i^2 = (x_{\mathrm{CRDP}} - x_i)^2 + (y_{\mathrm{CRDP}} - y_i)^2 \tag{2.1}$$

where S denotes the area of the triangle and E_i denotes the amount of remaining energy at each CH_i.

To minimize coverage overlapping, the objective function accounts for the energy of each CH that composes the Delaunay triangle. As an NLP method for solving Eq. (2.1), a "limited memory BFGS (L-BFGS) method" is used; this is a very efficient NLP method for solving the unconstraint optimization problem. Note that all but the angular values can be transmitted to the sink. CHs should, however, know these, which are computed at the sink node using the second law of cosine. The sink node eventually obtains the energy state and position of each CH.

Algorithm 1 Pseudo Code of DC-CTO scheme

```
loop
  //Expire timer: t
  Initialize(t); //t=0;
  t- -; //Reducing the expire time
  if (event == IDENTIFIED) then
    //If event is occurred in the WSNs
    numEvt = numEvt + 1; //numEvt: recognition counter
    if (numEvt > Threshold) then
      //Threshold: threshold for number of identification
      //Operate DC-CTO scheme: NLP-based approach
      Initialize(numEvt); //numEvt=0;
      Function Call: Initial Phase
      Function Call: DCC Phase
      Function Call: TPA Phase
    end if
    if (t==0) then
      //When expire timer is expired
      //Operate DC-CTO scheme: NLP-based approach
      Initialize(numEvt); //numEvt=0;
      Function Call: Initial Phase
      Function Call: DCC Phase
      Function Call: TPA Phase
    end if
  end if
end loop
```

FIGURE 2.6 The pseudo code of a DC-CTO scheme.

TPA Phase Having calculated the distance between a point and three neighbor CHs, we can now calculate the optimal assigned transmission power of each CH by using EIRP formula. The pseudo-code of a DC-CTO scheme is illustrated in Fig. 2.6.

2.5.5 Position or Location-Aided Routing Protocols (Iwata et al., 1999; Mauve et al., 2001; Jain et al., 2001; Karp and Kung, 2000; Lin and Wang, 1999; Stojmenovic and Lin, 2001)

Position- or geo-based routing protocols use the geographic position of nodes to make routing decisions. Geo-routing is commonly structured around two core functions: the *location estimation function* and the *geographic forwarding function* (Mauve et al., 2001). The former is used to determine the position of the intended destination (sink) node whereas the latter, based on the calculated location information, routes the call to its intended destination.

The specifics of location service and geographic forwarding are discussed in Section 2.5.5.1 and 2.5.5.3, respectively.

2.5.5.1 Methodologies for Location Estimation in WM^{2}Nets

Localization is the process of identifying the location of nodes. There are several ways to localize mesh nodes. Nodes can be localized at deployment time, for instance, using a global positioning system (GPS) receiver that is attached to the object or person deploying the nodes. With mesh devices needed to last for months or even years without battery replacement, traditional GPS-based localization techniques are, however, not suited for these requirements. Running GPS (www.navcen.uscg.gov/gps) on each device is costly and energy prohibitive for a number of applications, not sufficiently robust to jamming for military applications, and limited to outdoor applications. Besides, the receivers at the lowest end give poor accuracy, with inaccuracies of tens of meters possible (Garmin's eTrex, see at www.garmin.com/products/etrex/spec.html). Receivers suitable for operation in WM^{2}Net must have an accuracy of submeter. These, however, come with a price of more than $5,000.

The problem can be remedied, however, by using non-GPS techniques as proposed by Capkun et al. (2001). A survey-grade device can be used to localize nodes after deployment (Girod et al., 2006). A closed-loop system that is equipped with a pan/tilt laser can provide similar accuracy without human intervention (Stoleru et al., 2005; Romer, 2003). Besides, coarse locations can be obtained by simply placing beacon nodes with known positions throughout the deployment area. Nodes can thus estimate their positions based on the beacons within their radio range. Each of these localization techniques achieves a different balance of human effort prior to deployment, node effort after deployment, and localization accuracy.

The following paragraphs discuss techniques for localization using the received signal from one or more reference nodes.

Received Signal Strength It is commonly known that the signal strength of a radio message decreases as the distance from the transmitter increases. The received signal strength (RSS) can be converted to a distance estimate given that there exists a mapping from these RSS indicator (RSSI) values to distances. For RF systems (Bahl and Padmanabhan, 2000; Hightower et al., 2000), problems such as multipath fading, background interference, and irregular signal propagation characteristics make range estimates, however, inaccurate. It remains thus questionable whether the distance can be accurately determined based on signal strength, propagation patterns, and fading models.

On the other hand, RSSI is the cheapest and simplest option available to measure distance as RSSI values come for free with all radio devices.

Time-of-Arrival Time-of-arrival (ToA) is the reception time of a signal (RF, acoustic, or other); that is, the time of transmission plus a propagation-induced time delay. The cornerstone of time-based techniques is the receiver's ability to accurately estimate the arrival time of the line-of-sight (LoS) signal. This estimation is though hampered both by additive noise and multipath signals.

Angle-of-Arrival Angle-of-arrival (AoA) provides information about the direction to neighboring nodes rather than the distance to neighboring nodes. AoA is calculated using RSS and ToA-based methods; thus, it provides complementary information to the ToA and RSS measurements.

There are two common methods that nodes measure AoA (see Fig. 2.7). The first is to use a node array and employ the so-called array signal processing techniques at mesh nodes. In this case, each node embeds a four-element Y-shaped microphone array, as shown in Fig. 2.7a. The AoA is estimated from the differences of arrival times for a transmitted signal in each array element. The estimation is similar to time-delay estimation, as discussed above, but generalized to the case of more than two array elements. When

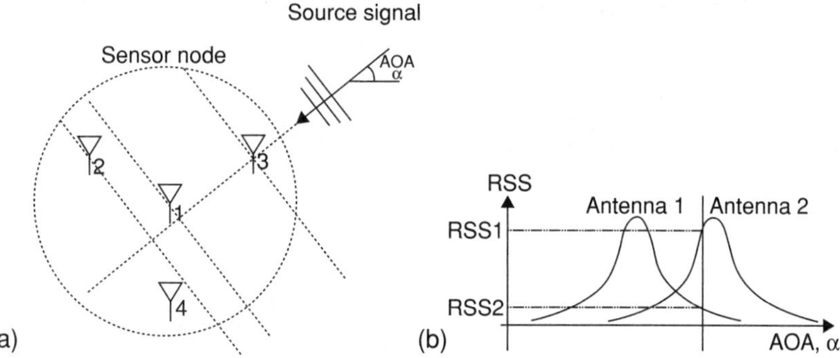

FIGURE 2.7 AoA estimation methods. (a) AoA is estimated from the ToA differences among sensor elements embedded in the node; a 4-element Y-shaped array is shown. (b) AoA can also be estimated from the RSS ratio RSS1/RSS2 between directional antennas.

the impinging signal is narrowband (i.e., its bandwidth is much less than its center frequency), then a time delay τ relates to a phase delay φ with $\varphi = 2\pi f_c \tau$ where f_c is the center frequency. Narrowband AoA estimators are often formulated based on the phase delay.

The second approach to AoA estimation takes advantage of the RSS ratio between two (or more) directional antennas mounted on mesh node (see Fig. 2.7b). Two directional antennas pointed in different directions, such that their main beams overlap, can be used to estimate the AoA using the ratio of their individual RSS values.

Both approaches require the use of multiple antenna elements. This requirement drives up the mesh device cost and size. The reader is referred to the work by Van Veen and Buckley (1988), Stoica and Moses (1997), Ottersten et al. (1993) for further discussions on AoA estimation algorithms and their properties.

Lateration In practice, it may not be feasible for all nodes to be equipped with special purpose location determination hardware, but rather a small fraction of the population. Such nodes, called "anchor nodes," can act as reference points for location information. Mesh nodes use the information from anchor nodes to estimate their own position. A common technique for locating objects using other objects whose position is known is called *lateration.*

In lateration, each node possesses information about the estimated distances (d_i) along with the locations ($x_i\, y_i$) of a series of anchors. This produces the following set of equations (with (x, y) being the location of the node, which is initially unknown):

$$\begin{gathered}(x_1 - x)^2 + (y_1 - y)^2 = d_1^2 \\ \vdots \\ (x_n - x)^2 + (y_n - y)^2 = d_n^2\end{gathered} \tag{2.2}$$

This can be linearized by subtracting the last equation from all others, which results in:

$$\begin{gathered}x_1^2 - x_n^2 - 2(x_1 - x_n)x - \; y_1^2 - y_n^2 - 2(y_1 - y_n)y \; = \; d_1^2 - d_n^2, \\ \vdots \\ x_{n-1}^2 - x_n^2 - 2(x_{n-a} - x_n)x - \; y_{n-1}^2 - y_n^2 - 2(y_{n-1} - y_n)y \; = \; d_{n-1}^2 - d_n^2,\end{gathered} \tag{2.3}$$

This can be reordered to a standard system of linear equations: $Ax = b$:

$$A = \begin{bmatrix} 2(x_1 - x_n) & 2(y_1 - y_n) \\ \vdots & \vdots \\ 2(x_{n-1} - x_n) & 2(y_{n-1} - y_n) \end{bmatrix}$$

$$b = \begin{bmatrix} x_1^2 - x_n^2 + y_1^2 - y_n^2 + d_n^2 - d_1^2 \\ \vdots \\ x_{n-1}^2 - x_n^2 + y_{n-1}^2 - y_n^2 + d_n^2 - d_{n-1}^2 \end{bmatrix} \tag{2.4}$$

This system can then be solved with standard least-squares calculations:

$$\hat{x} = (A^T A)^{-1} A^T b \tag{2.5}$$

The location estimate $\hat{x}$ can be verified by calculating the total residue of the given distances (d_i) and the distances to location estimate $\hat{x}$:

$$residue = \frac{\sum_{i=1}^{n} \sqrt{(x_i - \hat{x})^2 + (y_i - \hat{y})^2} - d_i}{n} \tag{2.6}$$

This residue should be as small as possible and at least smaller than the known radio range.

Min-Max As shown above, lateration requires quite a lot of calculations. Mesh nodes, however, are relatively limited in processing power, as they do not feature a mathematical coprocessor. Min-max is a simpler approach; it does not work with circles but with square bounding boxes around each anchor with the length of a side being twice the distance estimate to that anchor. A node can now combine two bounding boxes of two anchors; the intersection of those two produces its own location.

This process continues for all known anchors and finally the position for the node is calculated as the center of the resulting intersection. All calculations required are only a few simple additions, subtractions, minimum or maximum. Simplicity makes Min-max a very interesting option for running on mesh nodes.

Ad Hoc Positioning Niculescu and Nath (2001) present a new ad hoc positioning method, called the ad hoc positioning system (APS) that extends the capabilities of GPS to non-GPS enabled nodes in a hop-by-hop fashion. Positioning is based on a hybrid method combining distance vector like propagation and GPS triangulation to estimate location in presence of signal strength measurement errors.

Ad hoc positioning is based on the following observation: if a node receives anchor information through multiple other nodes (N_1 and N_2 in Fig. 2.8) with known distance estimates (a and b, respectively) it calculates its own distance to that anchor. Because it is distance a away from N_1 and distance b away from node N_2 there can only be two possible locations for this node. See Fig. 2.8 for more detail. In this respect, APS assumes at least three nodes (called landmarks) that are GPS enhanced, or know their position by some other means.

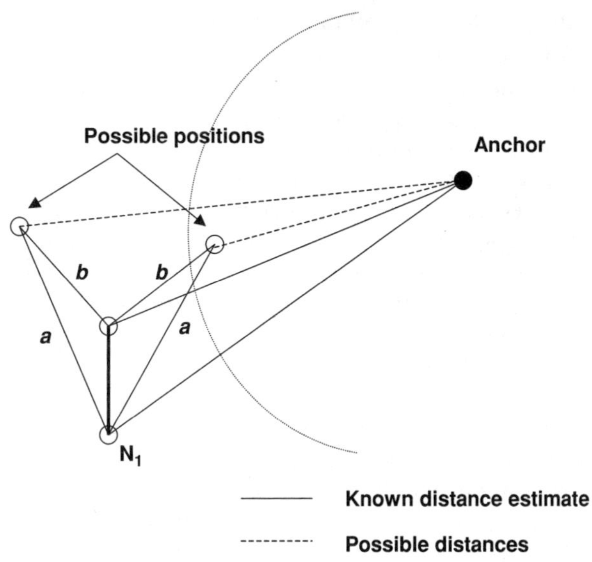

FIGURE 2.8 Ad hoc positioning visualized.

More accurate calculations can be achieved with a higher number of neighbors passing information about this anchor. With a third neighbor (N_3) connected to either one of the previous neighbors, two circles can be drawn again with one other neighbor; this results in again a maximum of two intersections. With two pairs of two intersections, the same intersection should be in both pairs, which is then the correct intersection and that gives the node the distance to the anchor. If the third neighbor is connected to both previous neighbors the node should be able to determine whether or not it is on the same side as this new neighbor of the line $N_1 \leftrightarrow N_2$.

For the second phase, that is, determining location, Ad hoc positioning uses the lateration approach.

2.5.5.2 Using Directional Antennas for Location Estimation

Whereas omnidirectional antennas radiate isotropically in space, directional antennas focus (direct) the transmitting energy in the desired direction. Directionality is commonly achieved through a phased array. A prerequisite is that the elements of the phased array be an appreciable fraction of a wavelength apart. This is not feasible in electrically small form factor microdevices. However, limited directionality can be cheaply integrated using standard patch arrangements with high dielectric constant antennas.

The following paragraphs provide three different techniques to the problem of location estimation with directional antennas using one or more nodes with a known location.

Directional Antenna Model One of the simplest semi-directional antennas is the patch antenna. The ideal patch radiation model is a hemispherical radiator, which allows for semi-directional radiation. The typical gain of a patch antenna is on the order of 3.5–6 dBi, depending on the dielectric substrate used in the design. A representative angular variation of the gain for a typical microstrip antenna is in the range of $\cos^2(\frac{\beta l}{2}\sin(\theta)) < G(\theta) < \cos(\frac{\beta l}{2}\sin(\theta))$, where β is the free-space constant and l is the longest length of patch, assuming the lowest order mode of operation (Clarricoats et al., 1989). The gain is defined as the ratio of the intensity, in a given direction, to the radiation intensity that would be obtained if the power accepted by the antenna were radiated isotropically.

In the E-plane cut, the antenna's radiated e-field from a standard patch radiator is ideally $E = \cos(\frac{\beta l}{2}\sin(\theta))$. This pattern dependence is in relation to a coordinate system with the z-axis perpendicular to the microstrip patch radiator. This is the ideal solution for a patch antenna with an infinite ground plane and is only slightly altered using a finite size ground plane. The ground plane is used to shield the radiating field from the rest of the circuitry and the other radiators. Unshielded radiators, such as those that are standard with the Motes (Scott, 2004), are susceptible to parasitic radiating currents, which result in asymmetric patterns.

The received power at an antenna is given by $P_r = \frac{P_t G_t(\Theta_t) G_r(\Theta_r)}{r^2}\left(\frac{\lambda}{4\pi}\right)^2 \sum$, where Θ_t and Θ_r are the transmitting and the receiving angles, respectively, and r is the distance between the transmitter and the receiver. λ is the RF wavelength of the carrier frequency.

Since $(\lambda/4\pi)^2$ is a constant, we will exclude it from future expressions.

Aligned antennas In a number of practical applications it is reasonable to expect that meshes will be manually deployed. For instance, a WM²Net composed of mesh devices, which are set up to monitor a bridge's health, have to be placed by construction workers on the bridge. In such scenarios even though it may not be possible to know the precise location of the mesh node, it is possible to place these nodes in a predetermined orientation.

If the antennas of target nodes are aligned, then we can use the power received at multiple receiving antennas of the target from a single transmitting antenna on an anchor for position estimation. Without loss of generality, consider that an anchor node is placed to the southeast of the target node as shown in Fig. 2.9.

The size of the mesh node would usually be much smaller than the transmission distance. So $d/r = \Theta_c$. Then the received power at the two receiving antennas of the target node is given by Eqs. (2.7) and (2.8) in two variables Θ_1 and r. Since these are nonlinear equations it is difficult to get a closed form solution for Θ_1 and r in terms of

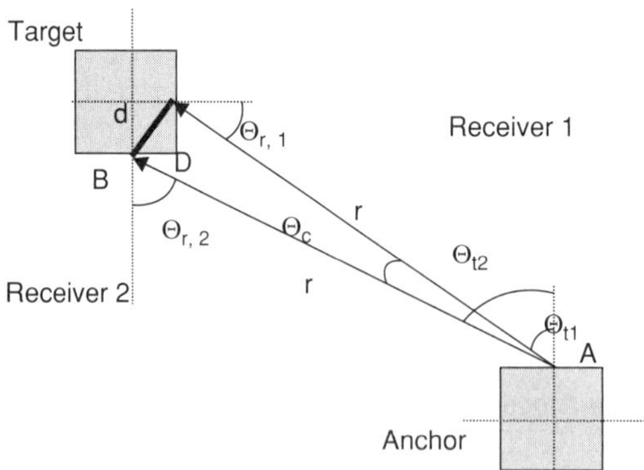

FIGURE 2.9 Location determination with aligned nodes.

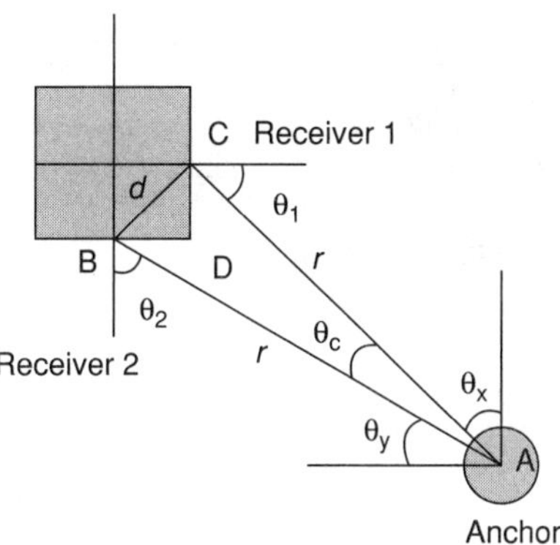

FIGURE 2.10 Location determination with an omnidirectional transmitter and directional receiver.

the input variables $P_{r,1}$ and $P_{r,2}$. However, these equations can be numerically solved by standard methods to obtain Θ_1 and r:

$$P_{r,1} = \frac{P_t}{r^2} G_t(\Theta_{t,1})\, G_r(\Theta_{r,1}) = \frac{P_t}{r^2} G_t\left(\frac{\pi}{2} - \Theta_1\right) G_r(\Theta_1) \tag{2.7}$$

$$\begin{aligned} P_{r,2} &= \frac{P_t}{r^2} G_t(\Theta_{t,2})\, G_r(\Theta_{r,2}) = \frac{P_t}{r^2} G_t\,(\Theta_2)\, G_r(\Theta_2) \\ &= \frac{P_t}{r^2} G_t\left(\frac{\pi}{2} + \frac{d}{r} - \Theta_1\right) G_r\left(\frac{\pi}{2} + \frac{d}{r} - \Theta_1\right) \end{aligned} \tag{2.8}$$

where $\Theta_1 = \Theta_{r,1}$ and $\Theta_2 = \Theta_{r,2}$.

Alternatively, if the orientations of these meshes are not perfect, Θ_1 in Eqs. (2.7) and (2.8) can be replaced by $\Theta_1' = \Theta_1 - \Phi$ unaligned, where Φ unaligned can be obtained from a digital compass (*DMC-SX Digital Magnetic Compass—Operator Manual*, Leica Vectronix AG, Switzerland) or some other simple algorithms (Fang et al., 2005). A possible approach is to mount an omnidirectional antenna with the four directional antennas on the same node and estimating Φ unaligned from the difference of the received power strength between the directional antennas and the omnidirectional antenna.

A baseline experiment for this is with the anchor node having omnidirectional dipole antennas. In this case, the gain of the transmitter, $G_t(\Theta)$, is constant over all Θ and denoted G_{omni}. Figure 2.10 shows this configuration. Now, since we know the distance as well as the relative direction of the target with respect to the anchor, we can estimate its position.

The estimates from multiple anchors can be averaged to obtain a better estimate of the position. Alternatively, the information about Θ_1 could be discarded and the range measurements (r) can be used to triangulate the position of the node in a least squares manner. Both these strategies have been evaluated using computer simulations. The results showed that the averaging strategy yields better results.

Generalization to Unaligned Antennas In cases where it is not possible to ensure a global orientation of all nodes of a network, additional measurements can be used to estimate position. Received power at two different antennas of the target node from two

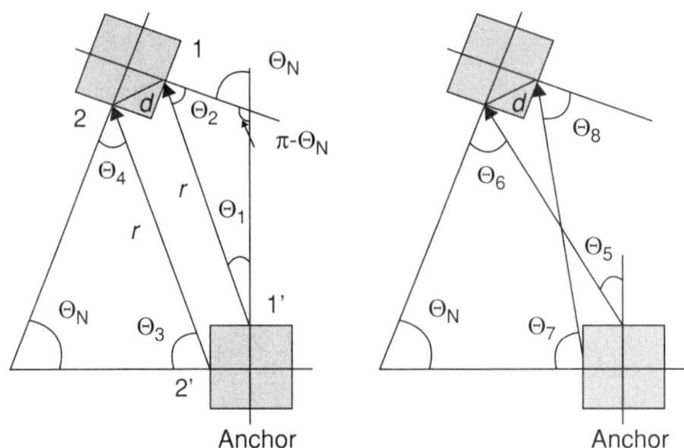

FIGURE 2.11 Location determination for unaligned antennas.

transmitting antennas of the anchor node is measured. Such an arrangement is shown in Fig. 2.11.

Geometric relations between the various transmission and receiving angles can be derived from Fig. 2.11.

$$\Theta_2 + \Theta_6 = \Theta_4 + \Theta_8 = \Theta_3 + \Theta_5 = \Theta_1 + \Theta_7 = \frac{\pi}{2} + \frac{d}{r}$$

$$\Theta_1 + \Theta_2 + \Theta_3 + \Theta_4 = \pi \tag{2.9}$$

$$P_{r,11'} = \frac{P_t^* \, G_t\,(\pi - \Theta_2 - \Theta_3 - \Theta_4)^* \, G_r\,(\Theta_2)}{r^2}$$

$$P_{r,21'} = \frac{P_t^* \, G_t\left(\frac{\pi}{2} + \frac{d}{r} - \Theta_3\right)^* G_r\left(\frac{\pi}{2} + \frac{d}{r} - \Theta_2\right)}{r^2} \tag{2.10}$$

Let $P_{r,ij}$ denote the power received by antenna i on the target node when antenna j is transmitting on the anchor node. We can use these equations to simplify the received power equations as follows.

$$P_{r,12'} = \frac{P_t^* \, G_t\left(\frac{d}{r} + \Theta_2 + \Theta_3 + \Theta_4 - \frac{\pi}{2}\right)^* G_r\left(\frac{\pi}{2} + \frac{d}{r} - \Theta_4\right)}{r^2} \tag{2.11}$$

$$P_{r,22'} = \frac{P_t^* \, G_t\,(\Theta_3)^* \, G_r\,(\Theta_4)}{r^2} \tag{2.12}$$

Equations (2.9) to (2.12) in the four variables Θ_2, Θ_3, Θ_4, and r can again be numerically solved to estimate the location of the target node.

This scheme requires that two target antennas be able to simultaneously receive transmissions from two anchor antennas. This would require a transmitter beam width of 180°. This is nonoptimal for four antennas covering a 360° plane but is a tradeoff for increased degrees of freedom in the orientation of nodes. Besides, the increased beamwidth will lend greater fault tolerance to the system by providing greater redundancy in the areas reached by multiple transmitting antennas. It will also make the antenna design easier since high directionality, that is, narrow beam width is not needed.

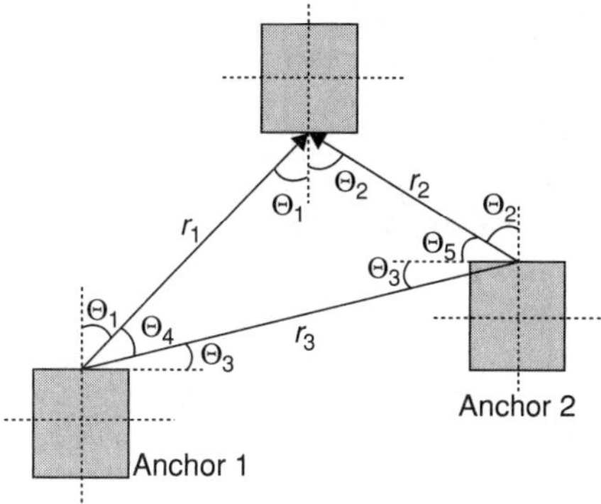

Figure 2.12 Location determination using measurements from two anchors.

Aligned Antennas with Two Anchors The two location determination methods described above rely on the difference in power received at two antennas of a node from the antennas on the same anchor node. The error in the power received can become correlated due to the proximity of the two antennas, even if they are pointed in separate directions. In a real life scenario the correlation can significantly reduce the accuracy of the location estimate, especially for a very small mesh node. To investigate the performance of the location determination with increasingly uncorrelated channels, two transmitted signals were sent from two nodes substantially positioned far enough from each other. This scheme is also useful in situations where more than one directional antenna would not fit on a single mote. The arrangement is shown in Fig. 2.12.

Since the location of the two anchors is known, the parameters r_3 and Θ_3 can be determined. Using geometric properties of the system, we get the following relations between the various angles

$$\Theta_5 = \frac{\pi}{2} - \Theta_2 \qquad \Theta_1 + \Theta_3 + \Theta_4 = \frac{\pi}{2}$$

The equations for the received power are given by

$$P_{r,1} = \frac{P_t}{r_1^2} G_t(\Theta_1) G_r(\Theta_1) \tag{2.13}$$

$$P_{r,2} = \frac{P_t}{r_2^2} G_t(\Theta_2) G_r(\Theta_2) \tag{2.14}$$

Using the law of sines along with relations between the angles derived earlier we get two more equations

$$\frac{r_2}{\cos(\Theta_1 + \Theta_3)} = \frac{r_3}{\sin(\Theta_1 + \Theta_2)} \tag{2.15}$$

$$\frac{r_1}{\cos(\Theta_2 - \Theta_3)} = \frac{r_3}{\sin(\Theta_1 + \Theta_2)} \tag{2.16}$$

This gives us four equations with four unknowns: $r_1, r_2, \Theta_1,$ and Θ_2. Thus, the distance and the angle with respect to each of the two anchors are determined. The mesh node's location can be estimated using either distance or angle pair or the final location that is estimated averaging out all estimates.

2.5.5.3 Methodologies for Geographic Forwarding

There exist three common strategies for geo-forwarding: **greedy forwarding, directed flooding**, and **hierarchical routing**. For the first two strategies, a node with a packet to relay forwards it to one (greedy forwarding) or more (directed flooding) one-hop neighbors that are located closer to the destination than the forwarding node itself. The position of the neighboring nodes is typically learned through a periodic beaconing scheme. That is, all nodes periodically broadcast their one-hop beacons that contain their latest position information. Using this information, the recipient nodes update their own positions, and so on. Each node thus maintains a table with the positions of all its direct neighboring nodes. The third forwarding strategy aims at structuring mesh nodes into a hierarchy. Hierarchical mechanisms may use different types of routing protocols at different levels of the hierarchy (e.g., a nonposition-based routing protocol at one level and a position-based protocol at a different level).

The position information calculated from the location service is included in the header of the packet for forwarding decisions. Intermediate nodes may not need to consult their location table to obtain a more accurate position of the destination, but simply route the packet using the location information carried in its header. However, if an intermediate node maintains a more accurate position for the destination, it may well choose to update the position information in the header of the packet prior to forwarding it.

2.5.5.4 Contention-Based Geographic Forwarding: A Communication Paradigm for Efficient Data Delivery in WM^{2}SNets[2]

Contention-Based Geographic Forwarding (CGF)

Traditional Geographic Forwarding Traditional geographic forwarding encapsulates two modes of operation: *greedy forwarding* and *void handling*. In the greedy forwarding mode, the forwarding decision is based on the location of the sender, the location of the destination, and the locations of the sender's neighboring nodes. The location of the destination is attached in the header of the packet. An intermediate node that receives the packet checks the location of the destination in the header of the packet prior to forwarding it. In case the node maintains more accurate location information for the destination in its database, it is then free to update the location information in the header of the packet. The location of the neighboring nodes is typically learned through a periodic beaconing scheme. That is, all nodes periodically update their own locations and broadcast their one-hop beacons that contain their latest location information. The beaconing scheme is a proactive component and independent of actual data traffic and network dynamics. Nodes maintain a table with the location of their one-hop neighbors as well as of distant nodes. If a sender cannot locate a next-hop node with a positive progress towards the destination, called *a communication void*, it switches to the void handling mode in order to route the packet to around the void.

---

[2] Excerpt from the invited article "Contention-Based Geographic Forwarding: A communication paradigm for efficient data delivery in wireless mesh networks," Dazhi Chen and Pramod K. Varshney, Department of Electrical Engineering and Computer Science, Syracuse University Syracuse, New York, NY 13244, USA.

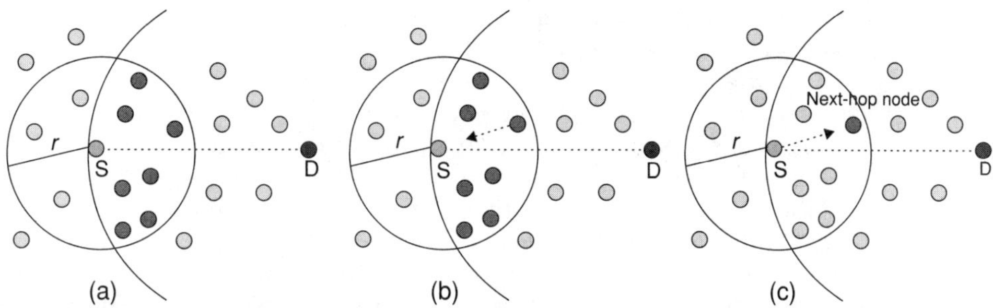

Figure 2.13 A high-level description of CGF. (a) The sender broadcasts a request to its one-hop neighboring nodes. (b) The winner from potential next-hop candidate nodes in the forwarding area confirms with the sender. (c) The sender forwards actual data packet to the established next-hop node.

The proactive nature of the beaconing mechanism induces several limitations, which are observed from three perspectives. First, the periodic transmission of beacons consumes a considerable amount of energy and wireless bandwidth resources unnecessarily. When no data are exchanged for long periods of time, beacons are sent unnecessarily. Second, in a highly dynamic network, a neighboring node selected as a next-hop node may no longer be available before beacons update such information. This leads to significant packet losses and hence to a very high communication overhead due to MAC layer packet retransmissions resulting in increased delivery ratios. In order to obtain more accurate position information, it is likely to increase the beaconing rate. However, this would result to significant increase of communication overhead. On the other hand, in a relatively static network, where the network infrastructure remains unchanged for long periods of time, beacons are still sent out at periodic intervals.

Contention-Based Geographic Forwarding To address the aforementioned performance concerns, a new geographic forwarding paradigm, called CGF[3] (Chen et al., 2005), is proposed aiming to eliminate the beaconing scheme. In contrast to traditional geographic forwarding, a distributed contention-based mechanism is used in CGF to select a next-hop node among a group of neighboring nodes.

Figure 2.13 illustrates the basic procedure of the CGF paradigm. To forward a data packet, a sender broadcasts a control packet requesting its one-hop neighboring nodes to relay its data. The neighboring nodes that are eligible to forward the data packet, for example, they are within a forwarding area which is expected to make a positive geographic progress towards the destination, are called next-hop candidate nodes. The sender then unicasts the actual data packet to the next-hop node that won the contention. If a communication void occurs, that is, if no next-hop node appears in the forwarding area, a void handling scheme is employed to route the packet around the void. This process is repeated until the packet is successfully delivered to its destination.

Based on the high-level description, CGF mainly consists of the following components:

[3] It is also called random forwarding, implicit forwarding, volunteer forwarding, and contention-based forwarding (Zorzi and Rao 2003; Blum et al. 2003; Chen et al., 2005a; Füßler et al., 2003; Heissenbüttel et al., 2004, Ferrara et al. 2005, Xu et al., 2005).

(1) **A predefined forwarding area.** A forwarding area is a geographical area with respect to the sender and the destination of the packet. Nodes that lie in the area are eligible next-hop candidates and contend with one another for the data-forwarding task. A common requirement for a forwarding area in CGF is that each transmission should make a positive progress towards the destination, unless a void occurs. Forwarding areas of different sizes, shapes, and geographic locations are defined in existing CGF protocols, including *maximum forwarding area (MFA)*, *maximum communication area (MCA)*, *60-degree radian area (DRA)*, and *Reuleaux triangle area (RTA)* (Chen et al., 2005; Zorzi and Rao, 2003; Blum et al., 2003; Chen et al., 2005a; Füßler et al., 2003; Heissenbüttel et al., 2004; Ferrara et al., 2005; Xu et al., 2005).

(2) **A distributed next-hop node selection scheme.** A distributed selection scheme is used to effectively establish a *single* next-hop node among a local set of next-hop candidate nodes in the forwarding area. In CGF, distributed selection is implemented via distributed contention arbitration and resolution among next-hop candidate nodes. Since the sender does not maintain the information of neighboring nodes, such information must be accessed instantly when needed. However, if all candidate nodes send back their latest information to the sender upon request, for the purpose of a centralized selection, it would be rather difficult to control communication overhead, provided that a large number of next-hop candidate nodes exist in the forwarding area. Thus, the establishment of a next-hop node should be accomplished either in a fully distributed manner (i.e., entirely by receivers themselves) or in a partially distributed manner (i.e., with partial assistance from the sender). During the contention phase, all candidate nodes independently make their own decision according to their current state; eventually only a single next-hop node is established. Note that distributed next-hop node selection in CGF is an instance for a more general local leader election problem (Chen et al., 2005) in distributed computing, where only a local leader needs to be elected from a small group of entities that can directly or indirectly communicate with each other. Given that packet duplication in wireless networks may count as a benefit to defend against packet loss, as pointed out by Chen et al. (2005), a solution for leader election in wireless networks may not require a strict restriction, that is, one and only one leader is elected.

(3) **Next-hop node selection criteria.** Geographic criteria, similar to those used in traditional geographic forwarding with the centralized selection of a next-hop node at the sender, can be applied to CGF with a distributed selection of the next-hop node among receivers as well. These criteria include *most forward within r (MFR)*, *nearest with forward progress (NFP)*, *compass routing*, and *random selection* (Mauve et al., 2001). Nongeographic criteria such as *node energy reserve* and *link reliability* can also be exploited especially if CGF is used for data delivery in extremely resource-constrained wireless networks, such as sensor networks (Blum et al., 2003; Chen et al., 2005a; Ferrara et al., 2005).

(4) **A void handling scheme.** Communication voids occur when greedy forwarding fails to locate a next-hop node in the forwarding area, even though one topologically valid route still exists. Prior to initiating a void handling scheme, a node should first determine whether the void is temporary due to packet collisions or unavailability of next-hop nodes. The problem of temporary voids will be solved as nodes wake up, collisions are resolved, or mobile nodes move in. Such

Component of CGF	Design Choices
Forwarding area	• MFA • MCA • DRA • RTA
Distributed node selection	• Fully distributed contention • Partially distributed contention
Node selection criterion	• MFR • NFP • Compass routing • Random selection • Node energy reserve • Link reliability • Hybrid
Void handling	• Passive participation • Active exploration • Void avoidance • One-hop flooding • Other void handling schemes provided by Chen and Varshney (2007)

Table 2.2 Design Choices for CGF

temporary voids only affect average packet delay. The void is not temporary; a void handling scheme is invoked. Most void handling solutions proposed for traditional geographic forwarding, such as *perimeter routing* in GPSR (Karp and Kung, 2000), *BOUNDHOLE* (Fang et al., 2004), and *cost-based forwarding* in PAGER-M (Zou et al., 2004), require nodes to collect local topology information periodically via beacons. These solutions are not a good design choice for the CGF paradigm, as a periodic beaconing scheme is used. Accordingly, some approaches that are reactive in nature and do not rely on a periodic beaconing scheme, such as *passive participation* (Zorzi and Rao, 2003; Chen et al., 2005a), *active exploration* (Chen et al., 2005a; Ferrara et al., 2005), *void avoidance* (He et al., 2003), and *one-hop flooding* (Stojmenovic and Lin, 2001a), are a better fit for the CGF paradigm. More design choices of a void handling scheme are available in the work by Chen and Varshney (2007).

The design choices of the above CGF components are summarized in Table 2.2. As demonstrated, there exist multiple design choices for each individual component. An optimum choice depends on the desirable features of a CGF protocol as well as on the unique characteristics of targeted wireless networks and applications.

On-Demand Geographic Forwarding: An Example for CGF Protocol Design A practical CGF protocol, called on-demand geographic forwarding (Chen and Varshney 2006), is proposed for efficient data delivery in large-scale, resource-constrained static wireless mobile mesh sensor networks (WM^{2}SNets) with unreliable sensors. In such networks, network dynamics is not due to node mobility but is mostly induced instead from unreliable sensors, that is, sensor failures, deaths, and additions.

Similar to CGF, the design objectives of on-demand geographic forwarding (OGF) include simplicity, energy-efficiency, scalability, and robustness to unreliable nodes. The main components of OGF are described below.

Forwarding Area The MFA to ensure a positive progress is a design choice in OGF. It is defined as the overlap region of two circular areas: the transmission circle of the sender and the circle that is centered at the data sink with a radius equal to the distance between the sender and the sink, shown as the shaded region in Fig. 2.14. The size of this area depends on the transmission radius of the sender and the distance between the sender and the data sink. Thus, it is variable. Any mesh node within this forwarding area has a shorter distance than that from the sender S to the data sink. Any active mesh

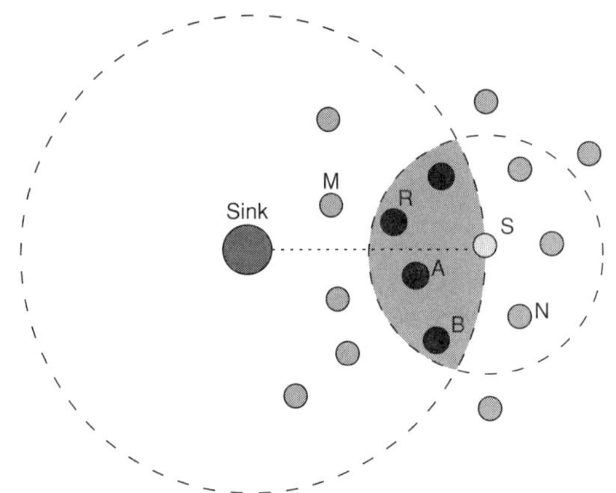

FIGURE 2.14 MFA.

node within this area is a next-hop candidate node with respect to the sender S. A group of contending nodes consists of nodes located within this area, and the number of nodes in this group varies with time, location, and node deployment. Since all nodes know the distance between themselves and the data sink, they can easily determine whether they resides in this region or not using the position information contained in the sender's request-to-forward (RTF) control packet.

Distributed Next-Hop Node Selection Due to the broadcast nature of wireless channels, the transmission of a packet from the sender is simultaneously received by a group of contending nodes. In this situation, if each node records the time instant at which such a packet is received, then the nodes in this group that received the same packet are implicitly synchronized at these time instants. These implicit synchronization points are valuable because, without the use of any costly time synchronization protocols, we can propose a timer-based scheme to establish a next-hop node in a distributed fashion. In OGF, the task of distributed selection is integrated into a handshake at the MAC layer. Thus, OGF is a cross-layer protocol merging tasks across both routing and MAC layers. This timer-based scheme is a partially distributed scheme, which includes a contention arbitration strategy entirely among receivers and a contention resolution strategy with assistance from the sender.

As shown in Fig. 2.15, whenever the sender S has a packet to forward, it senses the medium physically and virtually. If the medium is determined to be free for an interhandshake space time, which must be larger than an intrahandshake space time, the sender broadcasts an RTF to all the neighboring nodes in its transmission range. Otherwise, it defers from transmitting and backs off. Either the position information or the distance-to-sink of the sender S is carried in the RTF packet. The nodes receiving this RTF packet compare their distance-to-sink with the sender's announced value. Those nodes with smaller distances-to-sink such as nodes R, A, and B in Fig. 2.14 automatically become next-hop candidate nodes. Each candidate node sets a contention response timer, which should elapse before it is allowed to reply a confirm-to-forward (CTF) packet corresponding to the RTF packet. In fact, the contention response time assigns different priority to each candidate node. Details of the contention response timer are discussed in

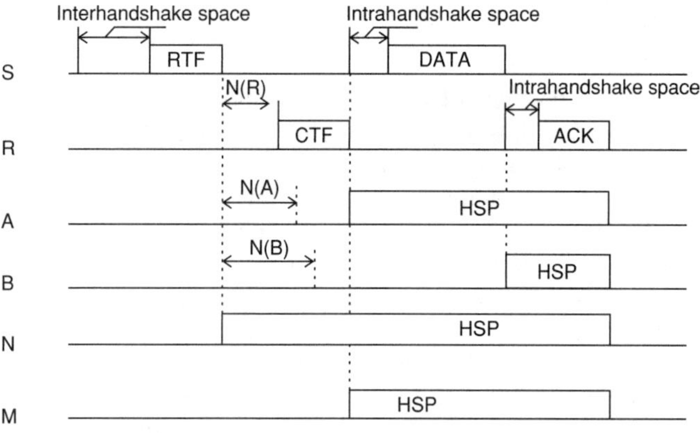

FIGURE 2.15 Distributed node selection through MAC handshake.

the following paragraphs. Other nodes that are outside this forwarding area but within the communication range of S, for example, node N, set their handshake silence period (HSP) values appropriately.[4] The node with the earliest timeout, for example, node R in Fig. 2.15, responds to the sender S with a CTF, which confirms that it would like to forward the data packet from the sender. Then the sender S forwards its data packet directly to the next-hop node R established previously. After receiving the data packet, node R replies with an ACK to indicate the end of this handshake.

In OGF, the next-hop candidate nodes monitor the channel for all transmissions during their contention response times. Whenever a CTF is heard, they conclude that another node with higher priority in the forwarding area has sent out a CTF response. They cancel their CTF responses and quit the contention. OGF, however, cannot guarantee that all candidate node pairs in the forwarding area can hear each other. Hence, only when a transmission carrier is sensed, they will assume that another node with higher priority exist in communication proximity. They will then defer their CTF replies. If a following data packet is heard, they will quit from contention. Note that the maximum distance between any two-candidate nodes in the forwarding area is less than twice the value of the transmission range. The correct operation of OGF on the contention arbitration can be guaranteed by adjusting the carrier sensing range to be at least twice the transmission range in node radios.

Furthermore, it is possible for two CTF responses to collide when two or more candidate nodes have the same response time delay. For this reason, an explicit contention resolution strategy is employed to handle the collisions of CTF response. Note that the sender can find out if a response collision has occurred, by checking the current MAC state. If a collision occurs, then the sender retransmits the RTF packet. The new RTF packet carries an additional bit of information to signal the contending nodes that there was a response collision. Only those candidate nodes that just sent out the CTF responses are allowed to participate in the new contention. The other candidate nodes keep silent since, in reception of the RTF packet, they are informed that there are other candidate

[4] HSP is similar to network allocation vector (NAV) used in virtual carrier sensing in IEEE 802.11.

nodes in the neighborhood with higher priority. The contention process reiterates until the collision is resolved and the sender establishes a single next-hop node.

Next-Hop Node Selection Criteria Multiple node selection criteria, including MFR, random selection, and node energy reserve, are employed together to determine a next-hop node in OGF. In OGF, these criteria are incorporated into the calculation of the contention response time, which is defined as the amount of time for which a next-hop candidate node should wait before it replies to the RTF packet (with an CTF packet). The response time is closely related to the priority level of a next-hop candidate node in the contention. Here, we introduce the function used in OGF, although a subtler function can be designed for further optimization. The basic idea of this example function is to force the nodes that are closer to the data sink and have more residual energy to take the forwarding responsibility with a higher probability. The introduction of a random value is to further disperse the system workload. The function is presented in detail below:

$$C_{\text{abs}} = \left[W_{\text{d}} \times \left(1 - \frac{L}{T}\right) + W_{\text{e}} \times \left(1 - \frac{R_{\text{e}}}{E}\right) + W_{\text{r}} \times V \right] \times M,$$

$$N_{\text{slot}} = \left\lfloor C_{\text{abs}} \big/ t_{\text{slot}} \right\rfloor,$$

where

M = the interhandshake space time (i.e., maximum contention response time);
L = distance-to-sink of the sender—distance-to-sink of a forwarding node;
T = transmission range of the sender;
R_{e} = remaining energy reserve;
E = maximum energy reserve;
V = random value in (0,1);
$W_{\text{d}}, W_{\text{e}}, W_{\text{r}}$ = weights assigned to distance, energy, and random value;
$W_{\text{d}} + W_{\text{e}} + W_{\text{r}}$ = 1;
C_{abs} = absolute contention response time;
t_{slot} = slot duration; and
N_{slot} = number of time slots in a contention response time.

Note that a CTF response packet is transmitted only at the beginning of a time slot and the value of slot duration is a protocol parameter in OGF.

Void Handling An on-demand and state-less void handling scheme, called *partial source routing* (PSR) is designed for OGF. PSR enables a void node (i.e., a node that encounters a void) to forward a data packet to either the data sink through its next-hop node or to a node with a shorter distance to the data sink than itself. The method consists of two phases: partial path discovery and source packet forwarding.

The partial path discovery phase at a void node uses a method somewhat similar to *expanded ring search* (Aggélou, 2004). This phase identifies a partial path to the data sink. In the first run, the time-to-live (TTL) field in a partial path request packet is set to two, that is, the request packet can only be broadcast to nodes that are two hops away from the void node. If this discovery fails, the void node initiates a partial path request with the TTL value increased by one, that is, it extends the flooding range by one more

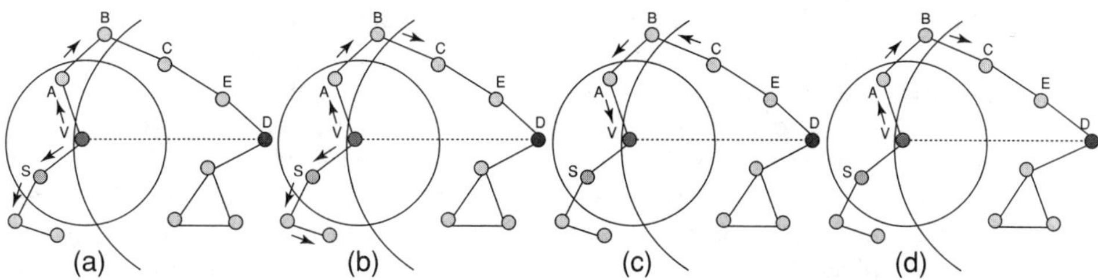

Figure 2.16 An illustrative example to demonstrate how PSR handles a void in OGF. (a) A partial path request packet is initiated by V with TTL = 2, this discovery fails. (b) Another partial path request packet is initiated with TTL = 3. (c) A partial path request packet is sent back from node C, which is closer to data sink D. (d) The stuck data packet at V is then forwarded to node C by source packet forwarding. Greed forwarding is reactivated for the packet at this time.

hop than the previous run. The maximum number of actual runs is a protocol parameter. If the void node receives one or more path reply packets, then the void node knows at least one partial path to the data sink or to another node with a better path to the sink. The node then selects the shortest path and goes to the source-forwarding phase.

In the source-forwarding phase, the void node attaches the full path in the data packet's header and forwards the packet to its next-hop node. The intermediate next-hop node scans the specified source path and forwards the packet over the specified hop. Should the packet be delivered to the data sink directly by PSR, the forwarding phase is then resumed. If not, a void is resolved. After the packet is delivered to a better-positioned node, that node will again start to forward the data packet using greedy forwarding until the packet is delivered to the data sink or the packet faces another communication void. If another void is encountered, the above process reiterates. Figure 2.16 illustrates an example how to handle a void using PSR.

Forwarding Table: Another Design Consideration In WM^{2}Snet context for which OGF is designed, the rate of topology changes due to unreliable nodes is at the time-scale of minutes, hours, and days, which in most situations is much smaller than the data transmission rate in WM^{2}Snet applications. Thus, the information about a previously established next-hop node may only need to be updated after a fairly large number of data packet transmissions. A forwarding table of small size is employed to store previously learned next-hop node information for data sinks currently in use, as shown in Table 2.3. The idea is similar to a routing cache used in on-demand routing protocols, such as the DSR (Johnson and Maltz, 1996); it basically stores the routing information carried in route discovery packets.

ID	Position	Next-hop ID
Data sink ID 1	(X_1, Y_1)	WM^{2}Snet node identifier
⋮	⋮	⋮
Data sink ID n	(X_n, Y_n)	WM^{2}Snet node identifier

Table 2.3 A Forwarding Table in OGF

Initially, there is no information available to reach the data sink. Thus, a distributed selection via contention is initiated and next-hop node information is learned. This is stored in the forwarding table. A timer is associated to control the lifetime of the entry. If a subsequent data packet at the sender needs to be delivered to the same data sink, the sender can exploit the next-hop information in the forwarding table to directly unicast the data packet. If the unicast is successful, the timer related to the entry is restarted. If a unicast fails, OGF assumes that the next-hop information is stale and initiates the contention phase to discover a new next-hop node to forward its data packets. When the timer expires and no data packet needs to be delivered to that data sink, the entry in the forwarding table is deleted. The expiration time of a timer is a protocol parameter.

2.5.5.5 On the Critical Connectivity Radii in WM^2Snets[5]

Related Background on Modeling WM^2Snets using Graph Theory A graph theoretical model of WM^2Snets is first introduced. Although most of the issues and theories discussed throughout this work can be extended to 3D ($\Re^3$), we focus on 2D ($\Re^2$) problems to simplify the illustration of the basic concepts.

A graphical model of a WM^2Snet can be built by using a vertex in the graph to represent every node and an edge in the graph to represent every node pair for which the two nodes can directly communicate with each other. The resulting graph G = (V, E), where V is the **vertex set** and E is the **edge set**, is called the **underlying graph** of the network. Let | X | denote the number of elements in a set X; then | V | (| E |) represents the number of vertices (edges) in the graph G = (V, E).

A widely used graphical model of a WM^2Snet is the **unit disk graph**. Suppose in the WM^2Snet any two nodes can directly communicate with each other if and only if their Euclidean distance is smaller than a given threshold r, known as the **connectivity radius** of the WM^2Snet. Then the associated graphical model is known as a unit disk graph.

Another important concept in modeling WM^2Snets by graphs rests on the fact that in many WM^2Snets, nodes are effectively randomly located. Suppose that nodes are independently and randomly distributed in a d-dimensional space (d equals 2 unless otherwise specified) following either a uniform distribution, parameterized by the number of nodes in a defined region, typically the unit square, or a Poisson distribution, parameterized by a node density per unit area. Any two nodes i and j can communicate directly with each other if and only if their Euclidean distance is smaller than a given threshold r. The underlying graph of that WM^2Snet is called a **random geometric graph** (Penrose, 2003), which is denoted by $G_n(r)$, or $G_n(r(n))$ when we need to emphasize the dependence of the connectivity radius r on n, as will occur later. The parameter n denotes either the total number of nodes in the case of a uniform distribution or the node density per unit area in the case of a Poisson distribution.

In the following section we provide answers to the above questions using properties of graph theory.

---

[5]Excerpt from the invited article *"On the critical connectivity radii in wireless mesh sensor networks," †Guoqiang Mao, ‡Brian D. O. Anderson and Barıs Fidan, §Jia Fang and A. Stephen Morse (*The work of G. Mao, B. D. O. Anderson and B. Fidan was supported by National ICT Australia. The work of J. Fang and A. S. Morse was supported in part, by grants from the U.S. Army Research Office and the U.S. National Science Foundation and the Xerox Corporation; †University of Sydney and National ICT Australia, Sydney, Australia; ‡Australian National University and National ICT Australia, Canberra, Australia; §Yale University).

Using Graph Theory to Model WM²Snets Whether or not one is working with random networks, a number of WM²Snet problems can be studied in the framework of graph theory. For example, in power control, a critical problem is the minimum transmitter power required to achieve a connected network (Gupta and Kumar, 1998) in which each node in the network has a path to all nodes in the network. If the transmitter power is too large, it may result in excessive interference and power consumption. On the other hand, a small transmitter power may result in the network degenerating into isolated components. Because of the relation between the transmitter power and the connectivity radius, the problem can be cast into graph theory as establishing the minimum r required for a set of nodes for which the underlying unit disk graph is a connected graph. When the graph is random, the question becomes one of determining the minimum r required to guarantee with a prescribed probability, usually $1 - \varepsilon$ for some prescribed small ε, that $G_n(r)$ is connected; alternatively the value $r(n)$ is sought for which, as $n \to \infty$, $G_n(r(n))$ is connected, while $G_n(r(n) - \varepsilon)$ is not connected for any $\varepsilon > 0$. In routing, a problem of concern is that there exist k independent paths between any two nodes for some positive integer k. For a random graph, the problem can be formulated as given the connectivity radius r, determining the probability that $G_n(r)$ is a k-connected graph, or what is virtually equivalent, given a probability that $G_n(r)$ is k-connected, determining the value of r that could ensure this.

Another important WM²Snet problem that can be studied in the framework of graph theory is localization. In particular, there is a large class of distance-based localization algorithms that estimate the Euclidean positions of all nodes in a network given the knowledge of the Euclidean positions of some nodes (i.e., anchors) as well as internode distances between certain pairs of nodes (Mao et al., 2006).

The distance-based localization problem can be formulated using graph theory as follows. Consider a WM²Snet with a set V of nodes, a set D of known distances d_{ij} between certain pairs of nodes $v_i, v_j \in \mathrm{V}\,(i, j \in \{1, \ldots, |\mathrm{V}|\}, i \neq j)$, and a set V_a of anchors v_{a_i} $(a_i \in \{1, \ldots, |\mathrm{V}_a|, \mathrm{V}_a \subset \mathrm{V})$ whose Euclidean coordinates $p(a_i)$ are known. Note that the distances between anchors are given implicitly and hence are known values. The localization problem is one of finding a mapping $p : \mathrm{V} \to \Re^{\mathrm{d}}$ which assigns coordinates $p(i) \in \Re^{\mathrm{d}}$ to each node v_i such that $\|p(i) - p(j)\| = d_{ij}$ holds for all pairs i, j for which d_{ij} is given, and the assignment is consistent with the known anchor positions. The greatest interest is in the case when there is only a single assignment $p(i)$ possible for every i, that is, the case when the network is uniquely localizable. This is illustrated in Fig. 2.17.

The WM²Snet localization problem can be split into two fundamental problems: an analytic existence/solvability problem and an algorithmic problem. The analytic existence/solvability problem is concerned with the properties of the network required to uniquely localize all nodes. The algorithmic problem is concerned with the design of efficient algorithms to solve the localization problem, the computational complexity of the localization algorithms, and the properties of the WM²Snet that can be exploited to simplify the computational complexity of the localization algorithms.

In the following section we show that answers to the earlier questions can be found in the framework of graph theory.

Graphical Properties of Uniquely Localizable Networks A fundamental problem in localization is whether every node in a WM²Snet with a given set of nodes and internode distances is uniquely localizable. Unique localizability is a fundamental property of the WM²Snet, which does not depend on specific localization algorithms. The unique

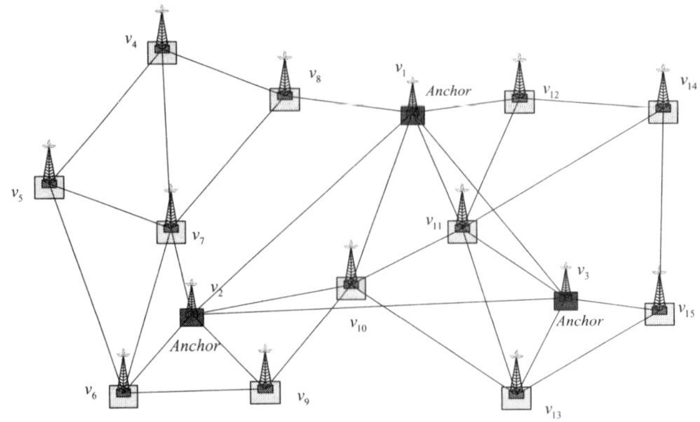

FIGURE 2.17 An illustration of the 2D WM2Snet localization problem. Absolute positions of three anchors v_1, v_2, and v_3 are known. The distances between can be calculated. The distances, d_{ij}, between each node pair v_i and v_j connected with a solid line are known. The goal is to estimate the position of every node v_i in the network that is consistent with the known distances and anchor positions.

localizability problem has been studied in the framework of graph theory (Goldenberg et al., 2005; Eren et al., 2004; Aspnes et al., 2006; Anderson et al., 2006). In this section, we shall investigate the graphical properties of a uniquely localizable WM^{2}Snet.

A graphical model G(V, E) of a WM^{2}Snet can be built using the procedure described in the last section. A d-dimensional **framework** G(V, p) is a graph G(V, E) together with a mapping $p : V \rightarrow \Re^d$. A framework is called a **realization** if the associated mapping p satisfies $\| p(i) - p(j) \| = d_{ij}$ for all pairs of $i, j \in V$ where there is an edge between i and j. Two frameworks G(V, p) and $G(V, q)$ are **equivalent** if $\| p(i) - p(j) \| = \| q(i) - q(j) \|$ holds for every pair $i, j \in V$ connected by an edge. Two frameworks G(V, E) and $G(V, q)$ are **congruent** if $\| p(i) - p(j) \| = \| q(i) - q(j) \|$ holds for every pair $i, j \in V$ no matter whether there is an edge between them. For two congruent frameworks, one can be obtained from the other by applying one or more of a translation, rotation and reflection.

A framework G(V, p) is called **generic** if the set containing the coordinates of all its points is algebraically independent over the rationales. A framework G(V, p) is called **rigid** if there exists a sufficiently small positive constant ε such that if G(V, p) is equivalent to G(V, q) and $\| p(i) - q(i) \| < \varepsilon$ for all $i \in V$, then G(V, q) is congruent to G(V, p). The closeness qualification produced by ε is critical in the definition, which is to be contrasted with the definition below of global rigidity. A graph G(V, E) is called **rigid** if there is an associated framework G(V, p) that is generic and rigid. Intuitively, if the underlying graph of a WM^{2}Snet (with at least three anchors in generic positions in $\Re^2$) is rigid, there can only be a finite number of solutions to the localization problem and there is no continuous deformation that can move a node (nodes) to a different position (positions) while satisfying the distance constraints. If the underlying graph is nonrigid, there are an infinite number of solutions to the localization problem.

A framework G(V, p) is **globally rigid** if every framework equivalent to G(V, p) is also congruent to G(V, p). A graph G(V, E) is called **globally rigid** if there is an associated framework G(V, p) that is generic and globally rigid. If the underlying graph of a WM^{2}Snet (with at least three anchors in generic positions in $\Re^2$) is globally rigid and the associated framework is generic, there can only be one solution to the localization problem, that is, every node in the network is uniquely localizable. Note that rigidity is a generic property in $\Re^2$. In other words, provided the mapping p is generic, the graph G alone determines the rigidity of the framework. One can speak of a rigid graph: this would be a graph such that a generic framework corresponding to the graph is rigid. Global rigidity is also known to be a generic property in $\Re^2$.

For a WM^{2}Snet containing three or more anchors in generic positions in $\Re^2$ and whose underlying graph is globally rigid, any generic framework associated with the graph is also globally rigid, that is, the WM^{2}Snet is unique localizable; and many nongeneric frameworks associated with the graph may also be globally rigid.

For a WM^{2}Snet containing three or more anchors in generic positions in $\Re^2$, and whose underlying graph is a unit disk graph, global rigidity of the graph is only a sufficient condition for unique localization and rigidity of the graph is a necessary condition for unique localization. Only in **generic situations** where a priori information is not helpful, global rigidity is both a sufficient condition and a necessary condition. An example of a nongeneric situation can arise in a WM^{2}Snet whose underlying graph is a unit disk graph. For such a WM^{2}Snet, even when its underlying graph is a rigid graph but not a globally rigid graph, it may still be uniquely localizable because the ambiguities associated with the nonglobally rigid nature of the underlying graph may sometimes be eliminated using the unit disk graph properties. This is illustrated in Fig. 2.18.

The following theorem summarizes the aforemetioned discussions:

Theorem 3.1 (Eren et al., 2004; Aspnes et al., 2006): Consider a 2-dimensional WM^{2}Snet with at least three anchors generically positioned in $\Re^2$. In generic situations where a priori information is not helpful, the necessary and sufficient condition for unique localization of the WM^{2}Snet is that its underlying graph is globally rigid. Otherwise,

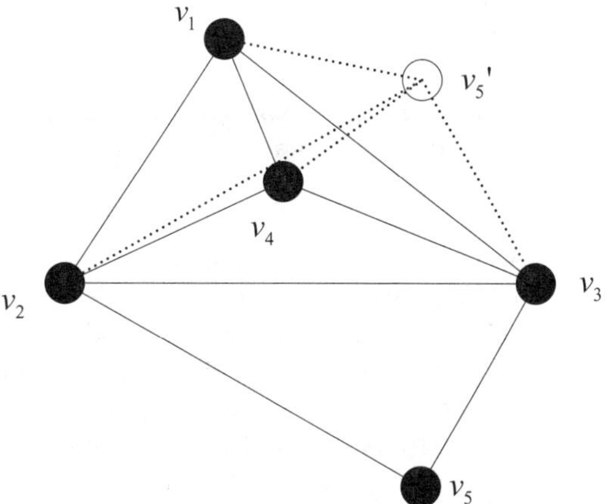

Figure 2.18 An example showing that a WM2Snet with a nonglobally rigid unit-disk graph in $\Re^2$ can still be possibly uniquely localized. There are five nodes and the solid lines represent the known internode distances. The anchors are v_1, v_2, and v_3. The graph is not globally rigid and the position of the fifth node can be either at v_5 or while satisfying the distance constraints. However, using the knowledge that the graph is a unit disk graph, if the fifth node is at v'_5, it will become the neighbor of nodes 1 and 4, a fact which is inconsistent with the knowledge of internode distances. Therefore, the fifth node can be uniquely localized at v_5 only.

if the underlying graph has the property of being a unit disk graph, then a sufficient condition for unique localization is that the underlying graph is globally rigid and a necessary condition is that the underlying graph is rigid.

In practice, given that a wireless node has a limited transmission range, a WM²Snet can be regarded as a unit disk graph. Therefore, global rigidity is only a sufficient condition for unique localization of a WM²Snet. Determining the necessary conditions for unique localization remains a challenging research issue.

Test for Uniquely Localizable Networks In the previous paragraphs, we have established the necessary and sufficient conditions for a uniquely localizable WM²Snet. A question of practical significance is: How do we test the underlying graph of a given WM²Snet (with the associated internode distances) for the property of rigidity or of global rigidity?

Generally, rigidity can be tested by examining the rank of a matrix whose entries are formed from the coordinates of the vertices (Eren et al., 2004). In $\Re^2$, an alternative well-known combinatorial necessary and sufficient condition for rigidity is given in Laman's theorem (Laman, 1970):

Theorem 3.2 (Laman's Theorem): A graph G(V, E) in $\Re^2$, where $|V| > 1$, is rigid if and only if there exists a subset $E' \subseteq E$ satisfying the following two conditions:

1. $|E'| = 2|V| - 3$;
2. For any nonempty subset $E'' \subseteq E'$, $|E''| \leq 2|V(E'')| - 3$, where $V(E'')$ denotes the set of all end vertices of the edges in E''.

Currently, no such combinatorial necessary and sufficient condition is available in $\Re^3$.

In $\Re^2$, an elegant necessary and sufficient condition for the global rigidity of a graph is given in the following:

Theorem 3.3 (Hendrickson, 1992; Jackson and Jordan, 2005): A graph G(V, E) in $\Re^2$ with $|V| \geq 4$ vertices is globally rigid if and only if it is ***redundantly rigid*** and ***3-connected*** (see below).

A graph is **redundantly rigid** if it remains rigid with the removal of any edge in the graph. In $\Re^2$, there is a variant of Laman' theorem called the pebble game (Jacobs and Hendrickson, 1997) for simply testing rigidity and redundant rigidity (Goldenberg, 2006). A graph is k**-connected** if there exist at least k paths which have no edge or vertices in common (apart from the end vertices) between any two vertices, or equivalently, it is not possible to find $k-1$ vertices whose removal would render the graph disconnected. Testing of 2-connectivity and 3-connectivity is discussed below.

In addition, for random graphs, there is a probabilistic but asymptotic result linking k-connectedness and the minimum node degree, as given in the following theorem.

Theorem 3.4. (Penrose, 2003, 1999): Consider a WM²Snet whose underlying graph is a random geometric graph, represented by $G_n(r)$, with the nodes of the WM²Snet uniformly randomly distributed within a unit square in $\Re^2$. Let r_n denotes the minimum r that makes $G_n(r)$ k-connected and let σ_n denotes the minimum r for which vertices in $G_n(r)$ have a minimum node degree k. Then $\Pr(\sigma_n = r_n) \to 1$ as $n \to \infty$.

An intuitive explanation of Theorem 3.4 is at a large value of n, with a very high probability the same connectivity radius which makes the graph $G_n(r)$ have a minimum node degree of k also makes the graph k-connected. We are of course interested in finite n behavior, and this is taken up in Theorem 4.3 below. When the number of nodes n is large, the difference between a network whose nodes are uniformly randomly distributed in a defined region D and one whose nodes are Poisson distributed in D and the mean number of vertices is n is small (Gupta and Kumar, 1998), and likewise for the associated graphs. Therefore, Theorem 3.4 may be extended to a graph whose vertices are Poisson distributed. Similarly, Theorem 4.2 in the next section may also be extended to a graph whose vertices are Poisson distributed.

Generation of Uniquely Localizable Networks Given a WM^{2}Snet that has been determined as not uniquely localizable, an interesting question is how we transform this WM^{2}Snet into one that is uniquely localizable by adding additional information about the network, that is, adding extra internode distances. Before we delve into technical discussions, some notations need to be introduced. Let $G = (V, E)$ be a graph. Then the graph G^2 is defined as $(V, E \cup E^2)$ where for any $v_i \neq v_j \in V$, $(v_i, v_j) \in E^2$ if and only if there exists a $v_k \in V$ such that $(v_i, v_k) \in E$ and $(v_k, v_j) \in E$. Thus G^2 is obtained from G by adding edges between the vertex pairs of G which are separated by exactly one intermediate vertex. Analogously, one can also obtain a graph $G^3 = (V, E \cup E^2 \cup E^3)$ where for any $v_i \neq v_j \in V$, $(v_i, v_j) \in E^3$ if and only if there exist two vertices $v_k, v_m \in V$ such that $(v_i, v_k), (v_k, v_m), (v_m, v_j) \in E$. A main result of this section is the following theorem:

Theorem 3.5 (Anderson et al., 2006; Cheung and Whiteley, 2005): Let $G = (V, E)$ be an **edge 2-connected** graph in $\Re^2$. Then $G^2 = (V, E \cup E^2)$ is globally rigid.

A graph is **edge k-connected** if there exists k paths between any two vertices that have no edge in common between any two vertices. It is obvious that a k-connected graph is also an edge k-connected graph but the converse is not necessarily true. There is a simple algorithm, known as the "ear decomposition" (Ramachandran, 1992), that can be used to test edge 2-connectivity, 2-connectivity, and 3-connectivity.

Based on Theorem 3.5, consider a WM^{2}Snet whose underlying graph is a unit disk graph with connectivity radius r; if the underlying graph is not globally rigid but edge 2-connected, by doubling the connectivity radius of the WM^{2}Snet, the graph becomes globally rigid and the associated mesh network becomes uniquely localizable. Doubling the connectivity radius can be achieved by an increase in the transmitter power. Note that this upward adjustment in connectivity radius needs to be performed only once, or at least occasionally. During the rest of the time, a lower transmitter power suffices to maintain a topology with a simple connectivity-based property, or with other desirable properties. Other alternatives to increasing the connectivity radius include increasing the density of nodes or employing additional measurements, for example, bearing measurements (Anderson et al., 2006).

The edge-2 connectedness condition in Theorem 3.5 can be mildly relaxed (Cheung and Whiteley, 2005). Specifically, if $G = (V, E)$ is connected and if the removal of any edge in E which disconnects G results in one of the two components being a single vertex, then G^2 is globally rigid.

Graphical Properties of Connected Networks and Easily Localizable Networks In the last section, we discussed the analytic existence/solvability problem in WM^{2}Snet localization.

An equally important problem is the computational complexity of WM^{2}Snet localization algorithms. The computational complexity of WM^{2}Snet localization algorithms has been studied in the literature. Generally, the computational complexity of localization algorithms is NP-hard and probably exponential in the number of vertices (Saxe, 1979). In this section, we show that for a mesh network whose underlying graph has certain properties, for example, **bilateration graph** and **trilateration graph**, in addition to global rigidity, the localization algorithm can be greatly simplified.

A graph $G = (V, E)$ is called a **bilateration graph** with seeds v_1, v_2, and v_3 if its vertices can be ordered as $v_1, v_2, \ldots, v_{|V|}$ such that $(v_1, v_2) \in E$ and each vertex $v_i, i = 3, 4, \ldots, |V|$ is adjacent to at least two of the vertices in $v_1, v_2, \ldots, v_{i-1}$. The ordering of the vertices $v_1, v_2, \ldots, v_{|V|}$ is called a **bilaterative ordering**. A network is called a **bilateration network** if its underlying graph is a bilateration graph. Similarly, a graph $G = (V, E)$ is called a **trilateration graph** if there exists an ordering of the vertices $v_1, v_2, \ldots, v_{|V|}$ such that the vertices v_1, v_2, v_3 induce a complete graph and each vertex $v_i, i = 4, 5, \ldots, |V|$ is adjacent to at least three of the vertices in $v_1, v_2, \ldots, v_{i-1}$. The ordering of the vertices $v_1, v_2, \ldots, v_{|V|}$ is called a **trilaterative ordering**. A network is called a **trilateration network** if its underlying graph is a trilateration graph. Given a graph known to be a bilateration (trilateration) graph, there can be more than one bilaterative (trilaterative) ordering that is consistent with the bilateration (trilateration) property.

A bilateration graph can be obtained from a connected graph by doubling the connectivity radius of a network with a connected underlying graph:

Theorem 4.1: (Anderson et al., 2006): Let G be a connected graph in $\Re^2$. Then $G^2 = (V, E \cup E^2)$ is a bilateration graph.

An important result about connected graphs is given in the following. It is relevant for bilateration graphs as well, since node radius doubling in a network with a connected graph results in a bilateration graph.

Theorem 4.2: (Gupta and Kumar, 1998): Let $G_n(r(n))$ be a random geometric graph derived from a network whose nodes are uniformly randomly placed a unit disk area in $\Re^2$. If the connectivity radius $r(n)$ of the network satisfies $r(n) = \sqrt{\frac{\log n + c(n)}{n\pi}}$, then the graph $G_n(r(n))$ is connected with probability one as $n \to \infty$ if and only if $c(n) \to +\infty$.

When $\lim_{n\to\infty} \sup c(n) < +\infty$, the graph is disconnected with a positive probability (Gupta and Kumar, 1998).

Connectivity is a **monotone property** of a graph. A graphical property is called an **increasing property**, if the property is preserved when edges are added to the graph. A graph property A is **monotone** if either A or A^c is increasing. In random geometric graphs, a monotone property is known to have a sharp threshold (Goel et al., 2004). The following Theorem in conjunction with Theorem 4.2 sheds more light on the connectivity of random geometric graphs with a finite (but large) number of vertices.

Theorem 4.3: (Goel et al., 2004): If A is a monotone property, then for $0 < \varepsilon < \frac{1}{2}$, let $r(n, \varepsilon) = \inf\{r > 0 : \Pr\{G_n(r) \in A \geq \varepsilon\}\}$. Define further $\delta(n, \varepsilon) \overset{\Delta}{=} r(n, 1 - \varepsilon) - r(n, \varepsilon)$. Then, for every monotone property in $\Re^2$ it is $\delta(n, \varepsilon) = O\left(\frac{\log^{3/4} n}{\sqrt{n}}\right)$.

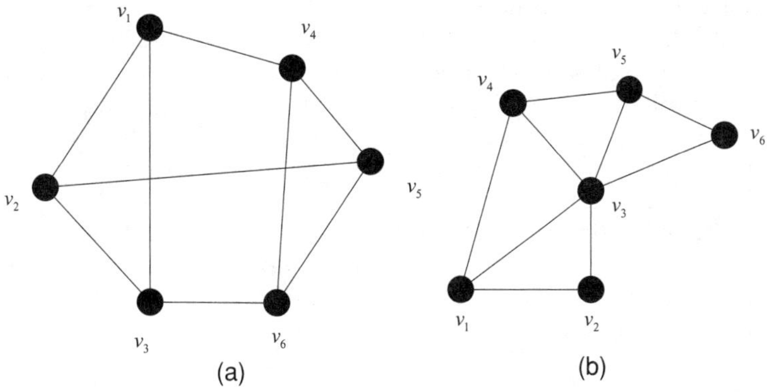

FIGURE 2.19 Relation between a bilateration graph and a globally rigid graph. (a) A globally rigid graph but not a bilateration graph; (b) A bilateration graph but not a globally rigid graph.

We examine now the style of algorithms that can be used to localize nodes where the underlying graph is a bilateration graph. Note first though that a bilateration graph is a rigid graph but is not necessarily globally rigid nor is a globally rigid graph necessarily a bilateration graph. This is illustrated in Fig. 2.19.

If the underlying graph of a network is both globally rigid and a bilateration graph, a sequential localization algorithm called "sweep" can be developed (Fang et al., 2006; Fang et al., 2006a) and (Goldenberg et al., 2006a) to efficiently localize all nodes in the network. The principle of the sweep algorithm is illustrated through the special case in Fig. 2.20. Consider a graph G that is both globally rigid and a bilateration graph with a bilaterative ordering $v_1, v_2, \ldots, v_k$ ($k = 8$ in Fig. 2.20). With some abuse of notation, let us also regard $v_1, v_2, \ldots, v_k$ as nodes in the WM^{2}Snet (and thus, there is associated with v_i a position).

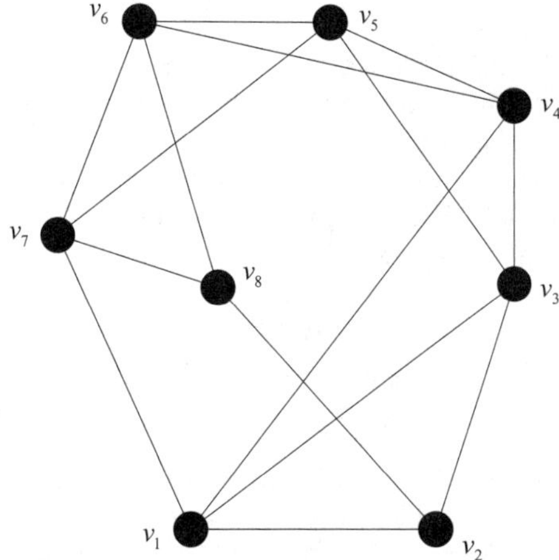

FIGURE 2.20 An illustration of the sequential sweep localization procedure for a network with a globally rigid bilateration graph. Solid lines represent the known internode distances and the nodes are localized sequentially, that is, node $v_i, i > 3$ is localized after node $v_1, \ldots, v_{i-1}$ and using the estimated (or known) positions of nodes $v_1, \ldots, v_{i-1}$ and known distances between v_i and nodes in $v_1, \ldots, v_{i-1}$.

Starting with three seed nodes v_1, v_2 and v_3 with known coordinates, without loss of genericity the coordinates of the three nodes are assumed to be $v_1 = (0, 0)$, $v_2 = (a, 0)$ and $v_3 = (b, c)$ with $a, c > 0$. Then knowledge of the internode distances v_1v_4 and v_3v_4 gives the position of v_4 with a binary ambiguity, that is, v_4 has two possible positions. Consider v_5 with known internode distances v_4v_5 and v_3v_5 to v_4 and v_3 respectively, for each possible position of v_4, v_5 has two possible positions that makes the total number of possible positions for v_5 four. Successively, we obtain the positions of v_6, v_7, v_8 with 2^3, 2^4 and 2^5 ambiguities. However, v_8 is also connected to v_2 (in the example shown in Fig. 2.20, the edge v_2v_8 is necessary to make a globally rigid graph), knowledge of distance to v_2 resolves the ambiguities in the position of v_8 and also reduces the ambiguities in the positions of other vertices. Finally, knowledge of internode distance v_1v_7 and position of v_8 sequentially removes all ambiguities in $v_7, v_6, \ldots, v_4$ and the unique localization of the network is achieved.

There are $\frac{1}{6}|V|(|V| - 1)(|V| - 2)$ different choices of three nodes from V. Therefore, the computational complexity involved in searching for the seed nodes is $O(|V|^3)$, that is, at most polynomial in the number of nodes. If one knows the seed but does not know the ordering, the computational complexity in choosing the ordering is $O(|V| + |E|)$. Another potential problem with the sweep algorithm is that the number of possible positions grows exponentially with the number of nodes. This exponential growth in the number of possible positions can be mitigated by considering all known distances of the newly swept nodes to the already swept nodes, instead of two distances only. When a node with more than two known distances to the already swept nodes is added to the set of swept nodes, the additional distance information will not only eliminate some of its own possible positions, but it also removes some of the possible positions of already swept nodes (Goldenberg et al., 2006a). The number of possible positions also reduces dramatically when the sweep process meets an anchor. Simulation using 250 nodes randomly uniformly distributed in a square region showed that when the average node degree equals six, the number of possible positions reaches its maximum value of 250 (Goldenberg et al., 2006a). Generally, the running time of the sweep algorithm grows linearly in the number of nodes (Goldenberg et al., 2006a). Details on the design of the algorithm can be found in the work by Goldenberg et al. (2006a) where extensions of the basic sweep algorithm to handle noisy distance measurements as well as distance and angle measurements are reported.

In the previous paragraphs, we have shown that if the underlying graph of a network is a globally rigid bilateration graph, an efficient localization algorithm can be designed whose computational complexity is at most polynomial in the number of nodes provided that the ambiguities are bounded. An even more efficient algorithm can be designed if the underlying graph has the property of a trilateration graph. A trilateration graph can be obtained from a connected graph by tripling the connectivity radius.

Theorem 4.4 (Anderson et al., 2006): Let $G = (V, E)$ be a connected graph with n vertices, and let $v_1, v_2, \ldots, v_n$ be an ordering of the vertices of G such that for all $m > 1$, the subgraph of G induced by the vertex set $V_m = \{v_1, v_2, \ldots, v_m\}$, denoted by G_m, is connected. Then $G^3 = (V, E \cup E^2 \cup E^3)$ is a trilateration graph with a trilaterative ordering $v_1, v_2, \ldots, v_n$.

In $\Re^2$, a trilateration graph is globally rigid but a globally rigid graph is not necessarily a trilateration graph.

In $\Re^2$, if the underlying graph of a mesh network is a trilateration graph and at least three of the meshes are anchors, then the whole network can be easily and uniquely localized. The computational complexity involved in searching for the seed nodes and a trilaterative ordering is the same as those involved in searching for the seed nodes and a bilaterative ordering in a bilateration network. These are $O(|V|^3)$ and $O(|V| + |E|)$ respectively. Now considering one has found the seed nodes and the corresponding trilaterative ordering and assuming temporarily the three seed nodes v_1, v_2 and v_3 have the known coordinates, $v_1 = (0, 0)$, $v_2 = (a, 0)$ and $v_3 = (b, c)$ with $a, c > 0$, the values of a, b and c can be derived from the known internode distances between the seed nodes. Then, it is obvious that other nodes can be localized relative to the three seed nodes sequentially, in a single sweep and in time $O(|V|)$. The estimated positions differ from the true positions by a translation, rotation and possible reflection. At this stage, knowledge of the anchor positions has not been used. By aligning the estimated positions of anchors with their true positions, the required translation, rotation and reflection to transform the estimated coordinates into coordinates in which anchors positions are consistent with their true positions can be obtained. The new estimated positions follow for the rest of the nodes through application of the same translation, rotation and reflection.

Though the computational complexity involved in searching for the seed nodes and the vertex ordering is the same as that involved in a bilateration network, the computational complexity involved in estimating a node's position is much simpler than that in a bilateration network and there is no growth in the number of possible positions. Therefore, localization algorithms in a trilateration network can be more efficient than those in a bilateration network. However, a trilateration network requires knowledge of a higher number of edges (or equivalently a larger connectivity radius) than a bilateration network.

Simulation Results Simulations were conducted to evaluate the critical connectivity radii required for a connected, a 2-connected network, a 3-connected network, a rigid network, a redundantly rigid network, a globally rigid network, a globally rigid bilateration network and a trilateration network respectively. Compared with the simulation results by Anderson et al. (2006), more extensive results, including the critical connectivity radii for a globally rigid bilateration network and a trilateration network, were obtained. We generate 50 instances of test network each with 200 nodes uniformly distributed in a unit square area.

Simulation results are shown in Fig. 2.21. Comparing r_1, r_2, r_g, r_{tri} it can be readily concluded that $r_g \leq 2r_2$ and $r_g \leq r_{tri} \leq 3r_1$. This is an expected result from Theorem 3.5 and Theorem 4.4. Moreover, it is observed that $r_g = r_{g,bi}$ in all simulations. As shown in Fig. 2.19, a bilateration graph is not necessarily globally rigid nor is a globally rigid graph necessarily a bilateration graph. However, simulation results seem to suggest that the scenario shown in Fig. 2.19 rarely occurs and with a very high probability, a globally rigid graph is also a bilateration graph. It is also noted that the difference between r_g and r_{tri} is usually small which indicates that a small increase in the connectivity radius will transform a globally rigid graph into a trilateration graph.

2.5.6 Multicasting in Mobile Wireless Networks

Background A number of WM^{2}Net scenarios require some degree of one-to-many or many-to-many interactions, such that a node forwards its data to multiple destinations.

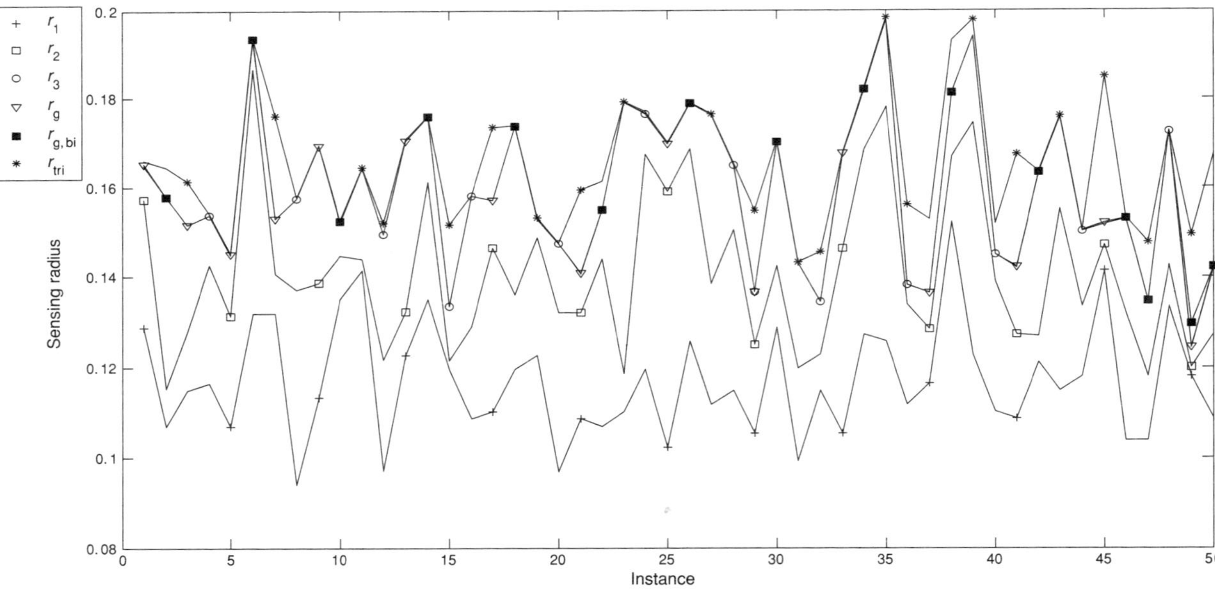

FIGURE 2.21 Critical connectivity radii to achieve a connected, a 2-connected network, a 3-connected network, a rigid network, a redundantly rigid network, a globally rigid network, a globally rigid bilateration network and a globally rigid trilateration network. r_k denotes the critical connectivity radius for a *k*-connected graph. r_{bi}, r_g, $r_{g,bi}$ and r_{tri} denote the critical connectivity radius for a bilateration graph, for a globally rigid graph, for a globally rigid bilateration graph and for a globally rigid trilateration graph, respectively.

Delivering multicast packets over a large network is a complex process. There are several components of this process that must take place to successfully establish multicast communications. The first step is the identification of the receivers. All of the hosts that want to receive a particular stream of multicast information must identify themselves to the network. This is called the *multicast registration process*, and it is facilitated with a unique set of IP addresses (called Class D addresses) that are reserved specifically for multicast communications. The receivers register with a particular group (e.g., to attend an individual webinar)—the concept of "groups" is central to multicast data delivery.

Once receivers join their respective groups (it is likely that a single receiver will join several groups), the network must deliver the multicast traffic to the correct end stations. To this end, the transmissions from the data source must be replicated at some point, so that the information can be received in multiple locations simultaneously. The delivery process is facilitated by a multicast routing protocol.

Most multicast data transmissions are unidirectional. Typically, a single host will transmit information (such as stock ticker updates or the content for a seminar) that will be received by multiple end stations.

2.5.6.1 Multicast Registration Process

The first step towards multicast communications is the identification of the receivers. This is accomplished via the Internet group management protocol (IGMP), which takes place between hosts and a local router.[6] Historically, IGMP has been an integral part of the Internet protocol (IP) suite. In fact, IGMP has a Protocol ID of 2, indicating that it is one of the earliest extensions of IP (ICMP is "1"). IGMP first appeared in RFC 988, in 1986 – 22 years ago. Since then, it has been revised several times. The latest edition is known as version 3 and was codified in RFC 3376 in 2002.

IGMP allows users to announce their intention to join particular multicast groups. These groups are identified by their unique Class D IP addresses. When a host wants to participate in a multicast group, it sends an IGMP "join" message to its local router. If multiple routers exist on a single segment, they can mutually elect a "designated router" (DR) that will manage all of the IGMP messages for the segment.

After a router receives one or more "joins" for a specific group, it will forward any packets destined for that particular group to the appropriate interface(s). The router will forward only one copy of the data packets per interface. If there are multiple receivers on a single interface, they will all receive the information by monitoring common multicast MAC and IP addresses.

IGMP is a stateful protocol. The router regularly verifies that hosts want to continue to participate in the multicast groups by sending periodic "queries" to the receivers. These queries are transmitted to a well-known multicast address (224.0.0.1) that is monitored by all systems. If the receivers are still interested in that particular multicast group, they will respond with a "membership report" message. When the router stops seeing responses to queries, it will delete the appropriate group from its forwarding table.

IGMP version 2 (from 1997) adds the ability for the receivers to gracefully exit from a multicast group. This is accomplished by creating a new "leave" message that a host will

---

[6] In a WM^{2}Net context this may well be the mesh node itself or a tailored "net-box" with routing capabilities.

use when it needs to depart from a particular group. This allows the router to immediately update its forwarding tables, without waiting for the expiration of the query timer.

Multicast protocols were initially designed based on the assumption that communications streams consist of a single source transmitting to multiple receivers. Nothing in the protocols prevented the inclusion of multiple sources transmitting to the same group of receivers. Several video-conferencing applications are designed to support this type of configuration. In this scenario, most receivers also act as sources. However, in other environments, receiving unwanted packets from multiple ad hoc sources could adversely impact some of the receivers' applications. Additionally, this could create security problems if an unauthorized source transmits invalid or malicious information to a large group of receivers (consider the stock quote example). IGMP version 3 (the current version) addresses this potential risk.

IGMPv3 adds the ability to specify the source(s) that a receiver is willing to listen to. Sources can be stipulated with "include" filters in the "join" and "report" messages, or sources can be specifically rejected with "exclude" filters. Overall, these filters greatly enhance the security and performance of multicast communications. These filters also add a new dimension to the tables for the participating routers, since they must now keep track of the acceptable sources for every multicast group.

2.5.6.2 IPv6

Multicast support was built into the IPv6 protocol at its inception. The first three bits of an IPv6 address identify the format prefix (FP) of the overall 128-bit address. If the FP is set to binary 111, then it refers a multicast address (similar to an IPv4 Class D address).

Instead of using IGMP, IPv6 has its own registration protocol called Multicast Listener Discovery (MLD). The first version of MLD (contained in RFC 2710) is similar to IGMP version 2—it supports the multicast "join" and "leave" functions. MLD version 2 is a proposed draft (not yet adopted by the IETF as an official RFC) that adds the "include" and "exclude" filter functionality as in IGMP version 3.

2.5.6.3 Multicast Routing

The registration process described above is used by receivers to join existing multicast groups. The local router uses this process to determine where it should deliver multicast data. When a router acts in this manner, it is called the Last Hop Router (LHR) as is located by the receivers and is placed at the end of the communications path. Conversely, the First Hop Router (FHR) is the nearest router to the source of the multicast traffic. Multicast routing is responsible for delivering efficiently the data from the first hop router to all of the participating last hop routers.

2.5.7 Multicasting in WM^{2}Nets

Given the limited available resources (bandwidth and power) of WM^{2}Nets, multicasting is a very alluring approach to deliver the same data packet to multiple destinations. The broadcast nature of the wireless medium turns multicasting in wireless networks a fundamentally different problem to multicasting in wireline networks. Traditional wireline multicast protocols such as the protocol independent multicast-sparse mode (PIM-SM) (Deering et al., 1996), the core based tree (CBT) (Ballardie et al., 1993), and the distance vector multicast routing protocol (DVMRP) (Waitzman et al., 1998) do not work well in a wireless multihop context. The frequent tree reorganization induced from these traditional multicast protocols can cause significant signaling overhead and scalability concerns; as shown by Ching-Chuan et al. (1997a, 1998, 1999) and Ching-Chuan and Mario (1997), these are prohibitive factors in using modified versions of wire-line multicast protocols in resource-limited WM^{2}Nets.

In what follows, a brief survey of the three common groups of multicast protocol categories is reviewed:

1. **Tree-based multicasting:** Tree-based multicast is applicable to networks with low-loss links and minimal mobility. Obraczka et al. (2001) argue that these protocols do not perform well due to their "fragile" forwarding structure owing to the increased node mobility. This is, however, not as much of a problem in predominantly static networks like a preconfigured wireless sensor network (WSNet). Examples of tree-based protocols are the following (early) protocols devised for operation in ad hoc networks: AMRIS (Wu and Tay, 1999), MAODV (Royer and Perkins, 2000) and LAM (Ji and Corson, 1998).
2. **Mesh-based multicasting:** Mesh-based multicast generates a more redundant forwarding structure where multiple paths exist between a source-destination pair. Existing studies (Ju Lee et al., 2002) show that mesh-based protocols seem to perform better in highly mobile networks than tree-based structures with the cost, however, of higher control and maintenance overhead. ODMRP (Ju Lee et al., 2002), CAMP (Garcia-Luna-Aceves and Madruga, 1999) and FGMP (Chiang et al., 1998) are examples of mesh-based protocols. In order to reduce the transmission overhead in mesh-based forwarding, the work by Zhou et al. (2006) studies the problem of constructing the optimal (minimum number of transmissions) forward mesh while ensuring that at least two disjoint paths from the source to each multicast recipients exist.
3. **Hybrid-structure-based multicasting:** Hybrid-structure-based multicasting combines the resilience of mesh-based protocols with the simplicity of tree-based protocols. AMROUTE (Xie et al., 2002) is an example of hybrid-structure multicast protocol. It maintains multiple virtual mesh links to ensure that the multicast-forwarding tree remains unchanged when topology changes. The major drawback of this scheme is that loop-freedom as well as optimal trees in case of nonstatic hosts is not guaranteed.

2.5.7.1 Multicasting in Multiradio Multirate Multichannel (MR^2-MC) WM^{2}Nets[7]

The use of multiple radios on a single node can significantly improve the spatial channel reuse and the system's capacity as such. The use of multirate media access control (MAC) protocols on the other hand can significantly improve throughput and fairness. However, multicasting in multirate WM^{2}Net is a rather complicated practice comparing to single-rate WM^{2}Net. This is so for the following two reasons. Different rates use different modulation schemes, which in turn produce different transmission ranges and neighbor-sets as such. As a consequence, the common implicit assumption that a single-rate multicast node reaches all its neighbors on a single broadcast transmission does not

[7]Excerpt from the invited article "Multicasting in multi-rate wireless mesh networks," [†]Junaid Qadir and Chun Tung Chou, [‡]Archan Misra ([†]School of Computer Science and Engineering, University of New South Wales, Australia; [‡]IBM T. J. Watson Research Center, Hawthorne, New York, NY, USA; E-mails: {junaidq, ctchou}@cse.unsw.edu.au; archan@us.ibm.com).

always hold true as noted by Chou et al. (2006). An extra degree-of-freedom allows a node to perform multiple "distinct-rate transmissions" to reach different subset of neighboring nodes at different rates. This sequel highlights the specifics of MR^2-MC, exposing their advantages as well as potential limitations. More specifically, as shown, broadcast performance in WM^{2}Nets can be improved by exploiting the inherent interface, channel, and rate diversity of multiradio multirate.

Single-Radio Single-Channel Multiple-Rate Multicast A number of centralized multicast algorithms for multirate WM^{2}Nets exist in the literature that aim to to construct *low-latency trees*. The problem of constructing a low-latency tree is referred to as the *minimum latency broadcasting*[8] (MLB) problem. A broadcast heuristic called *broadcast incremental bandwidth* (**BIB**) was first proposed by Chou and Misra (2005). The BIB algorithm is very similar to the *broadcast incremental power* algorithm (BIP) (Jeffrey et al., 2002) in that both use a modified version of Prim's algorithm (a minimum spanning tree (MST) algorithm) to add links to an existing tree whereas minimizing a certain cost metric. However, while BIP focuses on the development of low-energy packet distribution trees, BIB primarily aims to choose high-rate links since the transmission of a transmitter to its neighbors is constrained by the slowest of the point-to-point links between the transmitter node and each individual neighbor. Subsequently, an improved heuristic based on the concept of *weighted connected dominating set* (**WCDS**) was proposed by Chou et al. (2006). We studied the use of CDS to perform multicast; WCDS is the multirate version of a *minimum CDS* (MCDS). The WCDS problem aims to find a subset Y in V, and also the link-layer multicast rate w_i for each node $y_i \in Y$ such that firstly, every element of V is in Y or in the neighborhood of Y; secondly, the set Y is connected; and lastly, the weighted sum $\sum_{\forall y_i \in Y} \frac{1}{w_i}$ is minimal. It must be noted that the WCDS reduces to the problem of finding MCDS when there is only one transmission rate.

Whilst BIB and WCDS both are centralized algorithms, a *localized* and *distributed* CDS-based approach called *multirate delayed-pruning WuLi* (**MDW**) has been proposed by (Qadir et al., 2007) for constructing low-latency broadcast trees in multirate WM^{2}Nets. This approach is similar to the approach adopted by Wu and Li (1999) to construct distributed CDS in single-rate WM^{2}Nets. It includes an additional stage though, called Rate-Maximization whose aim is to increase the transmission rate of all nodes above the lowest rate. The results reported by Chou et al. (2006) indicate that centralized rate-aware multicasting algorithms (e.g., WCDS) give results approximately two times the optimal, whereas (Qadir et al., 2007) reports distributed rate-aware algorithms (e.g., MDW) perform approximately two times the performance of centralized rate-aware multicasting algorithms.

Multiple-Radio Multiple-Channel Multiple-Rate Multicast The problem of efficient multicast/broadcast in multiradio multirate WM^{2}Net is especially challenging since interface-diversity of multiradio WM^{2}Net must be exploited. Four centralized broadcast/multicast *rate-aware* algorithms were presented by Qadir et al. (2006) and (Qadir et al., 2006) for the MLB problem in MR^2-MC WM^{2}Net. The first algorithm called *multiple-radio shortest-path-tree* (**MSPT**) is a simple extension of Dijkstra's SPT algorithm; since it does not exploit WBA, it was the worst performing of the four algorithms provided by Qadir et al.

[8] It is assumed that the multicast tree can be built from pruning the broadcast tree (see also Chou et al., 2006; Qadir et al., 2006; Qadir et al., 2006a).

(2006). The second algorithm called *multiple-radio weighted-connected-dominating-set tree* (**MWT**) exploited WBA but not the availability of multiple radio interfaces on the same node. The other two algorithms, *l*ocally parallelized *m*ultiradio WCDS *t*ree (**LMT**) *and* (*p*arallelized, *a*pproximate-shortest, *m*ultiradio WCDS *t*ree (**PAMT**) differed in how they exploited both the WBA and the interface diversity on individual nodes.

We will briefly discuss the three best performing algorithms presented by Qadir et al. (2006). Firstly, the **MWT** algorithm is an extension to the WCDS algorithm, which is designed for the MLB problem for SR-SC multirate networks (Chou et al., 2006). In SR-SC multirate WM^{2}Net, WCDS performs creditably against other low-latency broadcast heuristics, because WCDS considers both the multirate nature of the network and the WBA of the underlying wireless medium. The MWT, like WCDS, is a greedy heuristic algorithm that decides the optimal transmission in each round from a set of eligible transmissions. However, as we shall see, *MWT does not consider the availability of multiple interfaces on each node, and thus fails to exploit the potential advantage of parallel transmissions at any intermediate node.* The second heuristic algorithm for MLB problem in MR^2-MC WM^{2}Net is the **LMT**, whose development was motivated by the observation that MWT, while taking into account the WBA and multirate nature of the underlying medium, does not '*as readily*' exploit the interface diversity on individual nodes. This observation can be explained by noting that MWT is inherently biased by its priority metric (used to decide the optimal transmission in each round) to include transmissions that cover greater number of uncovered nodes. This metric tends to work well when the number of radio interfaces/channels is small. However, it fails to exploit the increased opportunities for parallel faster transmissions (on different orthogonal channels) when the number of interfaces is higher. Accordingly, the LMT algorithm is based on the observation that a node m covered by a transmission may also be covered by a transmission *of the same node at a lower rate on an orthogonal channel*. Thus, we may be able to cover node m on an orthogonal channel without exceeding the nominal delay, by considering node m as a covered node of the latter transmission. The third heuristic for the MLB problem in MR^2-MC is the **PAMT** algorithm, which like the LMT algorithm is adapted from the MWT algorithm, and is designed to be *adaptive* to number of radio interfaces and channels available. The PAMT algorithm is intended as an improvement over the LMT algorithm. The LMT algorithm, during any particular round, might decide to cover some nodes with a transmission that has a longer latency path to s (the source node) compared to other eligible transmissions (*by currently unused interfaces on other intermediate nodes*) that can possibly take place on an alternative, noninterfering channel in "*parallel.*" The trees constructed from PAMT algorithm, adapts to the number of available radio resources (by parallelizing or funneling the transmissions over the available radio interfaces). Thus, the tree constructed by the PAMT algorithm resembles the WCDS for SR-SC WM^{2}Net (which is the best performing tree for SR-SC WM^{2}Net), and the SPT for MR^2-MC WM^{2}Nets with infinite radio resources (shortest-path-tree with infinite radio-resources is the optimal low-latency tree).

A key finding of the MLB research, for SR-SC multirate WM^{2}Net by Chou and Misra (2005) and Chou et al. (2006), and for MR^2-MC WM^{2}Net by Qadir et al. (2006, 2006a), is that the multicast efficiency of a transmission rate can be predicted reasonably from the *Rate-Area Product* (RAP) product of the transmission rate and its transmission coverage area (Chou and Misra, 2005). This finding was exploited by Qadir et al. (2006a) to choose a single rate from the multiple transmission rates available *for all link-layer multicasts*. This framework was referred to as "*single best-rate multicast*" (**SBM**) framework, in contrast to the fully multirate multicast (**FMM**) framework used by Chou and Misra (2005),

Chou et al. (2006), and Qadir et al. (2006, 2006a), and an algorithm called *parallelized connecting dominating set* (**PCDS**) using the SBM framework was proposed by Qadir et al. (2006a). The results by Qadir et al. (2006, 2006a) indicate that by incorporating interface-diversity, MR^2-MC multirate multicast algorithms (e.g., PAMT and PCDS) improve the performance of SR-SC multirate algorithms (e.g., WCDS and MDW) by approximately 70–80% in MR^2-MC WM^2Nets.

2.5.8 Network Coding

In traditional routing, an intermediate network node replicates a packet from the incoming to a suitable outgoing link or a set of outgoing links. The set of more than one outgoing link is selected when the packet is multicast/broadcast. Network coding instead generalizes the function of the intermediate nodes, such that each packet sent out from an outgoing link is a product of the packets coming from different incoming links. In effect, with network coding, relay nodes combine the two communication flows coming from two or more nodes.

Although the network coding has initially exhibited its benefits for multicast in wireline packet networks, the unreliability and the broadcast nature of the wireless setting appear to set a fertile ground for developing network-coding solutions (Lun et al., 2005; Fragouli et al., 2006). One particular scenario in which the network coding in multihop wireless networks (shortly: *wireless network coding*) is promising to bring significant throughput/delay improvements is the case of bidirectional or two-way traffic (e.g., interactive multimedia communications with voice and video).

Let us illustrate network coding in operation using the example depicted in Fig. 2.22. As illustrated, node A has a packet P_{AC} destined to node C and, vice versa, C has a packet $\mathbf{P}_{CA}$ destined to node A. For simplicity, assume that all packets have identical length and the length is equivalent to the duration of a time slot. The conventional relaying is depicted on Fig. 2.22a where it takes 4 time slots to transmit the 2 packets, P_{AC} and P_{CA}. Figure 2.22b depicts the two-way relaying through network coding with decode-and-forward (NCDF) proposed by Wu et al. (2004), Larsson et al. (2005), and Popovski and Yomo (2006).

With NCDF, in the first two slots the relay node B receives and decodes the packets $\mathbf{P}_{AC}$ and $\mathbf{P}_{CA}$, respectively. Then B creates the packet $\mathbf{P}_B = \mathbf{P}_{AC} \oplus \mathbf{P}_{CA}$, where $\oplus$ denotes the bitwise XOR operation, and broadcasts the packet $\mathbf{P}_B$ to A and C. Since node A knows a priori the packet $\mathbf{P}_{AC}$, it can extract the desired packet $\mathbf{P}_{CA}$ as $\mathbf{P}_{CA} = \mathbf{P}_{AC} \oplus \mathbf{P}_B$. Similarly, node C extracts the packet $\mathbf{P}_{AC} = \mathbf{P}_{CA} \oplus \mathbf{P}_B$. If we ignore data re-transmissions (due to channel impairments), NCDF takes only 3 slots to transport the same amount of data from the source (A) to the destination (C). This is a significant improvement on network resource utilization and system's performance as such.

2.5.8.1 Cross-layered Transport with Network Coding for Bidirectional Traffic in Multihop Wireless Networks[9]

Physical-Layer Wireless Network Coding in Two-Way Relay Channels The physical-layer network coding schemes considered in this work (Popovski and Yomo, 2006; Popovski and

[9] Excerpt from the invited article "Cross-layered transport with network coding for bi-directional traffic in multi-hop wireless networks," Petar Popovski and Hiroyuki Yomo, Department of Electronic Systems, Aalborg University, Niels Jernes Vej 12, DK-9220 Aalborg, Denmark, Email: {petarp, yomo}@kom.aau.dk.

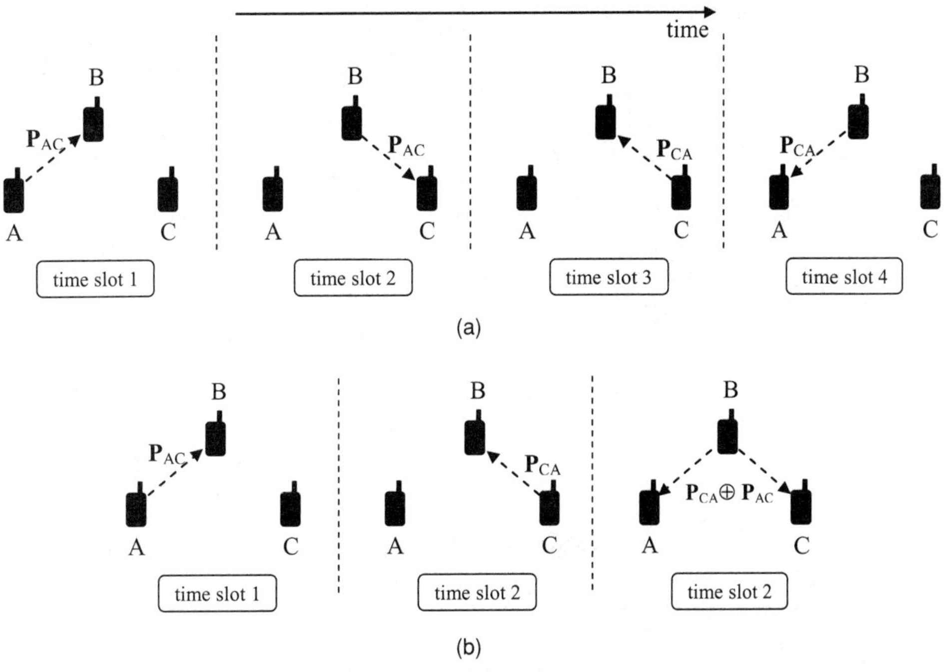

Figure 2.22 Conventional two-way relaying versus two-way relaying with NCDF. (a) Conventional two-way relaying. (b) Two-way relaying with NCDF.

Yomo, 2006a; Rankov and Wittneben, 2005) consist of two phases; namely, *the multiple access* (MA) and *the broadcast phase*. Let us consider the communication scenario illustrated in Fig. 2.23 where node A exchanges data with node C through B.

The **MA phase** has a duration equal to N transmission symbols. During this phase, both nodes A and C are transmitting. The N-dimensional complex baseband signal

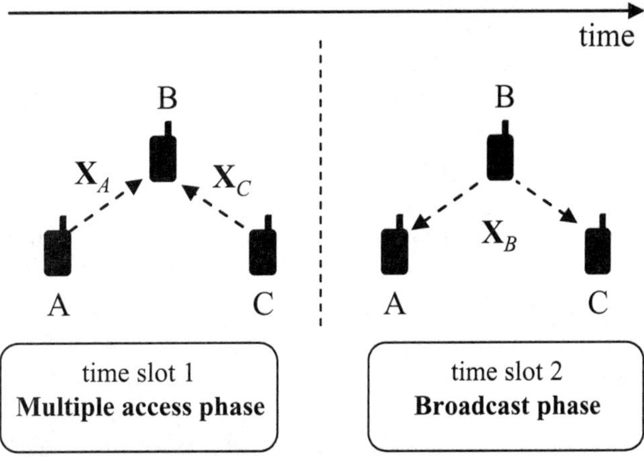

Figure 2.23 Physical-layer wireless network coding scheme that consists of two phases.

received at node B is given as

$$\mathbf{Y}_\mathrm{B} = h_\mathrm{AB}\mathbf{X}_\mathrm{A} + h_\mathrm{CB}\mathbf{X}_\mathrm{C} + \mathbf{Z}_\mathrm{B} \tag{2.17}$$

where $\mathbf{X}_\mathrm{A}$ ($\mathbf{X}_\mathrm{C}$) is a N-dimensional baseband vector that represents a data packet $\mathbf{P}_\mathrm{AC}$ ($\mathbf{P}_\mathrm{CA}$) which needs to be transported from A to C (from C to A). The complex value h_AB (h_CB) denotes the channel coefficient between nodes A and B (nodes C and B). We assume that channel gains h_AB and h_CB remain constant during the MA and broadcast phases. In addition, we assume that h_AB and h_CB are known to all three nodes A, B, and C. The transmission rate of the signal $\mathbf{X}_\mathrm{A}$ ($\mathbf{X}_\mathrm{C}$) is denoted by $R_\mathrm{A}(R_\mathrm{C})$, expressed in bits per symbol. The N-dimensional vector $\mathbf{Z}_\mathrm{B}$ represents the additive noise at the receiver front of B and each component of this vector is a complex Gaussian-distributed value with a variance σ^2. Further on, we assume the transmission power of A and C is normalized, such that each component x of the vectors $\mathbf{X}_\mathrm{A}$, $\mathbf{X}_\mathrm{C}$ has average energy of 1, that is, $E[|x|^2] = 1$. Hence, the signal-to-noise ratio (SNR) at B for the signal transmitted by A in the absence of the signal transmitted by C is given by $\gamma_\mathrm{AB} = |h_\mathrm{AB}|^2/\sigma^2$ and, in analogy, $\gamma_\mathrm{CB} = |h_\mathrm{CB}|^2/\sigma^2$. Finally, we assume that the signals of A and C are synchronized at B.

In the **broadcast (BC) phase**, node B transmits a signal whose M-dimensional baseband complex representation is denoted by $\mathbf{X}_\mathrm{B}$. In general, the duration of the broadcast phase need not be equal to the duration of the MA phase that is, $M \neq N$. The information content of $\mathbf{X}_\mathrm{B}$ is determined by the signal $\mathbf{Y}_\mathrm{B}$, which is received in the previous MA phase, and the decoding/processing operation performed at node B. Upon the reception of signal $\mathbf{Y}_\mathrm{B}$, node B switches to either of the following three modes of physical-layer wireless network coding: *network coding with amplify-and-forward (NCAF), network coding with joint decode-and-forward (NCJDF),* and *network coding with denoise-and-forward (NCDNF).* Due to the lack of space, in the sequel we describe NCAF and NCJDF, while a detailed description of NCDNF can be found (Popovski and Yomo, 2006a).

Before delving into the details of each of the two schemes, let us quickly assess the throughput gain brought by the physical-layer network coding scheme. For that purpose, let us assume that the MA phase and the BC phase have identical duration and the channel is perfect (noiseless). In this context, node A (C) can perfectly recover the packet $\mathbf{P}_\mathrm{CA}$ ($\mathbf{P}_\mathrm{AC}$) after receiving the signal $\mathbf{X}_\mathrm{B}$. Thus, it takes only two slots to transmit the two packets $\mathbf{P}_\mathrm{CA}$ and $\mathbf{P}_\mathrm{AC}$ from the source to the destination, which means a 100% throughput improvement over the conventional relaying.

If B operates solely in the NCAF mode as $\mathbf{X}_\mathrm{B}$ is obtained from:

$$\mathbf{X}_\mathrm{B} = \beta\mathbf{Y}_\mathrm{B} \tag{2.18}$$

where β is a real coefficient to account for the amplification done at B. The amplification coefficient is chosen such that each component of vector $x_\mathrm{B} \in \mathbf{X}_\mathrm{B}$ has a normalized energy $E[|x_\mathrm{B}|^2] = 1$, which results in $\beta = \sqrt{\frac{1}{|h_\mathrm{AB}|^2+|h_\mathrm{CB}|^2+\sigma^2}}$. Clearly, in this case $\mathbf{X}_\mathrm{B}$ has the same dimension as $\mathbf{Y}_\mathrm{B}$, such that the MA phase has identical duration with the BC phase. The signals received by A and C after the broadcast phase are:

$$\begin{aligned}\mathbf{Y}_\mathrm{A} &= h_\mathrm{AB}\mathbf{X}_\mathrm{B} + \mathbf{Z}_\mathrm{A} = h_\mathrm{AB}\beta\mathbf{Y}_\mathrm{B} + \mathbf{Z}_\mathrm{A} = \beta h_\mathrm{AB}^2\mathbf{X}_\mathrm{A} + \beta h_\mathrm{AB}h_\mathrm{CB}\mathbf{X}_\mathrm{C} + \beta h_\mathrm{AB}\mathbf{Z}_\mathrm{B} + \mathbf{Z}_\mathrm{A}\\ \mathbf{Y}_\mathrm{C} &= h_\mathrm{CB}\mathbf{X}_\mathrm{B} + \mathbf{Z}_\mathrm{C} = h_\mathrm{CB}\beta\mathbf{Y}_\mathrm{B} + \mathbf{Z}_\mathrm{C} = \beta h_\mathrm{CB}^2\mathbf{X}_\mathrm{C} + \beta h_\mathrm{AB}h_\mathrm{CB}\mathbf{X}_\mathrm{A} + \beta h_\mathrm{CB}\mathbf{Z}_\mathrm{B} + \mathbf{Z}_\mathrm{C}\end{aligned} \tag{2.19}$$

where $\mathbf{Z}_A$ and $\mathbf{Z}_C$ denote the complex-valued noise at A and C, respectively, and is assumed to have identical per-component variance of σ^2 as the noise at node B. Assuming that A and C are aware of the values β, h_{AB}, h_{CB}, then they subtract the a priori signals from Eq. (2.19) in order to obtain the signal ($\mathbf{D}_A$ and $\mathbf{D}_C$, respectively) to decode the desired packet:

$$\mathbf{D}_A = \beta h_{AB} h_{CB} \mathbf{X}_C + \beta h_{AB} \mathbf{Z}_B + \mathbf{Z}_A \quad \mathbf{D}_C = \beta h_{AB} h_{CB} \mathbf{X}_A + \beta h_{CB} \mathbf{Z}_B + \mathbf{Z}_C \tag{2.20}$$

From Eq. (2.20), it follows that A (C) observes the signal from C (A) through an equivalent Gaussian channel. The SNRs available for A to decode $\mathbf{X}_C$ and for C to decode $\mathbf{X}_C$ are denoted by $\gamma_A^{(AF)}$ and $\gamma_C^{(AF)}$, respectively, and are expressed as:

$$\gamma_A^{(AF)} = \frac{\gamma_{AB}\gamma_{CB}}{2\gamma_{AB} + \gamma_{CB} + 1} \quad \gamma_C^{(AF)} = \frac{\gamma_{AB}\gamma_{CB}}{\gamma_{AB} + 2\gamma_{CB} + 1} \tag{2.21}$$

A detailed analysis of NCAF is given by Popovski and Yomo (2006).

If B applies NCJDF, it attempts to decode signals $\mathbf{X}_A$ and $\mathbf{X}_C$ from signal $\mathbf{Y}_B$. This can be achieved, for example, by a successive interference cancellation. Theoretically, the values γ_{AB} and γ_{CB} can be used to determine the rate pairs (R_A, R_C), which can be successfully decoded at B. However, those rates are achieved by considering infinite code words and impractical coding schemes (Cover and Thomas, 1991). Instead, here we assume that nodes A and C are using finite-length packets with practical modulation/coding methods. In addition, we assume that A and C are using identical modulation/coding procedures, which implies that $R_A = R_C$. This condition is not mandatory for the operation of NCJDF, but is rather convenient in order to specify a simple protocol that can be used to compare the performance of NCJDF with the other two schemes for physical-layer wireless network coding. We specify the operation of NCJDF to be as follows:

This scheme combines joint decoding with AF. If transmission is accomplished according to **Step 5**, then the performance is improved as compared to NCAF. This is because the relay node does not amplify the noise. Note that this procedure can be generalized to the case when the transmission rates in the MA phase are different. For example, if $R_A > R_C$ then the packet size of $\mathbf{P}_{AC}$ is larger than the size of $\mathbf{P}_{CA}$, such that $\mathbf{P}_{CA}$ should be padded with zeros in order to have a consistent XOR operation in **Step 5**.

Figure 2.24 illustrates the throughput measurements for the two way-relaying scenario (Popovski and Yomo, 2006). To obtain these results, BPSK-modulated packets are used, each consisting of 100 bits. For given SNR values γ_{AB}, γ_{CB} the channel coefficients h_{AB}, h_{CB} are fixed to be both real, as this creates the least favorable situation for NCJDF. Notably, the other schemes are not affected by such a choice. Figure 2.24a shows the normalized throughput in bits/symbol when $\gamma_{AB} = \gamma_{CB}$. In this case, both flows at the multiple access channel have identical strengths, which prohibit successive interference cancellation, and the NCJDF scheme always operates as NCAF. However, when $\gamma_{CB} = \gamma_{AB}/4$ on Fig. 2.24b, the successive interference cancellation at B takes effect and NCJDF is better than NCAF.

In both figures, NCDF outperforms the conventional relaying. It also outperforms NCSF and NCJDF at low SNRs, where the noise amplification in NCAF/NCJDF is exacerbated. Nevertheless, at high SNRs, NCAF/NCJDF almost doubles the throughput as compared to the conventional relaying.

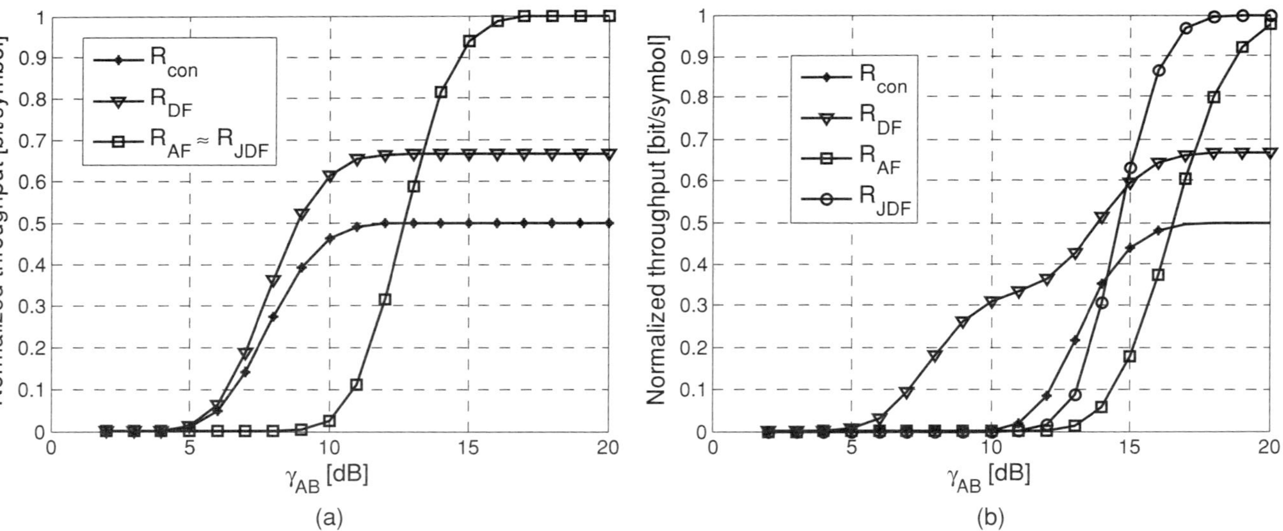

FIGURE 2.24 Comparison of the throughput achieved with different relaying schemes for BPSK-modulated data. R_{con} is the throughput for conventional relaying, while R_x is the throughput, if NCx is applied where x is DF, AF, or JDF. (a) $\gamma_{CB} = \gamma_{AB}$. (b) $\gamma_{CB} = \frac{\gamma_{AB}}{4}$.

Using Network Coding in Wireless Multihop Networks It is interesting to see how the aforementioned schemes for wireless network coding can be applied when the communication flows are hopping data over more than two hops. The multihop operation of the NCDF mode is closer to the current design of the communication protocols, as the relay nodes are always decoding the packets. A description of NCDF can be found in the work by Wu et al. (2004). Here a distributed implementation of the NCAF scheme is described using the scenario illustrated in Fig. 2.26. In this example, there is a two-way traffic between nodes A and E. The traffic is relayed via nodes B, C, and D. All nodes are symbol-synchronized to a common clock. In the odd slots the nodes A, C, and E transmit, while B and D receive. The situation is opposite in the even slots. Here ai and ej denote the baseband representations of i-th packet from A and j-th packet from E, respectively. Thus, in slot 5, when A transmits a3 and C transmits a2 + e1, the node B receives a3 + a2 + e1.

All packets have identical length of N symbols. Each node k has a buffer S_k to store N baseband symbols and operates running the algorithm, illustrated in Fig. 2.25. For each node and each slot on Fig. 2.26 the value of the buffer at the end of the slot. For example, at the beginning of slot 6, the buffer of C is $S_C = a2 + e1$, and the received signal is a3 + e1 + a2 + e2, such that the buffer value at the end of the slot is $S_C = a3 + e2$. To preserve the consistency of the algorithm, we assume that a virtual node transmits a new packet to node A (E). For example, in slot 4 the virtual node of A transmits a3, the value of new_received at A is a3 + a2, the buffer at the slot start is $S_A = a2$ and the buffer at the end of the slot is $S_A = a3$. The packets ai (ej) are representing the sink_data for node E (A). In slot 4 the node E receives the sink_data a1 and, in slot 5, after E broadcasts a1 + e2, it removes the sink_data a1 and the buffer value remains $S_E = e2$.

Elaborating upon Fig. 2.26, one can observe that in a stationary regime, A and E can inject two new packets in the network each two slots, which results in throughput of 1 packet per slot.

2.5.8.2 The Effect of Multimode Operations on Routing and Scheduling for WM^{2}Snets[10]

Formulating the Maximum Data Retrieval Problem Retrieving data from sensors in a network involves determining which path each set of data will take and when each node will transmit. That involves a joint optimization of scheduling and routing decisions. As illustrated in Table 3.1, the difference in the power consumption between the idle and the receive mode is marginal. Therefore, the maximum data retrieval (MDR) problem models the three basic operational modes: sleep, receive, and transmit, where the receive mode refers to the case where the transceiver is on but not active (transmitting), regardless of whether it is actually receiving or is in the idle mode. When switching between these three operational modes, the RF transceiver is likely to incur an energy spike (Wang et al., 2001). Two energy spikes are considered: one incurred when turning on the transceiver from the sleep mode, and the other incurred due to powering up the transmitter to actually transmit.

Let us consider a network of N WM^{2}Snet nodes with $N_t(i)$ the set of nodes that can successfully transmit data to i as long as no more than one node from within the set is

[10] Excerpt from the invited article "The effect of multi-mode operations on routing and scheduling for wireless mesh sensor networks," †Shanchieh Jay Yang and ‡Moises Sudit (†Department of Computer Engineering, Rochester Institute of Technology, ‡Industrial and Systems Engineering, University of Buffalo).

```
if node_k_state=transmit
        broadcast Sk;
        if node_k_type=sink
                remove sink_data from Sk;
else
        receive new_received;
        Sk=new_received- Sk;
end
```

Figure 2.25 Pseudocode of the algorithm run by node k for NCAF over multiple hops.

transmitting at the same time. Similarly, $N_r(i)$ is referred to as the set of receiving nodes of i. For simplicity, it is assumed that $N_t(i)$'s and $N_r(i)$ are time invariant sets.

Let b_i denote the buffer size of each node, λ_i the average sensing rate, e_i^{tot} the total initial energy, p_i^s the power levels to operate in the sleep mode, p_i^r the additional power needed to have the RF transceiver on, p_i^t the additional power needed to transmit at a rate p_{ij} to neighbor j, and the energy spikes incurred when transitioning from the sleep to the receive mode (e_i^{sr}) and that from the receive to the transmit mode (e_i^{rt}). Based on these, the power consumed while operating in the sleep, receive, and transmit modes for node i is p_i^s, $p_i^s + p_i^r$, and $p_i^s + p_i^r + p_i^t$, respectively.

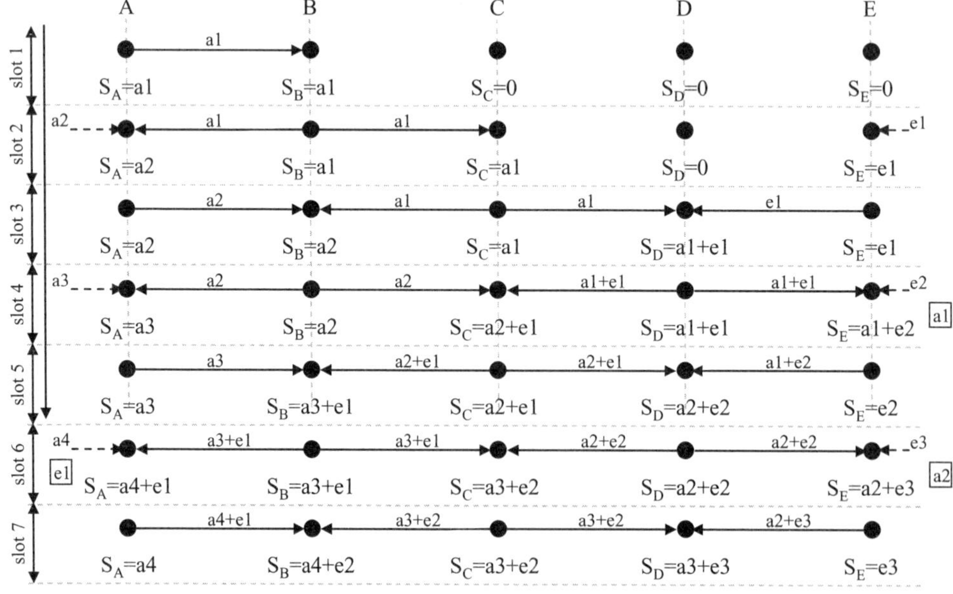

Figure 2.26 Example operation of NCAF over multiple hops.

Based on these definitions, a mixed-integer programming model is developed to maximize the total data retrieved by the sinks until one of the WM^{2}Snet nodes dies due to energy exhaustion.

$$\max \sum_{i \in S} \sum_{j \in N_t(i)} \sum_{k=1}^{K} F_{jik}\, c_{ji} \tag{2.22}$$

$$\textit{subject to}\;\; T_k - M(1 - x_{ijk}) \le \min(T_k, M \cdot X_{ijk})\;\; \forall i \in N,\; j \in N \cup S,\; 1 \le k \le K$$

$$T_k - M(1 - Y_{ik}) \le G_{ik} \le \min(T_k, M \cdot Y_{ijk})\;\; \forall i \in N,\; j \in N \cup S,\; 1 \le k \le K \tag{2.23}$$

$$\sum_{j \in N_t(i)} \sum_{l \in N_t(j)-1} x_{ljk} - M\left(1 - \sum_{j \in N_r(i)} x_{ijk}\right) \le 0\;\; \forall i \in N,\; 1 \le k \le K \tag{2.24a}$$

$$\sum_{j \in N_t(i)} \sum_{l \in N_r(j)-1} x_{jlk} - M\left(1 - \sum_{j \in N_t(i)} x_{jik}\right) \le 0\;\; \forall i \in N,\; 1 \le k \le K \tag{2.24b}$$

$$Y_{ik} - \left(\sum_{j \in N_t(i)} x_{ijk} + \sum_{j \in N_t(i)} x_{jik}\right) \ge 0\quad \forall i \in N,\; 1 \le k \le K \tag{2.25a}$$

$$Z_{ik} - \left(\sum_{j \in N_r(i)} x_{ijk} - \sum_{j \in N_r(i)} x_{jik-1}\right) \ge 0\quad \forall i \in N,\; 2 \le k \le K \tag{2.25b}$$

$$w_{ik} - (Y_{ik} - Y_{ik-1}) \ge 0\;\; \forall i \in N,\; 2 \le k \le K \tag{2.25c}$$

$$0 \le \sum_{i=1}^{k} \sum_{j \in N_t(i)} c_{ji}\, F_{jil} + \sum_{l=1}^{k} \lambda_{il}\, T_l - \sum_{i=1}^{k} \sum_{j \in N_r(i)} c_{ij}\, F_{ijl} \le b_i \quad \forall i \in N,\; 1 \le k \le K \tag{2.26}$$

$$e_i^{\text{sr}} \sum_{k=1}^{K} W_{ik} + e_i^{\text{rt}} \sum_{k=1}^{K} Z_{ik} + p_i^{\text{s}} \sum_{k=1}^{K} T_k + p_i^{\text{r}} \sum_{k=1}^{K} G_{ik} + p_i^{\text{t}} \sum_{k=1}^{K} \sum_{j \in N_r(i)} F_{ijk} \le e_i^{\text{tot}}\;\; \forall i \in N \tag{2.27}$$

The model essentially determines the time intervals T_k, during which a subset of WM^{2}Snet nodes transmit data to their neighbors. Constraint (Eq. 2.22) determines whether WM^{2}Snet node i transmits to node j at time slot k (X_{ijk}), as well as the duration of the transmission (F_{ijk}). Similarly, constraint (Eq. 2.23) determines if the communication module of node i is in the receive (idle) mode at time slot k (Y_{ik}) along with the amount of time the module is in the receive mode (G_{ik}).

The MDR model accounts for the spatial reuse on the network, the energy spent for switching between the operational modes (sleep-to-receive and receive-to-transmit), as well as the store-and-forward capability of the WM^{2}Snet nodes over time. Constraints (Eq. 2.24a and 2.24b) prevent the neighbors of each transmitter from receiving, and the neighbors of each receiver from transmitting. Note that by altering $N_t(i)$'s and $N_r(i)$'s, one can change the spatial reuse constraints to account for a different wireless signal interference model. Constraint (Eq. 2.25a) forces the communication module to be on, if the corresponding node at the time interval k is either transmitting or receiving (but not both). It is important to note that a communication module could still be on, even if the node is neither receiving nor transmitting. This occurs when the expenditure of energy for a time period is less than the cost of switching the transceiver from off to on. Constraints (Eq. 2.25b and 2.25c) allow the model to track the energy spikes due to switching from sleep to receive (W_{ik}) and from receive to transmit (Z_{ik}), respectively.

Constraint (Eq. (2.26)) prevents the buffer from overflowing and ensures that the amount of transmitted data does not go beyond what has been sensed and received. Finally, Constraint (Eq. 2.27) decrements the energy accordingly due to the operational modes in each time interval, and guarantees that the amount of energy spent over time does not exceed the total energy available. M and K are two large numbers used for the model.

Observations from Optimal Data Retrieval

Observation 1: Store-and-Forward Operation for Scheduling Observations from the optimal solutions for sample cases of MDR problems reveal that the operational modes of most individual WM^{2}Snet nodes follow a "store-and-forward" (S&F) cycle as shown in Fig. 2.27. This S&F cycle dictates the routes that data takes to reach the sinks.

Consider the same notation defined above, but ignore the use of subindex i for that we shall illustrate the optimal S&F cycle for an individual WM^{2}Snet node case. To distinguish between data arriving at the wireless node via the sensing unit and data arriving via the

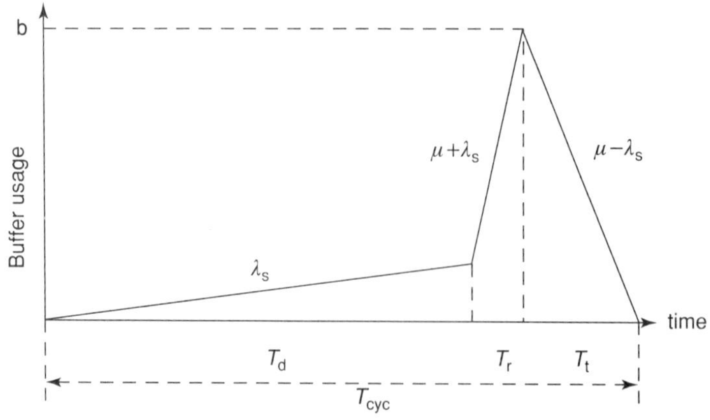

FIGURE 2.27 Buffer usage of a wireless WM2Snet node over an ideal busy cycle.

RF receiver, let λ_s and λ_r be the average sensing rate and the average receiving rate, respectively, and μ be the transmission capacity. Note that this analysis assumes a flow model (instead of a packet model). This assumption is appropriate as long as the operation (sleep, receive, and transmit) intervals are much larger than the time scales to perform packet transmission and event sensing.

The S&F cycle shown in Fig. 2.27 starts with the WM^{2}Snet node collecting data only via sensing with a rate of λ_s for T_d amount of time. Following up the sensing period, the buffer usage increases at a rate of $\mu + \lambda_s$ by performing both receiving and sensing at the same time—recall that the data is received at rate μ. After T_r amount of time, the buffer is full and node starts to transmit what has been stored and that continuously arrives via sensing. In this transmission phase, the buffer usage decreases at a rate of $\mu - \lambda_s$. Note that during the first T_d time interval, the WM^{2}Snet node may choose to be in the sleep mode, or to stay in the receive mode to avoid the penalty of incurring an energy spike e_{sr}.

Consider the stability condition for this S&F cycle. A periodical operation with the S&F cycle is stable, if the buffer usage does not go beyond the buffer size b and will go down to zero at the end of the cycle. Based on these two constraints and the fact that transmission and receiving cannot happen simultaneously, we derive the following stability condition, which is more restricted than '$\lambda_s + \lambda_r \leq \mu$' for traditional flow models where transmission and receiving can happen simultaneously.

Lemma 1: The S&F cycle is stable if and only if $\lambda_s + 2\lambda_r \leq \mu$.

Proof: The stability condition follows by solving the following equality and inequalities.

$$
\begin{aligned}
T_{cyc} &= T_d + T_r + T_t \\
T_{cyc}\lambda_r &= T_r\mu \\
T_{cyc}\lambda_s + T_r\mu - b &\leq T_t\mu \leq T_{cyc}\lambda_s + T_r\mu
\end{aligned}
\qquad \blacksquare
$$

Satisfying the stability condition, we may now determine the amount of data transmitted (R_{cyc}) and the amount of energy used (E_{cyc}) within an S&F cycle, and the time interval of the S&F cycle (T_{cyc}). First, the total data that can be transmitted in an S&F cycle is what can be transmitted in an interval of $b/\mu - \lambda_s$ with transmission rate of μ. This gives

$$R_{cyc} = {b \cdot \mu}\Big/{\mu - \lambda_s} \tag{2.28}$$

Since the total data arrival rate at the WM^{2}Snet node, either via sensing or receiving, is $\lambda_s + \lambda_r$, and the total transmitted data in the time interval T_{cyc} is R_{cyc}, we have

$$T_{cyc} = {b \cdot \mu}\Big/{(\mu - \lambda_s)(\lambda_s + \lambda_r)} \tag{2.29}$$

Furthermore, notice that the sensing power p^s will be consumed during the entire T_{cyc} interval, additional power p^r will be consumed during the T_r time interval and possibly the T_d interval if e^{sr} is large, and the transmit power p^t will be consumed during the T_t

interval. This gives the closed form expression for the energy used during an S&F cycle as shown below.

$$E_{\text{cyc}} = \frac{b \cdot p^s \cdot \mu}{(\mu - \lambda_s)(\lambda_s + \lambda_r)} + \frac{b \cdot p^r \cdot (\lambda_s + 2\lambda_r)}{(\mu - \lambda_s)(\lambda_s + \lambda_r)} + \frac{b \cdot p^t}{(\mu - \lambda_s)} + \min\left(e^{sr}, \frac{b \cdot p^r \cdot (\mu - \lambda_s - 2\lambda_r)}{(\mu - \lambda_s)(\lambda_s + \lambda_r)}\right) + e^{rt} \tag{2.30}$$

The above equations allow one to analyze the lifetime of a WM^{2}Snet node given the total initial energy e^{tot} available to the node. Notice that Eqs. (2.28) to (2.30) correspond to the case where most amounts of data are transmitted using least amount of energy in a busy S&F cycle. This observation leads to the theorem below that finds the upper bounds for the lifetime of the WM^{2}Snet node and the total amount of data that can be transmitted during this lifetime.

Theorem 1: The lifetime (T_{life}) and total amount of data transmitted (R_{life}) of an individual WM^{2}Snet node with initial energy $e^{\text{tot}} \gg E_{\text{cyc}}$ will be upper bounded based on the following inequalities, respectively:

$$T_{\text{life}} \leq \frac{e^{\text{tot}} \cdot T_{\text{cyc}}}{E_{\text{cyc}}} := T^*_{\text{life}} \tag{2.31}$$

$$R_{\text{life}} \leq \frac{e^{\text{tot}} \cdot R_{\text{cyc}}}{E_{\text{cyc}}} := R^*_{\text{life}} \tag{2.32}$$

The proof of Theorem 1 lies on the fact that $R_{\text{cyc}}/E_{\text{cyc}}$ calculates the most energy efficient bits per joule transmitted by an individual WM^{2}Snet node. With a total available energy e^{tot} that is much larger than E_{cyc}, one cannot transmit more than $R_{\text{cyc}} \cdot e^{\text{tot}}/E_{\text{cyc}}$ amount of data. The lifetime then follows. Note that the inequalities shown in Theorem 1 become equal if E_{cyc} exactly divides e^{tot}. Since we assume $e^{\text{tot}} \gg E_{\text{cyc}}$, we can approximate a WM^{2}Snet node's lifetime and the total amount of data transmitted over the lifetime as T_{life} and R_{life}, respectively.

Observation 2: WM^{2}Snet Node Role Assignment for Routing Though optimal, not all WM^{2}Snet nodes in a general network can operate in periodical S&F cycles. The S&F models developed in the previous section, however, help define the role each node may play for the entire network data retrieval operation.

Lifetime Critical Nodes are the nodes whose battery energy is first exhausted and consequently determine the lifetime of the entire network. Recall Eqs. (2.28) to (2.30), which suggest that the lifetimes of WM^{2}Snet nodes with the same physical make-up depend solely on λ_r and λ_s. Let us assume that λ_s are about the same for all WM^{2}Snet nodes. Thus, λ_r is the catalytic factor of a node's lifetime; the larger it is the shorter the lifetime. Consequently, nodes that are one hop away from the sinks (i.e., $n \in N_t(S)$) will first exhaust their energy, since all sensed data exchange flows through them; this translates to a very large λ_r.

Since the nodes in $N_t(S)$ are critical to the network lifetime, it makes sense to have them operate in the optimal S&F cycles, while letting others transition between operational modes more often. This implies that a possible approximated solution to the MDR problem may be found by determining the flow distribution over the entire network (*i.e.*,finding $\lambda_{r,i}$, $\forall i \in N$) that minimizes the lifetime of the nodes in $N_t(S)$.

Flow Dispersing Nodes are those placed far away from the sinks. Although they are not critical to the lifetime of the entire network, they use energy less efficiently than those closer to the sink. This general statement can be explained by considering a chain of n WM^{2}Snet nodes, indexed as $1, 2, \ldots, n$ to indicate the number of hops each node is away from the sink. In this chain network, nodes may operate in such way that node $i+1$'s transmission schedule matches node i's receiving schedule, and so on. This operation, however, makes the effective buffer size for S&F diminish as the node is located farther away from the sink. Assuming that the physical buffer size is b, node 1 in the chain network can use the entire buffer size, $b_1^* = b$, but the subsequent nodes can use only buffer size b_i^* as given below.

$$b_i^* = b_{i-1}^* \times \frac{(\lambda_{\mathrm{r},i} + \lambda_{\mathrm{s},i})(\mu + \lambda_{\mathrm{s},i})}{(\lambda_{\mathrm{r},i-1} + \lambda_{\mathrm{s},i-1})(\mu + \lambda_{\mathrm{s},i-1})}, \qquad \forall i = 2, \ldots, n.$$

This is due to the fact that the nodes closer to the sink receive more data volume, which is the aggregation from all receiving nodes. Because of the diminishing buffer size, nodes that are far away from the sink cannot store much data before forwarding it. Consequently, the energy spikes become the predominant portion of the energy spent in an S&F cycle (recall E_{cyc} from Eq. 2.30).

A solution to alleviate the energy inefficiency problem is to have the far-away nodes store data until the buffer is almost full and 'disperse' the flow with multiple upstream paths. Note that a properly designed flow dispersion scheme should also help balance traffic arriving to the lifetime critical nodes.

Sensing Effective Nodes. Previous discussions consider the uniform sensing case where the WM^{2}Snet nodes in a network have about the same λ_{s}. In the case of nonuniform sensing rates, we find that higher λ_{r} may not lead to increasing total data transmitted by that WM^{2}Snet node over its lifetime—recall Eq. (2.26). By taking a closer look at the models derived above, we show the following corollary, which can be proved by deriving from the conditions $\frac{dT_{\mathrm{life}}}{d\lambda} < 0$ and $\frac{dR_{\mathrm{life}}}{d\lambda} < 0$, respectively.

Corollary 1: As λ_{r} increases and all other parameters are fixed, a WM^{2}Snet node will have (1) T_{life}^* monotonically decreases regardless of other parameter values, and (2) R_{life}^* decreases if and only if $\lambda_{\mathrm{s}}/\mu > p^{\mathrm{s}}/p^{\mathrm{r}}$.

The above suggests that although WM^{2}Snet node lifetime decreases with the number of flows that are directed to it, the total amount of data transmitted (or forwarded) by the node may or may not go down as the lifetime does. In fact, the total transmission will go up as λ_{r} increases if and only if it is more energy efficient to receive a bit of data from one of the neighbors rather than to sense a bit of data as the condition: $\lambda_{\mathrm{s}}/\mu > p^{\mathrm{s}}/p^{\mathrm{r}}$ suggests. Figure 2.28 depicts the numerical results and show for the cases of $\lambda_{\mathrm{s}}/\mu = 1.0 \cdot p^{\mathrm{s}}/p^{\mathrm{r}}, 1.0 \cdot p^{\mathrm{s}}/p^{\mathrm{r}}$ and $1.5 \cdot p^{\mathrm{s}}/p^{\mathrm{r}}$, respectively how T_{life}^* and R_{life}^* vary with λ_{r}. The results demonstrate that maximizing the WM^{2}Snet node lifetime does not necessarily mean maximizing the total data transmitted by that node. Also interesting from Fig. 2.28 is that, regardless of whether sensing or receiving is more energy efficient, as $\lambda_{\mathrm{r}} \to \infty$, the total transmission

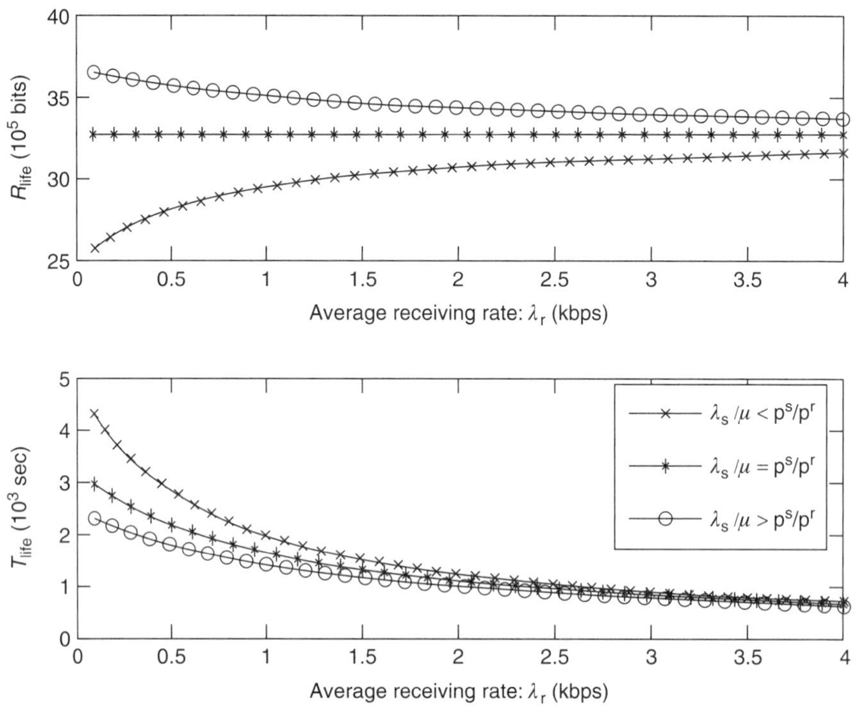

FIGURE 2.28 The changes of T^*_{life} and R^*_{life} as λ_r increases for the different cases of λ_r/μ.

approaches to the case where they are equally efficient. Of course λ_r can never go to infinity, which is due to the stability condition. This observation suggests that it may be wise to distribute the least flow possible towards nodes that potentially collect most data via sensing. In other words, the sensing nodes should be focus on sensing while the other nodes forward the data.

2.5.8.3 Analysis of Retransmission, Streaming and Opportunistic Data Delivery Methods in WM^{2}Snets[11]

Synopsis of Data-Link Layer (DLL) protocols Let us first present three commonly used data link protocols, namely the automatic repeat request (ARQ), the streaming communication model, and the ExOR. ARQ (Gnawali et al., 2004; Karl and Willig, 2005) is the simplest protocol for ensuring reliable communications. It has several versions that differ in retransmission strategies. In this work, we only consider the Stop-and-Wait version. The transmitter sends one packet at a time and sets a timer. The receiver either receives the packet and sends an acknowledgement back to the transmitter or receives nothing and

[11] Excerpt from the invited article "Analysis of retransmission, streaming and opportunistic data delivery methods in wireless mesh sensor networks," Jingbo Sun and Rachel Cardell-Oliver, School of Computer Science & Software Engineering and Cooperative Research Centre for Plant-Based Management of Dryland Salinity, University of Western Australia, M002, 35 Stirling Highway, Crawley, WA 6009, Australia, Email: {jingbo,rachel}@csse.uwa.edu.au

keeps quiet. If the transmitter receives an acknowledgement from the receiver, the last transmitted packet is deleted from its buffer and the next packet is transmitted. Otherwise, the transmitter retransmits the packet one or more times, up to some threshold, before the packet is declared lost.

The streaming communication model (Cao et al., 2006) assumes no timeout mechanisms at the transmitter side but sends packets in sequence. If a gap is found in the received packets, the receiver then sends retransmission request packets (RRPs) to inform the transmitter. As a response, the transmitter retransmits the lost packet(s). To maintain the correct sequence of packets, all new packets received are buffered during the recovery period.

Finally, ExOR is an integrated routing and MAC protocol whose aim is to increase the throughput of large unicast transfers in multihop wireless networks by using more than one forwarder (Morris and Biswas, 2005). Only one node sends packets at a time. The source broadcasts a batch of packets to its neighbors, including a list of candidate forwarders in each packet, prioritized by the estimated cost to the destination. In addition, the sender also sends a batch map indicating the condition of packets received. Receivers buffer successfully received packets and await their time slot. The forwarder with the highest priority broadcasts its buffered packets. The remaining forwarders transmit the packets, which are not acknowledged in the batch map of higher priority nodes.

Efficiency Analysis Link efficiency is the ratio of data bytes sent and the total bytes transmitted over a single radio link between two nodes. Path efficiency is the accumulated link efficiency over a multihop path. Link and path efficiency are indicators for the amount of energy consumed in sending a fixed number of packets. Let λ be the packet length ratio between acknowledgement and data packets.

ARQ protocols Efficiency Analysis As illustrated in Fig. 2.29, suppose that sender A sends M packets to receiver B with the forward and backward link reception probability $p$and q, respectively. Thus, M/pq is the upper bound on A's transmissions for M packets successfully sent. The receiver is expected to send $M/pq \times p = M/q$ acknowledgement packets. The link efficiency over a single hop link is then (Cao et al., 2006):

$$\eta_{\text{ARQ}} = \frac{pq}{1 + p\lambda} \tag{2.33}$$

Streaming Protocol Efficiency Analysis For the streaming protocol, the successful transmission ratio of the sender A for one packet is p of and only the lost packets $(1 - p)$ need a RRP sent by receiver B. Therefore, the successful ratio of RRP is pq whereas the receiver B is expected to send $1/pq$ times to guarantee that the RRP is correctly received by the sender A. On the other hand, the sender A retransmits the lost packet when it receives

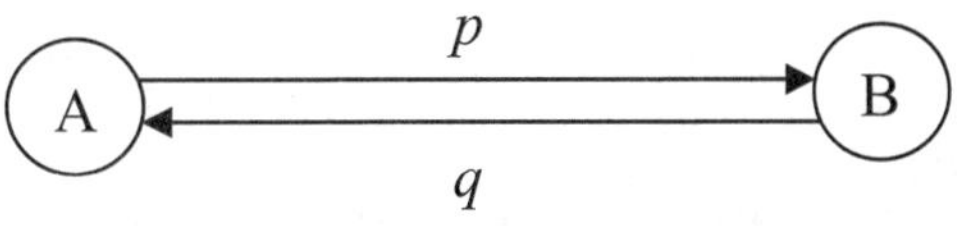

Figure 2.29 ARQ and streaming network architecture.

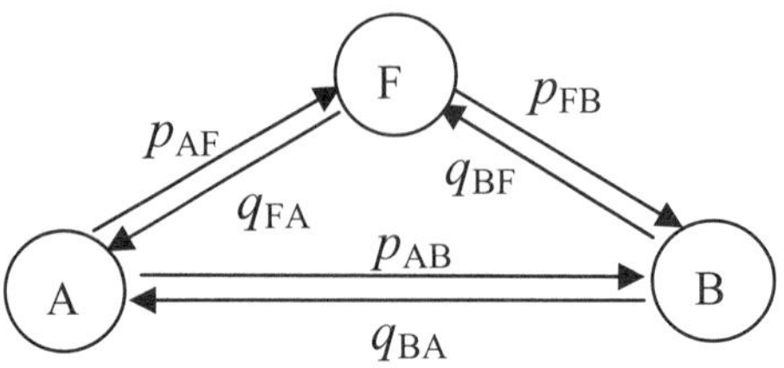

FIGURE 2.30 ExOR network architecture.

the RRP sent by B. Thus, the number of retransmissions for lost packets is $1/p$. Therefore, the link efficiency over a single hop link is (Cao et al., 2006):

$$\eta_{\text{streaming}} = \frac{M}{M + (1-p)M(1/p + \lambda/pq)} = \frac{pq}{q + (1-p)\lambda} \tag{2.34}$$

Equations (2.33) and (2.34) tell us that the link quality is the only factor that affects link efficiency. In the following, the link efficiency of ARQ, streaming, and ExOR is examined.

ExOR Efficiency Analysis For the ExOR protocol, we consider the case with one forwarder shown in Fig. 2.30. Source node A sends packets to destination node B which has first priority, using second priority forwarder F as necessary. The forward packet and backward reception probabilities for each link between nodes are given by p and q respectively. Suppose node A sends M packets to node B, then node B receives Mp_{ab} and node F receives Mp_{af}. Then node B sends K ($K = 10$ provided in Morris and Biswas, 2005) acknowledgment packets, containing a batch map that lists the packets received by B. K > 1 is usually chosen to ensure that at least one acknowledgement is received. Suppose node F and node A receive at least one of these packets from node B, then node F forwards all packets that it has received but node B has not. There will be $Mp_{\text{af}} - Mp_{\text{ab}}p_{\text{af}}$ of these. So the total number of packets sent in the first round is $\mathit{TotalNum}_{1\text{st}} = M + K + M(p_{\text{af}} - p_{\text{ab}}p_{\text{af}}) = MQ_1 + K$, where $Q_1 = 1 + p_{\text{af}} - p_{\text{ab}}p_{\text{af}}$.

Suppose node A receives at least one packet from node F in the first round. In the second round, node A sends thus $\mathit{ASend}_{2nd} = M - Mp_{\text{ab}} - M(p_{\text{af}} - p_{\text{ab}}p_{\text{af}}) = MQ_2$, where $Q_2 = 1 - p_{\text{ab}} - p_{\text{af}} + p_{\text{ab}}p_{\text{af}}$. Therefore, the total number of packets that should be sent in the second round is $\mathit{TotalNum}_{2\text{nd}} = MQ_2Q_1 + K$. Hence, in the nth round, the total number of packets that should be sent is $\mathit{TotalNum}_{n\text{th}} = MQ_2^{n-1}Q_1 + K$. Therefore, the total number of packets sent in n rounds is $\mathit{TotalNum} = M\sum_{i=1}^{n} Q_2^{i-1}Q_1 + nK$.

Suppose nodes resume from sending packets when $MQ_2^{n_a-1} < 1$, which is the packet number sent by node A, and $MQ_2^{n_f-1}(p_{\text{af}} - p_{\text{ab}}p_{\text{af}}) < 1$, which is the packet number sent by node F. So there is $n_a > 1 - \lg M/\lg Q_2$ and $n_{\text{f}} > 1 - \lg M(p_{\text{af}} - p_{\text{ab}}p_{\text{af}})/\lg Q_2$. When the packet reception rate is equal to 1, then $n_{\text{f}} = 0$. This is because forwarder needs not forward any packets to the destination node. Here, we use n_a and n_f to indicate sending times of node A and F. Thus the total sending time is $n > 2 - \lg M/\lg$

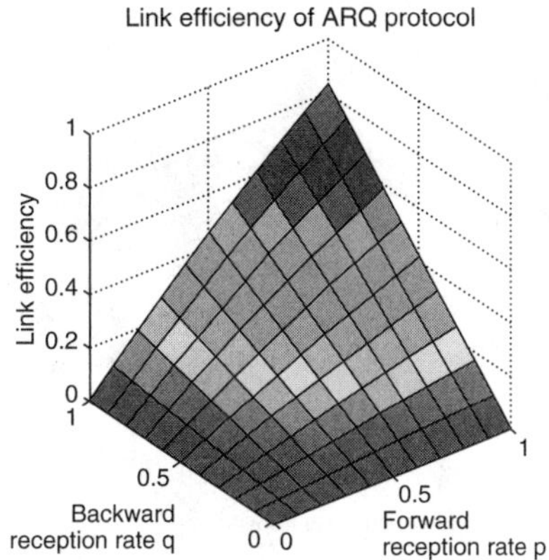

FIGURE 2.31 Estimated link efficiency of ARQ.

$Q_2 - \lg M(p_{af} - p_{ab}p_{af})/\lg Q_2$ and the estimated link efficiency of the ExOR protocol is:

$$\eta_{\text{ExOR}} = \frac{M}{\frac{M(1-Q_2^n)Q_1}{1-Q_2} + n\lambda K} = \frac{1}{\frac{Q_1(1-Q_2^n)}{1-Q_2} + \frac{n\lambda K}{M}} \tag{2.35}$$

where n is the minimum integer satisfying Eq. (2.35).

Based on Eq. (2.35), we found that not only the packet reception probability can impact the link efficiency, but the number of packets sent in each batch too. To increase link efficiency, a relatively large number of packets per batch should be sent.

Figures 2.31–2.34 show the link efficiency of ARQ, streaming, and ExOR protocols based on the Eqs. (2.33) to (2.35) for different forward and backward link reliability λ is set to 0.2. We observed that ARQ's maximum link efficiency is above 0.8. The streaming protocol can achieve link efficiency of 1. The maximum efficiency for ExOR is nearly 1 and above 0.8 with the batches sizes of 100 and 10, respectively. So streaming is a good

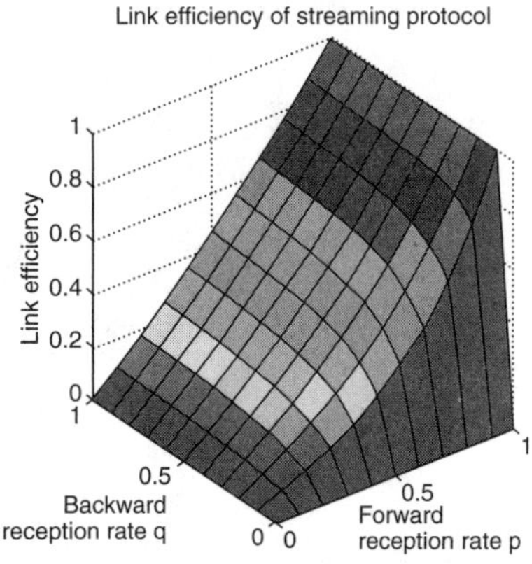

FIGURE 2.32 Estimated link efficiency of streaming.

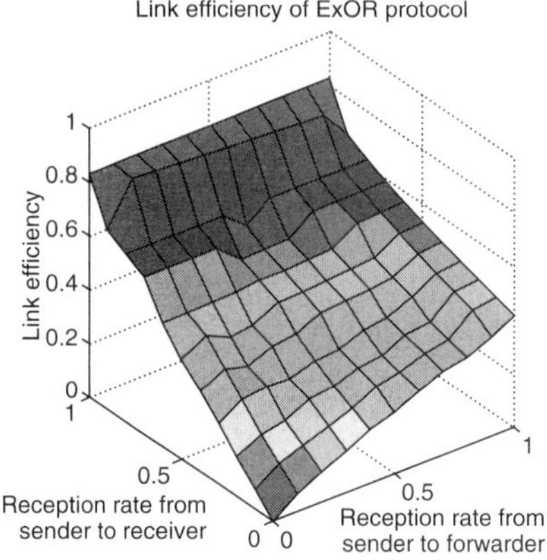

Figure 2.33 Estimated link efficiency of ExOR (batch size = 10).

choice when the link quality is good. We also found that only when both link directions are good can ARQ achieve good performance. However, streaming can work very well as long as the forward packet reception rate is good. For ExOR, when the link quality is low, the use of a forwarder node can significantly improve the link efficiency.

Buffer Requirement Analysis In this section we consider the buffer requirements of ARQ, Streaming, and ExOR protocols. ARQ can provide loss-free, duplicate-free and in-sequence delivery with only one buffer, as long as the sender's timeout is larger than the maximum time to send a data packet and an acknowledgement (Karl and Willig, 2005).

It is difficult to estimate the buffer size for the streaming protocol, as this depends on how long it takes for both sender and receiver to successfully recover from a lost packet (Cao et al., 2006). The size of the buffer should be chosen in such a way so that the probability of the sender detecting that a packet is lost before it deletes that packet from its buffer is sufficiently small. Furthermore, the buffer should be sufficiently large so that enough time is allowed to recover lost packets. Cao et al. (2006) derived equations are

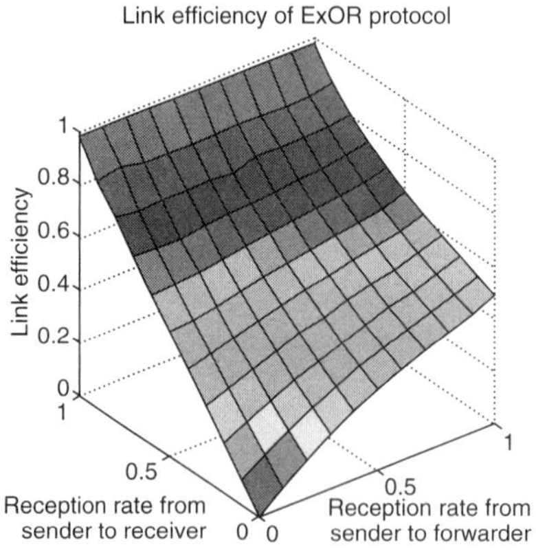

Figure 2.34 Estimated link efficiency of ExOR (batch size = 100).

for both cases. As shown, for a link with reception probability (0.5, 0.5), the buffer size of 16 is sufficient.

The buffer size of ExOR is equal to the batch size. However, since the batch size directly impacts the link efficiency, the buffer should be as large as possible to maximize link efficiency.

End-to-End Delay Analysis End-to-end delay is the total time taken for a packet to traverse from the source to the destination. Low end-to-end delay means nodes can have more sleeping time and thus saves more energy. In this section, we will compute and compare the link end-to-end delay of ARQ, streaming, and ExOR. We assume link reception probability as (p, q).

In the ARQ protocol, the sender's timer expires at T_a. For a packet to be successfully received, the source is expected to send it *1/pq* times. Let the delivery latency at one hop be T_{dl} and $T_{dl} \ll T_a$. Therefore, the end-to-end delay of the ARQ in a single link is $Delay_{ARQ} = T_{dl} + (1-p)\frac{1}{pq}T_a$.

For streaming protocol, let T_{dl} denote the per-hop delivery latency. The timeout of RRPs is T_{na} and the source generates data packet at a fixed interval T_{dp}, where $T_{dl} \ll T_{na}$ and $T_{dl} \ll T_{dp}$. In order to detect the lost packet, the receiver has to wait for a successful transmission. After the lost packet is detected, it is expected to take $1/pq$ RRPs to recover the packet. And it takes on average $1/p$ times to have a packet successfully received. Therefore, based on the analysis provide by Cao et al. (2006), the total end-to-end delay over one hop can be estimated as

$$Delay_{streaming} = T_{dl} + (1-p)(T_{dp}/p + T_{na}/pq)$$

One hop link ExOR is similar to the ARQ protocol. The difference is that instead of sending a single packet, the source node sends a batch of packets. Suppose the source node sends packets at a fixed interval T_{df} and the destination node sends acknowledgment back at the interval $\frac{1}{3}T_{df}$. Let the packet delivery latency be T_{dl}, where $T_{dl} \ll T_{df}$ and the batch size be M. The end-to-end delay over one hop can thus be estimated as $Delay_{ExOR} = [T_{dl} + (1-p)\frac{1}{pq}T_{df}]/M$.

Graphing these equations shows that the end-to-end delay of all three protocols is similar when links are reliable in both directions. However, for low quality links, the streaming protocol performs very badly, taking over twice as long as ARQ protocol to deliver a packet. ARQ is seriously affected when the backward link quality is worse than the forward link, a situation which is common in WM^{2}Nets. The use of a forwarder node in ExOR gives it the lowest end-to-end delay, even for very poor quality links. The worst case performance ratios for Stream:ARQ:ExOR are 15:7:2.

Implementation Experiments We implemented the ARQ, Streaming, and ExOR protocols using TinyOS/nesC (http://www.tinyos.net/) and Fleck motes (Sikka et al., 2004) and ran 45 one-hour experiments with different link qualities. To achieve low link reliability we took off the antenna from the sender and receiver and put two flecks SMA to SMA at a distance of about 5 cm. Data packets were transmitted every 5 s and, for ExOR, a forwarder node is added into the network to forward the sender's packets as necessary. ExOR batch size is 10 and the destination node sends 10 acknowledgments. The λ parameters were set to $\lambda_{ARQ} = 2/3$, $\lambda_{streaming} = 17/26$ and $\lambda_{ExOR} = 1$. A sample of measured link efficiencies is shown in the table below, together with the predicted efficiency for each using Eqs. (2.33) to (2.35), up to noisiness in estimating link reliability.

Protocol	Link Quality (A,B) (A,F) (F,B)	Measured Link Efficiency	Predicted Link Efficiency
ARQ	(0.98,0.98)	0.57	0.58
ARQ	(0.25,0.23)	0.05	0.048
Streaming	(0.99,1)	0.98	0.98
Streaming	(0.27,0.32)	0.0242	0.107
ExOR	(0.99,0.26) (0.91,1) (1,0.97)	0.48	0.50
ExOR	(0.075,0.13) (0.33,0.998) (0.93,0.90)	0.142	0.136

2.6 Hazardous Operation in WM^{2}Nets

A *hazard* is the out-of-order execution of queries and commands resulting from a lack of coordination between WM^{2}Net nodes and their associated AP (or CH in clustered configurations). Let us address the hazardous operation problem with an example of a WM^{2}Snet composed of sensor devices. Consider a battlefield application with two types of actors—tanks and ambulances, and one type of sensor—image sensors. Further, consider the case where the image sensors have detected the presence of an enemy target, and this message has been received by the sink. The sink issues two sequential commands: the first is sent to activate the tanks to attack the enemy targets whereas the second is sent to ambulances to rescue wounded soldiers. Now, if the order the commands are executed is reversed, the ambulances will arrive prior to the elimination of the enemy soldiers, resulting in a catastrophic situation where the rescue workers are placed at risk of being attacked.

Three different types of hazards are identified.

- **A command-after-command (CAC) hazard**—A CAC hazard occurs when the order of two sequential commands is not guaranteed. If two sequential commands, Command 1 and Command 2, are issued to two different actors in the event region, a CAC hazard occurs, if Command 2 is executed prior to Command 1. The battlefield application scenario described above is an example of CAC hazard.

  A CAC hazard can be formally defined as follows: Consider a set of n directives, $I_1, I_2, \ldots, I_n$. Let I_k and I_{k+1} be two dependent sequential commands sent to two actors in the event region, A_x and A_y. Let the execution of the command I_k from actor A_x resumes at time, T_1, and the execution of command I_{k+1} from actor A_y starts at time T_2. A CAC hazard occurs when $T_2 < T_1$.
- **A query-after-command (QAC) hazard**—A QAC hazard occurs when a query is issued upon the reception of a command but gets executed prior to the execution of the command. If two sequential directives, a command, Command 1, and a query, Query 2, are sent to an event region and the response to Query 2 is initiated prior to the execution of Command 1, a QAC hazard is then occurred. Formally, a

QAC hazard can be defined as follows: Let I_k and I_{k+1} be two sequential directives, with I_k being a command sent to an event region where an actor, A_x, sits and I_{k+1} being a query sent to the event region where a sensor, S_y, sits. Let the execution of the command I_k from the actor A_x resume at time, T_1, and the response to query I_{k+1} from node S_y starts at time, T_2. A QAC hazard occurs when $T_2 < T_1$.

- **A command-after-query (CAQ) Hazard**—A CAQ hazard is the opposite of a QAC hazard; it occurs when a query is executed prior to the command. Consider two sequential directives, say Query 1 and Command 2, issued to the event region in that order. If Command 1 gets executed prior to Query 1 for this event region, a CAQ hazard occurs. Let I_k and I_{k+1} be two related, sequential directives, with I_k being a query sent to the event region where a sensor, S_x, sits and I_{k+1} being a command sent to the event region where an actor, A_y, sits. Let $R_{k,x}$ denote the response to query I_k from node S_x, which is initiated at time, T_1 and $E_{k+1,y}$ denote the execution of command I_{k+1} from actor A_y starting at time, T_2. A CAQ hazard occurs when $T_2 < T_1$.

2.6.1 A Practical Approach to Addressing Hazards in WM²SAnets[12]

2.6.1.1 Hazard Modeling and Hazard-free Operation

Design Assumptions In this section, a formal description for avoiding the three hazards addressed in Section 2.6 is presented. The requirements for hazard-free operation are then illustrated.

To make the problem more tractable, the following simplifying assumptions are made:

- *Network Model*: We assume a WM²SAnet where sensors and actors are both static and are randomly distributed in the field.
- *Location Information*: We assume that sensors and actors are aware of their location (see Section 2.5 for related background and works).
- *Sensing, Acting, and Communication Ranges*: We assume that the sensing range (R_s) is equal to the communication range for sensors (T_s), and the acting range (R_a) is equal to the communication range for actors (T_a).
- *Routing Model*: We also assume that a reliable WM²Net routing protocol exists for delivering directives and gathering responses (Park et al., 2004).

The generic hazard-free operation goal can be described as follows:

Settings: Consider a WM²SAnet network where the sink issues a set of directives to a set of sensors or actors. Directives issued by the sink are subject to a set of dependencies determined by the sink. Let Ω denote the set of directives, Δ the set of entities, and Γ the dependency set. Each element, λ_m (λ_m in Γ), defines the dependency of two directives I_i and I_j, where $I_i, I_j \in \Omega$. If I_i is required to be executed before I_j, we use $I_i \rightarrow I_j$ to denote this dependency requirement.

[12] Excerpt from the invited article "A practical approach to addressing hazards in wireless sensor and actor networks," Ramanuja Vedantham, DSPS R&D Center, Texas Instruments, TX, USA, E-mail: r-vedantham@ti.com

To prevent all three hazards, any two directives that are dependent should be executed sequentially according to the dependency.

Observations: Consider any two entities, D_x and D_y, in Δ within the event region that are required to execute two directives, I_i and I_j, with dependency requirement $I_i \rightarrow I_j$. Let $t_{Ii \cdot Dx}$ denote the time that an instruction, I_i, is executed from some entity, D_x. To prevent a hazard, it is required that $I_i \rightarrow I_j$ is valid for both entities. Let $A(D_x)$ and $A(D_y)$ denote the sensing/acting region of D_x and D_y respectively. For hazard-free operation, the following four equations need be satisfied: (i) $t_{Ii \cdot Dx} < t_{Ij \cdot Dx}$, (ii) $t_{Ii \cdot Dy} < t_{Ij \cdot Dx}$, (iii) $t_{Ii \cdot Dx} < t_{Ij \cdot Dy}$, and (iv) $t_{Ii \cdot Dy} < t_{Ij \cdot Dy}$.

It can be shown that for equations (i)–(iv) to be satisfied, directives $I_i \rightarrow I_j$ need be ordered in the region $A(D_x) U A(D_y)$ (Vedantham et al., 2006). Based on this result, the following two inferences can be made:

- Any pair of dependent directives issued to entities that do not have any overlapping execution regions can be executed concurrently across the two entities, even though the relative ordering must be preserved within each entity.
- Any pair of dependent directives issued to entities with overlapping execution regions needs be ordered in the union of the two regions.

In this regard, applying these rules, for a given entity, D_x, pair-wise with any other entity in the event region, we can define a region in the neighborhood of D_x called the *dependency region*, where perfect ordering is necessary. The dependency region of a WM^2SAnet node can be defined as the region with radius equal to the sum of the sensing and acting ranges, whereas the dependency region of an actor is the region with radius twice the acting range.

Throughout the following paragraphs one such hazard-free approach, called the neighborhood clock (NC) approach, is described.

NC introduces the notion of a NC on every WM^2SAnet sensor and actor for ordering the directives within all dependency regions (see below for details). The NC is used to enforce synchronization between WM^2SAnet sensors and actors within a dependency region. It does not enforce synchronization beyond the dependency region of any WM^2SAnet sensor or actor, thereby allowing the other nodes in the event region to execute the directives concurrently.

When the sink learns about an event, the sink creates a reference clock for that event and initializes this clock to a unique start value, denoted by NC_0. This reference clock is used to indicate the progression of directives sent by the sink. This information is flooded throughout the event region. When a WM^2SAnet sensor or actor in the event region receives the message, it initializes its NC using the initial reference clock value. In this fashion, all nodes can synchronize their initial NC values. The reference clock is incremented whenever the sink sends a new directive. The reference clock of the sink, RC_i, is piggybacked with the *i*-th directive. Since the RC values increase linearly for every directive sent, the NC is ordered according to the sequence in which the directives were issued.

Each entity, D_x, maintains its own view of the progress in the network, based on its NC identifier, NC(x), where the view number is set to be NC(x) + 1. NC manages to synchronize the views on all WM^2SAnet sensors and actors within the dependency region. This is accomplished with the synchronization of the NC values of all NCs. A WM^2SAnet sensor (actor) moves to the next view only after all other WM^2SAnet sensors

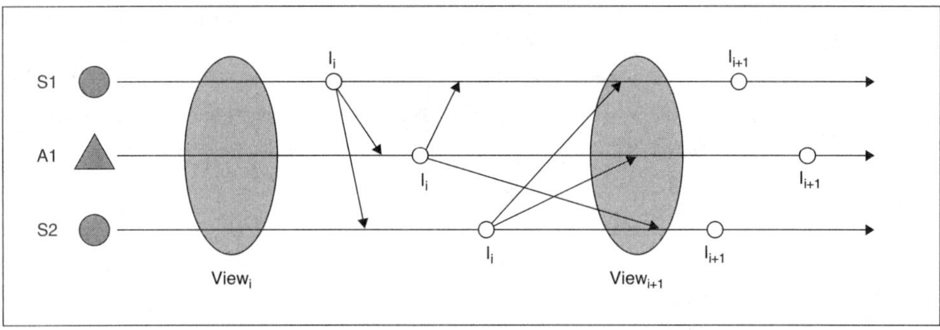

FIGURE 2.35 View movement in NC approach.

(actors) have moved into its current view. For the case when one or more node failures occur within the dependency region of a node, the node waits for the timeout value to expire before incrementing the NC clock. Thus, the difference between the views of any two nodes within the dependency region will be at most 1.

Any entity, D_x, can execute a directive only if the NC value piggybacked is the same as the current view number. That is, if an entity is in $view_i$, it is allowed to execute the directive with an NC value equal to i. Applying this scheme, NC ensures the atomic execution of all directives from all entities. Once an entity executes a directive, it notifies all other entities in dependency region for the completion of the directive. The progress of views within a dependency region, which consists of two WM^{2}SAnet sensors and one actor, is illustrated in Fig. 2.35. As illustrated in the left ellipse area, at a certain time all the entities have moved into $view_i$. So, all these entities are allowed to execute the directive with $NC = i$. After the execution, each entity notifies the other entities about the execution of the directive. Whenever an entity receives notifications from all other entities, it will move to the next view, $view_{i+1}$, as illustrated in the right ellipse area.

It is evident that the NC mechanism ensures hazard-free operation within the dependency region of any entity. As illustrated by Vedantham et al. (2006), the vector NC that extends the idea of a scalar NC is to include an array of clocks based on the number of dependency lists between directives for each type of event where a dependency list is defined by the chain of dependencies for a specific directive. We do not discuss the vector NC in detail due to space constraints.

Simulation Studies The performance of the NC approach with two basic strategies (Vedantham et al., 2006), namely the Wait-For-All (WFA) and the Bounded Delay (BD), is evaluated and discussed. The following performance metrics are considered:

- The *directive-execution throughput*, which is defined as the number of directives executed per second;
- The *correctness*, which is the probability of hazard occurrence; and
- The *total traffic per directive*.

Simulation Setup For all simulations, 2000 WM^{2}SAnet sensors and actors are randomly placed on a 2,000 m × 2,000 m square area. The sensing and communication range of

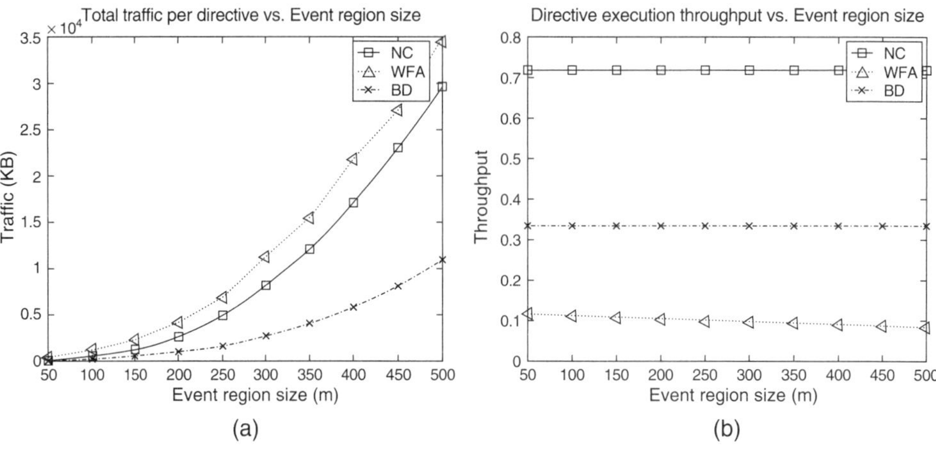

FIGURE 2.36 Performance for different event region sizes.

sensors is set to 30 m, and the acting and communication range of actors is set to 60 m. When an event is detected, a minimum sensor (Gupta et al., 2003) and actor coverage is formed. The CSMA/CA is used for the provision of MAC functions.

The events considered in the simulation are regional events with varying radii ranging from 50 m to 500 m. In the following results, if not specified explicitly, the distance from the sink to the event center is 1,000 m, the radius of event region is 200 m and the loss rate per hop is 10%. Amongst the directives issued by the sink, the probabilities of queries and commands are both 50%. A default value for the event-processing time, $T_{EP} = 2$ s, is assumed. The correctness of NC and WFA are both 100%.

Varying the Event Region Size Figure 2.36 shows the performance results of the three approaches under varying event region sizes. As shown in Fig. 2.36a, the traffic per directive of all three approaches increases when the event region size increases. In BD, this is mainly attributed to the increase of nodes in the event region. While BD shows a very good performance in terms of overhead, it is only at the expense of correctness and throughput. For NC, aside from this reason, since each node has to receive notifications from all other nodes within its dependency region, the overhead is relatively large. For the WFA strategy, the traffic measured is very large, which is attributed to the acknowledgement packets that nodes send to the sink in response to directives. Figure 2.36b shows the throughput variation for increasing event size. NC has the largest directive-execution throughput when compared to the basic approaches. This is because the dependency region is just the 2-hop neighborhood region. Both NC and BD demonstrate constant throughput values since their mechanisms are not affected by the region sizes, wheres WFA's throughput drops slightly due to the fact that it must wait for more time to receive all the acknowledgements before issuing the next directive.

Varying the Distance from the Sink to the Event Center Figure 2.37 shows the performance results of the three approaches for varying sink-to-event distances. We can see that WFA

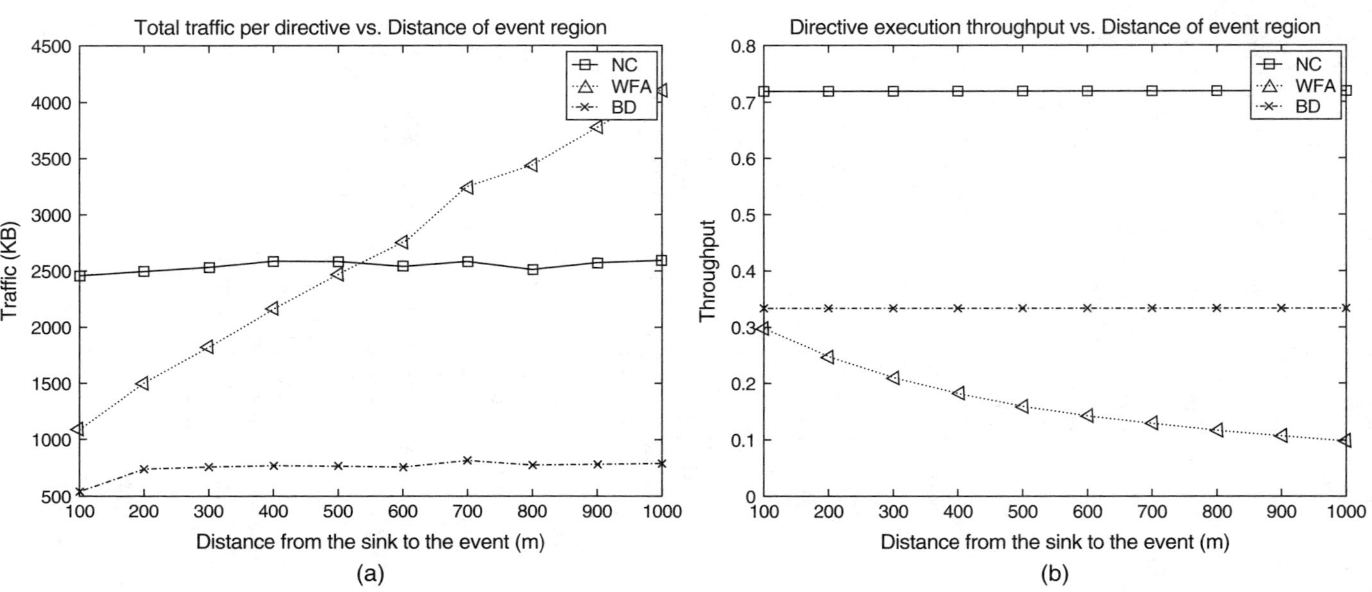

FIGURE 2.37 Performance for different sink-to-event distance.

induces much higher overhead when dealing with far-away events. This is because all WM^{2}SAnet sensors and actors in the event region are required to respond back to the sink. As shown in Fig. 2.37a, both BD and NC have (almost) constant traffic, which only increases slightly with increasing sink-to-event distance. This is because the average traffic in delivering the directive is almost a constant. Additionally in NC, the traffic generated within the dependency region will always be a constant. Figure 2.37b shows that NC has largest throughput. The throughput of WFA drops because the waiting time for issuing a directive increases with increasing sink-to-event distance. Unlike WFA, the throughputs of NC and BD do not change with the sink-to-event distance, since the latency between the executions of successive directives does not depend on the distance of the event from the sink.

Testbed Implementation The feasibility of the NC approach is also demonstrated using a WM^{2}SAnet testbed implementation. The goal of the testbed implementation is two-fold: (i) to demonstrate the functionality of a real-life WM^{2}SAnet testbed, and (ii) to show the feasibility of the NC approach in resolving hazards.

Testbed Setting The testbed consists of two parts: (i) a WM^{2}SAnet sensor network connected to the sink; (ii) a wireless actor network from the sink to the actors. The testbed comprises of twelve sensor devices and two actor heads, each of which controls four different actors—fans, rotating pedestal and two LED-array light sources. The sensor devices are based on Crossbow MICA2 and MICA2DOT motes, from two MOTE-KIT5040 professional kits (CrossBow Technology Inc., Wireless sensor networks: Product reference guide: http://www.xbow.com). Six of the MICA2 processor/radio boards were mounted with MTS310 sensor boards; the latter were equipped with acceleration, magnetic, light, temperature, acoustic, and sounder sensors. The MICA2 and MICA2DOT motes are set to the lowest power setting with an effective transmission range of 2–6 feet. We use the Surge Reliable application for the different sensors to communicate with the sink in a multihop fashion, either using other sensors or relays. The application transmission rate is fixed to 1 packet transmission every 8 s, where the packet size is approximately 128 bytes.

The two actor heads are composed of a combination of a Dell laptop, attached to a Devasys USB I2C/IO board. The laptop, Devasys USB I2C/IO board combination acts as a wireless transceiver and microcontroller to turn on or off the actor devices attached to the configurable output pins. We consider 4 different actor devices attached to the each of the actor heads—(1) 2 LED-array light sources, (2) 1 mini-fan, and (3) 1 rotating pedestal. For the actor network, the terminal computer and the laptop-switch combination are equipped with 802.11b cards. To send commands between the sink and the two actor heads, we establish a reliable connection using TCP sockets.

Demonstration for a WM^{2}SAnet To address the first goal, the testbed is set up as shown in Fig. 2.38. The 6 sensor devices communicated in a single or multihop fashion (using other sensors or relays) to the sink. In the testbed, associated with sensor nodes, S1, S2, S5, and S6, there are corresponding LED array actors, A1, A2, A5, and A6. LED array actors A1 and A2 are controlled by the actor-head 1 (laptop, Devasys USB I2C/IO switch), and similarly, actor-head 2 controls the LED array actors, A5 and A6. The sink issues commands to the actors based on the data received from the corresponding sensors at

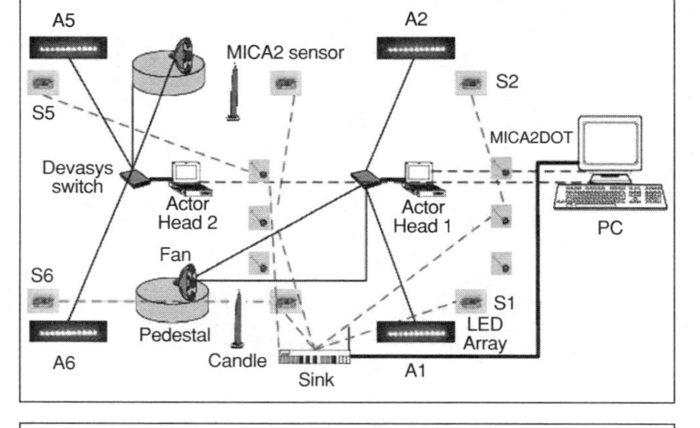

FIGURE 2.38 WM^2SAnet testbed for addressing hazards. (a) Logical representation of the WM^2SAnet testbed. (a) Illustration of the WM^2SAnet testbed.

the rate of 1 command packet every 16 s (half that of the sensor-data reporting rate). The rate is set to such a low value in order to ensure that there is at least one 1 sensor-data packet received at the sink from each sensor. We refer to this duration of 16 s as an epoch. The sink issues a command to a LED array actor if the average light reading from all the packets received in the previous epoch from the corresponding sensor is below a minimum threshold. Similarly, a command is issued to turn off an actor if the average reading from the corresponding light sensor is above a maximum threshold. This command issue to turn on or off an actor is sent to the corresponding actor-head using the wireless channel. The actor-head controls the output pin of the switch resulting in the desired output.

Demonstration of a CAC hazard and the basic NC Approach The second goal of the testbed implementation is to demonstrate the occurrence of CAC hazards. In this context, we consider the lighting of the candle as an external event. The goal is to ensure that the candle remains lit while the temperature in the region does not increase.

This can be accomplished by issuing two sequential commands in the following order: (1) command C1 is issued to the rotating pedestal, to ensure that when the fan is on, the direction of air flow is not concentrated only in the direction of candle ensuring the candle from being put out and (2) command C2 is issued to turn on the fan, so that the temperature in the region does not increase noticeably. This would correspond to the desired sequence of two commands.

When a CAC hazard occurs, the sequence of command execution is reversed: (1) command C2, which is the command issued to turn on the fan, gets executed first and (2) command C1, which is the command to turn on the rotating pedestal, gets executed only after the execution of C2. We artificially introduced the occurrence of a CAC hazard by probabilistically introducing a delay prior to the issue of each command at the sink. Thus, the candle is put out prior to the execution of the command to turn on the pedestal (C1). This resulted in an undesirable consequence, that is, the candle being extinguished.

To address the CAC hazard, the notion of sequence numbers is introduced and is associated with each directive issued by the sink. The sequence number acts as the scalar virtual clock as described in the NC mechanism. A command is executed only if the NC clock associated with the actor is exactly 1 less than the sequence number associated with the command issued by the sink. If the sequence number piggybacked is twice or higher the NC value, the command is queued and is executed only when the first condition is met (NC = Seq._No. − 1). After the execution of each command, the NC clock value is incremented. In the scenario considered, the command C1 has sequence number 1, and that of C2 is 2.

2.7 WM^{2}Net Testbeds and Prototypes

To improve WM^{2}Net performance, it is important to understand the fundamental characteristics of the network and its behavior. Such an understanding allows network operators to understand what applications are suitable to the environment and what kind of adaptations may be needed for different applications to operate over the network. Our goal here is to delve into the details of realistic testbeds and develop an intuitive feel for the induced complexity, configuration parameters, network node properties, associations and services. We can thus see how a few simple techniques work and could be practically used in a WM^{2}Net context.

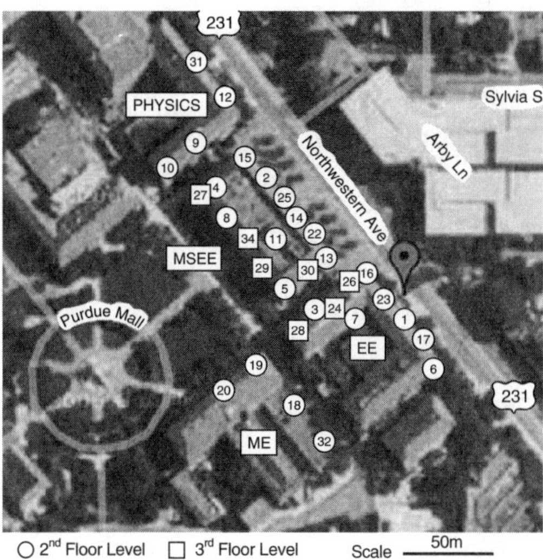

FIGURE 2.39 MAP testbed.

2.7.1 Measurement-Based Characterization of a Wireless Mesh Network[13]

2.7.1.1 Setup and Methodology

For the measurement studies we used the MAP testbed (Mesh@Purdue. http://www.engineering.purdue.edu/MESH), shown in Fig. 2.39. MAP consists of 32 mesh routers spread out across four academic buildings on the Purdue University campus. Each router has two radios. For this study, we used one of them: the Atheros 5212 based 802.11a/b/g wireless card. Each radio is attached to a 2 dBi rubber duck omnidirectional antenna with a low loss pigtail to provide flexibility in antenna placement. Each mesh router runs Mandrake Linux 10.1 and the open-source *madwifi*. IP addresses are statically assigned. To minimize interference from other 802.11b networks and/or other interfering sources, such as microwaves, the measurements were performed during night hours.

2.7.1.2 Latency Measurements

Methodology For this set of measurements the **optimized link state routing** (OLSR) protocol (Clausen et al., 2003) was used in MAP, enhanced with ETX routing metric. To avoid the control packets generated from OLSR interfering with the actual measurements, we ran OLSR daemon for appropriate time to ensure path stability and availability between all pairs of nodes. At this point, that is, when routing stability is reached, control messages are not considered further up. Hence, the same routing paths have been used for all the experiments. The methodology adopted for measuring packet latencies was as follows: Each node generates and sends 100 1470-byte ping packets to each other node. In this way, we measured the round-trip-time (RTT) between any pair of nodes for the paths calculated by the OLSR protocol. The packet size used is typical for the majority

[13] Excerpt from the invited article "Measurement-based characterization of a wireless mesh network," Saumitra M. Das, Dimitrios Koutsonikolas, Y. Charlie Hu. School of ECE, Purdue University, West Lafayette, USA, E-mail: {smdas, dkoutson, ychu}@purdue.edu

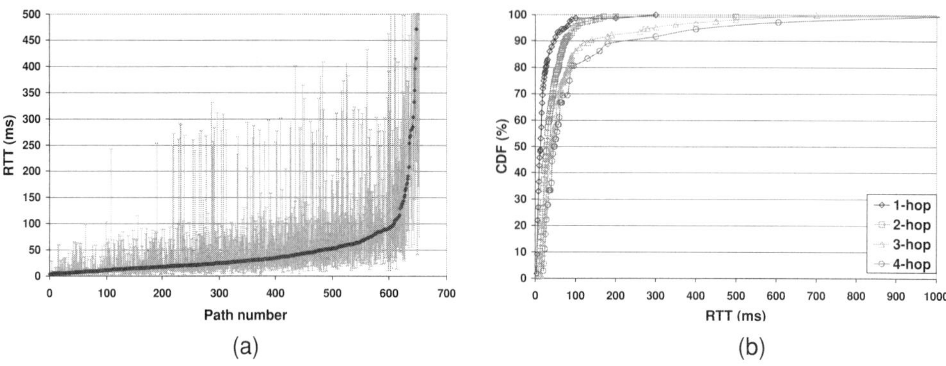

FIGURE 2.40 (a) RTT. (b) CDFs of RTTs.

of Internet applications. The packet interarrival time was 0.01 s, which is equivalent to a sending rate of about 1.1 Mbps.

Results The results of the latency measurements are shown in Fig. 2.40. Figure 2.40a plots the average (over 100 packets), maximum and minimum RTT values for each of the 654 paths used by OLSR (i.e., for each of the 654 pairs of nodes that were well connected to each other).[14] Figure 2.40a plots the CDF of RTT values for 1-hop, 2-hop, 3-hop, and 4-hop paths.

From Fig. 2.40a we observe that RTTs span over a large range of values between 3.45 and 1374 ms (for clarity, we kept the maximum value of the y-axis in the graph at 500 ms). Also, for many paths, the variations are very large; this mainly occurs for maximum values. For example, paths with average RTT below 50 ms have maximum values larger than 250 ms. This shows that network conditions experience steep changes over time and there exist periods of very poor connectivity. This can be a significant problem for applications such as VoIP or video streaming, which require constant jitter. On the other hand, minimum values (the true RTT) are not in general much smaller than the average ones, which shows that the quality of the paths is not very unstable. One more encouraging observation is that for the majority of the paths (600 out of 654 paths, about 91% of all paths), the average RTTs are lower than 100 ms, which is tolerable. Figure 2.40b confirms these observations, showing that a large fraction of paths have low RTTs. About 99% of the 1-hop paths and 95% of the 2-hop paths have RTTs lower than 100 ms. Also, 50% of them have RTTs lower than 50 ms. The majority of longer paths also experience low RTT values—86% of the 3-hop paths and 82% of the 4-hop paths have RTTs lower than 100 ms. The problem with paths longer than 1 hop is that their CDFs have very long tails. For example, there is one 2-hop path, from node 12 to node 17, which had an average RTT of 1374 ms.

2.7.1.3 Loss Measurements

Methodology For the loss measurements, we used the same ping messages used in the previous section and measured the average loss rate over 100 packets for each path. The results are shown in Fig. 2.41. Figure 2.41a shows the average loss rate for each of the 654

[14] The remaining paths were found with one or more bad links. These were not considered.

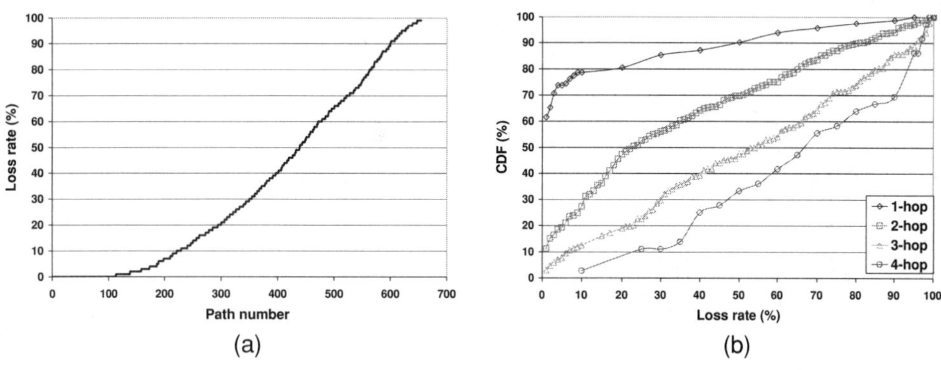

FIGURE 2.41 (a) Loss rate. (b) CDFs of loss rates.

paths where Fig. 2.41b the CDFs of the loss rates for the 1-hop, 2-hop, 3-hop, and 4-hop paths.

Results From Fig. 2.41a we observe that loss rates cover the whole range between 0 and 100%. The results are not as encouraging as in the case of RTT measurements. More specifically, only 300 out of 654 paths (46%) have loss rates lower than 20%. On the other hand, for 33% of the paths loss rates are higher than 50%.

From Fig. 2.41b we can see the correlation between loss rate and the number of hops. The majority of 1-hop paths are paths with very low loss rate. 78.6% of them have loss rates lower than 10%. For paths longer than 1 hop, the loss rates increase rapidly. Only 47% of the 2-hop paths, 19% of the 3-hop paths and 8% of the 4-hop paths have low loss rates of 20% or less. In general, with this set of measurements we observed that the majority of paths show satisfactory average RTTs, but high loss rates. In the next session, we demonstrate transport layer issues. The aforementioned link conditions are assumed.

2.7.1.4 Transport Measurements

Methodology In this section, we characterize the performance of UDP and TCP transport layer protocols. The methodology followed is as follows: Nodes initiate a UDP session to each, one at a time. A 5 s of idle time interval between any two sessions is used. The sending rate is 5 Mbps. The *iperf* tool (dast.nlanr.net/Projects/Iperf) was used for this experiment. In the second part of this experiment, we used *netperf* (www.netperf.org/) to initiate TCP sessions between any two pairs of nodes. Again, a single TCP session was always active. The routing paths where either *iperf* or *netperf* was unable to initiate a connection were not included in the results. The results are shown in Figs. 2.42 and 2.43.

Results Figure 2.42 shows that UDP outperforms TCP significantly. Figure 2.42a shows that 50% of the paths have UDP throughput higher than 1,700 kbps, while from Fig. 2.42b the percentage of paths that have TCP throughput higher than 1,700 kbps is only 30%. Moreover, for half of the paths, TCP throughput is lower than 600 kbps. Even worse, for 31% of the paths, TCP throughput is less than 200 kbps; on the other hand, the minimum UDP throughput observed is 298 kbps. This result shows that TCP performs poorly in multihop wireless networks. From *tcpdump* and *tcptrace* we found the reason for this to be the lossy environment, which prevents TCP from increasing its window.

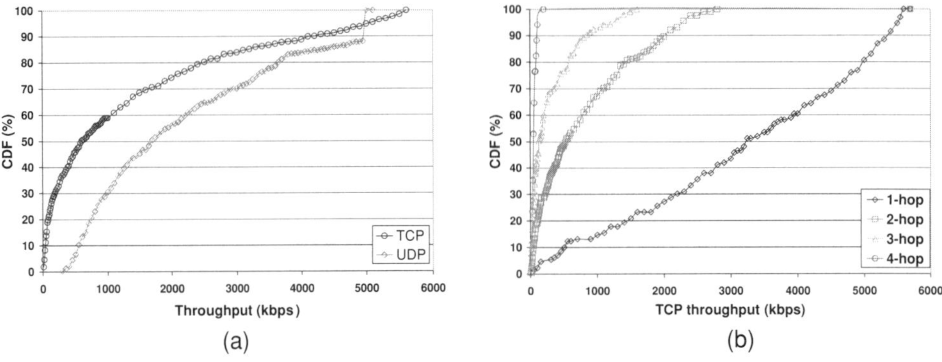

FIGURE 2.42 (a) CDF of TCP/UDP throughputs. (b) CDF of TCP throughputs for different hop count paths.

2.7.1.5 Interference Measurements

Methodology The 32 nodes of the testbed can form up to 992 (directional) links. However, many of these links may not always be available; this may occur, for instance, if two nodes are far away from each other such that the measured BER at the receiving side is below some critical threshold. To calculate the exact number of links as well as to produce estimates on their quality, we first transferred a series of 100 ping messages between each pair of nodes. Note that in this case no routing protocol was used, since we are interested in links and not in multihop paths. Hence, we could not use the ping messages as in previous sections. If all 100-ping messages between a pair of nodes were lost, this link is classified as nonexisting. This experiment assures that the network is not partitioned. As expected, the link quality and node distances do not directly relate. For example, we observed a perfect link (0% loss) between nodes that are placed in different buildings (e.g., 1 and 13—see Fig. 2.39). Although in different buildings, however, nodes are close to windows and signal is propagated in the low-loss outdoor environment. On the other hand, there is no link between nodes 19 and 20, although their distance is half compared to the distance of link 1–13. In total, we found 257 links with loss rate less than 100%.

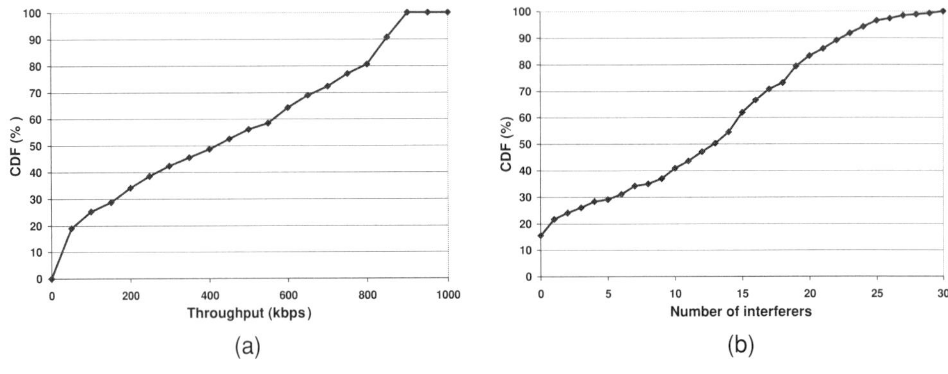

FIGURE 2.43 (a) CDF of individual throughputs. (b) number of interferers for the 257 links of the testbed.

To measure interference, we executed the following series of experiments. In the first set, each node broadcast 1470-byte packets with its highest transmission speed for 30 s. All other nodes that could receive packets measured the throughput in these 30 s. This experiment produced the throughput of all existing links. Figure 2.43a shows the cumulative distribution function (CDF) of the throughput for the 257 existing links.

In the second series of experiments, following a methodology similar to (Padhye et al., 2005), we measured the pairwise interference among all nodes. For this experiment, we had each *pair* of nodes broadcast 1,470-byte packets *together* for 30 s. In each 30-s interval, all other nodes except for the two senders measured the throughput from the two senders. At the end of this experiment, for each pair of nodes, one viewed as the sender S and the other the interferer I (and vice versa), we calculated the receiver throughput ratio (RTR) at each one of the rest 30 nodes (*R*) as follows:

$$RTR_{\mathrm{R}}^{\mathrm{S,I}} = Throughput_{\mathrm{R}}^{\mathrm{S,I}} \,/\, Throughput_{\mathrm{R}}^{\mathrm{S}}$$

where $RTR_{\mathrm{R}}^{\mathrm{S,I}}$ is the receiver throughput ratio at receiver R when S is the sender and I the interferer node, $Throughput_{\mathrm{R}}^{\mathrm{S}}$ is the throughput at R from node S when only S transmits and $Throughput_{\mathrm{R}}^{\mathrm{S,I}}$ is the throughput at R from node S when both S and I transmit simultaneously. If $RTR_{\mathrm{R}}^{\mathrm{S,I}} < 0.9$, we consider that node I is an interferer for link S→R. In this way, we found out all nodes that are not interferers for a particular link S→R according to the pairwise interferer model.

In the third set of experiments, we measured multiway interference[15] as follows. We considered each of the 257 links in turn, along with their noninterferers (say *n*). For each of these links we had again the sender broadcast packets for 30 s along with 1, 2, 3, ... up to *n* interferers simultaneously. At every single point of time, we measured the sender's throughput at the receiver and calculated the receiver's throughput ratio.

Results

Pairwise Interference Figure 2.43b shows the CDF of the number of interferer nodes for the 257 existing links. We observe that the number of interferers varies but, in general, pairwise interference is a widespread phenomenon. About 15% of the links have no interferers. On the other hand, 60% of the links have 10 or more interferers and 50% have 13 or more interferers. Finally, the last 17% of the links have more than 20 interferers, and, as we mentioned before, we removed these heavily affected links from the remaining experiments.

According to (Padhye et al., 2005), the pairwise interference model is accurate in predicting interference in the majority of practical cases. The results seem to verify this. However, we still wanted to study further, What happens to those links that are not affected by pairwise interference? This is the subject of the following paragraphs.

Does Multiway Interference Occur? Figure 2.44 shows three examples used in the testbed. In these figures, S is the sender, R is the receiver, and I_1, I_2, I_3, I_4 are nodes that were found not to interfere according to the pairwise model, but they might interfere if two, three, or four of them transmit simultaneously. Table 2.4 shows the receiver throughput

[15] That is, interference from the cumulative effect of multiple transmitters on a given link. Such interference cannot be measured using pair-wise measurements as these only consider two transmitters at a time.

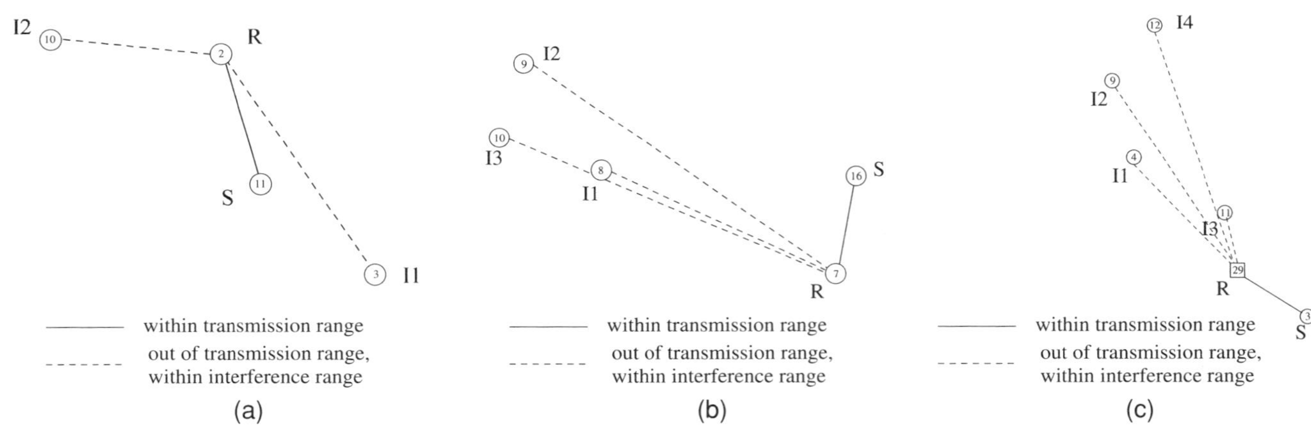

FIGURE 2.44 Three examples showing that pairwise interference is not always enough.

Example	Throughput Ratio			
	Min 1 intf.	Min 2 intf.	Min 3 intf.	4 intf.
a.	0.92	0.74	—	—
b.	1	0.88	0.74	—
c.	0.93	0.96	0.91	0.86

Table 2.4 Three Examples Demonstrating the Impact of Pairwise Interference on Throughput Ratio

ratio for different number of interferers in the three examples of Fig. 2.44. The second column shows the minimum ratio at the receiver when interferers are considered one at a time. The third column shows the minimum ratio when interferers are considered in pairs, whereas the fourth shows the minimum ratio when interferers are considered three at a time. Finally, the last column shows the ratio when all four interferers transmit simultaneously. As we observe, the ratio for the second column is above 0.9 in all the three examples (no interference according to the pairwise model).

In the first example, node 11 is the sender, node 2 is the receiver, and nodes 3, 10 the interferers. As Table 2.4 shows, if we allow both nodes 3 and 10 to transmit simultaneously with node 11, the receiver throughout ratio is reduced to 0.74, quite below the threshold of 0.9 and the quality of link 11→2 worsens significantly. Note again that the distance of the nodes is not directly mapped to the quality of the links. For example, nodes 11 and 10 are almost equal distance away from node 2, however, only link 11→2 exists.

In the second example, node 16 is the sender, node 7 is the receiver, and nodes 8, 9, 10 the interferers. In this example, all the three interferers are very far from link 16→7 and they cannot affect it (throughput ratio is 1 when each of these three nodes transmits simultaneously with the sender). Even when two of these nodes transmit together, the ratio is not significantly reduced—it only drops to 0.88, very close to the threshold limit. But when all three nodes 8, 9, 10 transmit simultaneously, the ratio becomes 0.74, which makes the link quality unacceptable.

Finally, the third example shows that in some cases even 5-way interference has to be considered. In this example, node 3 is the sender, node 29 is the receiver and nodes 4, 9, 11, and 12 the interferers. As Table 2.4 shows, the throughput ratio remains above the threshold when two or three of the interferers transmit simultaneously (with minimum values equal to 0.96 and 0.91, respectively) but it drops to 0.86 when all four interferers transmit simultaneously.

How Widespread is Multiway Interference? Out of 257 links, we found 16 of these where pairwise interference model could not accurately predict interference. Out of these 16 links, 4 had throughput between 200 and 400 kbps, 9 had throughput between 600 and 800 kbps, and the rest three had throughput higher than 800 kbps without any interferers present. Some general observations are as follows. As we observe in Fig. 2.44a, about 20% of the links have throughput lower than 50 kbps. For those links, no observation can be made about interference, since the quality is so bad that adding more interferers cannot make it worse. In fact, when we repeated the experiments, some even showed zero throughput in some cases. Such links will probably not be selected by a routing protocol that uses a link-quality based routing metric (e.g., ETX or SPP). Hence, in the rest of this study, we ignore these links. Similar methodology is followed by Padhye et al. (2005) where low quality links are rejected using an ETX-based threshold.

We did not find any case of 3-way interference for the 36 links with throughput between 50 and 200 kbps (about 14% of the total 257 links). For these links, throughput is still very low, although nonzero. For many of them we observed large variations in throughput when the number of interferers changed. For example, in many cases throughput was increased when we added interferers, compared to the case where only the sender transmitted. For the rest of them one interferer was enough to reduce the throughput ratio to very low levels, hence the pairwise model was enough. Note that a reasonable routing algorithm should also avoid most of these links. For the 77 links of medium quality with throughput between 200 and 600 kbps (30% of the total 257), the pairwise interference model was successful in predicting interference in almost all cases. For those links, we did not observe the strange variations described in the previous paragraph, but in most cases, one interferer was enough to change the link quality from medium to low and reduce the throughput ratio below acceptable levels. Only 4 links remained unaffected by single interferers, but were affected when two or three interferers transmitted simultaneously.

The majority of instances (9 out of 16 links), where the pairwise interference model was not sufficient, occured for the 41 links (16% of the total 257) of medium to high quality and with throughput varying between 600 and 800 kbps. Since throughput is high enough for these links, there could be margin for gradual decrease by adding more interferers. Hence, we had cases where one interferer reduced the ratio only slightly but without crossing the 0.9 threshold, the second interferer sent the ratio close to the threshold, and three or more interferers resulted in large throughput reduction. Finally, for the 51 high quality links (20% of the total 257) with throughput higher than 800 kbps, the common case is that if such a link is not affected by other nodes, when they are considered one at a time, it is also not affected when the other nodes are considered more than one at a time. Hence, again the pairwise interference model gives the correct answer in most cases. But we still found three cases where 3-way interference should be considered. Thus, while multiway interference does occur and, when it does, it significantly affects throughput, we found that the phenomenon is not widespread and depends on the initial link quality.

2.7.2 MeshDVNet: A Fully Functional IPv6 Wireless Mesh Network Testbed[16]

2.7.2.1 Introduction

The design of MeshDVNet leverages upon two core tasks of WMRs: mesh backbone self-organization and end-to-end user connectivity. For the former, the MeshDV software uses WMRs to self-organize and establishes the mesh backbone in a proactive way. To achieve this, enhanced cross-layer techniques are used. For the latter task (end-user connectivity), MeshDVNet maintains a standard WLAN access mode. This means that all users with a device running any standard operating system and equipped with an IEEE 802.11 wireless card can get connected without the need to install any additional software. Clients connected to the MeshDVNet platform are all logically placed in the same subnetwork (Fig. 2.45), regardless of the WMR they are associated to. The MeshDV daemon, which runs on each WMR, hides the multihop mesh backbone part to clients.

[16] Excerpt from the invited article "MeshDVNet: A fully functional IPv6 wireless mesh network testbed," Luigi Iannone. LIP6/CNRS—Université Pierre et Marie Curie (Paris VI), Paris, France, E-mail: luigi.iannone@lip6.fr

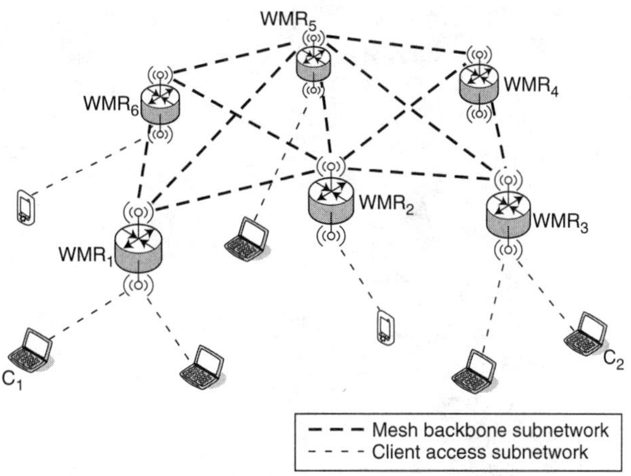

FIGURE 2.45 Logical architecture and example of wireless mesh network deployment.

Even if clients are associated to different WMRs far apart, thanks to MeshDV, they can communicate to each other as if they where on the same LAN segment. Communication between clients is set up in an on-demand fashion. On-demand connectivity (Aggélou, 2004) prevents the network from distributing useless information to all the nodes, reduces the routing setup signaling overhead, and eases the management of mobile clients.

2.7.2.2 The MeshDVBox and the MeshDVNet Platform

The MeshDVNet (LIP6-UPMC RNRT-InfRadio Project: http://rnrt-infradio.lip6.fr/indexEnglish.html) testbed is composed of custom wireless mesh routers that we call MeshDVBox. The WMR is not a proprietary solution but is based on off-the-shelf components. In particular, a WMR is bundled in a Soekris net4521 box (Soekris Engineering: http://www.soekris.com/net4521.htm). It has a PC-like architecture with a 133 MHz AMD ElanSC520 processor, 64 MByte SDRAM and equipped with a 1 GByte Microdrive. It has also two PCCard/Cardbus slots and one mini-PCI socket. Figure 2.46 shows a picture of a MeshDVBox.

Each MeshDVBox uses two wireless cards, each having their its external antenna. The first one is the client interface that is based on the IEEE 802.11 b/g (Institute of Electrical and Electronics Engineers (IEEE), 1999a) Technology and functions as access point for client connections. This interface is a Proxim 8470-WD b/g card (Proxim wireless: http://www.proxim.com/products/wifi/client/11bgpccard/) occupying one of the Cardbus slots (see left hand side of Fig. 2.46). The second card is the mesh interface that is based on the IEEE 802.11a (IEEE 802 LAN/MAN Standards Committee, 1999) technology and operates in ad hoc mode to form the mesh backbone with the other WMRs' peer interfaces. This interface is a NetGate 5354 MP Plus Aries2 4G a/b/g card (Netgate: http://www.netgate.com) occupying the mini-PCI socket (see right hand side of Fig. 2.46). Since the two technologies used (802.11 a and 802.11 b/g) work in separate frequency bands, the mesh backbone and the client access subnetworks are physically independent. All wireless interfaces are IPv6 configured. Each MeshDVBox runs NetBSD 4 (the development branch of the NetBSD Project (The NetBSD Project: http://www.netbsd.org). Figure 2.47 depicts a snapshot of the network testbed deployed.

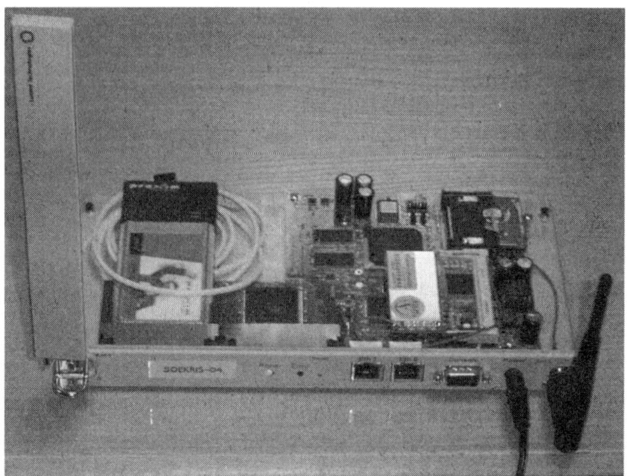

FIGURE 2.46 The MeshDVBox wireless mesh router.

The MeshDVBoxes are named Sxx, with xx ranging from 01 to 12. The device named C01S is an access point used to connect the testbed to the wireline LAN through the S04 MeshDVBox that acts as a GW.

On the wireline LAN, an IPv6 DNS together with an IPv6-to-IPv4-proxy server is deployed. In the same picture, the backbone's routes that are established using the routing daemon are also depicted. Figure 2.47 is a simplified version of a snapshot of the supervision web page, used to monitor if MeshDVBoxes are running and reachable. This page is publicly available (IPv6 only) at http://www.infradio-jussieu.lip6.fr/supervision/supervisionmesh-scott.html. MeshDVNet offers IPv6 connectivity on the entire building floor and can be used without any particular setup on PCs, PDAs, or laptops. Common Internet usage, including basic services like Web browsing, e-mailing, MHh and so on, can be performed without any particular problem.

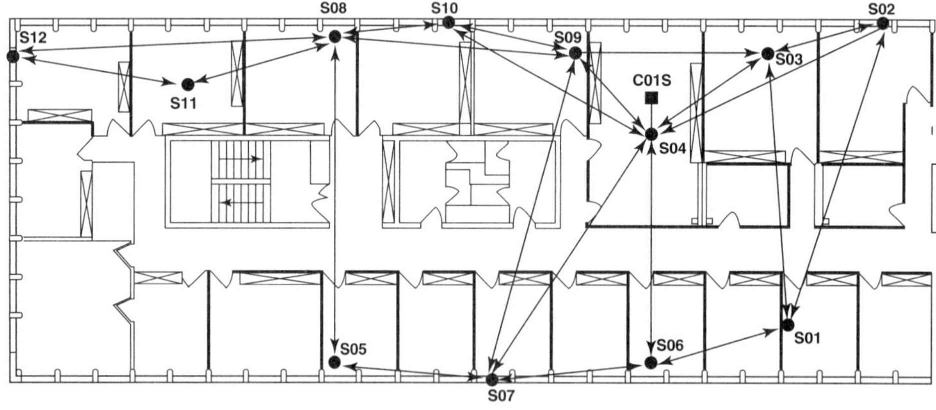

FIGURE 2.47 The MeshDVNet wireless mesh network testbed deployed at the LIP6 Laboratory. Nodes named from S01 to S12 are MeshDVBoxes. Arrows are the corresponding backbone links.

2.7.2.3 The MeshDV Architecture

As already stated, the MeshDVNet platform is based on the MeshDV daemon. MeshDV is an integrated software platform that is able to perform the following tasks:

- Proactive route building inside the mesh subnetwork using cross-layer metrics,
- GW advertisement,
- DNS advertisement,
- Address prefix advertisement,
- On-demand communication setup between clients associated to different WMRs,
- Clients' mobility management.

The alluring property of MeshDV is that all these tasks are performed in a totally transparent manner. Clients associated to the same MeshDVBox communicate as in traditional WLAN networks. Even more, MeshDV is able to facilitate clients of different WMRs communicate as if they where associated to the same WMR.

The functional architecture of MeshDV is depicted in Fig. 2.48. MeshDV has been designed in order to take advantage of the two-tier architecture of WM^{2}Nets and is composed of the following four main modules (see Fig. 2.48): (a) *Client Manager Module*, (b) *NDP Proxy Module*, (c) *IPv6 Forwarder Module*, (d) *Enhanced DV Module*. The *Client Manager Module* implements on-demand mechanisms necessary to discover which MeshDVBox a client is associated to. Details of this mechanism are provided below.

MeshDV has a specific module that allows it to act as Neighbor Discovery Protocol proxy. This module manages ICMPv6 requests issued by local clients in order to set up a communication to remote clients. The advantage is that clients do not have to perform any particular operation, since only standard ICMPv6 messages are exchanged between clients and MeshDVBoxes. In this way MeshDV behaves totally transparent to the users.

The *NDP-Proxy Module* and the *Client Manager module* provide a lookup mechanism and a standard interface to clients that synergetically facilitate the establishment of a communication between clients associated to different MeshDVBoxes. Once the communication is set up, data packets are transported through the mesh backbone using an

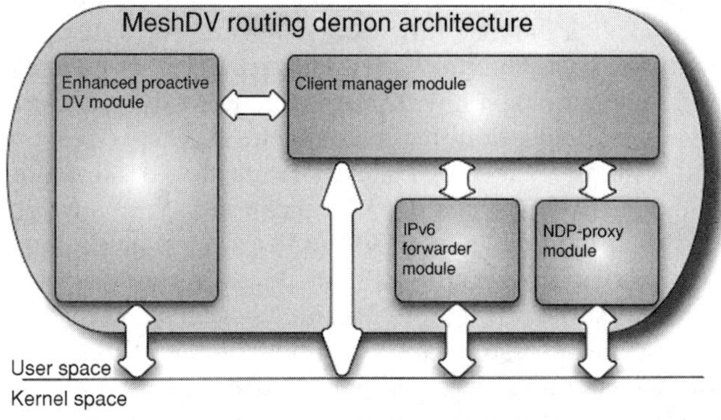

FIGURE 2.48 Functional architecture of the MeshDV daemon.

IPv6-in-IPv6 tunneling approach. The *IPv6 Forwarder Module* performs this task. Such approach introduces a certain amount of overhead, which is due to the size of the IPv6 header. Its major advantage on the other hand is that MeshDVBoxes along a communications path keep state information about the ongoing communication. Only MeshDVBoxes at the edges of the communication path, that is their *Client Manager module*, need be aware of the ongoing communication.

The last module is the *Enhanced DV Module*, whose task is to maintain proactively a mesh of routes among all MeshDVBoxes in the network. This is a fundamental task, since the abovementioned modules and mechanisms rely on the existence of a mesh backbone. In the next subsection, we describe the routing metrics used from the *Enhanced DV Module*.

Cross-Layer Routing on the Mesh Backbone The core of the *enhanced DV module* is an IPv6 implementation of a modified version of the DSDV (Perkins and Bhagwat, 1994) routing protocol. Destination-sequenced distance-vector (DSDV) routing is a simple distance vector routing protocol based on the Bellman-Ford algorithm. As the name suggests, a sequence number is attached in each update message to ensure fresh and loop-free paths. We implemented two different versions of the Enhanced DV module.

The first version uses the hop-count metric; thus, shortest-path routes are constructed. Each link (hop) is assigned a cost equal to one and hence routes are selected from minimizing the following cost function: $C_{\mathrm{HC}}(Path) = \sum_{\forall (i,j) \in Path} 1$, where *Path* is the set of all links from the source to the destination.

The second version uses the raw transmission rate as a metric. The intention is to implement the cross-layer metrics as proposed by Iannone and Fdida (2005). Iannone and Fdida (2005) coupled physical transmission rate, interference, and packet error rate with transmission power control in order to use them as routing metrics and improve the transport capacity of the $\mathrm{WM^2Nets}$' backbone. Compared to the hop-count metric, the raw transmission rate metric selects links of high transmission rates. In IEEE 802.11a, which the technology used in MeshDV backbone, the possible values are 54, 48, 36, 24, 18, 12, 9, and 6 Mbps. MeshDV obtains the rate a WMR uses to talk to each direct neighbor. The cost that can be associated to a link in terms of transmission rate R is the inverse of the rate itself; consequently, routes are chosen from minimizing the following cost function: $C_{\mathrm{CL}}(Path) = \max_{\forall (i,j) \in Path} [1/R_{ij}]$, where R_{ij} is the raw data transmission rate on the link from i to j and *Path* is the set of all links of the communication path. The composition rule of the rate metric (*i.e.* the max function) is introduced to discard congested paths with low transmission rates. The traditional layered approach relegates the rate adaptation mechanism to the MAC layer, with no interaction with the routing layer. This layered design choice, while elegant from an architectural perspective, under-exploits the wireless medium. Rate-aware routing instead prefers links that offer high transmission rates, calculating thus high throughput communication paths, as we showed in Iannone and Fdida (2005a) and Iannone et al. (2006). As an example, we show in Fig. 2.49 the TCP throughput obtained between the MeshDVBoxes S02 and S12 (see Fig. 2.47) when using cross-layer routing. The throughput obtained has an average of almost 600 kbps, which may appear as a modest result, but has a nonnegligible value compared to the almost zero throughput that can be observed when using the hop-count metric (see Iannone et al., 2006). This result is the reason why during normal use on MeshDVNet we use cross-layer rate-aware routing.

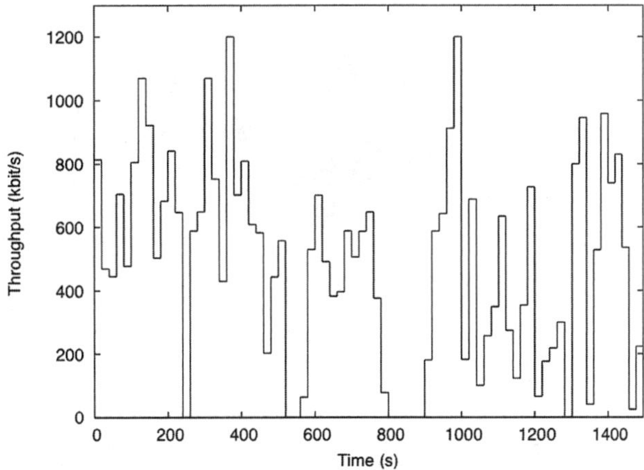

Figure 2.49 Throughput obtained on the MeshDVNet testbed between the S02 and the S12 MeshDVBox.

Client Auto-Configuration Once the mesh backbone is built, the other important task is the assignment of correct auto-configuration parameters to clients. Since the platform is based on IPv6, the stateless auto-configuration features of this protocol can be used (Thomson and Narten, 1998). More specifically, once MeshDV receives the address prefix from a GW it configures and initiates the router advertisement daemon (*rtadvd*). This daemon broadcasts periodically a routing advertisement (RA) message on the client interface. The result is two-fold: first, each associated client receives, and second, an address prefix that is used to build a unique IPv6 address following the rules of the standard; each associated client sets the WMR that issued the RA message as its default route.

Besides, this approach has one more strength: it also works when there are no GWs in the WM^{2}Net cloud (*i.e.* no address prefix announced from the GW). Indeed, in this case MeshDV configures the *rtadvd* daemon to announce the standard prefix for the IPv6 site-local addresses. Using this standard prefix, clients can still build an address, even if its uniqueness is guaranteed only inside the mesh (site) cloud. This allows clients to have an IP address and communicate each other even if the WM^{2}Net cloud is not connected to the Internet. If a GW appears after this configuration, site-local and global unique addresses can coexist. Thus, the only change is the fact that clients obtain a global unique address that can be used to establish a connection towards the Internet.

In the presence of a GW, clients have a globally unique IP address and know how to communicate outside the local subnetwork. Lastly, a list of DNS servers needs be assigned to the clients to resolve domain names. For this purpose, once MeshDV receives a valid DNS entry it configures the DHCPv6. Clients that have configured their address and default route are able to send DHCP queries and obtain the DNS list.

Once the above steps are performed, a client has all the means to access the Internet or to communicate with other clients using the standard IP protocol stack. No particular manual configuration is necessary. Also, no additional software is installed in clients.

Clients Communication Setup As previously mentioned, all clients that access MeshD-VNet have the same prefix, no matter where they are associated to. In this way when

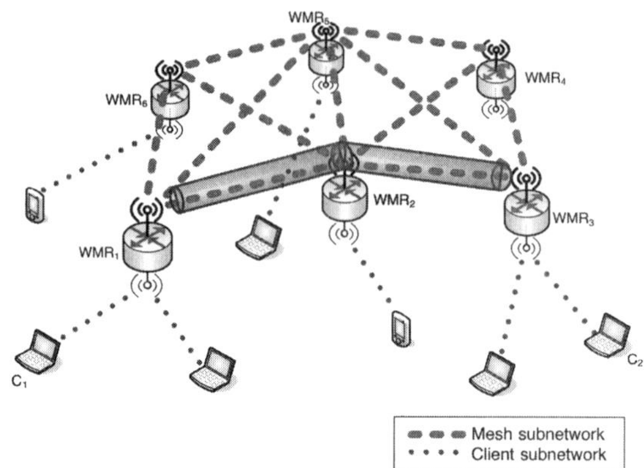

FIGURE 2.50 Example of two communicating clients associated to different WMRs.

two clients need to communicate, they perform the same steps no matter to which WMR they are associated. If the two clients are associated to the same WMR no particular action needs be performed from the MeshDV daemon. As an example, let us consider the scenario where client C_1 intends to send a packet to client C_2 and both are associated to the same WMR.[17] The following steps show how the communication between C_1 and C_2 is set up.

1. To set up a communication, C_1 sends a multicast ICMP Neighbor Solicitation packet asking "Who is C_2?"
2. C_2 receives the message and replies to C_1 with an ICMP Neighbor Advertisement packet "I am C_2".
3. C_1 receives the reply and stores C_2's MAC address.
4. C_1 is ready now to send a data packet to it and prepares an IEEE 802.11 frame and sends it through.
5. C_2 receives the data packet.

Let us now suppose that the two clients are associated to two different WMRs, as in Fig. 2.50. The following steps show how the communication between C_1 and C_2 is set up, according to MeshDV rules. The corresponding temporal diagram is depicted in Figure 2.51.

1. In order to setup a communication, C_1 sends a multicast ICMP Neighbor Solicitation packet asking "Who is C_2?"
2. The neighbor discovery protocol proxy (NDP-Proxy) module of WMR_1 receives the packet from C_1 and sends a message to the Client Manager module, requesting to discover which WMR C_2 is associated to.

[17] For simplicity we will simply use C_X to indicate both the name and the IP address of a client.

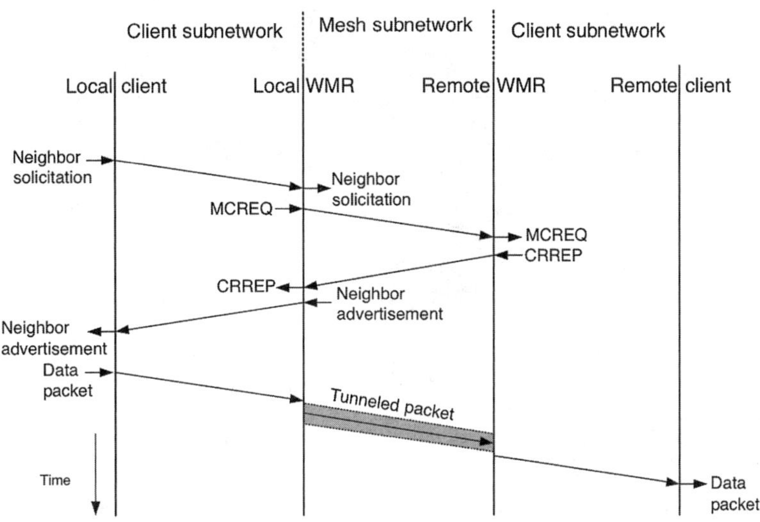

FIGURE 2.51 Temporal diagram of communication setup between clients associated to different WMRs.

3. The Client Manager module of WMR_1 checks whether C_2 is locally associated. Since it is not, it sends a multicast client request (MCREQ) on its mesh interface, asking the other WMRs "Who is managing C_2?"
4. The Client Manager module of WMR_3 receives the multicast request and checks its own associated client list. It knows C_2, thus it sends a unicast client request reply (CRREP), answering, "I am managing C_2."
5. The Client Manager module of WMR_1 receives the unicast reply. It stores the fact that C_2 is associated to WMR_3 and sends a message to the NDP-Proxy module to announce that it knows to which WMR the client is associated.
6. The NDP-Proxy module of WMR_1 sends to C_1 an ICMP Neighbor Advertisement packet associating the IP address of C_2 to the Ethernet address of the client interface of WMR_1.
7. C_1 receives the ICMP packet and sends the data packet to C_2 in an 802.11 frame destined to WMR_1.
8. The data packet is captured from the Forwarder module of WMR_1, which encapsulates it in another packet destined to WMR_3 and sends it.
9. The kernel of WMR_3 receives the tunneled packet, decapsulates it and automatically forwards it on the local client interface where C_2 is associated to.
10. C_2 receives the data packet.

The above steps not only show how to set up a communication but also the role of the MeshDV's modules depicted on the right hand side of Fig. 2.48. The following observations can be made. First, the *client manager module* is aware of the clients associated locally to the WMR from directly accessing the card driver. That is why at step 4, on WMR_3, the Client Manager Module does not generate any query to C_2; it is already aware of its

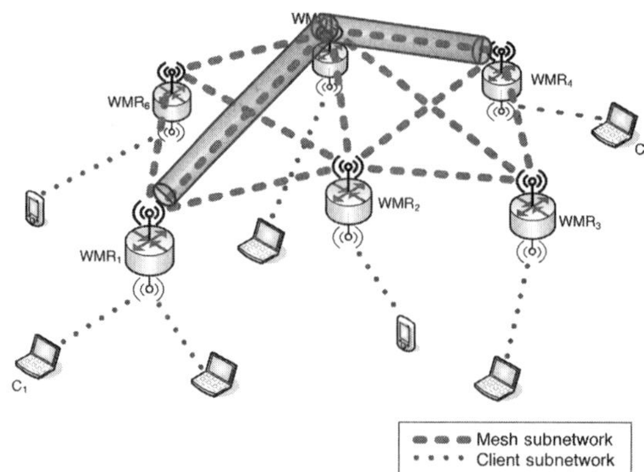

FIGURE 2.52 Example of moving client and the corresponding adaptation of MeshDV.

presence. Second, compared to the first example, no changes have occurred to C_1. The only change is the increased delay between the query for C_2 and reception of the reply. Finally, one can remark that in step 8, the packet issued from C_1 is encapsulated into another packet. As already mentioned, this tunneling approach (see Fig. 2.50) is adopted to avoid having information maintained on the ongoing communications. Indeed, in this particular example (see Fig. 2.50), WMR_2 is totally unaware of the ongoing communication of the two clients. Nevertheless, it is able to forward packets coming from WMR_1 and destined to WMR_3, containing the original data packet, which WMR_3 is able to correctly deliver to C_2.

2.7.2.4 Mobility Management

The abovementioned tunneling approach has also the nice property of allowing simple management of moving clients. Starting from the end of the previous example, let us suppose that after a while client C_2 moves from WMR_3 and associates itself to WMR_4, as shown in Fig. 2.52. WMR_3 readily knows about the departure of C_2, since the *client manager module* has acquired this information from the driver. At the same time, the module knows that a client associated to WMR_1 has asked for C_2 (see steps 4 in the previous example). Thus, WMR_3 sends a message to WMR_1 announcing that C_2 is no longer associated to it. WMR_1 simply acknowledges the message and repeats steps 3 to 5 of the previous example, in order to discover who is managing C_2 now. The only difference is that it is WMR_4 that answers now. Then, all subsequent packets will be tunneled towards WMR_4, as shown in Fig. 2.52.

The interesting point is that since all WMRs advertise the same network prefix, the client does not change address when changing association. Furthermore, the simple tunneling mechanism avoids network-wide updates. Indeed, only some updates at the edges of the tunnel are necessary (WMR_1, WMR_3, and WMR_4 in the previous example), whereas routers in the middle of the path are totally unaware of the change (see Fig. 2.52).

An example of mobility measurement performed on the MeshDVNet testbed is showed in Fig. 2.53. In particular, once the client changes its association, there is an interruption of the traffic that in the average lasts approximately 10 s. While this gap may appear large, this is the result of measurements pursued on the first prototype of

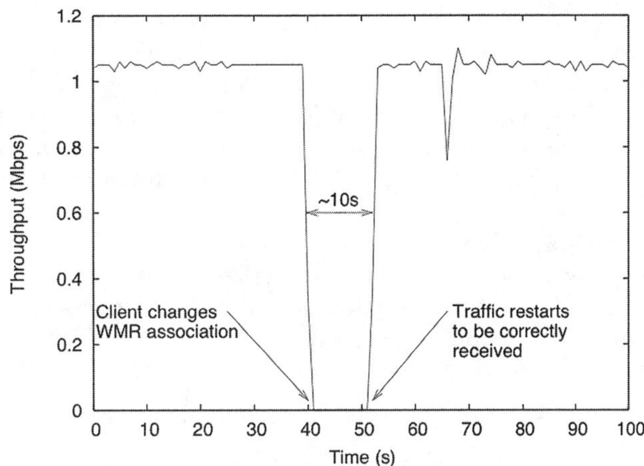

FIGURE 2.53 Traffic evolution between two clients, with one moving to associate to a different WMR.

MeshDV. Furthermore, several different factors lead to this value. MeshDV is placed above the network layer and the mobility management is totally distributed. This means that forcedly there is a certain amount of delay attributed to communication (propagation) delay and layer (protocol conversion) latency. In particular, IEEE 802.11 standard does not define a handover mechanism; the lookup time for a new access point thus depends on the operating system used. Moreover, all the advertisements need to get through the output queue of several interfaces, which may yet contain stale packets. The sum of these produces a nonnegligible (order of seconds) amount of communication delay. Nevertheless, MeshDV is able to manage clients' mobility and re-route traffic without breaking active TCP connections.

2.7.3 OntoSensor: An Ontology for WM^{2}Snet Application Development, Deployment, and Management[18]

2.7.3.1 Ontological Engineering Basics

Ontological engineering has been a subject of computing science since the 1980s, particularly in knowledge representation and automated reasoning endeavors. The work in ontologies centers on developing shared, machine readable conceptualizations of knowledge that intelligent computer systems can use. Representative work concerning ontologies includes (Lenat and Guha 1989; Skuce and Monarch 1990; Chandrasekaran et al., 1999; Uschold and Gruninger, 1996), and more recent work in the context of the Semantic Web includes (McGuinness, 2003). The Semantic Web effort (Berners-Lee et al., 2001) focuses on defining and moving ontologies and knowledge representation standards from traditional stand-alone systems to the highly-distributed World Wide Web. Semantic Web infrastructure is being leveraged along with commercially available sensors in the

[18] Excerpt from the invited article "OntoSensor: An ontology for wireless mesh sensor network application development, deployment, and management," David J. Russomanno and J. Caleb Goodwin, Department of Electrical and Computer Engineering, The University of Memphis, Memphis, TN 38152, USA.

design, implementation, and test of the laboratory environment. This approach appears to be consistent with Department of Defense initiatives. For example, the program manager of DARPA's Advanced Technology Office states (DeBeasi, 2006): "Military networks shall converge to civil technologies." Since networked WM^2Snet sensors are the focus, the evolving infrastructure being standardized for the Semantic Web was viewed as an appropriate means for developing OntoSensor and the supporting infrastructure in the laboratory environment.

The term ontology has been used in a variety of contexts. This study adopts a conceptualization of declarative knowledge as described by Genesereth and Nilsson (Genesereth and Nilsson, 1987) to define an ontology and includes, but is not necessarily limited to, the following (Russomanno et al., 2005a):

- The classes to which objects belong (e.g., sensor types)
- The class hierarchy or taxonomic structure (e.g., set of radiant sensors is a subset of all sensors)
- The relational basis set among the classes (e.g., a sensing element is part of a sensor)
- The functional basis set among the objects (e.g., bandwidth ("JERS SAR") = 1.275 GHz)
- The capability for executing sophisticated programs or procedures for evaluating the truth of literals or other properties (e.g., procedural attachment)

Once a basic ontology is defined, a programming language can be used to express knowledge using concepts defined in the ontology. The Web Ontology Language (OWL) (Smith, et al., 2003) is rapidly gaining popularity for expressing ontologies in the context of the Semantic Web.

The OntoSensor platform, an Ontology-based for Sensor Network Application Development, Deployment, and Management, uses Protégé 2000 (Stanford Medical Informatics, 2004) whereas it extends the IEEE Suggested Upper Merged Ontology (Niles and Pease, (2001). OntoSensor is based in part upon a preliminary SensorML (Botts, 2004) specification. Figure 2.54 depicts a snapshot of OntoSensor through the Protégé 2000 interface.

SensorML provides a framework that consists of a series of UML class diagrams for defining relevant associations and properties to sensors. OntoSensor deviates from SensorML, which lacks the semantic richness of specific sensor classes, properties, and associations among classes using axiomatic-grounded terms. More sophisticated semantics than currently specified by SensorML may be necessary to support automated reasoners. OntoSensor includes knowledge models for a variety of data acquisition boards, sensors, and processor/radio units common in commercial wireless sensing environments. It also includes preliminary definitions for a variety of wireline imaging sensors such as a Sony XCD-SX910 camera, FLIR Merlin InSb configurable as a mid-wave or short-wave infrared (IR) camera, and a FLIR ThermoVision A40 long-wave IR camera. OntoSensor comprises a hierarchy of sensor classes, the knowledge model for a given sensor type metadata, such as sensitivity and other performance parameters for the sensing elements, as well as physical characteristics such as mass, radio frequencies, dimensions, and power supply information for wireless motes.

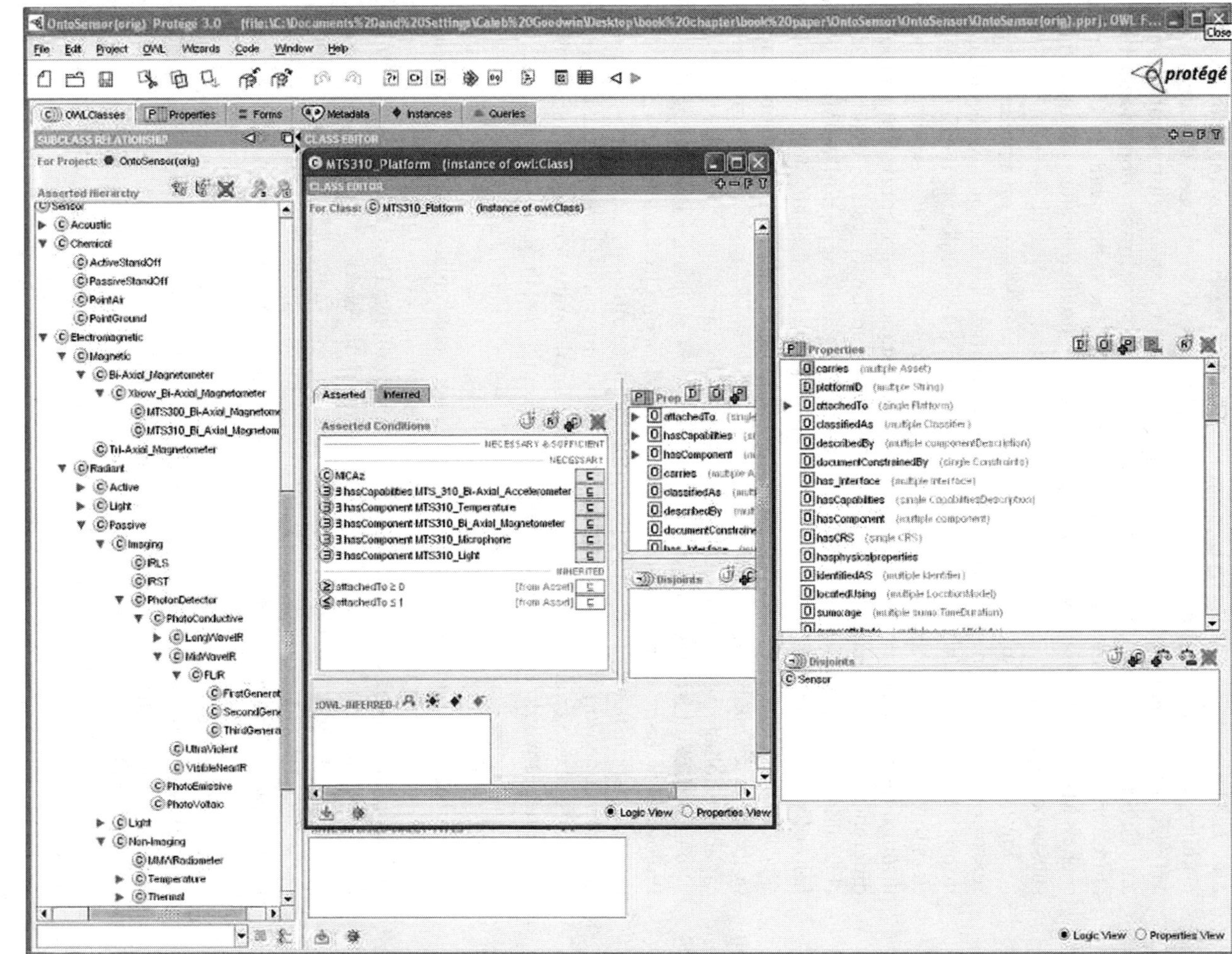

FIGURE 2.54 Screen shot of the OntoSensor using Protégé 2000.

The contents of OntoSensor can be exported to OWL format and processed by any software with OWL parsing capability. An excerpt of the OWL file is illustrated in Fig. 2.55. The excerpt shows the OWL constructs that capture the following knowledge: The resource MTS420_1 is an instance of the OntoSensor *MTS420* class. This instance of a MTS420 sensor is appropriate for collecting temperature, barometric pressure, humidity, and GPS readings, has day and night operability, has maximum operational temperature range of 60 °C, etc. Certain other properties of the components of the platform, such as MTS_420_Ambient and MTS420_Biaxial_Accelerometer, can be obtained through subsequent queries.

Figure 2.56 shows an excerpt of a class diagram that is based on a preliminary draft SensorML specification of OntoSensor. The capabilities of a sensor have been captured through the class *CapabilitiesDescription*. This class in turn is linked to the *GenericProperty* class through two associations. The sensitivity and resolution of a sensor can be derived through the *performanceProperty* association. For instance, noise equivalent temperature difference (NETD), which is one gross measure of sensitivity, can be determined from obtaining the values of certain sensor parameters like focal length, horizontal and vertical field of view, frame rate, overscan ratio and so on (Driggers et al., 1999). Through the *supportedApplications* association, queries can determine the types of sensors to attempt to locate and task based on some application criteria like resolution, night operation capability, foliage penetration, all weather capability, etc.

Figure 2.57 illustrates simple test queries of the OntoSensor knowledge base using Prolog. All of the query results have been truncated for brevity. Line one describes a query that takes an instance of an MEP510 sensor and determines its capabilities. The second query finds sensors in the knowledge base with temperature sensing capabilities. Line three describes a query that retrieves the operational parameters for a MEP510 instance. Such queries would comprise subgoals of an agent's overall query.

2.7.3.2 Laboratory Environment

The primary focus of the laboratory environment is to create ontological information for low-cost magnetic, passive IR, light, temperature, humidity, barometric pressure, and acoustic sensors common in WSNet configurations. These plain sensors typically provide scalar samples rather than high bit-rate data sources such as video sensors or IR cameras. In addition, capturing ontological information for these plain sensors and motes facilitates the rapid development of a prototype environment to experiment with novel approaches for the design of software agents and the utilization of Semantic Web infrastructure.

Currently, the wireless environment consists of two sensor networks, each having multiple sensor nodes built from parts available from commercial vendors such as Crossbow Technology, Inc. Sensing elements are, for example, MEP410, MEP510, MTS420, and MTS310 models available from Crossbow (CrossBow Technology Inc., Wireless sensor networks: Product reference guide, available at http://www.xbow.com). The MEP410 includes sensors for ambient light, barometric pressure, photosensitive light, relative humidity, and temperature. The MEP510 senses relative humidity and temperature. A MIB510 GW is used for the aggregation of data obtained from MEP410 and MEP510 sensor nodes. The MTS420 includes a biaxial accelerometer and also senses ambient light, barometric pressure, temperature, GPS, photosensitive light, and relative humidity. The MTS310 has a biaxial accelerometer and magnetometer along with acoustic and photosensitive light sensing elements. Also, the MTS420 and MTS310 sensors are coupled with

```
<MTS420_Platform rdf:ID="MTS420_1">
      <hasComponent rdf:ID="#MTS420_Bi-Axial_Accelerometer"/>
      <hasComponent rdf:ID="#GPS"/>
      <hasComponent rdf:ID="#MTS_420_Ambient"/>
      <hasComponent rdf:ID="#MTS420_Photosynthetic_Radiation"/>
      <hasComponent rdf:ID="MTS420_Relative_Humidity"/>
      <hasComponent rdf:ID="MTS420_BarometricPressure"/>
      <hasComponent rdf:ID="#MTS420_Temperature"/>
      <hasComponent rdf:ID="#MICAz"/>
<hasCapabilities>
   <PlatformCapabilities rdf:ID="OntoSensor_Individual_226">
         <supportedApplication rdf:resource="Motion Detection"/>
         <supportedApplication rdf:resource="Day Night Operation"/>
         <supportedApplication rdf:resource="Location Awareness"/>
         <supportedApplication rdf:resource="Ambient Light Measurement"/>
         <supportedApplication rdf:resource="Humidity Measurement"/>
         <supportedApplication rdf:resource="Photosynthetic Light Measurement"/>
         <supportedApplication rdf:resource="Temperature Measurement"/>
         <supportedApplication rdf:resource="Barometric Pressure Measurement"/>
         <performanceProperty>
          <Generic_Property rdf:ID="OntoSensor_Individual_234">
         <Attr_Name rdf:datatype="http://www.w3.org/2001/XMLSchema#string">
               Operating Temperature Range Max</Attr_Name>
         <Attr_Value rdf:datatype="http://www.w3.org/2001/XMLSchema#string">60</Attr_Value>
          <Generic_Property rdf:ID="OntoSensor_Individual_235">
         <Attr_Name rdf:datatype="http://www.w3.org/2001/XMLSchema#string">
               Operating Temperature Range Min</Attr_Name>
         <Attr_Value rdf:datatype="http://www.w3.org/2001/XMLSchema#string">-10</Attr_Value>
         </performanceProperty>
   </PlatformCapabilites>
</hasCapabilities>
</MTS420_Platform>
```

Figure 2.55 Excerpt of OntoSensor instance.

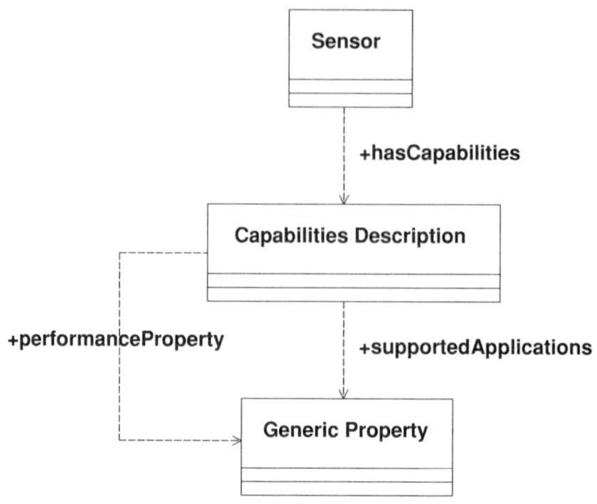

Figure 2.56 Excerpt of OntoSensor model.

```
1                                                  ?-
sensor_capability('MEP510_1',ListOfCapabilities).
ListOfCapabilities = [
'Day/Night Operation',
'Humidity',
'Temperature']

2                                                  ?-
sensor_capability(SensorInstance,'Temperature').
SensorInstance = 'MEP510_3' ;
SensorInstance = 'MEP510_1' ;
SensorInstance = 'MEP510_2' ;
SensorInstance = 'MTS310_1'

3                                                  ?-
sensor_parameters('MEP510_3',ListOfParameters).
ListOfParameters = [
'ACCURACY'='+-0.5 C',
'ACCURACY'='+-2% RH',
'BATTERY LIFE'='0.5 yr',
'FREQUENCY'='433 MHz',
'POWER'='<-20dBm +5dBm>',
'RANGE'='<-40C +123C>'
'RANGE'='>500 ft',
'RANGE'='<0RH 100RH>',
'RESOLUTION'='0.01 C',
'RESOLUTION'='0.03% RH']
```

Figure 2.57 Example query result.

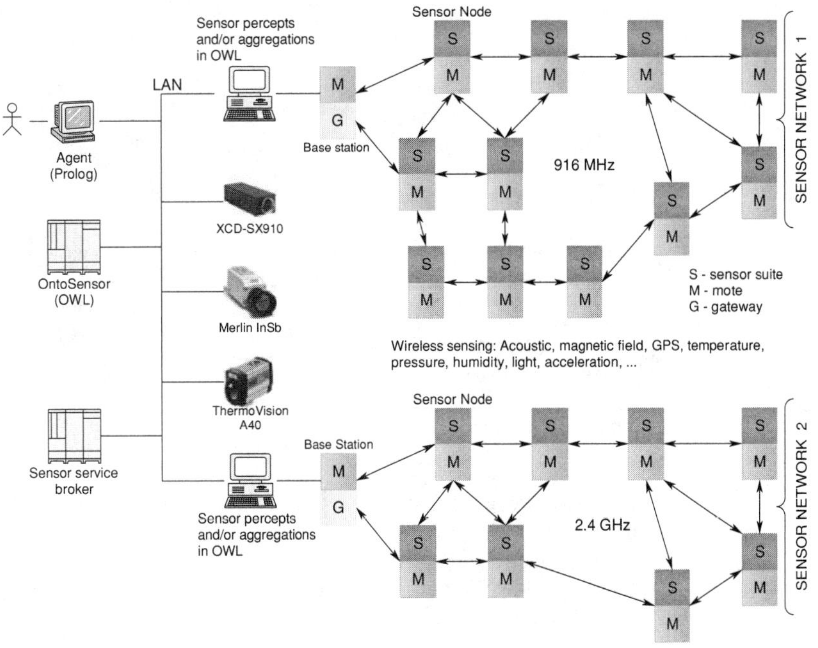

Figure 2.58 Laboratory environment (abridged).

MICAz motes. The MTS420 and MTS310 data acquisition boards and motes use a MIB520 network GW for data aggregation. The motes operate at 916 MHz or 2.4 GHz and handle the data processing and communication needs for each sensor node. Crossbow's XMesh software is used for the dynamic formation of wireless communication links between the nodes.

The prototype network is composed of two computers that serve as base stations and store the data collected from the MIB510 and MIB520 network GWs. The base stations run Crossbow's MoteView application to retrieve data from the network GWs and the data is stored in a PostGRE database. In addition, each base station executes custom software, which was developed specifically for the laboratory environment to extract sensor data and metadata into OWL repositories, which OntoSensor references. The software is a preliminary implementation of a Web service that is evoked at the base stations to store selective sensor data into OWL repositories within the networked environment. In addition, base stations provide physical storage aggregation for sensor data and metadata, which are used in response to queries on sensors' behalf. Figure 2.58 depicts an abridged view of the laboratory environment.

On-the-fly discovery of sensors that may satisfy a user or agent's high-level needs is critical in a network-centric environment. In general, service-oriented architectures (SOAs) strive to establish principles and standards for description, discovery, connection, and communication of Web services (Singh and Huhns, 2005). In the application domain of sensors, an elementary SOA (see Fig. 2.59) consists of the following logical/physical entities: (1) individual sensors or a network of sensors that register/publish descriptions of services that they provide; (2) sensor service brokers that maintain a registry of the

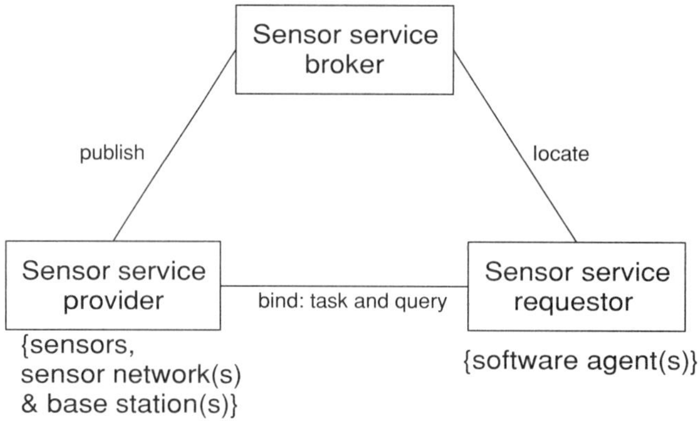

FIGURE 2.59 Generic model for sensor services.

published services; and (3) users or service requesters who search the registry offered by the broker to locate sensors that offer the required services.

Universal description, discovery, and integration (UDDI) is a de-facto registry standard for the SOA. However, UDDI alone is not capable of supporting semantic searching of services. This is because UDDI has limited utility in applications where the requestor issues imprecise queries; that is, queries that are either too specific or too general for an exact match of the services published in the registry.

To provide more accurate results, an understanding of the semantic relationships between the various sensor concepts, instances, and their properties is required. OntoSensor enhances the matching process of finding the appropriate sensor services in response to a given agent's query by using its taxonomic hierarchy to determine the type of sensor instance that is listed in a repository. For example, an agent may attempt to locate camera services in a geographic location of interest; however, only specific models of IR cameras have been registered in the area of interest. Instead of returning a failure of matching the agent's query, knowledge from OntoSensor enables the sensor service broker to return information to the agent where IR cameras are available. In the event the agent is concerned for these alternative sensor services, additional metadata can be provided by the broker.

Sun Microsystems (Horan, 2005) and Microsoft (Liu and Zhao, 2005; Liu et al., 2005a; Woo et al., 2006) have pursued research on sensor services. In addition, the Naval Research Laboratory (NRL) (Luo et al., 2005) defines an approach to support semantic service description and matchmaking with registries that use an existing UDDI specification. The laboratory environment focuses on experimenting with the semantic matching process of finding the appropriate sensors in response to a high-level query using knowledge-based support, which, however, is not originally part of the sensor service broker's registry. The NRL (Luo et al., 2005) outlines an approach to bulk-load semantic data into UDDI tModels before queries are issued. TModels provide a mechanism to extend UDDI repositories with external information. To enable semantic searching, the activity develops a method that associates a subset of OntoSensor with a UDDI registry of sensor services. The UDDI repository may not initially have pointers to ontological schema information in the tModels about the specific sensors that have been registered. In order to respond more effectively to subsequent queries that require semantic

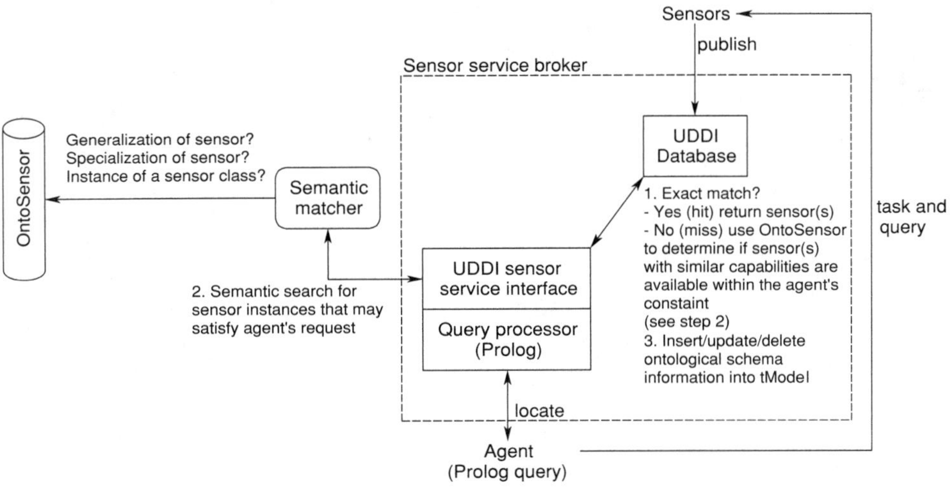

FIGURE 2.60 Sensor semantic service description and match architecture.

matchmaking, there is a need to augment the specific facts in the UDDI repository with relevant ontological schema information. For discovery performance matters, it is important to extract from the UDDI repository only a subset of OntoSensor that is relevant to the sensor service instances.

A simplification of the approach we experimented with is shown in Fig. 2.60. In this scheme, the UDDI database can be viewed as a sensor services registry cache. If the query results in hits, that is, sensor services that match the user's request are directly specified in the UDDI registry, then these sensors' bindings (specifications to query and/or task the sensor) are returned to the agent; otherwise, the UDDI sensor service interface will attempt to find generalizations or specializations via OntoSensor that may semantically match (to some degree) the agent's query. If such matches are found, the ontological information from OntoSensor pertinent to the specific sensor facts that are published in the registry is inserted/updated to the UDDI tModels, much like a conventional cache, for use in subsequent queries.

CHAPTER 3

Energy-Aware WM²Net Communications

3.1 Introduction

Wireless mobile mesh network (WM²Net) devices are battery-powered devices with on-chip integrated capabilities, including communications, computing, and processing. The battery capacity of WM devices is a very important issue in the design of WM²Nets as it restricts the operation time of a wireless radio. In some application scenarios, replenishment of power resources might be impossible. Consequently, carefully scheduling and budgeting battery power in wireless mobile networks has become a critical issue in WM²Net design.

The **power consumption** of a wireless device can be categorized in five parts (Li et al., 2001): (1) the power spent for the transmission of messages (P_{Txm} mode), (2) the power spent for the reception of messages (P_{Rec} mode), (3) the power spent for the processing of messages (P_{Prc} mode), (4) the power spent while the system is idle (P_{Idle} mode), and (5) the power spent while the system is in sleep mode (P_{Sleep} mode).

When a wireless node is in the sleep mode, it is basically shut down except that a low-power timer remains on to wake itself up at a later time (Deng et al., 2004; 2004a). Therefore, it consumes a tiny, but nonnegligible fraction of the energy consumed in the active (Txm, Rec, and Prc) modes (Kumar et al., 2004a; Sinha and Chandrakasan, 2001). As illustrated in Table 3.1, for RangeLAN2, the power consumption for the doze mode is 2 mA.

Table 3.1 provides an indicative list of power consumption for different WNICs (Range LAN: http://www.proxim.com/products/rl2/7410.shtml; Feeney and Nilsson 2001; Adcon telemetetry: http://www.adcon.com).

It is to be noted that there may be some additional circuitry for data encoding and decoding. Application-specific integrated circuits may also be used in some cases. In all these scenarios, the design of WM²Net algorithms and protocols is affected by the corresponding power expenditures, in addition to those that have been discussed. Mixers, frequency synthesizers, voltage control oscillators, phase-locked loops (PLL) and power amplifiers all consume valuable power in the transceiver circuitry. It is important that power consumption considers not only the active power but also the startup power consumption in the transceiver circuitry. The startup time, being of the order of hundreds of microseconds, makes the startup power a nonnegligible factor. This high value for the startup time can be attributed to the lock time of the PLL. As the transmission packet

Card	Txm (mA)	Rec (mA)	Idle (mA)	Sleep (mA)	Power Supply (V)
RangeLAN2-7410	265	130	*NA	2	5
WaveLAN (11Mbps)	284	190	156	10	4.74
Smart Spread	150	80	*NA	5	5

*NA = not available.

Table 3.1 Power Consumption Comparison Among Different Wireless LAN Cards (Range LAN: http://www.proxim.com/products/rl2/7410.shtml; Feeney and Nilsson, 2001; http://www.adcon.com)

size is reduced, the startup power consumption starts to dominate the active power consumption. As a result, it is inefficient to turn the transceiver ON and OFF, because a large amount of power is spent in turning the transceiver back ON each time.

3.2 Related Background on Power Consumption

Shih et al. (2001) present a formula for the radio power consumption as

$$P_c = N_T[P_{Txm}(T_{on} + T_{st}) + P_{out}(T_{on})] + N_R[P_{Rec}(R_{on} + R_{st})] \quad (3.1)$$

where

P_{out} = the output power of the transmitter
T_{on} = the transmitter on time
R_{on} = the receiver on time
T_{st} = the transmitter startup time
R_{st} = the receiver startup time
N_T = the number of times transmitter is switched on per unit time
N_R = the number of times receiver is switched on per unit time

The latter depends on the task and medium access control (MAC) scheme used. T_{on} can further be rewritten as L/R, where L is the packet size and R the data rate. Today's state-of-the-art, low-power radio transceiver has typical P_{Txm} and P_{Rec} values around 20 dbm and P_{out} close to 0 dbm (National Semiconductor Corporation, 2000). Note that PicoRadio aims at a P_c value of −20 dbm.

The design of a small-sized, low-cost, ultra low-power transceiver is discussed by Porret et al. (2000). A direct conversion architecture is proposed for the transceiver circuitry. Based on their results, authors present a power budget and estimate the power consumption to be at least an order of magnitude less than the values given above for P_{Txm} and P_{Rec} values.

Energy expenditure in data processing (Prc mode) is much less compared to data communication (Txm and Rec modes). The example described in the work of Pottie and Kaiser (2000) effectively illustrates this disparity. Assuming Rayleigh fading and fourth power distance loss, the energy cost of transmitting 1 KB a distance of 100 m is approximately the same as that for executing 3 million instructions by a 100 million instructions per second (MIPS)/W processor. Local data processing is therefore crucial in minimizing power consumption in a multihop WM²Net.

Further limitations of cost and size led engineers to complementary metal oxide semiconductor (CMOS) technology for the microprocessor. However, a CMOS transistor pair draws power when it is switched. This switching power is proportional to the switching frequency, device capacitance (which further depends on the area) and square of the voltage swing. Reducing the supply voltage is hence an effective means of lowering power consumption in the active state. Dynamic voltage scaling, explored by Min et al. (1995) and Pering et al. (1998), aims to adapt processor power supply and operating frequency to match workloads. When a microprocessor handles time-varying computational loads, a linear decrease in power consumption is observed when low operating frequencies are used during periods of reduced activity, whereas reducing the operating voltage produces quadratic gains. On the other hand, this compromises the peak performance of the processor, which remarkably is not always desired. Significant energy gains can then be obtained from adapting the processor's operating voltage and frequency to instantaneous processing requirements.

Sinha and Chandrakasan (2001) propose a workload prediction scheme based on adaptive filtering of the past workload profile and analyze several filtering schemes. Other low-power CPU organization strategies are discussed in the work of Govil et al. (1995), Lorch and Smith (1996), and Weiser et al. (1994). The power consumption in data processing (P_{Prc}) can be formulated as follows:

$$P_{\mathrm{P}} = C V_{\mathrm{dd}}^2 f + V_{\mathrm{dd}} I_o e^{V_{\mathrm{dd}}/n V_{\mathrm{T}}}$$

where C is the total switching capacitance; V_{dd}, the voltage swing; and f, the switching frequency. The second term indicates the power loss due to leakage currents (Sinha and Chandrakasan, 2001). The lowering of threshold voltage to satisfy performance requirements results in high subthreshold leakage currents. Coupled with the low duty cycle operation of the microprocessor in a WM²Net node, the associated power loss becomes significant (Shih et al., 2001).

3.3 Issues of Power-Aware Communications

In wireless mesh networks, the lack of a centralized authority dictates that the wireless stations must always stay awake. Hence, when a message is broadcast through the system, all nodes will receive and process the message, either directly or through relaying. Power is thus consumed unnecessarily, which is principally due to overhearing other nodes' transmissions.

Energy efficiency may be the most important design criterion for mobile networks in general—the operation time of a mobile radio is basically restricted by its battery capacity. Limited battery power restricts the communications (transmission) range as well as the duration of the active period for the nodes. Below some critical threshold for battery power, a node will not be able to function as a router, thus immediately affecting the network connectivity, possibly isolating one or more segments of the network. Fewer routers almost always mean fewer routes and therefore increased likelihood of degraded performance in the network. In fact, communication becomes meaningless if a node is not able to communicate owing to low battery power. Since exchange of messages necessarily means power consumption, network protocols should be designed with battery power consumption in mind (for related works, see: Aggélou, 2004; Wang et al., 1997; Toh, 2001; Singh and Raghavendra, 1998; Rodoplu and Meng, 1999; Chang and Tassiulas, 2000; Bahl et al., 2001; Li et al., 2001; Xue and Li, 2001).

The wireless link-only communication paradigm in WM^{2}Nets, however, makes energy savings difficult to achieve. Given that in a WM^{2}Net, nodes themselves collectively form a network infrastructure for communication, the corresponding reduction of nodes' lifetime directly affects the network lifetime. *This is the primary design principle of energy-aware communications: to equally balance energy expenditure among WM^{2}Net nodes to prolong network lifetime, while at the same time conserving overall power consumption as much as possible.*

To illustrate the effectiveness of energy-aware communications in WM^{2}Nets, let us consider the multihop communication scenario in Fig. 3.1. Based on the hop-count metric, host A selects the shortest route A-F-E to reach E. Similarly, B chooses route B-F-D to reach D. Both communication pairs go through F, which may quickly drain off F's battery energy. This may make F die earlier, forcing the sink node (G) to become isolated from the network. Consequently, energy cost should be taken into account to lengthen the network lifetime.

On the other hand, decreasing the transmitting power in wireless networks is paramount to keep interference low and make thus more efficient use of the RF spectrum (Adler and Scheideler, 2000; Li et al., 2001; Vikas Kawadia and Kumar, 2003). In a system with base stations (or access points, likewise), the advantages seem obvious. In wireless peer-to-peer networks, this advantage does not seem that obvious though. Data paths in WM^{2}Nets are composed of wireless nodes where links have to be fashioned out of the

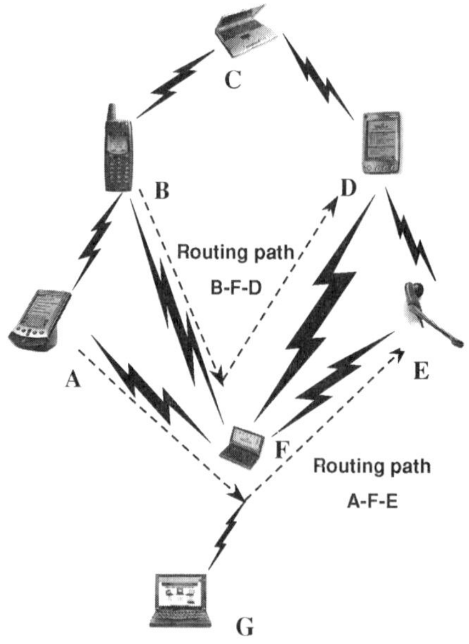

Figure 3.1 A scenario that may quickly drain off the energy of host F, an articulation point in the network.

radio by nodes choosing the power levels at which they transmit. Transmit power control can reduce (or increase) the local scope (one-hop neighbors) of nodes, and as a result the channel contention at the MAC layer (Vikas Kawadia and Kumar, 2003). Decreasing the transmitted power, however, means increasing the number of hops between a pair of nodes involved in the average data path. This further implies that more choke points are potentially introduced in the network.

Besides, there is a mutual dependence of power control and communication protocols: power control impacts on the routes employed as the power level dictates what links are available for routing and, vice versa, the power control protocol needs connectivity information which is provided by the communication protocols. This mutual dependency motivates the need for a joint solution for power control and communication protocols, or else *power-aware communications*.

There are three major issues involved in power/energy-aware communication protocols. First, the goal is to find the path that either minimizes or balances the power consumed. Balanced energy consumption does not necessarily lead to minimized energy consumption; it simply prevents a certain node from being overloaded, thus ensuring longer network lifetime. Second, energy awareness can be implemented either at purely routing layer or routing layer with help from other layers such as MAC or application layer. For example, information from the MAC layer, such as interference levels, is beneficial as MAC often supports power saving features that the routing protocol can exploit to provide better energy efficiency. Third, some routing protocols assume that the transmission power is controllable and nodes' location information is available (e.g., via GPS). With these observations in mind, the problem of finding a path with the least consumed power becomes a conventional optimization problem on a graph where the weighted link cost corresponds to the transmission power required for transmitting a packet between the two ends of the link.

Four major requirements to the power efficiency issue are identified in battery-operated WM^{2}Nets: The first requirement is *autonomy*. Mesh nodes may be placed out in the field, unattended, for months or years. Once a WM^{2}Net is in place, its lifetime must last as long as possible based on the initially provided amount of energy.

The second requirement is *scalability*. In order to cover large spatial extents or monitor the extents at a high temporal resolution, mesh networks are required to retain their power efficiency against a large number of mesh nodes and a large amount of data generated.

The third requirement is *adaptability*. Mesh nodes are required to vary their transmitting power as well as their data transmission paths to their sink. The data aggregation and transmission path adjustment can significantly reduce power consumption.

The fourth requirement is *simplicity*. Due to limited resource availability, mesh control software needs to be simple in its design and lightweight in its footprint in order to minimize power consumption.

The difficulty of power control for wireless multihop mesh networks is that it infringes on several layers in the layered hierarchy. Clearly, power control impacts on the physical layer due to the need for maintaining link quality. However, power control also impacts the network layer, as shown in Fig. 3.2a. If all nodes are transmitting at 1 mW, then the route from node N_1 to node N_5 is $N_1 \rightarrow N_2 \rightarrow N_3 \rightarrow N_4 \rightarrow N_5$. However, if all nodes transmit at 30 mW, then one can choose the route $N_1 \rightarrow N_3 \rightarrow N_5$. Transmit power thus also affects routing latency as choosing low power levels longer (in hops) routes are constructed.

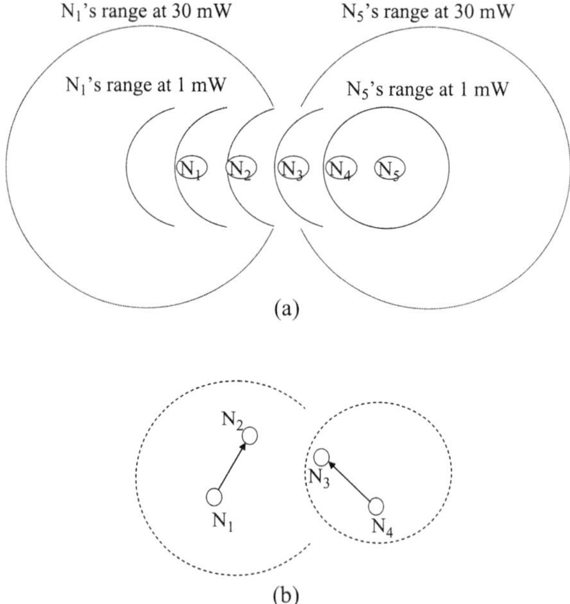

FIGURE 3.2 Power control implications on the physical, routing, and transport layers. (a) Routing. (b) Transport.

Furthermore, power control also impacts on the transport layer. In Fig. 3.2b, transmissions from node N_1 to N_2 with increased power would increase the interference levels at the transmission of N_4 to N_3. This would result to a loss of several packets on the communication link from N_4 to N_3.

In summary, choosing an excessively high power level causes excessive interference as seen in Fig. 3.3a. This reduces the traffic carrying capacity of the network in addition to reducing battery life. On the other hand, in Fig. 3.3b, having a very small power level results in fewer links and hence network partitioning effects may exhibit. When the power level is just right, the network is still connected and there is no excessive interference as shown in Fig. 3.3c.

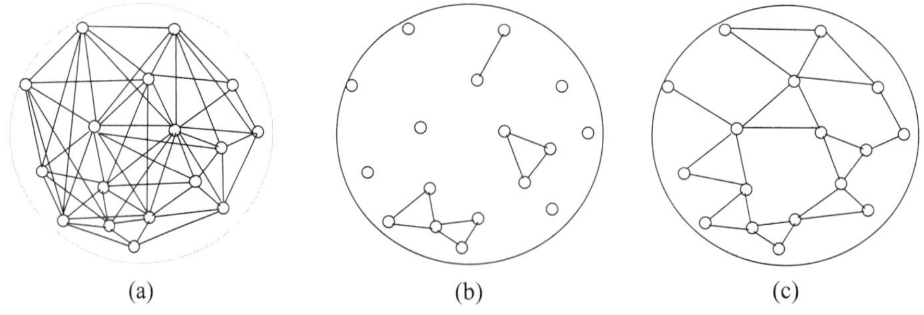

FIGURE 3.3 Choice of power level affects network connectivity and level of interference. (a) High. (b) Low. (c) Just right.

It should be made clear though that reducing transmitter power does not imply a decrease of receiver's sensitivity. By reducing output power, you will reduce the area over which the transmitter will interfere with other transmissions. However, reducing sensitivity will only add to the problem of data collisions. In addition, reduced sensitivity requires higher transmit power as well as offering less protection against Raleigh fading, multipath, etc.

3.4 Power-Aware Network Categories

Based on the aforementioned requirements, wireless networks can be broadly classified into two categories: simple or fixed-power radio networks, where mobile hosts use fixed transmission powers, and power-controlled or variable-power radio networks, where mobiles are able to vary their transmission power (Adler and Scheideler, 2000).

In the **variable-power case**, nodes can dynamically vary their transmitter power levels; the chosen transmission power depends on the distance between the transmitter and receiver nodes. Banerjee and Misra (2002) observed that the choice between a path with many short-range hops and a path with fewer long-range hops is nontrivial, but involves a trade-off between the reduction in the transmission energy for a single packet and the potential increase in the frequency of retransmissions. Even if all links have identical error rates, it is not always true that splitting a large-distance (high-power) hop into multiple small-distance (low-power) hops results in overall energy savings. Indeed, the analysis of Banerjee and Misra (2002) shows that if the number of hops exceeds an optimal value, the rise in the overall error probability negates any apparent reduction in the transmission power.

The idea of allowing mobile hosts to vary their transmission power is already used in power-controlled code division multiple access (CDMA) systems, where the base station can direct mobiles to reduce their power so that to reduce the system's interference and thus allow more users on the system (Jacobsmeyer, 1996; Gibson, 1996).

For the **fixed-power case**, the transmission power of a node is chosen independent of the distance or quality of the link (Malkin, 1998; Moy, 1998). If communication links are assumed to be error-free, conventional minimum-hop routing could fit in constant power/energy-efficient routing. If this is the case, the minimum-hop path may not be the most energy-efficient approach, since an alternative path with more hops may prove to be better, given that its overall error rate is sufficiently low. Similarly, for the variable-power case, a path with a greater number of smaller hops may not always be better; the savings achieved in the individual transmission energies (given by Equation 3 in the work of Banerjee and Misra, 2002) may be nullified by a larger increase in link errors and consequently retransmissions.

The latter issue, that of retransmissions, is of paramount importance: wireless links typically employ link-layer frame recovery mechanisms (e.g., link-layer retransmissions, or forward error correcting codes) to recover from packet losses. Additionally, protocols such as TCP or SCTP employ additional source initiated retransmission mechanisms to ensure a reliable transport layer. The energy-efficiency associated with a candidate path should thus reflect not merely the

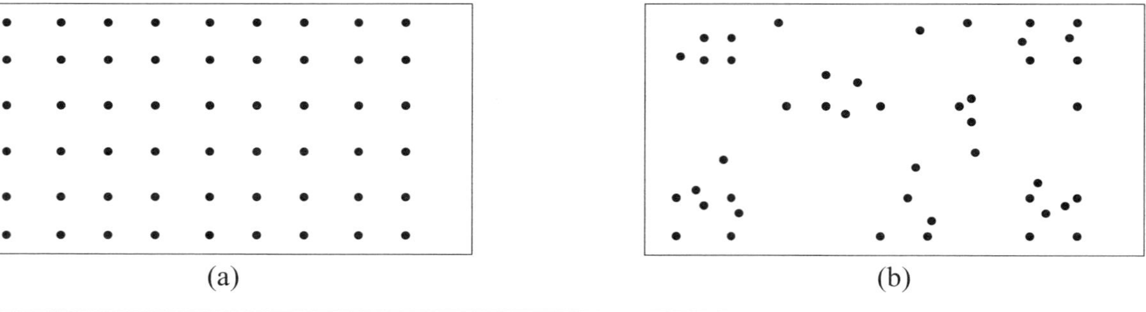

FIGURE 3.4 Homogeneous vs. Clustered networks. (a) Homogeneous spatial dispersion of nodes. (b) Nonhomogeneously dispersed.

energy spent in just transmitting a single packet across a link, but rather the "total effective energy" spent in packet delivery, which includes the energy spent in potential retransmissions as well.[1] It is thus important to consider the link's error rate as part of the *route selection* algorithm, since the choice of links with relatively high error rates can significantly increase the effective energy spent in reliably transmitting a single packet.

Given the respective intricacies of mobile multihop radio networks and the semantics of power-aware routing, the question that naturally arises then is: What is the smallest common power level at which the entire network becomes connected? Or else, how can the power control problem be implemented in a distributed asynchronous fashion by the nodes participating in the network to optimally resolve multihop radio connectivity?

Ideally, energy-aware routing protocols manage to operate all nodes at a common power level, which is chosen to be the smallest power level at which the network is connected (Kawadia and Kumar, 2003). A common power level, however, does not guarantee a common signal-to-noise plus interference ratio (SINR) at all receivers (Kawadia and Kumar, 2003; Banerjee and Misra, 2002). In cellular systems, there exists a feasible choice of power levels that ensure equal SINR for all the transmitters to a single receiver (i.e., the base station). However, in WM^{2}Nets there is no fixed power level that could ensure a common SINR for all the transmitter-receiver pairs. It all turns on the density of system's hosts and the patterns of communications. If there exist a very large number of hosts in a rather small geographical area, that would inherently generate increased contention for the RF channel. Lowering the transmitting power, the potential for interference among hosts on the same channel is greatly reduced, but this in significantly different geographical locations.

Gupta and Kumar (2000) show that when nodes are homogeneously dispersed in space, which is the case when a large number of nodes are uniformly distributed (see Fig. 3.4a), then the choice of a common transmit power level has several appealing features and properties.

However, when nodes are nonhomogeneously dispersed, as in Fig. 3.4b, power control becomes more complicated. For example, let us consider that the nodes in a WM^{2}Net are organized in clusters, which are about 2 km apart. Within a cluster, each node is reasonably within a few meters of at least one other node. When communication is within a cluster, it would be prudent therefore to limit the transmit power to that level necessary to carry out communications within that cluster (see Fig. 3.5), but not high enough to cause interference between clusters. On the other hand, when the communication is between different clusters, a higher power level would be needed.

Obviously, the lowest common power level that assures network connectivity will imply that the common power level is dictated by outlying nodes: those that are far from others (that is, nodes X and Y in Fig. 3.5). All nodes within clusters A and B are mutually reachable at 1 mW except X and Y. Without loss of generality, we can say that these nodes form a 1 mW cluster. Nodes X and Y can be mutually reached by using a higher power level of, say, 150 mW. A power-aware network layer protocol, which converges to the lowest power level such that the network is connected, will in this case converge to 150 mW.

[1] This is especially relevant in multi-hop wireless networks, where variable channel conditions often cause packet error rates as high as 15%–25% (Banerjee and Misra, 2002).

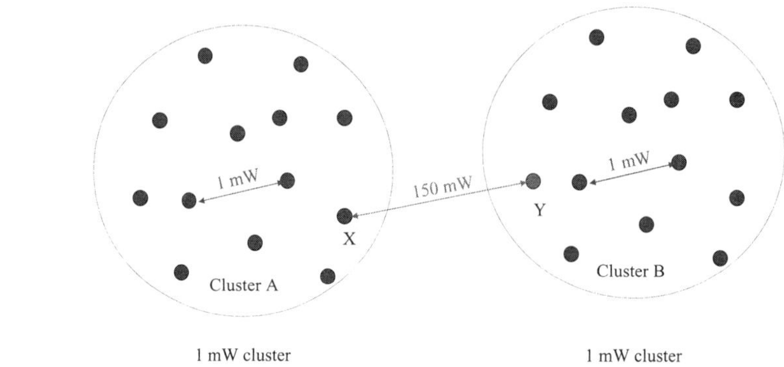

Figure 3.5 A common power level is not appropriate for nonhomogeneous networks.

Thus, every node in the network will be forced to use 150 mW even though 1 mW is enough for most communications.

3.4.1 Battery-Aware Routing (BAR) for Streaming Data Transmissions in Wireless Mesh Networks[2]

3.4.1.1 Related Background on Battery Models

Battery and Discharging The most commonly used batteries in wireless devices are nickel—cadmium and lithium—ion batteries. In general, a battery consists of cells arranged in series, parallel, or a combination of both. Two electrodes—an anode and a cathode, separated by an electrolyte—constitute the active material of each cell. When the cell is connected to a load, a reduction-oxidation reaction transfers electrons from the anode to the cathode. To illustrate this phenomenon, Fig. 3.6 shows a simplified symmetric *electrochemical* cell. In a fully charged cell (Fig. 3.6a), the electrode surface contains the maximum concentration of active species. When the cell is connected to a load, an electrical current flows through the external circuit. Active species are consumed at the electrode surface and replenished by diffusion from the bulk of the electrolyte. However, this diffusion process cannot keep up with the consumption, and a concentration gradient builds up across the electrolyte (Fig. 3.6b). A higher load electrical current I results in a higher concentration gradient and thus a lower concentration of active species at the electrode surface (Doyle et al., 1993). When this concentration falls, the battery voltage drops. When the voltage is below a certain cut-off threshold, the electrochemical reaction can no longer be sustained at the electrode surface, and the battery stops working (Fig. 3.6e). The electroactive species that have not yet reached the electrode are not used. The unused charge is referred to as discharging loss. However, the discharging loss is not physically "lost," but simply unavailable due to the lag between the reaction and the diffusion

[2] Excerpt from the invited article "Battery-aware routing for streaming data transmissions in wireless mesh networks," *Chi Ma and †Yuanyuan Yang (*Department of Computer Science, State University of New York, Stony Brook, NY 11794, USA; †Department of Electrical and Computer Engineering, State University of New York, Stony Brook, NY 11794, USA).

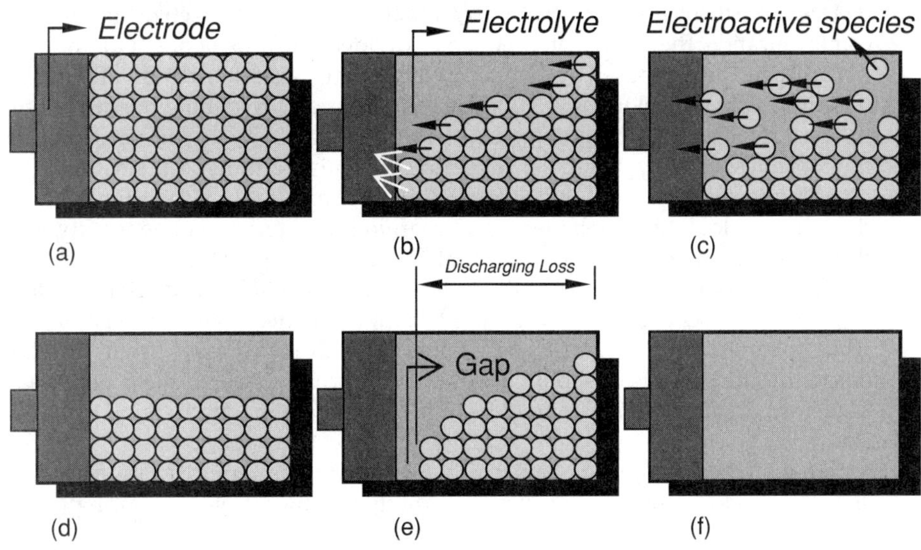

FIGURE 3.6 Battery operation at different states. (a) Fully charged status. (b) In discharging. (c) In recovery. (d) After recovery. (e) Discharging loss. (f) Power fully consumed.

rates. Before the battery dies, if the battery current I is reduced to zero or a very small value, that is, in the battery recovery regime (Fig. 3.6c), the concentration gradient flattens out after a sufficiently long time, reaching equilibrium again. The concentration of active species near the electrode surface following this recovery period makes unused charge available again for extraction (Fig. 3.6d). Effectively recovering the battery can reduce the concentration gradient and recover discharging loss, hence prolong the lifetime of the battery (Fig. 3.6f). Experiments on nickel—cadmium and lithium-ion batteries show that the discharging loss might take up to 30% of the total battery capacity (Rakhmatov and Vrudhula, 2003). Hence, precisely modeling battery behavior is essential for optimizing system performance WM²Nets.

Mathematical models that can capture the battery discharging behaviors have been developed (Panigrahi et al., 2001; Rakhmatov and Vrudhula, 2003). These models are independent of battery chemistry. Panigrahi et al. (2001) provided an abstract model to describe battery recovery behavior. This model treats discharge and recovery as a negative exponential function and represents them as a transient stochastic process. However, as pointed out in the work of Rao et al. (2003), this method uses large lookup tables that require considerable effort to configure, and its accuracy and computational complexity are barely acceptable. Therefore, it has limited utility for the implementation in power-sensitive wireless networks. Rakhmatov and Vrudhula (2003) proposed an analytical battery model, which can accurately estimate the battery behavior. This model combines a high-level representation of the battery with analytical expressions, which are based on physical laws. The high-level representation is determined by experimental data. This model can effectively capture the effect of battery discharge and recovery. However, it requires long computing time and large pre-computed look up tables. In addition, due to their complex computations and large lookup tables, all these battery models are off-line computed models.

On-Line Computable Discrete Time Battery Model Due to the packetized nature of network communication, the battery lifetime can be divided into a sequence of discrete time slots. According to the proposed on-line computable battery model, battery lifetime is divided into time slots with slot length δ. We use I_n, α_n, and ζ_n to denote the discharge current through the battery, the dissipated battery energy, and the discharging loss in the n-th time slot, respectively. We use T to denote the entire lifetime of the battery. In the n-th time slot, the battery is either discharging ($I_n > 0$) or idle ($I_n = 0$). A time slot is called a discharging slot when the battery is discharging. Without loss of generality, we assume that the discharge current I is a constant in a discharging slot.

The condition of a battery at the n-th slot is measured by its discharging loss at that time. A high discharging loss indicates a "fatigue" battery, which needs some recovery, while a battery with low discharging loss is "well recovered." Intuitively, an energy efficient routing protocol should always choose well-recovered mesh nodes as routers. Therefore, a battery model should be able to calculate the discharging loss at any time slot.

An analytical model can be used to compute the battery discharging loss at a time slot. The model computes the energy α_n dissipated by the battery during the n-th time slot $[n\delta, (n+1)\delta]$

$$\alpha_n = I_n \times F(T, n\delta, (n+1)\delta, \beta) \tag{3.2}$$

where $F(T, n\delta, (n+1)\delta, \beta) = \delta + \pi^2/3\beta^2[e^{-\beta^2(T-(n+1)\delta)} - e^{-\beta^2(T-n\delta)}]$. This model can be interpreted as follows. The dissipated energy α_n in Eq. (3.2) comprises two components. The first, $I_n \times \delta$, is simply the energy consumed in device during $[n\delta, (n+1)\delta]$ whereas the second, $I_n \times \pi^2/3\beta^2[e^{-\beta^2(T-(n+1)\delta)} - e^{-\beta^2(T-n\delta)}]$, is the amount of battery discharging loss in the duration. It can be seen that the discharging loss decreases as the lifetime T increases. β (>0) is a constant, which is an experimental chemical parameter and may vary from battery to battery. The larger the β, the faster the battery diffusion rate, hence the less the discharging loss.

Experiments have been conducted to evaluate the accuracy of this model. The results indicate that for a battery with α and β, the model can estimate the battery lifetime T for various I with less than 4.5% error rates. For example, for a battery charged with $I = 912$ mA for the first 25 min and recovering for the next 10 min, the model can accurately describe the battery behavior during discharging and recovery. More details on the battery model can be found in the studies of Yang and Ma (2005) and Ma and Yang (2005).

The model is also very simple for on-line computation. It does not require lookup tables or complex computations. Next, we show how this model can be used to calculate the discharging loss at each time slot. As defined earlier, ζ_n is consumed in the n-th slot and recovered step by step in the following $n+1, n+2, \ldots,$ slots until the battery dies. ζ_n^k is defined as the ζ_n after κ following slots ($\kappa \geq 1$). That is,

$$\zeta_n^{\kappa} = I_n \times \frac{\pi^2}{3\beta^2}[e^{-\beta^2\kappa\delta} - e^{-\beta^2(\kappa+1)\delta}] \tag{3.3}$$

It should be mentioned that the discharging loss ζ_n is only a potential type of energy. For example, in Fig. 3.7, if the battery dies at time t_0, then energies ζ_1^4, ζ_2^3, and ζ_3^2 do not have a chance to be recovered. Thus, the battery permanently loses them. This model

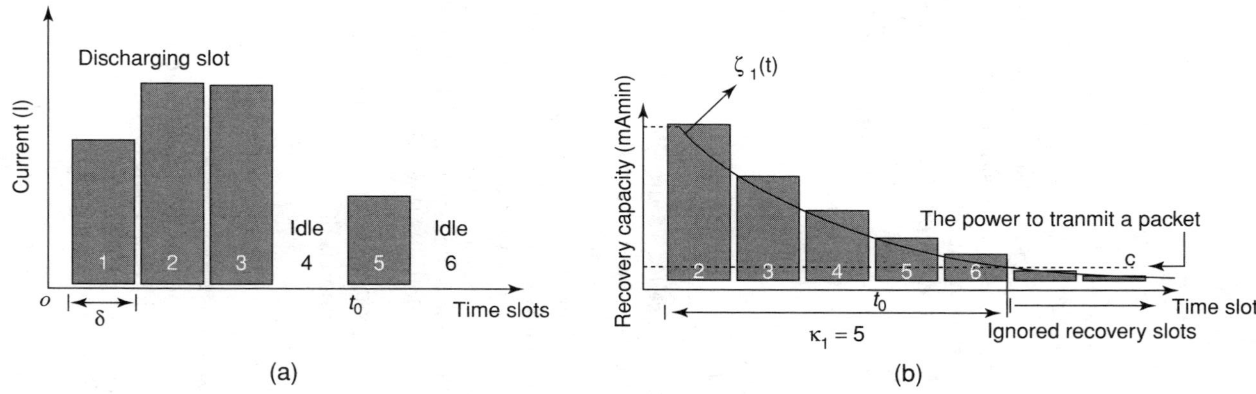

FIGURE 3.7 Battery discharging model. (a) Slots 1, 2, 3, and 5 are discharging slots. (b) The over-charged capacity at slot 1 is recovered gradually in the following slots.

also gives a way to further simplify the computation of ζ_n. Assume that the energy for transmitting a single packet is c. By observing Eq. (3.3), we know that, if ζ_n^κ is less than c, then the recovery of ζ_n after κ slots can be ignored. From $I_n \times \frac{\pi^2}{3\beta^2}[e^{-\beta^2\kappa\delta} - e^{-\beta^2(\kappa+1)\delta}] < c$, we obtain

$$\kappa > \frac{1}{\beta^2\delta} \log \frac{\pi^2(1 - e^{-\beta^2\delta})}{3\beta^2 c/I_n} \tag{3.4}$$

where κ is referred to as the recovery length of ζ_n. Thus, we can truncate recovery slots and only look ahead for limited $\kappa = \frac{1}{\beta^2\delta} \log \frac{\pi^2(1-e^{-\beta^2\delta})}{3\beta^2 c/I_n}$ time slots to calculate the approximate value of ζ_n. In the example (Fig. 3.7b), the effective recovery length of slot 1 is from slot 2 to slot 6 ($\kappa = 5$). The recovery of ζ_1 after slot 6 is ignored.

In summary, in this section we introduce an on-line computable discrete time battery model, and compare it with previous off-line battery models. The model can reduce the computational complexity and make it possible to be implemented in wireless networks on-line. The model also makes on-time ζ_n computation feasible by truncating the recovery slots to κ. Next, we discuss how this model can be applied to energy efficient routing in WM^{2}Nets.

3.4.1.2 BAR in WM^{2}Nets

Streaming data (e.g., multimedia transmissions) is usually modeled as streaming packets from a source to its corresponding destination (Gerla and Xu, 2003). The source and its corresponding destination are called a source-destination pair. Once a source-destination pair is setup, packet transmissions will continue for a long period. Therefore, a critical issue in such transmission is how to maximize the communication lifetime between a source-destination pair. The main idea of the BAR is to choose the "well recovered" mesh nodes as routers, and leave "fatigue" nodes for recovery. By dynamically scheduling routing paths to efficiently recover the node battery capacity, we can minimize the discharging loss on mesh nodes and maximize the lifetime and data throughput between a source-destination pair. In this section, we first study the power metric of energy efficient routing and then provide a BAR protocol.

Metric of Battery-Aware Energy Efficiency To transmit a packet from node a to node b in a WM^{2}Net, the energy consumption is often modeled as

$$E = e + \gamma d_{(a,b)}^n + \zeta \tag{3.5}$$

where ζ is the discharging loss. To carefully schedule the battery activity and budget energy consumption, a node should capture its battery status. Since the battery model introduced in the previous section is suitable for on-line computation, battery-awareness now can be easily implemented in WM^{2}Net routing. In the next subsection, we present a BAR protocol, which finds the routing path by using the metric in Eq. (3.5).

BAR Protocol Here, we present the BAR protocol based on the battery model. The BAR algorithm is used to set up a routing path between a source and a destination. We assume that nodes are randomly deployed. Each node knows its geographic position. A battery with parameter β powers a node. We also assume that the source node transmits a stream of packets to the destination node. This models the applications such as video monitoring

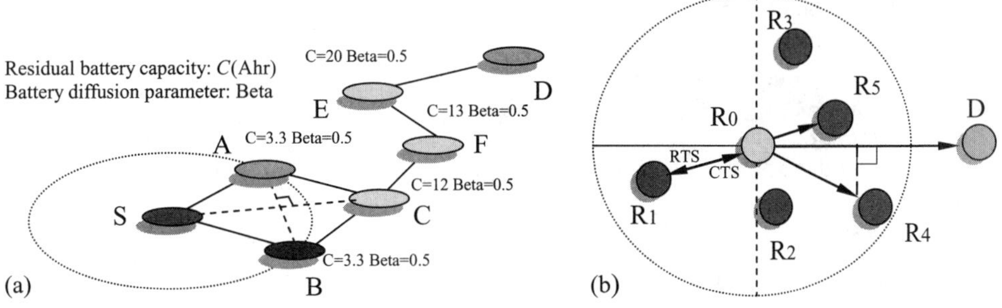

FIGURE 3.8 Battery-aware routing. (a) By alternating between node A and node B, the network achieves longer lifetime. The current at each node is I = 3.5 A. (b) Next routing hop selection.

or multimedia transmissions where transmission is viewed as a stream. A node is called a routing node if it is on the routing path from the source to the destination. In each time slot, a routing node can be assigned for a task or in idle. A task may be a routing activity, video displaying, software execution or any other power-consuming function at this node. Multiple tasks may be assigned in the same slot. Since the current I_n of each task is known, κ_n can be computed from Eq. (3.4).

We first give an example in Fig. 3.8a to illustrate that by considering the battery behavior, the network lifetime can be prolonged significantly. In this WM²Net, source node S will transmit packets to destination node D. The battery residual capacity C and parameter β are indicated in Fig. 3.8. We compare the following two approaches. In the first approach, S sends packets to D through multihop path S-A-C-F-E-D. After 45.35 min, node A uses up its energy. After that, the routing path changes to S-B-C-F-E-D. The total connection lasts 90.70 min. However, the lifetime can be extended in a simple way by alternating between the above two paths. In the second approach, node A and node B alternate each other as the router. Node A recovers its battery while node B is routing, and so on. In this way, the total lifetime is 113.15 min. It is increased by 24.8%.

In the battery-aware energy efficient routing protocol, we alternatively recover batteries to extend node lifetime. The basic idea is that we always choose the most fully recovered nodes as routing nodes. A general comparison between existing routing protocols and BAR is provided in Table 3.2. From Table 3.2 we can see that BAR adopts the power metric in Eq. (3.5) and helps to select those "well recovered" nodes as routers. BAR protocol complements existing protocols rather than replaces them. For example, the greedy routing protocol, most forward within radius (MFR) (Takagi and Kleinrock, 1984), finds the next routing hop with the farthest communication distance. Then in BAR-MFR, which is the BAR algorithm employing the MFR strategy, the next routing hop is found with the farthest communication distance among those nodes whose discharging loss is the lowest. BAR can also be employed in conjunction with other existing routing protocols to improve their energy efficiency through battery-awareness.

In the following, the details of the BAR protocol are further analyzed. As shown in Table 3.2 to find a next hop, one node has to collect the battery discharging loss status from all neighbors. In order to reduce the communication overhead, in turn to dissipate less power during routing path setup, BAR adopts κ instead of ζ to measure the battery discharging loss. Since the larger the recovery length κ, the higher the discharging loss

Given: Source S and destination D; **Find**: A routing path from S to D.
Existing Routing Protocols: $X := S$; **Repeat** Find the next hop Y of X by the metric in (4); $X := Y$; **Until** $Y = D$ or Y do not exist. *Battery Aware Routing*: $X := S$; **Repeat** Collect battery discharging loss status from one-hop neighbors of X; Find the next hop Y of X by the metric in (5); $X := Y$; **Until** $Y = D$ or Y do not exist.

Table 3.2 Comparison of Routing Path Setup in Existing Routing Protocols and BAR

ζ, and the less the battery is recovered. Also, adopting κ can reduce the packet size as ζ is a floating number while κ is an integer.

We introduce two vectors $\vec{R}$ and $\vec{C}$ to record the battery status. $\vec{R}$ and $\vec{C}$ are the recovery status and the residual capacity in each time slot, respectively. Each routing node maintains the two vectors. Figure 3.9 gives an example of $\vec{R}$ and $\vec{C}$ on a routing node. The battery is assigned for four discharging tasks, each with a different recovery length $\kappa_1, \kappa_2, \ldots, \kappa_4$. The number of recovering tasks is put into $\vec{R}$. In Fig. 3.9, $\vec{R}$ = [0,2,3,3,2,1].

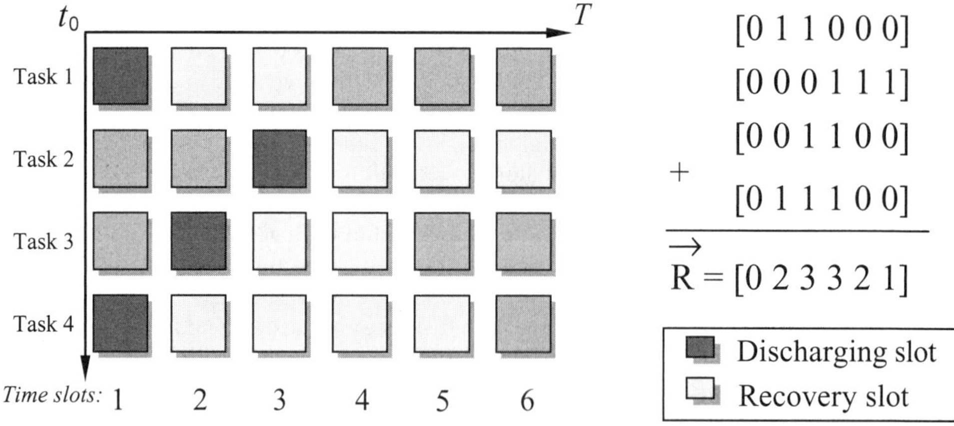

Figure 3.9 The battery status at a node. Current time is 0. The battery is assigned for four tasks. Recovery vector is calculated by summing over all recovering slots. Recovery vector is [0,2,3,3,2,1]. Capacity vector is the battery residual power in each time slot. In this case, it is [500,420,375,390,405,410] (mAhr).

To calculate $\vec{C}$, we simply subtract the "battery capacity" from the "power dissipated" in each slot. The source also needs two vectors $\vec{R}'$ and $\vec{C}'$, which are the recovery vector and power needed for this transmission, respectively. For example, if a transmission takes 2δ time, the recovery length $\kappa = 4$ and battery power consumption is 50 mAhr per time slot, then we have $\vec{R}' = [0,1,1,1,1,0] + [0,0,1,1,1,1] = [0,1,2,2,2,1]$ and $\vec{C}' = [50,50,0,0,0,0]$.

Along the routing path from the source to the destination, each hop is a sender and its next hop is the receiver. To select the next hop for a sender, the BAR protocol is employed to select the most fully recovered node. On receiving a packet called request-to-send (RTS), available receivers reply a packet called clear-to-send (CTS). Their $\vec{R}$ and $\vec{C}$ are included in CTS. The sender receives CTS from all its n_1 one-hop neighbors. By checking the following rules for each node $j = 1 \ldots n_1$, it selects the best available node:

1. $\vec{C_j} - \vec{C}' > 0$ must be satisfied in order to ensure that the next hop will not use up its battery during the transmission.
2. $\vec{R_j} + \vec{R}' > 0$ is computed for each receiver, and the receiver with minimum recovery demand $\vec{R_j} + \vec{R}'$ is selected as the next hop.

Note that the minimal $\vec{R_j} + \vec{R}'$ may not be unique. We can select the next hop when on a tie by different strategies as shown in the description of the BAR protocol in Table 3.3. In the NextHopSelect procedure we use several existing routing algorithms to show how BAR adopts existing routing strategies. These protocols are MFR, nearest with forward progress (NFP), compass routing, and geographic distance routing (GEDIR). MFR (Takagi and Kleinrock, 1984) greedily chooses the next routing hop farthest away within communication distance, while NFP (Hou and Li 1986) attempts to choose the nearest node as next forwarding node to minimize the energy required per routing task. The compass routing method, also referred to as DIR (Melodia et al., 2004), aims to route the packet to the neighbor with the closest angle to destination. GEDIR (Melodia et al., 2004) selects a routing hop that is the closest to the destination. These routing algorithms are commonly used in today's wireless networks. As will be seen in the simulation results, BAR complements them and achieves better performance for streaming data transmissions in WM²Nets. However, it should be pointed out that BAR is not restricted to these routing algorithms.

Figure 3.8b gives an example of applying the BAR protocol. The packets are to be transmitted from source R_0 to destination D. R_0 has one-hop neighbors: R_1, R_2, R_3, and R_4. Their recovery vectors and capacity vectors are shown in Table 3.4. R_0 receives $\vec{R_j}$ and $\vec{C_j}$ $(j = 1 \ldots 4)$ from neighbors. Since $\vec{C_2} - \vec{C}' < 0$, R_2 is not selected. R_1 is not in the forwarding direction. Thus, only R_3, R_4, and R_5 are considered. Among them $\vec{R_4} + \vec{R}' > 0$ and $\vec{R_5} + \vec{R}' > 0$ are both minimum. On this tie, we have the following choices: we can forward the packet to the nearest node (BAR-NFP) by selecting R_5, to the most progress node (BAR-MFR) by selecting R_4, to the closest node to the destination (BAR-GEDIR) by selecting R_4, or to the compass router (BAR-DIR) by selecting R_4.

The communication complexity and time complexity of the BAR protocol is linear to the diameter of the network. Thus, its complexity is $O(\sqrt{n})$, where n is the number of nodes. Note that the overhead of RTS and CTS communications in path setup here is negligible. Unlike in regular packet routing where a routing path is setup to transmit

Receiver BAR procedure:
begin
  while power not used up **do**
    Update $\vec{R}$ and $\vec{C}$;
    Receive RTS from sender;
    Reply CTS with $\vec{R}$ and $\vec{C}$;
    Receive $\vec{R}'$ and $\vec{C}'$ from sender;
    $\vec{R} = \vec{R} + \vec{R}'$;
    $\vec{C} = \vec{C} - \vec{C}'$;
    Call {*Sender BAR procedure*};
    // Selecting the receiver's next hop, and so on
  end while
end

Sender BAR procedure:
begin
  Hop Queue = {};
  Broadcast RTS;
  Receive $\vec{R_1}, \vec{R_2}, \ldots, \vec{R_{n1}}, \vec{C_1}, \vec{C_2}, \ldots, \vec{C_{n1}}$
    from n_1 neighboring node s;
  for $j = 1 \ldots n_1$ **do**
    if $\vec{C_j} - \vec{C}' > 0$ **then**
      if node j is in forward direction **then**
        Hop Queue = Hop Queue$\cup\{j\}$;
      end if
    end if
  end for
  Call {*BAR NextHopSelect procedure*};
end
Send $\vec{R}'$ and $\vec{C}'$ to the selected nexthop;

BAR NextHopSelect procedure:
  begin
    for each node j in Hop Queue **do**
      nexthop = j with min $\{\|\vec{R_j} + \vec{R}'\|\}$;
      the minimum is not unique **then**
      // Adopting different strategies:
        if under *BAR-MFR* algorithm **then**
          nexthop = j with max {distance for j};
        if under *BAR-NFP* algorithm **then**
          nexthop = j with min {distance from j};
        if under *BAR-GEDIR* algorithm **then**
          nexthop = j
         with minimum {distance of j and destination};
        if under *BAR-DIR* algorithm **then**
          nexthop = j
         with min {angle of j and destination};
      end if
  end for
end

Table 3.3 BAR Protocol

Recovery Vector	Capacity Vector ($\times 10^{-2}$ Ahr)
$\vec{R}' = [0,1,2,2,2,1]$	$\vec{C}' = [5,5,0,0,0,0]$
$\vec{R}_1 = [0,4,7,3,5,1]$	$\vec{C}_1 = [70,63,51,20,23,12]$
$\vec{R}_2 = [1,2,3,1,0,0]$	$\vec{C}_2 = [3,4,6,8,9,2]$
$\vec{R}_3 = [2,3,4,3,3,2]$	$\vec{C}_3 = [32,30,18,22,25,19]$
$\vec{R}_4 = [1,2,3,2,1,0]$	$\vec{C}_4 = [20,22,15,17,14,12]$
$\vec{R}_5 = [1,2,3,2,1,0]$	$\vec{C}_5 = [10,11,13,16,18,15]$

Table 3.4 Recovery Vectors and Capacity Vectors for the Example in Fig. 3.8b

a single packet, in multimedia transmissions a routing path is setup for a stream of continuous packets from the source to its destination. Once the path is set up, packets will be continuously transmitted for a period. Therefore, the path set up period is a very small fraction of the entire transmission lifetime. Also, in some applications nodes are mobile. To take the mobility of nodes into consideration, we can let the source node update the routing path every t time, where t depends on the mobile speed of the nodes and is a multiple of δ. The slower the speed, the larger the t is. In the simulations we use $t = 20$ min. As will be seen from the simulations results in Section 3.4.1.3, the BAR protocol achieves good performance in prolonging network lifetime, increasing data throughput and reducing power dissipation.

3.4.1.3 Performance Evaluations

In this section we evaluate the performance of the BAR protocol and apply it to improve the performance of existing routing protocols: MFR, NFP, GEDIR, and DIR, for streaming data transmissions in WM²Nets.

Simulation Setup The network is set up in a 150×150 field. Two hundred and fifty mesh nodes are randomly uniformly deployed in this area. The radius of each node is 15. Nodes are mobile. The nodes are assumed to move at a random direction at the speed of 0.1/min. The moving direction of each node is also randomly chosen every minute. The length of a time slot δ is set to 10 min. There are 25 source-destination pairs randomly distributed in this field. Each source transmits packets to its corresponding destination, and updates its routing path every 20 min (i.e., $t = 20$ min). A source defers from transmitting when there is no routing path between the source and its corresponding destination. To model real-world applications, the battery performance of MICAz sensor (Ma et al., 2004; Crossbow datasheet on MicaZ: http://www.xbow.com/Products/Product pdf files/Wireless_pdf/MicaZ_Datasheet.pdf;) is used throughout the simulations. Node battery has a full battery power of 40×10^3 mAmin. The discharging diffusion parameter is $\beta = 0.5$. Fig. 3.10 shows the field of randomly distributed nodes in the simulations.

Simulation Results The BAR protocol is implemented in this network and its performance is compared with aforementioned protocols. The performance metrics evaluated are network lifetime, power dissipation and data throughput.

Lifetime We first evaluate the lifetime of WM²Net under different protocols. Because the BAR protocol makes nodes use up their battery power gradually, the lifetime of

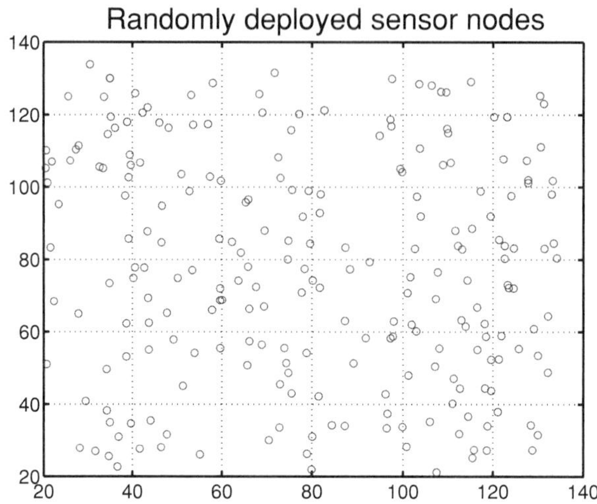

Figure 3.10 Simulation setup: 250 or nodes randomly deployed in the field.

WM²Net is extended. Figure 3.11 shows that the number of the nodes in the network decreases as transmissions go on. We can see that since the battery-aware protocol is sensitive to battery status and carefully recovers node battery, the decrease under BAR is slower than existing protocols. Also, note that the lifetime of the entire network is extended as well. The network lifetime can be extended by up to 41.6% compared with existing protocols. Among these protocols, BAR-NFP has the shortest lifetime, because the NFP protocol has more nodes along a routing path, and in turn consumes more energy for each transmission. Thus, its lifetime is the shortest.

Power Dissipation Under previous routing protocols, a routing node tends to use up its battery power without recovery. Then the routing path switches to another alive node. Therefore, firstly, the residual battery power on routing nodes is very different. Some nodes are almost using up their power, while leaving other nodes with full battery power. Secondly, the residual power at mesh nodes is very low because discharging losses of batteries are not recovered. Thirdly, routing paths in the network tend to be re-setup more frequently, because a path is disconnected even if only a single router on it uses up its power. The operations of detecting disconnected paths and re-setup a transmission consume a lot of extra power. Such operations also lead longer delay in transmissions. This will greatly affect the quality of real-time communications such as video transmissions. As can be seen, under the BAR protocol, the power consumed at nodes is more gradual: the residual power at each node is almost at the same level. Figure 3.12 shows the power distribution of the nodes under BAR protocol in the middle of network transmission (at 45th minute). The x-axis and y-axis show the geographic positions of nodes in the network. The z-axis stands for the residual battery power of each node. It can be seen that by adopting the BAR protocol, mesh nodes can preserve higher battery energy. The results show that the average battery power on a node can be increased by up to 43% compared with existing protocols. It also shows that alive nodes are distributed more uniformly under the BAR protocol.

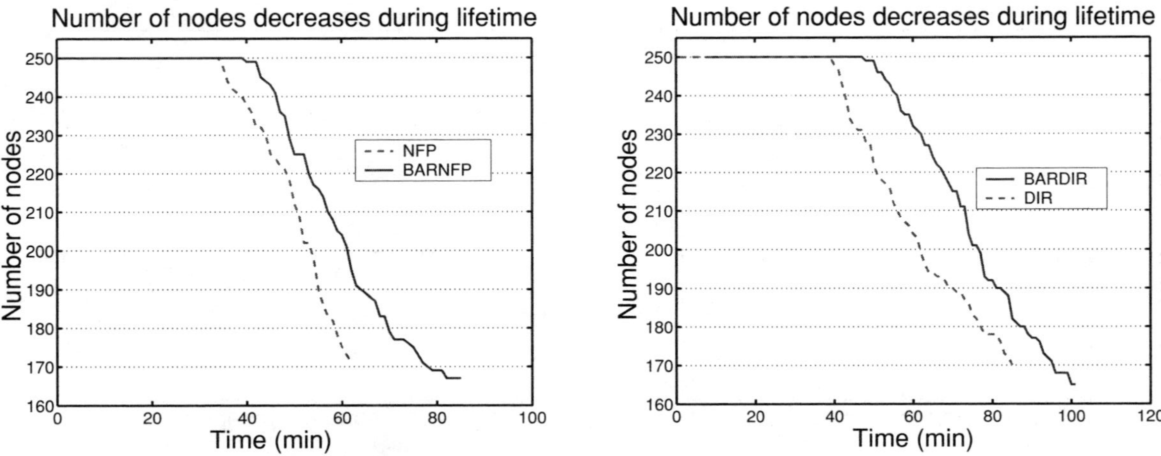

FIGURE 3.11 The number of nodes decreases during the lifetime of the network. (*continued*)

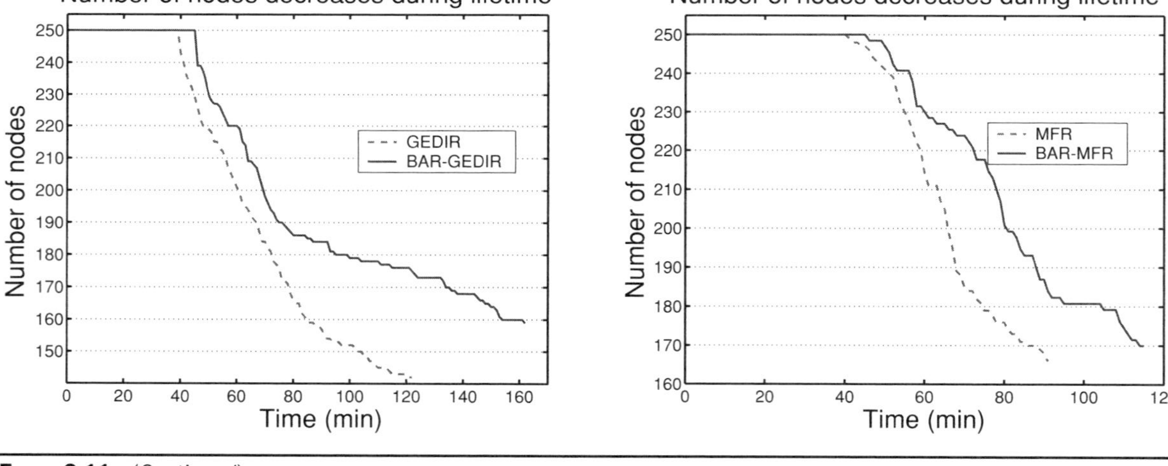

FIGURE 3.11 *(Continued)*

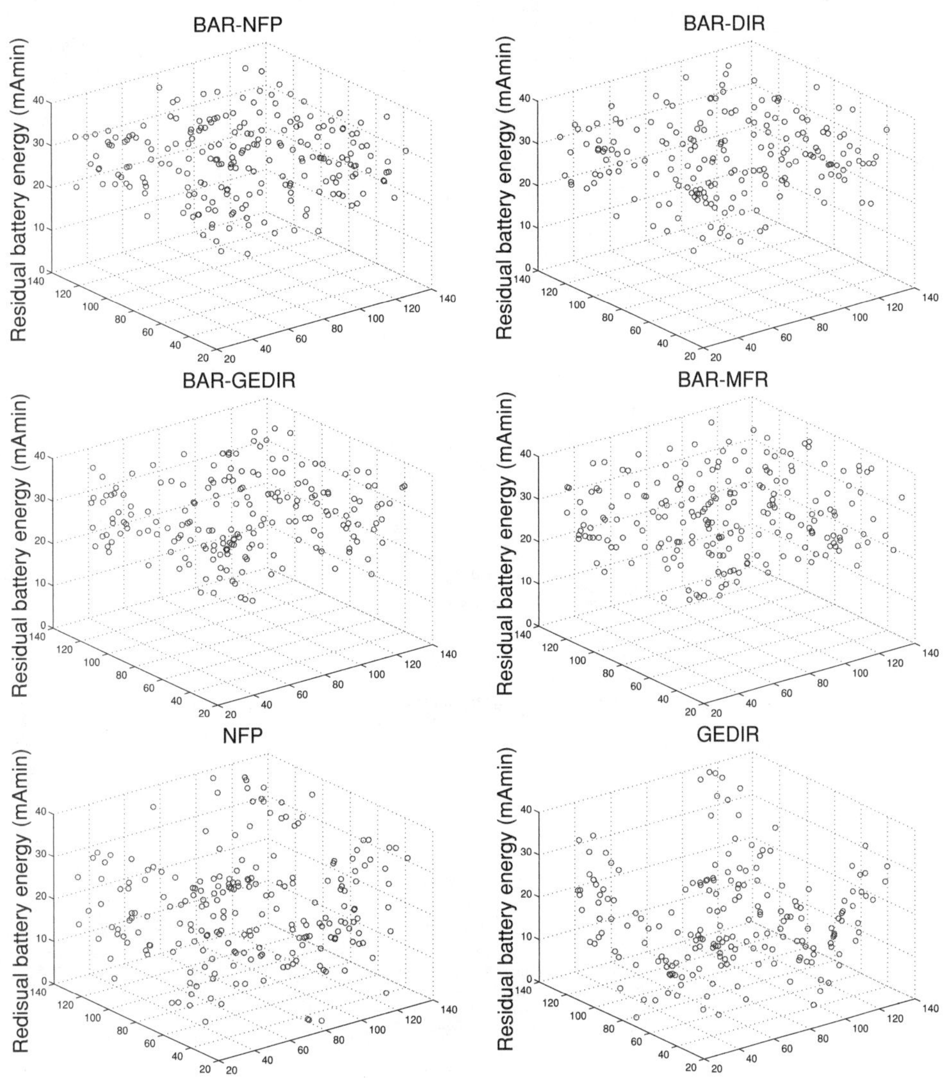

FIGURE 3.12 The residual battery power of mesh nodes under different protocols at the 45th minute. *(continued)*

Data Throughput The BAR protocol improves the data throughput as well. Firstly, the lifetime is prolonged under BAR, which in turn increases the data throughput. Secondly, under the BAR protocol a higher number of nodes exist in the network. Therefore, it is more likely to successfully set up routing paths between source-destination pairs. Figure 3.13 compares the total data throughput under different protocols in the entire network lifetime. The y-axis is the total number of the successfully transmitted packets during the network lifetime. BAR achieves better performance. The data throughput under BAR is improved by up to 32% compared with existing protocols.

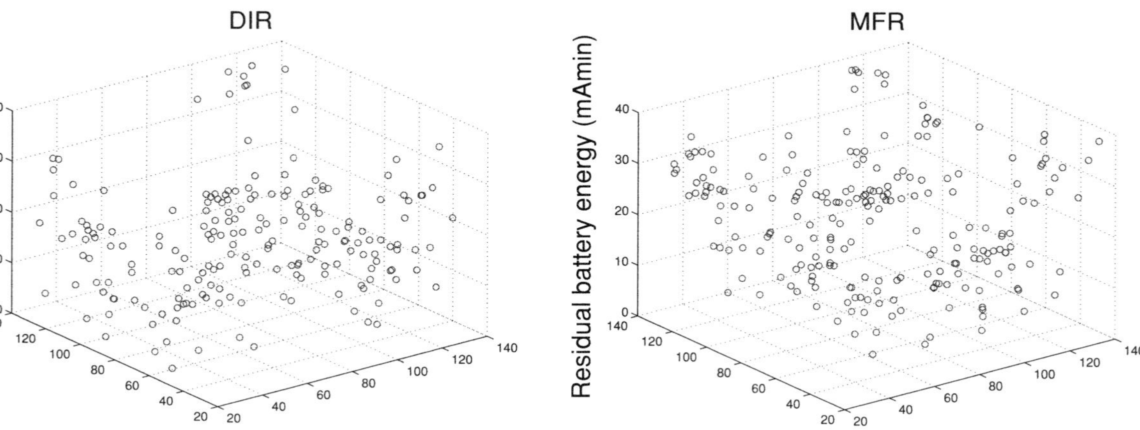

FIGURE 3.12 *(Continued)*

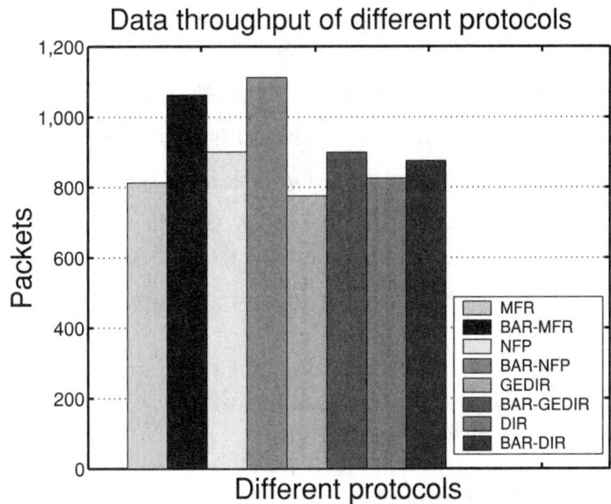

FIGURE 3.13 The total data throughput of different protocols during their network lifetime.

3.4.2 Power-Aware Algorithm for Heterogeneous Wireless Mesh Network[3]

3.4.2.1 System Model

A *probabilistic data-flow graph* (PDFG) models a mesh network application. A PDFG G = <U,ED,T,M> is a directed acyclic graph (DAG), where U = $<u_1, u_2, \ldots, u_N>$ is the set of nodes; ED $\subseteq$ U $\times$ U is the edge set that defines the precedence relations among nodes in U. M = $<M_1, M_2, \ldots, M_R>$ is a mode set. There is a timing constraint L that PDFG must satisfy.

Let the random variable $T_{Mj}(u)(1 \leq j \leq R)$ represent the execution times of each node $u \in$ U from node j. $P_{Mj}(u)$ and $E_{Mj}(u)$ denote the probability and energy consumption, respectively. Let us then define an assignment A to be a function from domain U to range M, where U is the node set and M is the mode set. For a node $u \in$ U, A(u) denotes the selected mode of node u.

Definition : We define the mode assignment with probability (MAP) problem as follows: Given R different modes: $M_1, M_2, \ldots, M_R$, a PDFG G = <U,ED> with $T_{Mj}(u)$, $P_{Mj}(u)$, and $E_{Mj}(u)$ for each node $u \in$ U executed on each mode M_j, a timing constraint L and a confidence probability P, find the mode for each node that gives the minimum total energy consumption E with confidence probability P under timing constraint L.

[3] Excerpt from the invited article "Power aware algorithm for heterogeneous wireless mesh network," *Meikang Qiu, †Edwin H.-M. Sha, ‡Hung-Chung Huang, §Wenyuan Li, and ¶Jiande Wu (*Department of Electrical Engineering, University of New Orleans, New Orleans, LA 70148, E-mail: mqiu@uno.edu; †Department of Computer Science, University of Texas at Dallas, Richardson, Texas 75083, E-mail: edsha@utdallas.edu; ‡Department of Systems Biology and Translational Medicine, Texas A&M Health Science Center, E-mail: hc.jhuang@tamu.edu; §Molecular and Computational Biology, University of Southern California, E-mail: wel@usc.edu; ¶Department of Electrical Engineering, University of New Orleans, New Orleans, LA 70148, E-mail: jw3@uno.edu).

Motivational Example

Our model relies upon the following considerations:

- The execution time (T) of a task is as a random variable.
- The energy consumption (E) depends on mode (M).
- Depending on the mode used, a task assumes different energy consumptions.
- The execution time of a node in active mode is less than that in vulnerable mode, whereas they both are less than the execution time in sleep mode.
- The relations of energy consumption are just the reverse.

The following study shows how to assign a proper mode to each node of a PDFG such that the total energy consumption is minimized while satisfying the timing constraint with a guaranteed confidence probability.

An typical PDFG is shown in Fig. 3.14a. Each node can select one of the three different modes: M_1 (active), M_2 (vulnerable), and M_3 (sleep). The execution times (T), corresponding probabilities (P), and energy consumption (E) of each node under different modes are shown in Fig. 3.14b. The input DAG has five nodes. Node 1 is a multi-child node, which has three children: 2, 3, and 5. Node 5 is a multi-parent node, and it has three parents: 1, 3, and 4. The execution time T of each node is modeled as a random variable.

Referring to Fig. 3.14, the minimum total energy consumptions with computed confidence probabilities under the timing constraint are shown in Table 3.5. A newly proposed algorithm, called MAP_Opt, generates the results. The entries with probability that is equal to 1 (see the entries in boldface) actually give the results for the hard real-time problem, which shows the worst-case scenario of the MAP problem. For each row of the table, the E in each (P,E) pair gives the minimum total energy consumption with confidence probability P under timing constraint L. For example, using the proposed algorithm, at timing constraint 8, we can get a (0.81,27) pair.

The assignments are shown in Table 3.6. Assignment A(u) represents the voltage selection of each node u. We change the mode of nodes 2 and 3 to be M_3 and node 1 to be M_2. Hence, we find the way to achieve minimum total energy consumption 27 with probability 0.81 satisfying timing constraint 8. When the heuristic algorithm

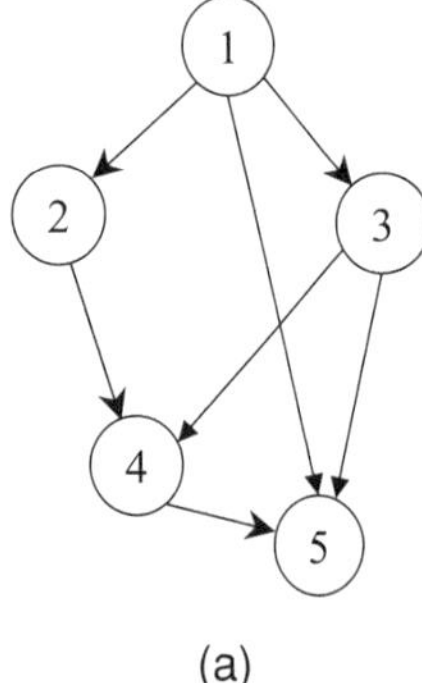

Nodes	M_1			M_2			M_3		
	T	P	E	T	P	E	T	P	E
1	1	0.7	8	2	0.9	4	5	0.9	2
	3	0.3		4	0.1		6	0.1	
2	1	1.0	8	2	1.0	6	3	1.0	2
3	1	1.0	8	2	1.0	6	3	1.0	2
4	1	0.9	9	3	0.9	5	5	0.6	2
	2	0.1		4	0.1		6	0.4	
5	1	0.9	12	3	0.9	6	5	0.8	1
	2	0.1		4	0.1		6	0.2	

(b)

FIGURE 3.14 (a) A mesh network topology. (b) The times, probabilities, and energy consumptions of its nodes in different modes.

T	(*P*,*E*)	(*P*,*E*)	(*P*,*E*)	(*P*,*E*)
4	(0.57,43)			
5	(0.57,39)	(0.73,43)		
6	(0.57,31)	(0.73,35)	(0.81,43)	
7	(0.73,27)	(0.81,35)	(0.90,43)	
8	(0.57,25)	(0.81,27)	(0.90,35)	(**1.00,47**)
9	(0.73,21)	(0.90,27)	(**1.00,39**)	
10	(0.50,20)	(0.81,21)	(0.90,27)	(**1.00,31**)
11	(0.65,16)	(0.73,17)	(0.90,21)	(**1.00,27**)
12	(0.81,16)	(0.90,21)	(**1.00,25**)	
13	(0.65,12)	(0.90,16)	(**1.00,21**)	
14	(0.81,12)	(0.90,16)	(**1.00,20**)	
15	(0.43,9)	(0.90,12)	(**1.00,16**)	
16	(0.72,9)	(0.90,12)	(**1.00,16**)	
17	(0.90,9)	(**1.00,12**)		
18	(0.43,7)	(0.90,9)	(**1.00,12**)	
19	(0.72,7)	(**1.00,9**)		
20	(0.90,7)	(**1.00,9**)		
21	(**1.00,7**)			

TABLE 3.5 Minimum Total Energy Consumptions with Computed Confidence Probabilities Under Various Timing Constraints

		Node id	***T***	**M**	***P***	***E***
MAP_Opt	A(*u*)	1	2	2	0.90	4
		2	3	3	1.00	1
		3	3	3	1.00	1
		4	1	1	0.90	9
		5	2	1	1.00	
	Total		**8**		**0.81**	**27**
MAP_Greedy	A(*u*)	1	3	1	1.00	8
		2	1	1	1.00	9
		3	1	1	1.00	9
		4	2	1	1.00	9
		5	2	1	1.00	12
	Total		**8**		**1.00**	**47**

TABLE 3.6 The Mode Assignment of MAP_Opt and MAP_Greedy Algorithms with Timing Constraint 8

MAP_Greedy (Shao et al., 2005) is used, the total energy consumption obtained is 47 and all nodes are in M_1. We observe a 42.5% savings on energy when using MAP_Op.

3.4.2.2 The Algorithms for MAP Problem

In this section, the adaptive online energy-saving (AOES) algorithm is introduced. The basic idea is to use a time interval (window) and obtain the best mode assignment for each node in this time window. The mode of each node is then adjusted online. In this algorithm, the MAP_Opt subalgorithm is used to calculate the optimum mode for each node.

Adaptive Online Energy-Saving Algorithm The AOES algorithm is shown in Algorithm IV.1. In the AOES algorithm, we use the adaptive model to solve the energy-saving problem for heterogeneous WM^2Nets. The adaptive approach includes three steps: First, collect updated information of each node and predict the probability distributive function (PDF) of execution time of each node in a time interval (window). Second, in each time interval (window), obtain the best mode assignment for each node during the time interval (window) to minimize the energy consumption while satisfying timing constraint with guaranteed probability. Third, use an on-line architecture adaptation control policy. Since the design considers nonstationary environments, the control policy varies with the environment but is stationary within a time interval (window).

Algorithm IV.1 Adaptive Online Energy-Saving Algorithm

Input: A mesh network, R different mode types, and the timing constraint L.
Output: A mode assignment to minimize energy E while satisfying L for the WM^2Net.

1. Collect data and predict the PDF of execution time of each node in a time interval (window).
2. Obtain the best mode assignment A by using MAP_Opt for each node in the time interval (window).
3. Output results: A and $E_{min.}$
4. Use online architectural adaptation to reduce energy consumption while satisfying timing constraints with guaranteed probability.
5. Repeat the above steps.

Definition: To solve the MAP problem, we use a dynamic programing method traveling the graph in a bottom-up fashion. For the ease of explanation, we will index the nodes based on bottom-up sequence. For example, Fig. 3.15a shows nodes indexed in a bottom-up sequence after topological sorting the Fig. 3.14a. Given the timing constraint L, a PDFG G, and an assignment A, we first give several definitions as follows:

(1) G^i: The subgraph rooted at node u_i, containing all the nodes reached by node u_i. According to the proposed algorithm, during each step a node is added, which becomes the root of its subgraph. For example, G^3 is the graph containing nodes 1, 2, and 3 in Fig. 3.15a.

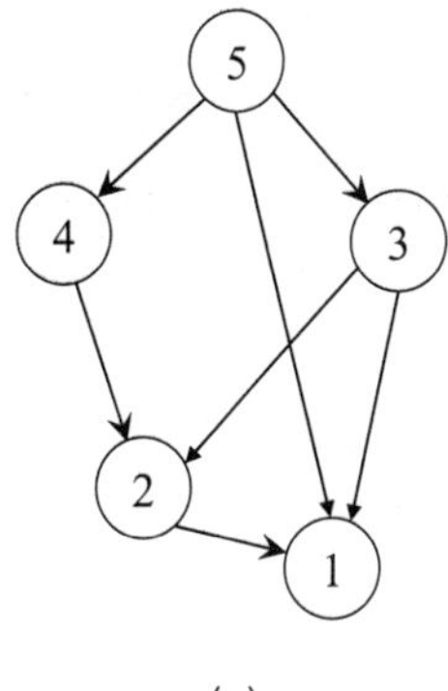

(a)

Nodes	M_1			M_2			M_3		
	T	P	E	T	P	E	T	P	E
5	1	0.7	8	2	0.9	4	5	0.9	2
	3	0.3		4	0.1		6	0.1	
4	1	1.0	8	2	1.0	6	3	1.0	2
3	1	1.0	8	2	1.0	6	3	1.0	2
2	1	0.9	9	3	0.9	5	5	0.6	2
	2	0.1		4	0.1		6	0.4	
1	1	0.9	12	3	0.9	6	5	0.8	1
	2	0.1		4	0.1		6	0.2	

(b)

FIGURE 3.15 (a) The resulted DAG of Fig. 3.14 (a) Being topologically sorted. (b) The times, probabilities, and energy consumptions of its nodes in different modes.

(2) $E_A(G^i)$ and $T_A(G^i)$: The total energy consumption and total execution time of G^i under the assignment A. According to the proposed algorithm, each step will achieve the minimum total energy consumption of G^i with computed confidence probabilities under various timing constraints.

(3) The proposed algorithm constructs table $D_{i,j}$. Each entry of table $D_{i,j}$ will store a linked list of (probability, energy) pairs sorted by probability in an ascending order. Here we define the **(probability, energy) pair** $(P_{i,j}, E_{i,j})$ as follows: $E_{i,j}$ is the minimum energy consumption of $E_A(G^i)$ computed by all assignments A satisfying $T_A(G^i) \leq j$ with probability $\geq P_{i,j}$.

In every step, an additional node is included for consideration. The information of this node is stored in local table $B_{i,j}$, which is similar to table $D_{i,j}$, but with accumulative probabilities only on node u_i. A local table stores only information such as probabilities and energy consumptions of a node itself. Table $B_{i,j}$ is the local table storing only the data of node u_i. In more detail, $B_{i,j}$ is a local table of linked lists that store pair $(P_{i,j}, E_{i,j})$ sorted by $P_{i,j}$ in an ascending order; $E_{i,j}$ is the energy consumption only for node u_i with timing constraint L, and $P_{i,j}$ is the *cumulative distributive function* (CDF) $F(j)$. The building procedures of $B_{i,j}$ are as follows. First, sort the execution time variations in an ascending order for each R. Then, compute the CDF under each R. Finally, let $L_{i,j}$ be the linked list in each entry of $B_{i,j}$, insert $L_{i,j}$ into $L_{i,j+1}$ while redundant pairs canceled out based on Lemma 4.1.

We introduce the operator "⊕" in this study. For two (probability, consumption) pairs H_1 and H_2, if H_1 is $(P^1_{i,j}, E^1_{i,j})$ and H_2 is $(P^2_{i,j}, E^2_{i,j})$, then after applying the ⊕ operation between H_1 and H_2, we get pair (P', E'), where $P' = P^1_{i,j} + P^2_{i,j}$, and $E' = E^1_{i,j} + E^2_{i,j}$. We denote this operation as "$H_1 \oplus H_2$." This is the key operation of the proposed algorithm. The meaning is that when two task nodes add together, the total cost is computed by adding the costs of all nodes together and the probability corresponding to the total cost is computed by multiplying the probabilities of all nodes based on the basic properties of probability and cost of a PDFG. For two independent events A and B, $P(\text{A and B}) = P(\text{A})^*P(\text{B})$, and $E(\text{A and B}) = E(\text{A}) + E(\text{B})$.

$D_{i,j}$ is the table in which each entry has a linked list that stores pair $(P_{i,j}, E_{i,j})$ sorted by $P_{i,j}$ in an ascending order. Here, i represents a node number, and j represents time. For example, a linked list can be (0.1,2) → (0.3,3) → (0.8,6) → (1.0,12). Usually, there are redundant pairs in a linked list. We give the redundant-pair removal Lemma in Lemma 4.1.

Lemma 4.1: Given $(P^1_{i,j}, E^1_{i,j})$ and $(P^2_{i,j}, E^2_{i,j})$ in the same list:

(1) If $P^1_{i,j} = P^2_{i,j}$, then the pair $E_{i,j}$ with minimum $E_{i,j}$ is selected to be kept.

(2) If $P^1_{i,j} < P^2_{i,j}$ and $E^1_{i,j} > E^2_{i,j}$, then $E^2_{i,j}$ is selected to be kept.

For example, we have a list with pairs (0.1,2) → (0.3,3) → (0.5,3) → (0.3,4), we remove the redundant-pair as follows: First, sort the list according $P_{i,j}$ in an ascending order. This list becomes to (0.1,2) → (0.3,3) → (0.3,4) → (0.5,3). Second, cancel redundant pairs. Comparing (0.1,2) and (0.3,3), we keep both. For the two pairs (0.3,3) and (0.3,4), we cancel pair (0.3,4) since the energy 4 is bigger than 3 in pair (0.3,3). Comparing (0.3,3) and (0.5,3), we cancel (0.3,3) since 0.3 < 0.5 while 3 = 3. The probability 0.3 is already covered by probability 0.5 while the costs are same. There is no information lost in redundant-pair removal.

Using Lemma 4.1, we can cancel many redundant-pairs $(P_{i,j}, E_{i,j})$ whenever we find conflicting pairs in a list during a computation.

The MAP_Greedy Algorithm

Algorithm IV.2 Heuristic Algorithm for the MAP Problem When the PDFG Is DAG (MAP_Greedy)

Input: R different mode types, a DAG, and the timing constraint L.
Output: A mode assignment to minimize energy while satisfying L.

1: Assign the lowest energy type to each node and mark the type as assigned.
2: Find a critical path (CP) that has the maximum execution time among all possible paths based on the current assigned types for the DAG.
3: For every node u_i in CP,
4: for every unmarked type p,
5: change its type to p,
6: calculate r = cost_increase / time_reduce
7: select the minimum r
8: if ($T > L$)
9: continue
10: else
11: exit /* This is the best assignment */

The MAP_Greedy algorithm is shown in Algorithm IV.2. A CP of a DAG is a path from source to its destination. To be a legal assignment for a data flow graph (DFG), the execution time for any CP should be less than or equal to the given timing constraint. In algorithm MAP_Greedy, we consider only the hard execution time of each node, that is, the case when the probability of the random variable T equals 1. This is a heuristic solution for hard real-time systems. We find the CP with minimized energy consumption first, then adjust the energy of the nodes in CP until the total execution time is less than or equal to L.

The MAP_Opt Algorithm

Algorithm IV.3 Optimal Algorithm MAP_Opt

Input: R different modes, a DAG, and the timing constraint L.
Output: An optimal mode assignment.

1. Topological sort all the nodes, and get a sequence A.
2. Count the number of multi-parent nodes t_{mp} and the number of multi-child nodes t_{mc}. If $t_{mp} < t_{mc}$, use bottom-up approach; Otherwise, use top-down approach.
3. For bottom-up approach, use the following algorithm. For top-down approach, just reverse the sequence. $|V| \leftarrow N$, where $|V|$ is the number of nodes.
4. If the total number of nodes with multi-parent is t, and there are a maximum of K variations for the execution times of all nodes, then we will give each of these t nodes a fixed assignment.
5. For each of the K^t possible fixed assignments, assume the sequence after topological sorting is $u_1 \rightarrow u_2 \rightarrow \ldots \rightarrow u_N$ in bottom-up fashion. Let $D_{1,j} = B_{1,j}$. Assume $D'_{i,j}$ is the table that stored minimum total energy consumption with computed confidence probabilities under the timing constraint L for the subgraph rooted on u_i except u_i. Nodes $u_{i1}, u_{i2}, \ldots, u_{iw}$ are all child nodes of node u_i and w is the number of child nodes of node u_i, then:

$$\text{If } w = 0,\ D'_{i,j} = (0,0);\quad \text{If } w = 1,\ D'_{i,j} = D_{i1,j};\qquad (3.6)$$
$$\text{If } w \geq 1,\ D'_{i,j} = D_{i1,j} \oplus \ldots \oplus D_{iw,j}$$

6. Then, for each k in $B_{1,k}$

$$D'_{i,j} = D'_{i,j-k} \oplus B_{i,k} \qquad (3.7)$$

7. For each possible fixed assignment, we get a $D_{N,j}$. Merge the (probability, energy) pairs in all the possible $D_{N,j}$ together, and sort them in ascending sequence according probability.
8. Then use the Lemma 4.1 to remove redundant pairs. Finally get $D_{N,j}$.

In algorithm MAP_Opt, we exhaust all the possible assignments of multi-parent or multi-child nodes. Without loss of generality, assume using the bottom-up approach. If the total number of nodes with multi-parent is t, and there are at most K variations for the execution times of all nodes, then we will give each of these t nodes a fixed assignment. We will exhaust all of the K^t possible fixed assignments. Algorithm MAP_Opt gives the optimal solution when the given PDFG is a DAG. In Eq. (3.6), $D_{i1,j} \oplus D_{i2,j}$ is computed as follows. Let G′ be the union of all nodes in the graphs rooted at nodes u_{i1} and u_{i2}. Travel all the graphs rooted at nodes $u_{i1}$and u_{i2}, For each node a in G′, we add the energy consumption of a and multiply the probability of a to $D'_{i,j}$ for only once, because each node can only have one assignment and there is no assignment conflict. The final $D_{N,j}$

we get is the table in which each entry has the minimum energy consumption with a guaranteed confidence probability under the timing constraint L. In the following, the Theorem 4.1 and Theorem 4.2 are presented.

Theorem 4.1: In each possible fixed assignment, for each pair $(P_{i,j}, E_{i,j})$ in $D_{i,j}(1 \leq i \leq N)$ obtained by algorithm MAP_Opt, $E_{i,j}$ is the minimum total energy consumption for the graph G^i with confidence probability $P_{i,j}$ under timing constraint j.

Theorem 4.2: For each pair $(P_{i,j}, E_{i,j})$ in $D_{N,j}(1 \leq j \leq L)$ obtained by algorithm MAP_Opt, $E_{i,j}$ is the minimum total energy consumption for the given DAG G with confidence probability $P_{i,j}$ under timing constraint j.

In algorithm MAP_Opt, there are K^t loops and each loop needs $O(|V|^2 * L * R * K)$ running time. The complexity of Algorithm MAP_Opt is $O(K^{t+1} * |V|^2 * L * R)$. Since t_{mp} is the number of nodes with multi-parent, and t_{mc} is the number of nodes with multi-child, then $t = \min(t_{mp}, t_{mc})$. $|V|$ is the number of nodes, L is the given timing constraint, R is the maximum number of modes for each node, and K is the maximum amount of execution time variation for each node. The experiments show that algorithm MAP_Opt runs efficiently.

3.4.2.3 Experiments

This section presents the experimental results. We conduct experiments on a set of DAGs. Three different modes, M_1, M_2, and M_3, are used in the system, in which a node with mode M_1 (active) is the quickest with the highest energy consumption and a node with type M_3 (sleep) is the slowest with the lowest energy consumption. The distribution of execution times of each node is Gaussian. We compare two methods base on algorithms MAP_Greedy and MAP_Opt. Method 1: Algorithm AOES with subalgorithm MAP_Greedy. Method 2: Algorithm AOES with subalgorithm MAP_Opt. The experiments are performed on a Dell PC with a P4 2.1G processor and 512 MB memory running Red Hat Linux 9.

Figure 3.16 shows a DAG with 17 nodes. This is for exp1 out of the six experiments conducted. We assume this is the topology of a WM^{2}Net. S is the source and D is the

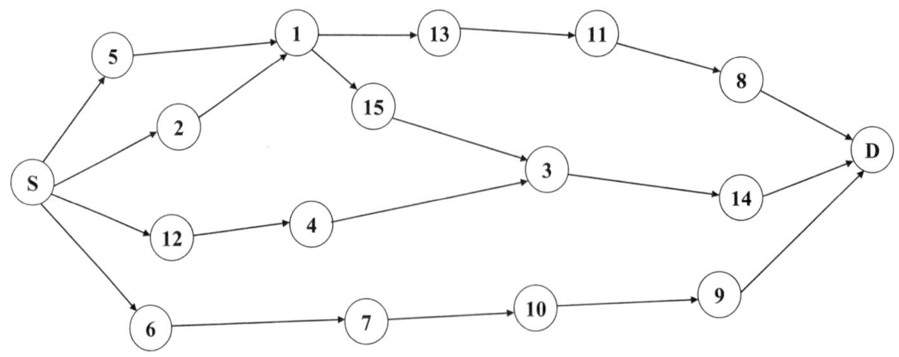

FIGURE 3.16 The PDFG of exp1.

destination. Each node has three modes with different execution times and energy consumptions. The collected data need to go through the topology to the destination within a timing constraint. Exclude the source and destination node; this DAG has 3 multi-child nodes and 5 multi-parent nodes. Using a top-down approach, we implemented all $3^3 = 27$ possibilities. The experimental results for exp1 are shown in Table 3.7. The first column of Table 3.7 stands for the timing constraint of the DAG. Column "Saving (% 1)" shows the percentage of reduction on system energy consumption, comparing the results of soft real-time with those of hard real-time, in other words, comparing the results of Method 2 with those of Method 1. The average percentage reduction is shown in the last row of Table 3.7 as "average saving." The entry with "X" means no solution available. Under timing constraint 50 in Table 3.7, there is no solution for hard real-time for Method 2 using the MAP_Greedy algorithm. However, we can find solution 6138 with probability 0.9 that guarantees the total execution time of the DFG is less than or equal to the timing constraint 50.

The experimental results of all six experiments are shown in Table 3.8. "Exp" is the name of the experiments, and "N" stands for the number of nodes of a DAG. Table 3.8 shows that the proposed algorithms can greatly reduce the total energy consumption while having a guaranteed confidence probability satisfying timing constraints. On average, Method 1 gives an energy reduction of 33.3% with confidence probability 0.9 under timing constraints, and an energy reduction of 41.4% and 48.0% with 0.8 and 0.7 confidence probabilities satisfying timing constraints, respectively. The experiments using MAP_Opt on these DAGs are finished within several minutes.

The advantages of Method 2 over Method 1 are summarized as follows. First, the proposed algorithms are efficient and provide overview of all possible variations of minimum costs comparing with the worst-case scenario generated by the MAP_Greedy. Although using the probabilistic approach the MAP problem becomes very complicated, the proposed algorithm gives very good results. First, it is possible to greatly reduce the system total energy consumption while have a very high confidence probability under different timing constraints. Second, given an assignment, we are able to get minimum total energy consumption with different confidence probabilities under each timing constraint.

3.4.3 Cross-Layer Energy Optimization in Multihop Wireless Mesh Networks[4]

3.4.3.1 Transmission Energy Minimization: A Source Coding Perspective

There have been several techniques proposed to reduce the energy consumption in WM^{2}Nets as well as to maintain a satisfactory signal quality of reception. Generally speaking, cross-layer optimization is necessary in order to take into account several interacting techniques and protocols such as sleep mode control and distributed source coding. In this study, we introduce a minimum energy (ME) source coding algorithm and perceive the benefits while using this scheme in a large WM^{2}Net.

Minimum Energy Coding for a WM^{2}Net Node Transceiver Minimum energy (ME) coding is a source coding algorithm that aims to reduce the transmission energy for a RF transmitter

[4] Excerpt from the invited article "Cross-layer energy optimization in multihop wireless mesh networks," Chun-Hung Liu, Department of Electrical and Computer Engineering, University of Texas at Austin, Austin TX, USA, E-mail: chliu@mail.utexas.edu

TC	Method 1	Method 2					
		0.7		0.8		0.9	
	Energy	Energy	Saving (% 1)	Energy	Saving (% 1)	Energy	Saving (% 1)
50	X	6,138		X		X	
60	X	6,124		6,125		6,178	
70	X	5570		5575		6173	
80	6,223	4,502	27.7	4,992	19.8	4,754	23.6
90	6,216	3,307	46.8	3,890	37.4	4,267	31.3
100	2,874	1,302	54.7	1,890	34.2	2,465	14.2
120	3,735	1,302	65.1	1,886	49.5	1,795	51.9
140	3,610	1,302	63.9	1,302	63.9	1,795	50.3
165	1807	1302	27.9%	1302	27.9%	1239	31.4%
166	1,302	1,302		1,302		1,302	
Average saving (%)			47.7		38.8		33.8

Table 3.7 Experimental Results of Method 1 and Method 2 for exp1

		Method 1	Method 2					
			0.7		0.8		0.9	
Exp	*N*	Energy	Energy	Saving (% 1)	Energy	Saving (% 1)	Energy	Saving (% 1)
exp1	24	2,982	1,558	47.8	1,850	38.0	2,008	32.5
exp2	35	4,215	2,283	45.8	2,378	43.6	2,687	36.3
exp3	43	5,176	2,572	50.3	2,953	43.0	3,513	32.1
exp4	62	6,402	3,348	47.7	3,812	40.5	4,410	31.1
exp5	78	9736	5073	47.9%	5736	41.1%	6427	34.0%
exp6	89	11865	6120	48.4%	6815	42.6%	7827	33.9%
Average saving (%)				48.0		41.4		33.3

TABLE 3.8 Experimental Results of Algorithms Method 1 and Method 2 for Different DAGS

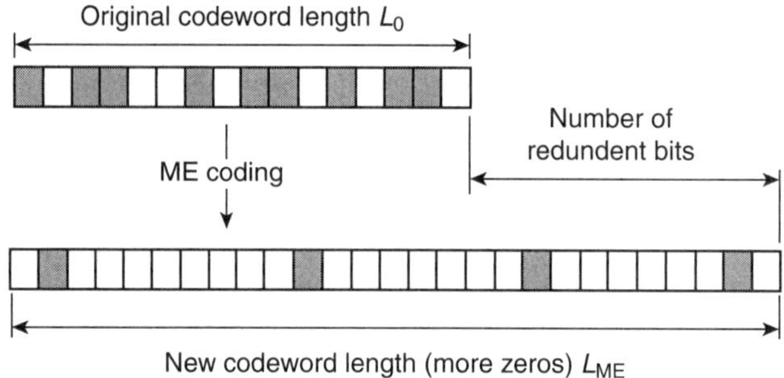

FIGURE 3.17 Principle of minimum energy coding.

that uses on–off keying (OOK) signal modulation (Erin and Asada, 1999). In on–off keying, higher power is consumed when one (high) bit is transmitted. Therefore, the number of high bits transmitted dominates the total power consumption. When a few redundant bits are added to the original codeword, as shown in Fig. 3.17, one can use a set of codewords that contain fewer high bits to represent the same number of source symbols. We can thus use those redundant bits for saving transmission power. ME coding is generated in two steps: codebook optimality and coding optimality. The former determines a set of codewords, named a codebook, with the fewest number of total high bits, whereas the latter assigns codewords to source symbols. While the fixed length ME coding is the simplest, it is most useful in practice. Figure 3.18 illustrates the entire set of usable codewords of length L sorted by the number of high bits. The i-th column lists all the codewords having i-1 high bits. The first column possesses only the $C_1^0 = 1$ codeword with zero high bits; the second column possesses $C_L^1 = l$ codewords with one high bit; the third column possesses C_L^2 containing two high bits and so on. All the codewords are numbered 1 through 2^L in the ascending order of the number of high bits, $W_0 = \{w_1, \ldots, w_{2^L}\}$. The last codeword w_{2^L} consists of all high bits, which consumes the largest amount of power to transmit. In order to save power, we do not use a codeword containing a large number of high bits, but represent all the q source symbols with the first q codewords in the

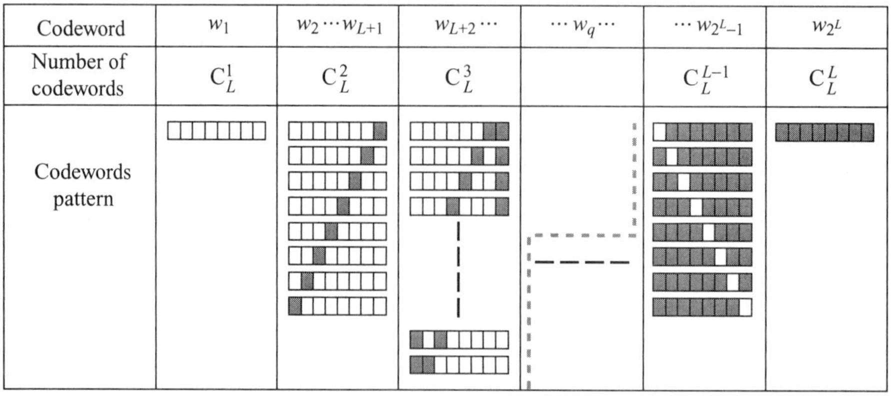

Codeword	w_1	$w_2 \cdots w_{L+1}$	$w_{L+2} \cdots$	$\cdots w_q \cdots$	$\cdots w_{2^L-1}$	w_{2^L}
Number of codewords	C_L^1	C_L^2	C_L^3		C_L^{L-1}	C_L^L
Codewords pattern						

FIGURE 3.18 Fixed-length ME codewords.

list. The remaining codewords are abandoned, as shown by broken lines in Fig. 3.18. This forms the optimal codebook. Use of this codebook is the necessary condition for minimizing the transmission power.

MAI Reduction in DS-CDMA WM²Nets by ME Source Coding The ME coding described above is useful for the reduction of multiple access interference (MAI) when using DS-CDMA (Liu and Asada). This is attained due to the low probability of overlap between different signals. Since ME coding intends to decrease the number of high bits in a codeword, it decreases the probability that two or more nodes transmit a high bit simultaneously, thus resulting in a large MAI reduction on transmission.

Several methods for reducing MAI have been proposed in the literature. Two conventional methods for reducing MAI at the transmitter side are to increase the system processing gain and to boost the signal power. Increasing the processing gain often entails a longer pseudo-random sequence for spreading the signal. This leads to the increase in memory size, computational complexity, and difficulty in designing spreading sequences. Boosting the signal power is not desirable for mobile applications and others where available power is limited. High output power is often prohibited by regulations in some countries as well. To reduce MAI at the receiver side sophisticated correlation filters must be used which, however, increase the complexity and cost of design. The salient characteristic of ME coding is that the more power it saves, the less MAI it achieves, however, at the expense of sacrificing transmission rate.

3.4.3.2 Energy Consumption Model

The energy saving coding technique in Section I deals with minimizing transmission energy. Focusing on minimizing transmission energy seems favorable in the traditional wireless link where the transmission distance could be large up to 100 m. In this scenario, the transmission energy dominates the total energy consumption. However, in WM²Nets the distribution of nodes is usually very dense (see Fig. 1.2) so that circuit energy consumption becomes more comparable or even higher than the transmission energy in the total energy consumption. Accordingly, there arises a fundamental question: is multihop transmission more energy-efficient than single-hop transmission in a WM²Net? Before investigating this question, we first have to understand the energy consumption model of a single-hop case in hardware and transmission. Then we can extend that model to the case with an n-hop transmission path in a WM²Net as depicted in Fig. 3.19.

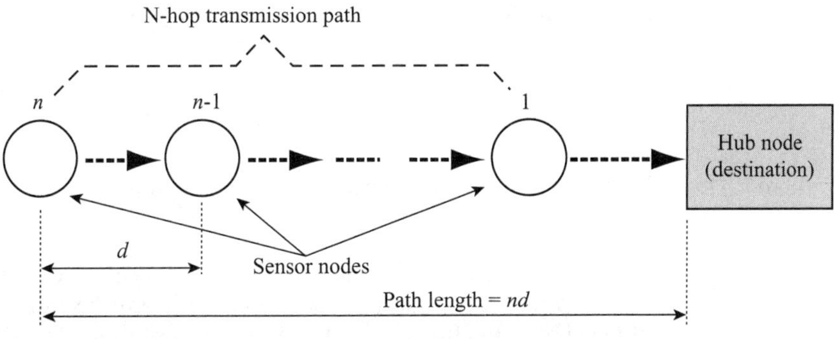

Figure 3.19 An n-hop transmission path in a multihop WM²Net.

Energy Consumption Model for Single Hop Transmission The power consumption model for a wireless transceiver must consider both transceiver and startup power consumption along with an accurate model for the amplifier. The latter becomes dominant with small packet sizes and long transition times. This is due to the frequency synthesizer settle down time. According to (Sankarasubramaniam et al., 2003), the energy consumption model for a bit E_b is given as

$$E_b = E_t + E_r + \frac{E_d}{l} \tag{3.8}$$

where E_t and E_r are the bit energy consumption for a transmitter and a receiver, respectively. E_d denotes the energy required to decode a packet whereas l is the number of bits per packet. The energy consumption of encoding data is assumed to be negligible in this case. This model describes the energy required to transmit a packet from a transmitting to a receiving node. This model was used for single-hop transmission to optimize the packet sizes and coding techniques while we will extend this energy model for a multihop scenario in the following subsection.

E_b in Eq. (3.8) with optimal power control can be written as

$$E_t = E_t^c + E_t^a d^\mu \tag{3.9}$$

where E_t^c is the energy consumption per bit in transmitter circuitry, E_t^a is the energy consumption per bit of the transmit amplifier, d is the transmission distance, and μ is the path loss exponent. An explicit formula for E_t^a can be found in the work by Chen and Callaway (2002) as follows:

$$E_t^a = \frac{\text{SNR} \cdot N_F \cdot N_0 \cdot B_w \cdot \left(\dfrac{4\pi}{\lambda}\right)^\mu}{g_a \cdot \eta_t \cdot R_b} \tag{3.10}$$

where SNR is the signal to noise ratio at the receiver side, N_F denotes the receiver noise figure, N_0 is the power spectrum density of the thermal noise, B_w is the channel bandwidth, λ is the carrier wavelength in meters, g_a is the antenna gain, η_t is the transmitter efficiency, and R_b is the transmission rate.

Thus, if we consider that all WM^{2}Net nodes communicate with the hub node via a single hop, we can calculate the total energy consumption E_Σ^{sh} as below

$$E_\Sigma^{sh} = \sum_{j=1}^{n} \left(m_b \left(E_c^t + E_c^a (jd)^\mu\right) + E_w^t\right) \tag{3.11}$$

where E_w^t is the bit energy required for startup and m_b is the number of bits transmitted in an n-hob transmission.

Energy Consumption Model of Multihop Transmission The analytical model for multihop transmission can be deduced from the previous analysis. Here the multihop energy analysis can be achieved by extending Eq. (3.8) to the linear multihop scenario assuming optimal power control (Min et al., 2002). According to Fig. 3.19, we can derive the bit energy for n-hop transmission from Eq. (3.8) as

$$E_b^n = \left(n\left(E_t^c + E_t^a d^{\mu}\right) + (n-1)\,E_r^c\right)\left(1 + \frac{\beta+\tau}{l}\right) + \frac{nE_w^t + (n-1)(E_w^r + E_d)}{l} \tag{3.12}$$

where E_c^r is the bit energy required in receiver circuitry, E_w^t and E_w^r are the bit energy needed for startup, E_t^a is the bit energy required for a successful transmission over one meter, μ is the path loss exponent, β is the preamble length, and τ is the coding overhead. The total energy consumed for a transmitting node for the n-hop case is

$$E_s^n = n\left(m_b\left(E_c^t + E_c^r d^{\mu}\right) + E_w^t\right) + (n-1)\left(m_b E_c^r + E_w^r + E_d\right) \tag{3.13}$$

The energy analysis of multihop transmission so far is based on the assumption that there is a single node transmitting information that is n hops away from the hub node. However, in reality almost all users will communicate with the hub node at the same time. In this context, the total energy consumption in the WM²Net can be expressed as follows:

$$E_{\Sigma}^{sh} = \frac{n(n+1)}{2}\left(m_b\left(E_c^t + E_t^a d^{\mu}\right) + E_w^t\right) + \frac{n(n-1)}{2}\left(m_b E_c^r + E_w^r + E_d\right) \tag{3.14}$$

3.4.3.3 Joint Energy Minimization for Physical, MAC, and Network Layers

The energy consumption model in a multihop WM²Net is described in Section 3.2. According to this model, we attempt to optimize a joint design across hardware, physical, MAC, and network layers. The optimization problem of joint design we formulate in this section is similar to the formulation approach provided by Cui et al. (2005). Let us consider a mesh node i and assume that N_i is the set of nodes transmitting to it and M_i the set of nodes receiving data from node i. In order to simplify the formulation, the energy a mesh node consumes during the transient mode is neglected. Let us assume that the transmitting energy that satisfies a target probability of bit error from node i to node j is denoted as $E_{i,j}^t = E_0 d_{i,j}^{\mu}$ in which E_0 is the required energy for a 1 m transmission distance. Since the transceiver of a mesh node cannot transmit and receive signals simultaneously, let $\theta_{i,j}$ and θ_{ji} denote the fraction of time that node i stays in the mode of transmitting to node j and in the mode of receiving from node j, respectively. Therefore, the average bit energy a node i consumes during its active mode is given as

$$\overline{E}_i = E_{c,i}^r \sum_{j \in \mathrm{N}_i} \theta_{ji} + \sum_{j \in \mathrm{M}_i} \theta_{i,j}\left(E_{c,i}^t + E_{t,ij}^a\right) \tag{3.15}$$

For a N-node network, we then formulate an energy optimization problem that maximizes the network lifetime:

$$\begin{aligned} &\text{minimize} \sum_{i=1}^{N} \overline{E}_i \\ &\text{subject to} \sum_{i=1}^{N} \sum_{j \in \mathrm{M}_i} \theta_{i,j} \le 1 \\ &\sum_{j \in \mathrm{M}_i} \theta_{i,j} R_{b,i} - \sum_{j \in \mathrm{N}_i} \theta_{ji} R_{b,j} = \overline{R}_{b,i}, \quad i = 1, \ldots, N \end{aligned} \tag{3.16}$$

where the first constraint is the time-sharing constraint by definition and the second constraint is coming from the conservation law of traffic flow in a node. The latter ensures that the difference between the total outgoing traffic and the total incoming traffic at each WM^{2}Net node is equal to the traffic generated by the node itself. $R_{b,i}$ and $R_{b,j}$ are the outgoing and incoming traffic rates, respectively, and $\overline{R}_{b,i}$ denotes the net traffic rate a node i generates. In particular, if node i is a relay node $\overline{R}_{b,i}$ must be equal to zero. Optimization problem Eq. (3.16) can be effectively solved by the linear programing techniques since the objective function and the constraints are all linear (Boyd and Vandenberghe, 2004).

This optimization problem considers the joint design of network (routing) and MAC (scheduling) layers. The solutions to problem Eq. (3.16) tell us then when multihop transmission is more energy-efficient than single-hop transmission. Since all of the communication links are time varying, the energy efficiency in the network can be further improved when adding the link adaptation into the model given in Eq. (3.16). The bit energy consumption for link (i, j) with a given target probability of bit error P_b should be bounded as

$$E_{i,j}^{\text{Link}} \le 2N_F N_0 E_{i,j}^t \left(\ln \frac{2}{P_b}\right)\left(\frac{2^{S_{i,j}-1}}{S_{i,j}}\right) + \frac{E_{c,i}^t + E_{c,j}^r}{S_{i,j}} \tag{3.17}$$

It is assumed that uncoded multiple quadrature amplitude modulation (MQAM) with constellation size $S_{i,j}$ (Cui et al., 2005a) is used. Therefore a new optimization problem that considers a joint design of routing, scheduling and link adaptation can be formulated as follows

$$\begin{aligned} \text{minimize} \quad & \sum_{i=1}^{N}\left(\overline{E}_i + \sum_{j\in M_i} E_{i,j}^{Link}\right) \\ \text{subject to} \quad & \sum_{i=1}^{N}\sum_{j\in M_i} \theta_{i,j} \le 1 \\ & \sum_{j\in M_i} \theta_{ij} R_{b,i} - \sum_{j\in N_i} \theta_{ji} R_{b,j} = \overline{R}_{b,i},\ i = 1, \ldots, N \\ & 2 \le S_{i,j} \le C_{i,j},\ \{i, j\} = 1, \ldots, N \end{aligned} \tag{3.18}$$

where $C_{i,j}$ is the link capacity per hertz between node i and node j, and $E_{i,j}^{Link}$ can be obtained by using the upper bound in Eq. (3.17).

3.4.4 The Network Size Impact on the Network Lifetime in Wireless Mesh Network[5]

3.4.4.1 Problem Statement

System Model Let us consider a circular supervised area with radius R as illustrated in Fig. 3.20. Let R_c and r^* denote the coverage range and transmission range, respectively.

[5] Excerpt from the invited article "The network size impact on the network lifetime in wireless mesh network," *Moez Esseghir, †Nizar Bouabdallah, and ‡Guy Pujolle (*LIP6 Laboratory, Pierre & Marie Curie

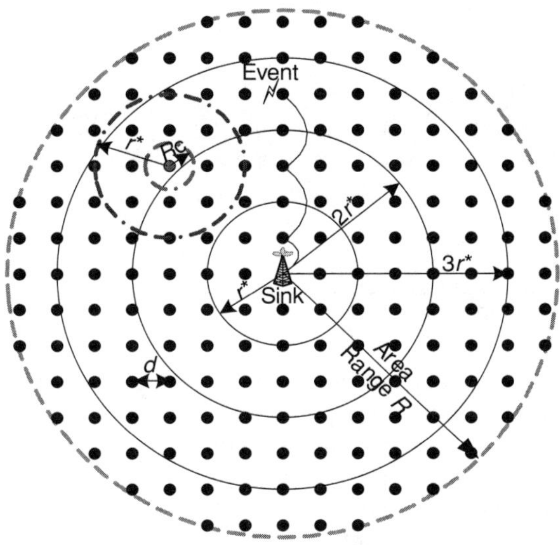

FIGURE 3.20 System model.

Let us assume that each node enters a sleep mode once there is no on-going transmission. A wake-up scheme is, thus, required to bring nodes to the active communication mode when necessary. In this study, we consider a perfect wake-up scheme where only the intended relaying node listens to the transmission, as specified in the study of Zhao and Tong (2005), thus eliminating unnecessary energy consumption. According to Zhao and Tong (2005), the transmission range r^* that minimizes the total energy consumed to report an event with a perfect wake-up scheme is given by:

$$r^* = \left(\frac{E_{\text{rx}} + e_{\text{tx}}}{e_{\text{out}}(\alpha - 1)} \right)^{\frac{1}{\alpha}} \tag{3.19}$$

where E_{rx} is the energy consumed in reception, e_{tx} is the energy consumed by the circuitry, e_{out} is the antenna output energy, and α is the path attenuation factor.

Energy Consumption Model To minimize the energy consumption, an event-reporting scheme is considered in which each event is reported by the closest node to the sink (AP) within the event radius. To illustrate the efficiency of this scheme, let us consider the example shown in Fig. 3.21. In this example, WM^{2}Net nodes C2 and C3 detect the event P since they are within the event radius R_c. Source data can be therefore communicated to the sink either by C2 or by C3 through intermediate nodes $\{C4, \ldots, C10\}$. The goal is to minimize the number of hops (or relaying nodes) between the reporting node and the sink in order to minimize the required energy to report P. To do so, node C3, being closer to the sink than C2, has to handle the reporting operation.

University. E-mail: moez.esseghir@lip6.fr; [†]Inria, France. E-mail: guy.pujolle@lip6.fr; [‡]LIP6 Laboratory, Pierre & Marie Curie University. E-mail: nizar.bouabdallah@inria.fr).

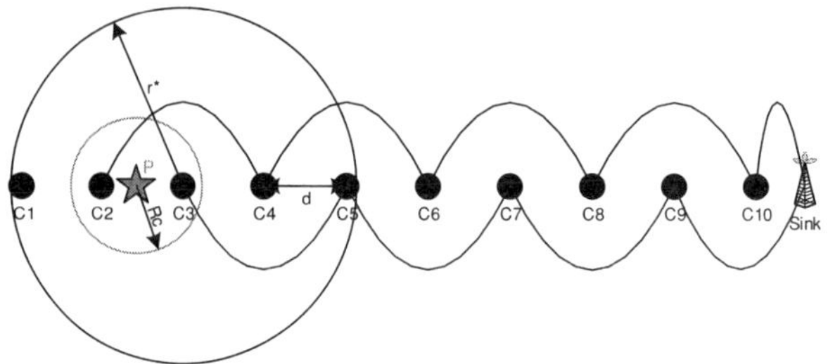

FIGURE 3.21 Event reporting in WM²Net.

Likewise, at each hop level, data must be relayed by the closest node to the sink within the transmission radius r^* of the previous relaying node. For instance, the message sent by C3 is relayed at the first hop by C5 instead of C4 and so on and so forth. In this case, the shortest path is {C3, C5, C7, C9}, which is composed of 4 hops. In contrast, C2 messages require 5 hops to reach the sink.

Let x denote the number of hops between the reporting node of an event P and the sink, and N the network size. Let us further assume for simplicity that the total energy spent in each hop is the sum of that spent in a single transmission-reception. It is then:

$$\begin{aligned} e(x) &= (E_{tx}(r^*) + E_{rx})x \\ &= (e_{tx} + (e_{out} \times (r^*)^{\alpha}) + E_{rx})x \end{aligned} \tag{3.20}$$

where, $E_{\text{tx}}(r^*)$ is the transmission energy required by a node that covers a neighborhood of radius r^*. Therefore, the average amount of energy required to report an event randomly occurring in the supervised area is:

$$e = (E_{\text{tx}}(r^*) + E_{\text{rx}})E[x] \tag{3.21}$$

where $E[x]$ denotes the average number of hops required to report an event P. According to Eq. (3.21), we notice that e behaves similarly to $E[x]$. Considering the system model of Fig. 3.20, $E[x]$ can be written as follows:

$$E[x] = \frac{\sum_{k=1}^{\lceil \frac{R}{r^*} \rceil} k\mathrm{S}_k}{\sum_{k=1}^{\lceil \frac{R}{r^*} \rceil} \mathrm{S}_k} = \frac{\sum_{k=1}^{\lceil \frac{R}{r^*} \rceil} k\mathrm{S}_k}{\mathrm{S}} \tag{3.22}$$

where S_k denotes the segment of S where each occurring event needs k hops to reach the sink. For instance, let us consider the example of Fig. 3.22, where the radius of the monitored is $R = 2r^*$. That is, an event P needs at most 2 hops to reach the sink. In this example, $\mathrm{S}_1(\mathrm{S}_2)$ denotes the portion of S where the number of hops required to report an event is 1 (2 respectively). Note that S_1 (blank surface of Fig. 3.22a) is simply a concatenation of the coverage areas of all nodes situated in zone Z_1 (i.e., within r^* from the sink: blank surface of Fig. 3.22b). On the other side, S_2 is the remaining surface of the monitored disk. We recall the difference between Z_k and S_k significations: S_k denotes the

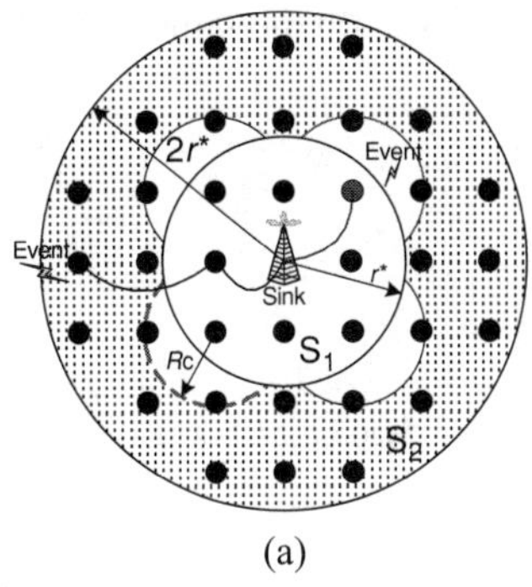

(a)

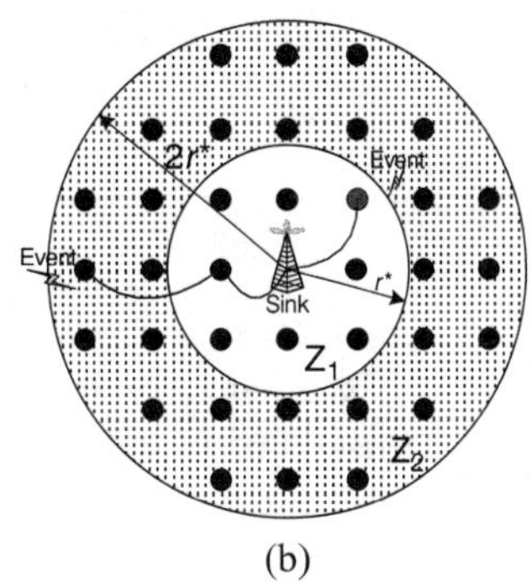

(b)

FIGURE 3.22 Decomposition of the monitored area in zones.

surface, where occurring events require k hops to reach the sink (Fig. 3.22a), whereas Z_k denotes the surface where nodes need k hops to report an event to the sink (Fig. 3.22b).

Energy Distribution Model Typically, in WM^{2}Nets, nodes are provided with the same initial amount of energy. However, considering the system model described above, it is easy to see that nodes in Z_1 (i.e., at 1 hop from the sink) support the entire reporting traffic. In contrast, the remaining nodes of the network handle only the reporting traffic of events that occur in S_2 (see Fig. 3.22). Doing so, nodes in Z_1 will be discharged prior to the remaining nodes of the network. Hence, this uniform distribution of the energy leads to energy wasting. Specifically, the network lifetime expires (loss of connectivity to the sink due to Z_1 nodes energy expiration), while Z_2 nodes still contain energy. To avoid this, the total provided energy E_t has to be properly distributed between zones $(Z_k)_{1\leq k\leq n}$. Note that n is the maximal number of hops required from a node to report an event. To achieve this, we propose a new energy distribution model. Recall that the total provided energy E_t depends on the network size N. In fact, E_t should be proportional to N. Thus, we assume in this study that E_t is given by:

$$E_t = N \times Es \tag{3.23}$$

where Es is a given amount of energy.

The basic idea is to provide nodes that belong to different zones $(Z_k)_{1\leq k\leq n}$ with different initial amounts of energy. In other words, the total initial energy E_t in the network is distributed between zones $(Z_k)_{1\leq k\leq n}$ according to their relative positions with respect to the sink. Specifically, each zone $(Z_k)_{1\leq k\leq n}$ is provided by $(E_k)_{1\leq k\leq n}$ initial amount of energy. Moreover, E_k is equally distributed among WM^{2}Net nodes belonging to Z_k. Then, $(E_k)_{1\leq k\leq n}$ can be written as follows:

$$E_k = \alpha_k{}^* E_t \textbf{ and} \sum_{k=1}^{n} E_k = E_t \tag{3.24}$$

where $(\alpha_k)_{1\leq k\leq n}$ are the normalized weights for $(Z_k)_{1\leq k\leq n}$.

Since nodes that belong to the same zone Z_k are provided with the same initial amount of energy $(e_k)_{1 \leq k \leq n}$, we have:

$$e_k = \frac{E_k}{N_k} \tag{3.25}$$

where $(N_k)_{1 \leq k \leq n}$ is the total number of nodes belonging to Z_k.

Normalized weights construction: To derive $(\alpha_k)_{1 \leq k \leq n}$, we first construct the non-normalized weight U_k for each zone Z_k. These weights can be calculated recursively as follows:

(1) First, let us consider zone Z_1. Nodes in Z_1 report events occurring in S_1 and relay reporting messages for events that occur in the remaining surface of the monitored area. So, U_1 can be defined as follows:

$$U_1 = 1$$

(2) Let us consider now Z_2. As for Z_1, nodes in Z_2 report events occurring in S_2 and relay reporting messages associated with events that occur in $(S_k)_{3 \leq k \leq n}$ (i.e., the remaining surface except S_1). Thus, U_2 can be written as:

$$U_2 = 1 - \frac{S_1}{S}$$

where S is the total surface of the supervised area.

(3) The other weights are defined on the same bases, and $(U_k)_{2 \leq k \leq n}$ can be written as follows:

$$U_k = 1 - \frac{S_1 + S_2 + \ldots + S_{k-1}}{S} \text{ and } U_1 = 1 \tag{3.26}$$

Then, the normalized weights α_k are given by:

$$\alpha_k = \frac{U_k}{\sum_{j=1}^{n} U_j} \tag{3.27}$$

Analytical Model In this section, we study analytically how the network lifetime T scales with the network size N. The energy distribution model described above is assumed. Uniform node distribution is also assumed. We first show the impact of network size on the average amount of energy consumed to report an event. Then, we prove that the speed of increasing the network lifetime increases faster than that of the number of nodes.

The Network Size Impact We now examine how e (see Eq. 3.21), the average amount of energy required to report an event and which will be denoted henceforth by $e(N)$, scales with the network size N. Note that $e(N)$ is calculated through Eq. (3.21) and using Monte Carlo method (Robert and Casella, 2004). Figure 3.23a shows that $e(N)$ is globally

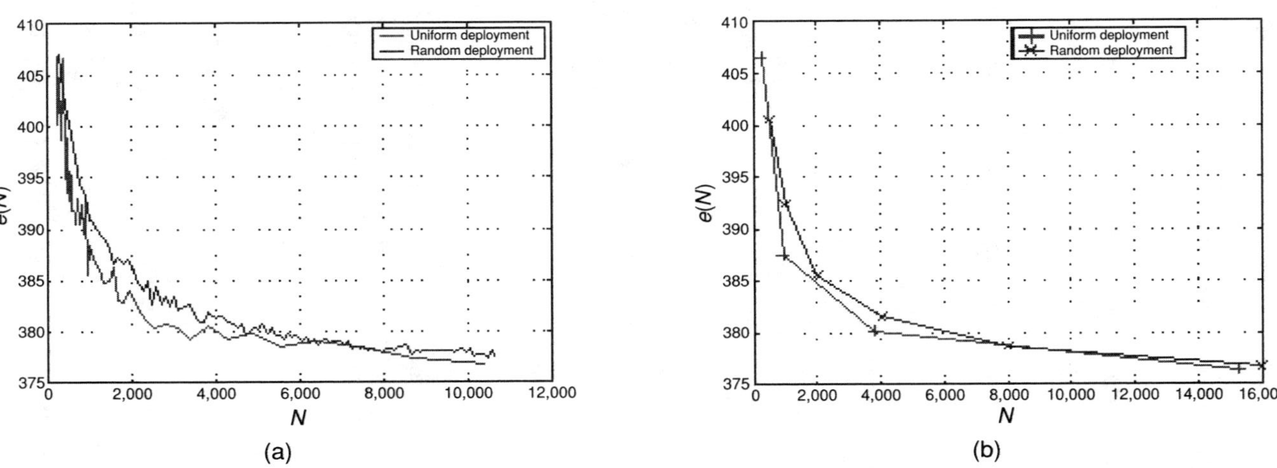

FIGURE 3.23 Average amount of energy required to report an event: $e(N)$ ($E_{rx} = 50$ nJ, $\alpha = 3$, $e_{out} = 0.1$ nJ, $e_{tx} = 81$ nJ).

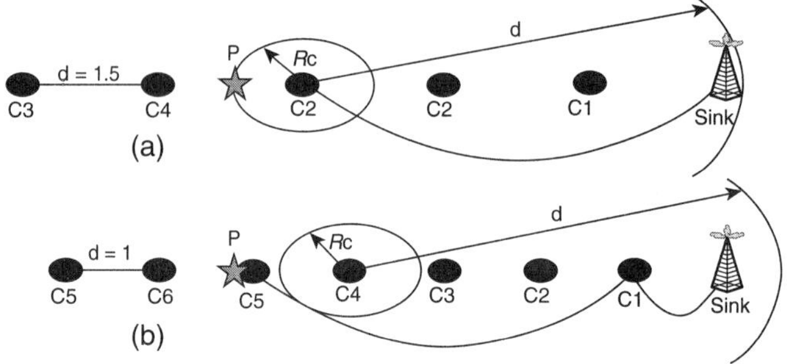

FIGURE 3.24 Illustration of a particular case where $e(N)$ increases with the network size N.

decreasing when N increases. Recall that the closest node C to the sink within the event radius always reports an event P. When N increases, the number of nodes within the event radius increases. So, the new reporting node C′ is likely to be closer to the sink than C. In view of this, both $E[x]$ and $e(N)$ decrease when N increases.

Nonetheless, Fig. 3.23a shows that $e(N)$, for some particular cases, can slightly increase when N increases. To understand this behavior, let us consider the example of Fig. 3.24. We suppose that the distance between the occurring event P and the sink is equal to 5.1 units distance (ud), $r^* = 4.6$ ud and $R_c = 1$ ud. First, the distance d between each two adjacent nodes is set equal to 1.5 ud (Fig. 3.24a). Recall that in regular topology, decreasing the distance d increases the network size or density. In this case ($d = 1.5$ ud), the event is reported by C3, which is at 1 hop from the sink. On the other side, when $d = 1$ (Fig. 3.24b) the event is reported by C5, which is at 2 hops from the sink. Hence, in this particular case, the required energy to report P increases although N increases. This explains the serrated shape of the curve in Fig. 3.23a.

To avoid this anomaly, the network size has to be increased while preserving the placement of the already deployed nodes. This is achieved by adding new nodes to the already deployed network; for instance, by inserting a new node between each two adjacent nodes, as depicted in Fig. 3.25. In this case, d is divided by factor 2. Doing so, we ensure that the number of hops required to report an event P does not increase with N. Results obtained according to this placement strategy are drawn in Fig. 3.23b, where we

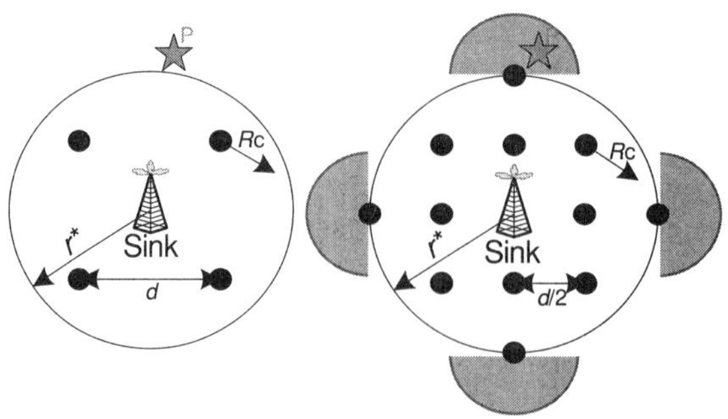

FIGURE 3.25 Keeping nodes' placement while increasing N.

can see that $e(N)$ always decreases with N (i.e., the serrated shape seen in Fig. 3.23a is avoided). The remainder of this analysis considers this placement strategy.

Recall that when N tends to infinity d tends to zero. This means that nodes are superposed to each other. In that case, the WM²Net lifetime tends to infinity.

The Network Lifetime as a Function of the Network Size We define the network lifetime T as the time spent from the deployment until the loss of coverage or connectivity. T depends on the network size, N, and the initially provided amount of energy for the network E_t.

Let $P(N)$ denote the total number of events that can be detected by a network consisting of N nodes. $P(N)$ is given by:

$$P(N) = \frac{E_t}{e(N)} = \frac{N \times Es}{e(N)} = \frac{\sum_{k=1}^{n} E_k}{e(N)} \tag{3.28}$$

Let now M be the average number of events occurring by unit of time. Then, the lifetime of a network consisting of N nodes is given by:

$$T(N) = \frac{P(N)}{M} \tag{3.29}$$

Theorem 1: $T(N)/N$ is an increasing function.

Proof: Let N_1 and N_2 be two network sizes with $N_1 < N_2$. As stated before, the average amount of energy consumed to report an event to the sink decreases with N. So, we have:

$$e(N_1) \geq e(N_2) \tag{3.30}$$

Let $T(N_1)$ and $T(N_2)$ denote the lifetimes for networks supervised by respectively N_1 and N_2 nodes.

$$\frac{T(N_2)}{T(N_1)} = \frac{P(N_2)}{P(N_1)} = \frac{N_2}{e(N_2)}\frac{e(N_1)}{N_1} \geq \frac{N_2}{N_1} > 1 \tag{3.31}$$

$T(N)/N$ is thus an increasing function. ■

Two properties can be derived from Theorem 1. First, the network lifetime $T(N)$ increases with N, which is a logical result (see Fig. 3.27a). Second, the speed of increasing $T(N)$ goes up faster than that of N (see Fig. 3.27b). In other words, the network lifetime $T(N)$ is not a cumulative function.

Corollary 1: $T(N_1 + N_2) \geq T(N_1) + T(N_2)$, where N_1 and N_2 are two integers.

Proof: Let N_1 and N_2 be two network sizes (integers). Based on Theorem 1, we have:

$$\frac{T(N_1)}{T(N_1+N_2)} \leq \frac{N_1}{N_1+N_2} \text{ and } \frac{T(N_2)}{T(N_1+N_2)} \leq \frac{N_2}{N_1+N_2}$$

```
Begin
(1) Deploy nodes on the supervised area,
(2) Charge nodes,
    Do
        (3.1) Generate a randomly event occurring in the
              monitored area,
        (3.2) Select a reporting node,
        (3.3) Find the path between the reporting node and
              the sink,
        (3.4) Decrease the provided amount of energy for
              each node all over the path,
    While event can be reported.
End.
```

FIGURE 3.26 Pseudo code for the developed simulator.

So, by a simple addition, we obtain:

$$\frac{T(N_1) + T(N_2)}{T(N_1 + N_2)} \leq 1$$

■

Experimental Results In this section, we present simulation results to assess the accuracy of the proposed analytical model. The developed simulator calculates the average number of events that can be detected by a network consisting of N nodes. Based on Eq. (3.29), we derive the average network lifetime $T(N)$.

The adopted algorithm is described in Fig. 3.26. The simulation resumes when the network is unable to report an event. This may occur due to the following two reasons: due to loss of either coverage or connectivity. In the first case, nodes that could detect P, and which are eligible for reporting P, have not enough energy to report the event to the sink. In the second case, one of the nodes belonging to the path that connects the reporting node to the sink has not enough energy to relay the reporting message.

Figure 3.27 reports the network lifetime $T(N)$ and $T(N)/N$ as a function of the network size N for both analytic and simulation results. Fig. 3.27a shows that the network lifetime $T(N)$ increases with N and Fig. 3.27b shows that the speed of increasing $T(N)$ increases faster than that of N.

Note that the network lifetime $T(N)$ is not a linear function of N since $T(N)/N$ is not constant for all the values of N. $T(N)$ behaves, in fact, as $\frac{aN^2}{N+b}$ functions. In Fig 3.27a, $T(N)$ seems to grow linearly with N due to the small number of observation points.

The simulation results confirm the analytic ones. Indeed, Fig. 3.27 reveals a perfect match between the simulation and the analytic results, which demonstrates the accuracy of the developed analytical model.

Optimum WM²Net Nodal Population In this section, the problem is tackled from a different perspective. Assuming that the network lifetime is known, the goal is to optimize the required number of nodes deployed. Below, we bring out the minimal number of nodes N_{opt} that ensures supervising an area A during a given period T.

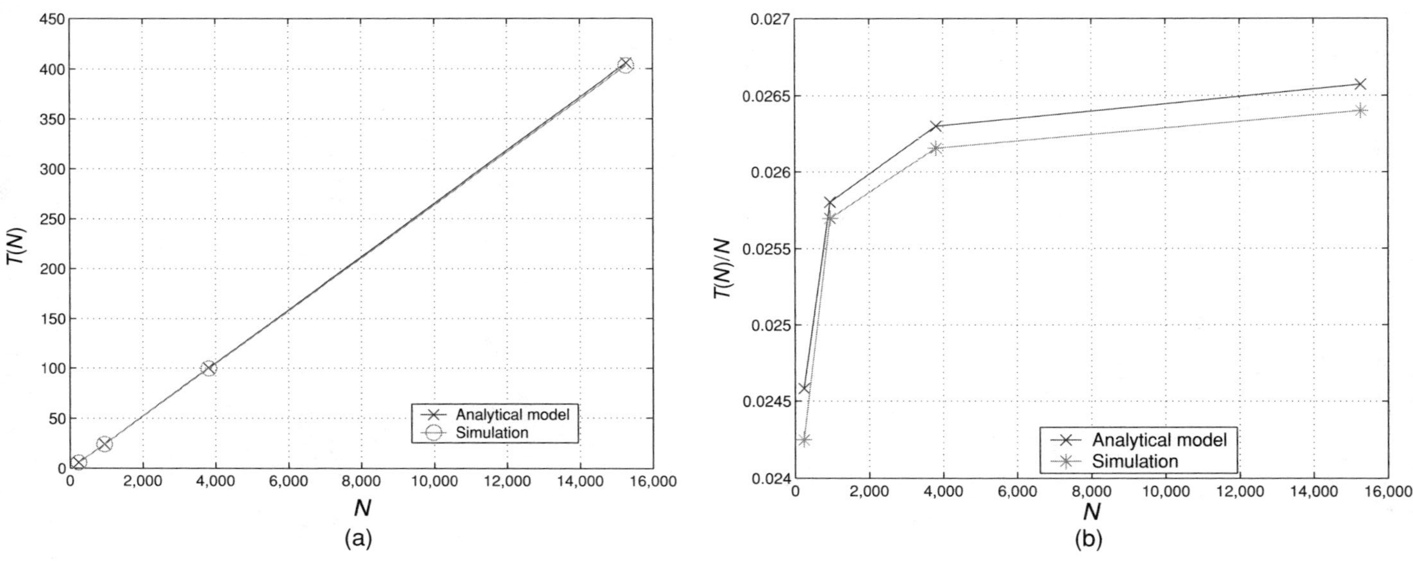

FIGURE 3.27 (a) Network lifetime $T(N)$. (b) $T(N)/N$ as a function of N.

Assume M to be the average number of events occurring in unit of time. The number of events P^* to be handled by the WM^{2}Net during T is given by:

$$P^* = T \times M \tag{3.32}$$

Theorem 2: The minimal number of nodes N_{opt} to supervise A during T is:

$$N_{opt} = \max(N^*, N_{\min}) \tag{3.33}$$

$$\text{where } N^* \text{ is given by } \frac{E_t}{e(N^*)} = P^* \tag{3.34}$$

and $N_{\min}$ is the minimum number of nodes that ensures network coverage as well as connectivity.

In addition, the N_{opt} nodes must be deployed at the same time (i.e., within the same deployment phase).

Proof: Let N^* be defined by:

$$\frac{E_t}{e(N^*)} = P^*$$

It is obvious that N_{opt} has to be greater than $N_{\min}$, otherwise connectivity and coverage are lost. Hereafter, we consider that $N^* \geq N_{\min}$.

(1) Let us first prove that N^* is minimal.

Suppose that it exists $N' < N^*$ that enables the detection of P^* events. The N' nodes can be deployed either at the same time or over multiple deployment phases. So, N' can be written as: $N' = \sum_i N_i$, where N_i is the number of nodes deployed at the step i and $N_i \geq N_{\min}$.

As $N_i < N^*$ and based on Theorem 1, we have:

$$\frac{T(N_i)}{T} = \frac{T(N_i)}{T(N^*)} = \frac{P(N_i)}{P^*} \leq \frac{N_i}{N^*}$$

Thus, $\frac{\sum_i P(N_i)}{P^*} \leq \frac{\sum_i N_i}{N^*}$ which gives: $1 \leq \frac{N'}{N^*}$: Absurd.

(2) Let us now demonstrate that the N^* nodes must be deployed at the same time to ensure the required network lifetime T. N^* can be written as follows:

$$N^* = \sum_{i=1}^{n} N_i$$

where N_i is the number of nodes deployed at the step i and $N_i \geq N_{\min}$. ■

Based on Corollary 1, the network lifetime is not a cumulative function. Hence, $T = T(N^*) \geq \sum_{i=1}^{n} T(N_i)$. So, in order to ensure that the network lifetime is T, N^* number of nodes have to be deployed within the same phase. Figure 3.28 reports the optimal number of required nodes N_{opt} as a function of the network lifetime T. Note that we have the

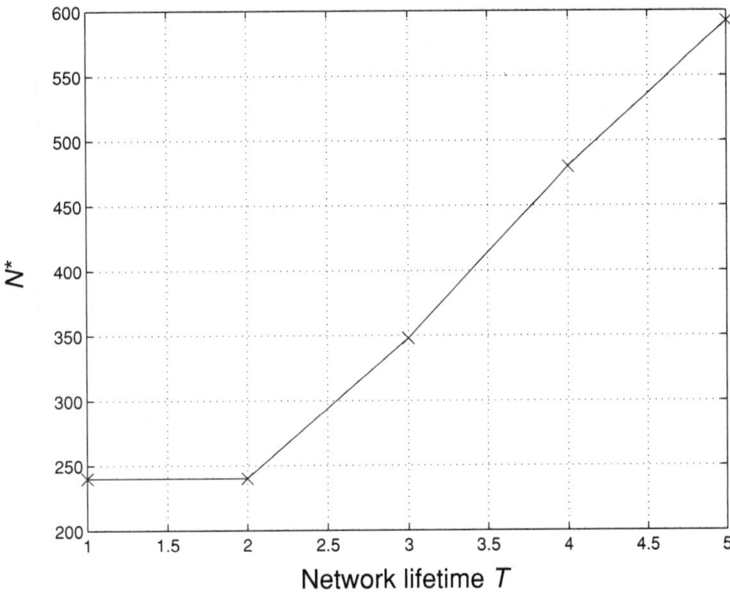

FIGURE 3.28 Minimal number of required nodes N_{opt}.

same optimal number N_{opt} of nodes for both $T = 1$ unit time (ut) and $T = 2$ ut. This illustrates the coverage and connectivity constraint on the optimal nodes to be deployed (see Eq. 3.33). So, for both $T = 1$ ut and $T = 2$ ut, N^* is larger than N_{min} and thus $N_{opt} = N_{min}$.

Extension to Random Deployment In this section, we extend the results presented above to the random deployment of nodes. We are especially interested in how the energy consumption scales with the network size. In fact, Theorem 1 and its corollary are based only on the principle that energy consumption decreases while increasing the network size N. We also give in this section simulation results to corroborate the proposed analytical model.

The Network Size Impact As for uniform distributed nodes, we use the same energy distribution model and examine how $e(N)$, the average amount of energy required to report an event, scales with the network size N.

For the same reasons as in Section B.1 above, we have the serrated shape in Fig. 3.29a and likewise, we avoid this anomaly by increasing the network size while preserving the placement of the already deployment nodes. In other words, at each step we add to the previously deployed nodes, in a randomly manner, the same number of nodes as already deployed. Thus, we almost reproduce the method used in Section B.1, which consists in dividing by 2 the distance between uniformly distributed nodes to increase the network size. Moreover, we assume that for the first step of the deploying process we have already the minimal number of nodes that ensures coverage.

Results obtained according to this placement strategy are drawn in Fig. 3.29b, where we can see that $e(N)$ always decreases with N (i.e., the serrated shape seen in Fig. 3.29a is avoided).

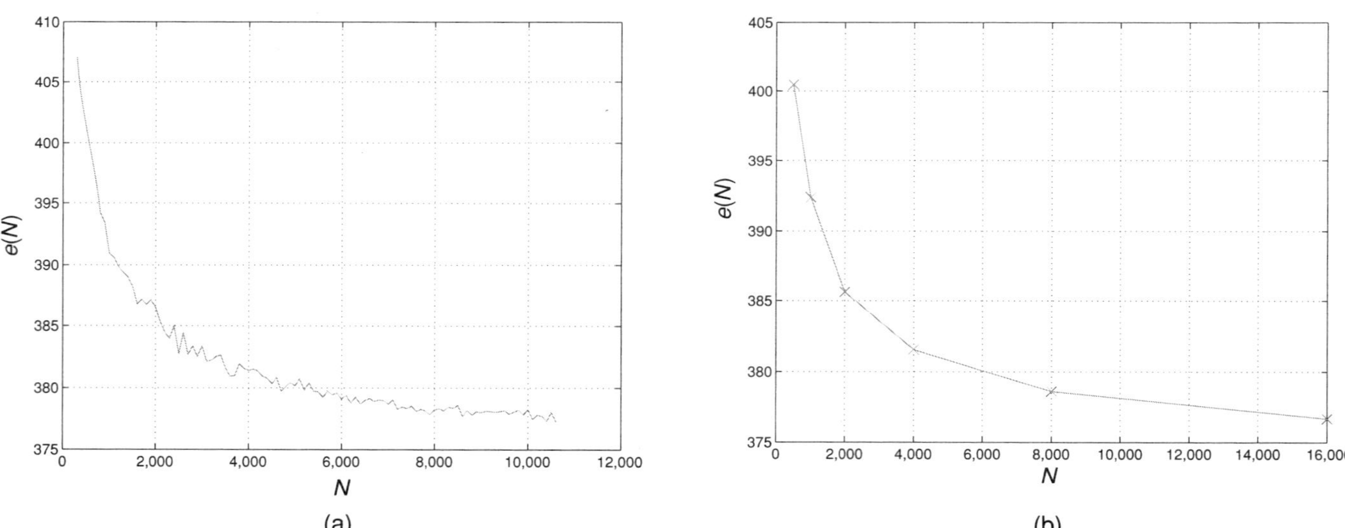

FIGURE 3.29 Average amount of energy required to report an event: $e(N)$ ($E_{rx} = 50$ nJ, $\alpha = 3$, $e_{out} = 0.1$ nJ, $e_{tx} = 81$ nJ).

The Network Lifetime as a Function of the Network Size As the average energy consumed to report an event scales in a similar fashion as in uniform nodal distribution, the proposed analytical model in the case of uniform nodes distribution is still valid in random deployment too. So, based on this analytical model and on the same simulation algorithm (see Fig. 3.26), we plot the network lifetime $T(N)$ and $T(N)/N$ as function of the network size N. Results are reported in Fig. 3.30. Figure 3.30a shows that the network lifetime $T(N)$ increases with N and Fig. 3.30b shows that the speed of increasing $T(N)$ goes up faster than that of N.

As for uniform deployment, the simulation results confirm the analytic ones in random deployment too. Indeed, Fig. 3.30 reveals a perfect match between the simulation and the analytic results, which demonstrate the accuracy of the developed analytical model.

3.4.5 Energy-Efficient Packet Relaying in Sparse Mobile Mesh Networks[6]

3.4.5.1 Related Background on Store-and-Forward Routing

A number of store-and-forward routing strategies exist in the literature for networks that share the properties of sparse and mobile with WM^{2}Nets (e.g., intermittently connected mobile networks (Vahdat and Becker, 2003; Lindgreny et al., 2003) or delay tolerant networks (Small and Hass 2005)). Epidemic routing (Vahdat and Becker, 2003) performs packet relaying in a way that is reminiscent of the concept of flooding. In epidemic routing, when two nodes meet they exchange copies of packets that were not received previously. Copies of a packet can thus be gradually spread to every node in the network including its destination. In practice, epidemic routing suffers from high usage of energy and bandwidth resources. A simple way to reduce the excessive resource consumed by epidemic routing is to replicate packets only to nodes that have higher delivery probabilities (Lindgreny et al., 2003) or to those with similar mobility patterns to the destination (Leguay et al., 2006). A more aggressive way to cut down resource usage is to use single-copy routing (Spyropoulos et al., 2004), where each intermediate node relays at most one copy per packet. Although these relaying strategies can substantially reduce the resource overhead per delivery, they increase latency of packet delivery. To address the latency-overhead dilemma, multi-copy relaying strategies are proposed by Small and Hass (2005) (Spyropoulos et al., 2004a) to limit the delivery overhead by replicating only exactly R (the replication factor) copies of each packet for the expected delivery delay. The R packet copies can be distributed through a process that can be represented as a balanced binary-tree (Spyropoulos et al., 2004a) or a locally-optimal tree (Small and Hass, 2005). The EBEC scheme (Liao et al., 2006) is a more complex version of the multi-copy routing schema. EBEC utilizes erasure coding to generate $R \times K$ message blocks per packet for enhanced redundancy. These message blocks are selectively distributed among nodes according to the joint probability of reaching the destination and with the smallest data delivery delay.

A common problem of conventional multi-copy routing protocols (Small and Hass, 2005; Spyropoulos et al., 2004a; Liao et al., 2006) is that the relaying process is heavily

---

[6] Excerpt from the invited article "Energy-efficient packet relaying in sparse mobile mesh networks," Zhuoqun Li, Lingfen Sun, and Emmanuel C. Ifeachor. University of Plymouth, Drake Circus, Plymouth, PL4 8AA, UK,. E-mail: zhuoqun.li@plymouth.ac.uk

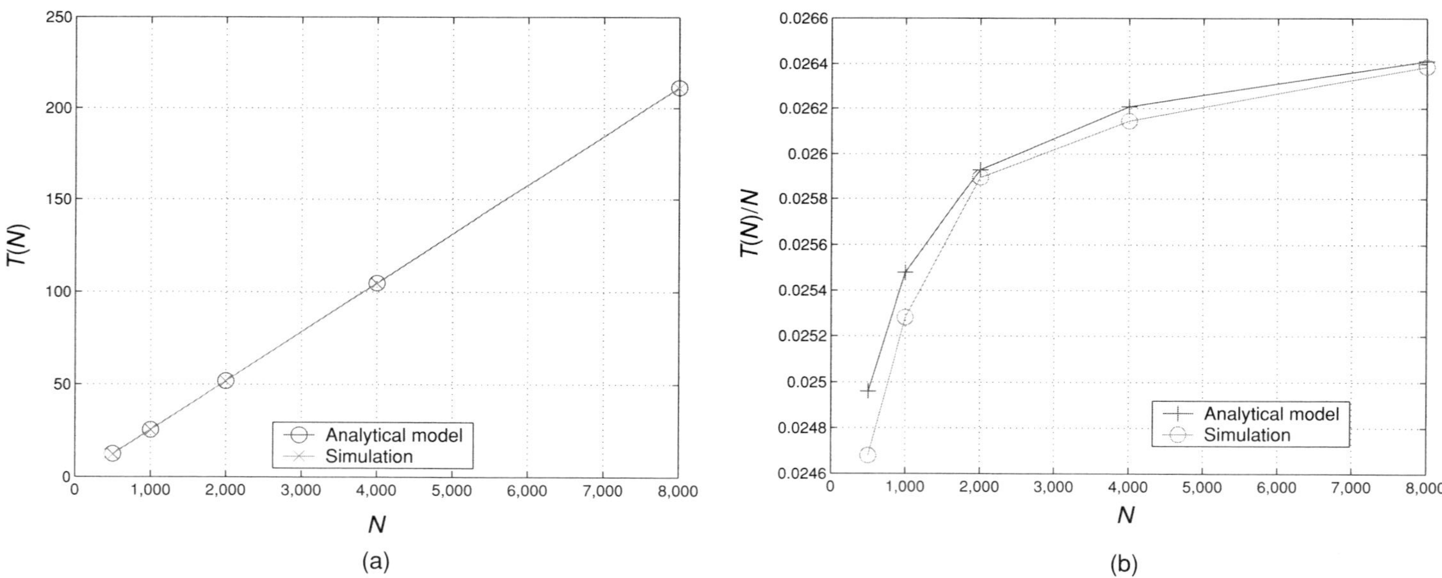

FIGURE 3.30 (a) Network lifetime $T(N)$. (b) $T(N)/N$ as a function of N.

controlled from the source node. More specifically, a source controls the relaying process by setting the replication factor before it initiates the process. However, it is rather difficult and complicate for the source to estimate a suitable R based on its limited knowledge about the network. If network conditions (e.g., network mobility or delivery probability) change during the relaying process, the predefined replication factor would no longer be appropriate, leading to a waste of energy because of unnecessary transmissions or a degradation of performance because of insufficient packet replications.

3.4.5.2 Adaptive Multi-Copy Routing

This work focuses on energy-efficient packet delivery in a special type of WM^{2}Net, that is, Sparse WM^{2}Net. This class of networks is built from sparsely distributed mesh nodes attached to a group of moving objects, for example, those that can be worn by the elderly for *e*-healthcare. Because of the sparse topology, connected end-to-end paths that are required by traditional routing protocols are seldom present in such networks. To address this, store-and-forward based packet delivery is proposed. However, existing store-and-forward strategies (e.g., multi-copy relaying) do not adapt to the variations in the network conditions due to their reliance on the source to determine the relaying process. In this study, a new store-and-forward based scheme, called adaptive multi-copy routing (AMR), is proposed for packet delivery in sparse WM^{2}Nets.

The relaying process in AMR is partially inspired by the binary Spray and Wait routing (Spyropoulos et al., 2004a). Figure 3.31 illustrates the packet relaying process using AMR in a WM^{2}Net built on a group of mobile meshes. As shown, the relaying process of a packet in AMR starts from the source (say s) and terminates at the destination (say d).

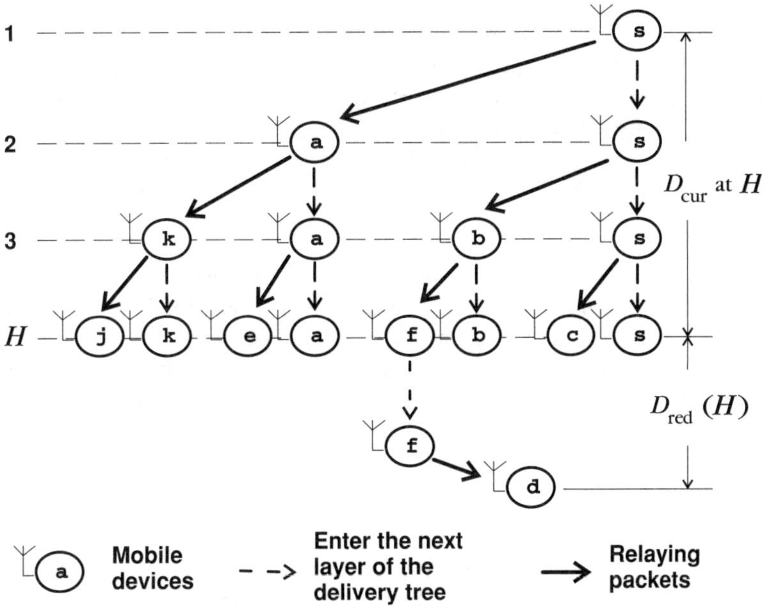

FIGURE 3.31 The binary delivery tree.

When node s meets[7] a relay node (say a), a replica of the packet is forwarded to a. A copy of the packet is sent to each mobile relay along the forwarding path. Upon receipt of a copy of the packet nodes would further disseminate it to those that they meet on their way until either they think that they have spread enough packet copies or they reach the destination. Such a replication process of a packet can be represented as a binary tree with the source of the packet as its root. The depth of the binary delivery tree determines the delay and the energy/traffic overhead spent for delivering a packet.

Unlike the conventional multi-copy protocols that use a source-defined replication factor R to control the depth of the delivery tree, in the AMR scheme, the depth of the relaying process of a packet is decided by intermediate relay nodes. In order to reduce energy consumptions in delivering a packet, relay nodes in the AMR scheme attempt to minimize the depth of the delivery tree as long as the resulting end-to-end delivery delay, that is, the sum of the elapsed delay D_{ecor} and the residual delay D_{ried} (see Fig. 3.31) does not exceed the given delay budget. It is evident that intermediate relay nodes tend to have a more up-to-date knowledge of the networking conditions than the source, the AMR scheme is thus able to set a more appropriate depth for the delivery tree than the source-based strategies. The details of the mechanisms of the AMR scheme are given in the following subsections.

3.4.5.3 Distributed Estimation of the Residual Delivery Delay

Distributed estimation of the residual delivery delay D_{ared} of a packet is a key mechanism of the AMR scheme. D_{red} is a function of both the number of copies of a packet existing in the network (e.g., R) and the mean intermeeting time[8] (denoted as $\overline{T}_{\mathrm{int}}$). The study of (Spyropoulos et al., 2004a) suggests that $\frac{\overline{T}_{\mathrm{int}}}{R}$ is a good approximation of the residual delivery delay when all nodes in the network perform IID random walks. However, the estimation of the mean intermeeting time $\overline{T}_{\mathrm{int}}$ in (Spyropoulos et al., 2004a) does not reflect the impact of network mobility. Groenevelt (Small and Hass, 2005) has proposed mathematic models to estimate $\overline{T}_{\mathrm{int}}$ at a specific level of nodal mobility for both random waypoint (RWP) and random direction (RD) models at their steady states. Suppose that all nodes in the network are constant moving in the RWP model, the mean intermeeting time is given by (Groenevelt, 2005):

$$\overline{T}_{\mathrm{int}} = \frac{LW}{2\,\omega\, r\, \overline{v}} \tag{3.35}$$

where $\omega \approx 1{:}3683$ is a specific constant for the RWP model, r is the radio radius, $\overline{v}$ is the mean internode relative speed, and $L \times W$ is the size of the network area.

If a node does not have the knowledge of one or more parameters listed in Eq. (3.35), it can collect its own intermeeting time with other nodes and use the average as an estimate of $\overline{T}_{\mathrm{int}}$. In practice, the size of the network area, the radio radius, and the model of the nodal movements are normally fixed and known a priori to a node. The only unknown parameter is the time-varying $\overline{v}$. Considering that a node can record its intermeeting

---

[7] By a meets b, we mean $d_{\mathrm{ab}} < r$, where d_{ab} is the Euclidean distance between a and b and r is the radio range of a mobile node. a(b) is referred to as the contact of b(a).

[8] Defined as the time elapsed from when a node meets another node for the first time and the next time when they meet again.

time with others only when it meets those that it has met before, it is easier for a node to collect its relative speeds to its contacts for estimating $\overline{T}_{\text{int}}$. $\overline{v}$ can be estimated by:

$$\overline{v} = \frac{1}{V^{\tau}} \sum_{\hat{v}_i \in V^{\tau}} \hat{v}_i \qquad (3.36)$$

where $\hat{v}_i$ is the relative speed sample that a node measured from a passing contact i and V^{τ} is the set of speed samples that a node collected during past τ seconds. τ should be designed to enable a node to collect enough samples (e.g., over 30 (Montgomery and Runger, 1999)) for an accurate estimation of $\overline{v}$ while emphasizing recent changes in the network mobility.

Given that the relaying process in the AMR scheme can be characterized from a binary delivery tree (see Fig. 3.31), a node can estimate the number of existing replicates of a packet as 2^{H-1}, provided that the current depth H of the delivery tree is known. Therefore, a relay node could estimate the residual delivery delay D_{red} when the depth of the delivery tree is H, by:

$$D_{red}(H) = \frac{\overline{T}_{\text{int}}}{2^{H-1}} \qquad (3.37)$$

3.4.5.4 Estimating the Delivery Probability

The estimation of D_{red} in Eq. (3.37) is based on the assumption that a mobile node could deliver a packet to its contacts with a probability of 1. However, the probability of successful one-hop delivery in reality is susceptible to a variety of factors (e.g. signal-to-noise ratio (SNR), contention, or link duration). As packet losses due to low SNR are normally addressed in the link layer and contention rarely happens in the sparse topology, in this study we discuss the delivery probability only under limited link duration, that is, the probability that the transmission of a packet ends before the link breaks.

As illustrated in Fig. 3.32, node j is passing by its contact d with a relative speed $\hat{v}_{\text{jd}}$. Let x denote the closest distance between j's trajectory and node d and T_{lk} be the link duration. We have $T_{lk} = \frac{2\sqrt{r^2-x^2}}{\hat{v}_{jd}}$. Let φ_i be the time required for transmitting the i-th packet to d and φ_{i-} be the amount of time that packet i has to wait until its transmission. We have $\varphi_{i-} = \sum_{k=0}^{i-1} \varphi_k + \varphi_{\text{ctrl}}$ $(i > 0)$ and $\varphi_{i-} = \varphi_{\text{ctrl}}$ $(i = 0)$, where φ_{ctrl} is the constant time for setting up a connection between the two nodes. Therefore, for a given φ_i and a relative speed estimate $\overline{v}$ the delivery probability of packet i (denoted as P_i) is the probability that x is small enough to make $T_{\text{lk}} > \varphi_i + \varphi_{i-}$, that is, $P_i(\varphi_i) = P(x \le \sqrt{r^2 - \left(\frac{(\varphi_i + \varphi_{i-})\overline{v}}{2}\right)^2})$. As x is uniformly distributed over $[0, r)$ when the nodal movements reach the steady state (assuming a RWP mobility model) (Navidi and Camp, 2004), the delivery probability of the i-th packet, P_i, can be estimated by:

$$P_i(\varphi_i) = \int_0^{\sqrt{r^2-\left(\frac{(\varphi_i+\varphi_{i-})\overline{v}}{2}\right)^2}} \frac{1}{r} dx = \begin{cases} \sqrt{1 - \left(\frac{(\varphi_i + \varphi_{i-})\overline{v}}{2}\right)^2} & \varphi_i \le \frac{2r}{\overline{v}} - \varphi_{i-} \\ 0 & \varphi_i > \frac{2r}{\overline{v}} - \varphi_{i-} \end{cases} \qquad (3.38)$$

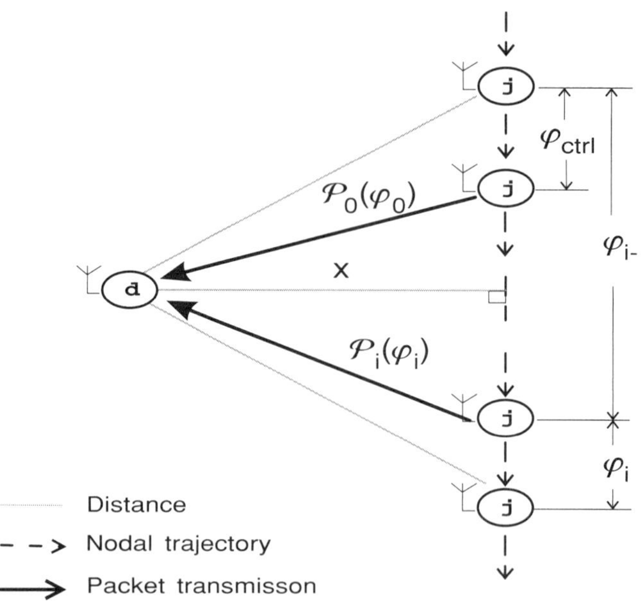

FIGURE 3.32 Delivery probability over one hop with limited link duration.

Equation (3.38) is based on the assumption that the mean relative speed $\overline{v}$ is known. For the estimation of the distribution of link durations with unknown relative speed except of its distribution, we refer the readers to (Cho and Hayes, 2005) or (Han et al., 2004). Knowing the time φ_i for transmitting the i-th packet to a contact and the current depth H of the delivery tree of the packet, a relay can now estimate the residual delivery delay of packet i as:

$$D_{red}(H, \varphi_i) = \frac{\overline{T}_{\text{int}}}{2^{H-1} P_i(\varphi_i)} \tag{3.39}$$

3.4.5.5 Distributed Adaptation for Energy-efficient Relaying

To allow intermediate relay nodes to independently adapt the relaying process, the packet header is extended with two new fields, namely, the current depth of the binary delivery tree H and the delivery delay target D_{tg}. D_{tg} is specified at the source according to the application layer requirement. H is increased by 1 at each relay node before the packet is forwarded to the next relay. The pseudo-code for the implementation of the adaptation mechanisms at relay nodes is given in Fig. 3.33. In the implementation, a relay can estimate the delivery probability of a packet to a contact using its knowledge of the average value of packet size, the available bandwidth, and the number of packets forwarded to a contact. From Fig. 3.33 we can see that, based on the information embedded in a packet's header, a mobile node can make its own decision on whether to forward this packet to a contact or to hold it until it meets the destination. On average, relay nodes in the same layer of the binary delivery tree of a packet would receive the packet around the same time (e.g. in a N-node network the mean time that a packet spent to reach layer $H+1$ of the binary delivery tree from layer H is about $\frac{\overline{T}_{\text{int}}}{2^{H-1}(N-2^{H-1})}$). Assuming both the buffer size

```
/* input:  Packet_Header pkt, Contact_ID   cID */
Function: AMR_Relay ( pkt, cID) ;

  if ( cID == pkt->dest ) {
              Relay_a_ Packet ( pkt, cID ) ;
               return ;
  }

  if ( pkt->H  >= DEPTH_LIMIT ) return ;

  D_cur = CURRENT_TIME – pkt ->time_stamp;

  P_delivery = P( pkt->size, Mean_Relative_Speed ) ;

  if ( D_cur + D_red ( pkt->H - 1,  P_delivery ) <= pkt-> D_tg ) {
              pkt->H = DEPTH_LIMIT ;
              return ;
  }

  else {
              pkt->H ++ ;
              copy_packet ( pkt, copy_of _pkt) ;
              Relay_a_ Packet  ( copy_of_pkt, cID ) ;
              return ;
  }
```

FIGURE 3.33 Pseudo-code for the distributed adaptation of the relaying process of a packet.

and the traffic load are constant, the average communication traffic between contacts is also constant. Therefore, the H-th layer relay nodes will draw the same conclusion on the delivery probability for the same packet and make the same decision on whether to forward it. To minimize the amount of energy spent on relaying packets, the forwarding process of a packet would be halted by relay nodes if the number (e.g., 2^{H-1}) of the packet's existing replicates is sufficient for achieving an end-to-end delivery delay that does not exceed the given budget.

3.4.5.6 Performance Evaluation

The simulated WM²Net consists of $N = 40$ mobile nodes randomly distributed in an area of 1200 m × 600 m size. Each node in the network is configured with a low power radio of a fixed range of 50 m. The data rate of the radio is 38.4 kbps and the transmission power is 0.075 W, as commonly seen from off-the-shelf products (CrossBow Technology Inc.,

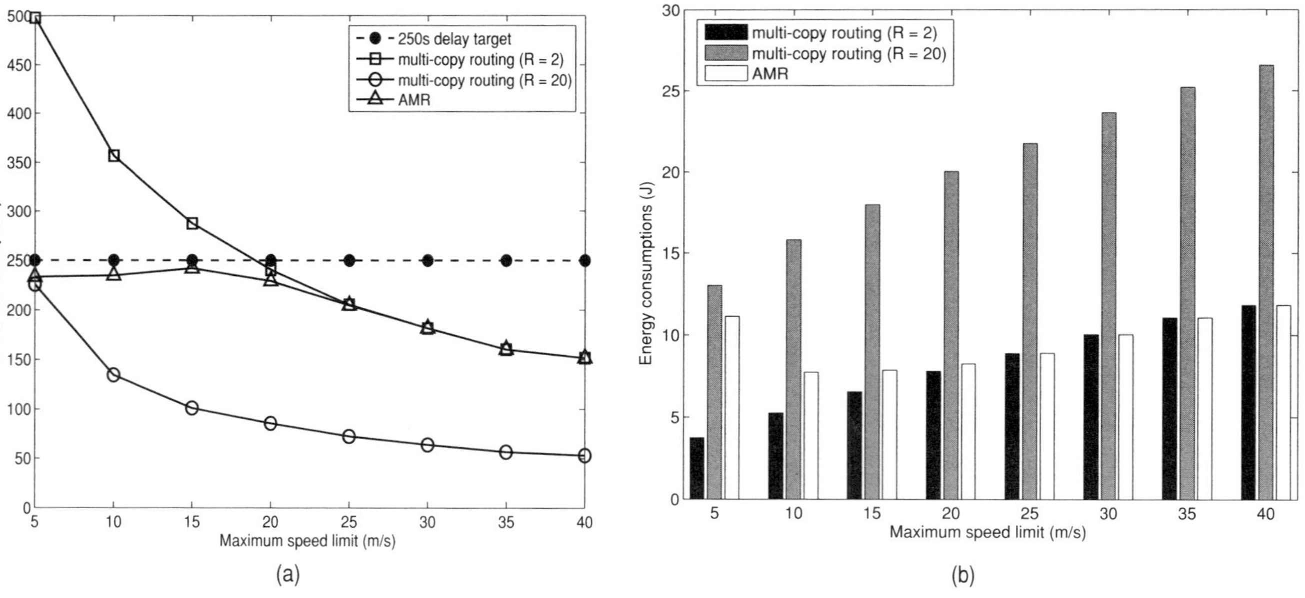

FIGURE 3.34 Performance comparison with delay target 250 s. (a) Delivery delay. (b) Total energy consumptions for packet transmissions.

Wireless sensor networks: Product reference guide: http://www.xbow.com). The network built on the short range radios has a relatively sparse topology (the node density is about 0.43). All nodes are mobile and constantly moving around (the pause time is kept at 0) according to the RWP model. The speed of nodes is uniformly distributed over $[v_{\min}, v_{\max}]$, where $v_{\min}$ is fixed at 2.5 m/s and $v_{\max}$ is varied from 5 m/s to 40 m/s. Every node in the network is acting as the packet source, relay, and receiver at the same time. A packet has the size S of 160 bits. Transmitting a packet over one-hop consumes $P_t \frac{S}{BW}$ joules of energy, where P_t denotes the transmit power of radios and BW for the data rate. A simulation starts with an initialization period of 1,100 s to allow the network mobility to reach its steady state. After that, each node generates a packet destined to each of the rest of the network every 0.5 s for a period of 500 s. The simulation then continues for another 900 s to allow most of the packets to reach their destination. Each simulation is repeated for 30 times with different random seeds to provide smoother results.

The performance of the AMR scheme (in terms of delay and total energy consumptions of packet transmissions) is compared with that of two multi-copy strategies of different R factor (e.g., 2 and 20) in Fig. 3.34 for an ideal delivery probability of 1 and a delay target of 250 s. The end-to-end latency of packet delivery of the three schemes is given in Fig. 3.34a. The energy consumed by these schemes for packet transmissions is given in Fig. 3.34b. We can see from Fig. 3.34a that within the AMR scheme, relay nodes keep adapting packet delivery to the varying network mobility. When the network mobility is low (e.g., $v_{\max} \leq 20$ m/s) the AMR scheme is able to set a right depth H for the delivery process resulting in end-to-end delays just below the specified 250 s budget. When the network mobility climbs up, the AMR scheme further constrained the delivery process making H to be 2. We can observe that in these cases the performance of the AMR scheme is similar to that of the multi-copy scheme with $R = 2$. Although from Fig. 3.34a we can see that the multi-copy strategy with$R = 20$ achieved the delivery target in all the scenarios, the energy consumptions of the scheme shown in Fig. 3.34b are up to 3 times of that of the AMR scheme in scenarios of high mobility. Figure 3.34b also demonstrates that the 2-copy multi-copy scheme used less energy than the AMR scheme in scenarios of $v_{\max} \leq 20$ m/s. But the extra energy is actually utilized by the AMR scheme in the low mobility situations to achieve close-to-target delivery delays for an overall optimal balance between the constraints of delay and energy consumptions.

The performance of the AMR scheme in comparison to the 2-copy and the 20-copy multi-copy schemes against varying delivery probabilities is shown in Fig. 3.35a for end-to-end latency and Fig. 3.35b for energy consumptions. The target of delivery delay is still kept as 250 s and the speed limit $v_{\max}$ is fixed at 20 m/s. The 2-copy multi-copy scheme has been shown in Fig. 3.34a to exhibit similar performance with the AMR scheme when the delivery probability is 1 and $v_{\max}$ is 20 m/s. However, Fig 3.35 shows that, due to lack of adaptability, the scheme missed the 250 s delay target for all the less-than-one delivery probabilities despite the fact that it consumes the least amount of energy for packet transmissions. On the contrary, we can observe from Fig. 3.35 that the adaptability of the AMR scheme enables it to spend some extra energy according to the present delivery probability so that the end-to-end delay target can be achieved. The delivery delay of the multi-copy scheme can be enhanced by increasing R, that is, with $R = 20$ in Fig. 3.35a the scheme can achieve the delay target regardless of the delivery probabilities. However, the downside of increasing R to more than needed is, again, the waste of energy in unnecessary transmissions, as demonstrated in Fig. 3.35b.

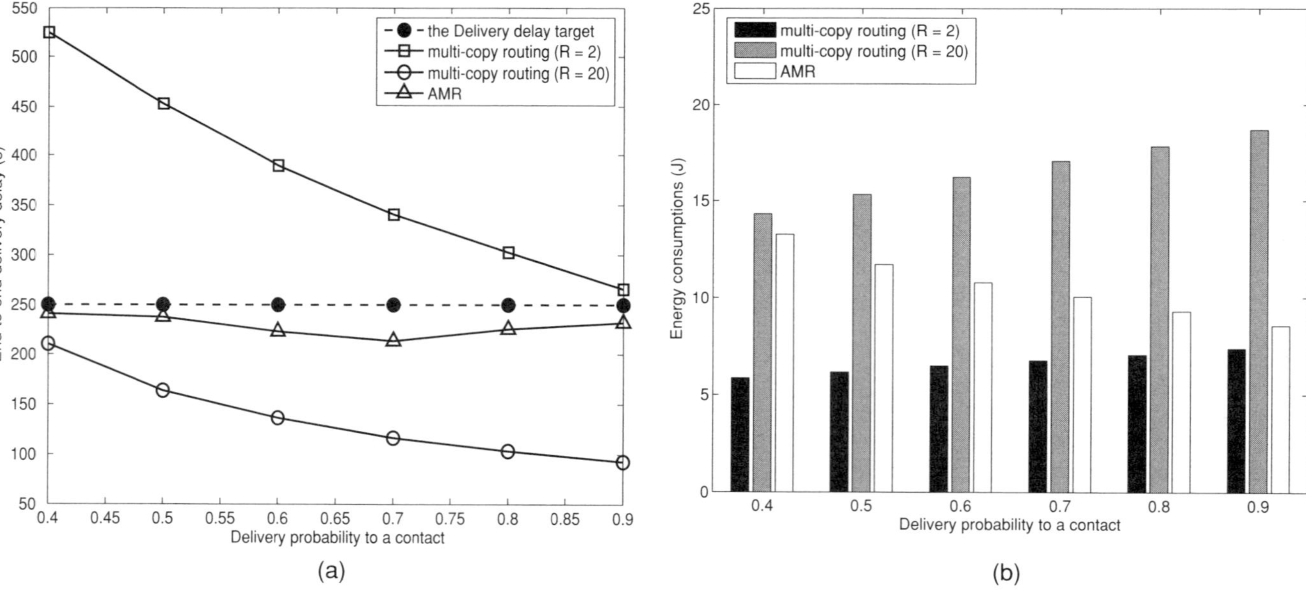

FIGURE 3.35 Performance comparison against one-hop delivery probabilities. (a) Delivery delay. (b) Total energy consumptions for packet transmissions.

The delivery ratios of the AMR scheme and those of other strategies being compared are not presented here. As we have designed a big enough relay buffer at each node and allocated enough time for the multihop packet relay, all the protocols investigated achieved a delivery ratio of over 98% in all scenarios.

3.4.6 Energy-Efficient Geographic Unicast and Multicast Routing in Mesh Networks[9]

3.4.6.1 Network and Energy Model

Throughout this study, a WM²Net is represented as an undirected graph $G = (V, E, \omega)$ where V is the set of vertices, E is the set of edges, and ω: $E \rightarrow R^+$ is a nonnegative cost function associated to edges. The packet reception rate between two nodes $u, v \in V$, $prr(u, v)$ is defined as the probability for v to receive a packet from u. An edge $u, v \in E \Leftrightarrow prr(u, v) > 1\%$. In real-life conditions, $prr(\cdot,\cdot)$ depends on the distance between nodes. In this study, the attenuation properties of wireless links are modeled using data derived from measurements by Zhao and Govindan (2003).

Furthermore, the energy model proposed by Rodoplu and Meng (1999) is used as a reference model for cost-assignment to network edges. In this model, it is assumed that the energy consumption associated with the transmission of a fixed size message at distance d is proportional to:

$$E(d) = d^{\alpha} + C \tag{3.40}$$

where α is the medium attenuation factor with $2 \leq \alpha \leq 6$, and C is a constant that represents the power spent to process the radio signal. As given by Rodoplu and Meng (1999), $\alpha = 4$ and $C = 10$.

Upon reception of a data packet, the receiver sends an ACK message to confirm the reception of the messages. If the data sender does not receive the ACK within a specific timeframe, it retransmits the data packet. We define T as the maximum number of retransmissions. The mean number of retransmissions needed to achieve a successful transmission of a data packet is calculated using the $prr(\cdot,\cdot)$ function. With these in mind, the following equation that estimates the energy, E^*, needed to send a message between two nodes $(u, v) \in V$ at distance $d_{u,v}$, is derived:

$$E^*(d_{uv}) = E(d_{uv}) \min\left(\frac{1}{prr(u, v)^2}, T\right) \tag{3.41}$$

where $E(d_{u,v})$ is the energy spent transmitting the message from u to v and $prr(u, v)$ the probability of v to receive a packet from u.

Using E^* to label the edges of the graph G, we then apply the Dijkstra algorithm to compute the energy-efficient path between two nodes in the presence of errors. Figure 3.36 illustrates the difference between the energy spent to transmit a single message and that to send the same message through the same distance with its correct

[9] Excerpt from the invited article "Energy-efficient geographic unicast and multicast routing in mesh networks," Juan A. Sánchez and Pedro M. Ruiz, Department of Information and Communications Engineering, University of Murcia, Faculty of Informatica, Campus de Espinardo s/n, E-30071, Espinardo, Murcia, Spain. E-mail: pedrom@dif.um.es

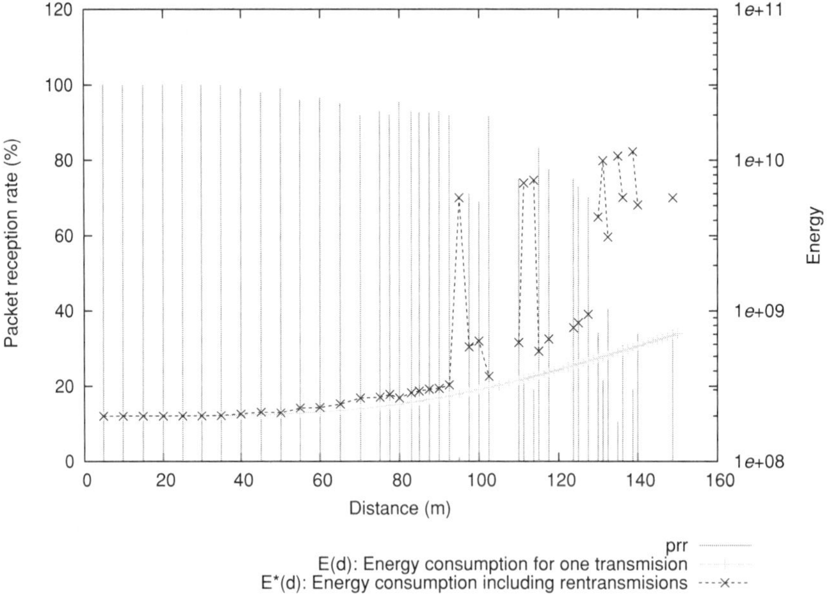

FIGURE 3.36 Packet reception rate and energy consumption.

reception guaranteed through the use of ACK messages, that is, $E(d)$ versus $E^*(d)$. The value of the empirically built *prr* table that is being used to compute the value of $E^*(d)$ is also illustrated in Fig. 3.36.

3.4.6.2 Adapting the Localized Energy-Efficient Multicast Algorithm (LEMA) (Sanchez and Ruiz, 2006a)

One-to-many communications in WM^{2}Net are commonly used to send messages to small groups of nodes. A list with the recipient nodes might be included in the message itself. The authors of LEMA (Sanchez and Ruiz, 2006a) define the Euclidean enclosure as a *connected* graph (i.e., there exist edges between each two nodes) whose member nodes are those in the destination list. The edges are labeled with the Euclidean distance between the linked nodes. When a node running LEMA receives a message, it uses that list to compute the Minimum Spanning Tree of the Euclidean Enclosure. The resulting tree is the one used for routing decisions. If the current node only has one descendant in the resulting tree, no splitting occurs. If the number of descendants is two or greater, the current node calculates a new path to all direct descendants. Each new path delivers a message to a subset of the original destination list. This partition is made using the same resulting tree. Calling first the descendants to the nodes directly connected to the root in the resulting tree, all the nodes being descendant of the first descendant share the same path.

The last part of the algorithm is the delivery of the message to the selected neighbors. When only one neighbor is selected to be the next forwarder, the current node computes an energy-efficient path to reach it using the Dijkstra algorithm over the local graph. That is, the graph is built using only neighbor information. The message is sent through the computed path until a node that provides advance towards the destination is found. The

Source Routing technique is used to force the message to follow the path (Sanchez and Ruiz, 2006a). The links of the local graph are labeled according to Eq. (3.40). To adapt LEMA for dealing with error-prone networks, we propose to label the links using the function E^* defined in Eq. (3.41).

On the other hand, when a node decides that its next forwarder must be a set of more than one neighbor, the shortest path tree (SPT) (Aggélou, 2004) algorithm is then used to determine the most optimum way to reach those neighbors. We apply the same modification to that part of the algorithm. The SPT algorithm is applied over the local graph whose edges are labeled using the function E^*. In that way, the paths and the resulting tree consider that some links are worse than others in terms of energy consumption.

Finally, when a node has no neighbors towards any of the destinations attached in the header of the message, a recovery mechanism is applied. This is based on the GPSR (Karp and Kung, 2000) algorithm. In fact, that algorithm is applied separately for every destination. Once the next forwarder is selected, the most energy-efficient path is computed. Therefore, we also modified the way in which the edges of the locally-planarized graph are labeled. In this case we also use the function E^* is used to label the links.

3.4.6.3 Experimental Results

Performance evaluations of the LEMA protocol are conducted by means of simulations. The basic scenario is a square of 250 m × 250 m with the source and destination nodes randomly placed. All nodes have the same maximum radio range set to 50 m. The two metrics evaluated in the simulations are the mean energy consumption and the packet delivery ratio. The focus of this study is the impact of the mean density in the performance of the protocol; the tests have thus been made on sets of graphs with different mean numbers of neighbors per node. (Due to space restrictions only the results for scenarios with 10 destinations are shown.) To increase the reliability of results, each algorithm has been simulated 50 times in each of the 50 different graphs for each combination of mean density and number of destinations. The results are processed and the mean of each parameter is used in the figures.

The simulated algorithms are two centralized (SPT and minimum incremental power (MIP) (Banerjee et al., 2003)) and a distributed multi-unicast routing algorithm, called energy efficient multi unicast (EEMU). MIP is a heuristic of the MIP (Banerjee et al., 2003). The SPT algorithm applies the Dijkstra algorithm over the complete topology. Both algorithms consider as edge weights the value of $E^*(d)$, with d being the distance between nodes, whereas EEMU is based on IPOW (Stojmenovic and Lin, 2001). IPOW is a geographic routing protocol that calculates energy-efficient paths.

3.4.6.4 Nodal Density versus Energy Consumption

Figure 3.37 shows the total energy consumption of each protocol for a spectrum of densities.

As expected, multicast oriented protocols perform better than EEMU. LEMA's performance is 65% better than EEMU even at low densities. The centralized algorithms, MIP and SPT, do not improve significantly with higher densities. This is mainly attributed to the higher probability of finding void zones (local optimum) when very low densities are deployed and routing is in the greedy mode. Thus, the performance of the greedy part of LEMA's algorithm improves as density increases. For densities higher than 12, the energy consumption is mostly due to the work of the greedy part of the algorithm.

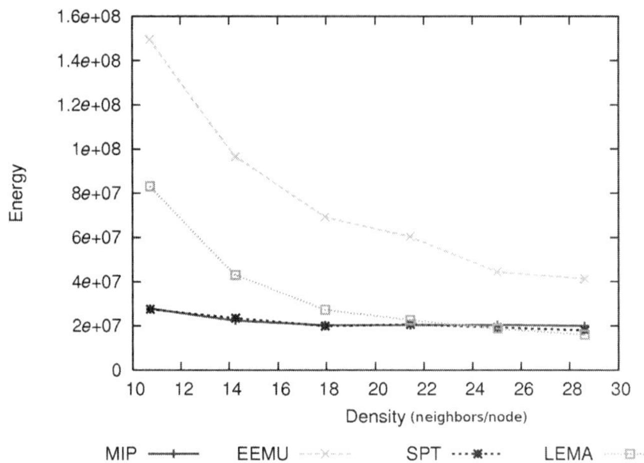

FIGURE 3.37 Total energy spent (10 destinations).

For that reason, MIP and SPT give better results comparing to the proposed protocol for densities lower than 12. For densities higher than 12, however, results show similar behaviors with LEMA being slightly better than MIP and SPT.

3.4.6.5 Delivery Ratio

As WM^2Nets are error-prone networks, building multicast trees with a high percentage of successful transmissions, and delivery ratios as such, is crucial. The energy estimation function must fulfill these two objectives allowing the protocol to determine reliable and at the same time energy-efficient paths.

Figure 3.38 depicts the delivery ratios that each protocol can achieve. As illustrated, the higher the nodal density, the higher the delivery ratios for all protocols. The delivery ratios achieved by LEMA in specific is greater than 82% in all the scenarios tested. Notably, EEMU shows better results because it is a multi-unicast algorithm and in the face of a link failure only a single destination fails to receive the message. In case multicast was used, all the descendants of that subtree would fail.

3.4.7 QoS-Constrained Optimal Energy-Management Minimizing Download-Times over Multichannel Wireless Links[10]

3.4.7.1 System Model and Problem Setup

The application scenario we consider is composed of an energy-limited (e.g., battery-powered) wireless mesh router (WMR) that serves $P \geq 1$ fixed (or, at most, nomadic) clients via a slow-faded downlink composed by $M \geq 2$ orthogonal subchannels. These subchannels may be orthogonal in any domain.

[10] Excerpt from the invited article "QoS-constrained optimal energy-management minimizing download-times over multi-channel wireless links," Enzo Baccarelli, *Mauro Biagi, Nicola Cordeschi, and Cristian Pelizzoni (*INFO-COM Department, Engineering Faculty, University of Rome, "La Sapienza," via Eudossiana 18, 00184, Rome, Italy).

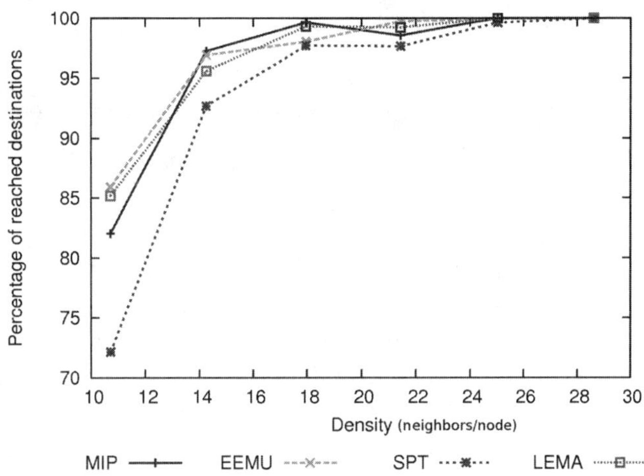

FIGURE 3.38 Delivery ratio (10 destinations).

Before proceeding, let us first introduce the adopted notation. Underlined bold lowercase symbols indicate vectors, bold lowercase characters denote scalar random variables (r.v.s), whereas the corresponding no bold (lowercase) symbols indicate their outcomes. E$\{\cdot\}$ is the expectation operator, $\Re_0^+$ is the set of the nonnegative real numbers, and $\equiv$ means "equal by definition." Finally, $P(A)$ is the probability of the event A, $p(\underline{\sigma})$ is the probability density function (pdf) of the (vector) r.v.$\underline{\sigma}$.

We assume that the traffic generated from the mesh-router is "elastic" (such as web browsing, e-mail, or FTP applications) (see Chapter 3 in the work of Kumar et al., 2004) so that the mesh-router acts as a wireless proxy-server (Triantafillou and Ackaterinidis, 2003). Specifically, we assume that the mesh-router shall transmit to the P clients $\Delta \equiv \sum_{m=1}^{P} \Delta_m$ information units (IUs), Δ_m being the number of IUs for the m-th client. The downlink is assumed slotted whereas the fading effects impairing the link are assumed constant over all slots. Let $\sigma_i(t) \in \Re_0^+,\ 1 \leq i \leq M$ be the (real-nonnegative) state of the i-th subchannel over slot t. Thus, to capture the random effects of the fading phenomena, the vector-state of the overall multichannel link is modeled as a real nonnegative vector random variable (r.v.). The corresponding random sequence $\left\{\underline{\sigma}(t) \equiv [\sigma_1(t), \ldots, \sigma_M(t)]^T \in \left(\Re_0^+\right)^M, t \geq 1\right\}$ of the link-states is assumed an independent identically distributed (i.i.d.) r.v. that follows a time-invariant probability density function $p(\underline{\sigma}) \equiv p(\sigma_1, \ldots, \sigma_M)$.

Setup of the Afforded Optimization Problem The constrained optimization problem we focus on deals with the optimal allocation of the overall available energy E_{tot} (in Joules) over both time-slots and subchannels aiming to *minimize* the resulting average download-time $\overline{\text{T}}$ (in multiples of the slot-time) that is required to transfer a total of Δ IUs. Since the average download-time $\overline{\text{T}}$ may be a critical system-performance parameter for *large* Δ, in the sequel we focus on application scenarios characterized by large Δ values. Thus, as provided in the work of Fu (2003), it is reasonable to assume that the overall energy $\mathcal{E}_{\text{tot}}$ available for downloading is *proportional* to the size Δ of IUs to be forwarded by the mesh router. That is,

$$\mathcal{E}_{tot} = K\Delta$$

where the constant K (in J/IU) determines the total average energy available for the download of a single IU. Thus, let $\mathcal{E}(t) \equiv \sum_{i=1}^{M} \mathcal{E}_i(t) \equiv \underline{1}^T \underline{\mathcal{E}}(t)$ be the total energy radiated by the mesh router over slot t, where $\underline{\mathcal{E}}(t) \equiv [\mathcal{E}_1(t),\ \mathcal{E}_2(t), \ldots, \mathcal{E}_M(t)]^T$ is the M-dimensional vector collecting the energies radiated over the M subchannels and $\underline{1}$ is the M-dimensional (column) vector with all unit entries. In general, $E_i(t)$ may depend on the *overall* link-state $\underline{\sigma}(t)$, on the residual energy $E^{(r)}(t) \equiv E_{\text{tot}} - \sum_{k=1}^{M} \sum_{i=1}^{t-1} E_k(i)$ still available at the beginning of the t-th slot, as well as on the residual number of IUs $\Delta^{(r)}(t) \equiv \Delta - \sum_{i=1}^{t-1} IU_i(t)$ that remain for downloading. Specifically, $IU_i(t)$ indicates the number of IU downloaded at slot t over the i-th subchannel and is linked to the overall number of downloaded IUs via the relationship $IU(t) \equiv \sum_{i=1}^{M} IU_i(t)$. Hence, $IU_i(t)$ may be expressed as $IU_i(t) \equiv \mathcal{R}(\mathcal{E}_i(t); \sigma_i(t))$, where $R(.;.)$ is the rate-function adopted to measure the "goodput" per slot transferred over each subchannel. Its behavior and analytical properties may depend on several system features, such as the requested QoS, the adopted modulation/coding policies, and the disturbing phenomena affecting the received signals. In the sequel, we assume that $\mathcal{R}(\mathcal{E}; \sigma)$ vanishes at $E = 0$ and $\sigma = 0$, that is, $R(E = 0; \sigma) = R(E; \sigma = 0) = 0$. In addition, $\mathcal{R}(\mathcal{E}; \sigma)$ is assumed to be nondecreasing both for $E \geq 0$ and $\sigma \geq 0$. Finally, for any assigned $\sigma \neq 0$, the function $\mathcal{R}(\mathcal{E}; \sigma)$ is assumed to be strictly concave over $E \geq 0$, with first-order derivative $\mathcal{R}_{\mathcal{E}}(\mathcal{E}; \sigma) \equiv \partial\mathcal{R}(\mathcal{E}; \sigma)/\partial\mathcal{E}$ nondecreasing over the σ-variable. We can thus write

$$E_i(t) \equiv \varepsilon_i\left(\underline{\sigma}(t); \mathrm{E}^{(r)}(t); \Delta^{(r)}(t)\right), \quad 1 \leq i \leq M,\ t \geq 1 \tag{3.42}$$

where the energy-allocation function $\varepsilon_i : \left(\Re_0^+\right)^M \times [0, \mathcal{E}_{tot}] \times [0, \Delta] \rightarrow \Re_0^+$ at the right-hand side (r.h.s.) of Eq. (3.45) is a real nonnegative function that aims to *minimize* the above mentioned average download-time $\overline{\mathrm{T}}$. We explicitly note that, in general, the energy $E_i(t)$ radiated over the corresponding i-th subchannel depends on the *overall* link-state $\underline{\sigma}(t)$ (see Baccarelli et al., 2007). Therefore, since the download-time $\mathbf{T}$ is a r.v. whose outcomes also depend on the link-states, it can be proved (Baccarelli et al., 2007) that for large values of Δ the optimization problem can be formally stated as follows:

$$\max_{\varepsilon_1(\underline{\sigma}),\ldots\varepsilon_M(\underline{\sigma})} \sum_{i=1}^{M} \int \mathcal{R}(\varepsilon_i(\underline{\sigma}); \sigma_i)\, p(\underline{\sigma}) d\underline{\sigma}, \tag{3.43}$$

$$\text{s.t.:} \quad \sum_{i=1}^{M} \int \varepsilon_i(\underline{\sigma}) p(\underline{\sigma}) d\underline{\sigma} \leq K \sum_{i=1}^{M} \int \mathcal{R}(\varepsilon_i(\underline{\sigma}); \sigma_i)\, p(\underline{\sigma}) d\underline{\sigma}, \tag{3.43a}$$

$$\sum_{i=1}^{M} \varepsilon_i(\underline{\sigma}) \leq \mathcal{E}_{\max} \tag{3.43b}$$

$$\varepsilon_i(\underline{\sigma}) \geq \mathcal{E}_i^{\min}, \quad 1 \leq i \leq M, \tag{3.43c}$$

where $\mathcal{E}_{\max}$ (J) is the maximum (e.g., peak) available energy per slot and $\mathcal{E}_i^{\min}$ the minimum energy to be radiated over the i-th subchannel.

3.4.7.2 The Optimal Energy Allocation Policy for Large Data to Download

Let us denote with $\overline{\mathrm{T}_0} \equiv \Delta / r_0$ the average download-time, where r_0 the average number of IUs downloaded over a slot-time with

$$r_0 \equiv \sum_{i=1}^{M} \int \mathcal{R}\left(\varepsilon_i^0(\underline{\sigma}); \sigma_i\right) p(\underline{\sigma}) d\underline{\sigma} (\mathrm{IU/slot}), \tag{3.44}$$

and $\underline{\varepsilon}^0(\underline{\sigma})$ the *optimal* energy-allocation policy achieving the constrained maximum in Eq. (3.43). Thus, since constant K in Eq. (3.43a) fixes the available average energy per IU, it is reasonable to believe that the analytical form assumed by the optimal policy $\underline{\varepsilon}^0(\underline{\sigma})$ heavily relies upon the value assumed for K.

If $r_{\min} \equiv \sum_{i=1}^{M} \int \mathcal{R}\left(\mathcal{E}_i^{\min}; \sigma_i\right) p_i(\sigma_i) d\sigma_i$ denotes the minimum average rate conveyed from the system, it is proved (Baccarelli et al., 2007) in fact that when $K < K_{\min} \equiv \mathcal{E}^{\min}/r_{\min}$ the constrained optimization problem in Eq. (3.43) does not admit a feasible solution. However, for $K \geq K_{\min}$ it is proved (Baccarelli et al., 2007) that the optimal energy-allocation policy is given by

$$\varepsilon_i^0\left(\underline{\sigma}, \mu\right) = \left[\mathcal{R}_{\varepsilon}^{-1}\left(\sigma_i; 1/x\left(\underline{\sigma}, \mu\right)\right)\right]_{\varepsilon_i^{\min}}, \quad 1 \leq i \leq M, \tag{3.45}$$

where $\mu \geq 0$ is the dual-variable associated to the energy-constraint in Eq. (3.43a), $\mathcal{R}_{\varepsilon}^{-1}(.,.)$ denotes the inverse function of $\mathcal{R}_{\varepsilon}(.;.)$ done with respect to the $\mathcal{E}$-variable, while $[f(x)]_a$ stands for max $\{a; f(x)\}$. Furthermore, the real nonnegative statistic $x\left(\underline{\sigma}, \mu\right)$ in Eq. (3.45) is *not decreasing* (in the Pareto sense) over $\underline{\sigma}$ and its analytical expression is given by

$$x\left(\underline{\sigma}, \mu\right) \equiv \min\left\{\varphi(\underline{\sigma}); (K + 1/\mu)\right\}, \tag{3.46}$$

where $\varphi(\underline{\sigma})$is the (*unique*) solution of the following functional equation (Baccarelli et al., 2007):

$$\sum_{i=1}^{M} \left[\mathcal{R}_{\varepsilon}^{-1}\left(\sigma_i; 1/\varphi\left(\underline{\sigma}\right)\right)\right]_{\varepsilon_i^{\min}} - \mathcal{E}_{\max} = 0 \tag{3.47}$$

3.4.7.3 Applicable Examples and Numerical Results

We test the actual performance of the developed optimal energy-allocation policy in Eq. (3.45) on the broadcast mesh system of the WOMEN system illustrated in Fig. 3.39. As shown, the WOMEN system consists of three types of nodes, namely mesh base stations (MBSs), WMRs, and wireless mesh users (WMUs). Specifically, MBSs are stationary (e.g., not mobile) nodes providing gateway/bridge functionalities to allow the connection of the overall mesh network to existing wireline broadband core networks (see the upper part of Fig. 3.39). WMRs are the nodes constituting the *mobile* wireless backbone (see the middle part of Fig. 3.39). They constitute the core of the WOMEN system and provide the functionalities required to implement the self-organization and self-management

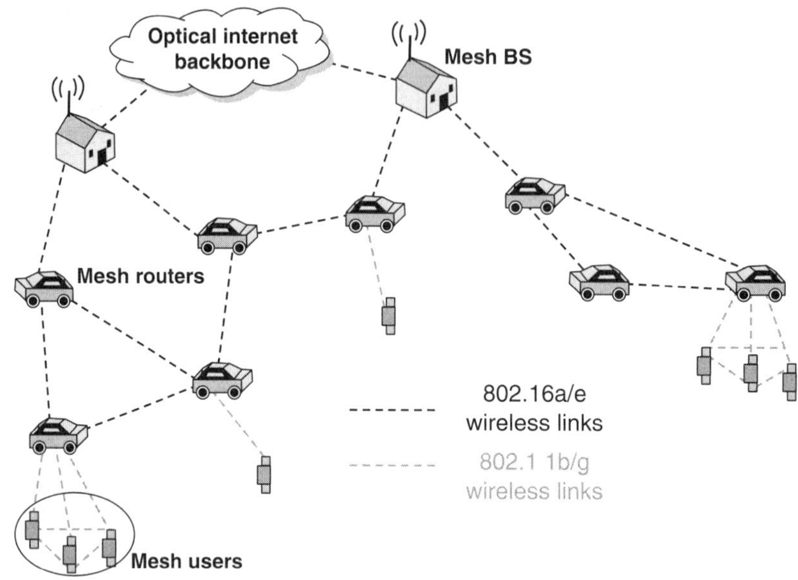

FIGURE 3.39 WOMEN system architecture.

procedures needed to build up and maintain the dynamic wireless mesh backbone, as, for example, the advanced routing protocols providing the multihop and the resource-management needed to implement QoS-guaranteed multimedia CDNs.

On the basis of the above considerations, in this contribution we focus on the "last-hop" of the wireless mesh network of Fig. 3.39, where a battery-powered (e.g., energy-limited) WMR acts as wireless proxy-server towards multiple clients that require the download of *huge-size* contents via a fading-affected shared downlink. Since the average size of multimedia objects may exceed several megabytes (Vakali and Pallis, 2003) while the size of conventional Web objects is typically of the order of 1–100 KB, current wireless proxy-servers fail to meet the (hard) QoS-requirements advanced by multimedia content-delivery applications (see special issue on Wireless Mesh Networks, IEEE Journal of Selected Areas in Communications (JSAC), No. 11, Vol. 11, November, 2006, for an updated overview). By fact, the typical request of the WMUs of Fig. 3.40 is for the *fast-download* of large-size multimedia objects cached at the WMR that, due to fading, may require *high* energy levels to be radiated by the (battery-powered) WMR. Thus, in the considered application scenario, the achievement of the right trade-off among radiated energy and download-time is a *still open crucial* question (www.womenproject.altervista.org).

Notably, we anticipate that the Shannon capacity of each subchannel is given by the following logarithmic rate-function (see Sections 5.3, 5.4, and 6.3 in the work of Paulraj et al., 2004):

$$\mathcal{R}(\mathcal{E};\sigma) \equiv \log(1+\mathcal{E}\sigma), \text{(nat/slot)}, \tag{3.48}$$

so that all the numerical results we present in the sequel refer to this rate-function. Furthermore, in the considered tests the vector link-state $\underline{\sigma}$ collects the instantaneous

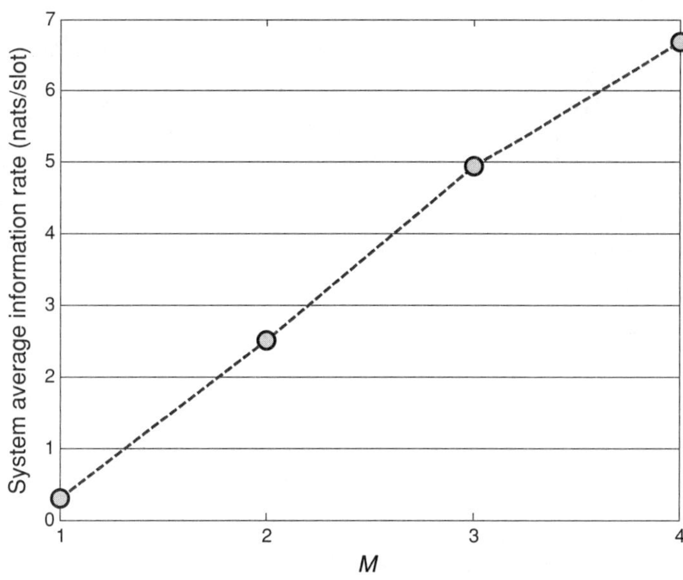

FIGURE 3.40 Average rate of the Multi-antenna broadcast system with $M = 1, 2, 3, 4$ and $K = 1.3$ J/nat.

(e.g., fading affected) SNRs measured at the output of the client-terminals of Fig. 3.39 when the energy radiated by the transmit mesh-router over a slot-time is unit.

Figure 3.40 depicts the increment in the information throughput induced by large values $M \equiv P$ of the served clients. Since the subchannels utilized by the clients are assumed to be orthogonal, large values of $M \equiv P$ increase the degrees of diversity of the downlink (Paulraj et al., 2004). Thus, the plot of Fig. 3.40 supports the conclusion that the optimal energy-allocation policy in Eq. (3.45) is capable to *effectively exploit* the degrees of diversity of the downlink, so to increase the aggregate throughput (e.g., the goodput) conveyed by the overall system in multiclient application scenarios.

To test the effects of finite Δ values, we compare the performance of the proposed scheduler with that attained by the corresponding optimal scheduler implemented according to the dynamic programing principle (Chapters 3 and 6 in the work of Sennot, 1999). Since the latter methodology explicitly accounts for the *finite* time-horizon, the performance of the implemented dynamic programing-based scheduler is the *optimal* one for any finite value of Δ. In order to carry out this performance comparison, we refer to the WOMEN system (see Fig. 3.39) with $P = M = 1$, $\mathcal{E}_1^{\min} = 0$, and $\mathcal{E}_{\max} = 8$ (J). Specifically, to facilitate the implementation of the dynamic programing algorithm over a finite time-horizon (Chapters 3 and 6 in the work of Sennot, 1999) the link-state $\sigma(t)$ is modeled as a *discrete scalar* r.v., that may assume the $N_\sigma = 16$ outcomes: $\{\sigma_i \equiv 0.1 + 0.5(i-1), i = 1, \ldots, 16\}$ according to the following (time-invariant) probabilities:

$$P(\sigma(t) = \sigma_i) = \frac{1}{C}\exp(-\sigma_i), \quad i = 1, \ldots, 16, \tag{3.49}$$

with $C = 2.2989$. Thus, for the considered system we have numerically computed the (optimal) dynamic-programing solution over finite time-horizon of the constrained

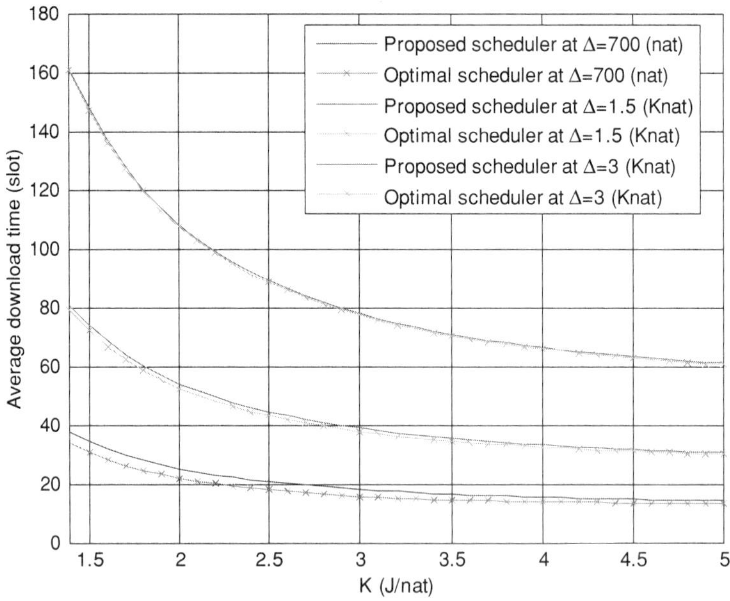

FIGURE 3.41 Average download time-vs-K for the dynamic programing-based optimal scheduler and the proposed one of Eq. (3.45).

optimization problem in Eq. (3.43) at $\Delta = 700$ (nat), 1.5 (Knat), and 3 (Knat). The performance curves we obtained are drawn in Fig. 3.41, where the corresponding performance plots of the proposed scheduler in Eq. (3.45) are also reported.

Although, in principle, the scheduler in Eq. (3.45) achieves optimality as Δ approaches infinite, nevertheless from Fig. 3.41 we observe a performance loss as high as 10% even at small values of Δ values. This loss virtually *vanishes as* Δ increases to *few Knats*.

3.4.8 Biologically Inspired Adaptive Power Management for WM^{2}Snets[11]

3.4.8.1 Design Principles for the Biologically-Inspired Architecture for Sensor NETworks (BiSNET) Agents

(1) **Decentralization:** Inspired by biological systems (e.g., bee colonies), there are no centralized entities in BiSNET to control and coordinate agents. Decentralization allows agents to be scalable and simple by avoiding a single point of performance bottlenecks and failures (Albert et al., 2000; Minar et al., 1999) and by avoiding any central coordination in deploying agents (Cabri et al., 2000).

(2) **Autonomy:** Similar to biological entities (e.g., bees), agents sense their local environments, and based on the sensed conditions, autonomously behave without intervention from/to other agents, platforms, base stations, and human administrators.

[11] Excerpt from the invited article "Biologically-inspired adaptive power management for wireless mesh sensor networks," Pruet Boonma and Junichi Suzuki, Department of Computer Science, University of Massachusetts, Boston, E-mail: {pruet, jxs}@cs.umb.edu

(3) **Food gathering and consumption:** Biological entities strive to seek and consume food for living. For example, bees gather nectar from flowers and digest it to produce honey. In BiSNET, agents (bees) read sensor data (nectar), and digest it to energy (honey).[12] (Energy gain is proportional to a change between the current and previous sensor data.)

(4) **Natural selection:** The abundance or scarcity of stored energy in agents affects their behaviors and triggers natural selection. For example, energy abundance indicates a significant change in sensor reading; thus, an agent emits a pheromone to stimulate replicating itself and its neighboring agents. A replicated agent migrates to a neighboring node to report sensor data to a base station. An energy scarcity (an indication of few changes in sensor reading) eventually causes the death of agents. Similar to biological natural selection where more favorable species in the environment becomes more abundant, the population of agents dynamically changes based on their energy levels (i.e., changes in their sensor readings).

3.4.8.2 BiSNET Agent

An agent consists of *attributes*, *body*, and *behaviors*. *Attributes* carry descriptive information on an agent. They include agent type (e.g., temperature sensing agent and CO sensing agent), energy level, sensor data to be reported to a base station, time stamp of the sensor data, and ID/location of a sensor node where the sensor data is captured. Application developers can define arbitrary attributes for their agents.

Body implements the functionality of an agent: collecting and processing sensor data. In each duty cycle, each agent gathers sensor data (as food) from the local sensor device, converts it to energy and processes it (e.g., discards it or reports it to a base station). Different types of agents collect different types of sensor data. *Behaviors* implement actions inherent to all agents. This study focuses on the following five behaviors:

- **Pheromone emission:** Agents may emit different types of pheromones (*replication pheromones* and *migration pheromones*) according to their local and surrounding network conditions. Agents emit replication pheromones in response to the abundance of stored energy (i.e., significant changes in their sensor readings). Different types of agents emit different types of replication pheromones, each of which carries sensor data. For example, temperature sensing agents emit temperature pheromones, which carry temperature data. CO sensing agents emit CO pheromones, which carry CO data. Replication pheromones stimulate the agents on the local and neighboring nodes to replicate themselves. Each replication pheromone can spread to one-hop away neighboring sensor nodes. On the other hand, agents emit migration pheromones on their local nodes when they migrate to neighboring nodes. Each replication and migration pheromone has its own concentration (or strength). The concentration decays by half at each duty cycle. A pheromone disappears when its concentration becomes zero.
- **Replication:** Agents may make a copy of themselves in response to the abundance of energy and replication pheromones. Each agent does not initiate

[12] The concept of energy in BiSNET is a logical concept that affects agent behaviors and does not necessarily represent the amount of physical battery.

replication until enough types of replication pheromones become available on the local node. For example, an agent may replicate itself only when both temperature pheromones and CO pheromones are available. A replicated (child) agent retains the same agent type as its parent's type, and aggregates multiple sensor data stored in multiple types of available replication pheromones. A child agent is placed on the node that its parent agent resides on, and it receives half of the parent's energy level. Each child agent is intended to move towards a base station to report (aggregated) sensor data.

- **Migration:** Agents may move from one sensor node to another in response to energy abundance (i.e., significant changes in their sensor readings). Migration is used to transmit agents (sensor data) to base stations. Each agent may implement one of or a combination of the following three migration policies:
  - **Directional walk:** Each agent may move to the nearest base station through the shortest path. Each base station periodically propagates *base station pheromones*, whose concentration decays on a hop-by-hop basis. Using base station pheromones, agents can sense where base stations exist approximately, and move towards the base stations by climbing pheromone gradients.
  - **Chemotaxis:** Agents may move to base stations by following migration pheromone traces on which many other agents travel. These traces can be the shortest paths to the base stations. When there are no migration pheromones on neighboring nodes, agents perform directional walk.
  - **Detour walk:** Each agent may go off a migration pheromone trace and follow another path to a base station when the concentration of migration pheromones is too high on the trace (i.e., when too many agents follow the same migration path). This avoids separating the network into islands. The network can be separated with the migration paths that too many agents follow, because the nodes on the paths consume more power than others and they go down earlier than others. In addition to the detour with migration pheromones, agents may avoid moving through the nodes where the concentration of replication pheromones is too high (i.e., where agents detect significant changes in their sensor readings). This detour walk distributes power consumption of agent migration over the nodes where agents detect no changes in their sensor readings, thereby avoiding the network to be separated.
- **Energy exchange:** Agents periodically deposit some of their energy units (honey) to their local platforms (hives), and keep the rest for living (i.e., for invoking their behaviors).
- **Death:** Agents die due to lack of energy when they cannot balance energy gain and expenditure. When an agent dies, the local platform removes the agent and releases all resources allocated to the agent.

Every agent expends a certain amount of energy to perform replication and migration behaviors. The energy costs to invoke these behaviors are constant for all agents.

Figure 3.42 shows a sequence of actions that each agent performs in each duty cycle. First, an agent reads sensor data (as nectar) with the underlying sensor device, and converts it to energy (honey). The energy intake (E_F) is calculated with Eq. (3.50). S represents the absolute difference between sensor data in the current and previous duty

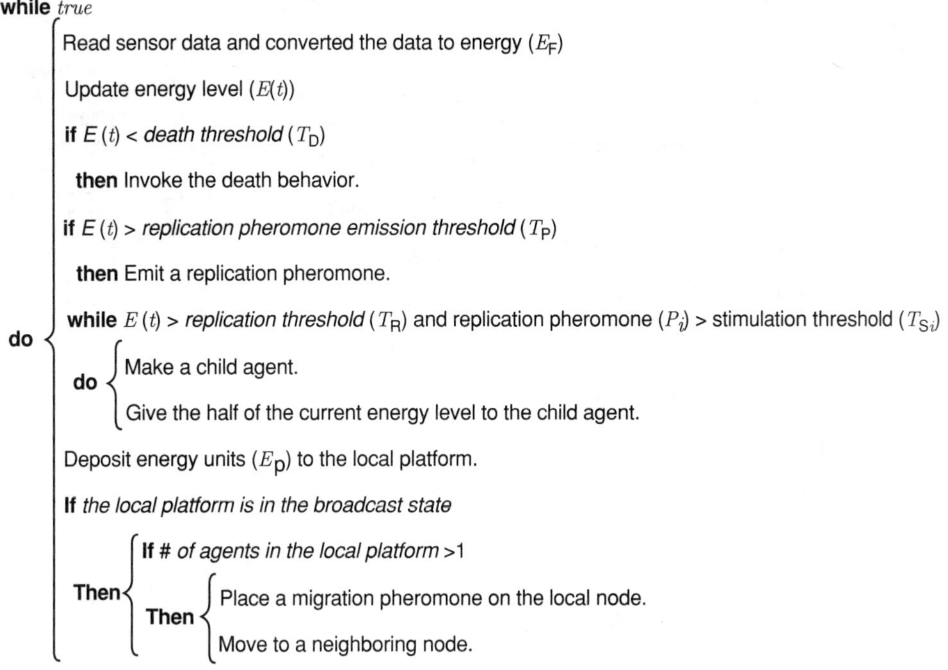

Figure 3.42 Agent actions in each duty cycle.

cycle. M is the metabolic rate, which is a constant value between 0 and 1.

$$E_F = S \cdot M \tag{3.50}$$

Different platforms may have different M values to prioritize particular types of sensor nodes. All agents on a platform follow the same M value that the platform has. The higher M value a platform has, the more often agents replicate and migrate on the platform because of higher energy intake.

Assuming that E_F is known, each agent updates its energy level as follows.

$$E(t) = E(t-1) + E(t) \tag{3.51}$$

$E(t)$ is the current energy level of the agent, and $E(t-1)$ is the agent's energy level in the previous duty cycle. t is incremented by one at each duty cycle.

If an agent's energy level ($E(t)$) becomes very low (below the death threshold: T_D), the agent dies due to energy starvation (see also Figs. 3.42 and 3.43).[13]

Then, an agent emits a replication pheromone if its energy level exceeds its replication pheromone emission threshold T_p (see Figs. 3.42 and 3.43). Agents continuously adjust

[13] If all agents on a node are dying at the same time, a randomly selected agent will survive. At least one agent runs on each node.

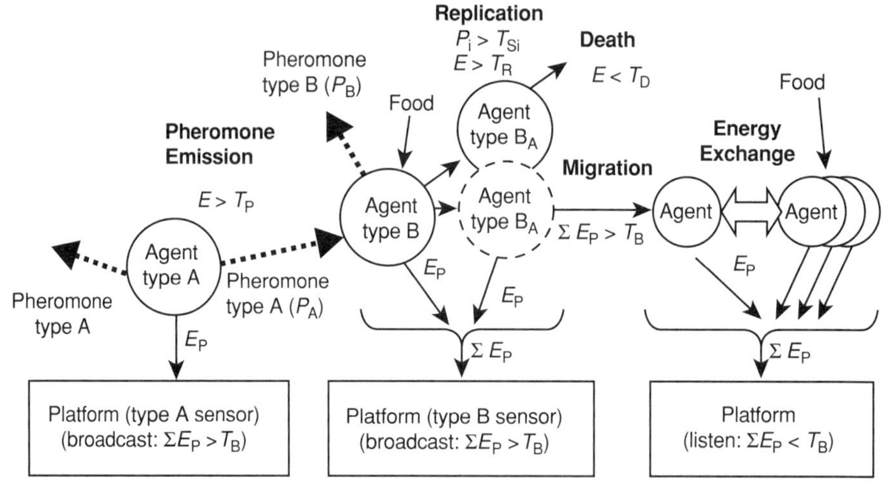

FIGURE 3.43 Agent behaviors.

their replication pheromone emission thresholds as the exponentially weighted moving average (EWMA) of their energy levels:

$$T_P(t) = (1 - \alpha)T_P(t - 1) + \alpha E(t) \tag{3.52}$$

$T_P(t)$ is the current replication pheromone emission threshold, and $T_P(t - 1)$ is the one in the previous duty cycle. EWMA is used to smooth out short-term minor oscillations in the data series of E (the energy level of an agent). It places more emphasis on the long-term transition trend of E; only significant changes in E have the effects to change T_P. The α value is a constant to control the sensitivity of T_P against the changes of E.

When a replication pheromone is emitted on a node, all the agents on the node can sense it. It may stimulate their replications. An agent replicates itself when it meets two conditions: (1) when the agent's energy level ($E(t)$) exceeds its replication threshold (T_R), and (2) when the concentration of each type of available replication pheromones (P_i)[14] exceeds its stimulation threshold T_{s_i} (see Figs. 3.42 and 3.43). Agents continuously adjust their replication thresholds as the EWMA of their energy levels (Eq. 3.53). The stimulation threshold of a replication pheromone changes as the EWMA of the pheromone's concentration (Eq. 3.54).

$$T_R(t) = (1 - \beta)T_R(t - 1) + \beta E(t) \tag{3.53}$$

$$T_{S_i}(t) = (1 - \gamma)T_{S_i}(t - 1) + \gamma P_i(t) \tag{3.54}$$

$T_R(t)$ is the current replication threshold, and $T_R(t - 1)$ is the one in the previous duty cycle. T_{s_i} is the current replication pheromone stimulation threshold for the replication

[14] P_i denotes the total concentration of replication pheromone type i, i indicates different types of replication pheromones available on the local node (e.g., temperature and CO pheromones).

pheromone type i, and $T_{s_i}(t-1)$ is the one in the previous duty cycle. The β and γ values are the constants to control the sensitivity of T_R and T_{s_i} against the changes of E and P_i, respectively. A replicating (parent) agent splits its energy units into halves ($\frac{E(t)-E_R}{2}$), gives a half to its child agent, and keeps the other half. E_R is the cost (energy units) for an agent to invoke the replication behavior. A replicated (child) agent aggregates the sensor data in the pheromones that stimulated its parent agent to perform a replication. A parent agent keeps replicating itself until its energy level becomes less than its replication threshold (T_R). Replicated agents may migrate to neighboring nodes when the local node is in broadcasting state (see Fig. 3.42).

As described above, agents replicate themselves only when they gain a large amount of energy on the local node and receive enough types of high-concentration pheromones from the local and neighboring nodes. This means that sensor data are aggregated and transmitted to base stations only when significant changes in sensor data are detected on the local and neighboring nodes. Agents do not respond to gradual changes in sensor readings (e.g., temperature changes during a day or between different seasons). This reduces power consumption in sensor nodes and expands the life of a wireless mobile mesh sensor network (WM²Snet) by avoiding unnecessary data transmission.

This adaptive data aggregation and transmission mechanism is designed with a self-healing capability in mind, which allows agents to detect and eliminate false positive sensor data. When a sensor node does not work properly owing to malfunctions or miscalibrations for example, each agent on the node emits the replication pheromones that contain false positive sensor data. A large number of false positive pheromones may be transmitted to neighboring nodes. However, they are discarded at the neighboring nodes because they are not aggregated with other types of pheromones (see Fig. 3.42). This means that false positive pheromones are not propagated more than two hops from a malfunctioning or miscalibrated node. Also, agents stop emitting false positive pheromones on the malfunctioning/miscalibrated node because their pheromone emission thresholds increase (see Eq. 3.52).

Each agent deposits a certain amount of energy (E_P) to its local platform (see Figs. 3.42 and 3.43):

$$E_P = \begin{cases} E(t) - E(t-1) & \text{if } E(t) \geq E(t-1) \\ 0 & \text{if } E(t) < E(t-1) \end{cases} \tag{3.55}$$

Each agent strives to keep its energy level ($E(t)$) close to the one in the previous duty cycle ($E(t-1)$).

When a platform's total energy gain ($\sum E_P$) is greater than a threshold (T_B), the platform changes its state to the broadcast state. This allows replicated agents and pheromones to move to neighboring nodes (see Figs. 3.42 and 3.43). Each agent implements one of or a combination of three migration policies (directional walk, chemotaxis, and detour walk) with the following equation:

$$WS_j = \sum_{t=1}^{3} w_t \frac{P_{t,j} - P_{t_{\min}}}{P_{t_{\max}} - P_{t_{\min}}} \tag{3.56}$$

Each agent calculates the weighted sum WS for each neighboring node j, and moves to a node that generates the highest weighted sum. t indicates the type of a pheromone;

$P_{1,j}$, $P_{2,j}$ and $P_{3,j}$ represent the concentration values of base station, migration or replication pheromones on a neighboring node j. $P_{t_{\max}}$ and $P_{t_{\min}}$ are the maximum and minimum concentration of P_t among neighboring nodes. w_t is used to determine which migration policies each agent performs. For example, w_2 and w_3 are zero for agents performing directional walk. w_2 is positive and negative for agents performing chemotaxis and the detour walk with migration pheromones, respectively. w_3 is negative for agents performing the detour walk with replication pheromones.

3.4.8.3 BiSNET Platform

BiSNET platform consists of two parts: *runtime services* and *state controller*. *Runtime services* hide lower-level computing and networking details (e.g., network I/O), and provide high-level services that agents use to read sensor data and perform behaviors. For example, the runtime services allow each agent to sense the type and concentration of each pheromone available on the local node.

State controller dynamically changes the state of a sensor node to control its duty cycle (sleep period). Each sensor node is in the *listen*, *broadcast*, or *sleep* state (Fig. 3.43). A platform and agents can work on a sensor node when its state is in the listen or broadcast state. In the listen state, a platform turns on a radio receiver to receive data (agents and pheromones) from neighboring sensor nodes. Each agent performs a series of actions described in Fig. 3.42. The listen state changes to the broadcast state if a platform gains energy more than the broadcast threshold ($\sum E_P > T_B$; see also Figs. 3.43 and 3.44). In the broadcast state, a platform turns on a radio transmitter to allow agents and pheromones to move to neighboring nodes.

When a platform gains no energy from agents ($\sum E_P = 0$), the platform goes into the sleep state (Fig. 4.43). The sleep period is determined as follows. P_{sleep} is a constant, and P_i is the concentration of each type of replication pheromones (the pheromone type i) available on the platform.

$$\text{sleep period} = \begin{cases} \dfrac{P_{\text{sleep}}}{\sum P_i} & \text{if } \sum P_i > 0 \\ P_{\text{sleep}} & \text{if } \sum P_i = 0 \end{cases} \tag{3.57}$$

The sleep period is reverse proportional to the total concentration of replication pheromones available on a platform ($\sum P_i$). This means that a platform increases its

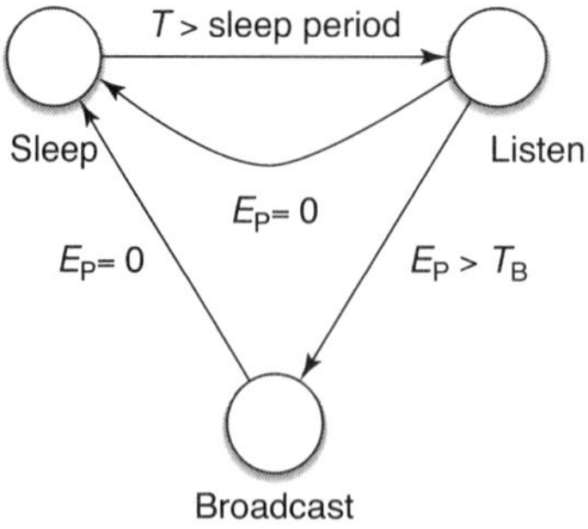

FIGURE 3.44 Platform state transition.

sleep period to reduce power consumption when agents find no significant changes in their sensor readings on the platform and its neighboring platforms.

This adaptive duty cycle management is designed with an inference capability in mind. When a platform receives replication pheromones from a neighboring node(s), it decreases its sleep period even if there is no change in the sensor reading on the local node (see Eq. 3.57). This way, agents can be more watchful on the node for a future potential change in their sensor readings so that they do not miss it during sleep period.

3.4.8.4 Simulation Results

This section shows a series of simulation results to evaluate how the biologically-inspired mechanisms in BiSNET impact the autonomy, adaptability, and simplicity of power management in WM^{2}Snets.

Simulation Configurations This simulation study emulates a WM^{2}Snet deployed in a forest to detect wildfires. The WM^{2}Snet consists of temperature sensor nodes and CO sensor nodes randomly deployed in a 25 × 24 grid topology (600 nodes); half of the nodes equip temperature sensors, and the other half equip CO sensors (Fig. 3.44). A wildfire moves from southeast to northwest. Simulations follow a model that describes wildfire spreading in nature (Drossel and Schwabl, 1992). In order to examine how different biologically-inspired mechanisms in BiSNET impact the operation of WM^{2}Snets, the following five BiSNET configurations are evaluated:

- **BiSNET-Mb:** Agents do not perform the replication behavior with replication pheromones. They replicate themselves when their energy levels exceed their replication thresholds (T_R), and do not perform data aggregation. Agents migrate to the base station with directional walk using base station pheromones.
- **BiSNET-RMb:** Agents perform the replication behavior with replication pheromones; they perform data aggregation. Agents migrate to the base station with directional walk using base station pheromones.
- **BiSNET-RMbm:** Agents perform the replication behavior (data aggregation) with replication pheromones. They migrate to the base station with base station and migration pheromones (directional walk, chemotaxis, and detour walk with migration pheromones). They perform chemotaxis by default and execute detour walk when the concentration of migration pheromones is too high on their local nodes.
- **BiSNET-RMbr:** Agents perform the replication behavior (data aggregation) with replication pheromones. They migrate to the base station with base station and replication pheromones (directional walk and detour walk with replication pheromones).
- **BiSNET-RMbmr:** Agents perform the replication behavior (data aggregation) with replication pheromones. They migrate to the base station with all of three pheromones (directional walk, chemotaxis, and detour walk with migration and replication pheromones).

In addition, BiSNET is compared with an existing WM^{2}Snet routing protocol, Gradient Based Routing (GBR) (Schurgers and Srivastava, 2001) and its three variants. In GBR, a base station periodically propagates a routing message to sensor nodes throughout the network. The routing message gradually assigns smaller *gradient height* values to nodes as it travels on a hop-by-hop basis. Given a gradient towards a base station, each node forwards sensor data to a neighboring node that has a higher gradient height. This routing protocol is similar to BiSNET-Mb. In GBR, sensor data are transmitted on the shortest paths to base stations; it is likely that network separations occur. To alleviate this problem, there are three variations of GBR (Schurgers and Srivastava, 2001):

- **GBR-R:** When a node finds multiple neighboring nodes that have the same gradient height (i.e., the same distance to a base station), it randomly selects one of them and forwards sensor data to the selected node.
- **GBR-P:** When the remaining amount of power becomes low on a node, the node increases its gradient height so that it does not receive sensor data from neighboring nodes.
- **GBR-S:** Nodes divert data transmission paths (or streams) to base stations. When a node receives sensor data from a neighboring node, it increases its gradient height for a while so that it does not receive sensor data from neighboring nodes. This routing protocol is similar to BiSNET-Mbm.

Success Rate and Latency of Data Transmission Table 3.9 shows the total number of sensor data that nodes collect and report to the base station throughout a simulation. BiSNET-RMb, -RMbm, -RMbr, and -RMbmr collect/report more sensor data than BiSNET-Mb because of data aggregation based on replication pheromones. The increase in the number of collected data is approximately 16%. Compared with GBR, BiSNET always operate in a higher temporal resolution because the inference mechanism in BiSNET allows nodes to collect more sensor data by reducing sleep periods. BiSNET-RMbm, -RMbr, and -RMbmr collect 22% more data than any GBR protocols do. Note that the success rate of data transmission of BiSNET and GBR is almost the same even if higher volumes of data are collected and reported to the base station.

Table 3.10 shows the average latency to transmit sensor data from nodes to the base station. With data aggregation enabled with replication pheromones, the latency is shorter in BiSNET-RMb than BiSNET-Mb because data aggregation reduces the number

	Number of Collected Data	Number of Reported Data	Success Rate (%)
BiSNET-Mb	420	400	95.24
BiSNET-RMb	485	460	94.85
BiSNET-RMbm	488	462	94.67
BiSNET-RMbr	488	462	94.67
BiSNET-RMbr	488	462	94.67
GBR	380	360	94.74
GBR-R	380	360	94.74
GBR-P	380	358	94.21
GBR-S	380	360	94.74

Table 3.9 The Total Number of Collected and Reported Sensor Data

	Latency (s)	Standard Deviation
BiSNET-Mb	35	4.5
BiSNET-RMb	33	4
BiSNET-RMbm	34	4
BiSNET-RMbr	37	4
BiSNET-RMbr	36	6
GBR	33	4
GBR-R	34	5
GBR-P	34	5
GBR-S	35	5

Table 3.10 Average Latency of Data Transmission

of migrating agents, and in turn, network traffic. In BiSNET-RMbm, -RMbr and -RMbmr, agents perform detour walk and do not always travel on the shortest path to the base station; however, the latency is not severely affected. The latency of BiSNET-RMbmr is only 1 s longer (or 3% longer) than that of BiSNET-Mb. Compared with GBR, the latency of BiSNET is 1–3 s longer (or 3–9% longer) because BiSNET transmits more data than GBR as shown in Table 3.8. The increase of 3–9% in latency is acceptable against the increase of 22% in data collection. Note that the standard deviation of latency is almost the same in BiSNET and GBR.

Power Consumption Table 3.11 shows the average power consumption of sensor nodes throughout a simulation. In BiSNET-RMb, -RMbm, -RMbr, and -RMbmr, with data aggregation enabled with replication pheromones, agents reduce the number of transmitted data to the base station via data aggregation. Compared with BiSNET-Mb, this contributes to 3–6% reduction in average power consumption. The power consumption in BiSNET-RMbm, -RMbr and -RMbmr is 2–3% higher than BiSNET-RMb because agents perform detour walk and do not always travel on the shortest path to the base station. Compared

	Power Consumption (mW)	Standard Deviation	Power Consumption per Reported Sensor Data (mW)
BiSNET-Mb	4,158	1,158	10.4
BiSNET-RMb	3,924	1,212	8.5
BiSNET-RMbm	4,041	984.5	8.8
BiSNET-RMbr	4,011	836	8.7
BiSNET-RMbr	4,044	745	8.6
GBR	3,930	1,015	10.9
GBR-R	3,660	910	10.2
GBR-P	3,480	850	9.7
GBR-S	3,480	840	9.7

Table 3.11 Average Power Consumption of Each Sensor Node

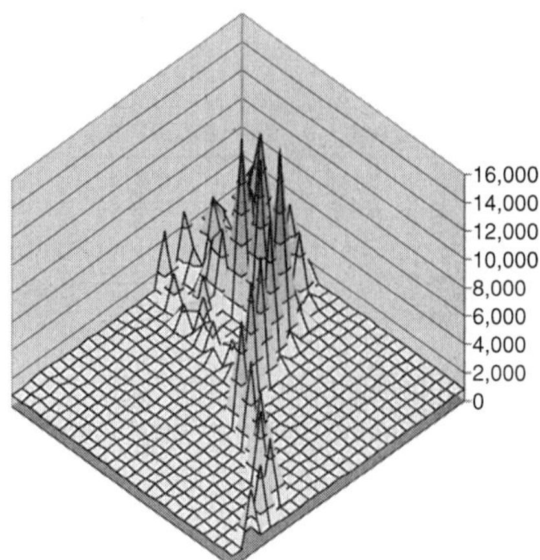

FIGURE 3.45 Power consumption in BiSNET-Mb.

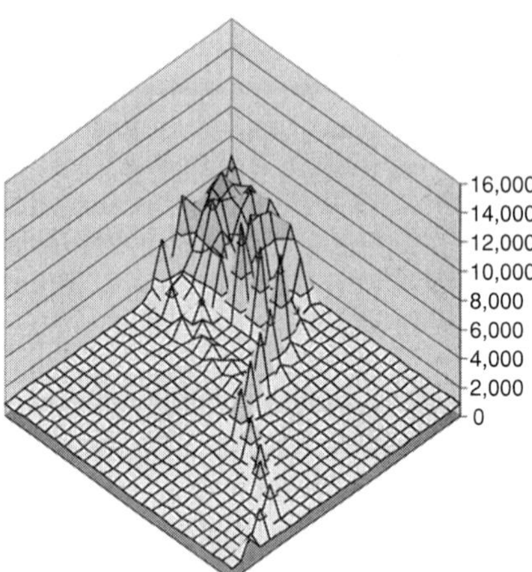

FIGURE 3.46 Power consumption in BiSNET-RMb.

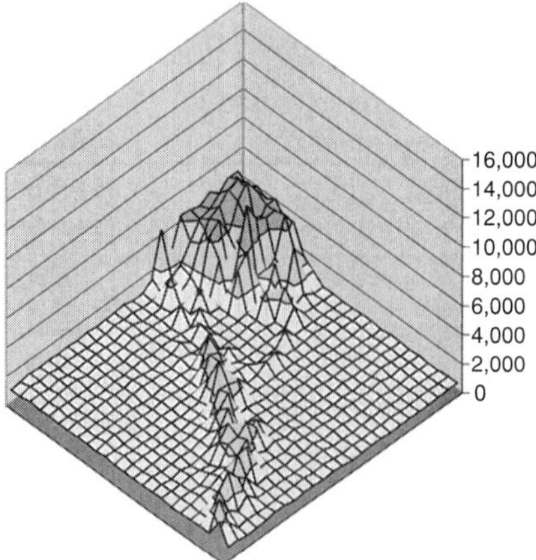

FIGURE 3.47 Power consumption in BiSNET-RMbm.

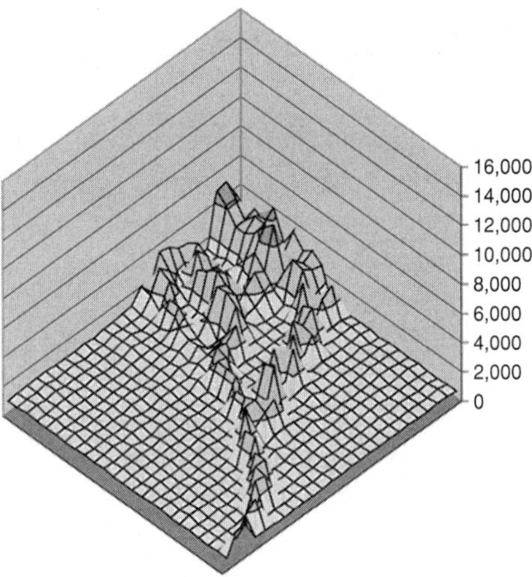

FIGURE 3.48 Power consumption in BiSNET-RMbr.

with GBR, BiSNET consumes 10–16% more power; this is so as it transmits more data as illustrated in Table 3.8. However, the power consumption per reported sensor data is consistently lower in BiSNET than GBR. For example, BiSNET-RMbmr consumes 12–18% less power per reported data than GBR-R, -P and -S. This means that BiSNET manages power consumption effectively while increasing the temporal resolution of data collection. Note also that the standard deviation of power consumption is consistently lower in BiSNET than GBR. This means BiSNET distributes power consumption over more nodes, thereby reducing a risk of network separation more effectively than GBR.

Figures 3.45 to 3.49 show how much power is consumed on 600 (25 × 24) nodes in five BiSNET configurations. In each figure, the base station is located at the upper corner

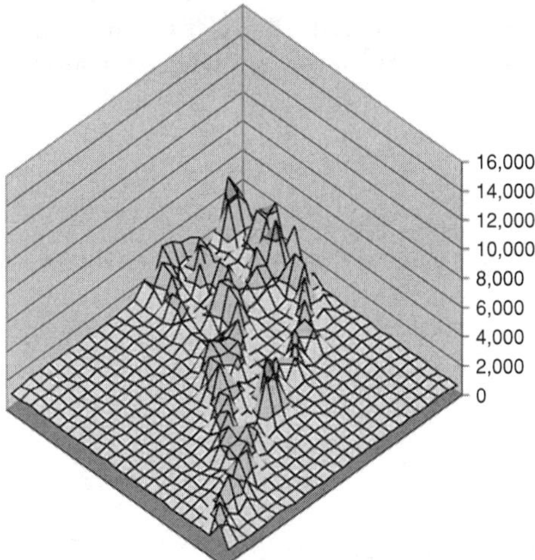

FIGURE 3.49 Power consumption in BiSNET-RMrbm.

	ROM (KB)	RAM (KB)
BiSNET	1.0	24
Blink	0.4	1.6
GBR	0.84	26
Agilla	3.59	41.6

TABLE 3.12 Memory Footprint in a MICA2 Mote

of the network. A wildfire starts at the bottom corner of the network, and move upward. These figures show that BiSNET reduces and distribute power consumption over nodes by leveraging data aggregation (with replication pheromones) and detour walk (with migration and replication pheromones). These mechanisms contribute to reduce a risk of network separation and expand the network lifetime.

Network Lifetime Figure 3.49 shows how soon sensor nodes go down due to lack of power in different BiSNET and GBR configurations. In BiSNET-Mb, 64 of 600 nodes (11% nodes) go down in 1,000 min. With data aggregation enabled with replication pheromones, 57 nodes (9.5% nodes) go down in 1,000 min. Moreover, with detour walk with migration and replication pheromones, the number of dead nodes decreases to 47 nodes (7.8% nodes) in BiSNET-RBbmr. Only one node goes down in 500 min. In contrast, in GBR-S, three nodes go down in 500 min and 52 nodes (8.7% nodes) in 1,000 min. The first death of a sensor node occurs in 482 min in BiSNET-RBbmr and in 350.5 min in GBR-S. BiSNET-RBbmr delays the first node death by 131.5 min (more than two hours). BiSNET successfully reduces and distributes power consumption over nodes and expand the network life.[15]

Simplicity: Memory Footprint In order to evaluate the simplicity of BiSNET, Table 3.12 shows the memory footprint of the BiSNET runtime in a MICA2 mote, and compares it with the footprint of Blink (an example program in TinyOS), which periodically turns on and off an LED, GBR, and Agilla, which is a mobile agent platform for WM^{2}Snets (Fok et al., 2005). As shown in Table 3.12, the BiSNET runtime is fairly lightweight in its footprint, and it can be deployed on sensor devises whose resource availability is severely limited.

[15] GBR outperforms the directed diffusion protocol (Intanagonwiwat et al., 2000) in terms of power efficiency (Faruque et al., 2005). Thus, it is fair to say that BiSNET-RBbmr also outperforms the directed diffusion protocol.

CHAPTER 4

Principles of Communications Coverage in WM²Nets

4.1 General Principles

Next to network lifetime, communication coverage is the primary performance indicator in wireless communications as it entails the level of communications availability in a certain area. The common premise is that real world propagation effects can be well related to the log and normal fading contributions to the lognormal fading model (Rappaport, 1996). Let us consider a typical point-to-multipoint (PMP) and a wireless mesh communication scenario (see Fig. 4.1), and highlight the clear advantages of mesh over PMP configurations. An urban communication setup is assumed. In this scenario, the majority of links show a high average loss and large variance. A grid of city streets is an extreme case of high shadowing and variance as such.

As shown in the PMP configuration, a node terminal with a poor link to base station (access point (AP)) needs to increase its transmission power to compensate for link losses whereas a node in a wireless mobile mesh network (WM^2Net) hands over its call to another terminal, which lies in a more advantageous position and can help as intermediate (relaying) node of communication between the node and AP. Nodes in a WM^2Net setting therefore synergetically hop around obstacles to establish low-power multihop communication links to reach AP.

Comparing these two configurations, it becomes evident that the PMP configuration (see Fig. 4.1a) can cope more efficiently with the *log fading* of an open environment, but in cluttered environments higher power levels must be used to cope with the *normal fading*. For meshes, the opposite holds true; their basic attribute to "skip around obstacles" allows meshes to deal well with cluttered environments, but not so well with distances.

Figure 4.2 shows the measured signal path loss of a typical cellular deployment around several city locations of Germany. The distribution may be described as lognormal. The constants n and σ have been drawn on the figure in a best-fit approximation to model the measurement distribution. n and σ describe the lognormal path loss following the well-known formula:

$$\text{Path loss} = (\text{a constant}) + 10^* n^* \log(\text{distance}) + X_\sigma$$

PMP approach:
Focus is on RF and deployment
Blast over and through obstacles

Mesh approach:
Focus is on smart software
Skip around obstacles

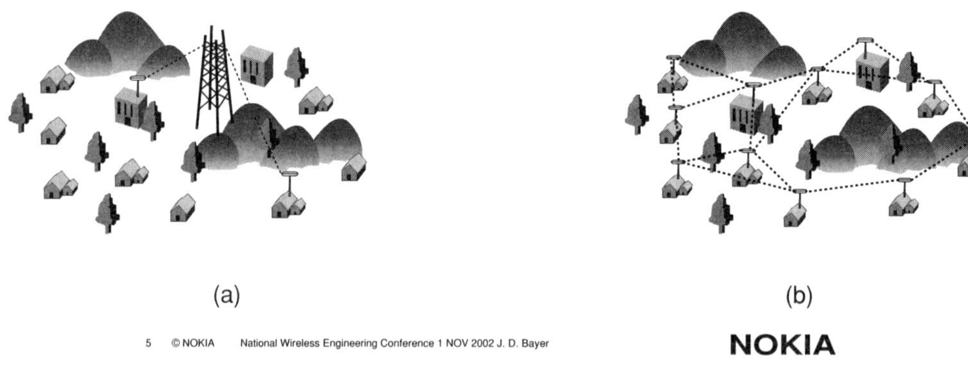

Figure 4.1 Distinction between PMP and mesh approach in the cluttered environment (Beyer, 2002).

n represents the average log fading attributable to distance and X_σ represents the variance of the normal distribution around this average caused by a distribution of cluttered environments.

Referring to Figs. 4.1 and 4.2 jointly, it may be deduced that for the PMP configuration, a large variance is bad.

RF path loss environment

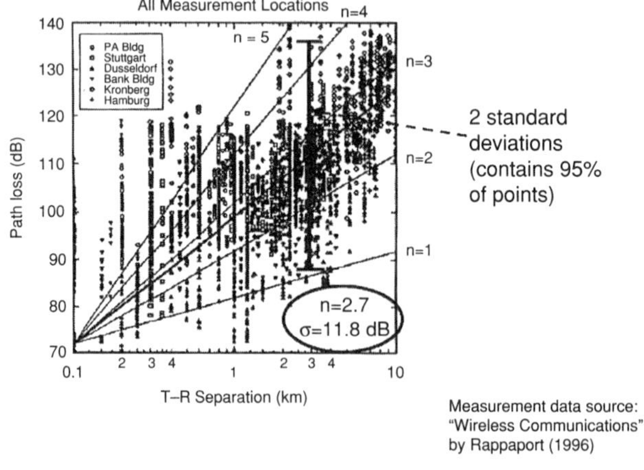

Figure 4.2 RF path loss, showing log and normal contributions (Beyer, 2002).

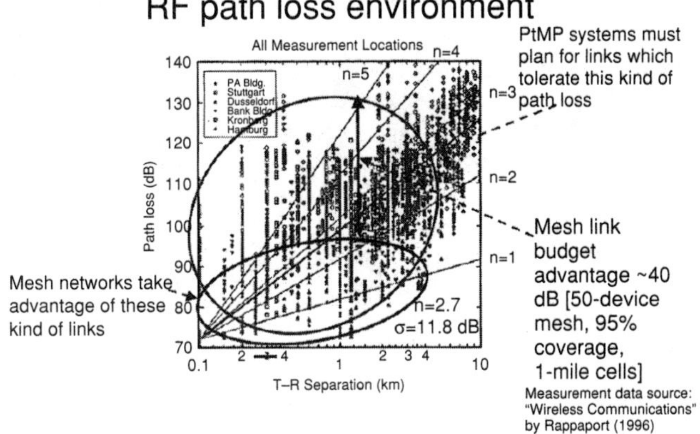

FIGURE 4.3 Sweet spot identification of mesh relative to PMP (Beyer, 2002).

In complete contrast for the mesh system, a large variance can be turned to advantage: nodes are enhanced with the necessary intelligence to pick the best link and, if this is not the direct node-to-AP link, communication "skips" around the obstacles over a distribution of RF paths with much less variance. This is called a restricted "sweet spot" of path losses depicted in the lower circle in Fig. 4.3. In the mesh case, therefore, lower RF powers are needed to cope with the lower power budget and a smaller interference footprint results.

4.2 Comparative Efficiency of Access Mesh and Cellular

As mentioned in previous section, an AP-enhanced wireless mesh configuration (i.e., access mesh) is more spectrally efficient comparing to a PMP cell. This is largely attributed to the lower average transmitting power, the localized channel contention as well as the lower interferences induced over a series of short hops, in a multihop WM^{2}Net context. Figure 4.4 illustrates the user throughput for a remote mesh terminal (RMT) that is sited around a corner from the AP, as depicted in the configuration scenario inside the graph.

The throughput performance is examined for the two possible communication schemes: a direct single-hop communication with distance AP-to-RMT of "d," and two-hop communications "hopping" around the corner via the forwarding mesh terminal (FMT) with distance AP-to-RMT of $d/\sqrt{2}$. As illustrated, the throughput performance is examined for different modulation schemes. The reader is referred to (Aggélou and Tafazolli, 2001) for a foundational work on enhancing Next Generation GSM cellular networks with relaying capabilities.

The aforementioned observations hold true only for idealized single-path scenarios. In practice, these are diminished by several factors, such as the dissimilar antenna gains of APs and mobiles, the overhead of relaying traffic from multiple users, the spectrum reuse contention within the mesh, and, of course, the routing overheads to combat node mobility.

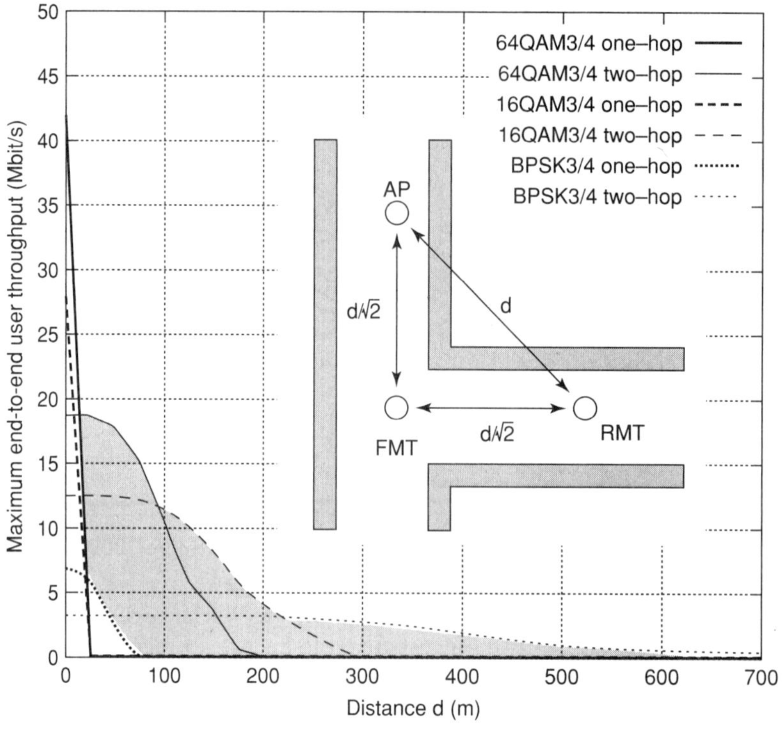

FIGURE 4.4 Use of a "forwarder" to "skip" around obstacles as in a mesh network (Esseling et al., 2002).

For the case of hopping between nodes of like type, consider node-to-node links in a mesh. If two hops of nearly equal length replace a single hop, as shown in Fig. 4.5, then:

- Only half the time-bandwidth product of spectral resource is available for each hop, and this acts to reduce the delivered data rate by a factor of 2
- But as each hop is half the length of the original link, the link budget is improved. This improvement can be exploited to improve spectral efficiency either

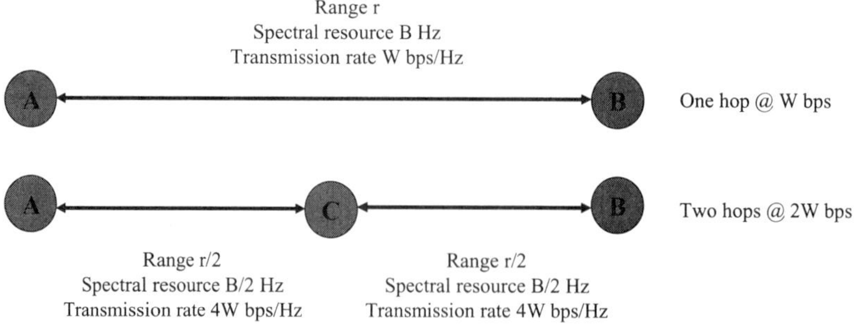

FIGURE 4.5 Two-hop vs. one-hop rate improvement between mesh nodes.

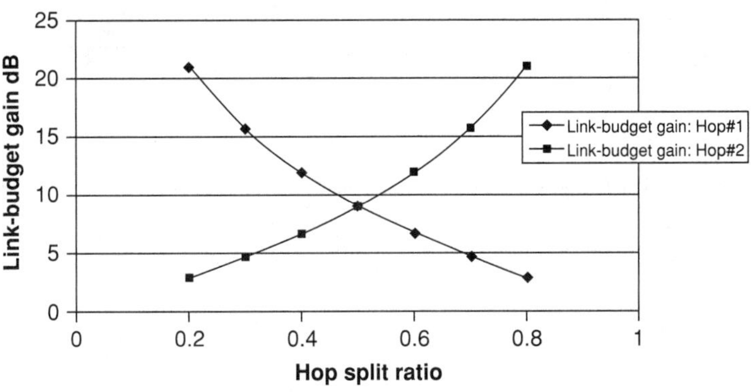

FIGURE 4.6 Two-hop link budget gain over single hop.

by increasing the transmission rate on each hop or reducing the transmit (Tx) power. For example, in a third-law propagation environment the link budget is improved by ×8 (~9 dB); this would permit a fourfold increase in transmission rate by changing from QPSK to QAM64. Alternatively, with spread-spectrum the coding gain could be reduced to realize a similar increase in transmission rate.

Overall, these two factors imply that twice as much data can be transferred using two shorter hops: that is, spectral efficiency is doubled. But, this prevails only when the path length is exactly halved. If instead there is asymmetry in the two-hop path lengths then the link-budget gain in the longer hop will diminish; higher rates become then unsupportable. This "sweet spot" in the path length split is illustrated in the link budget graph in Fig. 4.6. Taking the lower of the two lines demonstrates a peak at 0.5 (i.e., the center), which tails off away from the center in either direction, hence the term sweet spot. Overall, the two-hop chain is only as good as the weakest link.

But the comparative performance is further eroded for the case of multihopping into a mesh AP or cellular base station as represented in Fig. 4.7. The hop(s) between mobiles

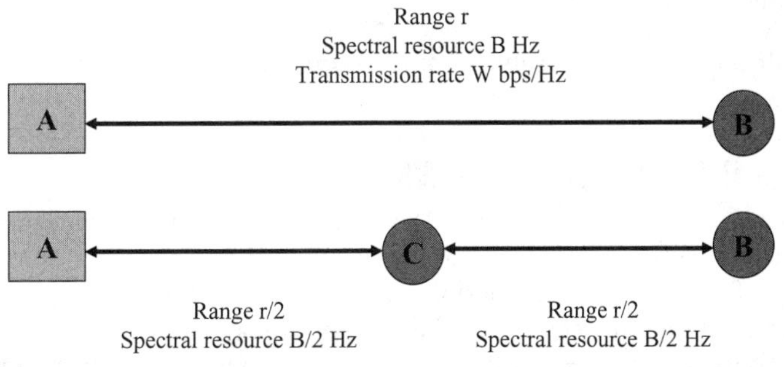

FIGURE 4.7 Two-hop vs. one-hop into high gain AP.

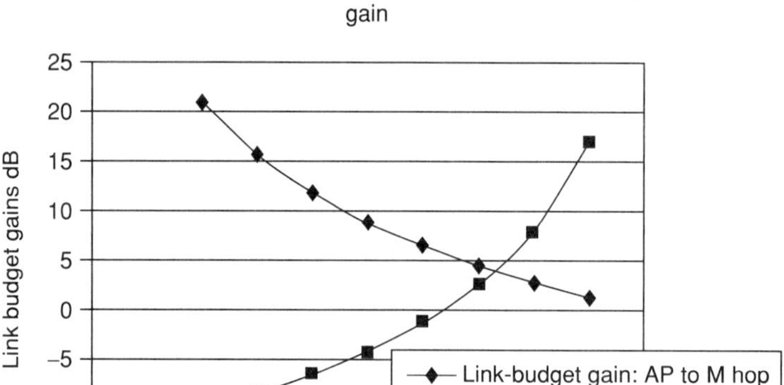

FIGURE 4.8 Two-hop vs. one-hop link budgets with high antenna gain.

lack the higher antenna gain and height of the link to the AP (node A in Fig. 4.7). Due to this imbalance, the "sweet spot" does not occur at the 50:50 path-length split.

The graph of Fig. 4.8 illustrates this for the case when the AP antenna gain is just 13 dB above the mobile nodes' gain—the "sweet spot" has moved to an approximately 75:25 path length ratio and the optimal link budgets on the two hops are only about 4 dB above the single-hop case. With this small link-budget gain the transmission rate might be little more than doubled. Thus, the best-case throughput rate of the two-hop route is the same as the single-hop.

Extending the analysis to a three-hop scenario, each hop would now be allocated one-third of the spectral resource. The "sweet spot" occurs at about 60:20:20 path length split and, at this point, the link budgets are about 7 dB above the single hop. This link budget gain might just support a tripling of transmission rate and so again achieve about the same throughput rate as the single-hop case. On the basis of the above, albeit highly simplified analysis we conclude that

Multihopping may rarely be much more spectrally efficient than a single hop.

4.3 Connectivity Principles in WM^{2}Nets

Key factors for a best-connected WM^{2}Net are first the development of optimal node placement strategies, and second, the number of meshes per unit of surface (i.e., nodal density). In reference to the nodal density, one tacit assumption on the WM^{2}Net nodal population is that the mesh node density should be high enough to ensure network-wide connectivity at any time. In this scope, the average degree of network connectivity increases with the node population. With regard to the former factor, deployment strategies are distinguished in those that account for network dynamics (e.g., attributed to mobility, the activation of nodes, and the switching between the "SLEEP" and "ACTIVE" mode at the medium access control (MAC) level), and these targeted for static environments. For the former category, when positions are static, the key deployment strategies are as follows:

- *k-coverage*: A region A is said to be k-covered if every point in A is within the communication range of k WM²Net nodes. According to this definition, to verify that an area is covered, one should enumerate all subregions resulting from the intersection of different WM²Net node regions and assert whether each of these is k-covered. Gage (1992), proposed a less complex technique, based on the following remark: a region is k-covered if for every communicating node the perimeter circle of its communication region is within the perimeters of at least k other nodes.
- *(k, R_t)-coverage* (Huang and Tseng, 2003): Given that N WM²Net nodes are deployed in an area A of radius R_t to track a target t, the (k, R_t)-coverage of the area is the probability that the random variable representing the number of WM²Net nodes including t within its range equals k for any target located in the deployment region. Huang and Tseng (2003) prove that the minimal nodal density corresponds to a maximum (k, R_t)-coverage that equals to:

$$\rho_s^{\min} = \frac{k}{\pi (R_s + R_t)^2} \tag{4.1}$$

  where R_s is the average communication range of the deployed WM²Net nodes.
- *k-connectivity*: The concept of *k-connectivity* is defined given that there are at least k-node distinct paths between every pair of nodes. In other terms, the network is *k-connected* if at least k nodes are within the transmission range of each WM²Net node. This coverage condition is often used throughout the literature to find a communication range assignment that ensures *k-connectivity*. For instance, Gupta and Kumar (1998) proved that if the average communication (transmission) range, denoted by R_s, verifies $R_c^2 = \frac{\log(n) + c(n)}{\pi n}$, the network is then asymptotically connected with high probability if and only if $\lim_{n\to\infty} c(n) = +\infty$.
- *Path-observability*: Path observation relies on measuring the exposure of a mobile target (having a specific trajectory) to the communication field. Exposure is defined as "the expected average ability of observing a target moving in the communication field" (Gosh and Das, 2006). Formally, the exposure during an interval $[t_1, t_2]$ has the following expression:

$$E(p(.), t_1, t_2) = \int_{t_1}^{t_2} I(p(t)) \left| \frac{dp(t)}{dt} \right| dt \tag{4.2}$$

  where $p(.)$ describes the target path, and $I(p(.))$ is a measure of sensitivity at a certain point along the path.

Notably, each of these deployment optimization approaches is associated with a specific reasoning. More specifically, k- and (k, R_t)-coverages are mainly used when the WM^{2}Net designer targets for maximum detection efficiency. On the opposite, k-connectivity is suitable primarily for enhancing the transmission and forwarding capabilities of nodes. Finally, path-observability corresponds to cases where the main objective is to track mobile targets in the monitored field.

For the latter category now, when positions are no longer static, the minimum number of meshes that theoretically could ensure optimal coverage is expected to be different to that of a deployment strategy that account for network dynamics. Let us assume that within a time interval $\Delta\Theta$, a mesh node can move at a distance ε from its initial position according to a given direction and speed. Hamdi et al. (2006) prove that the area A_0 that is no longer covered by a specific WM^{2}Net node after time interval $\Delta\theta$ is given by:

$$A_0 = \pi R^2 - 2\arccos\left(\frac{\varepsilon}{2R}\right) R^2 - \varepsilon\sqrt{R^2 - \left(\frac{\varepsilon}{2}\right)^2} \tag{4.3}$$

where R is the communication range.

Furthermore, as shown by Hamdi et al. (2006), if $s_1, \ldots, s_N$, are N_S mobile meshes; uniformly distributed in a region of area A, and s_i $(1 \leq i \leq N)$ meshes move at a distance ε during a time interval $\Delta\theta$ according to uniformly distributed random directions δi $[0, 2\pi]$, then k^θ denotes an integer such that A is $(k^\theta,\ R_t)$-covered at instant θ:

$$k^{\theta+\Delta\theta} = \left\lfloor \frac{k^\theta}{\pi R_0^2}\Psi(\varepsilon) \right\rfloor \tag{4.4}$$

where $\lfloor . \rfloor$ denotes the floor operator and R_0 is the sum of the target radius and communication range and

$$\Psi(\varepsilon) = \left\lfloor \pi(R_0 - \varepsilon)^2 + \pi\left(R_0^2 - (R_0 - \varepsilon)^2\right) I_1(\varepsilon) + \pi\left((R_0 + \varepsilon)^2 - R_0^2\right) I_2(\varepsilon) \right\rfloor$$

$$I_1(\varepsilon) = \int_{R_0-\varepsilon}^{R_0} \int_0^{2\pi} \left(2\pi - \arccos\left(\frac{R_0^2 - r^2 - \varepsilon^2}{2r\varepsilon}\right)\right) dr\, d\alpha$$

$$I_2(\varepsilon) = 2\pi - \int_{R_0-\varepsilon}^{R_0} \int_0^{2\pi} \arccos\left(\frac{r^2 + \varepsilon^2 - R_0^2}{2r\varepsilon}\right) dr\, d\alpha$$

This illustrates the mobility effect on the area coverage. In fact, in this particular situation (i.e., uniformly isotropic mobility), the function $\varphi(\theta) = \kappa^\theta$ is the root of the differential equation $\chi'(\theta) = \frac{\partial\Gamma}{\partial\varepsilon}(\theta, 0)\,\chi(\theta)$, where $\Gamma(\theta, \varepsilon) = \Psi(\varepsilon)$.

4.3.1 Robust Connectivity Energy-Aware Routing for Wireless Mesh Networks[1]

4.3.1.1 Introduction

It is a common practice for energy-aware routing algorithms to use the time elapsed until the first node in the network fails as the definition of network lifetime. In many practical WM^{2}Net applications, the death of the first node may not impact the overall communication task. In this view, the definition of the network lifetime is defined as the time elapsed until there is no route from any source to any destination. In other words, the network lifetime should be defined as the time until the network becomes disconnected/disintegrated. Using this definition for the network lifetime, the network connectivity becomes the basic criterion in pursuing routing decisions.

Throughout this study, we employ the notion of algebraic connectivity of a graph in the spectral graph theory to quantify the importance of the node. In particular, we propose to quantify the importance of a node by the Fiedler value (the second smallest eigenvalue of the network Laplacian matrix) of the remaining graph when that particular node dies. One property of the Fiedler value is that the larger it is, the more connected the graph will be. By considering the nodes' importance from the graph connectivity perspective in the routing design, the node with higher importance will be retained in the network, therefore the connectivity of the remaining network is maintained as long as possible. The proposed algorithm has several advantages; it is an online algorithm, efficient in maintaining the connectivity of the remaining network, and flexible to be used along with any other existing energy-aware routing algorithms that employ distributed Bellman-Ford/Dijkstra algorithms in their implementations.

4.3.1.2 Facts from Spectral Graph Theory

We briefly summarize some important facts from spectral graph theory (Fiedler, 1973; Chung, 1997) that will be used to prove properties of the proposed algorithms. We consider a simple finite graph. The following notations will be used throughout this study: $\mathrm{G} = (\mathrm{V}, \mathrm{E})$ is the graph with set of vertices V (of size $|\mathrm{V}| = n$) and set of edges E (of size $|\mathrm{E}| = m$). The Laplacian matrix associated with a graph is defined as

$$L(i, j) = \begin{cases} d_{v_i} & \textit{if } v_i = v_j \\ -1 & \textit{if } (v_i, v_j) \in \mathrm{E} \\ 0 & \textit{otherwise} \end{cases}$$

where $i, j \in \{1, \ldots, n\}$ are the nodes indices, d_{v_i} is the degree of node v_i. The eigenvalues of the Laplacian matrix, $L(\mathrm{G})$ $(\lambda_0(\mathrm{G}) \leq \cdots \leq \lambda_{n-1}(\mathrm{G}))$ are usually referred to as the *graph spectra*. The following lemma describes the relationship between the graph spectra and its connectivity (Chung, 1997).

Lemma 1: If G is connected, then $\lambda_1(\mathrm{G}) > 0$. Moreover, if $\lambda_i(\mathrm{G}) = 0$ and $\lambda_{i+1}(\mathrm{G}) \neq 0$, then G has exactly $i + 1$ disjoint connected components.

This lemma indicates that if G is strongly connected then $L(\mathrm{G})$ has the simple eigenvalue 0. If the eigenvalue 0 of the Laplacian matrix $L(\mathrm{G})$ has multiplicity n, then there are n

[1] Excerpt from the invited article "Robust connectivity energy-aware routing for wireless Mesh networks," Charles Pandana and K. J. Ray Liu, University of Maryland, College Park, MD 20742, E-mail: cpandana@glue.umd.edu

disconnected components. We focus on the second smallest eigenvalue of the Laplacian matrix, since we are dealing with a connected graph. The second smallest eigenvalue of the Laplacian matrix is also referred to as the algebraic connectivity of the graph G (Fiedler value) (Fiedler, 1973). Its properties are summarized in the following lemmas:

Lemma 2: The Fiedler value $\lambda_1(G)$ is non-decreasing for graphs with the same set of vertices, that is, $\lambda_1(G_1) \leq \lambda_1(G)$ if $G_1 = (V, E_1)$, $G = (V, E)$, and $E_1 \subseteq E$.

We observe that G and G_1 have the same number of vertices. Since G_1 has fewer edges compared to $G(E_1 \subseteq E)$, this implies that G_1 is less connected compared to G. From Lemma 2, we have $\lambda_1(G_1) \leq \lambda_1(G)$. It is in this sense that the Fiedler value represents the degree of connectivity in a graph. The following lemmas give the relation of Fiedler value for graph obtained from removing a vertex and all its adjacent edges, the upper, and lower bounds for the Fiedler value.

Lemma 3: Let G_1 be a graph obtained from removing 1 vertex and all the adjacent edges from G, then $\lambda_1(G_1) \geq \lambda_1(G) - 1$.

Lemma 4: Let d_{v_i} be the degree of node v_i, we have $\lambda_1(G_1) \leq \left(\frac{n}{n-1}\right) \min_{v_i} d_{v_i}$.

Lemma 5: Let $\varepsilon(G)$ be the edge connectivity of the graph G (the minimal number of edges whose removal results in losing connectivity). We have $\lambda_1(G_1) \geq 2\varepsilon(G)\left(1 - \cos\left({}^{\pi}/_{n}\right)\right)$.

4.3.1.3 Keep Connect Algorithms

In this section, we use the facts of spectral graph theory described in the previous section to develop an online algorithm that drives the routing algorithm to maximize the network lifetime with the connectivity criterion. We emphasize that the definition of network lifetime used in this study is the time until the network becomes disconnected. We first determine how to calculate the connectivity weight. From what is stated above, recall that the Fiedler value qualitatively represents the connectivity of a graph in the sense that the larger the Fiedler value is, the more connected the graph will be. The degree of connectivity of the remaining graph can be quantified by the Fiedler value of the graph obtained by removing that particular node and all the edges connected to that node from the original graph. We design the connectivity weight of each node by setting the weight of node v_i as $1/\lambda_1(G_{-v_i})$. In this way, the node that causes severe reduction in the remaining network connectivity will be avoided when doing the routing decision. The details of the algorithm for finding the connectivity weight using the Fiedler value are shown in Table 4.1.

The MTEKC(y) algorithm is obtained by embedding the connectivity weight $W(\cdot)$, to the original minimum total energy (MTE) (Toh, 2001) algorithm. The modified algorithm (see Table 4.2) employs $e_t(v_i, v_j)W(v_i)^y + e_r(v_i, v_j)W(v_j)^y$ as the link/edge cost between node v_i and node v_j, where $e_t(v_i, v_j)$ and $e_r(v_i, v_j)$ are the Tx and receive (Rx) energy for delivering a packet from node v_i to v_j. We note that in the MTEKC algorithm, the parameter y determines how significantly the connectivity weight should impact the weighted MTE.

Define graph $G_{-v_i} = (\{V - v_i\}, E_{-v_i})$ as the graph obtained by removing node v_i and all the connecting edges

1. Initialization: Set nodes' weights as zeros $W(v_i) = 0, \forall v_i \in V$.
2. For each node v_i:
   a. Form the Laplacian matrix $L(G_{-v_i})$ of graph $G_{-v_i} = (\{V - v_i\}, E_{-v_i})$ as (1).
   b. Find the Fiedler value and let denote the Fiedler value as $\lambda_1(G_{-v_i})$
   c. Set the weight of node as $W(v_i) = 1/\lambda_1(G_{-v_i})$.

End for

Table 4.1 Keep Connect Algorithm Using Fiedler Value

4.3.1.4 Upper Bound on the Energy Consumption

The works of Chang and Tassiulas (2004), Li et al. (2001), and Aslam et al. (2003) highlight the tradeoff between the energy consumed along a route and the bias towards recourse-rich nodes for maximizing lifetime and the number of packets delivered. Hence, it is important to show the bounds on the energy consumption of the proposed algorithm. For simplicity, we include only the Tx energy between two nodes in calculating the energy of the route (Feeney and Nilsson, 2001). We denote r^* as the MTE route connecting any fixed source node v_0 and destination node v_d. Equivalently, the MTE route is represented as $r^* = \arg\min_{r \in R(v_0,v_d)} \sum_{i=1}^{d(r)-1} e_t(v_i, v_{i+1})$, where $R(v_0, v_d)$ is the set of all routes connecting the source node v_0 and destination node v_d; $d(r)$ is the number of hops between in the route. Furthermore, we denote $r^\dagger$ as the MTEKC(y) route obtained using the Fiedler value and this route satisfies $r^\dagger = \arg\min_{r \in R(v_0,v_d)} \sum_{i=1}^{d(r)-1} \frac{e_t(v_i, v_{i+1})}{\lambda_1(G_{-v_i})^y}$. We first give some simple lemmas on the lower bound and upper bound of the metric in the keep connect algorithm. The proofs are straightforward, using Lemmas 1–5, hence they are omitted.

Lemma 6: (Lower bound of MTEKC(y)): The MTEKC(y) route has the following property $\sum_{i=1}^{d(r)-1} e_t(v_i, v_{i+1}) W(v_i)^y \geq \left(\frac{n-2}{(n-1)\min_{v_i} d_{v_i}}\right)^y \sum_{i=1}^{d(r)-1} e_t(v_i, v_{i+1}) \geq \left(\frac{(n-2)n}{2(n-1)m}\right)^y \sum_{i=1}^{d(r)-1} e_t(v_i, v_{i+1})$ where d_{v_i} is the degree of node v_i in the graph, n and m are the number of vertices and edges.

Lemma 7: (Upper bound of MTEKC(y)): The MTEKC(y) route satisfies the following inequality $\sum_{i=1}^{d(r)-1} e_t(v_i, v_{i+1}) W(v_i)^y \leq \left[\frac{1}{2(\varepsilon(G)-1)(1-\cos(\pi/_{n-1}))}\right]^y \sum_{i=1}^{d(r)-1} e_t(v_i, v_{i+1})$ where $\varepsilon(G)$ is the edge-cut or edge connectivity of the graph.

1. For any source-destination pairs, find the MTE path with edge cost as: $e_t(v_i, v_j) W(v_i)^y + e_r(v_i, v_j) W(v_j)^y$ for $v_i \in V$, $v_j \in S_{v_i}$, where $e_t(v_i, v_j)$ and $e_r(v_i, v_j)$ are the Tx and Rx energy for delivering a packet from node v_i to v_j. S_{v_i} denotes the neighbors of node v_i. $W(v_i)$ is the weight of node v_i.
2. If node fails, re-compute the "alive" nodes' weight using *Keep Connect* algorithm. Repeat Step 1.

Table 4.2 MTEKC(y)

Lemma 8: The MTEKC(y) route is exactly the same as the MTE route for a complete graph.

Proof: From the definitions of MTE route and MTEKC(y) with Fiedler value route, we have $\sum_{i=1}^{d(r^*)-1} e_t(v_i, v_{i+1}) \leq \sum_{i=1}^{d(r^\dagger)-1} e_t(v_i, v_{i+1})$ and $\sum_{i=1}^{d(r^\dagger)-1} \frac{e_t(v_i, v_{i+1})}{\lambda_1(G_{-v_i})^y} \leq \sum_{i=1}^{d(r^*)-1} \frac{e_t(v_i, v_{i+1})}{\lambda_1(G_{-v_i})^y}$. We note that removing one node from a complete graph with n nodes results in another complete graph with $n-1$ nodes. Therefore, we have $\lambda_1(G_{-v_i}) = n-1$. Simplifying and combining both expressions, we have $\sum_{i=1}^{d(r^*)-1} e_t(v_i, v_{i+1}) = \sum_{i=1}^{d(r^\dagger)-1} e_t(v_i, v_{i+1})$. Since it is less likely to have two different routes with the same total Tx energy in random network deployment for fixed source node v_0 and destination node v_d, we conclude that $r^* = r^\dagger$. ■

Using Lemmas 6–8, we can show:

Theorem 1: The energy consumed in the MTEKC(y) satisfies the following upper bound

$$\sum_{i=1}^{d(r^\dagger)-1} e_t(v_i, v_{i+1}) \leq \left[\frac{(n-1)\,m}{n\,(n-2)\,(\varepsilon\,(G)-1)\,(1-\cos\,(\pi/_{n-1}))}\right]^y \sum_{i=1}^{d(r^*)-1} e_t(v_i, v_{i+1}).$$

The following theorem gives the asymptotic bound on the ratio of energy consumed by MTEKC(y) using the Fiedler value.

Theorem 2: Suppose that the network satisfies $m = a_1 n$ and $\varepsilon(G) - 1 = a_2$, where a_1 and a_2 are some constants. Then the upper bound on the ratio of energy consumed can be presented as

$$\frac{\sum_{i=1}^{d(r^\dagger)-1} e_t(v_i, v_{i+1})}{\sum_{i=1}^{d(r^*)-1} e_t(v_i, v_{i+1})} = O(n^{2y})$$

Proof: Using the assumptions of Theorem 2 besides Theorem 1, it is straightforward to show that $\frac{\sum_{i=1}^{d(r^\dagger)-1} e_t(v_i,v_{i+1})}{\sum_{i=1}^{d(r^*)-1} e_t(v_i,v_{i+1})} \leq \left(\frac{a_1 n(n-1)}{a_2 n(n-2)(1-\cos(\pi/n-1))}\right)^y \leq C\left(\frac{n(n-1)(1+\cos(\pi/n-1))}{n(n-2)\sin^2(\pi/n-1)}\right)^y$, where $C = (a_1/a_2)^y$. As $n \to \infty$, the ratio is less than $C\left(\frac{(n-1)^2}{\pi^2}\right)^y$; this shows the theorem. We note that we have used the small angle approximation in the sinusoidal function $(\sin(\theta) \approx \theta, \theta \geq 1)$. ■

From this theorem, we see that the ratio of energy can be easily controlled from parameter$y$. If $y = 1/2$, for instance, the ratio of consumed energy increases then as a linear function of the number of nodes. In the extreme case, setting $y = O(1/n)$ makes the proposed algorithm approaching to MTE as $n \to \infty$.

4.3.1.5 Simulation Results

We simulate the routing algorithms in a discrete-event simulator. The simulator initially deploys nodes in the network. All events are time-stamped and queued. The most current event will be de-queued and according to the type of the events certain task is triggered. There are three types of events: packet arrival events, reporting event, and sending events. In the packet arrival event, packets are injected from the sources node to destination node. We assume the packet arrival follows the Poisson arrival process with mean μ. The reporting event occurs periodically to retrieve the simulation parameters in this report interval. All events that are neither the packet arrival events nor reporting events are the sending events. In the sending event, a packet is sent to the next hop. The next hop is determined based on the routing algorithm used. Whenever a packet arrives at a node, it is queued in the node's buffer and will be sent in the next transmission time. Whenever a packet reaches its destination, the number of delivered packets is incremented and the event associated with that packet is freed. The channel between two adjacent nodes in the network is modeled as a fading process that attenuates the Tx signal proportionally to the distance $d^{-\alpha}$. This model is general enough to describe both the free space propagation and the two-ray ground propagation model, which is typically used in many ad hoc simulators (Xu and Saadawi, 2001). In all cases, we use the network lifetime (time before the remaining network becomes disconnected), packet delivery time, average Tx energy per packet, and total delivered packet before the remaining network becomes disconnected as performance metrics.

We generate five random networks with the parameters shown in Table 4.3. Figure 4.9 shows the simulation results for normalized network lifetime, routing time, transmission energy, and total successful delivered packets of the MTEKC ($y = 1$) algorithm with respect to the MTE algorithm. The x-axis of the figures represents network realization, which is numbered according to Table 4.3. From this figure, we see that the proposed algorithm can achieve from 10% to 53% increase in the network lifetime for broad network loads with an increase of less than 20% transmission energy. Notably, the proposed algorithm causes an increase of the overall routing time, which under certain circumstances may be a critical concern in a WM²Net context. Finally, the proposed algorithm achieves up to 36% improvement in the total delivered packets. In addition to having a longer lifetime and more packets delivered, the algorithm is more robust in terms of the network connectivity as shown in Fig. 4.10. Figure 4.10 shows the decrease in the algebraic network connectivity whenever a node fails. The x-axis is the number of dead nodes and the y-axis is the algebraic network connectivity. It is obvious that the proposed algorithm is more robust in terms of network connectivity. Next, we show the performance of the distributed implementation. We note that the distributed implementation is based on the distributed reinforcement learning scheme for network routing (Littman and Boyan, 1993; Pandana and Liu, 2005). We modify the algorithm by

Network	1	2	3	4	5
Nodes	102	101	115	124	86
Network connectivity	0.3994	0.5679	0.7282	0.5543	0.1141

* Five networks of 100 nodes are generated using 2D Poisson point process in the area of 100 m × 100 m. The Tx energy per packet between two nodes is equivalent to $3 \times 10^{-3}\, d^2$, where d is the distance between the nodes. The maximum Tx and Rx powers equal to 1.4 W and 0.7 W, respectively. This implies that the farthest node that can be reached by a node is about 21.6 m away. The initial energy of all nodes is equal to 10,000 units.

TABLE 4.3 Five Random Networks That Are Generated with the Parameters*

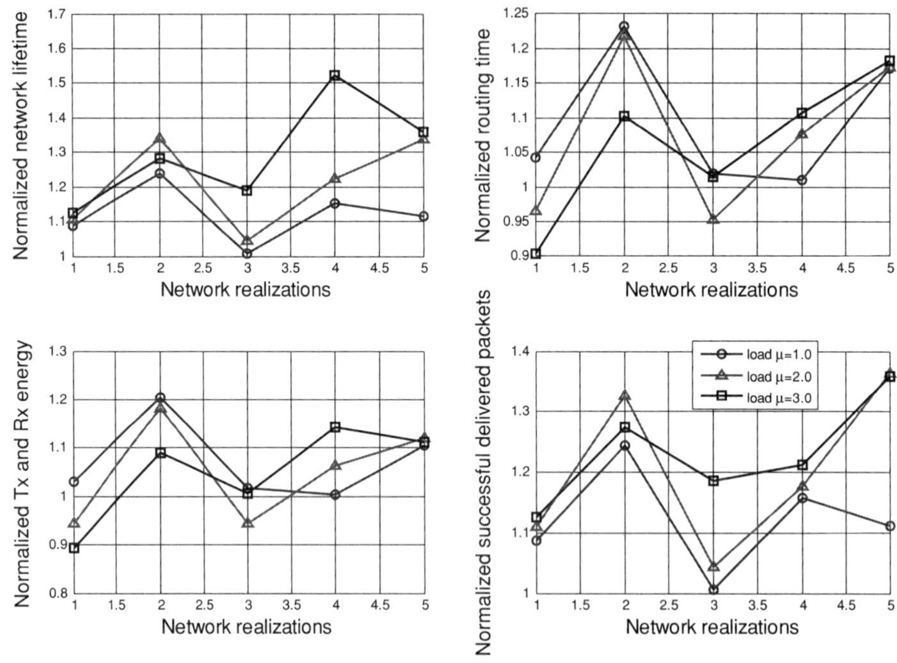

Figure 4.9 Normalized metrics for MTEKC (1) with respect to MTE for different packet arrival rates.

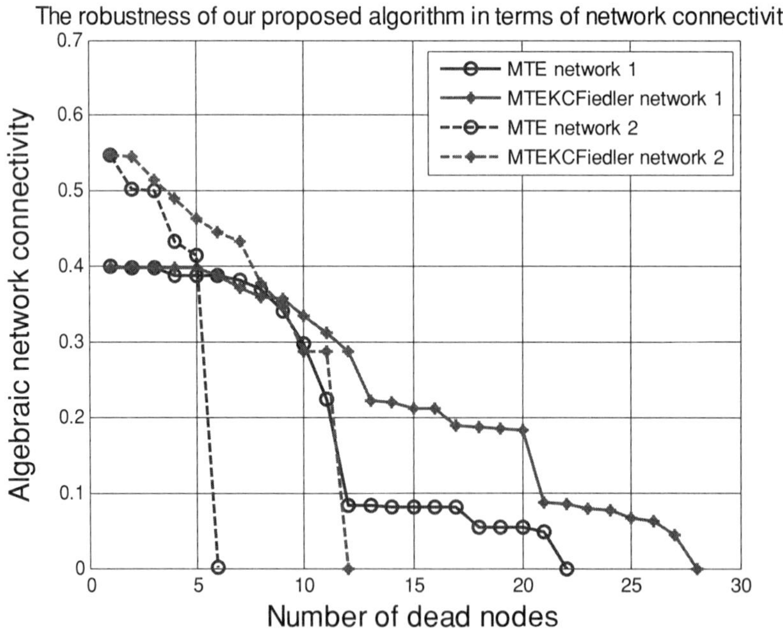

Figure 4.10 Robustness of the proposed algorithm in terms of network connectivity. Zero algebraic connectivity implies the disconnected network.

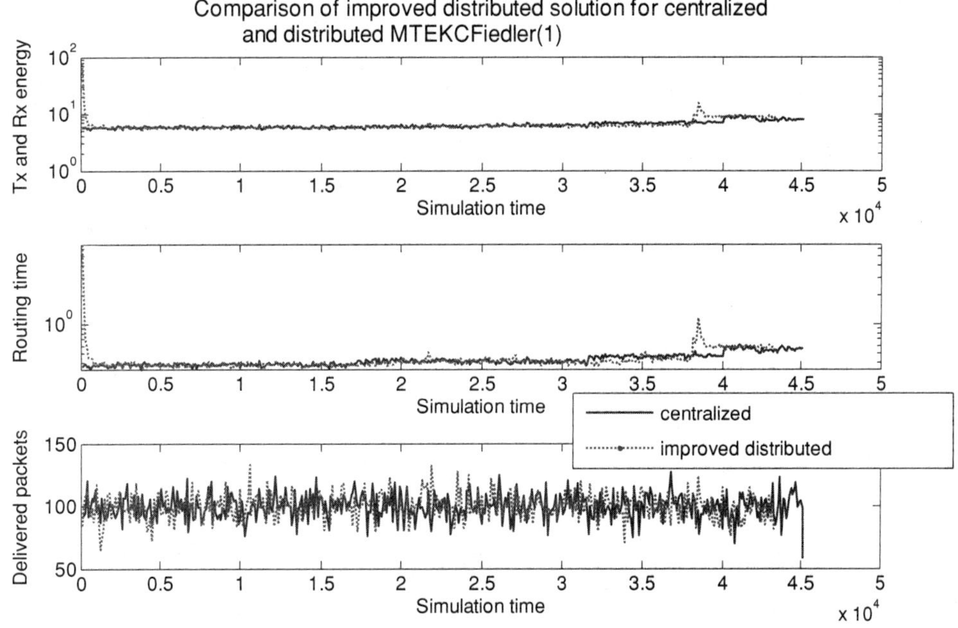

FIGURE 4.11 Tx and Rx energy, routing time, and the delivered packets for distributed solution.

periodically sending 10 small packets in between sending the actual data packets. This approach can better learn the routing state in the network and it achieves the solution of the centralized solution as shown in Fig. 4.11. This figure shows that the improved distributed solution can achieve the near-centralized solution.

Finally, we compare the performance of distributed algorithm for random source and destination generation for distributed MTEKC and MTE algorithms using the previously defined five networks (Table 4.3) in terms of normalized network lifetime, routing time, transmission energy, and total delivered packets. From Fig. 4.12, we see that the improved distributed solution along with the keep connect algorithm can improve the network lifetime and the total delivered packet before the network becomes disconnected.

4.3.2 Relay Placement for Topology Design of Wireless Mesh Networks[2]

4.3.2.1 Relay Placement in Homogeneous Networks

The basic relay placement algorithm for achieving connectivity in WM²Net was proposed by Lin and Wang (1999). The authors proved the problem to be NP-Hard, and presented an algorithm they proved to be a 5-approximation. The algorithm restricts the placement of relay nodes on straight lines joining two mesh nodes. They then form a complete graph between WM²Net nodes with weight of each edge as the number of relay nodes needed,

[2] Excerpt from the invited article "Relay placement for topology design of wireless mesh networks," Abhishek Kashyap, Department of Electrical and Computer Engineering, University of Maryland, E-mail: kashyap@glue.umd.edu

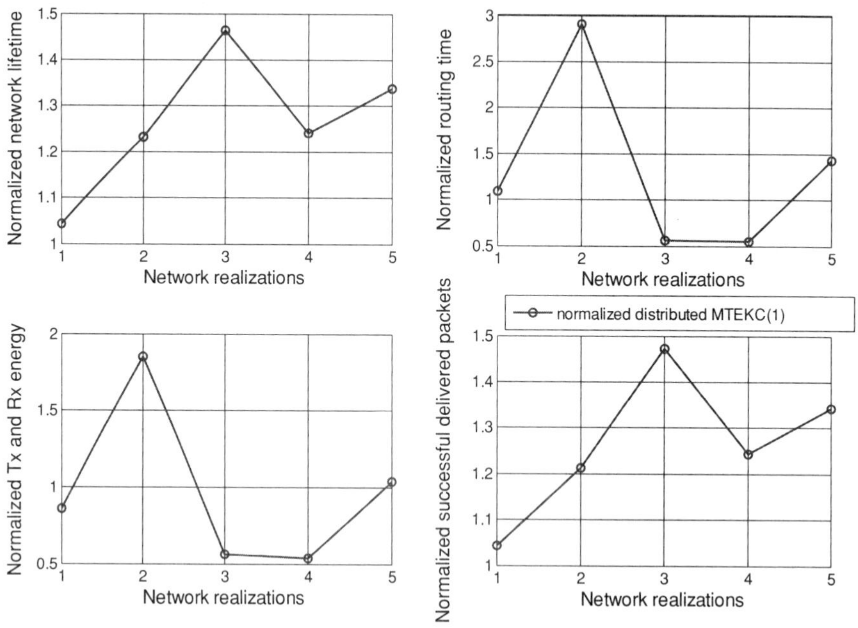

FIGURE 4.12 Performance of normalized metrics for distributed MTEKC (1).

and compute a minimum spanning tree (MST) of the resulting graph. The algorithm was later proved to be a 4-approximation for the Euclidean plane by Chen et al. (2001) and Măndoui and Zelikovsky (2000). The algorithms for constructing a fault-tolerant topology discussed throughout this study rely upon this algorithm.

Network Model and Problem Statement We model the network as a graph G = (V, E), where V is the set of mesh nodes, and E is the set of links between them. We assume each node has a limited transmission range, which we normalize to one. It is assumed that a node can connect to all nodes within its transmission range. A link $e = (x, y)$ belongs to E if nodes x and y are within unit distance of each other. The links can be omnidirectional RF, directional RF, or Free Space Optical (without obscuration).

The objective is to construct a network topology, that is, k-vertex or k-edge connected on the mesh nodes or on the mesh and relay nodes. We assume the relay nodes are identical to the mesh nodes in terms of their transmission range and type of links. The problems can be stated as follows:

Partial k-connectivity. Given a graph G = (V, E), find the minimum number of relay nodes (denoted by set R) needed (and their locations) such that the set of nodes V is k-vertex (edge) connected ($k \geq 2$) in the resulting graph $G' = (V + R, E')$, $E \subseteq E'$. The objective is to construct a graph such that for all $u, v \in V$, $\lambda(u, v) \geq k$; where $\lambda(u, v)$ is the number of internally vertex-disjoint (or edge-disjoint) paths between u and v in G′.

Full k-connectivity. Given a graph G = (V, E), find the minimum number of relay nodes (denoted by set R) needed (and their locations) such that the set of nodes

V is k-vertex (edge) connected ($k \geq 2$) in the resulting graph $G' = (V + R, E')$, $E \subseteq E'$. The objective is to construct a graph such that for all $u, v \in V + R, \lambda(u, v) \geq k$; where $\lambda(u,v)$ is the number of internally vertex-disjoint (or edge-disjoint) paths between u and v in G'.

Algorithms The algorithm for achieving partial k-vertex connectivity was first presented by (Bredin et al., 2005, Kashyap et al., 2006).

Algorithm B.1: Relay Placement for Partial k-Vertex Connectivity

1. Construct a complete graph $G_c = (V, E_c)$ by adding an edge between each pair of vertices of graph G.
2. Weight the edges in E_c as $c_e = \lceil |e| \rceil - 1$. Here, $|e|$ represents the length of edge e. The cost is the minimum number of relay nodes needed if they are placed on the straight line joining the two mesh nodes.
3. Compute an approximate minimum cost spanning k-vertex connected subgraph of G_c. We use the 2-approximation algorithm of Khuller and Raghavachari (1996) for $k = 2$, and the k-approximation algorithm of Kortsarz and Nutov (2003) for $k > 2$. For $k \leq 7$, we can use the improved approximation algorithms proposed by Auletta et al. (1999) and Dinitz and Nutov (1999). Let the resulting graph be G'_c.
4. Place relay nodes (number equal to the weight of the edge) on the edges in G'_c with link costs greater than zero.
5. For all pairs of nodes (including the relay nodes) in G'_c within each other's transmission range, form an edge.
6. For the relay nodes sorted arbitrarily, do the following (starting at $i = 1$):
   - Remove node i (and all adjacent edges).
   - Check for k-vertex connectivity between the mesh nodes. We can use the algorithm of Cheriyan and Thurimella (1993) for the purpose.
   - If the graph is k-vertex connected, repeat for $i = i + 1$, else put back the node i and corresponding edges, and repeat for $i = i + 1$.
   - Stop when all relay nodes have been considered.
7. Output the resulting graph.

Full k-vertex connectivity can be achieved by placing $k - 1$ additional relays at the position each added relay in the partially *k-vertex* connected graph, and at each WM²Net node that has an edge with at least one relay node incident on it (Bredin et al., 2005).

The algorithm for partial edge connectivity is similar to Algorithm B.1 (Kashyap, 2006). In Step 1 of Algorithm B.1, we form a complete multigraph with k edges between each pair of mesh nodes. The 2-approximation algorithm of Khuller and Vishkin (1994) is used for computation of a *k-edge* connected subgraph in Step 3 of the algorithm. The last change is that the graph is checked for *k-edge* connectivity using the algorithm of Matula (1987) in Step 6. For full *k-edge* connectivity, $\lceil k/2 \rceil - 1$ relays are added at the position of each added relay in the partially *k-edge* connected graph, and at each mesh node that has an edge with at least one relay node incident on it.

We now present the approximation results for these algorithms:

Theorem II.B.1 (Kashyap, 2006): If an optimal network uses s relay nodes so that a network in the Euclidean plane is partially k-vertex connected, Algorithm B.1 forms a network with maximum of $c(3\lceil k/2\rceil(\lceil k/2\rceil + 1) - 1)s$ relay nodes, such that the network is partially k-vertex connected. Here, c is the approximation ratio for the algorithm used to compute a k-vertex connected subgraph in Step 3.

Theorem II.B.2 (Kashyap, 2006): If an optimal network uses s relay nodes so that a network in the Euclidean plane is partially k-edge connected, Algorithm B.1, modified for edge connectivity, forms a network with maximum of $10\lceil k/2\rceil s$ relay nodes, such that the network is partially k-edge connected. The approximation ratios get multiplied by $3k$ for full k-vertex connectivity, and $3\lceil k/2\rceil$ for full k-edge connectivity.

Computational Complexity Let there be N mesh nodes in the network. Steps 1–5 of Algorithm B.1 for vertex connectivity take $O(k^2N^5)$ time. Let there be N' relay nodes in the network at Step 5 of the algorithm. Step 6 of the algorithm takes $O(k^3N'(N+N')^2)$ time. Thus, the algorithm takes $O(k^2N^5 + k^3N'(N+N')^2)$ time (Kashyap, 2006). Similarly, the edge connectivity algorithm takes $O((kN)^2 + kN'(N+N')^2)$ time (Kashyap, 2006).

4.3.2.2 Networks in Higher Dimensions and with Obstacles

We now present approximation ratio results for networks of higher dimensional metric spaces. In the following analysis, obstacles are also accounted for.

Let us first define the MST number (M): the MST number of a metric space is defined as the maximum node degree in a minimum-degree MST spanning points. The MST number for the Euclidean plane is 5 (Monma and Suri, 1992), three-dimensional Euclidean space is 12, and rectilinear plane (two-dimensional space with metric defined by the L_1 norm) is 4 (Robins and Salowe, 1995).

The connectivity algorithm of Lin and Wang (1999) was proved to have an approximation ratio of $M - 1$ by Măndoui and Zelikovsky (2000). We present the results for the algorithm of Section B for partial k-edge connectivity and 2-vertex connectivity:

Theorem III.1 (Kashyap, 2006): Algorithm B.1 has an approximation ratio of $2M$ for achieving partial 2-vertex connectivity.

Theorem III.2 (Kashyap, 2006): Algorithm B.1, with modifications for edge connectivity, has an approximation ratio of $2M\lceil k/2\rceil$ for achieving partial k-edge connectivity.

The results can be extended for full connectivity with the same multiplicative factor as for the Euclidean plane.

We now consider the case where obstacles are accounted for in the network setting. We assume that there exist polygonal regions in the network where relay nodes cannot be placed. The aforementioned algorithms are then modified since relay nodes cannot necessarily be placed on straight lines joining two mesh nodes. We use the algorithm proposed by Arkin et al. (2005) to calculate the minimum number of relays required to connect a pair of mesh nodes. The weight of each edge in Step 2 of Algorithm B.1 is set to the number of relays needed on each edge using the algorithm proposed by Arkin et al. (2005). Kashyap (2006) proves that the algorithms still show similar approximation guarantees as in the case where no obstacles exist.

4.3.2.3 Heterogeneous Networks

In this section, we consider networks with non-uniform transmission ranges for mesh and relay nodes. We assume mesh nodes with a transmission range in the range $[T_{\min}, T_{\max}]$, and relay nodes with a transmission range of T_{relay}. We normalize the transmission range, and let $T_{\min} = 1$, $T_{\max} = \alpha$, $T_{\text{relay}} = \gamma$. We assume two nodes can communicate if both are within each other's transmission range. That is, for nodes i and j to communicate, the distance between them should not be more than the minimum of their transmission ranges. We first present the algorithm to achieve partial k-vertex connectivity in such a network. The algorithm is similar to Algorithm B.1, and was proposed by Han et al. (2006). The only modification from Algorithm B.1 is the set of positions the relays are placed at, thus affecting the weight function in Step 2. In the algorithm for the heterogeneous model, if two nodes i, j with transmission ranges T_i, T_j are more than $\min\{T_i, T_j\}$ apart (let the distance be l_{ij}), we use Algorithm C.1 to place relays for connecting them. For each pair of mesh nodes, we assign the number of relays used as the edge weight that is used in Step 2 of Algorithm B.1.

Algorithm C.1: Relay placement in heterogeneous networks

1. Let $T_i \leq T_j$. Place one relay r at distance $\min\{T_i, \gamma\}$ from i.
2. If $l_{rj} > \min\{T_i, \gamma\}$, place one relay r' at distance $\min\{T_i, \gamma\}$ from j.
3. Place $\lceil l_{rr'}/\gamma \rceil - 1$ relays at equal spacing between the two relays r, r'.

Han et al. (2006) prove the algorithm to be a $c((8\gamma^2 + 1/4)k^2 + 3k/2 + 2)$-approximation for partial *k-vertex* connectivity in the Euclidean plane. Here, c is the approximation ratio of the best performing algorithm for computing a *k-vertex* connected subgraph of a graph. The authors also provide approximation analysis for nodes distributed in metric spaces with dimension d. They prove the algorithm for partial *k-vertex* connectivity is an $O((2\sqrt{d}\alpha)^d k^2)$-approximation. Authors also provide approximation results for full *k-vertex* connectivity and for the model where links can be unidirectional.

Kashyap (2006) analyzes the algorithms for 2-vertex and k-edge connectivity. We prove the algorithm to be a $2(5 + 11 \lceil \log_{\sqrt{3}} \min\{\alpha, \gamma\} \rceil + 5 I_{\lceil \log_{\sqrt{3}} \gamma \rceil > \lceil \log_{\sqrt{3}} \alpha \rceil})$-approximation for 2-vertex connectivity in the Euclidean plane. For k-edge connectivity, we prove the algorithm to be a $2(5 + 11 \lceil \log_{\sqrt{3}} \min\{\alpha, \gamma\} \rceil + 5 I_{\lceil \log_{\sqrt{3}} \gamma \rceil > \lceil \log_{\sqrt{3}} \alpha \rceil}) \lceil k/2 \rceil$-approximation. We present similar results for full connectivity as well. The bounds are much tighter than the bounds provided in the work of Han et al. (2006).

4.4 Ensuring Connectivity in Wireless Mobile Networks

A number of factors affect the network connectivity in WM²Nets (Aggélou, 2004). Antenna gain, antenna pattern, mobility behavior of mobile units, battery lifetime, transmitted power, receiver sensitivity, and nodal populations are some to name a few. To enable nodes to react prior to the occurrence of a network disconnection, mobiles shall be capable of determining their own connectivity in an ad hoc fashion. The simplest approach for achieving this goal is for mobile radios to periodically transmit a small (in size) packet, often called a "Beacon" (Aggélou, 2004). On its reception, this is used as an indication of connection existence. Specifically, within a specified time frame, nodes keep track of the number of packets that are successfully received out of a number of transmitted ones, and use that ratio to determine per-neighbor (link) or per-destination (end-to-end) connection quality. If this ratio is high enough in both directions, the link is then declared up or *good*; otherwise, it is marked down or *bad*. Links can also be marked bad if too many retransmissions are required to send a packet to a neighbor. Enhancements to this procedure include measuring signal strength or signal-to-noise (SNR) ratio, using overhead traffic, and so on. This tactic requires the packet radios to determine which nodes can hear them and which nodes they can hear. Evidently, for a highly populated network the control and maintenance of information on radio connectivity is a very tedious procedure, increasing processing overheads and battery usage, both being prohibiting factors in networks with battery-powered wireless units where battery scavenging is a critical design parameter.

To this avail, limited battery power restricts the communications coverage as well as the operation time of mobile devices. Below some critical threshold for remaining battery power, a node will not be able to function as a router[3], thus immediately affecting the network connectivity, possibly isolating one or more segments of the network. Fewer routers almost always mean fewer routes and therefore increased likelihood of degraded performance in the network. In fact, communication becomes meaningless if a node is not able to communicate owing to low battery power. Since exchange of messages necessarily means power consumption, many ad hoc networking mechanisms, especially routing and security protocols, explicitly include minimal power consumption as a design objective [see Chapter 4 in Aggélou, 2004 for a complete detailed description].

Indeed, power control impacts on the links/paths employed, since the power level dictates what links are available for routing and, vice versa, the power control protocol needs connectivity information, which is provided by the routing layer. However, an excessively high power level will cause excessive interference as seen in Fig. 13.3a (see also Adler and Scheideler, 2000; Li et al., 2001; Kawadia and Kumar, 2003). This reduces the traffic carrying capacity of the network in addition to reducing battery life. On the other hand, in Fig. 13.3b, having a very small power level results in fewer links and hence network partitioning effects may occur. When the power level is just right, the network is still connected and there is no excessive interference as shown in Fig. 13.3c. This mutual dependence of power control and connectivity motivates the need for a joint solution for power control and routing, or else *power-aware routing*.

---

[3] The limitations on power consumption imposed by portable wireless radios result in a node transmission range that is typically small relative to the span of the network. Consequently, similar to mobile ad hoc networks, in WM^{2}Nets mobile terminals do not always have direct radio links to all the radio terminals in the network; this implies that terminals must communicate with each other either directly or indirectly, using relaying stations via intermediate mobile hosts. Therefore, nodes are also acting as routers and dynamically establishing communications amongst themselves to form an infrastructure-less wireless network.

4.5 Network Partitioning versus Network Disconnection

Network partitioning is the physical or logical grouping of nodes into different physical or logical zones. Communication between two nodes of the same or different partitions is established either directly (one-hop) or indirectly (multihop). Figure 4.13a illustrates a network with a single partition. If all nodes can communicate with each other (via single- or multihop links), we can assume without loss of generality that the network has a single partition that itself contains several groups of nodes. This is illustrated in Fig. 4.13b where the network has a single partition (full network connectivity) that contains three groups of nodes. If some nodes or group of nodes cannot communicate with the rest of the network, either directly or indirectly, we refer to this situation as *network disconnection*. The main causes of network disconnection in wireless networks include the mobility behavior of nodes, their operational limitations (e.g., battery lifetime, mobile's antenna gain and transmission power), channel conditions, and others. With disconnected partitions being a reality in wireless mobile networks, upper layer routing and other applications, involving nodes in separate partitions, are severely disrupted and may terminate if the partitions do not merge in time. Such situation is unacceptable in mission-critical network applications where every node must receive a certain level of quality of service (QoS) or have constant access to an important information repository.

To provide therefore for QoS guarantees in WM^{2}Nets, it is imperative for communication protocols to be capable of predicting future network states, including the possibility of occurrence of network disconnection. Indeed, a communication protocol being aware of potential future disconnection could adapt the client and/or the server behaviors in order to ensure continuous operation. In addition, predicting *the timing* of the partitioning can further improve the efficiency and performance of the applications and system's per se.

To illustrate all these considerations with an example, let us consider a mission-critical service in ad hoc networks. The service can be a critical information database or a web server that must be accessible from all mobile nodes in the network. The service runs on a single mobile node referred to as the server node. To guarantee continuous service availability, instances of the service can be dynamically replicated on any other mobile unit. Figure 4.13 illustrates the progression of network partitioning. At time t, there is full (direct or indirect) connectivity in the network (single partition). The server node provides its services to all mobile users in the network. As time evolves, mobiles are freely moving about and at time $t + 1$, three partitions are formed (P_1, P_2, and P_3) with full (direct or indirect) connectivity.

At time $t + \tau$, however, partition P_3 disconnects from the rest of the network, as the physical distances of all nodes in P_3 to the closest node(s) in P_1 (node N_1) and P_2 (node N_2), that is, d_2' and d_2, respectively, are greater than the minimum communication threshold (say d_0) beyond which a transmitted signal is unsuccessfully received. In order to continue the service in partitions P_1 and P_2, the service must be replicated onto the departing partition at any time earlier than $t + \tau$ but also later than $t + 1$. The server node executes the partition prediction algorithm at periodical intervals and produces estimations on the time of separation. Clearly, if the estimated time is smaller than the actual separation time ($t + \tau$), the server node timely replicates the service to N_1 and N_2 and the network survives from a critical strike on upper layer protocols.

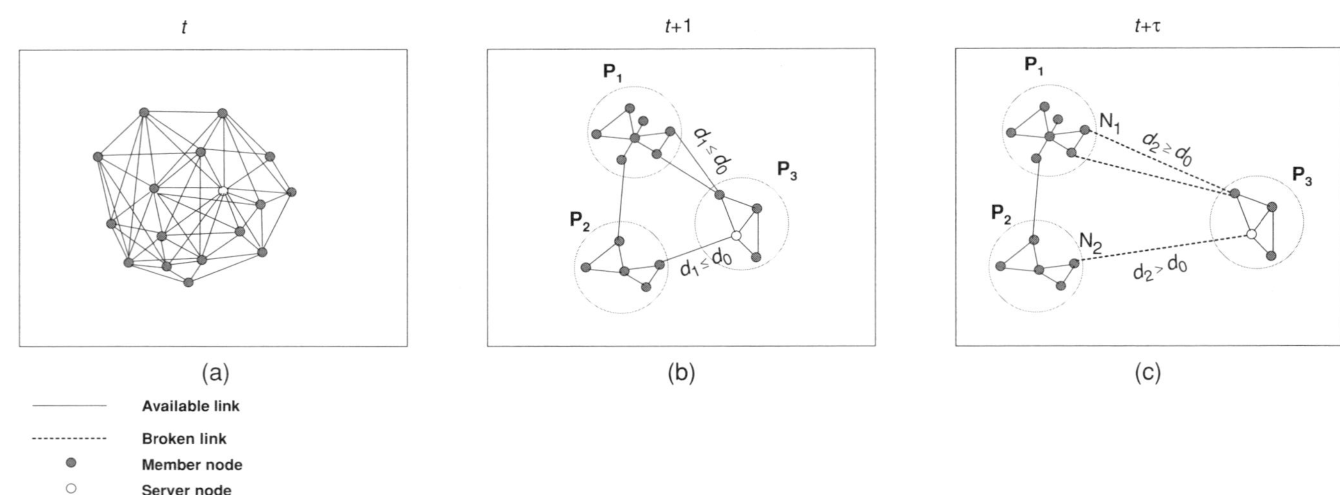

Figure 4.13 Progression of network partitioning. (a) Single partition—no disconnection. (b) Three partions—no disconnection. (c) Three partitions—disconnection.

4.5.1 Forecasting Network Disconnections in Mobile Wireless Mesh Networks[4] (Aggélou, 2005)

4.5.1.1 Preliminaries

To tackle the network disconnection problem, the author has recently proposed a methodology for estimating future connectivity states of a time-varying topology. The proposed methodology combines features from self-learning techniques inspired from pattern recognition practices and the theory of time-series. More specifically, the proposed mechanism is based on a two-level approach: first, identify and characterize the movements of mobile users; use then this characterization to quantitatively model the topology changes. Second, based on the topology-changing pattern, a time series-based classifier derives statistical data on future network connectivity states.

Applying the proposed mechanism to the application scenario depicted in Fig. 4.13, the server node executes the partition prediction algorithm at periodical intervals and produces estimations on the time of separation. Should the estimated time be smaller than the actual separation time $(t + \tau)$, the server node replicates timely the service to N_1 and N_2; hence, the network survives from a critical strike on upper layer protocols.

4.5.1.2 Related Work on Network Disconnection Prediction

Network disconnection is a wide-scale network problem where user-dependent as well as network-specific parameters are two catalytic factors in the formation of a partial or full network disconnection in WM²Nets. Node mobility and network partitioning make intermittent network disconnections a reality.

Park and Corson (1997) use a method to detect network partitions after they occur. The aim of this detection is to find the nodes that are no longer reachable and thus, erase the deprecated routes that lead to them. Shah et al. (2001)_aimed at enhancing data access in an ad hoc network by detecting partitions prior to their occurrence. Their proposed method is based essentially on a data replication mechanism. Every node embeds a positioning system (such as GPS) and by successive measures computes its velocity. Regularly, it spreads that information to the other nodes. Thus, each node knows the behavior of the other members of this group. So, they can *predict* when a node storing a particular data will leave the group before it effectively does so. At this point, the owner of the data elects a node of the group to be another host of the data and replicates it on this node. The main advantage of this method is that each node knows exactly *when* the partition occurs if node movements are almost regular. On the other hand, this method has two main disadvantages. First, it requires a positioning system, which is often expensive and bulky. Second, the exchange of position and speed information generates a continuous and relatively high network load. Moreover, as the trajectory of nodes is interpolated, this involves a non-negligible computation. Wang and Li (2002) propose a similar partition prediction, although their solution is more centralized. It is based on an extension of the reference point group mobility (RPGM) by Hong et al. (1999). They extend this model to handle the velocity of nodes and thus regroup nodes according to their position and speed. For achieving this, each node sends its position and velocity to a server. Then, this server runs a sequential clustering algorithm from the

[4] George Aggelou and *Nikos Argyreas, "Forecasting network disconnections in mobile wireless mesh networks" (*Institute of Informatics and Telecommunications, NCSR Demokritos Research Center, Athens, Greece).

field of pattern recognition (Friedman and Kandel, 1999) to regroup the nodes. Then, as the server knows the groups' position and velocity it informs nodes about a future network partition. This method has the same problems with those provided in the study of Shah et al. (2001) and additionally requires a centralized server.

Several other solutions that tackle the disconnected problem exist in the literature (e.g., Karumanchi et al., 1999; Wang and Li, 2002; Li and Rus, 2000, 2002; Goyal and Caffery, 2002; Davis et al., 2001; Chatzigiannakis et al., 2001). Karumanchi et al. (1999) utilized designate servers and a quorum-based scheme for updating and querying information. The work by Wang and Li (2002) presented run-time algorithms that dynamically place servers during the network partitioning (Wang and Li, 2002). Li and Rus (2000, 2002) intermediate nodes between the source and the destination of data and automatically change their trajectories to avoid unpredictable transmission delays. The approach minimizes the degree of changed trajectories for the mobile nodes. Goyal and Caffery (2002) used the depth first search approach to detecting critical links of the network and preventing degraded network performance. There is an essential issue appeared in these schemes. Forcing a mobile node to deviate its trajectory for reinforcing critical links for other connections could break more significant links. In order to detect network partitions, each node has to maintain global information. Davis et al. (2001) described a movement and communication model where data packets can be carried by the mobile nodes across the different partitions. Predefined infrastructure consists of the stationary nodes and the mobile nodes. Mobile nodes carry the packets sent by a stationary source and move to a stationary destination located in another partition. Each mobile node has a finite-sized buffer and drops packets based on their proposed dropping strategies. Chatzigiannakis et al. (2001) designed a communication strategy to enable a group of mobile nodes to move as a "snake" around the network. When some nodes in the group connect to the source, the source will deliver the packets to a member of the group. The group forwards the packets to the destination if the destination is reachable.

All the approaches illustrated above can be broadly classified into three categories. First, those where services or data are replicated to every network partition so the mobile nodes can obtain services even if the server nodes are not within the same network partition. Second, mobile nodes that have formed a critical link need to modify their trajectories for strengthen the link or reinforce the link by asking other nodes to change their locations. Third, mobile nodes can physically carry information across the network partitions. The nodes carry the data packets to the destination or move around the network in order to provide data to needed nodes. However, using the network internal capability to handle the disconnected problem could interfere with normal operations and network performance.

Although those algorithms demonstrate good performance results, their operation relies heavily upon either expensive and bulky hardware (GPS), or mobility prediction schemes, which however can predict link availability that is caused by *local* topology changes, but fall short to make any predictions on *global scale* topology changes such as *network partitionings*.

4.5.1.3 Introductory Concepts

The proposed framework assumes cluster-based network architectures, in which network nodes are partitioned into groups called *clusters* (see Section 2.5). Clustering is a technique commonly used in pattern recognition problems, where we want to train a machine to recognize something and to classify it somewhere.

The following paragraphs summarize several properties and definitions of clustering.

4.5.1.4 Clustering Criteria and Clustering Algorithms

For a given set of nodes (vectors), to determine the cluster to which it is more likely to belong, a cluster analysis is first applied. To develop a clustering task, two steps are essential:

- *Clustering criterion*. Different criteria may be used given the type of clusters that are expected to represent the data set. For example, a compact cluster of vectors (see Fig. 4.15) in the l-dimensional space may be sensible according to one criterion, whereas an elongated cluster may be sensible according to another. The clustering criterion may be expressed via a cost function or some other types of rules.
- *Clustering algorithms*. Having adopted a clustering criterion and a vector assignment method, often called proximity measure, this step refers to the choice of a specific algorithmic scheme that unravels the clustering structure of the data set.

Different choices of clustering criteria and clustering algorithms may lead to thoroughly different clustering results. To demonstrate this, let us consider the following example. Consider Fig. 4.14. How many "sensible" ways of clustering can we obtain for these points? The most "logical" answer seems to be two. The first clustering contains four clusters (surrounded by dashed circles). The second clustering contains two clusters (surrounded by solid circles). Which clustering is "correct"? It seems that there is no definite answer. Both clusterings are valid. Evidently, when additional constraints are considered, the system will be biased to the most sensible one.

Definition of Clustering Algorithm Let $X = \{x_1, \ldots, x_N\}$ be the vector (data) set. An *m-clustering* of X is defined as the partition of X into *m-sets* (*clusters*), $C_1, \ldots, C_m$, such that the following three conditions are met:

1. $C_i \neq \emptyset, i = 1, \ldots, m$
2. $\bigcup_{i=1}^{m} C_i = X$
3. $C_i \cap C_j = \emptyset, i \neq j, \quad i, j = 1, \ldots, m$

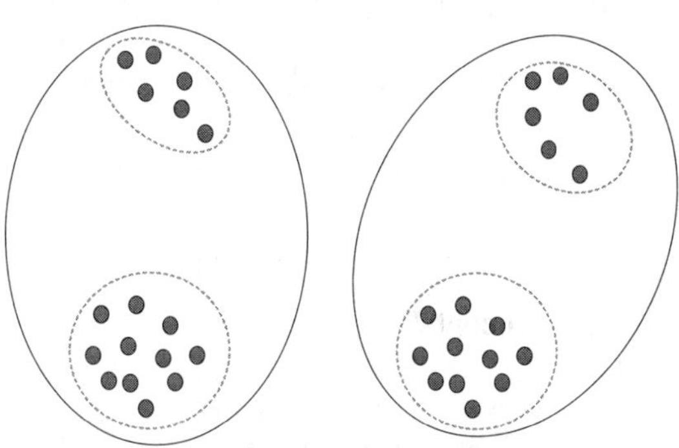

Figure 4.14 A coarse clustering of the data results in two clusters, whereas a finer one results in four clusters.

Under these definitions, each vector belongs to a single cluster. This type of clustering is sometimes called *hard* or *crisp*. An alternative definition is in terms of the *fuzzy sets*, introduced by Zadeh (1965). A fuzzy clustering of X into m clusters is characterized by m functions u_j where

$$u_j : X \rightarrow [0, 1], \quad j = 1, \ldots, m \tag{4.5}$$

and

$$\sum_{j=1}^{m} u_j(x_i) = 1, \quad i = 1, 2, \ldots, N, \quad 0 < \sum_{j=1}^{N} u_j(x_i) < N, \quad i = 1, 2, \ldots, m \tag{4.6}$$

These are called *membership functions*. The value of a fuzzy membership function is a mathematical characterization of a set; that is, a cluster in our case, which may not be precisely defined. That is, each vector x belongs to more than one cluster simultaneously "up to some degree," which is quantified by the corresponding value of u_j in the interval [0, 1]. Values close to unity show a high "grade of membership" in the corresponding cluster whereas values close to zero a low grade of membership. The values of these membership functions are indicative of the structure of the data set, in the sense that if a membership function has close to unity values for two vectors of X, that is, x_k, x_n, they are considered similar to each other.

The right condition in Eq. (4.6) guarantees that there are indeed cases where clusters exist with no vectors shares. This is analogous to the condition $C_i \neq \emptyset$ of the aforementioned definition. The definition of clustering into m distinct sets C_i can be recovered as a special case of the fuzzy clustering if we define the fuzzy membership functions u_j to take values in [0, 1], that is, to be either 1 or 0. In this case, each data vector belongs exclusively to one cluster and the membership functions are now called *characteristic functions* (Klir and Yuan, 1995).

In this study, the hard type of clustering is considered.

Definition of Proximity Measures

Proximity Measures Between a Vector and a Set In many clustering schemes, a vector x is assigned to a cluster C taking into account the proximity between x and $C, d(x, C)$. There are two general directions for the definition of $d(x, C)$. According to the first one, all points $y \in C$ contribute to $d(x, y)$. Typical examples of this case include:

- The *max proximity function:* $d_{\max}^{\text{ps}}(x, C) = \max_{y \in C} d(x, y)$
- The *min proximity function:* $d_{\min}^{\text{ps}}(x, C) = \min_{y \in C} d(x, y)$
- The *average proximity function:* $d_{\text{avg}}^{\text{ps}}(x, C) = \dfrac{1}{n_C} \sum_{y \in C} d(x, y)$

According to the second direction, a cluster C is equipped with a representative and the proximity between x and C is measured as the proximity between x and the representative of C. Many types of representatives are available in the literature. Among them, the point, the hyperplane, and the hypersphere are most commonly used. Point

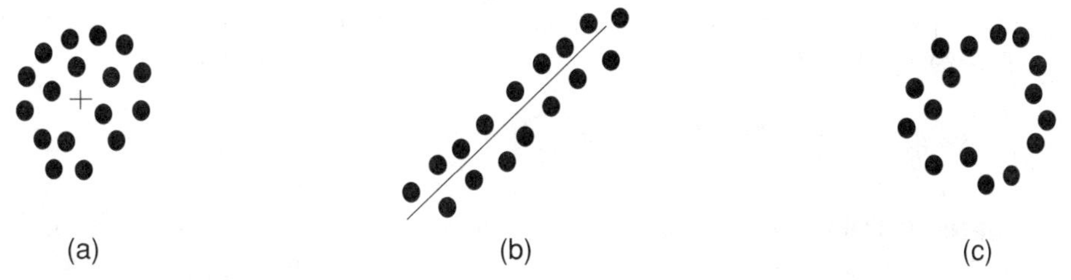

FIGURE 4.15 (a) Compact cluster; (b) hyperplanar (linear) cluster; (c) hyperspherical cluster. (Theodoridis and Koutroumbas, 1998.)

representatives are suitable for compact clusters (Fig. 4.15a) wheres hyperplane (hyperspherical) representatives may be used for clusters of linear shape (Fig. 4.15b) and hyperspherical shape (Fig. 4.15c).

For completeness sake, an overview of these candidate schemes is presented below. The reader can find a more in-depth description in the work (Chapter 14) of Theodoridis and Koutroumbas (1998).

Point Representatives Typical choices for a point representative of a cluster are:

- The *mean vector* $m_p \in C$ defined as

$$m_p = \frac{1}{n_c} \sum_{y \in C} y \tag{4.7}$$

  where n_c is the cardinality of C. This is the most common choice when point representatives are employed and we deal with data of a continuous space. However, it may not work well when we deal with points of a discrete space F^l. This is because it is possible for m_p to lie outside F^l. To cope with this problem, we may use the mean center m_c of C, which is defined next.
- The *mean center* $m_c \in C$ defined as

$$\sum_{y \in C} d(m_c, y) \leq \sum_{y \in C} d(z, y), \forall z \in C \tag{4.8}$$

  where d is a proximity measure between two points.

Hyperplane Representatives Linear shaped clusters (or hyperplanar in the general case) cannot be accurately represented by a single point. In such cases, we use lines (hyperplanes) as cluster representatives (e.g., Duda et al., 2000).

The general equation of a hyperplane H is

$$\sum_{j=1}^{l} a_j x_j + a_0 = a^T x + a_0 = 0 \tag{4.9}$$

where $x = [x_1, \ldots, x_1]^T$ and $a = [a_1, \ldots, a_1]^T$ is the weight vector of H. The distance of a point x from H is defined as

$$d(x, H) = \min_{z \in H} d(x, z) \tag{4.10}$$

Hyperspherical Representatives For hyperspherical clusters, the ideal representative is a circle (hypersphere). The general equation of a hypersphere Q is:

$$(x - c)^T (x - c) = r^2 \tag{4.11}$$

where c the center of the hypersphere and r its radius. The distance from a point x to Q is defined as

$$d(x, Q) = \min_{z \in Q} d(x, z) \tag{4.12}$$

Figure 4.16 provides geometric insight into this definition. However, other nongeometric distances $d(x, Q)$ have been used in the literature.

Proximity Functions Between Two Sets Most of the proximity functions d^{SS} used for the comparison of sets are based on proximity measures, d, between vectors. If D_i, D_j are two sets of vectors, the most common proximity functions are:

- The *max proximity function*: $d^{SS}_{\max}(D_i, D_j) = \max_{x \in D_i, y \in D_j} d(x, y)$
- The *min proximity function*: $d^{SS}_{\min}(D_i, D_j) = \min_{x \in D_i, y \in D_j} d(x, y)$
- The *average proximity function*: $d^{SS}_{\text{avg}}(D_i, D_j) = \frac{1}{n_{D_i}, n_{D_j}} \sum_{x \in D_i} \sum_{y \in D_j} d(x, y)$ where n_{Di} and n_{Dj} are the cardinalities of D_i and D_j, respectively.
- The *mean proximity function*: $d^{SS}_{\text{mean}}(D_i, D_j) = d(m_{D_i}, m_{D_j})$ where m_{Di} is the representative of D_i, $i = 1, 2$.

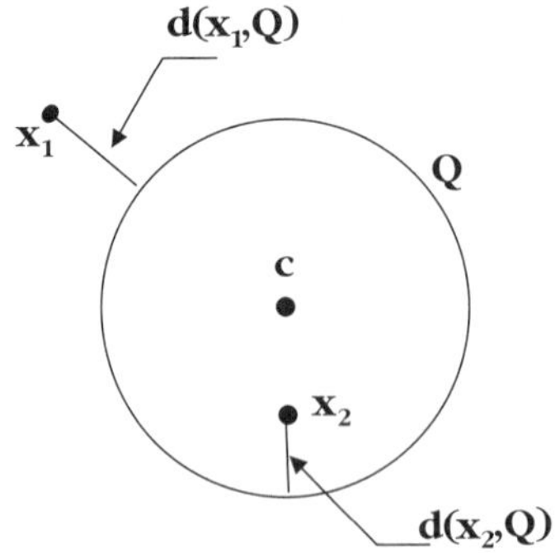

FIGURE 4.16 Distance between a point and hypersphere.

- Another proximity function is based on the mean proximity function and is defined as[5]

$$d_e^{SS}(D_i,\ D_j) = \sqrt{\frac{n_{D_i} n_{D_j}}{n_{D_i} + n_{D_j}}} d(m_{D_i},\ m_{D_j})$$

where m_{Di} is defined as in the previous case.

Intuitively, different choices of proximity functions between different data (vector) sets may lead to thoroughly different clustering results. Moreover, if we use the same proximity function between sets but different proximity measures between points, this will, in general, lead to different clustering results.

4.5.1.5 Proposed Framework for Estimating Future Network Connectivity States

The proposed framework for network disconnection prediction in a clustered ad hoc network comprises two generic elements: a clustering algorithm, called the wireless mesh sequential algorithmic scheme (WiMeSAS), and the core mechanism that produces the estimations on network disconnections, called the network disconnection prediction algorithm (NPDA). These are further analyzed in the following paragraphs.

The WiMeSAS In this section the WiMeSAS clustering algorithm is presented. WiMeSAS belongs to the family of sequential algorithms, which, given a certain proximity measure, tend to produce single compact and hyperspherically or hyperellipsoidally shaped clusters (Theodoridis and Koutroumbas, 1998). WiMeSAS is a generalization of the basic sequential algorithmic scheme (BSAS) presented in the work of Theodoridis and Koutroumbas (1998). Let $d(x, C)$ denote the distance between a vector x and a cluster C. This may be defined considering either all vectors of C or a representative vector of it. According to BSAS, the user-defined parameters required by the algorithmic scheme are the threshold distance Θ, which is the maximum distance for successful communication between two clusters, and the maximum allowable number of clusters, q.

The basic idea of the vector assignment in clusters according to the BSAS algorithm is as follows (see also Fig. 4.17a): each new vector is assigned either to an existing cluster or to a newly created one, depending on its distance from the already formed clusters. Therefore, a decision for the vector x is reached prior to the final cluster formation, which is determined after all vectors are presented.

A refinement of BSAS, called Modified BSAS (MBSAS), overcomes this drawback. The algorithmic scheme consists of two phases (see Fig. 4.17b): the first phase involves the determination of the clusters via the assignment of *some* of the vectors to them. During the second phase, the unassigned vectors are presented for a second time to the algorithm and are assigned to the appropriate cluster. The number of clusters is determined in the first phase. The decision taken during the second phase for each vector accounts all clusters. The cost paid in MBSAS is that all vectors need be accounted for twice. In addition, similar to BSAS, MBSAS is sensitive to the order in which the vectors are presented.

[5] This definition is a generalization of that given in Chapter 13 of the work by Theodoridis and Koutroumbas (1998).

Basic sequential algorithmic scheme (BSAS)

- $m = 1$
- $C_m = \{x_1\}$
- For $i = 2$ to N
  - Find $C_i : d(x_i, C_i\} = \min_{i<j<m} d(x_i, C_j\}$
  - If $d(x_i, C_k\} > \Theta)$ AND $(m < q)$ then
    - * $m = m+1$
    - * $C_m = \{x_i\}$
  - Else
    - * $C_i = C_i \cup \{x_i\}$
    - * Where necessary, update representatives $C_i = C_i \cup \{x_i\}$
  - End {if}
- End {For}

(a)

Modified basic sequential algorithmic scheme (MBSAS)

Cluster Determination

- $m = 1$
- $C_m = \{x_1\}$
- For $i = 2$ to N
  - Find $C_i : d(x_i, C_i\} = \min_{i<j<m} d(x_i, C_j\}$
  - If $d(x_i, C_k\} > \Theta)$ AND $(m < q)$ then
    - * $m = m+1$
    - * $C_m = \{x_i\}$
  - End {if}
- End {For}

Pattern Classification

- For $i = 1$ to N
  - If x_i has not been assigned to a cluster, then
    - * Find $C_k : d(x_i, C_k\} = \min_{i<j<m} d(x_i, C_j\}$
    - * $C_k = C_k \cup \{x_i\}$
    - * Where necessary, update representatives
  - End {if}
- End {For}

(b)

FIGURE 4.17 Pseudocodes for BSAS and MBSAS algorithms.

In contrast to BSAS and MBSAS, the proposed WiMeSAS algorithm overcomes these drawbacks. Moreover, in WiMeSAS, the number of clusters is not known a priori whereas new clusters are created as the algorithm evolves. In addition, the proximity measure, $d(x, C)$, in WiMeSAS is defined by taking into account all vectors of C and not a representative vector of it (see Definition 1 below).

The following definitions are necessary prior to describing WiMeSAS. In the following text, a two-dimensional vector set $\boldsymbol{S} = \{x_1, x_2, \ldots, x_n\}$ in space $\Re^D$ is assumed. **P(S, D)** denotes a partition (D) in vector space S, $\boldsymbol{C} = \boldsymbol{C(P)} = \{C_i, \ldots, C_m\}$ the set of clusters formed as a result of partition $P(S, D)$, and $\boldsymbol{M} = \boldsymbol{m[P(S, D)]}$ the cluster population of C as a result of partition $P(S, D)$.

Definition 1: Definition of Proximity Measure: We define as the proximity measure of a vector x from a cluster C, the minimum distance of x from any vector of C, such that $\mathrm{d}(x, C) = \min\{d(x, x_i), x_i \in C\}$

Definition 2: Definition of Distance Between Two Clusters: We define as the distance between two clusters, C_a and C_b, the minimum distance between any two vectors from C_a and C_b. That is, $d(C_a, C_b) = \min\{d(x_i, x_j), x_i \in C_a, x_j \in C_b\}$. Hence, for any D-partition and any two clusters, C_a and C_b , it holds $d(C_a, C_b) > D$.

Definition 3: Definition of m-Clustering: We define as an m-clustering of S, the partition of $S, P(S, D)$, into m subsets (clusters), $C_1, \ldots, C_m$, with $m \leq n$, where n is the population of vector set, such that the following three conditions are met:

D.4.1. $\quad C_i \neq \emptyset, i = 1, \ldots, m$

D.4.2. $\quad \bigcup_{i=1}^{m} C_i = S$

D.4.3. $\quad C_i \cap C_j = \emptyset \quad \forall i \neq j, \quad i, j = 1, \ldots, m$

D.4.4. $\quad \exists D \in \Re : \forall x_i \in C_k, \exists x_j \in C_k : d(x_i, x_j) \leq D, 1 \leq k \leq m,$

Conditions (D.4.1) to (D.4.3) constitute the classic definitions of an m-clustering formation (Theodoridis and Koutroumbas, 1998).

Condition (D.4.4) assures that the minimum distance between any two vectors of same cluster is smaller than or equal to D.

From condition (D.4.4), it holds that

D.4.5. $\quad x \notin C_k \Leftrightarrow \forall x_i \in C_k : d(x, x_i) > D$

D.4.6. $\quad \forall x \in C_i, \forall y \in Cj : d(x, y) > D$

Inequality (D.4.5) assures that the distance of any vector x in S to any vector of C_k is greater than D if and only if x is not a member of any cluster C_k. Similarly, condition (D.4.6) indicates that the proximity measure of any two vectors of different clusters is greater than D.

In the following, the WiMeSAS algorithm is illustrated.

The WiMeSAS Algorithm: Let C' denote the set of vectors in S that are not assigned to any of the existing m-clusters, such that $C' = \{x \in S, x \notin C_i, i = 1, \ldots, m\}$

```
m = 0
WHILE (S − C' ≠ ∅ )
   Choose a random vector x, such that x ∈ S − C
   C_m = {x}
   m = m + 1
   REPEAT
      FOR i = 1 to n
         IF x_i ∉ C AND d(x, C_m) ≤ D THEN
            C_m = C_m ∪ {x}
            ClusterChanged = True
         ELSE
            ClusterChanged = False
         END (IF)
      END (FOR)
   UNTIL (ClusterChanged = False)
END (WHILE)
```

Proof of Correctness: Conditions (D.4.1) to (D.4.3) are obviously true. We examine condition (D.4.4):

For $m = 0$ (creation of first cluster), condition (D.4.4) holds true. This is so because in this stage $C = C_1$.

Hence,

$\mathbf{x} \in S - C$ if and only if $d(x, C_1) > D$ and $x \in C_1$ if and only if $d(x, C_1) \leq D$.

Let us assume that condition (D.4.4) holds also true for $m = k$. We shall then prove that condition (D.4.4) holds also true for $m = k + 1$:

$x \in S - C$ if and only if $d(x, C_{k+1}) > D$ and $x \in C_{k+1}$ if and only if $d(x, C_{k+1}) \leq D$. ■

WiMeSAS Theorem 1: For a vector set S, let us assume two partitions of it, $P(S, d_1)$, $Q(S, d_2)$, with a set of clusters $C(P) = \{C_i, \ldots, C_m\}$ and $C(Q) = \{C'_i, \ldots, C'_m\}$, respectively. If it holds that $d_1 < d_2$, we then conclude that each cluster created from partition P will constitute a subgroup of some cluster created from Q

$$\forall C_a \in C(P), \exists C'_a \in C(Q) : C_a \subseteq C'_a$$

Proof: We will prove that condition $\forall C_a \in C(P), \forall C'_a \in C(Q) : C_a \not\subset C'_a$ does not hold true.

Assuming that this condition is true, this implies that $\exists x, y \in C_a, \exists C'_b, C'_c \in Q : x \in C'_b, y \in C'_c$. From this, it is implied that $d(x, y) > d_2 > d_1 \Rightarrow x, y \notin C_a$, which according to assumptions this cannot be true. Thus Theorem 1 holds true under all cases. ■

WiMeSAS Theorem 2: For a vector set S, let us assume two partitions of it $P(S, d_1)$ and $Q(S, d_2)$. If the populations of P and Q are m_1 and m_2, respectively, it holds that

$$d_1 < d_2 \Rightarrow m_1 \geq m_2$$

Proof: It is evident that the clusters $C(Q)$ from partition $Q(S, d_2)$ have a distance greater than d_2. From the assumption, it is thus true that they will also be at distance greater than d_1. That is,

$$\forall C_i, C_j \in C(Q) \Rightarrow d(C_i, C_j) > d_2, d_2 > d_1 \Rightarrow d(C_i, C_j) > d_1$$

Hence, $C_i, C_j \in C(P)P(S, d_1)$. Therefore, m_1 is at least equal to m_2. Hence, for each cluster $C_i \in C(Q)$ of $Q(S, d_2)$ is an individual set and can be partitioned into a number of subclusters from partition $P(S, d_1)$. It then holds that $m_2 \leq m_1$. ■

WiMeSAS Theorem 3: For a vector set S with two partitions $Q(S, d)$, $P(S, d)$, it holds $Q \equiv P$.

Proof: We consider two cases: when the vectors belong to the same cluster of partition P and when they do not belong to the same cluster of partition Q.

Let us assume that they belong to the same cluster of partition P. It is then true that $\exists C_a \in C(P) : x, y \in C_a \Rightarrow d(x, y) \leq d$. Let us now assume that they do not belong to the same cluster of partition Q. It is then true that $\exists C'_i, C'_j \in C(Q) : x \in C'_i, y \in C'_j, C'_i \neq C'_j$. This implied that $d(x, y) > d$, which can not be true. ■

General Remarks on WiMeSAS Algorithm It is evident that the WiMeSAS algorithm does not use point representatives as a proximity measure between two clusters. The reason the point representative method is not favored in WiMeSAS is twofold: (1) it turns out that sequential clustering algorithms with point cluster representatives favor compact clusters provided in the work (Chapter 12) of Theodoridis and Koutroumbas (1998) for discussion). Since there is no a priori evidence on what type of clusters will be eventually formed throughout an experimental study, given a certain network setup, the use of cluster representatives would confine the formation of different types of clusters, and (2) clustering algorithms with a single vector are known to be based on global clustering criteria whereas algorithms where all vectors are used for its representation are known to be based on local clustering criteria. Given that network disconnection is a wide-scale topology change, it is more appropriate to consider the global clustering criteria.

Furthermore, whereas the results of BSAS and MBSAS are strongly biased on the order in which the vectors are presented to the algorithm (see Fig. 4.17), it is easy to realize that the order in which the vectors are presented to the WiMeSAS algorithm (line 3 in WiMeSAS pseudocode) does not impact its clustering results in terms of the number and size of clusters. Definition 3 ensures this observation.

Network Disconnection Prediction of a Time-Varying Vector Set In general, the design of a protocol for use in wireless mobile networks is predicated upon two basic tenets: (1) optimality is inherently difficult or impossible to achieve in highly dynamic environments, given the restrictions imposed from the wireless physical media, and (2) to ensure acceptable levels of performance in these environments, efficiency is considered more important than optimality. Consequently, with disconnection being a reality in wireless mobile networks, the overriding design principle of any communication protocol would be its ability to evaluate network connectivity as a function of time whereas to the expense of potentially higher communication overhead and/or communication delays.

With respect to the design trade-off stressed above, two observations can be made for the estimation algorithm. First, no attempt is made to maintain or specify the criteria for optimal cluster organization. Essentially, any clustering algorithm could plug in and play in conjunction with the NPDA mechanism. Second, an estimation mechanism relies on a self-learning procedure whose task is to identify certain characteristics of the topology, which in fact underlie the data set.

Methodology Let $S = \{x_1, x_2, \ldots, x_n\}$ be a two-dimensional vector set in space $\Re^D$, whose position is time-varying. For n successive discrete time intervals $t_1, t_2, \ldots, t_n$, with $t_1 < t_2 < \cdots < t_n$, the vector set S is also time varying with $S(t_i)$ the vector set at time t_i. Similarly, $P(t_i, d) = (t_i, P(S, d)) \equiv P(S(t_i), d)$ is the d-partition of $S(t_i)$, and $C(t_i) = \{C_1(t_i), C_2(t_i), \ldots, C_M(t_i)\}$ is the set of clusters with population $m(t_i) = m[P(S(t_i), d)]$.

For the time-varying vector set $S(t_i)$ let us assume two partitions of it, $P(t_i, d_1) = (t_i, P(S, d_1)) \equiv P(S(t_i), d_1)$ and $Q(t_i, d_2) = (t_i, Q(S, d_2)) \equiv Q(S(t_i), d_2)$, with $d_1 < d_2$. Let us also assume that for a series of successive discrete time intervals $t_1, t_2, \ldots, t_{n-1}$, the cluster population of partition $Q(t_i, d_2)$ is unity, that is $m(t_i) = m[Q(S(t_i), d_2)] = 1$, with $1 \leq i < n$; also the network disconnects at time t_n for partition $P(t_n, d_i)$. From the latter assumption it is evident that $m(t_i) = m[P(S(t_i), d_i)] = k > 1$, for $i = n$. As a result of Theorems 1 and 2, it becomes clear then that:

(a) $m'(t_n) = \lambda > \kappa > 1$ with $m'(t_n) = m\,[P(S(t_n), d)]$, and
(b) A network disconnection is impossible to occur for partition $P(S(t_i))$ during any time interval $t_1, t_2, \ldots, t_n$. It holds true therefore that $m'(t_i) \geq m(t_i) \Rightarrow m'(t_i) \geq 1$.

In this context, we can specify an appropriate value for d, say d', with $d_1 < d' < d_2$, such that a $Q(t_i, d_2)$ network disconnection occurs at some earlier time interval from its actual time of occurrence. In other words, assuming a $Q(t_i, d')$ partition, network events are in fact $P(t_i, d_1)$ network events that are *shifted back in time by some time offset. Hence, we define a d_2' partition of smaller dimension to d_2 (i.e., $d_2' < d_2$)*, such that $Q(t_i, d_2)$ events occur at some time interval t_i, with $1 \leq i \leq n-1$. In this line of thoughts, it is possible for the algorithm to produce certain statistics related to the time of network disconnection and thus to evaluate the possibility of occurrence within a specific time window. This concept is illustrated in Fig. 4.18.

Given that $d_2' = f^* d_2 < d_2$, with $0 < f \leq 1$, the selection of an appropriate value for f demonstrates the following trade-off: for large values of f (i.e., $f \to 1$) the efficiency of the system, in terms of accuracy of prediction (prediction error), is expected to be higher, whereas the time offset for the system to react in the face of a network disconnection is relatively small. For small values of f (i.e., $f \to 0$) on the other hand, the efficiency of the system is expected to be lower (less reliable predictions), whereas the time offset for the system to react higher (earlier prediction). Let us designate this time offset as t_0.

In the following, we illustrate this trade-off with an example. Let us assume $d_0 = 120$ m. For $f = 0.991$ ($d = 119$ m), the forecast is very short-term: a distance of two vectors (or clusters) will likely increase in a very short time with no time left for the NPDA engine to react to an upcoming event at $d = 120$ m. For example, if a method can forecast an earthquake 1 s prior to its occurrence, this forecast has no effective result to the affected area.

On the other hand, for $f = 0.08$ ($d = 10$ m), the forecast is very long-term, which is desirable but, however, for two vectors (or clusters) at distance $d = 10$ m, there could be no guarantee that their distance will increase to 120 m and, if so, when will this occur. In other words, this forecast will have no accuracy. Using the earthquake example, if a

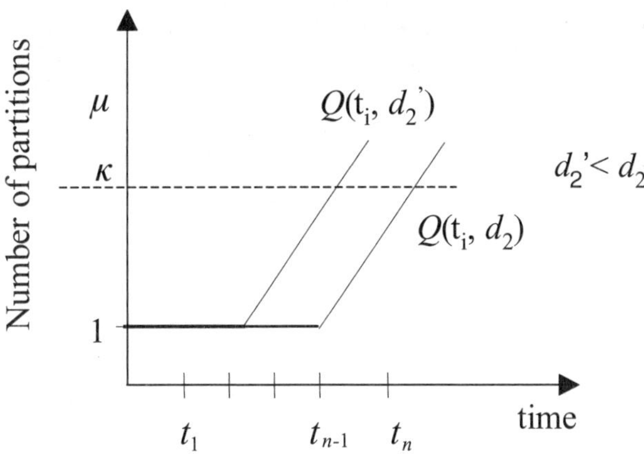

Figure 4.18 Illustration of the proposed method for network disconnection prediction.

method can forecast an earthquake 10 years prior to its occurrence, this forecast has again no effective result on the affected area.

If d' and d'' are the distance thresholds produced from the large and small values of f, respectively, the objective of the NPDA is twofold: (1) to specify an appropriate value for d, with $d' < d < d''$, that produces nearly optimum prediction figures—in other words, to find the optimum value for f—and (2) to bind the mean elapsed time offset between the estimate time of disconnection occurrence and the actual time of disconnection. Let us denote this time as $\bar{t}_0$, with

$$\bar{t}_0 = \frac{\sum_{i=1}^{n} t_0^i}{n}$$

where n is the number of events.

In essence, a finely tuned process is needed to produce prediction statistics that lie in a specific time window of events. For this purpose, a time series-based classifier is developed to produce statistics upon a certain event within a short time interval. The lower bound of the classifier is set to the estimated time, whereas its upper bound to the actual time of occurrence. In order to produce reliable statistics, the classifier engine shall be carefully fed with the appropriate data set. This is the initial data that is necessary for the training of the engine. This is further analyzed in the following paragraphs.

As aforementioned, given two partitions $P(t, d_1)$ and $Q(t, d_2)$, with $d_1 < d_2$, if a disconnection occurs for $Q(t_i, d_2) = (t_i, Q(S, d_2)) \equiv Q(S(t_i), d_2)$, a disconnection will also occur (see Theorems 1–3) for one or more partitions in $P(t_i, d_1) = (t_i, P(S, d_1)) \equiv P(S(t_i), d_1)$. In order to establish a reliable prediction, we should first register and study the *relative* formation of clusters in time. To accomplish this, the following data need be calculated for each cluster C_i in partition $P(t_i, d_1) = (t_i, P(S, d_1)) \equiv P(S(t_i), d_1)$

1. The distance (d_i) to its closest cluster; that is, $d_i = \min\{d(C_i, C_j), i \neq j\}$
2. The maximum distance ($d_{\max}$) among all d_i instances: $d_{\max} = \max\{d_i = \min\{d(C_i, C_j), i \neq j, \forall i, j\}$

In fact, this data shall comprise the initial data set to be used for the training of the classifier engine. Let us illustrate this methodology step-by-step using the example in the following table.

Timestep (t_i)	$m[P(S(t_i), d_0)]$ ($d_0 = 120$ m)	$m[P(S(t_i), d)]$ ($f = 0.9$)	Critical Distance
1	1	1	0.0
2	1	1	0.0
...	1	1	0.0
41	1	1	0.0
42	1	1	0.0
43	1	2	108.5
44	1	2	109.8
45	1	2	111.2
46	1	2	110.4
47	1	2	109.5
48	1	2	108.8
49	1	2	108.0
50	1	1	0.0
51	1	1	0.0
52	1	2	108.0
53	1	2	108.2
54	1	2	108.4
55	1	2	108.7
56	1	2	109.0
57	1	2	109.3
58	1	2	109.7
59	1	2	110.1
60	1	2	110.4
61	1	2	110.4
62	1	2	110.4
63	1	2	110.4
64	1	2	110.4
65	1	4	110.4
66	1	4	110.4
67	1	4	110.9
68	1	4	112.1
69	1	4	113.3
70	1	4	114.6
71	1	4	115.8

(*continued*)

Timestep (t_i)	$m[P(S(t_i), d_0)]$ ($d_0 = 120$ m)	$m[P(S(t_i), d)]$ ($f = 0.9$)	Critical Distance
72	1	4	117.1
73	1	4	118.3
74	1	4	119.6
75	2	2	120.9
76	2	2	122.1

The first column depicts time steps of events (0–76). The second column depicts the actual (real) partitioning of a $P(S(t_i), d_0)$ network. Distance threshold (d_0) is the distance beyond which the communication is lost (i.e., more than one clusters are produced with distance higher than d_0). As mentioned above, the value of d_0 directly affects the number of clusters formed by WiMeSAS. If d_0 is too small, unnecessary clusters will be created but the possibility of network disconnection also increases. On the other hand, if d_0 is too high, a smaller than appropriate number of clusters will be created. In both cases, improper choice of d_0 may lead to meaningless clustering results. We assume that d_0 is a system parameter and fixed.

The third column depicts the estimated partitioning of a $P(S(t_i), d)$ network using NPDA with $d = f^*d_0$. In order to forecast that communication between two points will eventually break, we observe how many clusters will be formed whose distance is smaller than 120 m, for example, 110 m. This is shown in the third column.

Whereas the information presented in the second column shows when a network disconnection will actually occur, no useful information on a potential future network disconnection can be extracted however from the third column. To this avail, a snapshot of a single run of NPDA/WiMeSAS algorithm is captured and depicted below. In this set-up, a number of sample-points were created and based on these; the clustering algorithm ran and created seven clusters.

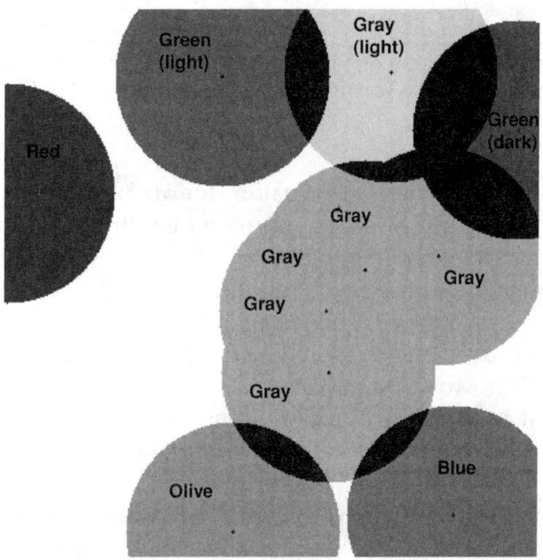

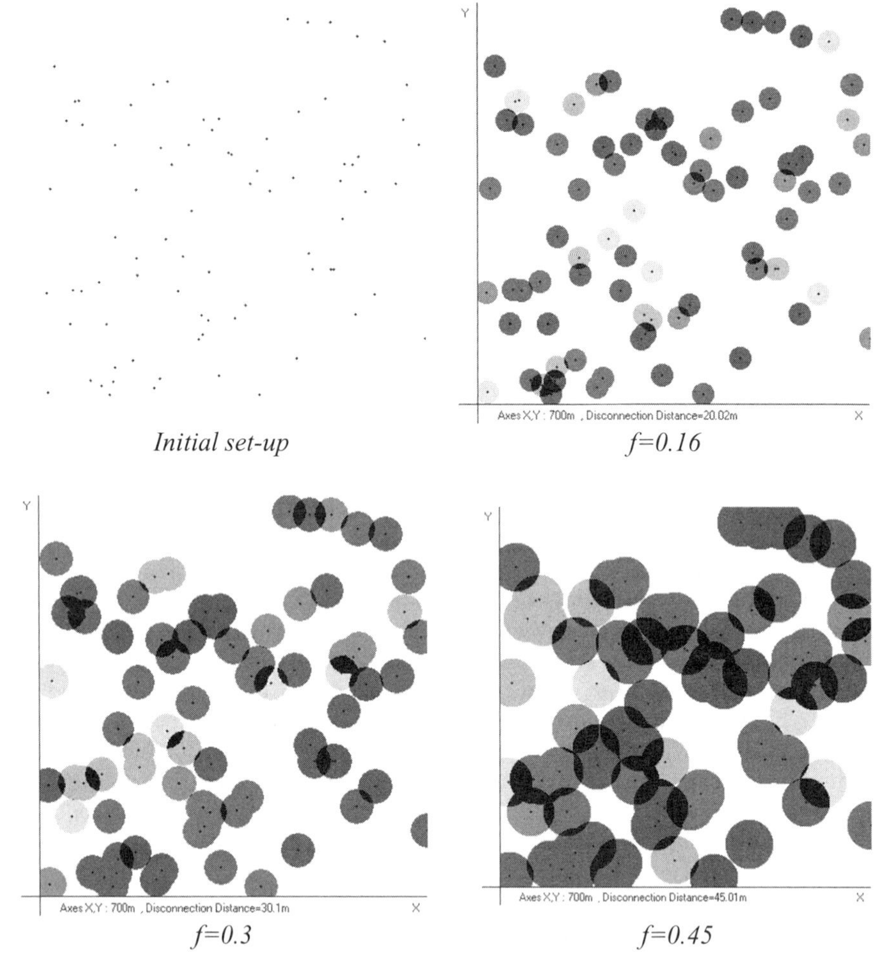

FIGURE 4.19 Evolution of clustering formation for an 80-node WM^{2}Net (MINPAUSE = 0 s, MAXPAUSE = 30 s, MINVEL = 0 m/s, MAXVEL = 3 m/s, MINMOVE = 30 s, MAXMOVE = 30 s).

The evolution of clustering formation for an 80-node simulated WM^{2}Net environment is illustrated in Fig. 4.19. To simulate mobility, a mobility manager chooses randomly whether a node should move or pause. If the node should pause, it stays stationary for a random period of time uniformly distributed over (MINPAUSE, MAXPAUSE). If the node decides to move, it will choose an arbitrary direction uniformly distributed over $(0, 2\pi)$ as well as speed and motion which are also uniformly distributed over (MINVEL, MAXVEL) and (MINMOVE, MAXMOVE), respectively. Direction and moving parameters remain constant only for the duration of the moving period.

Returning now to our previous reasoning and considering this clustering set-up, it becomes obvious that the on-line determination of the moving pattern of clusters is an extremely difficult practice. This information can be extracted from the data provided in the fourth column, which illustrates how *rapidly* the partitioning of the vector space

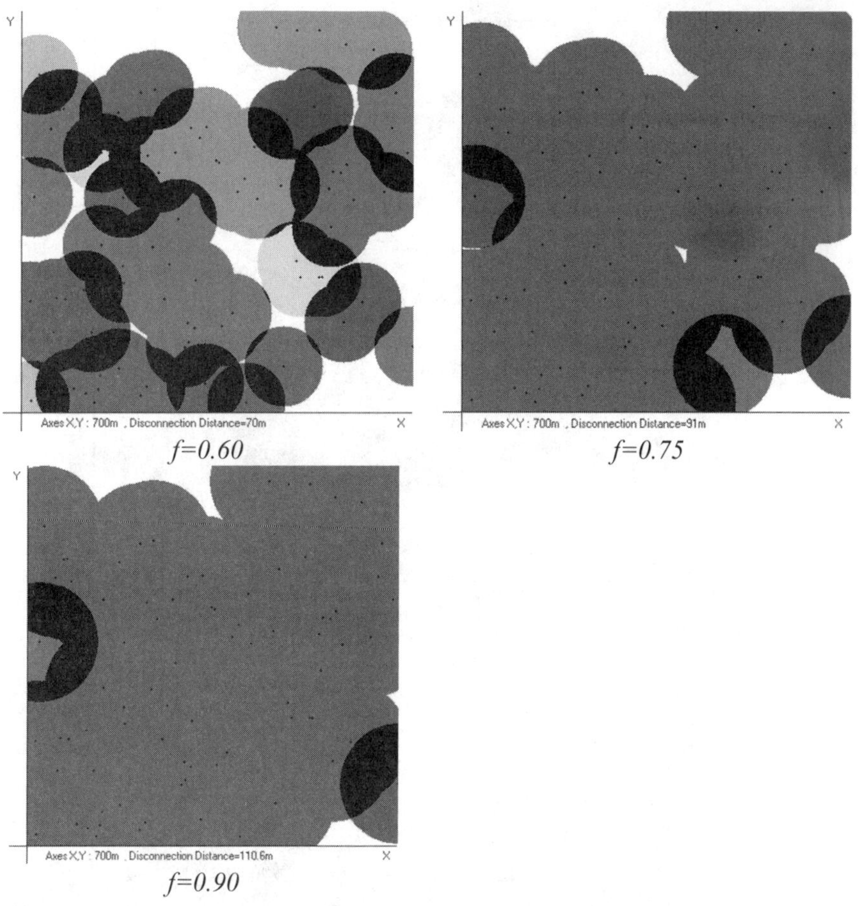

FIGURE 4.19 *(Continued)*

occurs as time evolves. *This data is the actual information needed for disconnection predictions.*

The fourth column in fact illustrates the critical distance. Critical distance is the maximum distance between two clusters beyond which communication fails. That is, given three formed clusters K1, K2, K3 with D(K1, K2) = 60 m, D(K1, K3) = 90 m, D(K2, K3) = 7 m, the critical distance is 90 m. Referring to the table above, we observe that up to time step 42 the critical distance is 0 (there is only one cluster formed). As a second cluster is formed, the critical distance starts to gradually increase, which implies that more than one clusters are formed and are slowly moving apart.

This is a good metric as it shows the tendency of points to move apart; in fact, this information constitutes the core of NPDA. As illustrated (Fig. 4.20), from time step 52 and beyond the critical distance is constantly increasing, which shows the tendency of the network to form disconnected partitions at some future state. The reader is referred to the work of Aggélou (2005) for a concise description of the experimental analysis of NPDA/WiMeSAS.

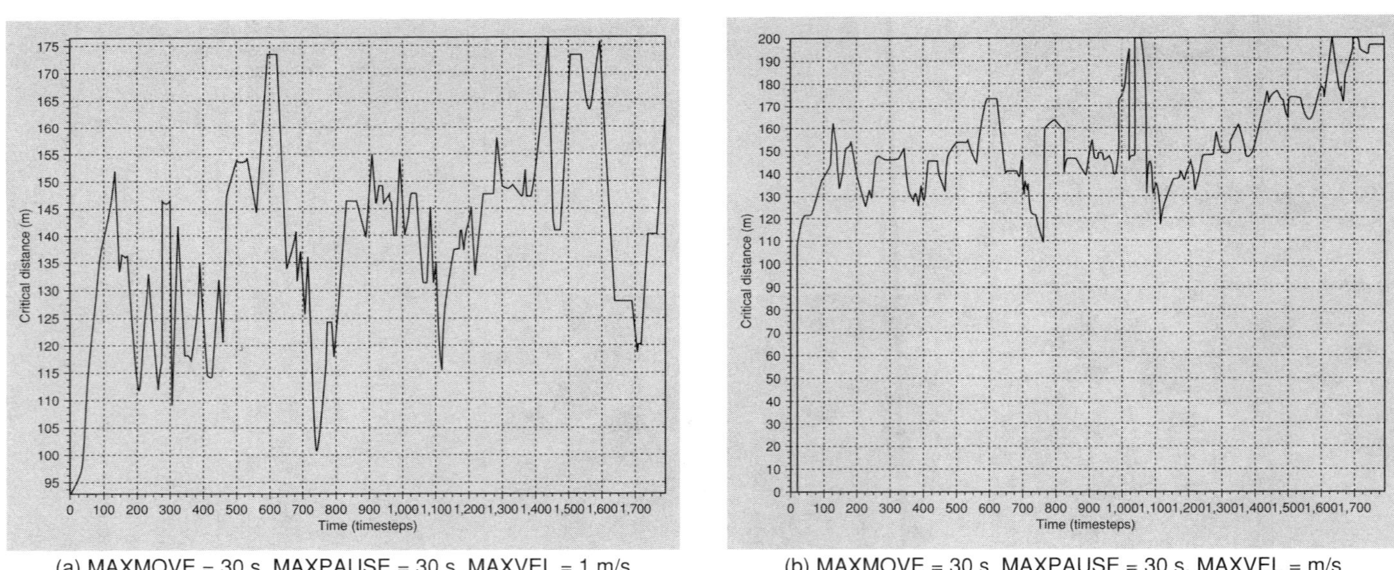

(a) MAXMOVE = 30 s, MAXPAUSE = 30 s, MAXVEL = 1 m/s

(b) MAXMOVE = 30 s, MAXPAUSE = 30 s, MAXVEL = m/s

FIGURE 4.20 Evolution of critical distance for different MinPause, MaxPause, MinMove, and MaxMove combinations. (a) $f = 0.4$. (b) $f = 0.9$. (c) $f = 0.4$. (d) $f = 0.9$. (e) $f = 0.4$. (f) $f = 0.9$. (g) $f = 0.4$. (h) $f = 0.9$. (*continued*)

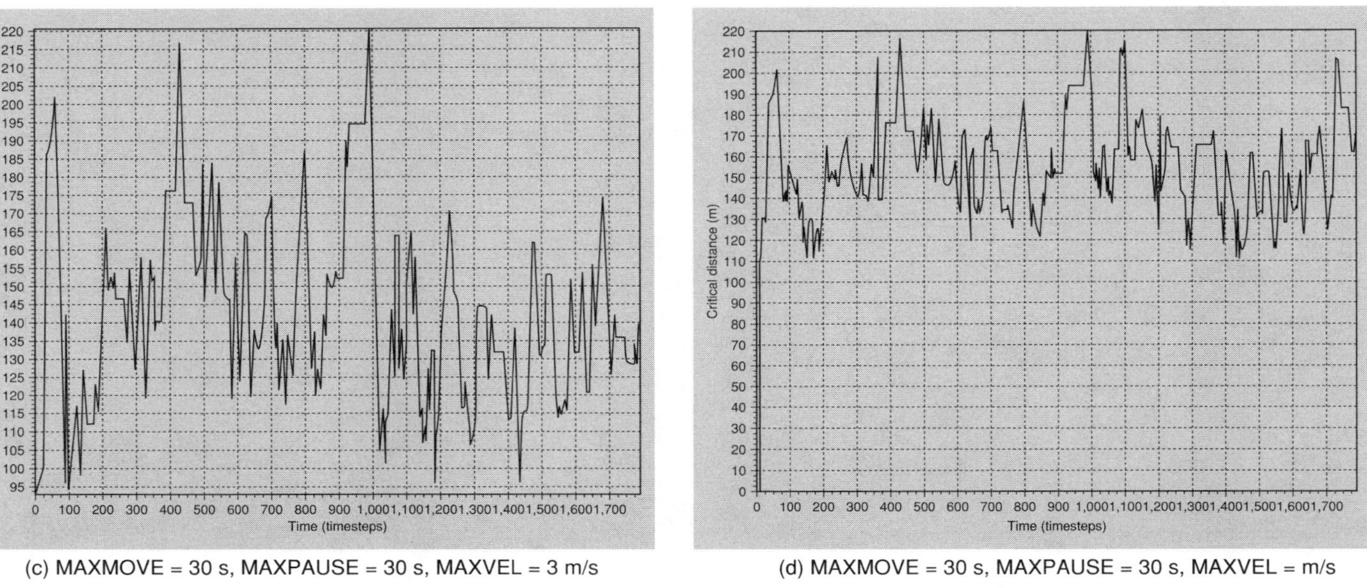

(c) MAXMOVE = 30 s, MAXPAUSE = 30 s, MAXVEL = 3 m/s

(d) MAXMOVE = 30 s, MAXPAUSE = 30 s, MAXVEL = m/s

FIGURE 4.20 *(Continued)*

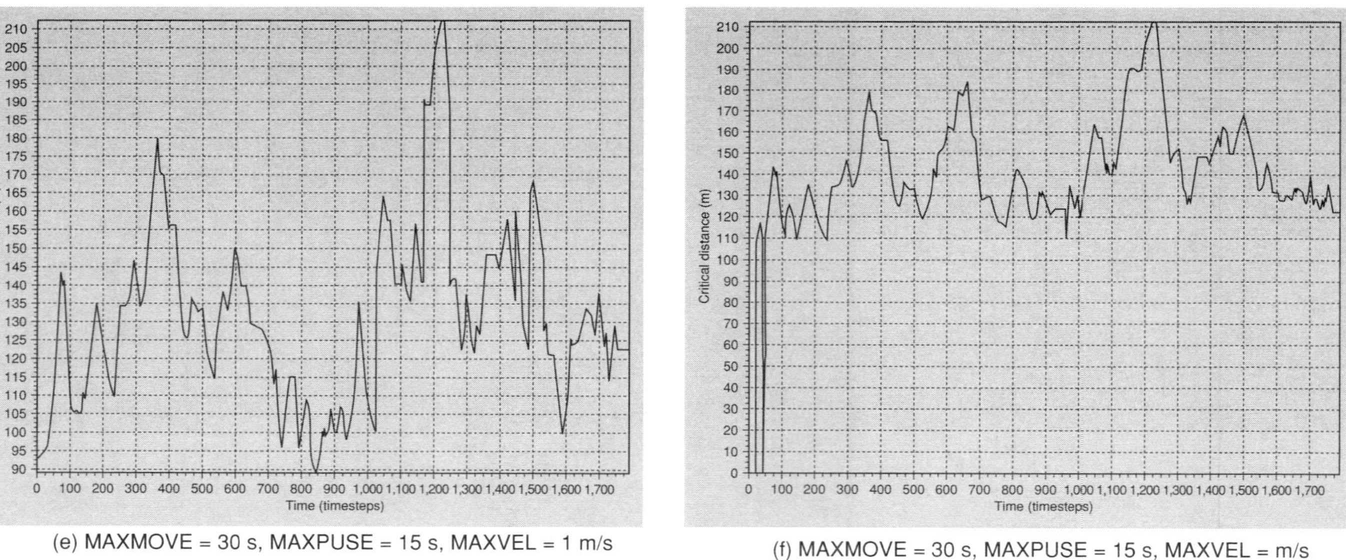

(e) MAXMOVE = 30 s, MAXPUSE = 15 s, MAXVEL = 1 m/s

(f) MAXMOVE = 30 s, MAXPUSE = 15 s, MAXVEL = m/s

FIGURE 4.20 *(Continued)*

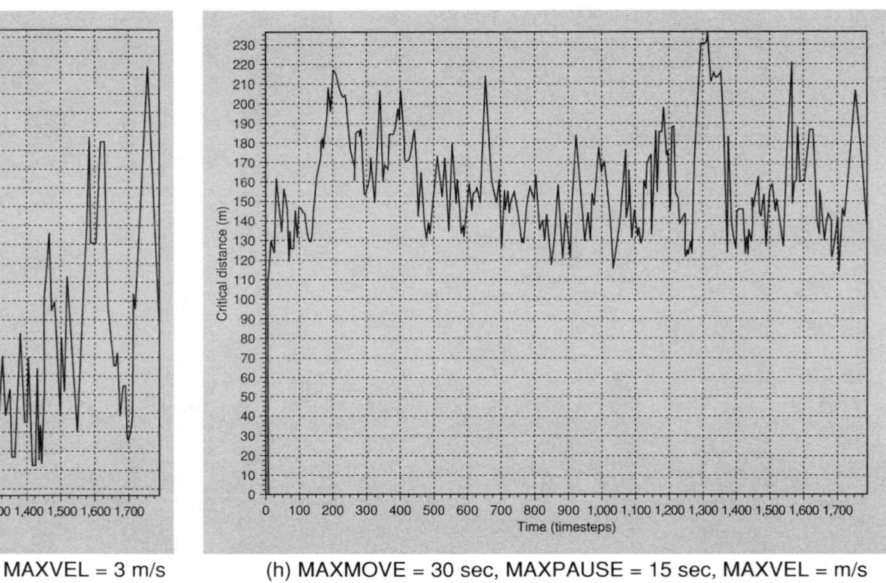

(g) MAXMOVE = 30 sec, MAXPAUSE = 15 sec, MAXVEL = 3 m/s

(h) MAXMOVE = 30 sec, MAXPAUSE = 15 sec, MAXVEL = m/s

FIGURE 4.20 *(Continued)*

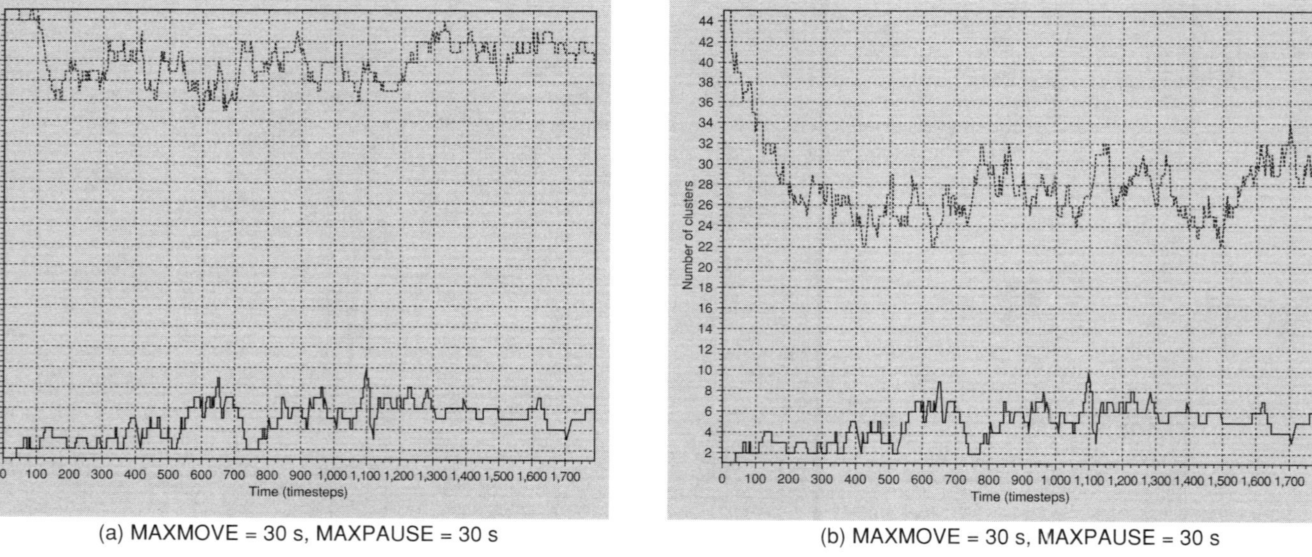

(a) MAXMOVE = 30 s, MAXPAUSE = 30 s

(b) MAXMOVE = 30 s, MAXPAUSE = 30 s

FIGURE 4.21 $I(m) = m(P(s, d_O))$ and $E(m) = m(P(s, d))$ Plots. 45 Nodes deployed. MAXVEL=3 m/s. *(continued)*

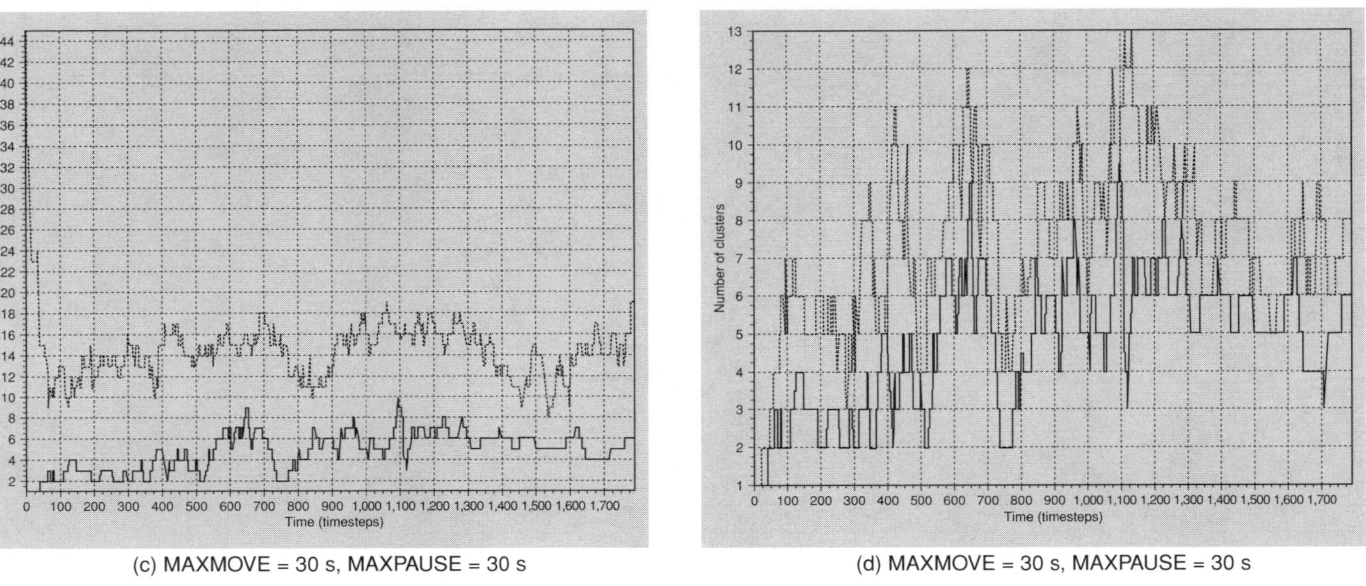

(c) MAXMOVE = 30 s, MAXPAUSE = 30 s

(d) MAXMOVE = 30 s, MAXPAUSE = 30 s

FIGURE 4.21 *(Continued)*

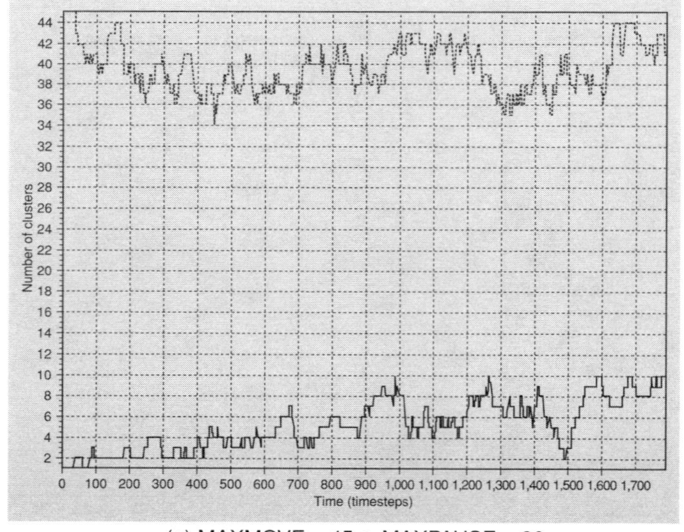

(e) MAXMOVE = 45 s, MAXPAUSE = 30 s

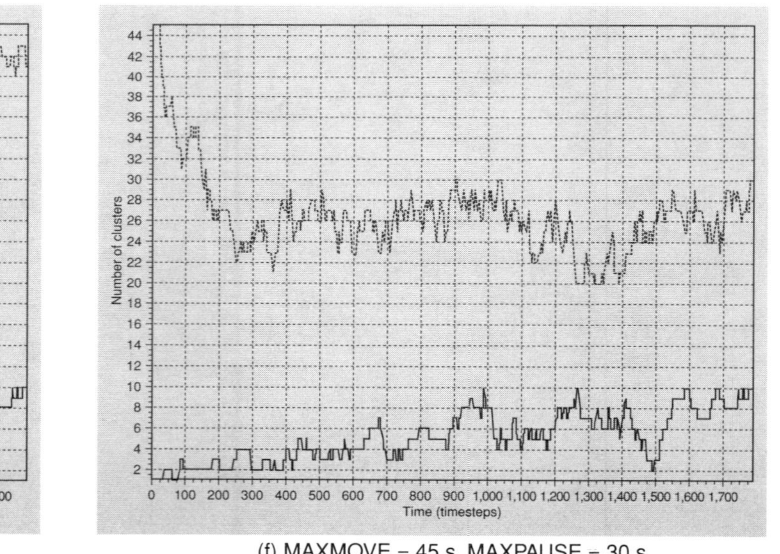

(f) MAXMOVE = 45 s, MAXPAUSE = 30 s

Figure 4.21 *(Continued)*

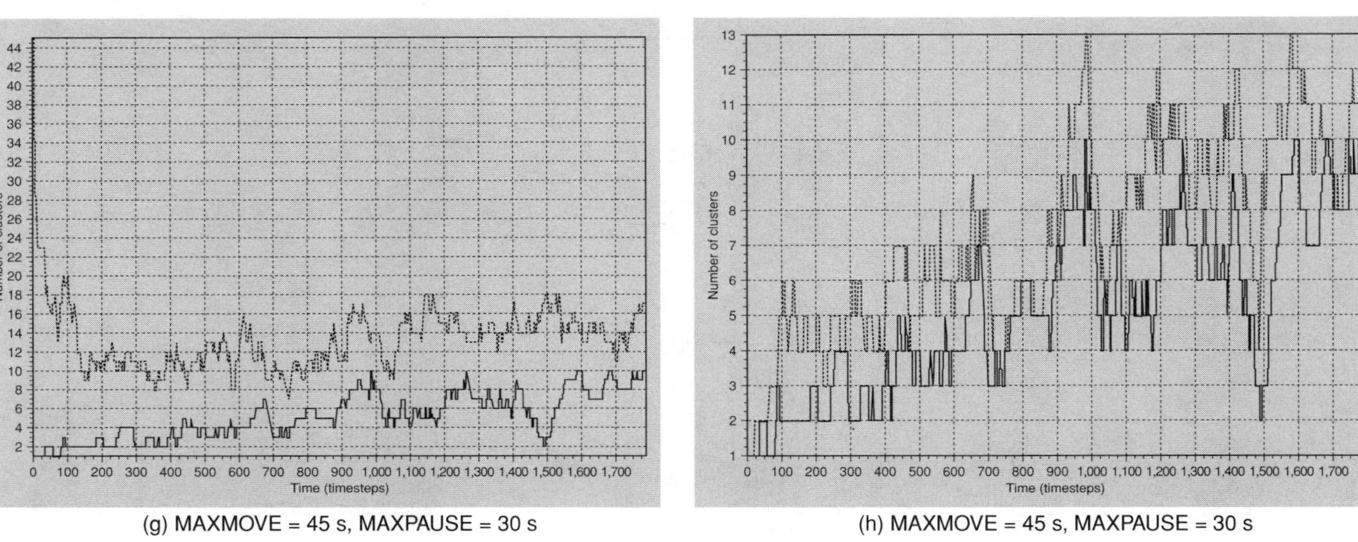

(g) MAXMOVE = 45 s, MAXPAUSE = 30 s

(h) MAXMOVE = 45 s, MAXPAUSE = 30 s

FIGURE 4.21 *(Continued)*

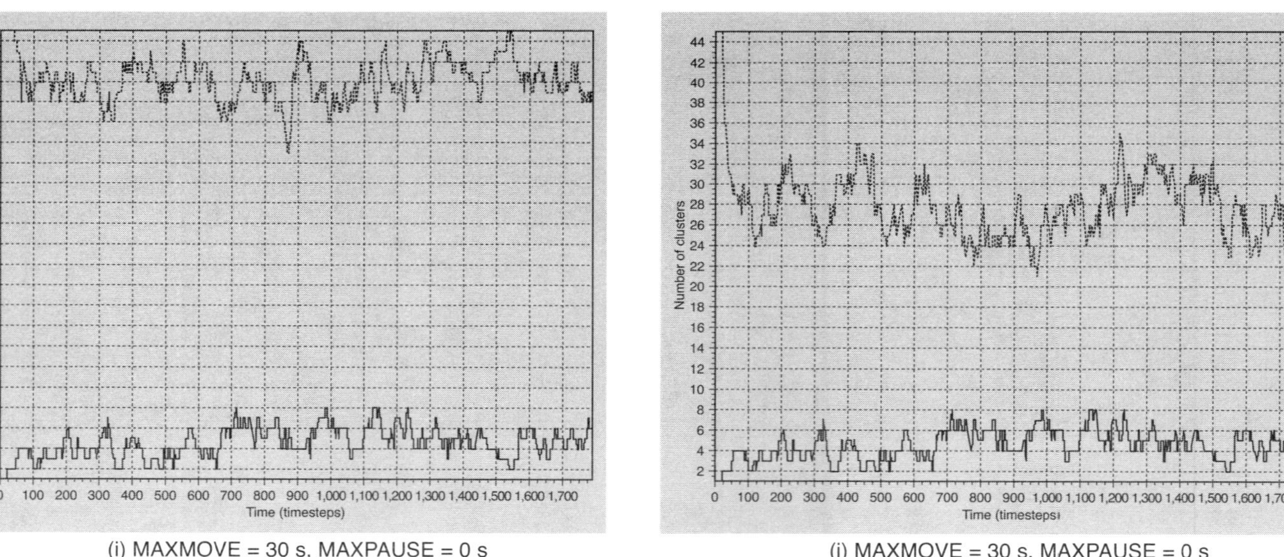

(i) MAXMOVE = 30 s, MAXPAUSE = 0 s

(j) MAXMOVE = 30 s, MAXPAUSE = 0 s

FIGURE 4.21 *(Continued)*

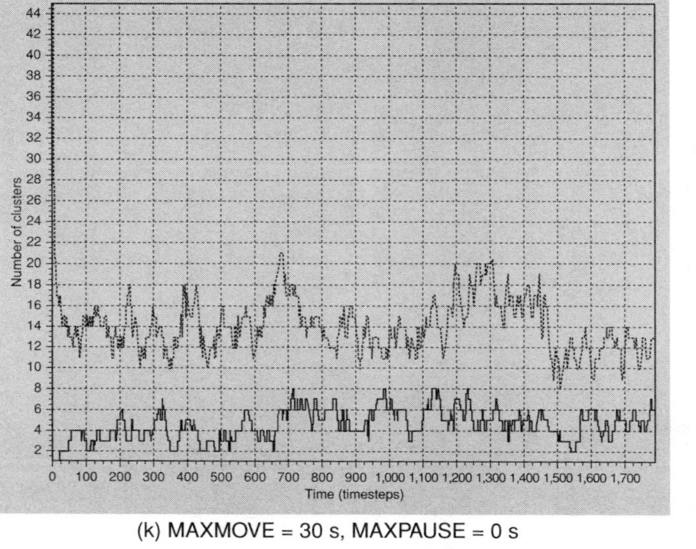

(k) MAXMOVE = 30 s, MAXPAUSE = 0 s

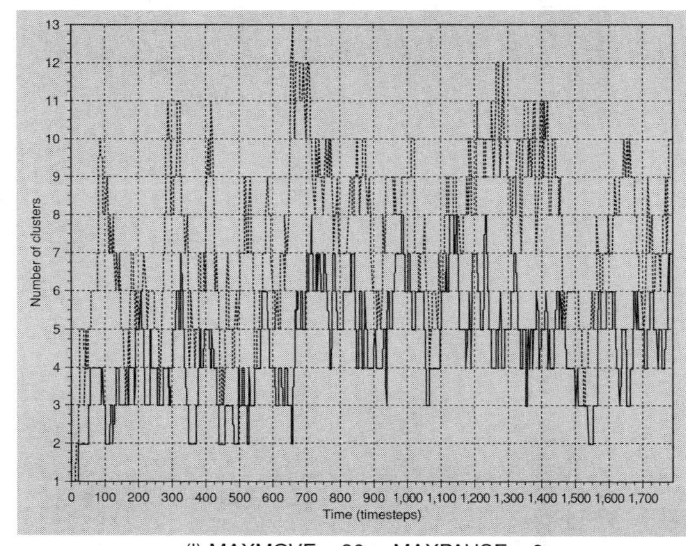

(l) MAXMOVE = 30 s, MAXPAUSE = 0 s

FIGURE 4.21 *(Continued)*

Figure plots the second and third column as a function of time. Let us designate the y-axis of the second and third column as $I_{t+t0}(m) = m\ (P(s, d_0))$ and $E_t(m) = m(P(s, d))$, respectively. The dotted line illustrates the cluster population for $d_0 = 120$ m ($I_{t+t0}(m)$), whereas the compact line the cluster population for $d_0 = 108$ m ($E_t(m)$). As illustrated, the two plots follow similar patterns. This is reasonable and, without loss of generality, we could state that E_t events are I_{t+t0} events deferred for time t_0.

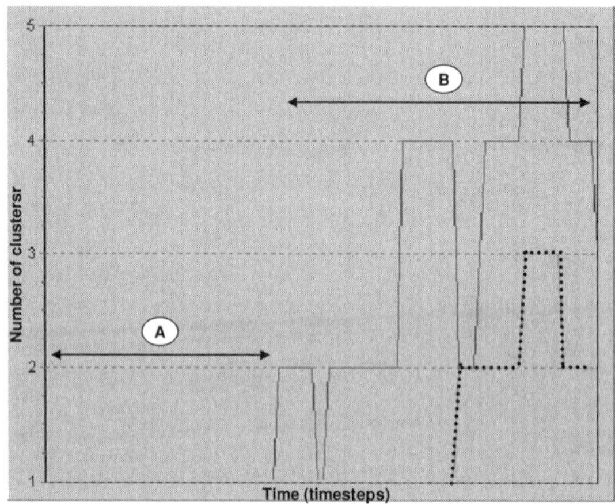

Referring to this figure, we can make the following observations:

- No disconnection is expected to occur during the "flat" time window, designated as "A" (that is, time $t = 1$ to 42—see table above); this is so because both I and E are valued at 1.
- Now, for the time window designated as "B":
  - Given a *partition space* of the form $[x, y]$, where $x \in A = m[P(S, d_0)]$ with $x \geq 1$, and $y \in B = m[P(S, d)]$ with $y > 1$, it is evident that a network disconnection will definitely occur when $A > 1$ (in this case, $B > 1$ is obviously true). For instance, it is certainly true that at time $t = 67$, where $A = 1$ and $B = 4$, no disconnection is expected to occur, whereas it is certainly true that at time $t = 75$ there will be at least two partitions disconnected when $x > 1$. It remains to evaluate the time offset (t_0) as well as the probability of occurrence within this time window.

The plots in Fig. 4.21 depict $I(m) = m(P(s, d_0))$ and $E(m) = m(P(s, d))$, for various values of f and network set-up configurations.

CHAPTER 5

Medium Access Control Principles in WM^{2}Nets

5.1 Introduction

In wireless communications, channel transmissions are overheard from all nodes in proximity to the transmitting node. A data packet collision occurs when more than two users are transmitting at the same time to the same node. The medium access control (MAC) protocol is responsible for controlling access to the physical medium as well as for accounting for the available resources. Efficient channel access during call setup is thus vital for minimizing the blocking/dropping rates.

The primary medium access mechanism underpinning multihop wireless networks is the IEEE 802.11 protocol, which features low cost, ease of setup, and high physical data rates (up to 54 Mbps). The most important feature of such IEEE 802.11-based mesh networks is that the radio links share the radio resources using a carrier sense multiple access (CSMA) based random access protocol. In CSMA (Kleinrock and Tobagi, 1975; Metcalfe and Boggs, 1976; Lam, 1980; Rom and Sidi, 1990), when a node wishes to send traffic, it first senses the radio carrier and proceeds with the transmission only if it considers the channel to be idle. The idea behind the CSMA paradigm is to reserve the transmission channel at the originator (source) by carrier sensing. In principle, CSMA-based systems have lower probability of collision. This is primarily attributed to the fact that when users sense other node's transmissions, they defer accessing the shared channel.

In extending CSMA to suit the needs of wireless mesh networks, a major complication that arises is that the carrier sensing operation must now cope with the following two forms of asymmetry:

(1) *Contention asymmetry*: Since a node has limited transmit power, it can communicate with only a subset of the nodes that form the network. Given the distribution of the nodes in the network, the set of nodes that a node can sense, or be sensed by, is quite different. This introduces asymmetry in the level of contention each link/node experiences.

(2) *Traffic asymmetry*: The rate at which a link-i contends for the radio channel is a direct function of the traffic, i it needs to carry. This in turn is a function of the topological location of the link. For example, links towards the interior of

the network multiplex the multihop traffic belonging to a number of end-to-end flows and as such, they typically carry more traffic than the links towards the periphery. This problem, though not unique to wireless networks, induces asymmetry in the amount of traffic the links need to carry.

Besides, similar to wireless ad hoc networks, where not all users in the network can hear each other, in wireless mobile mesh networks (WM^{2}Nets) the **hidden and exposed terminal problems** (Kleinrock and Tobagi, 1975; Tobagi and Kleinrock, 1975; Leiner et al., 1987) also arise. Hidden and exposed nodes are a phenomenon of carrier and packet sensing MAC protocols. Packet sensing ensures that received packets can be decoded only if their energy level is above a certain threshold. Carrier sensing, on the other hand, mandates a station that has data to send but hears a level of energy higher than a given threshold to defer and try to access the channel after some random time. Packet sensing is performed at the MAC layer, whereas carrier sensing is done at the *physical layer*, which is commonly abbreviated as PHY.

A "hidden node" cannot "hear" that a neighbor node has a communication session in place with another node, and thus does not defer from attempting to gain access. The reason for not hearing the transmission can be due to distance, an obstacle in the propagation path, being in the null of an antenna, or, in the case of direct sequence spread spectrum (DSSS), the use of a different code. When this station attempts to gain access, it interferes with the reception at the receiving node.

The exposed node problem is very different. An exposed node simultaneously hears multiple disjoint sections of a network and never gets the opportunity to contend. This is because the carrier is sensed busy for large period of time wherein the exposed node continuously defers from transmitting.

The example in Fig. 5.1 demonstrates the hidden/exposed terminal problem. Let us assume that stations A and B are hidden from each other and that both wish to transmit to a third station, named *Receiver*. When A is transmitting to *Receiver*, the carrier sensing of B cannot capture any transmission event, and thus B can immediately start a transmission to *Receiver* as well. Therefore, both stations A and B would transmit at the same time to *Receiver*.

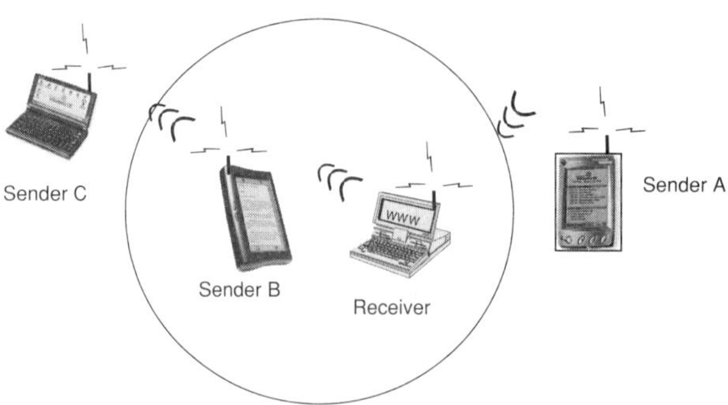

Figure 5.1 Illustration of the hidden and exposed terminal problem.

Now consider the case when A is transmitting to C. Since *Receiver* can "hear" A, *Receiver* cannot risk initiating a transmission to B for fear of causing a collision at A. In Figure 5.1, the *Receiver* is "exposed" to A.

5.2 Approaches to Mitigate the Hidden and Exposed Terminal Problem

The hidden and exposed stations phenomenon may occur in both infrastructured as well as peer-to-peer (ad hoc) WM^{2}Nets. Notably, the problem becomes more severe in peer-to-peer WM^{2}Nets where almost no coordination exists among the stations. The hidden terminal problem can significantly reduce the amount of traffic carried by the system. Some of this lost capacity can be regained through special mechanisms that allow receivers to control access to the channel. A few techniques are briefly discussed in below.

5.2.1 Tackling the Problem at the MAC Layer

In general, the shared wireless medium in WM^{2}Nets requires the use of appropriate MAC protocols to mitigate the medium contention issues as well as to allow for efficient use of the limited bandwidth. Specialized MAC protocols could also help alleviate the hidden/exposed terminal problem. For instance, to avoid the hidden and exposed terminal problem, the IEEE 802.11 basic CSMA access mechanism is extended with a virtual carrier sensing mechanism, called *request-to-send* (RTS)/*clear-to-send* (CTS). As pointed above, in the RTS/CTS mechanism, after access to the medium is gained and before transmission of a data packet begins, a short control packet, called RTS, is sent to the receiving station announcing the upcoming transmission. This message contains the destination address and the duration of the transmission. The receiver replies to this with a CTS packet to indicate readiness to receive the data. CTS packets also contain the projected length of the transmission. This information is stored by each active station in its NAV, the value of which becomes equal to the end of the channel busy period. Therefore, all stations within the range of at least one of the two stations (receiver and transmitter) know how long the channel will be used for this data transmission (see Fig. 5.2).

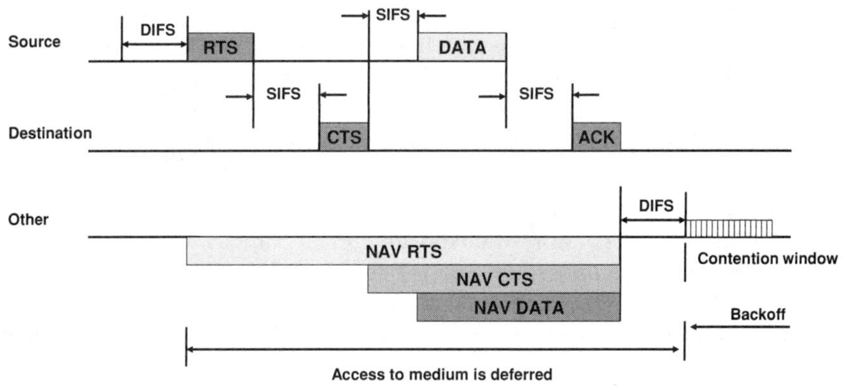

FIGURE 5.2 The RTS/CTS mechanism.

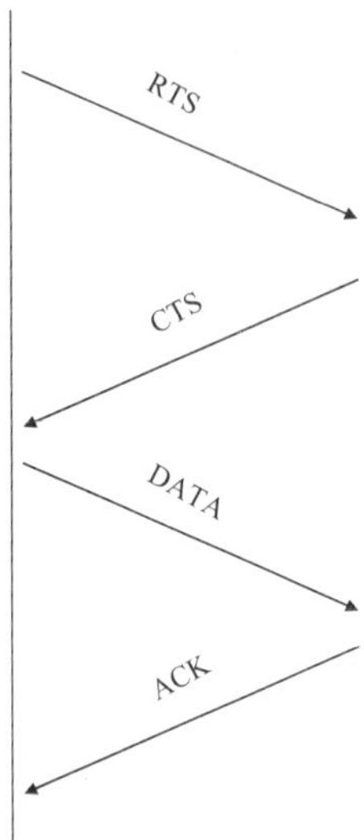

Figure 5.3 The 4-way handshake. coordination

Referring to the hidden and exposed terminal example of Fig. 5.1, when B wishes to transmit to *Receiver*, it first sends a RTS message to *Receiver*. In response, *Receiver* broadcasts a CTS message that is received by both A and B. Since B has received the CTS message unsolicited, A knows that *Receiver* is granting permission to send to a hidden terminal and hence refrains from transmitting. Upon receiving the CTS message from *Receiver* in response to its RTS message, B transmits its own message.

Not only does the above dialogue solve the hidden terminal problem but also solves the exposed terminal problem. Upon receiving an unsolicited CTS message, A refrains from transmitting. After an appropriate interval, determined by the attributes of the channel (i.e., duration of a time slot, etc.), A can send its own RTS message to C as the prelude to a message transmission.

The RTS/CTS mechanism, for the majority of the carrier-sensing protocols, resumes with an acknowledgement sent for each data packet. If an acknowledgement is not received, the MAC layer retransmits the data. This entire sequence is called the 4-way handshake as shown by Fig. 5.3.

5.2.2 RTS/CTS Collisions and Loss of State Information

Even though the RTS/CTS mechanism solves the hidden station problem *during the transmission of user data*, collisions may still occur *during the transmissions of control packets* (the small RTS and CTS packets), thus resulting in loss of RTS/CTS packets. In addition, the RTS-CTS solution proposed for solving the hidden terminal problem in various MAC schemes, like IEEE 802.11, MACA etc., assumes bidirectional links. Bidirectionality assures that the transmission of an RTS is always followed from a CTS exchange. In most wireless environments, however, due to asymmetrical propagation conditions, links are often unidirectional. To this avail, and if unidirectional links are assumed, the RTS/CTS exchange may indeed fail and MAC cannot do much about the exposed and hidden terminal problem. In this case, packet loss has to be handled at a different layer.

To illustrate how RTS/CTS information may be lost, consider the following example (for single channel operation):

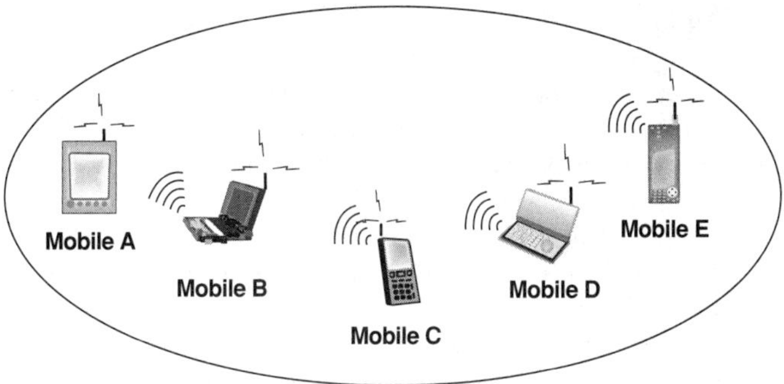

Assume that node D transmits a data packet to node E (using the RTS/CTS handshake). When the packet is in transit, node A decides to transmit to node B, also using the RTS/CTS handshake. Let us see now what node C experiences:

1. C receives node D's RTS.
2. C receives node D's DATA. (It becomes "jammed" by this transmission.)
3. C receives node B's CTS (sent in response of node A's RTS).
   Since node C is jammed by the DATA sent from D to E, it cannot successfully decode node B's CTS. Therefore, node C will not learn that node B is about to receive a packet.
4. After node D completes its transmission, node C is free to transmit. Node C does not know that B is receiving a packet and assumes that the channel is free. Let us say that node C decides to transmit to node D.
5. Node C sends an RTS to D, receives a CTS back and then starts sending DATA. At this point, node C's DATA (or even the RTS) collides (at node B) with node A's DATA.

The above example illustrates how a node can lose an RTS or CTS (or, typically, "state" information). Moreover, a node that loses state information may cause the loss of state information to other nodes as well. For example, assume in the previous example that in (4), node D decides to transmit to node C (instead of node C transmitting to D). Now, (5) will be that node C sends a CTS answering node D's RTS, and node C's CTS collides with the node A's DATA packet at node B. Note that at this point node B also loses state information (misses the CTS) and therefore it does not lose the A's DATA packet but it is also not aware of the D to C communication.

State information loss is a serious problem in IEEE 802.11-like protocols as this may result to collisions of on-going data transmissions of other stations.

Similar situations (although less often) can occur with multichannel systems, where the trigger of "loss of state" is not jamming of a node due to simultaneous communication but due to the fact that during normal reception (in a single transceiver system) a node can only listen to one channel at a time. Thus, a node that is listening in the DATA channel is not aware of the activities in the control channel, thus potentially causing him to lose state information.

5.2.3 Tackling the Problem at the PHY

When the density of nodes is very high, the usage of control signals (RTS/CTS) could cause congestion problems in the network. The wireless spectrum is admittedly a scarce resource and mobiles must consume bandwidth judiciously. With this in mind, the hidden and exposed node problems can also be tackled at the PHY, by using multiple codes (e.g., the transmitter-directed codes).

Using multiple codes implies that each node has a unique PN code in the network and transmits using its own PN code. The transmitter senses the channel for the known PN codes and if the receiver is idle (i.e., there are no data for this node) the node is then free to transmit. Since the PN codes are unique in the network, the hidden node problem is addressed. Also, since the node needs not be idle until the completion of the on-going data transfer of neighboring nodes, the exposed node problem is also addressed.

Another configuration of using a set of codes is for the sender to include in its RTS, which is sent over a common code, a preferred code (unused at that time) and for the receiver to send a CTS in response, if the code is acceptable. Then DATA-acknowledge (ACK) takes over the selected code.

A downside of using completely orthogonal multiple spreading codes is the requirement of a synchronous system. It is, however, very difficult to achieve perfect synchronization in a pure ad hoc system without the aid of sophisticated clocks or advanced and complex synchronization mechanisms.

5.2.4 Tackling the Problem with Smart Antennas

From the preceding discussions on hidden/exposed terminal problem, it becomes evident that the source of the problem is the propagation properties of omnidirectionally transmitted signals. In this regard, *smart directional or adaptive antennas* could be the right panacea. This type of antennas has the ability to point (direct) transmission energy towards a specific angular direction and thus cancel out most of the interfering signals. The sender can then focus the transmission energy in narrow regions thus reducing the effect of interferers. At the receiver side, the receiver runs an angle-of-arrival (AoA) algorithm to determine the direction of the maximum strength signal. It then places nulls in the other directions aiming to cancel the interfering signals. Adaptive antennas are described in more detail in Section 6.3.

In WM^{2}Nets, however, if both transmitter and receiver employ directional antennas, a lot of issues need working out, including pragmatic ones. In general, utilizing smaller antennas imply a higher operating frequency. This in turn implies poorer propagation characteristics. With smart antennas, one needs (typically) half-wavelength spacing between elements. At 2.4 GHz, a cylindrical 8-element array would have a radius of about 8 cm—making it quite unwieldy to carry on a PDA or a laptop. If the directional beams are not very narrow, using directional antennas does not make the problem go away altogether, just make it less severe. So, increase the operating frequency to 24 GHz ISM band and you get a mere 0.8 cm radius. But then, LoS operation is required.

Additionally, if the option to use beams exists, the implicit gain of the focused beam would imply that two nodes could communicate reliably over much greater distances than in the omnidirectional mode. For example, if both antennas show a 6 dB gain, the effective signal will be sixteen times as strong; the effective range could thus be from two to four times as great. If that is the case, then WM^{2}Net protocols that establish temporary point-to-point links rather than multihop paths will be needed. Of course, if the range is too large, there will also be a need to set up multihop paths, but each link could be of the form of vectors, rather than nodes. This may worth it, but sounds more complex.

On the other hand, even if one could construct an antenna capable of synthesizing perfectly power controlled pencil beams and these could be used for WM^{2}Net communications, there would still be a need for omnidirectional broadcasting. Beacons, broadcasts, and initial signal acquisition, all use omnidirectional coverage. Therefore, at least when starting up the link, the receiver can be expected to have an omni receiving antenna. Once a link is established should perform "like a piece of wire" at its best, or in RF parlance, similar to a motion tracking power-controlled pencil beam.

In addition, there is at least one other wrinkle in using omni-mode of operation. With omni-directional broadcasts, it is not always necessary to send an ACK. If A sends a packet to B for forwarding, as B rebroadcasts the packet, A hears it and knows all is well. An explicit ACK in this case is not needed. Directional antennas, on the other hand, do foster the need for an explicit ACK in almost all cases thus raising resource utilization concerns. Without loss of generality, these observations allow us to note that

> Any antenna system in WM^{2}Net might eventually be required to function both as omnidirectional as well as directional.

The analysis so far considers steered-beam or switched-beam directional antennas. An alternative technique is to steer a null in the direction of the dominant interfering signal. In fact, a preferred form of implementation is to steer the array so as to maximize the received signal-to-interference ratio, using such metrics as received symbol quality, training burst quality, per-packet error rate, etc.

Smart antenna technology is one step further in the evolutionary path towards capacity improvements and interference mitigation. The idea of smart antennas is to use base station antenna patterns that are adaptive to the radio conditions. This can be visualized as the antenna directing a beam towards the communication partner only. The difference between the fixed and the smart antenna concept is illustrated in Fig. 5.4.

To motivate the use of Smart antennas, let us enumerate their capabilities over and above those provided by directional antennas.

1. *Silencing interferers:* If a receiver knows that there are interfering transmitters in its neighborhood, it can form a directed beam towards the sender while simultaneously placing nulls in the direction of the other transmitters. A null effectively cancels the received signal power from a transmitter (even if the interferer is more powerful than the desired transmitter) and ensures a high SINR at the receiver.
2. *Enhanced neighbor discovery:* The identification of signal direction is necessary for transmission (to beamform appropriately) as well as for reception (to silence interferers). Some works typically use a form of sequential polling to identify the direction of one-hop neighbors (Jakllari et al., 2003). Thus, for a 45°-sectored antenna, there are eight directions in which a node will periodically poll neighbors. Adaptive antennas, on the other hand, can considerably ease the complexity of this task by running an AoA algorithm.

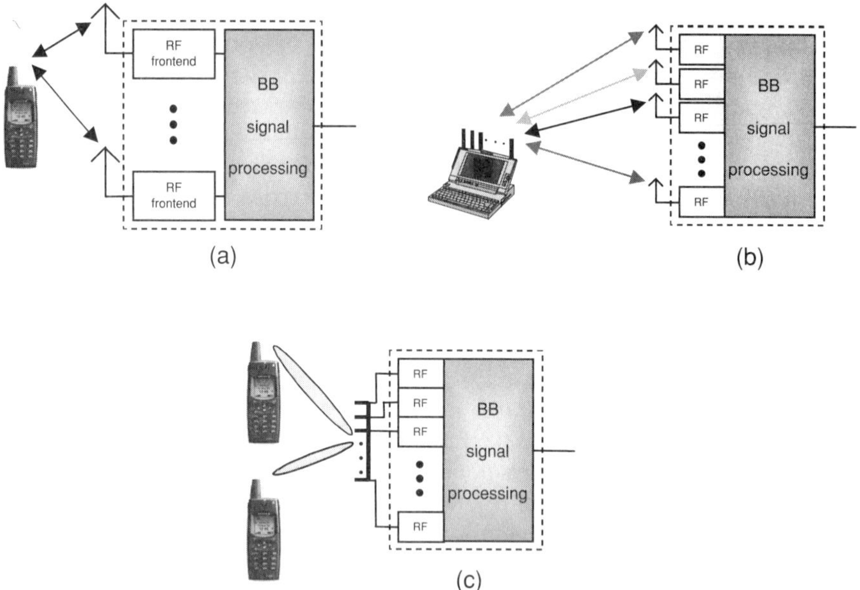

FIGURE 5.4 Illustration of the difference between a traditional base station radiation pattern and a smart antenna base station. (A) Diversity concepts. (b) Multiple antenna concepts. (c) Adaptive antenna concepts.

3. *Flexible beamforming:* A smart antenna system can be configured as an omnidirectional antenna or as a directional antenna with variable beamwidths (limited by the number of antenna elements) and with arbitrarily precise boresight.[1] This flexibility allows engineers to explore the protocol space with arbitrary combinations of beamwidths for collision avoidance (CA) and data transmission.

Nevertheless, as stretched in Section 6.3, the integration of smart antennas into mobiles is a difficult task. A major challenge faced by technology experts is the limited space in handsets. The lack of space in a mobile terminal forces violating the golden rule that the spacing between the array elements should not be smaller than half of the used wavelength of the transmitted or received radiation. With a wide spacing, a better directivity can be obtained.

[1] Sectored antennas, for example, are relatively inflexible in this regard as more packet collisions may occur for communicating nodes that lie outside the 3 dB beamwidth of the main beams.

In the past years, though, the miniturization of electronics have made it possible to produce antennae about the size of a grain of rice! The Micro Reach Xtend antenna from FRACTUS (http://www.fractus.com), for instance, measures 3.7 mm × 2 mm. Designed for the ISM 2.4 GHz band, the miniature antenna is designed using FRACTUS' space-saving fractal antenna technologies and developed especially for Bluetooth headsets and micromobile handsets. It supports all standards working at the 2.4 GHz ISM band, including Wi-Fi, Bluetooth, and Zigbee.

5.3 Overview of the IEEE 802.11 Protocol Specifications

In 1997, the Institute of Electrical and Electronics Engineers (IEEE) adopted the first digital wireless data transmitting standard, named IEEE 802.11, with data rates up to 2 Mbps (IEEE 802 LAN/MAN Standards Committee, 1999). Originally, IEEE 802.11 was conceived as part of the IEEE 802.4 token bus standard with a given name of 802.4L. In 1990, the 802.4L group was renamed to IEEE 802.11 W-LAN Project Committee, which created an independent 802 standard tasked with defining three PHY specifications and one common MAC layer for the lower portion of the Data-Link layer for W-LANs. The purpose of the IEEE 802.11 standard was to foster industry product compatibility between W-LAN product vendors (802.3 standard (Ethernet) (Metcalfe and Boggs, 1976)) which, consequently, led to the approval of the IEEE 802.11 standard on June 27, 1997 (IEEE 802 LAN/MAN Standards Committee, 1999).

Since then, two IEEE standards have been ratified to extend the data rate of W-LANs by enhancing the PHY specifications. These specifications are the IEEE 802.11a and the IEEE 802.11b, both ratified in 1999. The IEEE 802.11a task group created a standard for W-LAN operations in the 5 GHz Unlicensed National Information Infrastructure (UNII) band, with data rates up to 54 Mbps. The IEEE 802.11b task group produced a standard for W-LAN operations in the 2.4 GHz band, with data rates up to 11 Mbps. Both standards, IEEE 802.11a and b, share the same MAC specifications with the original IEEE 802.11 standard (IEEE 802 LAN/MAN Standards Committee, 1999). The differences are evident in newer PHY specifications, where IEEE 802.11a utilizes orthogonal frequency-division multiplexing (OFDM), while IEEE 802.11b utilizes complementary code keying (CCK).

Apart from these two standards, several other task groups (designated by letters) have been created to extend the IEEE 802.11 standard (IEEE 802 LAN/MAN Standards Committee, 1999). Among these, the IEEE 802.11e task group targets at supporting voice and video over IEEE 802.11 networks using an enhanced MAC layer with quality-of-service (QoS) features, and the IEEE 802.11g task group, which is working to develop a higher-speed extension to IEEE 802.11b while retaining compatibility, that is, it uses the 2.4 GHz frequency band. Figure 5.5 illustrates this evolutionary path of IEEE 802.11 standards.

The key features of IEEE 802.11 family of specifications are summarized in Table 5.1.

5.3.1 IEEE 802.11 Architecture

An IEEE 802.11 W-LAN can be implemented either with infrastructure or without infrastructure support (i.e., wireless ad hoc communications). In an infrastructure-based network, there is a centralized controller for each cell, often referred to as access point

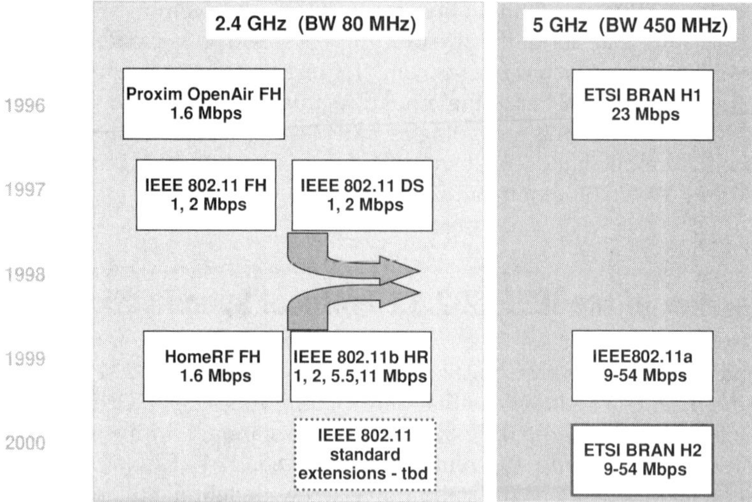

FIGURE 5.5 W-LAN standards evolution.

(AP). The AP is normally connected to the wireline backbone[2] network (Distribution System in IEEE 802.11 nomenclature) thus providing Internet access to mobile devices. All traffic goes through the AP, even when this is sent to a destination that belongs to the same cell. It may also act as a point coordinator (PC) to provide contention-free services to the associated stations (Matthias, 2001).

In the infrastructure-free network topology, a group of stations communicate directly with each other in an ad hoc fashion, independent of any infrastructure or base stations.

When two or more stations come together to communicate with each other they form a basic service set (BSS). The minimum BSS consists of two stations. A stand-alone BSS that is not connected to a base is called an independent basic service set (IBSS). The IBSS addresses the mobile peer-to-peer (ad hoc) configuration mode. Stations in IBSS-mode periodically broadcast beacons. The first station to instantiate an IBSS sets the Beacon interval, which is broadcast with every beacon or probe response Frame. Every station in the IBSS is therefore aware of the beacon intervals of the IBSS. For a station prior to transmitting its Beacon, it shall first calculate a random "beacon backoff" interval. If the station has not received a Beacon before its backoff interval expires, it proceeds and transmits its beacon frame. During the time of this beacon contention, data and announcement traffic indication message (ATIM)[3] transmissions are being halted. The algorithm makes sure that there will always be one beacon transmitted in the IBSS. The station last transmitting a beacon frame is elected to respond to probe requests. A new station in an IBSS does not transmit any beacon or probe response until it hears one from its IBSS.

[2] This backbone network is typically wireline, but can also be wireless. For the case of a wireless backbone, the IEEE 802.11 standard makes use of a special frame format that effectively tunnels the original frame over the IEEE 802.11 wireless network.

[3] ATIMs are used for power save mode.

Industry Standards	Roaming Support	Supported PHY Technology	Data Rate (in Mbps)	ISM Band (GHz)	UNII Band (in GHz)	Network Classification
IEEE 802.11	YES	DSSS, FHSS, Diffuse Ir	1, 2	2.4–2.48	*NA	W-LAN
IEEE 802.11a	YES	OFDM	6, 9, 12, 18, 24, 36, 48, 54	N/A	5.15–5.25 5.25–5.35 5.72–5.87	W-LAN
IEEE 802.11b	YES	DSSS	1, 2, 5.5, 11	2.4–2.48	*NA	W-LAN

*NA = not applicable.

Table 5.1 Key Features of IEEE 802.11 Standards

Furthermore, in contrast to the centralized polling-based mode, where the AP provides its own clock as common clock and handles the timing synchronization among the mobiles, synchronization among the IBSS stations is achieved by means of a distributed algorithm. The IEEE 802.11 uses two main functions for the synchronization of the stations in an IBSS:

1. *Synchronization acquisition*: This functionality is necessary for joining an existing IBSS. The discovery of existing IBSSs is the result of a scanning procedure of the wireless medium. During the scanning, the station receiver is tuned to different radio frequencies, searching for particular control frames. The station may initialize a new IBSS only if the scanning procedure does not result in finding any IBSS.
2. *Synchronization maintenance*: The system maintenance is implemented via a distributed algorithm, which is based on the transmission of beacon frames at a known nominal rate. The station that initialized the IBSS decides the beacon interval.

Creating large and complex networks using a number of BSSs leads us to the next level of hierarchy, the extended service set (ESS). The ESS shown in Fig. 5.6 consists of a series of overlapping BSSs (each containing an AP) connected together by means of the DS. The beauty of the ESS is that the entire network looks like an independent BSS to the link control layer. This means that stations within the ESS can communicate or

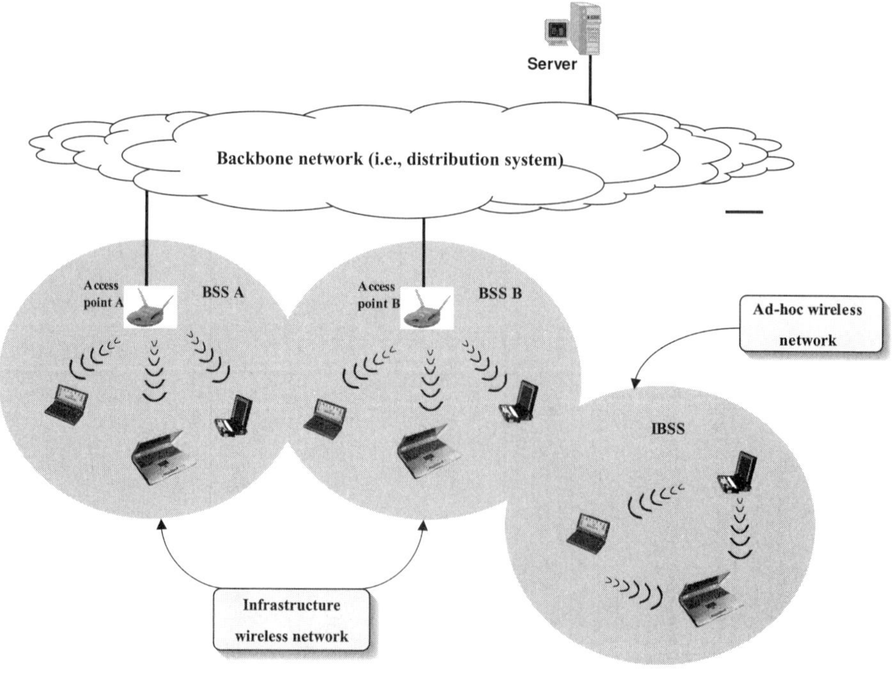

FIGURE 5.6 Infrastructure and ad hoc IEEE 802.11 W-LANs.

even move between BSSs transparently to the link control layer. One of the requirements of IEEE 802.11 is to be used with existing wireline networks. IEEE 802.11 solved this challenge with the use of a *Portal*. A portal is the logical integration between wireline LANs and IEEE 802.11. It also can serve as the AP to the DS. All data going to an IEEE 802.11 LAN from an 802.X LAN must pass through a portal. It thus functions as bridge between wireline and wireless.

Furthermore, the services that DS can support are divided into two sections: station services and distribution system services (DSS). There are five services provided by the DSS: Association, Reassociation, Disassociation, Distribution, and Integration. The first three services deal with station mobility. If a station is moving within its own BSS or is not moving, the station's mobility is termed No-transition. If a station moves between BSS's within the same ESS, its mobility is termed BSS-transition. If the station moves between BSS's of differing ESS's it is an ESS transition. For a station to use the LAN services, it must affiliate itself with the BSS infrastructure. This is done by Associating itself with an AP. Associations are dynamic in nature because stations move, turn on, or turn off. A station can be associated with only one AP. This ensures that the DS always knows where the station is. Association supports no-transition mobility but is not enough to support BSS-transition. BSS-transition is achieved through the Reassociation service, which allows the station to switch its association from one AP to another. Both association and reassociation are initiated by the station. Disassociation occurs when the association between the station and the AP is terminated. This can be initiated by either party. A disassociated station cannot send or receive data. Notice that it is not mentioned ESS-transition so far. That is because ESS-transition is not supported. A station can move to a new ESS but will have to reinitiate connections.

Distribution and Integration are the remaining DSS's services. The distribution service is simply forwarding the data from the sender to the intended receiver. The message is sent to the local AP (input AP), then distributed through the DS to the AP (output AP) that the recipient is associated with. If the sender and receiver are in the same BSS, the input and out AP's are the same. So the distribution service is logically invoked whether the data is going through the DS or not. Integration occurs when the output AP is a portal. Thus, 802.× LANs are integrated into the IEEE 802.11 DS.

SS are authentication, deauthentication, privacy, and MAC service data unit (MSDU) delivery. With a wireless system, the medium is not exactly bounded as with a wireline system. In order to control access to the network, before stations are allowed to converse, they must first pass a series of tests to ensure that they are legitimate users. That is really authentication all about. Once a station is authenticated, it may then associate itself. The authentication relationship may be between two stations inside an IBSS or to the AP of the BSS. Authentication outside of the BSS does not take place. There are two types of authentication services offered by IEEE 802.11. The first is open system authentication. This means that anyone who attempts to authenticate will receive authentication. The second type is Shared Key Authentication. In order to become authenticated the users must be in possession of a shared secret. The shared secret is implemented with the use a data encryption algorithm called the wired equivalent privacy (WEP)[4] algorithm to protect authorized stations from eavesdroppers. The shared secret is delivered to all stations ahead of time in some secure method (such as someone walking around and

---

[4] WEP algorithm is based on the RC4 PRNG algorithm developed by RSA Data Security, Inc.

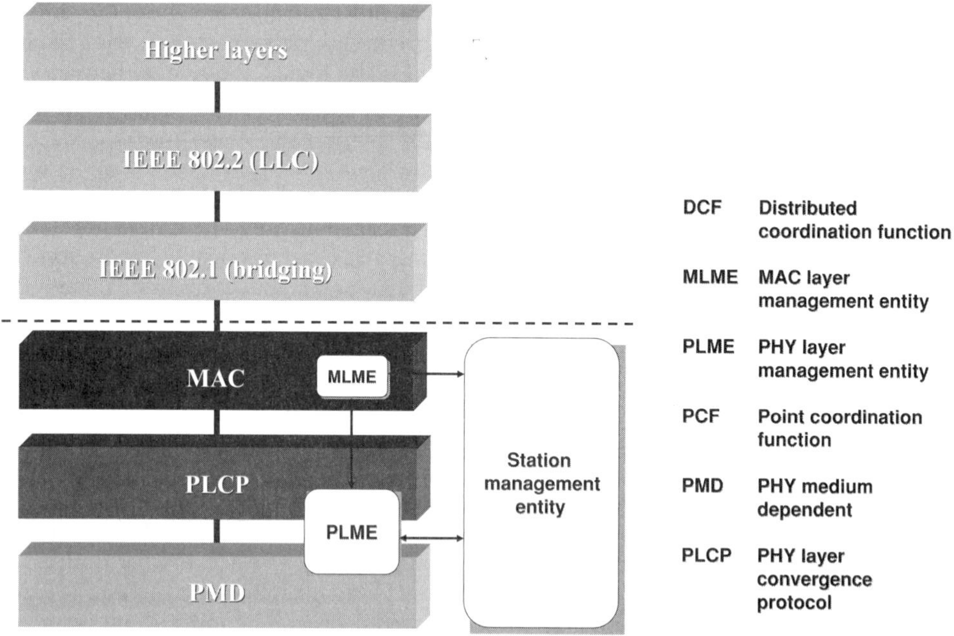

Figure 5.7 IEEE 802.11 reference model.

loading the secret onto each station). De-authentication occurs when either the station or AP wishes to terminate a station's authentication. When this happens the station is automatically disassociated. If WEP is not used, stations are "in the clear" or "in the red" status, meaning that their traffic is not encrypted. Data transmitted in the clear are called *plaintext*, whereas encrypted data are called *ciphertext*. All stations start "in the red" until they are authenticated.

Finally, with regard to addressing in IEEE 802.11 (see Fig. 5.7 for the IEEE 802.11 reference model), the authors of the IEEE 802.11 standard allowed for the possibility that the wireless media, distribution system, and wireline LAN infrastructure would all use different address spaces. IEEE 802.11 specifies addressing only for over the wireless medium, though it was intended specifically to facilitate integration with IEEE 802.3 wireline Ethernet LANs. IEEE802 48-bit addressing scheme was therefore adopted for IEEE 802.11, thereby maintaining address compatibility with the entire family of IEEE 802 standards. That is, each device owns a unique 48-bit station address. Further, each subnet has its unique identifier, which is the AP's MAC address (48 bit) in infrastructure networks or a random 48 bit IEEE 802 locally administered address (two bits + 46-bit random number) in infrastructureless networks. In the vast majority of installations, the distribution system is an IEEE 802 wireline LAN and all three logical addressing spaces are identical.

5.3.2 Key IEEE 802.11 MAC Layer Features

The IEEE 802.11 MAC layer provides a basic access mechanism that supports several characteristics such as clear channel assessment (CCA), both asynchronous and time critical data delivery, link setup, encryption and authentication, power management, and channel synchronization (Bray and Sturman, 2000). Further, link control and management, fragmentation and de-fragmentation (also known as fragmentation and reassembly), and roaming are situated in MAC layer entities (Matthias, 2001). Roaming, a key feature of W-LANs, allows mobile users to roam about BSS. Figure 5.6 illustrates how two APs are interconnected with the wireline backbone infrastructure while a mobile user seamlessly moves between two BSSs. This roaming capability is a key feature of the IEEE 802.11 family of specifications that other short-range wireless technologies, such as Bluetooth, do not support.

IEEE 802.11 MAC layer provides IEEE 802.11 users both contention-based and contention-free access control services for IEEE 802.11 W-LANs. The basic access method for peer-to-peer ad hoc networking in the IEEE 802.11 MAC protocol is the *distributed coordination function* (DCF), which incorporates a *CSMA with collision avoidance* (CSMA/CA) protocol as the basic medium access mechanism. On top of the DCF resides the point coordination function (PCF), which provides contention-free medium access in infrastructure-based network configurations (Matthias, 2001). The PCF operates similarly to a polling system (Conti et al., 1997), where a PC provides the transmission rights at a single station through a polling mechanism.

5.3.2.1 IEEE 802.11 MAC Framing

There are different types of MAC level messages. These are distinguished as data, management, and control. Data frames contain user data, management frames support the different MAC services (e.g., authentication frame), and control frames are meant to support delivery control of data and management frames. According to IEEE 802.11 standard, an IEEE 802.11 MAC frame consists of a MAC header, a frame body, and a CRC-32 frame check sequence (FCS). The basic structure of a MAC frame is depicted in Fig. 5.9 (IEEE 802 LAN/MAN Standards Committee, 1999).

The MAC header consists of seven fields and is 30 bytes long. These fields are frame control, duration, address 1, address 2, address 3, sequence control, and address 4.

The "Duration/ID" field is 2 bytes long. It contains the data on the duration value for each field and for control frames it carries the associated identity of the transmitting station. The "address" fields identify the BSS, the destination address, the source address, and the receiver and transmitter addresses. Each address field is 6 bytes long. The "sequence control" field is 2 bytes and is split into 2 subfields, fragment number and sequence number. Fragment number is 4 bits and tells how many fragments the MSDU is broken into. The sequence number field is 12 bits and indicates the sequence number of the MSDU. The "frame body" is a variable length field from 0 to 2312. This is the payload. The "FCS" is a 32-bit cyclic redundancy check (CRC), which ensures there are no errors in the frame. For the standard generator polynomial, see Section 5.2.3.

The "frame control" field, shown in Fig. 5.10, is 2 bytes long and is composed of the following fields: protocol version, Type, Subtype, To DS, From DS, More Flag, Retry, Pwr Mgt, More Flag, More Data, WEP, and Order.

Specifically, the "protocol version" field is 2 bits in length and will carry the version of the IEEE 802.11 standard. "Type" and "Subtype" fields are 2 and 4 bits, respectively. They work together hierarchically to determine the function of the frame. The remaining 8 fields are all 1 bit in length. The "To DS" field is set to 1 if the frame is destined for the distribution system. "From DS" field is set to 1 when frames exit the distribution system. Frames that stay within their BSS have both of these fields set to 0. The "More Flag" field is set to 1 if there is a following fragment of the current MSDU. "Retry" is set to 1 if this frame is a retransmission. "Power management" field indicates if a station is in power save mode (set to 1) or active (set to 0). "More data" field is set to 1 if there are MSDUs buffered for that station. The "WEP" field is set to 1 if the information in the frame body was processed with the WEP algorithm. The "Order" field is set to 1 if the frames must be strictly ordered.

5.3.2.2 IEEE 802.11 MAC Managements Entities

Both specified layers in IEEE 802.11, DCF and PCF, contain their own management entities, called management information bases (MIBs), which contain layer-specific information (Matthias, 2001). Specifically, there are three management entities namely the physical layer management entity (PLME), the MAC layer management entity (MLME), and the station management entity (SME).

The *PHY management entity* provides services to set the physical transmission channel, such as the hopping pattern in the FHSS system.

The *MAC sublayer managment entity (MLME)* is responsible for connection setup and maintenance, and power management. It provides the following services: power management, scan, synchronization, authenticate, deauthenticate, associate, reassociate, disassociate, reset, and start. Scanning for stations is used to find a station in communication range; the standard defines active or passive scanning. To synchronize all timers in the stations in a BSS and IBSS, the MLME provides the synchronization service. The synchronization information is provided either by the AP in a BSS (i.e., centralized) or via a distributed algorithm in an IBSS. In the centralized configuration, the AP sends a beacon frame containing BSS properties and the timing synchronization information in regular intervals.

Furthermore, a station must first authenticate itself before it associates itself with a BSS. Associated stations may use the distribution system (e.g., communicate with other stations in other subnets, requesting point coordination services), whereas nonassociated station must not. The reset service resets the MAC entity, and the start service is used to either start a BSS or an IBSS, if the station has not synchronized itself with a BSS.

The *SME* services. These are not fully specified in the standard. It is a layer-independent entity and would typically get and set values from the MLME and the PLME, to pass them to higher-level management functions. It can be considered as residing in a control plane.

5.3.2.3 The IEEE 802.11 PCF

In order to support time-bounded services, the IEEE 802.11 standard specifies the optional use of the PCF, in which a PC (or PCF station) has priority control of the medium. The PC resides in the AP. It schedules a contention-free period, which is announced by sending a beacon frame after the SIFS timer expires. That is, when PCF is active, the PCF station allows only a single station in each cell to have priority access to the medium at any one time.

This is implemented through the use of the previously mentioned PIFS and a beacon frame (see Fig. 5.11) that notifies all the stations in the cell to defer transmitting for the length of the contention free period (CFP). If one of the stations does not hear the expected beacon, it sets its NAV to a known maximum value for the length of the CFP. The length of the contention free period can vary within each CFP repetition interval according to the system load. Having silenced all of the stations, the PCF station can then allow a given station to have contention free access through the use of an (optional) polling frame that

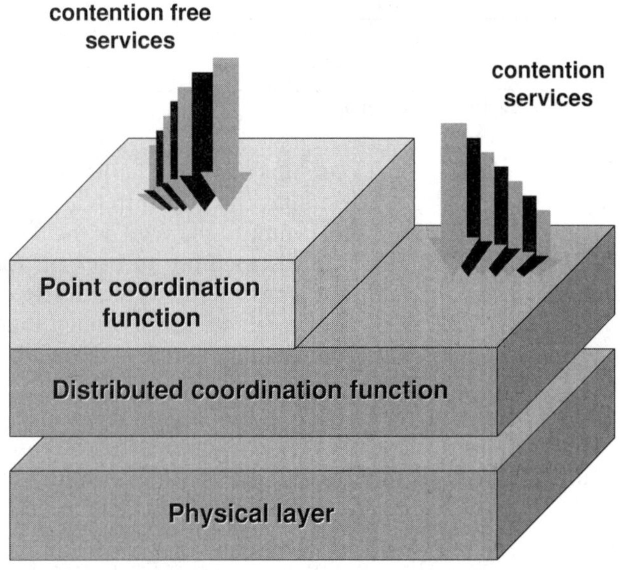

FIGURE 5.8 IEEE 802.11 architecture.

is sent by the PCF station. A station interprets a poll as a resource grant to transmit. QoS guarantees could be foreseen using this mechanism.

A typical wireless LAN installation would use different channels for adjacent cells so as to prevent two PCF stations (i.e., APs) from using the same channel during the contention-free period. This would allow coexistence with an ad hoc network that is using DCF only, even on the same channel (see Fig. 5.8).

5.3.2.4 The IEEE 802.11 DCF method for Medium Access in Infrastructureless Wireless Networks

The *DCF* specifies the use of the CSMA protocol with CA. The CSMA used in wireless networks is similar to the CSMA scheme used in wireline LANs. However, the collision detection (CD) technique for wireline LANs cannot be used effectively for wireless LANs since nodes cannot detect over-the-air collisions when they occur. The absence of detection is caused by the strong signals present at the transmitters that also serve to drown out other communicating signals (Wickelgren, 1996).

The CA property of the CSMA-based W-LANs helps them to reduce the number of over-the-air collisions. In short, the basic CSMA/CA medium access function allows for options that can minimize collisions by using RTS, CTS, data and ACK transmission

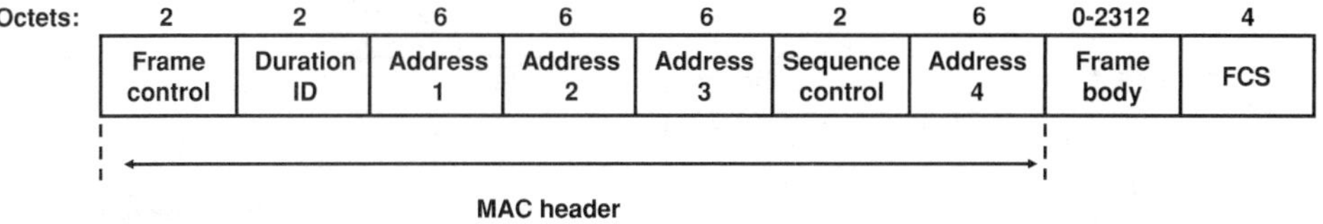

FIGURE 5.9 IEEE 802.11 frame.

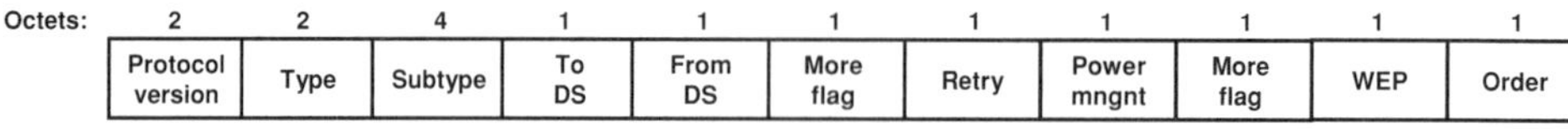

Figure 5.10 IEEE 802.11 MAC header.

frames, in a sequential fashion. Communication is established when one of the wireless nodes sends a short message RTS frame. The RTS frame includes the destination and the length of message. The message duration is known as the network allocation vector (NAV). The NAV alerts all others in the medium to back off for the duration of the transmission. The receiving station issues a CTS frame, which echoes the sender's address and the NAV. If the CTS frame is not received, it is assumed that a collision occurred, and the RTS process starts over. After the data frame is received, an ACK frame is sent back verifying a successful data transmission.

Before delving into the details of DCF, however, it is sensible to introduce the interframe space concept of IEEE 802.11, which is used to provide priority medium access. In general, this can be seen as a set of medium access timings for the different frame types, the random backoff procedure, the frame transfer procedures, the acknowledgement procedures, and the RTS/CTS procedures. More specific, each timer expires after a certain time period. According to which timer expires, certain frame types may or may not be sent, subject to their priority. These interframe spaces are summarized in Fig. 5.11.

To this end, when using the DCF access method (summarized in Fig. 5.12), before a station initiates a transmission, it senses the channel to determine whether another station is transmitting. If the medium is found to be idle for an interval that exceeds the *distributed interframe space* (DIFS), the station continues with its transmission.[5] The transmitted packet contains the projected length of the transmission. Each active station stores this information in a local variable, named NAV. NAV, therefore, conveys information about how long the medium will remain busy (see Fig. 5.12a). This mechanism prevents a station from listening to the channel during transmissions thus enabling the implementation of power-saving policies.

The CSMA/CA protocol does not rely on the capability of the stations to detect a collision by hearing their own transmissions. Hence, immediate positive acknowledgments are employed to ascertain the successful reception of each packet transmission.

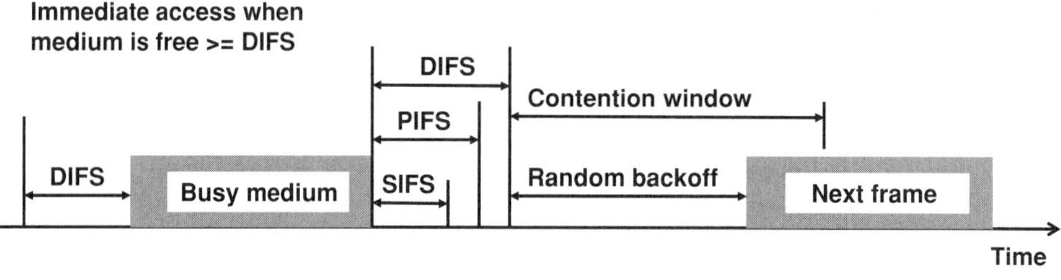

Figure 5.11 Primary access mechanism.

[5] To guarantee fair access to the shared medium, a station that has just transmitted a packet and has another packet ready for transmission must perform the backoff procedure before initiating the second transmission.

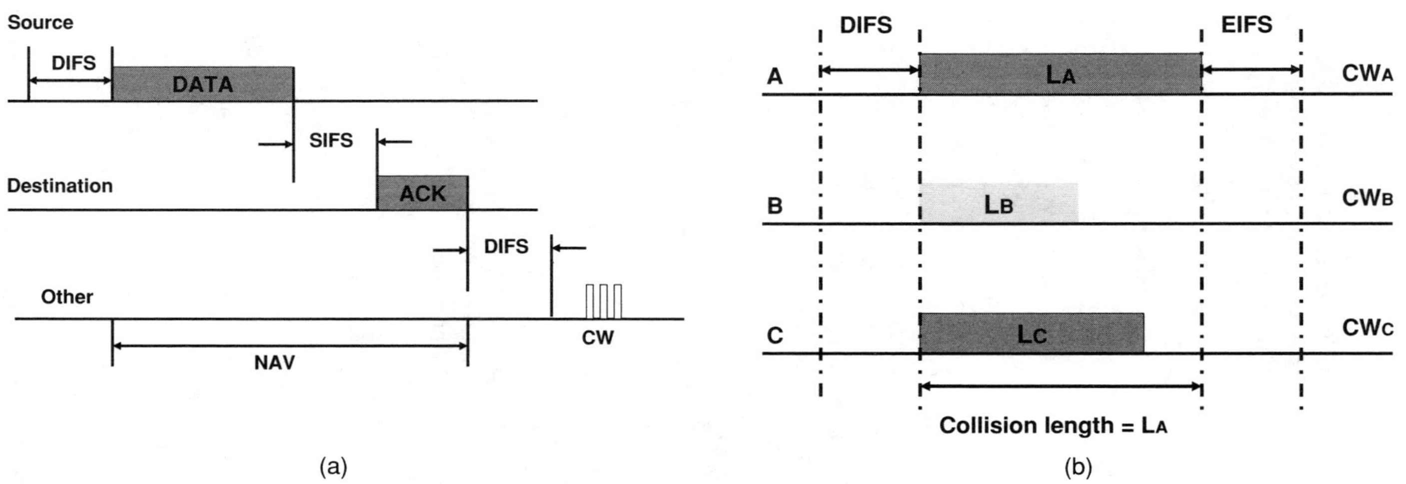

Figure 5.12 EEE 802.11 DCF. (a) A successful transmission. (b) A collision.

Specifically, the receiver after the reception of the data frame waits for a time interval, called the *short interframe space* (SIFS), which is less than the DIFS, and then initiates the transmission of an ACK frame. The ACK is not transmitted if the packet is corrupted or lost due to collisions. A CRC algorithm is adopted to discover transmission errors. Collisions among stations occur when two or more stations start transmitting at the same time (see Fig. 5.12b). If an acknowledgment is not received, the data frame is presumed to have been lost, and a retransmission is scheduled.

After an erroneous frame is detected (due to collisions or transmission errors), the channel must remain idle for at least an *extended interframe space* (EIFS) interval before the stations reactivate the backoff algorithm to schedule their transmissions (see Fig. 5.12b).

The CA scheme in CSMA/CA further provides a random backoff delay feature before a new transmission attempt is executed, which guarantees a time spreading of the transmissions. This random delay helps avoid collisions from simultaneous multi-user transmissions, since other wireless nodes could also be waiting to send data over the network (Wickelgren, 1996). When a station S, with a packet ready for transmission, observes a busy channel, it defers the transmission until the end of the ongoing transmission. At the end of the channel busy period, the station *S* initializes a counter (called the *backoff timer*) by selecting a random interval (*backoff interval*) for scheduling its transmission attempt. The backoff timer is decreased for as long as the channel is sensed as idle, stopped when a transmission is detected on the channel, and reactivated when the channel is sensed as idle again for more than a DIFS. The station transmits when the backoff timer reaches zero.

5.3.2.5 Power Saving

Since most IEEE 802.11 devices are wireless mobile (e.g., small handhelds and personal digital assistants), power consumption optimization is a critical matter. There are two power-saving modes included in the MAC protocol: *awake* and *doze*. In the *awake* mode, stations are fully powered all the time. In the *doze* mode, nodes must "wake up" periodically to listen for beacons, which indicate that AP has queued messages. Nodes must inform their associated AP before entering doze.

A station in the power saving mode (i.e., the doze state) cannot transmit or receive frames. The IEEE 802.11 standard defines power management procedures for cases with and without infrastructure. In the presence of infrastructure, a dozing station periodically wakes up and listens to selected time stamped beacons that are sent by the AP. If the beacon indicates that the AP has queued data for that station, the station sends a special poll frame that tells the AP to send the data.

In the absence of infrastructure, the policy adopted within an IBSS is completely distributed for preserving the self-organizing behavior. The power-conserving stations in the ad hoc cell wake up for only short predefined periods of time, which are announced in the ATIM window, to hear if they should remain on in order to receive a frame. All requests for transmissions are placed within the ATIM window. A station in sleep mode that receives one, wakes up for an interval, which equals at least the next beacon interval. The ATIM window is also determined by the first station to initialize the IBSS. If the time ATIM window is zero, the power save mode is not used.

5.3.2.6 THE IEEE 802.11 PHYs

The PHY in IEEE 802.11 consists of two sublayers: the physical medium dependent sublayer, and the PHY convergence sublayer (CS) on top. The physical medium dependent

Region	Allocated Spectrum (GHz)
US	2.400–2.4835
Europe	2.400–2.4835
Japan	2.471–2.497
France	2.4465–2.4835
Spain	2.445–2.475

TABLE 5.2 Global Spectrum Allocation at 2.4 GHz

sublayer specifies how to send and receive data over the wireless medium, whereas the CS maps the MAC frames to the physical medium dependent functions (Matthias, 2001). The original IEEE 802.11 standard specifies the use of three different PHY layers, any of which can utilize the same MAC layer. These PHY layers include two spread spectrum techniques at 2.4 GHz (Table 5.2) in the ISM band: frequency-hopping spread spectrum (FHSS) and DSSS. The third PHY layer is an optical technique utilizing diffuse infrared (DFIR).

In Europe, the same 2.4 GHz band (as the U.S. ISM band) has been allocated to allow wireless LAN operation, whereas in Japan only the frequencies from 2.471 to 2.497 GHz have been allocated, requiring thus special provisions in the IEEE 802.11 draft standard. The IEEE 802.11 committee allowed the definition of multiple PHY layers, in part, because the members of the committee had some interest in each of the aforementioned PHY layers and hence they sought to accommodate all of them. The benefit of this approach is that IEEE 802.11 compliant wireless LAN users can exploit the advantages of each of the PHYs deployed (e.g., see Goldberg, 1995). The downside of this approach is that in order to permit interoperability between two users, the users need to specify additionally the type and data rate of their wireless LAN system.

All IEEE 802.11 devices using one of these three technologies are required to operate at 1 Mbps data rate, with 2 Mbps as an option. The maximal size of an MSDU is 2312 bytes, excluding the MAC header and the physical preamble. There are several physical-layer dependent parameters that are relevant to the design of the MAC protocol. On one hand, there is the Rx/Tx (Receiver/Transmitter) turnaround time that varies from:

- 0 ms (infrared),
- 10 ms (direct sequence),
- to 19 ms (frequency hopping).

This time is contained in both the length of the interframe spaces and the length of the backoff slots in the contention window. The backoff slottime for the three layers is defined as:

- 6 ms for infrared
- 20 ms for direct sequence
- 50 ms for frequency hopping

In addition, each PHY introduces a physical preamble with different length, which is added to each packet. Infrared adds 92–112 timeslots of 250 $n + 32$ bits, direct sequence 192 bits whereas frequency hopping adds 122 bits.

Given therefore a specific PHY technique with different backoff slot times and physical preamble lengths, significantly different performance measures are observed at the MAC layer.

5.3.3 IEEE 802.11 as Ad Hoc Network

As previously mentioned, the IEEE 802.11 concept allows for infrastructureless networks. The IEEE 802.11 standard is a good platform to implement a single-hop local ad hoc network mainly because of its extreme simplicity. Multihop networks covering areas of several square kilometers could also be built by exploiting the IEEE 802.11 technology.

When referring to ad hoc configuration, IEEE 802.11 refers to the IBSS, the Independent BSS, in which all stations are in mutual communication range and communicate directly. The IBSS has no AP and thus cannot use the distribution system. Since there is no AP available, stations cannot associate to any BSS, hence only frames, which do not use the DS, are allowed.[6] Therefore, the stations make use of just the SS (authentication, MSDU delivery, privacy). The major changes occur in the MAC layer, in particular the MLME, which provides contention services (DCF services), authentication, and privacy to the higher layers.

As with the globally managed BSSID (in infrastructure mode), this is now locally assigned. It is an IEEE 802 locally administered address with the individual/group bit set to 0 and the universal/local bit set to 1. The remaining 46-bit number is chosen at random. The random algorithm should for similar seeds (adjacent IBSS for example, if local time is used as seed) return different numbers.

5.4 The Worldwide Interoperability for Microwave Access (WiMAX)[7]

The IEEE 802.16 standard (Wolnicki, 2005), also known as World Interoperability for Microwave Access (WiMAX), is designed to provide ubiquitous high-speed wireless access to customers, such as broadband home networks, community networks, and enterprise networking. Both point-to-multipoint (PMP) mode and mesh mode are defined in the IEEE 802.16-2004 released version (IEEE Standard for Local and metropolitan area networks, 2004). WiMAX intends to offer broadband access services to customers using fixed and mobile technologies. The fixed wireless access is guaranteed through the implementation of the IEEE 802.16d standard while the mobile access technology falls under the 802.16e WiMAX.

5.4.1 WiMAX Standards

IEEE 802.16d, which was ratified in July 2004, specifies the air interface for fixed broadband wireless access systems. It targets the 10–66 GHz licensed band with the line-of-sight (LoS) and the license-exempt frequencies below 11 GHz. The specification details the physical and MAC layers design. IEEE 802.16d supports both PMP and mesh communication modes.

---

[6] These frames are called Class 1 and Class 2 frames in the standard.

[7] Excerpt from the invited article "Worldwide interoperability for microwave access (WiMAX)," Neila Krichene and Noureddine Boudriga; *Mohammad S. Obaidat (*Computer Science Department, Monmouth University, W. Long Branch, NJ 07764, USA. E-mail: Obaidat@monnmouth.edu)

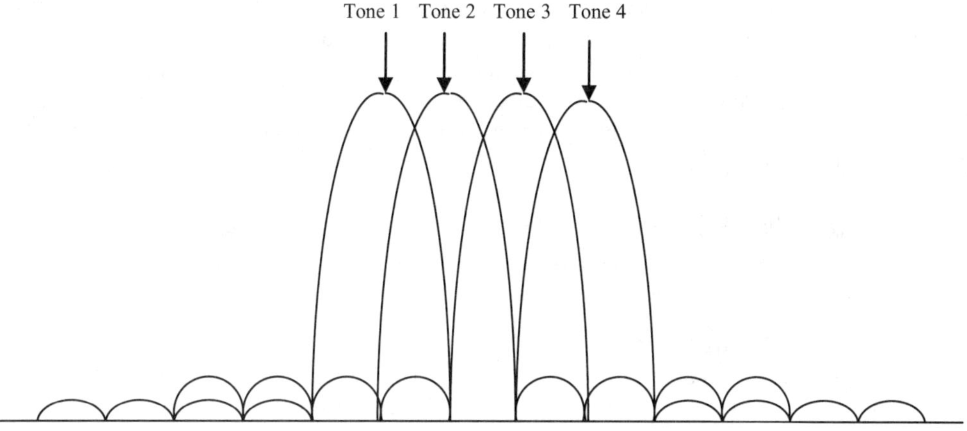

Figure 5.13 OFDM principle (Olexa, 2004).

WiMAX forum focuses on the OFDM variant. In OFDM, the subcarriers or "tones" overlap but are spaced apart at precise frequencies in order to guarantee orthogonality and prevent interference. In fact, Fig. 5.13 shows that the center of the modulated carrier coincides with the edge of the adjacent carrier. In this sense, the independent demodulators that perform a discrete Fourier transform see only their own frequencies.

WiMAX defines 256 subcarriers simultaneously transmitted with 192 data subcarriers, 8 pilot subcarriers and 56 nulls. Data are split into 192 parallel data streams each of which will be transmitted at 1/192 of the original rate. Pilot subcarriers guarantee a reference to minimize frequency and phase shifts that may occur during the transmission while null carriers are used for guard bands and the DC carrier.

The MAC layer is connection oriented. Every subscriber station (SS) is characterized by a 48-bit MAC address but the 16-bit connection identifier (CID) is used to uniquely identify each SS. MAC PDUs present a maximum length of 2048 bits; they are formed by a fixed-length header, a variable-length payload, and an optional CRC field. Three connection types are used: (a) the basic connection is used to transfer short, time-critical and radio link control messages, (b) the primary management connection carries long messages related with connection establishment and authentication, and (c) the secondary management connection transfers standard-based messages like DHCP and TFTP.

A SS first has to implement the network entry process in order to communicate with other nodes. During this phase, the SS synchronizes at the physical level by listening to a periodic frame preamble. It then processes an initial ranging and sends a capability request message indicating the supported modulation level, coding scheme, rates, and duplexing methods. Then an authentication phase followed by an IP address assignment and a connection creation is realized.

IEEE 802.16e WiMAX, on the other hand, is an amendment to the 802.16d WiMAX solution; it targets the 2–6 GHz licensed bands, implements mobility services, and supports subscribers moving at vehicular speed ranging from 75 to 93 miles/h (Fili, 2005). IEEE 802.16e adapts the scalable OFDMA (SOFDMA) technique at the PHY in order to improve the multiaccess capabilities while enhancing the MAC layer by addressing mobility issues and particularly handover.

SOFDMA modulation technique: To optimize the media access assignment, SOFDMA assigns a set of subcarriers (also referred to as subchannels) to different users. For instance, subcarriers 1, 3, and 7 may be assigned to user 1 while subchannels 2, 5, and 9 to user 2, and so on (Wolnicki, 2005). Mobility leads to a subchannel reassignment, which in turn is affected by the MS-to-BS distance. SOFDMA varies the size of FFT according to the available bandwidth.

5.4.2 The Air Interface Specifications for Fixed Access

The air interface specifications of 802.16d WiMAX provides fixed and nomadic access for both outdoor and indoor user scenarios.

5.4.2.1 Operation Modes

As aforementioned, WiMAX supports both PMP and mesh modes. In PMP, SS may communicate only with BS; meaning that the BS has to route data between communicating SSs within its coverage range. The BS broadcasts downlink transmissions to all stations within its coverage; the SSs listen to the broadcast subframe to verify whether they are concerned by a particular traffic by checking the CID in the received protocol data units (PDUs) (IEEE, 2004). Besides, the BS schedules the uplink transmission based on the SSs' requests. PMP also supports multicast and broadcast. Contention control is guaranteed by adopting either unsolicited bandwidth grants, polling, or contention-based procedures.

In the mesh mode, a traffic exchange may directly occur between two communicating SSs without having to deal with the BS. The BS does not necessarily route data; but a mesh BS is needed in order to provide connection to the backhaul. The adopted transmission algorithm is based on an adaptive scheduling mechanism that may be centralized or distributed. In distributed scheduling, all mesh nodes including the BS need to coordinate their transmissions in their two-hop neighborhood by broadcasting their available resources, requests and grants for downlink and uplink to all their neighbors (IEEE, 2004). In centralized scheduling, the mesh BS is responsible for collecting the requests coming from SSs within a certain range. It then determines granted resources for each link in both directions (e.g., uplink and downlink) and broadcasts those grants to all mesh SSs within the hop range. The current schedule is then deduced by the SSs using a particular algorithm.

5.4.2.2 CS Specification

The CS is responsible for performing packet classification and payload header suppression. Upon receipt of higher-layer packets, CS maps them to a particular service flow associated to certain QoS requirements. Packets are classified with respect to multiple criteria such as destination or source IP addresses. A packet matching a criterion is delivered to a MAC connection that has been associated to that criterion. As service flow may match multiple criteria, multiple classifiers may be defined for the same flow. Payload header suppression (PHS) consists of removing redundant information from higher layer packets headers by masking some of their bits. The receiving entity on the second connection's side unmasks the concerned bits and rebuilds the header before retransmitting the packet to the upper layers (IEEE, 2004).

5.4.2.3 MAC Layer Specifications

PDU Formats A MAC header and a payload form a MAC PDU. The payload may be formed using zero or more subheaders and zero or more SDUs or different PDU fragments. An optional cyclic redundant check (CRC) field may also exist. The MAC PDU size varies according to the content of the payload field; this allows MAC layer transmitting transparently different higher-level traffic without knowing the format or the handled bit patterns (IEEE, 2004). The IEEE 802.16d defines two header formats, which are the generic MAC header and the bandwidth request header. The first format starts each PDU containing MAC management messages or CS data. The second format is applied when requesting additional bandwidth.

Network Entry and Initialization The network entry process consists of associating a mobile subscriber with a base station. For this, a bit (Physical) and frame (MAC) synchronization is processed and an initial ranging is performed to adjust the optimized transmission power. Additional operations such as setting up the management connections, authenticating the SS, and performing higher-layer protocol interaction need also to take place to end the entry process (Olexa, 2004).

Scheduling Scheduling aims at coordinating transmissions between nodes. The 802.16d defines four services: the unsolicited grant service (UGS), the real-time polling service (rtPS), the Non-real-time polling service (nrtPS), and the best effort (BE) service. Particular QoS parameters need to be defined when a scheduling service is enabled for a service flow. UGS supports real-time communications occurring at periodic intervals such as VoIP. This service requires the definition of the maximum sustained traffic rate, the maximum latency, the tolerated jitter and the request/transmission policy. The rtPS supports periodic real-time variable-size transmissions such as MPEG video streams. Thus, this scheduling service defines the minimum reserved traffic rate, the maximum sustained traffic rate, the maximum latency, and the request/transmission Policy. The nrtPS scheduling service guarantees a minimum data rate for delay-tolerant applications, which generate variable-sized traffic such as FTP.

Bandwidth Allocation Three dedicated CIDs are assigned to each SS to exchange control messages and provide QoS. The amount of the requested bandwidth is determined by calculating the number of bytes needed to transmit the MAC header and payload while ignoring the PHY layer overhead. While an SS requests bandwidth for a connection, the BS allocates the resources with respect to an SS Basic Management Connection. As SSs need bandwidth in order to emit their bandwidth requests, three polling strategies can be identified: unicast, multicast, and broadcast.

Contention Resolution As the BS controls the uplink channels by means of UL-MAP messages, it is able to detect the mini-slots experiencing collisions, which may affect the Initial Ranging and the Request procedures. To resolve collisions, the standard proposes a procedure based on a truncated binary exponential backoff (BEB) algorithm. The initial backoff window and its maximum size are specified in the UCD message and determined by the BS. An SS with data to send enters the contention resolution process and sets its internal backoff window to the value specified in the UCD. It then randomly chooses a number within the backoff window referring to the number of contention transmission opportunities that should be deferred by the SS before transmitting. If the SS deduces that the sent data have experienced contention, it should increase its backoff window by a factor of two without exceeding the maximum window size. The procedure is repeated until the transmission occurs successfully or the threshold retries number is reached.

QoS Support QoS provision aims at associating MAC packets sent on a connection into a specific service flow identified by the CID. A service flow is a unidirectional packet flow belonging to a QoS class. It is characterized by a set of QoS parameters including: the maximum sustained traffic rate, minimum reserved traffic rate, maximum latency, request/transmission policy, traffic priority, and tolerated jitter. A service flow is also described by different attributes that determine the adopted packet scheduling and the method of requesting bandwidth. These attributes may be dynamically managed using MAC control messages to accommodate the dynamic service demand. They are: the service flow ID (SFID), CID, ProvisionedQoSParamSet, AdmittedQoSParamSet, ActiveQoSParamSet, and the authorization module (which is a logical function implemented at the BS in charge of accepting or rejecting changes to QoS parameters and classifiers). Three types of service flow are used: provisioned, admitted, and active.

5.4.2.4 PHY Overview

The IEEE 802.16d air interface specifications define multiple PHYs for different frequency bands and region-by-region frequency rules to cope with customers' demands and target diverse markets. WiMAX solutions represent a specific implementation of the IEEE 802.16d standard as they adopt common profiles in order to guarantee interoperability. The key features of the WiMAX 802.16d specifications are based on the 256-point FFT-OFDM modulation technique with a channel bandwidth up to 10 MHz and an optional uplink subchannelization. TDD and FDD multiplexing strategies are supported while an optional space timing coding may be adopted. QoS is guaranteed on a per-connection basis while Automatic Retransmission Request is used to address transmission errors. IEEE 802.16 standard defines five PHYs where AAS refers to adaptive antenna system, STC refers to Transmit Diversity Scheme and ARQ refers to Automatic Transmission Request.

Wireless MAN-SC The specifications of this PHY require LoS and support TDD and FDD multiplexing techniques to guarantee a flexible spectrum usage. Transmission parameters such as the adopted coding and modulation may be adjusted to each SS on a frame-by-frame basis. The use case (e.g., registration, contention user traffic) and the targeted performances determine the number of time slots assigned to the uplink channel. The downlink channel uses the TDM multiplexing technique as information of a SS is multiplexed into a single frame and received by all SSs within the same sector. To support half-duplex FDD, SS provision is also made for a TDMA portion of the downlink. Data bits coming from the Transmission CS on the downlink are randomized, FEC encoded than mapped to a QPSK, 16-quadrature amplitude modulation (QAM) or 64-QAM (optional) constellation depending on the signal-to-noise ratio (SNR) condition of the radio link. On the other hand, the uplink uses TDMA transmissions where TDMA bursts carrying variable-length MAC PDUs are mapped to a QPSK.

The TDD downlink subframe structure is made up of a Frame Start Preamble used by the SSs for synchronization and equalization purposes and a control section formed by the DL-MAP and the UL-MAP fields.

Wireless MAN-SCa This PHY uses the 2.5–11 GHz frequency band and enables NLOS transmissions. It is based on single-carrier technology while the allowed channel bandwidths should not be less than 1.25 MHz and are limited to provisioned bandwidth divided by any power of two. To guarantee more robustness to the air interface, additional features are used such as adaptive antennas with STC transmit diversity technique. This PHY also implements concatenated FEC using Reed6Solomon and pragmatic trellis coming with an interleaving possibility.

Transmitted burst sets include a preamble and one or more concatenated data bursts described by the corresponding burst profile. The transmitted bits are first randomized then FEC encoded in order to be directly mapped to QAM symbols using the appropriate Gray-coding map. The obtained symbols are then multiplexed into a duplex frame. A power control algorithm is also supported for the uplink channel in order to guarantee initial calibration and periodic adjustment procedures without leading to data losses.

Wireless MAN-OFDM This PHY operates in the frequency band below 11 GHz and designed for NLOS operations. The time structure of the OFDM symbol is formed by the useful symbol time Tb and a copy of the last Tg of the useful symbol period, termed Cyclic Prefix or CP, used to collect multipath while preserving the orthogonality of the tones. The CP overhead can be reduced by increasing the FFT size; however, this reduction may affect the sensitivity of the system to phase noise of the oscillators. The WiMAX forum adopted a 256 FFT OFDM modulation.

The SS shall hear all possible values of the CP during the initialization phase until that of BS is found. The same CP is then used on the uplink. The BS should select a permanent CP duration on the downlink; otherwise, all the SSs will be obliged to resynchronize with it. On the frequency domain, data subcarriers, pilot subcarriers, and null subcarriers form an OFDM symbol.

The OFDM frame in PMP mode (Fig. 5.14) is formed by the physical PDUs of BS and SSs, gaps and guard intervals. It is formed by a downlink subframe composed of downlink PHY PDUs and an uplink subframe consisting of contention intervals scheduled for initial ranging and bandwidth requests with one or more multiple uplink PHY PDUs, each transmitted by a different SS. Downlink or uplink PHY bursts are formed by an integer number of OFDM symbols carrying MAC messages. In an UL allocation, an SS with no data to transmit should emit an UL PHY burst message with a bandwidth request header attached on it. All SSs must transmit during their UL allocations; standard-padding mechanisms shall be used if necessary.

Wireless MAN-OFDMA The WirelessMAN-OFDMA PHY operates in frequency bands below 11 GHz and guarantees NLOS transmissions. Nominal channel bandwidths should not be less than 1 MHz. An OFDMA symbol time structure is the same as the OFDM type. When considering an OFDMA symbol in the frequency domain, we distinguish data subcarriers, pilot subcarriers, and null carriers. Active subcarriers are classified into subsets of subcarriers forming subchannels. An OFDMA physical slot, which is the minimum possible data allocation unit, needs a time and a subchannel dimension to be characterized. The WirelessMAN-OFDMA defined by the IEEE 802.16-2004 uses a 2048-point transfer function and multiple carriers to provide users with multiple channel access.

5.4.3 The Mobile Air Interface Specifications

5.4.3.1 SOFDMA Overview

The 802.16e adopts the SOFDMA, which enhances the traditional OFDM modulation. As described earlier, OFDM consists of dividing the bandwidth into orthogonal subcarriers.

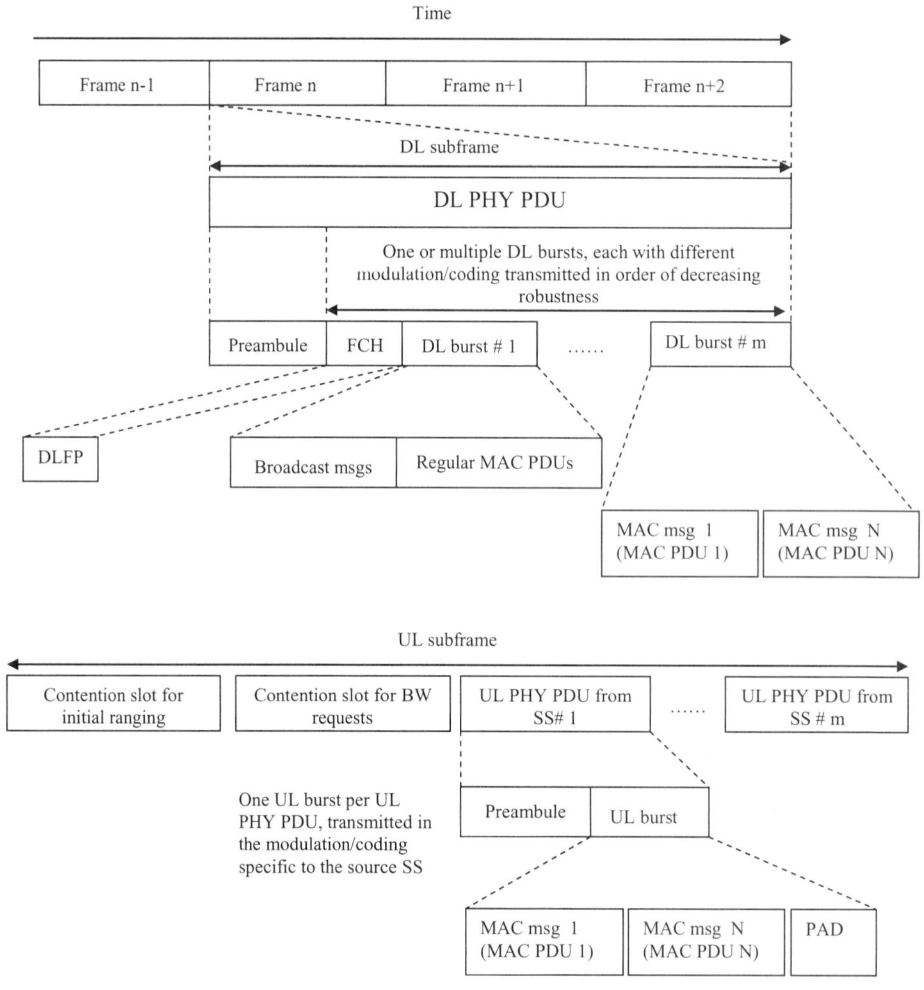

FIGURE 5.14 An OFDM frame in the FDD-PMP mode.

It enables data multiplexing from multiple users. The OFDMA technique consists of assigning a set of data, pilot and null subcarriers. The data and pilot subcarriers are grouped into subsets of subcarriers called subchannels. The minimal duration of a subchannelization is one slot during in which 48 data tones (subcarriers) are carried. Subcarriers are then permuted using either diversity or contiguous permutation. Diversity consists of pseudorandomly defining subcarriers in order to guarantee the frequency diversity and enable an intercell interference averaging. Clusters distributed throughout the subcarrier space form groups of clusters. Two clusters in a group form a subchannel, which groups 48 data subcarriers and 8 pilot subcarriers. The diversity permutation technique is used in the context of mobile applications ("Mobile WiMAX—Part I: A technical overview and performance evaluation," prepared on behalf of the WiMAX Forum,

February 21, 2006: http://www.wimaxforum.org). The contiguous permutation defines blocks of contiguous subcarriers. Eight contiguous data subcarriers and one pilot subcarrier in a symbol form a bin. A slot is defined as a set of bins following the rule $N*M = 6$ where N is the number of contiguous bins and M is the number of contiguous symbols. Contiguous permutations fit to fixed, portable, or low-mobility contexts.

Location Management Mobile customers need to access WiMAX services regardless of the time and their current position. Location management intends to locate mobile users despite the frequent changes in their network address in order to determine how to route the incoming calls and messages. Location management consists of the Location tracking and updating (registration) and the Location finding (paging).

Often, the location management schemes combine location tracking and location finding to minimize the overhead and paging delay. Generally speaking, a wireless network is always divided into location areas (LA) and paging areas (PA) that may overlap. LAs are relatively wide areas over which location updates take place while PAs are areas over which paging updates take place (Batayneh, 2006). In the mobile version of WiMAX, each set of BSs covering a varying important range forms a paging group that may contact the MS on the downlink by sending cyclic paging information without requiring that the MS send messages on the uplink. The MS synchronizes with the sent traffic after decoding a preferred BS downlink. The broadcast paging traffic may order a network entry or establish location and acknowledge messages. Location management updates are periodically performed in a secure or insecure manner (Lax and Dammander, 2006).

Homogeneous Handover Handover or handoff consists of transferring a communication in progress from one cell transmitter and receiver and frequency pair to another cell transmitter and receiver using a different frequency pair without interrupting it. More specifically, handover is executed when the mobile station moves out the coverage range of the BS or when the MS detects that a different BS can offer a better signal quality or QoS. The handover process has been developed for connection oriented speech systems like GSM and UMTS. However, it presents a new research domain that needs to be further investigated when it comes to applying it to packet-based networks such as IEEE 802.16e, IEEE 802.20, and 802.16e WiMAX (Lax and Dammander, 2006). A MS experiences homogeneous handoff when it is moving within networks presenting the same access technology. Basic handover operations consist of first collecting network information that is used to select the future serving base station; then executing the transfer of the communication.

IEEE 802.16e Handover Handover mechanisms are implemented at the MS and the BS levels. To be able to perform handover, the MS needs to collect network-related information by using the BS broadcast topology advertisements or scanning the neighboring BSs to localize them and sense the quality of their channels. The scanning process begins when the MS requests scanning intervals interleaved with normal operation periods from the serving BS. The BS assumes that the MS has entered the scanning mode and may buffer data destined to it. The end of the scanning period is assumed when the BS receives MS' PDUs again. The collected information is then saved for a certain period; it will guide the MS through the handoff process when needed (Lax and Dammander, 2006). To perform the handoff, six phases need to be executed:

- Cell reselection: During this phase, the MS collects network-related information. This phase will not necessarily be followed by a handoff execution.
- Handover decision and initiation: A handover decision may be assumed from the MS or the serving BS. A handover request is sent to the related entity then a set of handover messages will be exchanged.
- The synchronization to the target BS downlink: To associate with the target BS, the MS needs to synchronize with it. It will then be followed by the ranging procedure.
- The ranging: this is performed after synchronization. It enables the MS acquiring correct transmission parameters such as the time offset and the power level.
- The termination of service: this phase marks the end of the handover process, in fact, the serving BS ends all connections established with the MS and deletes information related with it.
- The handover cancellation: the handover process may be cancelled by the MS although it has already began; however, the cancellation request must be done before a certain time threshold.

WiMAX Handover WiMAX handover furnishes call handoff mechanisms at high levels. Layer-2 handoffs have been designed in order to last less than 50 ms (Mobile WiMAX—Part I: A technical overview and performance evaluation, prepared on behalf of the WiMAX Forum, February 21, 2006: http://www.wimaxforum.org). WiMAX handover addresses advanced procedures such as interaccess service network (ASN) handoff, roaming and seamless handoff at vehicular speed along with micro/macro mobility. An ASN is formed by at least a BS and a gateway. The BS manages the MSs in its coverage range while the ASN gateway relays data to the connectivity service network (CSN), which is defined as a network of Internet gateways, routers, servers, and proxies providing IP connectivity to WiMAX subscribers. The handoff process involves two types of entities: the serving entities and the target ones. More specifically, the serving BS is the BS managing the MS before the handover, while the serving ASN Gateway is the ASN gateway corresponding to the serving BS. The target BS is the one associated with the MS after handover, and the target ASN gateway is the gateway related with the target BS. We also distinguish the anchoring ASN gateway, which is the ASN GW that receives and relays the CSN data to the serving ASN gateway. This means that the MS mobility is completely transparent to the CSN and that is not necessary to frequently change the assigned IP address. The handoff process is implemented by the handoff function, the data path function, and the context function. The handoff function manages the signaling messages and the decisions triggered by the handover process, while the data path function establishes the path and manages data transmissions. Finally, the Context function handles the information and context related with the MS and their exchange in the backbone (Lax and Dammander, 2006).

Intra-ASN handover occurs when the target BSs belongs to the same ASN. The target BSs may be managed by the same ASN gateway that also manages the serving BS. Intra-ASN handoff is rapidly performed since all transitions are placed within the same network. Furthermore, there is no need for changing the MS's assigned IP as the handover will be transparent to external ASN entities. Inter-ASN handover ambiguously involves BSs belonging to different ASNs. This means that the ASN gateways will also belong to different networks and they must coordinate their operations in order to fluently process the handoff. Either anchoring or reanchoring is adopted. Contrary to reanchoring, anchoring does not change the established path and does not induce data redirection. The anchoring decision is taken by the target or anchor ASN gateway. IEEE 802.16e WiMAX defines two kinds of handover: the controlled and the uncontrolled type. Contrary to the uncontrolled handover, the controlled handover hardly guarantees QoS requirements.

5.4.4 The Security Sublayer

5.4.4.1 Architecture

This sublayer—also known as the MAC Privacy Sublayer—is located between the MAC and the PHYs. It employs an encapsulation protocol as well as a key management protocol. The encapsulation protocol defines data encryption and authentication algorithms and states the rules for applying them to the MAC PDU payload while the key management protocol or PKM manages the keys' distribution based on X.509 digital certificates, RSA public key encryption algorithms, and other strong encryption algorithms such as the 56-bit DES (Johnston and Walker, 2004).

The authentication procedure begins when a SS (PKM client) requests a key from the BS (PKM server). A hybrid cryptography scheme generates a shared secret that will be used to authenticate the SS as well as secure the future PKM exchanges.

5.4.4.2 PKM Protocol

The PKM protocol suite aims at defining authorization and data encryption keys for the requesting SSs. First, a SS needs be authorized from the BS. A periodic reauthorization procedure takes place to refresh the shared keys for more robustness. The authorization procedure begins when an SS sends an informative authentication information message to its BS including its digital certificate. In response, the SS sends an Authorization Request back asking for an authorization key and a security association (SA) identifier. In reception, the BS verifies the SS identity using the received certificate and generates an authentication key (AK). The AK is then encrypted using the SS's public key and sent to SS in an Authorization Reply message. After verifying the SS's identity, the BS determines the network services the SS is authorized to access. During reauthorization, all these steps are executed except for the Authentication Information message. When authorized, the SS needs to get a traffic encryption key (TEK) for each identified SA received within the authorization reply. When the PMP mode is adopted, these TEKs are periodically refreshed using periodic Key Request messages. They are also encrypted using a Key Encryption derived from the AK. When the mesh mode is adopted, all neighbors as well as SAs identified in the Authorization Reply run these steps.

SA Management An SA defines the security parameters of a connection. These primarily include the keys used and the supported encryption algorithms. An SA may be primary, static or dynamic. The primary SA is established between the SS and the managing BS during the initialization process. A set of static SAs may then be provisioned from the BS for the basic unicast service (Ribeiro, 2005). Finally, dynamic SAs are established on the fly in response to the initiation and termination of specific service flows. It is worth mentioning here that static and dynamic SAs may be shared by a group of SSs when multicast is used. A SA identifier (SAID) identifies each SA. All Transport connections should be mapped to an existing SA. However, the Basic and the Primary Management connections, which are created when a SS joins the network, should not be mapped to SAs.

Cryptographic Methods The cryptographic methods include the supported cryptographic algorithms and the key sizes that are used by the PKM protocol. The standard defines two data encryption schemes: one employs the DES in cipher bloc chaining (CBC) mode and the second uses AES in CCM mode. When the encryption algorithm identifier associated with an SA equals 0×01, data flow that will be transmitted on that connection can be encrypted using the CBC mode of the DES algorithm. When the length of the final block to encrypt is less than 64 bits, a residual termination block processing is used for encryption. When the encryption algorithm identifier related to an SA equals 0 × 02, data flows that will be transmitted on that connection will be encrypted using the CCM mode of the AES algorithm. The receiver of an encrypted PDU should decrypt and authenticate it according to the CCM specifications, then discard it if it is invalid.

Keys and Certificates

Key Usage Key information needs be synchronized so that the BS to maintain it for all SAs and client SSs. When a new AK is assigned to an SS, it will remain active until the specified Lifetime value set by the issuing BS expires. The Authorization Reply generated as a response to the Authorization Request should specify the remaining lifetimes of the AK. However, when a SS has not been able to reauthorize before the expiration of its AK, the BS should judge it as unauthorized and remove all the TEKs associated with that SS's primary SA. Finally, the BS needs to support two different and simultaneously active AKs for each SS; that is the reason why both keys present overlapping lifetimes.

Certification Management The IEEE 802.16d uses the X.509 Version 3 certificate format along with other certificate extensions. This format defines the tbsCertificate.version and the tbsCertificate.serialNumber field that together identify the certificate. These are assigned from the certification authority. We may also distinguish the tbsCertificate.signature field, which defines the algorithm processed to sign the certificate, along with the tbsCertificate.validity information, which determines when the certificate becomes active and when it expires. Finally, the X.509 format identifies the signature Value field, which contains the computed digital signature of the certificate, the tbsCertificate.subjectPublicKeyInfo field, which contains the public key, and the parameters along with the identifier of the algorithm with which the key may be used.

5.5 Enhancing Efficiency and Effectiveness of 802.11 MAC in Wireless Mesh Networks[8]

5.5.1 Increasing Parallelism by Power Control and Enhanced Carrier Sensing

The IEEE 802.11 MAC provides two CA mechanisms, the mandatory basic CSMA/CA and the optional virtual carrier-sensing scheme with RTS/CTS (IEEE 802 LAN/MAN

[8] Excerpt from the invited article "Enhancing efficiency and effectiveness of 802.11 MAC in wireless mesh networks," *Yuanzhu Peter Chen; †Jian Zhang and †Ivan Marsic (*Memorial University of Newfoundland; †Rutgers University).

Standards Committee, 1999). Under the basic scheme, a station refrains from medium access if it senses ongoing transmission(s) on the wireless channel. The mechanism to determine whether or not the channel is busy is called CCA. A prevalent CCA mode is known as carrier sense with energy detection. That is, the CCA decision is based on whether the energy of a detectable 802.11 signal exceeds a threshold, called *carrier sense threshold*. Given a carrier sense threshold, the corresponding *carrier sense range* is defined as the minimum distance allowed between two concurrent transmitters (Yang and Vaidya, 2005a). On the one hand, it may be true that the smaller the carrier sense range (or the higher the carrier sense threshold), the better the spatial reuse and the higher the efficiency. On the other hand, the interference level at a receiver can also increase as the carrier sense range decreases, that is, concurrent transmitters get closer; effectively this impairs the effectiveness of the CA mechanism. An interference model has been developed to describe the relationship between the transmission power, the carrier sense threshold, and the aggregate throughput. Using such a model, the optimal carrier sense threshold is specified to maximize the aggregate throughput for a regular topology, as described next.

5.5.2 Static Basic Carrier Sensing Based on Interference Model

Yang and Vaidya (2005a) and Kim et al. (2006) derived the worse case interference and signal-interference-noise ratio (SINR) at a receiver station as follows. The thermal noise is ignored for simplicity.

We denote the carrier sense threshold by T_{cs}, the corresponding carrier sense range by D, the transmission power by P_{tx}, and the transmission range by R. When a sender S_0 is transmitting, a concurrent transmitter must be at least a distance D away from S_0. Therefore, in the worst case there can be a total of 6 interferers distributed on the circle centered at the sender with radius D. This can be approximated using the Honey-grid model (Hekmat and Van Mieghem, 2002) as in Fig. 5.15.

As illustrated, the worst-case interference occurs when the distances between the receiver R_0 and the six interferers approximately equal $D - R$, $D - R$, $D - R/2$, $D + R/2$,

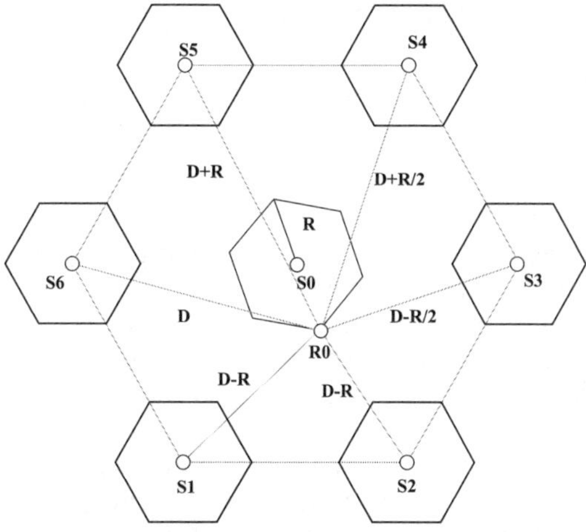

Figure 5.15 Worst case interference scenario.

$D + R$, and D, respectively. It can be shown that the interference contributed by other potential interferers in the network can be neglected (Hajek et al., 1997). Thus, the interference from these six interferers dominates the total interference at R_0. It can be expressed then as

$$I = \frac{2P_{tx}}{(D-R)^{\theta}} + \frac{P_{tx}}{\left(D - \frac{R}{2}\right)^{\theta}} + \frac{P_{tx}}{D^{\theta}} + \frac{P_{tx}}{\left(D + \frac{R}{2}\right)^{\theta}} + \frac{P_{tx}}{(D+R)^{\theta}} \tag{5.1}$$

where a path-loss radio propagation model with the path loss exponent θ is assumed. The corresponding SINR at R_0 can be expressed as an increasing function of D/R:

$$\begin{aligned} \text{SINR} = f\left(\frac{D}{R}\right) &= \frac{\frac{P_{tx}}{D^{\theta}}}{\frac{2P_{tx}}{(D-R)^{\theta}} + \frac{P_{tx}}{\left(D - \frac{R}{2}\right)^{\theta}} + \frac{P_{tx}}{D^{\theta}} + \frac{P_{tx}}{\left(D + \frac{R}{2}\right)^{\theta}} + \frac{P_{tx}}{(D+R)^{\theta}}} \\ &= \frac{1}{\frac{2}{\left(\frac{D}{R} - 1\right)^{\theta}} + \frac{1}{\left(\frac{D}{R} - \frac{1}{2}\right)^{\theta}} + \frac{1}{\left(\frac{D}{R}\right)^{\theta}} + \frac{1}{\left(\frac{D}{R} + \frac{1}{2}\right)^{\theta}} + \frac{1}{\left(\frac{D}{R} + 1\right)^{\theta}}} \end{aligned} \tag{5.2}$$

Using the Shannon capacity theorem, for a certain channel bandwidth W, the achievable channel rate is at most $\Gamma_c = W \cdot \log_2(1 + SINR)$. The total network capacity can be expressed then as $\Gamma_n = \Gamma_c \frac{U}{U_A}$, where U is the area of the network and U_A is the area "consumed" by each transmitter, that is, $\sqrt{3} \cdot D^2/2$. Thus, U/U_A is the total number of concurrent transmissions in the network. The carrier sense range D is simply $\left(\frac{P_{tx}}{T_{cs}}\right)^{1/\theta}$. So, the network capacity can be further expressed as

$$\Gamma_n = C_0 \left(\frac{T_{cs}}{P_{tx}}\right)^{2/\theta} \cdot \log_2\left(1 + f\left(\frac{1}{R} \cdot \left(\frac{P_{tx}}{T_{cs}}\right)^{1/\theta}\right)\right) \tag{5.3}$$

where C_0 is constant.

Using the network capacity function defined above, the highest aggregate throughput can be achieved by adjusting either the transmission power P_{tx} or the carrier sense threshold T_{cs}, or both. Some approaches use the above analytical model to determine an invariant optimal value of the carrier sense threshold for all the stations in the network given a fixed transmission power. Note that the above capacity is derived assuming that the network consists of dense and busy transmitters. In practice, however, it is not typical that all of the receivers in a network will experience the worse-case interference. Moreover, the locations of transmitters and their mutual interference in a network are not necessarily stationary or on a regular pattern. Therefore, instead of holding the carrier sense threshold or transmission power of all nodes constant all the time, a class of methods is proposed to adjust these parameters dynamically. These dynamic control methods are usually combined with the virtual carrier-sensing scheme as described in the following section.

5.5.3 Dynamic Schemes with Virtual Carrier Sensing

5.5.3.1 Virtual Carrier Sensing and Its Inefficiency and Ineffectiveness

As a complement to the basic CA scheme, virtual carrier sensing (Bharghavan et al., 1994) is dedicated to solving the collision problem due to hidden stations (Tobagi and Kleinrock, 1975a). The idea is to reserve the wireless channel by preceding the data frame transmission with an RTS/CTS handshake. The neighboring stations that receive the RTS/CTS frames are blocked from transmitting for a period of time specified in the frames. This is achieved by setting the NAV of an overhearing node's MAC agent; this counts down as the transmission progresses. Therefore, the transmission range of the RTS/CTS effectively determines the blocking area. The original design (Bharghavan et al., 1994) assumes that the stations are able to interfere with the upcoming DATA/ACK frames only if they can receive RTS/CTS, that is, that the transmission range of control frames equals the interference range. However, there commonly exists a disparity between the RTS/CTS transmission range and the interference range. Instead, it may result in one of the two opposite situations, that is, either the failure of CA or unnecessary false blocking, depending on which range is larger.

In our discussion, we define the *interference range* of a receiver R as the distance from R within which another transmitter may interfere with the current frame reception. Recall the two conditions needed for a receiver to receive a frame with an acceptable error rate: (1) the power of the received signal exceeds a threshold, called *receiver sensitivity*, denoted by P_{rth}, and (2) the SINR exceeds another threshold, called *capture threshold*, denoted by T_{cap}. The distance that a signal propagates before its power drops below P_{rth}, that is, the transmission range R, can be derived by solving

$$\frac{P_{\text{tx}}}{R^{\theta}} = P_{\text{rth}} \tag{5.4}$$

The interference range D_{I} is obtained by calculating the shortest distance between the receiver and an interferer so that the SINR on the receiver is right above the capture threshold when the sender and interferer transmission power levels for DATA frames are P_{tx} and P_{inf}, respectively. In other words, the following rule needs be satisfied:

$$\text{SINR} = \frac{P_{\text{tx}}}{r^{\theta}} \Big/ \frac{P_{\text{inf}}}{D_{\text{I}}} \geq T_{\text{cap}} \tag{5.5}$$

This shows that the interference range is not a fixed value in that it changes with the actual distance r ($r \leq R$) between the transmitter and the receiver, and with the capture threshold T_{cap} that is set by the modulation scheme used. Thus, it is common that the CTS transmission range does not necessarily match the current interference range. When the transmission range of CTS is smaller than the interference range, the CTS frame cannot be decoded correctly by all potential interferers, leading to collisions, referred to as the ineffectiveness of CA. On the other hand, a CTS with an excessively large transmission range may cause low spatial reuse, especially in wireless multihop networks, referred to as its inefficiency.

An example shown in Fig. 5.16a assumes that all nodes transmit RTS/CTS/DATA/ACK frames with the same power and modulation scheme. Although node X

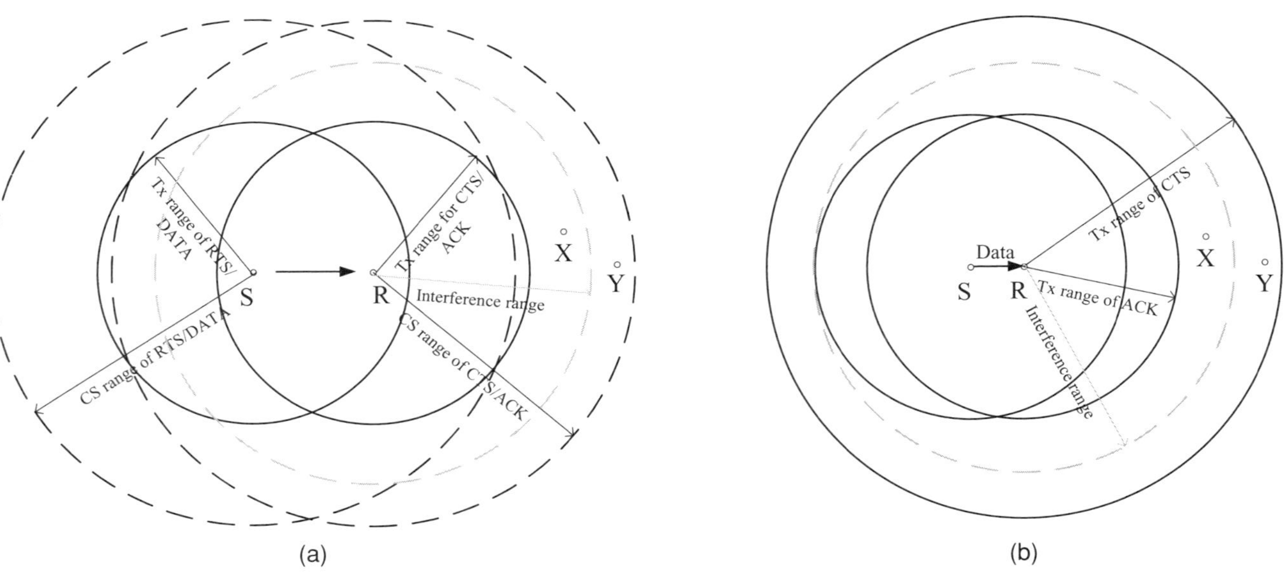

FIGURE 5.16 (a) Ineffectiveness of CA and (b) inefficiency of spatial reuse.

may sense node R's transmission since it is within R's carrier sense range, it cannot decode the CTS frame since it is outside of the transmission range of CTS of node R. Therefore, although node X will stay silent for the period of this CTS transmission, it may still transmit during the DATA frame from S to R since it failed to set its NAV based on the CTS frame. This may result in a DATA frame collision since node X is within the interference range of receiver R. This is the so-called hidden station problem, which still cannot be avoided by the original RTS/CTS scheme.

A solution for avoiding such collisions could be the increase of the RTS/CTS transmission range (i.e., increasing the RTS/CTS transmission power). For example, see the work by Gomez et al. (2001): the RTS and CTS are sent at the highest power level, and the data and ACK at a lower power level. However, it turns out that the above collision problem cannot be well solved by such a strategy. The reason is that by enlarging the CTS transmission range of receiver R to defer more potential interferers, at the same time we also increase the interference of RTS/CTS frames at the neighboring nodes due to the higher transmission power, that is, the interference range of receiver R is also increased due to a larger P_{inf} in Eq. (5.2). This paradox can be mitigated by multichannel schemes, for example, PCMA provided in the study of Monks et al. (2001), by transmitting RTS/CTS frames on a different channel than the DATA/ACK frames.

For single-channel networks, a way to enlarge the RTS/CTS range without increasing the transmission power is to use a lower-rate modulation scheme that requires lower receiver sensitivity P_{rth}. In Eq. (5.1), using the same P_{tx} but a smaller P_{rth}, the corresponding transmission range increases. Thus, as shown in Fig. 5.16b, the CTS frame of receiver R can now reach some potential interferers, such as node X. On the other hand, an excessively large transmission range of CTS may lead to inefficiency. As shown in Fig. 5.16b, node Y is unnecessarily blocked although its transmission would not interfere with the data reception of R (because it is beyond its interference range). Thus, as discussed next, the problem becomes: *how to improve the spatial reuse/efficiency without impairing the effectiveness of CA.*

5.5.3.2 Soft Blocking Schemes

The IA-MAC (Cesana et al., 2003) provides a single-channel solution. Its idea is similar to (Monks et al., 2001), but operating in single-channel networks. The idea, here referred to as "soft blocking," is to conditionally set the NAV of every node that overhears a CTS frame. Assume that a low-rate modulation scheme is selected for RTS/CTS frames and their transmission range is sufficiently large, as in Fig. 5.16b. To improve efficiency, if a node, say node Y, can tell that its transmission will not interfere with the reception at receiver, say R, Y may choose then not to set its NAV when overhearing a CTS. Node Y decides this by using the transmission power information carried explicitly and/or implicitly by RTS/CTS frames. The process is described below. Before and upon receiving an RTS from the sender, the receiver can measure the interference $P_{\text{I-current}}$ and the power of the received RTS as $P_{\text{rcv-RTS}}$, respectively. The minimum SINR should not drop below the capture threshold, which is

$$SINR = \frac{P_{\text{rcv-RTS}}}{P_{\text{I-current}} + P_{\text{I-add}}} \geq T_{\text{cap}} \tag{5.6}$$

To calculate the maximum additional interference $P_{\text{I-add}}$ that the system can tolerate, the receiver calculates Eq. (5.3). The receiver then inserts the result ($P_{\text{I-add}}$) in the CTS frame to advertise it to its neighbors. When a neighbor overhears this CTS frame, it first measures its power. Given the assumption of symmetry of the channel and equal transmission power for all nodes, the interference of a neighboring node at the receiver is about the same as the power that the neighboring node perceives from the receiver (via the CTS frame). If the perceived power of the CTS is higher than $P_{\text{I-add}}$, this neighbor sets its NAV according to the CTS and stays silent. Otherwise, it ignores the CTS frame presuming that its transmission will not disturb the current reception. Therefore, the parallelism/efficiency is improved by such a "soft blocking" scheme with virtual carrier sensing. Yet, the CA is still effective. The method is simple with no need for power control, its overhead on CTS is negligible, and the symmetry assumption is reasonable. Note that the collision may still occur if *aggregate interference* is considered. For example, in the worst interference case in Fig. 5.15, assume that the transmission will not be disturbed by single transmission from any of the six interferers. But the cumulative interference from the concurrent transmissions may be higher than the maximum additional interference. Since these interferers are out of the sensing range of each other, they may start their transmissions simultaneously, which leads to reception failures at receiver R_0 in Fig. 5.15.

5.5.3.3 Power Control Schemes

Power control in 802.11 MAC was originally proposed for the purpose of power saving (Gomez et al., 2001; Jung and Vaidya, 2002). It was first designed by Jung and Vaidya (2002), using a power control scheme, called POWMAC, to enhance spatial reuse and to manage interference in wireless multihop networks, aiming at improving the network throughput. The basic idea can be illustrated as follows. In Fig. 5.16b, node X is blocked since its transmission with regular power level disturbs the reception at R. However, if node X has a packet for a receiver nearby, say node Y, X may lower its "voice" (power) so that its interference is below the additional tolerable value for reception at R and yet its power is strong enough for reception on Y.

POWMAC considers the additional tolerable interference as a resource, which is shared with other concurrent transmissions. Like IA-MAC, power and interference information is exchanged via RTS/CTS handshakes. The process is as follows. When a sender i has a frame for a receiver j, it first calculates the maximum allowable transmission power (P_{MAP}) it can use without disturbing its neighbors:

$$P_{\text{MAP}}(i) = \min_{u}\{P_{\text{MTI}}(u)/G_{iu}, P_{\text{MAX}}\} \tag{5.7}$$

Here, P_{MTI} is the maximum tolerable interference (described below) of i's neighbor u and G_{iu} is the channel gain between nodes i and u that can be estimated if both the transmission power and received signal strength are known. Sender i then places P_{MAP} into its RTS frame and transmits it with the maximal power P_{MAX}. In addition, the sender also includes the estimated number N of future unintended transmitters that could interfere with the receiver, based on the current network load (Muqattash and Krunz, 2004). Upon receiving this RTS, the receiver j determines whether the regular transmission power P^{ij}_{load} of DATA frame is within the range $P^{ij}_{\min} \le P^{ij}_{\text{load}} \le P_{\text{MAP}}$, where $P^{ij}_{\min}$ is the minimum power required for DATA frame so that it can be decoded given the current interferences from existing transmissions. If P^{ij}_{load} does not fall within this range, the receiver sends a negative CTS back to sender i to reject the request. Otherwise, it calculates the maximum additional interference power $P_{\text{I-add}}$ that it can tolerate from N future unintended transmitters, in addition to the existing ones. The calculation of $P_{\text{I-add}}$ is similar to the related process described in IA-MAC. Unlike IA-MAC, a POWMAC receiver further splits the total tolerable $P_{\text{I-add}}$ across N potential interferers:

$$P_{\text{MTI}} = \frac{P_{\text{I-add}}}{N} \tag{5.8}$$

As Eq. (5.5) shows, the maximum tolerable interference for any single sender P_{MTI} is a fraction of the aggregate interference $P_{\text{I-add}}$. The calculated P_{MTI} is then broadcast with the CTS frame to neighboring potential transmitters so they can use it to properly set their maximum allowable transmission power P_{MAP}, Eq. (5.4).

As noted above, with more flexible allocation of transmission power and adaptive blocking area, a power control scheme for 802.11 MAC can further improve spatial reuse and the network throughput as such. In the soft-blocking scheme, the state of a neighboring node is either "on," that is, in the blocking range, or "off," that is, out of the range. In contrast to such a simple on-off control, dynamic power control schemes provide more flexible methods for dealing with various interference scenarios in wireless mesh networks. Note that the performance of POWMAC highly depends on the accuracy of the propagation model and the interference-error model described above. For implementation, it is imperative for the 802.11 products to measure and compare the power with the level of accuracy the POWMAC protocol (Abdesslem et al., 2006) requires. Moreover, for multirate wireless networks with rate-adaptive MAC (Holland et al., 2001), the throughput gain through power control may be ambiguous. That is because the rate adaptation mechanism may use the resource dedicated to tolerate additional interference to increase the link rate instead of increasing the number of concurrent transmissions, as in POWMAC.

5.5.3.4 Self-Learning Carrier Sensing

Compared to above schemes, the method of self-learning carrier sensing (Chen et al., 2006a) does not require any propagation modeling or power control. Here, the sender collects the historical RTS/CTS success ratio and the signal strength, and builds a black-box mapping model to describe their relationship. The update of the mapping curve is triggered by an access request event. Prior to an access attempt, the sender looks up the mapping curve indexing the current sensed signal strength and obtains the estimated success ratio. If the success ratio is lower than some threshold, the sender backs off and waits until it reckons the channel is clear. This method, although simple, is adaptive and easy to implement. On the other hand, this 2-D mapping can be flawed and inaccurate in the case when more media access behaviors and patterns are present.

5.5.4 Exploit Channel and/or Spatial Diversity with MAC-Layer Scheduling

5.5.4.1 Head-of-Line Blocking Problem

Another type of method (Jain and Das, 2005; Kim et al., 2006a; Wang et al., 2004; Zhang et al., 2006) for increasing concurrent transmissions and improving parallelism in mesh networks is to exploit the channel/spatial-diversity by rescheduling the frames in the sender's queue. In wireless mesh networks, some stations can be particularly overloaded. For example, a mesh network gateway (Aguayo et al., 2004) needs to deliver simultaneously multiple down-stream data flows (e.g., between the Internet and the wireless stations). Similarly, a mesh router may have to serve several neighbors by forwarding their packets along multihop paths. The efficiency of such stations is critical to the capacity of a mesh network. However, the performance of the regular 802.11 MAC protocol is susceptible to the head-of-line (HOL) blocking problem.

The HOL blocking problem occurs when the frame currently at the head of the queue in the sender's MAC layer cannot be transmitted successfully due to, say, the temporary unavailability of the receiver. As discussed in Section 5.3, in 802.11, each time a DATA or RTS transmission times out, the contention window is doubled. The frame will not leave the queue until the transmission is acknowledged or until the maximal number of retries is reached. This frame is thus blocking the subsequent frames from being transmitted although their receivers may be available at this time. Due to the exponentially-growing backoff time overhead, the HOL blocking problem can greatly lower channel utilization and network capacity. Simulations (Zhang et al., 2006) indicate that the MAC layer backoff time fraction at the sender may reach up to 70%. For a loaded mesh router or gateway, HOL blocking problem could result in a serious congestion. During the backoff process at a mesh gateway, more and more frames could arrive from wireline Internet connection and be blocked in the queue. For a loaded mesh router, the backoff forces the router to spend more time in receiving than transmitting. With more frames arriving and the head frame blocking the queue, the router's queue eventually overflows and it starts dropping packets. This may further trigger an upper layer (e.g., TCP) backoff, leading to further degradation of throughput performance. Thus, in order to improve the performance of multihop mesh networks, the HOL blocking problem must be addressed.

5.5.4.2 MRTS

A straightforward solution to the HOL problem is to reschedule the frames in the sender's queue based on the status of their next-hop nodes. For example, node B in Fig. 5.17 cannot receive traffic from A as it is blocked from another transmission. Instead of waiting for B, node A may first send its traffic queued to other nodes available, such as E. As a

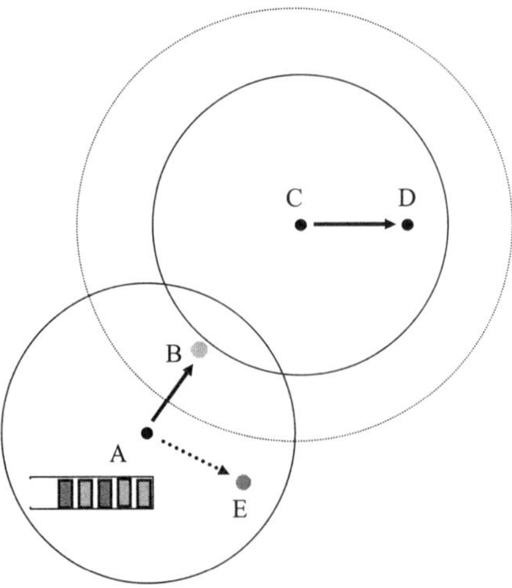

FIGURE 5.17 Rescheduling for HOL blocking problem.

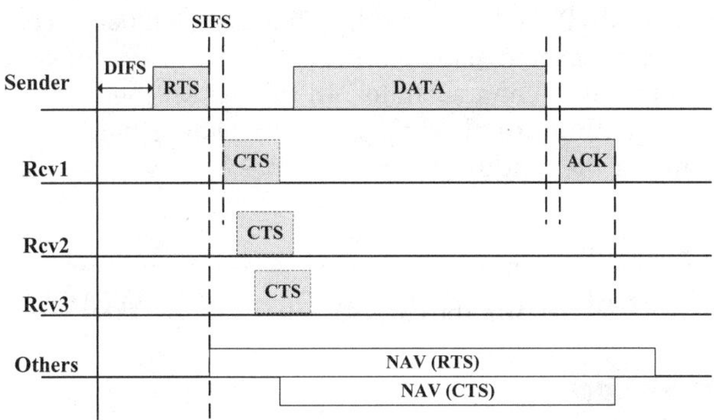

FIGURE 5.18 MRTS protocol timeline.

result, the backoff overhead is avoided whereas the channel utilization is also improved. In addition, the number of concurrent transmissions is increased.

To obtain the state information of the next-hop neighbors, a multicast RTS/CTS (MRTS) handshake is proposed by Jain and Das (2005) and Wang et al. (2004). An MRTS, in contrast to a unicast RTS in conventional RTS/CTS, is directed to a list of receivers. That is, an MRTS frame contains a list of next-hop receivers for which the sender has DATA packets currently queued. Each element of the list contains a receiver's address and the NAV of its corresponding packet. The priority among different receivers is decided by the order in which the receivers are arranged in the MRTS frame. That is, the earlier a receiver's address appears on the MRTS list, the sooner this receiver can return a CTS. This mechanism helps to avoid the collision of CTS frames returned by the receivers. Unless it is blocked by an ongoing transmission in its neighborhood, the first candidate receiver (Rcv1 in Fig. 5.18) that successfully receives the MRTS replies with a CTS. If a lower-priority candidate (Rcv2 or Rcv3) detects that all higher-priority candidates remained silent for certain period of time, it has the right to reply with a CTS. The lower a receiver's priority is, the longer its waiting time is. For example, the n-th receiver has to wait for SIFS + $(n - 1) \times$ slot_time. Such a right-to-reply is implicitly propagated down the chain until a nonblocked receiver sends a CTS or all receivers remain silent and the sender times out. The sender finds the responding receiver's address from the received CTS frame. Then, the sender retrieves the corresponding frame from its queue and transmits it to that receiver. The dialog ends with an ACK from the receiver if the transmission is successful. Since the MRTS probes the availability of multiple receivers almost simultaneously, the likelihood of MRTS failure, that is, no receiver available, is low. Hence, the idle time due to backoff on the loaded stations can be significantly lowered and their utilization is improved.

The multicast characteristic of MRTS provides another appealing feature. That is, it measures the channel conditions of multiple receivers almost simultaneously. Therefore, based on the observed MRTS responses, the sender can estimate the neighbors' channel states and their correlations, that is, how diverse/correlated the states of any two neighbors are. From this, it may also estimate their geographical relations since geographically proximal stations are likely to share similar channel states. The use of such information may enhance the channel-state diversity of the MRTS receiver list, and thus further improve the success ratio of MRTS. An extension of the adaptive channel-state-based scheduling with MRTS is developed by Zhang et al. (2006) to enable receiver candidate selection. The new scheme also determines the length of the MRTS list, which is adaptable to the candidates' channel states. In this sense, the extended scheme constructs a list of receivers with mutually diverse channel states based on historical observations. It also minimizes the length of MRTS frames, yet without jeopardizing their success ratio.

5.6 On the Effect of Optimal Power Control in WM²Nets[9]

5.6.1 Introduction

We consider a wireless mesh network with n nodes and m source-destination pairs (and a given offered traffic load for each pair), using a scheduling-based MAC protocol such as time division multiple access (TDMA), and a routing mechanism that may be unicast or multicast based, $m \leq n(n-1)$. Under a given set of nodal transmit power levels $P = (P_1, \ldots, P_n), 0 \leq P_i \leq P_{\max}, i = 1, \ldots, n$, we define the source-destination throughput vector $\lambda = (\lambda_1, \ldots, \lambda_m)$ to be achievable for the wireless mesh network if there exists an associated *temporal* (based on the channel sharing MAC protocol) *and spatial* (based on the underlying routing mechanism) *joint scheduling-routing scheme* (henceforth referred to as *joint scheduling and routing scheme*) that yields the throughput vector λ. Let $S(P)$ denote the set of all achievable source-destination throughput vectors under the power vector P.

In this study, we analyze the effect of nodal transmit power vector P on the maximum (or supremum) level of a general (real-valued) function of the source-destination throughput levels $\Omega(\lambda_1, \ldots, \lambda_m)$ subject to $\lambda \in S(P)$. We refer to the latter supreme level attained under power vector P as the *conditional* (with respect to P) *supreme value of the objective function* and represent this value by $\Omega^*(P)$. That is,

$$\Omega^*(P) = \operatorname*{Sup}_{(\lambda_1,\ldots,\lambda_m)} \{\Omega(\lambda_1, \ldots, \lambda_m) : (\lambda_1, \ldots, \lambda_m) \in S(P)\}. \tag{5.9}$$

Given a selected power vector P, the conditional supreme value of the objective function $\Omega^*(P)$ is achieved (in finite time or asymptotically in time) as the system designer selects an *optimal joint scheduling and routing scheme* over the underlying (finite or infinite) operational time period T. The objective is to characterize the key features of a power vector solution that maximizes the conditional supreme value of the objective function $\Omega^*(P)$ over the set of power vectors $P = (P_1, \ldots, P_n), 0 \leq P_i \leq P_{\max}, i = 1, \ldots, n$. We call such a power vector an *optimum power vector*, identify an associated optimal joint scheduling and routing scheme as an *optimum joint scheduling and routing scheme*, and denote the resulting value of the objective function as the *optimum objective function value.* Let Ω^* denote the optimum objective function value. Then, we have

$$\Omega^* = \operatorname*{Max}_{P}\{\Omega^*(P): P = (P_1, \ldots, P_n), 0 \leq P_i \leq P_{\max}, i = 1, \ldots, n\}, \tag{5.10}$$

[9] Excerpt from the invited article "On the effect of optimal power control in wireless mesh networks," Arash Behzad and Izhak Rubin, Electrical Engineering Department, University of California (UCLA), Los Angeles, CA 90095-1594, E-mail: {abehzad, rubin}@ee.ucla.edu

or equivalently,

$$\Omega^* = \operatorname*{Max}_{P}\left\{ \operatorname*{Sup}_{(\lambda_1,\ldots,\lambda_m)} \{\Omega(\lambda_1,\ldots,\lambda_m) : (\lambda_1,\ldots,\lambda_m) \in S(P),\ P = (P_1,\ldots,P_n), \right.$$

$$\left. 0 \le P_i \le P_{\max}, i = 1,\ldots,n\}\right\}. \tag{5.11}$$

Assuming that $\Omega(\lambda_1,\ldots,\lambda_m)$ is not directly a function of P (i.e., Ω is affected by P only through $S(P)$, so that power-related expenditures are not directly included in the objective function), we prove that, independent of nodal distribution, traffic pattern, and offered traffic load, $\Omega^*(P)$ is maximized (over the set of all nodal power vectors P) by properly increasing the nodal transmit power levels. Under the special case of this analysis for which the transmission power levels of all nodes are assumed to be identical (yet programmable), we prove that the power vector $P = (P_1 = P_{\max},\ldots,P_n = P_{\max})$ maximizes $\Omega^*(P)$, independent of nodal distribution, traffic pattern, and offered traffic load. For the latter special case, when the objective function $\Omega(\lambda_1,\ldots,\lambda_m)$ is defined as $\mathrm{Min}(\lambda_1,\ldots,\lambda_m)$, so that $\Omega^*(P)$ represents the throughput capacity under power vector P, the results imply that $P = (P_1 = P_{\max},\ldots,P_n = P_{\max})$ maximizes the throughput capacity (over the set of all permissible nodal power vectors P), independent of nodal distribution, traffic pattern, and offered traffic load.

5.6.2 System Model

We consider a wireless mesh network that consists of n nodes, which are located based upon any arbitrary distribution in a given area. Every node, when scheduled to access the communications channel, transmits at a fixed data rate of W bps, and variations in transmission power merely affect the transmission range. A single transmission may be intended for more than one receiver (i.e., link-layer multicasting is allowed). All nodes are equipped with identical half-duplex radios and omnidirectional antennas. A node can successfully receive from at most one other node in the same time instant. We assume node i to transmit at a fixed (yet programmable) transmission power P_i, $0 \le P_i \le P_{\max}, i = 1,\ldots,n$; assume a transmission to occupy the entire bandwidth of the system under consideration. Channel time is slotted into identical synchronized time slots. Slot duration τ is assumed to be equal to the transmission time of a packet plus some overhead duration that includes the maximum propagation delay.

The source-destination association can be selected based on an arbitrary traffic pattern. Source nodes may be attached to hosts that generate multicast messages that are destined to designate groups of hosts and thus have to be routed to several destination nodes. Other nodes may be attached to hosts that generate only unicast messages. Let us assume that a source node s generates traffic flows that are classified according to their disjoint destination sets as follows: an r-th class traffic flow (originating at node s) wishes to reach destination set $D_{s,r}$, $r = 1,\ldots,q_s$ so that each packet of the flow will be received by every nodal member of $D_{s,r}$. The offered traffic load for such multicast type traffic flows destined from node s to $D_{s,r}$ is denoted as $f(D_{s,r})$. Clearly, the total number of distinct source-destination nodal pairs can be calculated as

$$m = \sum_{s=1}^{n}\sum_{r=1}^{q_s} \| D_{s,r} \|, \tag{5.12}$$

where $\|A\|$ denotes the cardinality of set A. The notation used above applies also to unicast flows when the destination set consists of a single nodal member.

To distribute across a mesh network packets that have multiple destinations, the system might duplicate such a packet into multiple copies and then use a unicast route (characterized by a single source-destination nodal pair) for each copy. In turn, to increase link capacity utilization, multicast routing subgraphs (such as source-based multicast trees) can be used to route a multicast packet efficiently to multiple destinations. In a wireless network system, whereby each node may employ an omnidirectional antenna, one can achieve further efficiency by capitalizing on the broadcast character of the multiple access radio channels. The results presented in this study could apply with any one of these network layer distribution methods. In this vein, we assume a joint scheduling and routing scheme, which in its general context can be described as follows: each nodal member of the selected routing subgraph adds a distinct entry in its routing table that specifies the set of (link-layer) receivers to whom such a packet should be transmitted. This assumes in fact that the scheduling mechanism permits such packet transmission to occur at this time slot.

Definition: Let $P_{(M)} = (P_{i_1}, \ldots, P_{i_M})$ be an arbitrary power vector, whereby $0 < P_{i_k} \le P_{\max}$, $k = 1, \ldots, M$. Power vector $P'_{(M)} = (P'_{i_1}, \ldots, P'_{i_M})$ is said to be relatively maximized with respect to power vector $P_{(M)}$ if

$$P'_{(M)} = \alpha(P_{(M)})P_{(M)}, \tag{5.13}$$

where $\alpha(P_{(M)})$ is a real positive scalar defined as

$$\alpha(P_{(M)}) = \min_{k=1,\ldots,M}\{P_{\max}/P_{i_k}\}. \tag{5.14}$$

Moreover, a power vector is said to be relatively maximum if at least one of its components is equal to $P_{\max}$.

We define a communication link to be formed from node i to node j under power level P_i if the SNR at j is not less than a threshold γ_C (that dictates the need for a minimal received power level), that is,

$$G_{ij}P_i/N_j \ge \gamma_C, \tag{5.15}$$

in which G_{ij} is the propagation gain (incorporating the effects of link loss phenomena such as fading and shadowing) for direct transmission from node i to node j, and N_j is the thermal noise power at receiver j (Gupta and Kumar, 2000). We represent a direct (link-layer) multicast transmission from node i to the set of nodes in J (where there is a communication link from node i to each of the nodes in J, physically implemented through the broadcast of a message by node i that is assumed to be simultaneously received by all nodes in J) by $i \to J$. Let $(i \to J; s; D_{s,r})$ denote the transmission $i \to J$ whose source node and the associated set of destination nodes are s and $D_{s,r}$, respectively, $r = 1, \ldots, q_s$. A *transmission scenario* $S_{(M)} = \{(i_1 \to J(i_1); s(i_1); D_{s(i_1),r(i_1)}), \ldots, (i_M \to J(i_M); s(i_M); D_{s(i_M),r(i_M)})\}$ is defined as a candidate set of (link-layer) multicast transmissions that are considered to take place at the same time slot, where all transmitting and receiving nodes are distinct. Note that the distinction of transmitting and receiving nodes guarantees that i) a node is not an intended receiver of more than one transmission at any time slot, and ii) a node is not transmitting and receiving at any time slot (i.e., the half-duplexing constraint is satisfied).

For such a transmission scenario $S_{(M)}$ under nodal power vector $P_{(M)} = (P_{i_1}, \ldots, P_{i_M}), 0 \leq P_{i_k} \leq P_{\max}, k = 1, 2, \ldots, M$, we say that the transmission from i_k is *successfully received* at j, $j \in J(i_k)$, if the SINR at j is not less than the minimum required threshold γ (Rappaport, 1996), that is

$$G_{i_k j} P_{i_k} \Bigg/ \left(N_j + \sum_{\substack{r=1 \\ r \neq k}}^{M} G_{i_r j} P_{i_r} \right) \geq \gamma, \quad k = 1, \ldots, M. \tag{5.16}$$

We define the cardinality of the set of successful receptions in a transmission scenario $S_{(M)}$ employing transmit power vector $P_{(M)} = (P_{i_1}, \ldots, P_{i_M}), 0 \leq P_{i_k} \leq P_{\max}, k = 1, 2, \ldots, M$, as the *spatial reuse factor of the transmission scenario* $S_{(M)}$ with respect to $P_{(M)}$. We define a transmission scenario $S_{(M)}$ to be *feasible* under power vector $P_{(M)}$ if all the transmissions are successfully received at all of their intended receivers. Consequently, the spatial reuse factor of the feasible transmission scenario $S_{(M)}$ under power vector $P_{(M)}$ is equal to $\sum_{k=1}^{M} \| J(i_k) \|$. Clearly, every achievable throughput vector λ under power vector P can be achieved under a joint scheduling and routing scheme over the underlying operational time period that can be represented by a sequence of feasible transmission scenarios under power vector P allocated to consecutive time slots. We refer to such a sequence as a *scenario sequence* with respect to power vector P. The i-th scenario sequence with respect to power vector P and the associated value of the objective function are denoted as $SQ_i(P)$ and $\Omega_{SQ_i(P)}$, respectively. Furthermore, the set of all possible distinct scenario sequences, each operating under the same power vector P, is denoted as $X(P)$. Then, based on the definition of the scenario sequence, we can express the conditional optimal value of objective function also as follows:

$$\Omega^*(P) = \underset{i}{Sup}\{\Omega_{SQ_i(P)} : SQ_i(P) \in X(P)\}. \tag{5.17}$$

5.6.3 Mathematical Analysis

Lemma 1: Let $S_{(M)} = \{(i_1 \rightarrow J(i_1); s(i_1); D_{s(i_1),r(i_1)}), \ldots, (i_M \rightarrow J(i_M); s(i_M); D_{s(i_M),r(i_M)})\}$ be an arbitrary transmission scenario under power vector $P_{(M)} = (\beta P_{i_1}, \ldots, \beta P_{i_M})$, $0 < \beta P_{i_k} \leq P_{\max}, k = 1, \ldots, M$, where β is a real positive number. The spatial reuse factor of transmission scenario $S_{(M)}$ with respect to this power vector $P_{(M)}$ is a monotonically nondecreasing function of β in the interval $(0, \alpha(\beta^{-1} P_{(M)}))$, independent of nodal distribution, traffic pattern, and offered traffic load.

Proof: Let us consider an arbitrary transmission $(i_k \rightarrow J(i_k); s(i_k); D_{s(i_k),r(i_k)})$ in $S_{(M)}$, $k = 1, \ldots, M$. Based on relation Eq. (5.16), the transmission from i_k is *successfully received* at j, $j \in J_{i_k}$ if

$$G_{i_k j} \beta P_{i_k} \Bigg/ \left(N_j + \sum_{\substack{r=1 \\ r \neq k}}^{M} G_{i_r j} \beta P_{i_r} \right) \geq \gamma. \tag{5.18}$$

Independent of nodal distribution, traffic pattern, and offered traffic load, the derivative of the left-hand-side of relation (Eq. 5.18) with respect to β can be calculated as

$$\frac{\partial}{\partial \beta}\left(\frac{G_{i_k j}\beta P_{i_k}}{N_j + \sum_{\substack{r=1\\ r\neq k}}^{M} G_{i_r j}\beta P_{i_r}}\right) = \frac{G_{i_k j} P_{i_k} N_j}{\left(N_j + \sum_{\substack{r=1\\ r\neq k}}^{M} G_{i_r j}\beta P_{i_r}\right)^2}, \tag{5.19}$$

and is noted to be always nonnegative. Therefore, by increasing the value of β, the SINR at j remains constant (when $N_j = 0$) or increases. In fact, in the limit as $\beta \to \infty$, the SINR at j converges to a constant; that is,

$$\lim_{\beta\to\infty}\left(\frac{G_{i_k j}\beta P_{i_k}}{N_j + \sum_{\substack{r=1\\ r\neq k}}^{M} G_{i_r j}\beta P_{i_r}}\right) = \frac{G_{i_k j} P_{i_k}}{\sum_{\substack{r=1\\ r\neq k}}^{M} G_{i_r j} P_{i_r}}. \tag{5.20}$$

Similarly, the SINR at all other intended receivers increase as β increases. Therefore, the spatial reuse factor of the transmission scenario $S_{(M)}$ under $P_{(M)}$ is a monotonically nondecreasing function of β, $\beta \in [0, \alpha(\beta^{-1} P_{(M)})]$, independent of nodal distribution, traffic pattern, and offered traffic load. ■

While Lemma 1 corresponds to a single time slot, the following theorem is related to the entire operational period.

Theorem 1: If $P' = (P'_1, \ldots, P'_n)$ is relatively maximized with respect to $P = (P_1, \ldots, P_n)$, then $\Omega^*(P') \geq \Omega^*(P)$, independent of nodal distribution, traffic pattern, and offered traffic load.

Proof: Let P' be relatively maximized with respect to P. Based on Lemma 1, every feasible transmission scenario $S_{(M)}$ under power vector $P_{(M)} = (P_{i_1}, \ldots, P_{i_M})$ is also a feasible transmission scenario under power vector $P'_{(M)} = (P'_{i_1}, \ldots, P'_{i_M})$. Therefore, based on the definition of scenario sequence, every scenario sequence under P is also a scenario sequence under P'.

Now, let N_{i,P_i} represent the set of all nodes j in which there is a communication link from node i to node j under power level P_i, $i = 1, \ldots, n$. Since $P'_i \geq P_i$, based on relation Eq. (5.15) node i may attain some additional communication links under P'_i (due to the higher SNR), that is. $N_{i,p_i} \subseteq N_{i,p'_i}$, $i = 1, \ldots, n$. As a result, under power vector P', additional routes may be explored, which translates into supplementary scenario sequences. Therefore, $X(P) \subseteq X(P')$.

Assume the i-th scenario sequence under power vector P is the same as the j-th scenario sequence under power vector P', that is, $SQ_i(P) \equiv SQ_j(P')$. Since, every scenario sequence consists of a sequence of *feasible* transmission scenarios, we have $\Omega_{SQ_i(P)} = \Omega_{SQ_j(P')}$. Consequently, since $X(P) \subseteq X(P')$ and based on relation Eq. (5.17), we conclude that $\Omega^*(P') \geq \Omega^*(P)$, independent of nodal distribution, traffic pattern, and offered traffic load. ■

Lemma 2: An optimum power vector always exists.

Proof: Let Π_k represent the set of all power vectors $P = (P_1, \ldots, P_n), 0 \le P_i \le P_{\max}, i = 1, \ldots, n$, which results in the same conditional supreme value of the objective function $\Omega^{(k)}$ over the underlying finite or infinite operational period. Based on the definition of a transmission scenario, for a wireless mesh network with n half-duplex nodes and m source-destination pairs, there can be at most

$$N_S = \sum_{i=1}^{\lfloor n/2 \rfloor} \left[\binom{n}{i} m^i \sum_{k=i}^{n-i} \binom{n-i}{k} \sum_{\overline{n_i} \in A_{k,i}} \binom{k}{n_1, \ldots, n_i} \right] \tag{5.21}$$

$$= \sum_{i=1}^{\lfloor n/2 \rfloor} \sum_{k=i}^{n-i} \sum_{\overline{n_i} \in A_{k,i}} \binom{n}{i} \binom{n-i}{k} \binom{k}{n_1, \ldots, n_i} m^i \tag{5.22}$$

distinct transmission scenarios, where $\overline{n_i} = (n_1, \ldots, n_i)$, and $A_{k,i} = \{(n_1, \ldots, n_i) : \sum_{j=1}^{i} n_j = k, n_j \ge 1, n_j \in Z^+, j = 1, \ldots, i\}$. To elaborate on relation (15), we should make the following notes while considering the definition of a transmission scenario: First, there are $\binom{n}{i}$ distinct ways to select i transmitters out of n nodes, $i = 1, \ldots, \lfloor n/2 \rfloor$. Any such a set of i transmitters can be associated with transmissions corresponding to m^i distinct source-destination pairs. Secondly, the maximum and the minimum number of receivers is $n - i$ and i, respectively, when there are i simultaneous transmissions invoked in the network, $i = 1, \ldots, \lfloor n/2 \rfloor$. Thirdly, there are at most $\sum_{\overline{n_i} \in A_{k,i}} \binom{k}{n_1, \ldots, n_i}$ combinations of allocating k receivers to i transmitters.

In turn, for any finite ad hoc wireless network, the total number of distinct Π_k sets (Ψ) is finite and bounded by

$$\Psi \le \sum_{i=0}^{N_S} \binom{N_S}{i} \tag{5.23}$$

$$= 2^{N_S}. \tag{5.24}$$

Relation (Eq. 5.23) is valid due to the fact that the use of two power vectors $\hat{P} = (\hat{P}_1, \cdots, \hat{P}_n)$ and $\tilde{P} = (\tilde{P}_1, \cdots, \tilde{P}_n)$ results in the same conditional supreme value of the objective function if their associated sets of feasible transmission scenarios are identical. But, the total number of distinct feasible transmission scenarios is always limited by the upper bound for the total number of distinct transmission scenarios (N_S), yielding the right-hand side of relation (Eq. 5.27). Clearly, the finite cardinality of N_S (Eq. 5.22) yields the finite cardinality of Ψ.

Based on the above mentioned property, the set of all power vectors $P = (P_1, \ldots, P_n), P_{\min}(i) \le P_i \le P_{\max}(i), i = 1, \ldots, n$, can be partitioned into a *finite* positive number of sets (Π_k) so that for each set the conditional supreme value of the objective function is identical. Therefore, exactly one of the sets (Π^*) achieves the optimum objective function Ω^* in the underlying finite or infinite operational period. Consequently, any power vector in Π^* is an optimum power vector, which completes the proof. ■

Theorem 2: Independent of the underlying nodal distribution, traffic pattern, and offered traffic load, there exists a relatively maximum power vector that is optimum.

Proof: Let us assume that there is no optimum relatively maximum power vector. Then, based on Lemma 2, there exists an optimum power vector P^* that is not relatively maximum. Let P' denotes the relatively maximized power vector with respect to P^*. But, based on Theorem 1, $\Omega^* \leq \Omega(P')$. Clearly, the latter contradicts the suboptimality of every relatively maximum power vector, which completes the proof. ■

In general, an optimum power vector is a function of nodal distribution, traffic pattern, and offered traffic load. However, based on Theorem 2, independent of the underlying nodal distribution, traffic pattern, and offered traffic load, there exists an optimum power vector for which at least one of the components is equal to P_{max}. Intuitively, this is due to the fact that relative maximality provides a higher *combinatorial diversity* (i.e., higher degree of freedom) in terms of the optimization of the joint scheduling and routing scheme. In fact, as it has be numerically demonstrated for specific cases of the theoretical results (that involve a specific definition of the objective function, specific traffic pattern and offered traffic load, and merely unicast routing mechanisms), the latter property can lead to significant increase in the conditional supreme value of the objective function (Behzad and Rubin, 2004).

The following conclusions follow directly from the latter theorem.

Corollary 1: Under the special case that the maximum transmission power is sufficiently high, a fully connected topology is an optimum topology of a wireless mesh network, independent of nodal distribution, traffic pattern, and offered traffic load.

Corollary 2: Under the special case that the transmission power of all nodes is assumed to be identical (yet programmable), the power vector $P = (P_1 = P_{\text{max}}, \ldots, P_n = P_{\text{max}})$ is optimum, independent of nodal distribution, traffic pattern, and offered traffic load.[10]

5.7 Dynamic Sleep Scheduling in Rechargeable WM^{2}Nets[11]

5.7.1 Rechargeable Mesh Systems

The dynamic node activation or dynamic sleep scheduling problem involves determining when a node should be involved in data communication (activation) or should be put on the "standby" mode (deactivation).

[10] The result is concluded based on the fact that power vector $P = (P_1 = P_{\text{max}}, \ldots, P_n = P_{\text{max}})$ is the only relatively maximum power vector under the considered conditions.

[11] Excerpt from the invited article "Dynamic sleep scheduling in rechargeable wireless mesh networks," Neeraj Jaggi, Koushik Kar, and Ananth Krishnamurthy. Rensselaer Polytechnic Institute, E-mail: {jaggin,kark,krisha}@rpi.edu

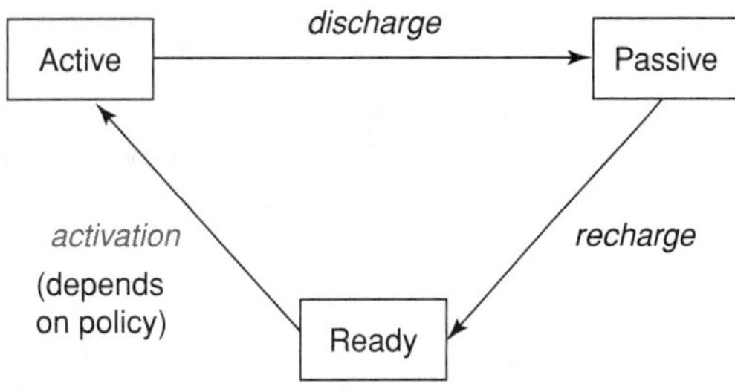

FIGURE 5.19 States and transitions under FA system model.

In this study, we distinguish wireless mesh systems in terms of activation (and deactivation) constraints on units. Two different system models are considered, namely, the full activation (FA) system model (Kar et al., 2006) and the partial activation (PA) system model (Jaggi, 2006; Jaggi et al., 2005). In the FA system model, mesh nodes get discharged after a certain duration of time, and need to be fully recharged. On the other hand, in PA system model, a node is modeled as an energy bucket of K quanta, and is available for activation as long as it has nonzero energy. Unlike FA model, in PA system model, a node activates (deactivates) itself even when it is not fully recharged (discharged).

The state transition diagram for a rechargeable node under FA system model is shown in Fig. 5.19. The discharge and recharge times under the FA model are modeled as exponentially distributed with means $1/\mu_1$ and $1/\mu_2$, respectively. Let $\rho = \frac{\mu_1}{\mu_2} \geq 1$, since the discharge rate typically exceeds the recharge rate. Spatial correlation among the nodes' recharge and discharge times is modeled by considering two different correlation models, namely, the independent lifetime (IL) and the correlated lifetime (CL) correlation models. In the IL model, the discharge and recharge times are independent, whereas in the CL model, the discharge times of all units that simultaneously enter the active state are the same. Similarly, the recharge times of all units that simultaneously enter the passive state are the same. The discharge (recharge) times of units that enter the active (passive) state at different times are independent of each other.

As shown in Fig. 5.20, a node in PA model is modeled as a finite-buffer quantum-queue with a buffer capacity of K. Note that the queue is discharged only when node is activated. The quanta arrival or recharge process at each node is modeled as Poisson with rate λ, and the discharge time of each quantum is exponentially distributed with mean $1/\mu$. Let $\gamma = \frac{\mu}{\lambda} \geq 1$, since average discharge rate is typically higher than the average recharge rate. Spatial correlation across the nodes' recharge and discharge processes

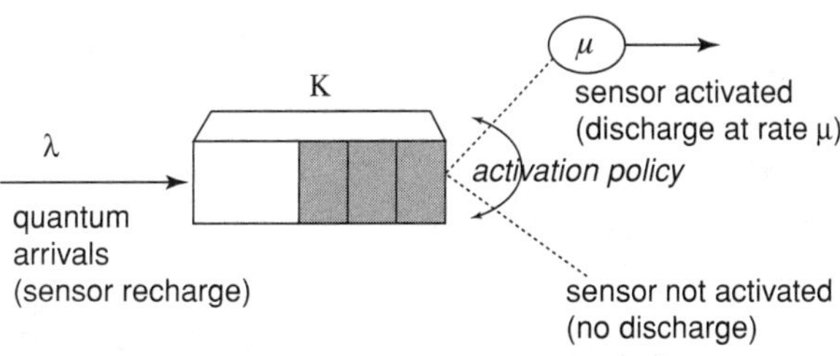

FIGURE 5.20 Finite buffer queuing model of a node under PA system model.

leads to two different correlation models, namely the independent discharge recharge (IDR) and the correlated discharge recharge (CDR) correlation model. In IDR model, the discharge as well as the recharge processes at different nodes are independent, whereas in CDR model, the discharge as well as the recharge processes at different nodes are completely correlated, that is, recharge quanta arrive at the same time at all nodes and get consumed, one by one, at the same time from all active nodes.

5.7.2 Performance Criteria

Consider a system of rechargeable nodes deployed over a certain region of interest. If a large number of units are deployed, it is likely that more units would remain charged at any given time. Thus, the overall system performance would typically improve with a more redundant node deployment. The performance of the rechargeable node system can be characterized by a continuous, nondecreasing, strictly concave function U satisfying $U(0) = 0$. More specifically, $U(n)$ represents the "utility" derived per unit area, per unit time, from n active nodes covering an area. As an example of a practical utility function, consider the scenario where each node can detect an event with probability p. If the utility function is defined as the probability for the system to detect an event, then $U(n) = 1 - (1 - p)^n$, where n is the number of active nodes. Figure 5.21 depicts the shape of this utility function for various values of detection probability p. The long-term system performance is represented by the time-average utility of the system. Let A denote the entire area in the physical space of interest. Let $n_\Pi(a, t)$ denote the number of active nodes that cover area element α at time t, under activation policy Π. Then the time-average utility under policy Π is given by

$$\lim_{t\to\infty} \frac{1}{t} \int_0^t \int_A U(n_\pi(a, t))\, da\, dt.$$

The decision problem is that of finding the sleep schedule or policy Π^*, such that the time-average utility is maximized.

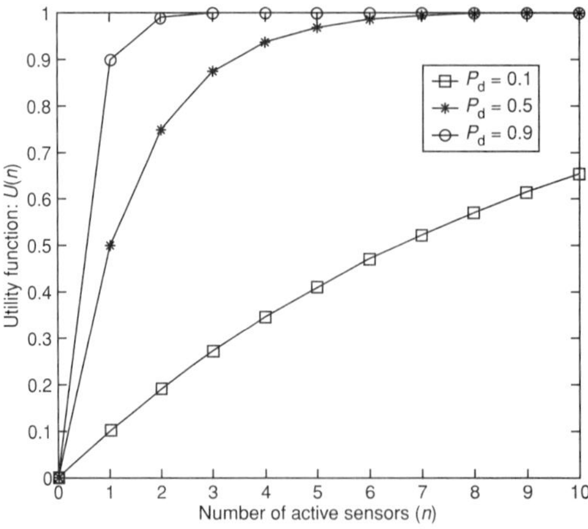

FIGURE 5.21 Utility function characteristics.

5.7.3 Threshold-Based Sleep Scheduling

Although the optimal sleep scheduling constraint can be formulated using the Markov decision process framework under specific cases, determining optimal policies can be nevertheless computationally prohibitive where global knowledge and coordination would be required. In practical scenarios, a communicating unit is required to take activation decisions in a distributed manner, based only on local topology and state information. This motivates us to study a class of simple threshold policies.

A threshold activation policy with parameter m is characterized as follows: At any decision time unit, a node s is activated if the number of active nodes does not exceed m. Otherwise s is kept in a ready state till the next decision time. In other words, a threshold policy with parameter m attempts to maintain the number of active nodes in the system as close to m as possible but never exceeding m. Activation decisions need to be taken only when the state of the overall system changes.

5.7.4 Performance Evaluation for Identical Coverage

Let us first examine the performance of threshold policies for a system of nodes that are deployed over an area A. In the identical coverage scenario, the performance of activation policy Π is given by $\lim_{t\to\infty} \frac{1}{t}\int_0^t n_\pi(t)\,dt$, where $n_\Pi(t)$ denotes the number of active nodes at time t when the system operates under policy Π. In this case, threshold policies achieve near-optimal performance in both FA and PA system models.

Theorem 1: The upper-bound to maximum achievable performance is given by (Kar et al., 2006; Jaggi, 2006):

(1) $U\left(\frac{N}{1+\rho}\right)$ under the FA system model.

(2) $U\left(\frac{N}{\gamma}\right)$ under the PA system model.

For FA system model, using Markov chain, the performance of threshold policies can be expressed in closed form, as a function of m. For PA system model, the performance of threshold policies can be expressed in terms of steady-state utilizations of equivalent queuing systems. Let $U^{\mathrm{T}}(m)$ denote the time-average utility achieved by threshold activation policy employing a threshold of m. Let us assume that N is divisible by both $(1+\rho)$ and by γ.

Theorem 2: Threshold policies provide the following performance bounds (Kar et al., 2006; Jaggi, 2006; Jaggi et al., 2005):

i. $U^T\left(\frac{N}{1+\rho}\right) \geq \frac{3}{4}U\left(\frac{N}{1+\rho}\right)$ under FA model, for both IL and CL correlation models.
ii. $U^T\left(\frac{N}{\gamma}\right) \geq \frac{K}{K+\gamma}U\left(\frac{N}{\gamma}\right)$ under PA model, for both IDR and CDR correlation models.

Note that the recommended threshold (m^*) is the energy-balancing threshold that arises from the equations $m\mu_1 = (N-m)\mu_2$ and $m\mu = N\lambda$ under the FA and PA system model, respectively. Also, in the latter case, the threshold policy achieves asymptotically optimal performance with respect to the mesh unit energy bucket size K.

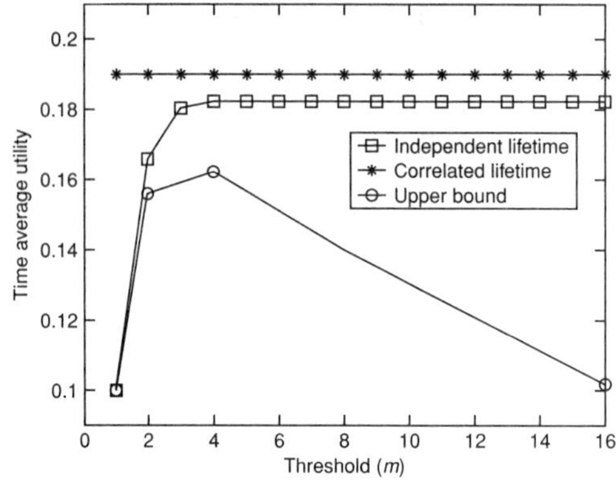

FIGURE 5.22 Performance for IL and CL correlation models ($N = 16$, $\rho = 7$, $p = 0.1$).

The threshold activation policies are of exceptional interest due to several reasons. Firstly, they are proven to achieve near-optimal performance under both system models. Secondly, they are robust in the presence of spatial correlations in the recharge and discharge times. Thirdly, they are computationally efficient, since nodes do not need to consider their current energy levels or the utility function while taking activation decisions. Finally, threshold policies can easily be extended to develop a distributed algorithm appropriate for more general network scenarios, as described in the next section.

Figures 5.22 and 5.23 plot the performance of threshold policies for various values of threshold parameter. For both the FA and PA system models, threshold policies satisfy the performance bounds as given in Theorem 2, and the bounds can be shown to be fairly tight. The presence of correlation in the discharge and recharge times deteriorates system performance, particularly at higher values of the threshold.

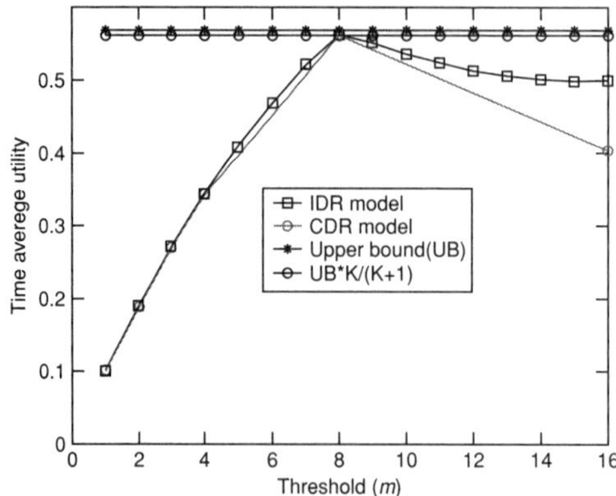

FIGURE 5.23 Performance for IDR and CDR correlation models ($N = 16$, $K = 100$, $\gamma = 2$).

5.7.5 Distributed Algorithm

In a realistic deployment scenario, mesh nodes would be deployed at random, and will typically cover different areas in the physical space of interest. In other words, the coverage areas of two nodes may overlap only partially, or may not overlap at all.

Let m_i denote the targeted threshold for node i. In other words, node i maintains a utility of $U(m_i)$ per unit area and time in its coverage area A_i. When the node is ready (available), then at any decision instant, it computes the current utility per unit time in its coverage area as follows. For a generic unit area element $A \in A_i$, let $n(A, t)$ denote the number of active nodes covering it at time t. Then the current utility per unit time at time t in the coverage area of node i is calculated as $\int_{A_i} U(n(A, t)) \cdot dA$. If the current utility is less than the targeted utility, then the node activates itself. Otherwise, the node remains in the ready (available) state until the next decision instant.

The targeted threshold may be different for different nodes depending upon the density of deployment of nodes in each node's coverage area. For each unit area element $A \in A_i$, let $N(A)$ denote the total number of nodes covering A. Then the node i would like to maintain a threshold of $\frac{N(A)}{1+\rho}$ in this area element, under the FA network model. The overall targeted utility per unit time in the node's coverage area A_i is given by $\int_{A_i} U\left(\frac{N(A)}{1+\rho}\right) \cdot dA$. Targeted utility can similarly be computed for PA network model.

In order to evaluate the network performance for a range of thresholds, let us introduce a local threshold parameter. If node i employs a local threshold parameter of α, its targeted utility is given by $\int_{A_i} U\left(\alpha \cdot \frac{N(A)}{1+\rho}\right) \cdot dA$, under the FA network model and similarly under the PA network model. Note that the value of $\alpha = 1$ corresponds to the recommended threshold given by Theorem 2. However, some of the invariants from the identical coverage case may not be satisfied in the general network scenario. For instance, even though all units employ a local threshold, α, it is possible that the target number of active units in some area, A, is exceeded.

5.7.6 Performance Evaluation for General Network

The maximum achievable utility in each generic area element in the network is summed up to obtain the upper bound on the achievable performance for the entire network. The discharge and recharge processes at nodes are triggered by the occurrence of discharge and recharge events respectively. An event drops at an area element in the network and affects nodes that lie in the vicinity of the area element. These events occur randomly at the area elements spanning the entire network. The amount of spatial correlation in the recharge and discharge times (processes) at nodes is modeled using the different event models, namely, independent events (IE) model and block-correlated event (BCE) model. Events occur according to a Poisson process, and are uniformly distributed in the area of interest. In IE model, an active node gets discharged by a quantum only when a discharge event occurs within its coverage area. The recharge process is modeled similar to the discharge process. Note here that the recharge and discharge processes at different nodes are not completely independent since nodes may have overlapping coverage areas. However, the degree of correlation is smaller in comparison to the BCE model. In BCE model, the network is divided into virtual blocks of equal sizes. A discharge (recharge) event occurring anywhere in the block affects all nodes located in this block

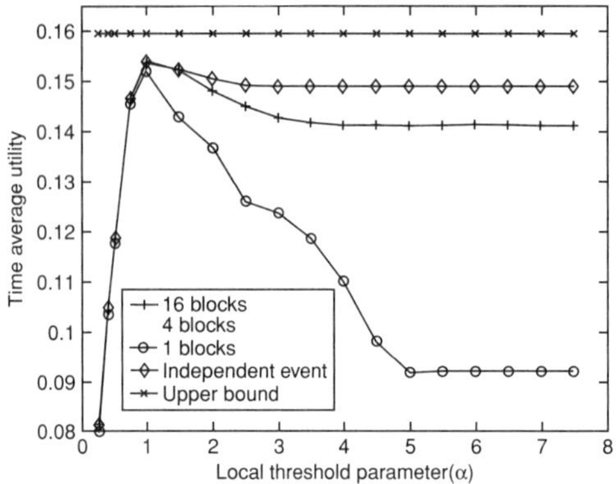

FIGURE 5.24 Network performance under FA system model.

in a similar manner. This introduces spatial correlation across the discharge (recharge) times (processes) of nodes.

The performance of the distributed node activation algorithm is evaluated using simulations for a wide range of system parameters, for different node system models. A total of $N = 52$ nodes, each having a circular coverage pattern of radius 12 units, are thrown uniformly at random in an area of size 50×50. With these parameters, the mean coverage of the network, defined as the average number of nodes covering any point in the deployment region, is observed to be approximately 9.1. The event detection probability for individual node $p = 0.1$. For FA network model $\rho = 3$, and for PA network model $\gamma = 2$. The node energy bucket size is $K = 100$.

Figures 5.24 and 5.25 plot the network performance for various values of local threshold parameter α, for FA and PA network models respectively. The performance for all the system models peaks at a value of α close to 1, and the peak performance is very close to the upper bound. Note that, as the number of blocks decrease, the size of each block increases which leads to an increase in the degree of spatial correlation. Therefore, the figures demonstrate that the performance of threshold policies degrade as the degree of spatial correlation increases. This performance drop is particularly significant

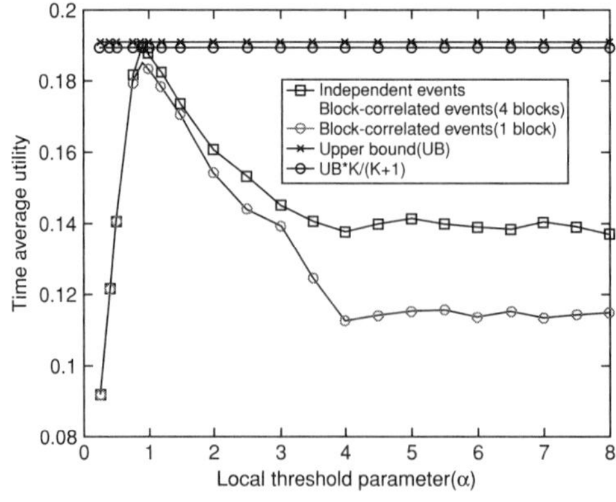

FIGURE 5.25 Network performance under PA system model.

at higher values of the threshold parameter α. Particularly, we observe that the performance bounds achieved for identical coverage case are also satisfied in the general network scenario.

5.8 On Improving Throughput and Fairness in Wireless Mesh Networks: A Listen-and-Learn Approach[12]

5.8.1 Introduction

The primary technology underpinning multihop wireless networks is the IEEE 802.11. It features low-cost, ease of setup, and high physical data rates (up to 54 Mbps). The most important feature of such IEEE 802.11-based mesh networks is that the different links of the mesh network share the radio resources using a CSMA-based random access protocol. Since a random access protocol does not require the scheduling overhead associated with coordinated access schemes such as TDMA, it greatly enables distributed network operation. In CSMA, when a node wishes to send traffic, it first senses the radio carrier and proceeds with the transmission only if it considers the channel to be idle.

In extending CSMA to suit the needs of WM^{2}Nets, a major complication that arises is that the carrier sensing operation must now cope with the following two forms of asymmetry:

1. *Contention asymmetry*: Since a node has limited transmit power, it can communicate with only a subset of the nodes that form the network. Given the distribution of the nodes in the network, the set of nodes that a node can sense, or, be sensed by, is quite variable. This introduces asymmetry in the level of contention each link/node experiences.
2. *Traffic asymmetry*: The rate at which a link-i contends for the radio channel is a direct function of the traffic λ_i it needs to carry. This in turn is a function of the topological location of the link. For example, links towards the interior of the network multiplex the multihop traffic belonging to a number of end-to-end flows and as such, they typically carry more traffic than the links towards the periphery. This problem, though not unique to wireless networks, induces asymmetry in the amount of traffic the links need to carry.

Indeed, the IEEE 802.11 protocol has been designed for networks where links are in carrier sensing range of one another. A direct application of IEEE 802.11 to multihop mode would result in undesirable performance such as some links receiving no throughput (Nandagopal et al., 2000; Medepalli and Tobagi, 2006), throughput instability (Xu and Saadawi, 2001), etc. The first goal of this study is to provide some new insights into the dependency between individual link throughputs and CSMA access parameters. We study the impact of IEEE 802.11 contention parameters (CWmin and CWmax) on link throughput and find that CWmin has a far greater impact than CWmax. The second goal

[12] Excerpt from the invited article "On improving throughput and fairness in wireless mesh networks: A listen-and-learn approach," Kamesh Medepalli and Fouad A. Tobagi, Department of Electrical Engineering, Stanford University, Stanford, CA 94305, Email: kmedepalli@gmail.com, tobagi@stanford.edu

of this study is to leverage upon these lessons and design a new distributed listen-and-learn algorithm that allows each link to dynamically adjust its contention parameters based on the traffic load. The algorithm exploits the structure of the optimal CSMA access policy (Medepalli and Tobagi, 2006a; Medepalli et al., 2006b), which unlike the Exponential Backoff in IEEE 802.11 uses a single backoff window that is a function of the number of contending nodes. As shown, the algorithm significantly improves link (and hence network) throughput while ensuring the disparity among the link throughputs to be small. Moreover, since it uses the structure of the optimal backoff policy, it significantly mitigates the TCP instability problem (Xu and Saadawi, 2001).

5.8.1.1 The Throughput Fairness Tradeoff

In this section, we illustrate the inherent throughput-fairness tradeoff in random access networks like IEEE 802.11 mesh networks. Due to space limitations, we will provide only some representative results. The reader is referred to (Medepalli and Tobagi, 2006; Medepalli, 2006c) for details on the mathematical analysis. We start by providing a brief description of the IEEE 802.11 random access protocol. A node wishing to transmit first senses the medium for a duration called DIFS. If the medium is idle, the node sets a backoff counter that is a uniformly distributed random number between 0 and W_n, where $W_n = \min\{2^n(W_0 + 1) - 1, W_L\}$ is the backoff window at the n-th retry of the packet, W_0 is the minimum contention window (referred to as CWmin), W_L is the maximum contention window (referred to as CWmax), and L is the number of stages in BEB. Thus, at each retry, the contention window is doubled according to the BEB process, until the maximum value of W_L is reached. In IEEE 802.11b, $W_0 = 31$, $W_L = 1023$, and $L = 5$. The node decrements the backoff counter for every idle slot on the channel and when the counter reaches zero, the node transmits the RTS control packet. If the destination receives this packet, it confirms by sending a CTS packet. Upon receiving the CTS, the sender will send the DATA packet. In reception of the DATA packet, the receiver sends an ACK to acknowledge the DATA packet.

With these in mind, let us now consider the 10-node random network shown in Fig. 5.26a where each node uses 0.2 W of transmit power. A dotted line between nodes indicates that the nodes can sense each other, while a solid line connects the communicating pairs. Figure 5.26b depicts the individual throughputs obtained when all links are saturated, that is, all links are busy. UDP traffic used for network traffic load throughout the simulation experiments.

We notice that even in the absence of multihop routing (all communications are local in this example) the throughput of each link is significantly biased from its topological location. For example, links between nodes 1 and 8 receive almost no throughput while links between nodes 2 and 3 receive very high throughput. The reason for this behavior is that the IEEE 802.11 standard proposes the same BEB parameters (CWmin, CWmax) for all links, regardless of the traffic situation in different parts of the network—thereby giving an unfair advantage to links with less traffic and less contention as such. The question that remains to be answered is the role the main contention parameters CWmin and CWmax play in this behavior. In Fig. 5.27a, the effect of CWmin is isolated by fixing the value of CWmax to 1023 slots whereas keeping Cwmin variable. Throughput and fairness are plotted for each value of CWmin. To quantify fairness, that is, the disparity in throughput achieved by different nodes, we use the Jain fairness index (Jain et al., 1984). The throughput-fairness tradeoff is clear—high throughput and high fairness are not achievable simultaneously. The throughput is monotonically decreasing with CWmin

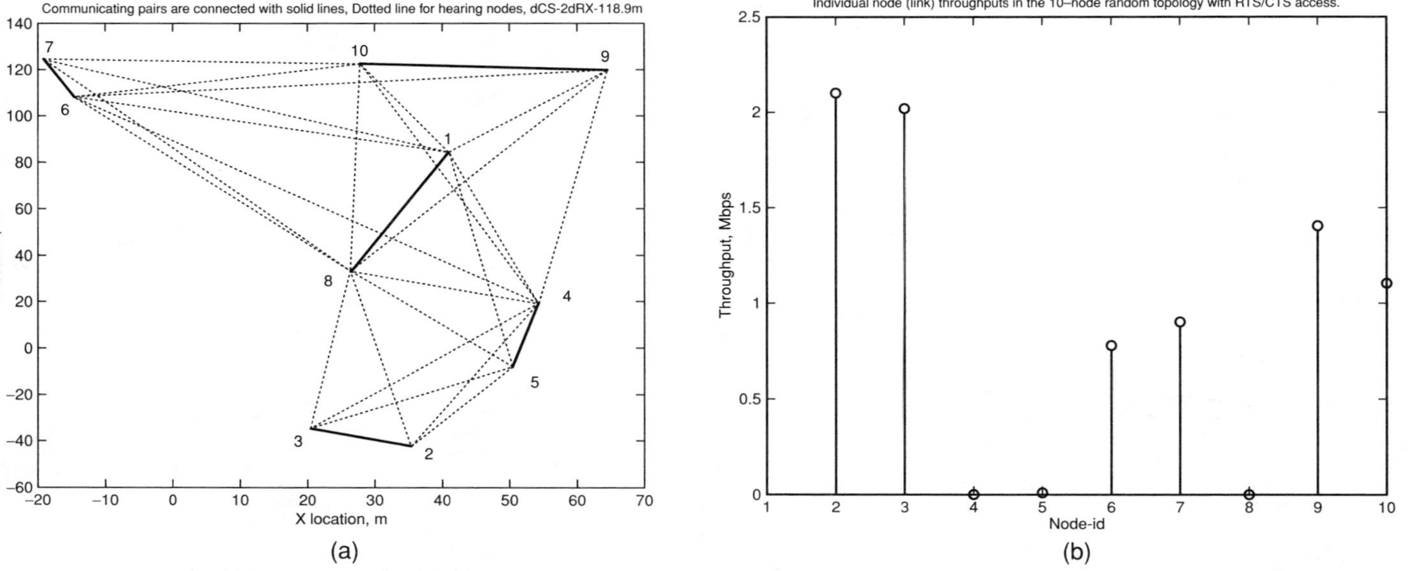

Figure 5.26 A random 10-node topology. Communicating pairs are shown using solid lines and dotted lines indicate nodes that are within carrier sensing range of each other. dCS = 2dRX = 118:9m (left). Individual link throughputs in the 10-node random topology (right). (a) A 10-node mesh network. (b) Individual link throughputs.

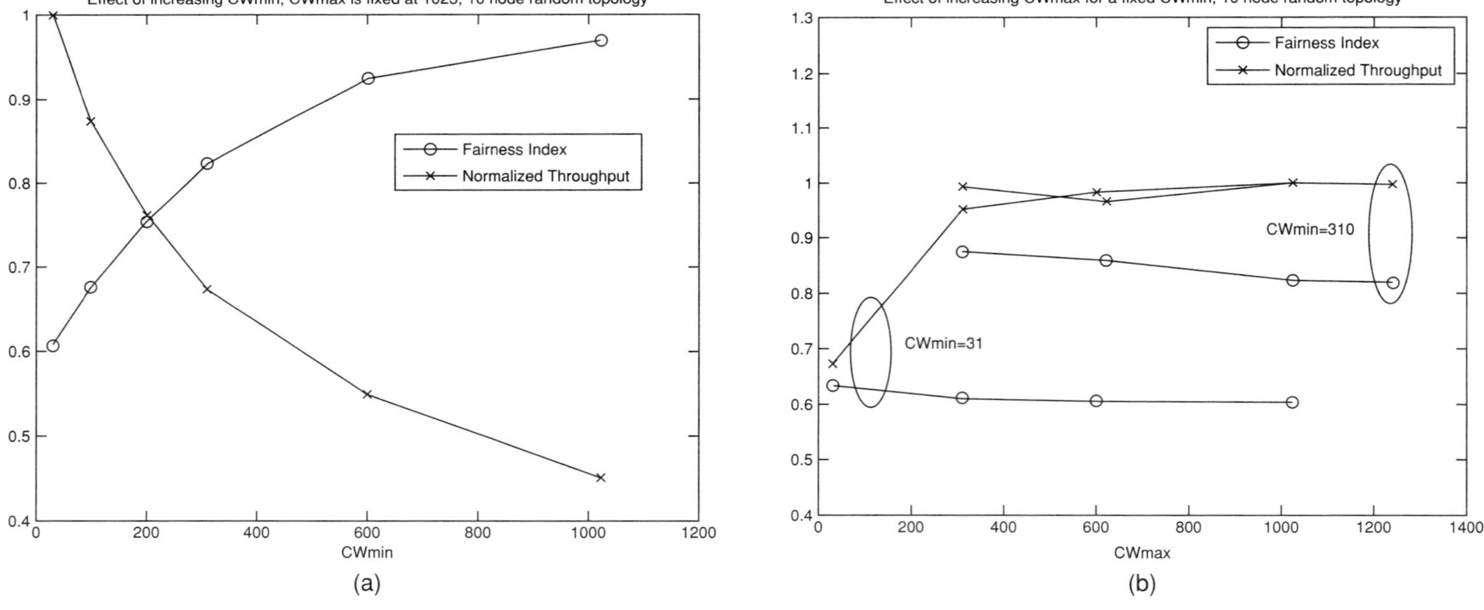

Figure 5.27 (Left) Effect of increasing CWmin while keeping CWmax fixed at 1023. (Right) Effect of increasing CWmax while keeping CWmin fixed. (a) Impact of CWmin. (b) Impact of CWmax.

whereas fairness is monotonically increasing with CWmin (recall that 802.11b standard specifies CWmin = 31 and CWmax = 1023). Note that increasing the CWmin has the affect of keeping the channel virtually idle. This has a remarkable effect on fairness because the nodes that were previously sensing the channel to be busy most of the time can now sense an idle channel with high probability. Moreover, the transmission is bound to be successful with high probability for large values of CWmin. In Fig. 5.27b, we have attempted to isolate the effect of CWmax on throughput and fairness by fixing the value of CWmin. We note that by increasing CWmax, we are aggravating the unfairness while improving the throughput. We can conclude that although increasing CWmax has the opposite affect of increasing CWmin, overall, it has a lesser effect than CWmin. We will use this result to set CWmax and CWmin to a common value according to the algorithm described in the next section.

Listen-and-Learn Approach for Improving Throughput and Fairness The results in the previous section depict that the contention parameters need to reflect the topological location of the node. To this end, we propose a simple heuristic contention metric called effective contention indicator (ECI), denoted $\sigma_{i,j}$, for the link-m between a source $i = s(m)$ and destination $j = d(m)$. It is taken to be the total number of links in the set $E_{i,j}$, which is the union of links k whose source $s(k)$ is in the carrier sensing range of $s(m)$ and whose destination $d(k)$ is in the interfering range of $d(m)$, that is, $\sigma_{i,j} = |E_{i,j}|$. We will shortly describe a distributed method called listen-and-learn, which allows links to independently compute $\sigma_{i,j}$. Recall that the objective of this stuy is not in merely coming up with a metric but rather to use it in an operational protocol for improving the overall network performance. For single-hop IEEE 802.11 networks whereby all nodes can hear each other, prior studies (Bianchi, 2000; Medepalli and Tobagi, 2006a) have showed that system throughput is maximized when the average contention window of users is made to be a *linear* function of the number of contending users N. In fact, we show that it is indeed necessary[13] for all users to operate using a single backoff window equal to the optimal average backoff window of $N\sqrt{2w_s}$ where w_s is the number of slots wasted on the channel when there is a collision. In a multihop network, not all nodes can directly communicate with each other, and the optimization of these random access networks remains an open problem in the field. The approach adopted is thus to exploit the fact that the optimal window size for single-hop networks is linear to the number of contending users; then set the contention window $W_{i,j}$ of link i-j using the metric $\sigma_{i,j}$ as follows:

$$W_{i,j} = \sqrt{2w_s}(f_N + b(f_N - \sigma_{i,j})) \tag{5.25}$$

where f_N is a network-wide constant that is approximately equal to twice the maximum edge degree in the network and b is a network-wide fairness control parameter that allows control over the relative priority of high contention and low contention links. The impact of this parameter will become clear when we discuss numerical results. Equation (5.25) suggests that for nodes to determine which window size to use, nodes need to determine the ECI for each link for which it is a source. This information is acquired

[13] That is to say that optimizing CWmin is not sufficient. It is necessary to disable BEB. Using additional backoff stages would penalize TCP throughput and performance in lossy channels. See the works of Medepalli and Tobagi (2006a) and Medepalli et al. (2006b) for relevant discussions.

through the use of a distributed algorithm called window-based adaptive backoff algorithm (WABA) (Medepalli, 2006c). The basic idea of the algorithm is that nodes maintain a small list of the nodes that it has heard from and includes this information (along with an estimate of the average packet transmission time, which dictates the channel cost due to collisions) in all its transmissions. The users who receive this packet will learn of the sender's neighborhood and update their ECI computation, should it be needed. Thus, all users learn the contention state of the channel in a distributed, asynchronous, and adaptive manner. They then use this information to modify the contention window size that they must use for their own transmission. If a link infers that its ECI has increased, it will automatically choose a smaller contention window to allow itself better success probability (see Eq. 5.25).

5.8.1.2 Performance Evaluation

We now evaluate the improvement in performance achieved when using the ECI contention metric.

Improving Throughput and Fairness In contrast to the single-hop case where maximizing the individual user throughput is equivalent to maximizing the system throughput, there is no such one-to-one correspondence between the user and the system objectives in a multihop network. Rather, one can consider the throughput-fairness tradeoff in a multihop scenario and evaluate how well a particular scheme improves the tradeoff. Figure 5.28 explores precisely that. In this figure, the throughput-fairness tradeoff

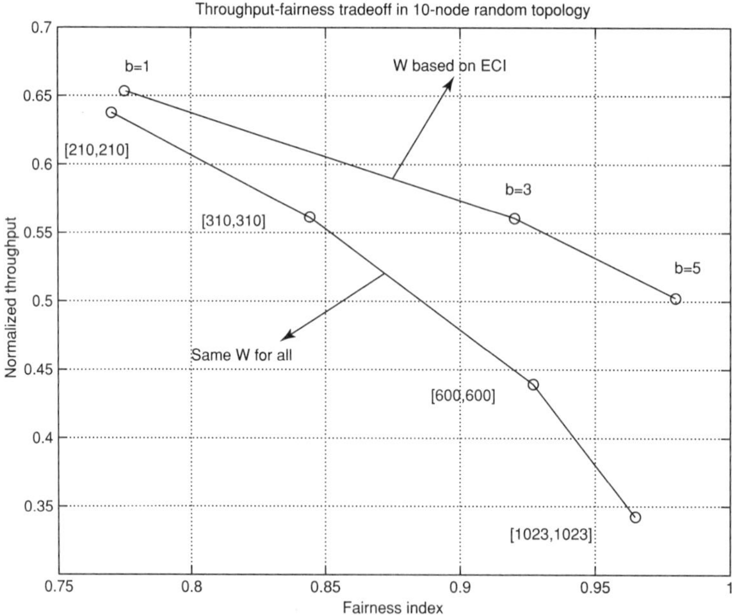

FIGURE 5.28 Improving throughput-fairness tradeoff in the 10-node random topology by setting the contention window size to be a function of the ECI.

for the random 10-node topology shown in Fig. 5.26a is considered. The x-axis of Fig. 5.26a denotes the Jain fairness index computed using the individual link throughputs. On the y-axis of Fig. 5.26a, we plot the sum of the throughput $\sum S_i$ normalized by the total throughput of the default IEEE 802.11 system which, as mentioned before, uses the same contention window parameters regardless of location-dependent contention.

The curve for IEEE 802.11 is obtained by modifying the Exponential Backoff parameters, that is, CWmin and CWmax, just as in Fig 5.27a and 5.27b except that we set CWmax = CWmin. Observe that the default system lies on the top left corner in Fig. 5.28 yielding the normalized throughput of 1.0 and fairness index of approximately 0.5. Since the initial contention window is a more influential control parameter than the maximum contention window, we modify the backoff parameters of 802.11 to set them to different values such as $[CW_{min}, CW_{max}] = [200, 200]$ and $[CW_{min}, CW_{max}] = [1023, 1023]$. It is important to clarify that all nodes still use the same backoff parameters; however, the common backoff parameters are what are varied. The same figure also shows results for the proposed algorithm, which infers ECI and then sets its contention window. Again, using different values of the fairness-knob b generated different points of the curve. We observe that when a high fairness index is desired (all links achieving similar throughput) the proposed scheme increases the commonly achievable throughput by about 60%. Thus, estimating contention and adjusting its contention window adaptively is a better way of sharing resources in multihop wireless networks than using 802.11 (as they have no inbuilt mechanism for estimating location-dependent contention).

Stabilizing TCP Performance in Multihop Networks Recall that the proposed approach involves two key features: (1) Estimate contention and (2) Use a single backoff window to suit the contention (this exploits the structure of the optimal backoff policy studied by Medepalli and Tobagi (2006a) and Medepalli et al. (2006b). While the results so far have focused on the first issue, we now show that using a single backoff window significantly reduces delay variability, promoting a better cross-layer interaction with TCP traffic. Consider the scenario shown in Fig. 5.29a where there are three TCP flows, each of different path length and each starting at different time. Flow 1 is from node 1 to node 10 and becomes active at time = 5 s. Flow 2 is from node 7 to node 4 and becomes active at time = 25 s and stays until time = 85 s. Finally, Flow 3 is a single-hop flow from node 5 to node 6 and is active from time = 40 s to 60 s. The goal is to study the throughput for each of these flows. For simplicity, we assume that all nodes are equally spaced apart and they all use the same transmission power of 0.4 W. We assume a simple path loss model with propagation exponent equal to 4. Carrier sensing threshold was set to −90 dBm while the receiver sensitivity was set to 12 dB above this for a data rate of 11 Mbps. The carrier sensing range and transmission range are thus as indicated in the figure. For TCP, we use TCP Reno and Delayed ACKs are enabled at the receiver with a delay of 100 ms.

Figure 5.29b illustrates the throughput for the three flows when IEEE 802.11 is used as well as when contention window is estimated via the proposed algorithm. When Flow 1 is the only flow (from time = 5 s to 25 s), we observe that with the default 802.11 system, the TCP throughout exhibits instability, yielding zero throughput at several time instants. The links in the middle contend more often as both TCP DATA and TCP ACKs need be transferred through in a timely manner. The significant variability in delays arising

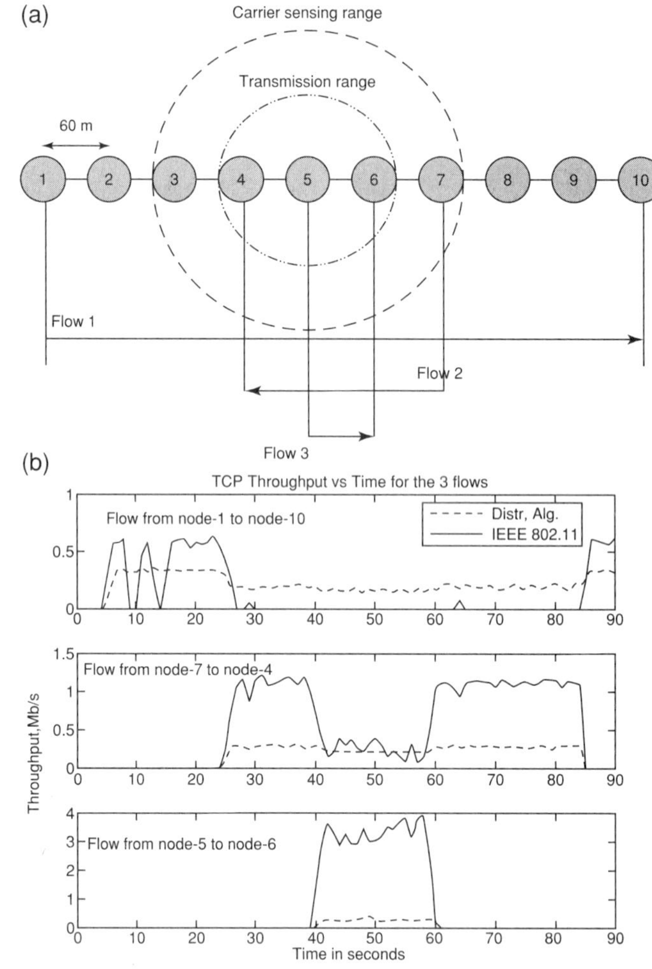

FIGURE 5.29 Three dynamic TCP flows are present in the above topology. Flow 1 becomes active at time = 5 s, flow 2 is active from time = 25 s to 85 s while Flow 3 is active from time = 40 s to 60 s (left). Improving stability and fairness of TCP throughput in a multihop network using ECI and distributed algorithm. TCP Reno with delayed ACKs was used. TCP packet size is 1,000 B and window size was set to 10 packets (for distributed algorithm $f_N = 15$ and $b = 8$) (right). (a) A linear topology with three flows. (b) Throughput vs. Time.

due to contention among the multiple hops and the TCP DATA/ACK packets results in instability of TCP (Xu and Saadawi, 2001). When Flow 2, being of shorter length and located topologically within Flow 1, becomes active, it completely captures the channel, leaving Flow 1 with zero throughput as seen in Fig. 5.29b. In contrast, adjusting the contention window size dynamically, the algorithm is able to enforce fair-share of the flows. Finally, when Flow 3 becomes active it shares a substantial portion of Flow 2's throughput. This is illustrated through the high disparity between Flow 2 and Flow 3 throughputs. In contrast, the distributed algorithm is able to maintain fairness among the throughputs of different flows.

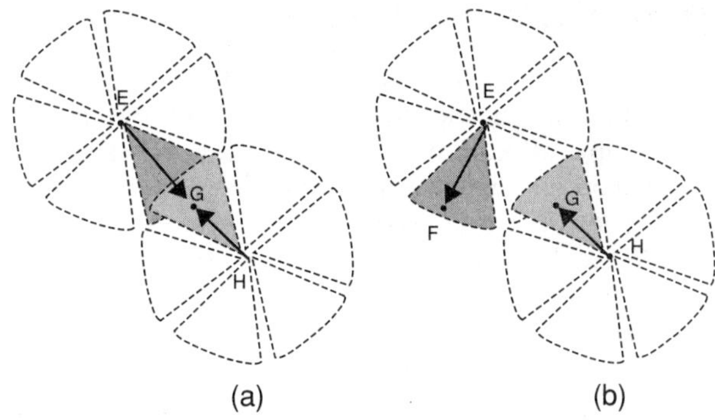

FIGURE 5.30 Region of interference.

5.9 Mathematical Modeling and Performance Evaluation for Centralized Scheduling in WiMAX Mesh Networks[14]

5.9.1 Preliminaries

Throughout this study, it is assumed that a routing tree rooted at BS is always available and constructed using shortest path routing.

According to the IEEE 802.16-2004 Standard, the Mesh mode supports only the time division mode (TDD). Furthermore, the MAC layer is assumed to schedule data to multiple access (TDMA) through a single carrier channel. Thus, as long as the bandwidth allocation result is calculated, frames in each link can be built in a simple structure based on the allocation. The following rules are defined:

Rule 1: A node cannot transmit and receive at the same time.
Rule 2: The relaying data traffic received by one SS cannot be transferred immediately to its neighbor in the same frame slot.
Rule 3: Spatial reuse is adopted to allow simultaneous transmission of two noninterfering traffic streams.

It is believed that interference is one of the most significant factors that limit the system throughput and scalability of wireless mesh networks. We assume a multibeam adaptive array (MBAA) as provided by Lichun and Garcia-Luna-Aceves (2002). Each BS and SS can successfully receive and transmit one or more overlapping packets at the same time by pointing their beams towards individual packet directions, while annulling all unwanted directions. As show in Fig. 5.30, the 360° horizon is divided into 6 sectors; the dotted region in Fig. 5.30a denotes the transmission range. When H transmits to G, it will interfere with G if G is receiving from E. The scheduled policy shall avoid this. Figure 5.30b shows the spatial reuse according to Rule 3, where node G can safely receive from node H, whereas a communication between E and F is in place.

[14] Excerpt from the invited article "Mathematical modelling and performance evaluation for centralized scheduling in WiMAX Mesh Networks," Jianfeng Chen Caixia Chi, Bell Laboratories, Alcatel Lucent Technologies, E-mail: {chenjf,chic}@alcatel-lucent.com

5.9.1.1 Mathematical Model and Scheduling Algorithm

Overview of 802.16 Centralized Scheduling Algorithm In the 802.16 centralized scheduling algorithm control messages and data packets are allocated in different mini-slots in a frame. Data subframe allocation is performed through the control message exchange; hence, there is no contention in the subframe.

The schedule control subframe is used to determine the amount of allocated resources for a link allocated within the data subframe. In the centralized scheme, Mesh SSs send resource requests to the Mesh BS. In response, the BS determines the amount of available resources for each link. Both the request and grant process use the mesh centralized scheduling (*MSH_CSCH*) message format. More specifically, SSs send *MSH_CSCH:Request* messages to the BS and the Mesh BS replies with the *MSH_CSCH:Grant* message.

The scheduler provides each node with bandwidth allocation such that traffic can reach its destination in the schedule validity. Usually, the longer the schedule validity the lower the bandwidth efficiency is. Minimizing the schedule validity becomes an objective of utmost importance in designing scheduling algorithms. In the following, we build an optimization model to study the minimal schedule validity that can be provided by a centralized scheduler with specified topology and traffic distribution. Although IEEE 802.16 standard lacks support for difference priority services in Mesh mode (BE is the only service type defined), it is still possible to classify different service priorities in the priority/class field in the Mesh CID construction. A generalized model in the following is proposed to support multipriority traffic.

Mathematical Model Let $T = (V, E)$ be an access tree where V is the set of APs, and E be the set of bidirectional wireless links between neighboring pairs of APs, $|V| = N$. All nodes in V are labeled with an integer and the root node is labeled with 0. The root node 0 is BS, and the other nodes $i \in V - \{0\}$ are SS. Each node $i \in V$ has a specified capacity P which is the data rate it can support. F_i, $F_i = \{j|(j, i) \in E\}$, is the neighboring parent nodes of node i. With T being an access tree, each node has at most one neighboring parent node, that is, $|F_i| = 1$. $N_i = \{j|(j, i) \in E\}$ be the neighboring children of node i. Each $j \in N_i$ is given a label $l, l = 1, \ldots, |N_i|$ and N_i^l is the lth neighboring child of node i. N_i^l and all its children form the lth branch of node i which is denoted as B_i^l. $C(i)$ represents all the children of node i, and $C(i) = \bigcup_{l=1}^{|Ni|} B_i^l$. $h(i)$ is the number of hops from root to node i. When $L \geq 1$, and $L = 1$ means traffic has no priority difference.

Suppose that traffic of each node has L priorities, and d_i^p is the uplink traffic request of node i with priority level p, $p = 1, \ldots, L$. d_{0i}^p is the downlink traffic from root node 0 to node i with priority level p, $p = 1, \ldots, L$. With all these inputs to the scheduling problem, we need to decide the uplink and downlink traffic with priority p of node i in frame k. Let $x_{i,k}^p$ be the uplink traffic of node i with priority p at frame k, and $y_{i,k}^{l,p}$ the downlink traffic of node i with priority p to branch l at frame k. The explanation of above notions is illustrated on the example access tree in Fig. 5.31. Let K^p represent the number of frame slots to carry the request of priority p to its destination. The relation between different K^p, $p = 1, \ldots, L$ is shown in Fig. 5.32. The scheduling problem is to find a scheme to minimize the total number of frame slots K^p. This problem can be formulated into a linear program problem as follows:

$$\min K^L \tag{5.26}$$

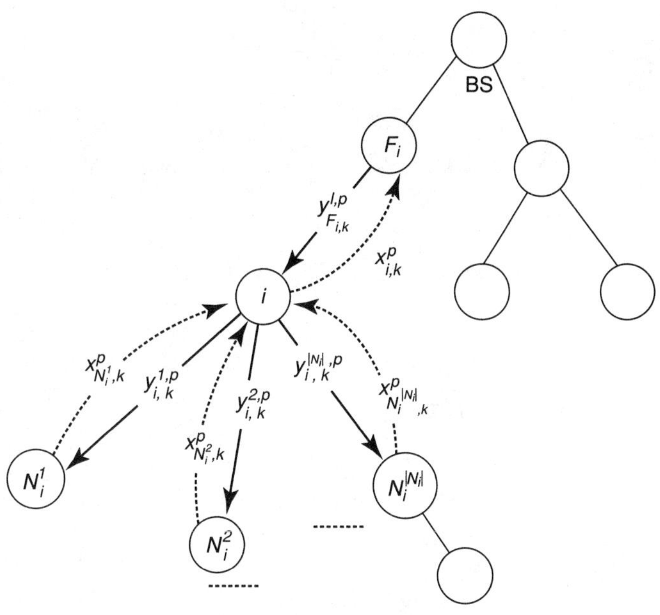

FIGURE 5.31 Access tree and its notions.

s.t.

$$\sum_{k=1}^{K^p+1-h(i)} x^p_{i,k} = d^p_i + \sum_{j\in C(i)} d^p_j, i = 1, \ldots, N, p = 1, \ldots, L. \quad (5.27)$$

$$\sum_{k=1}^{K^p} y^{l,p}_{F_i,k-1} = d^p_{0i} + \sum_{j\in C(i)} d^p_{0j}, i = 1, \ldots, N, i = N^l_{F_i}, l = 1 \cdots |N_{F_i}|, p = 1, \ldots, L. \quad (5.28)$$

$$\sum_{p=1}^{L} \left(x^p_{i,k} + \sum_{l=1}^{|N_i|} y^{l,p}_{i,k} + y^{l,p}_{F_i,k} + \sum_{l=1}^{|N_i|} x^p_{N^l_i,k} \right) \leq P, i = 0, \ldots, N, i = N^l_{F_i},$$

$$l = 1, \ldots, |N_i|, p = 1, \ldots, L. \quad (5.29)$$

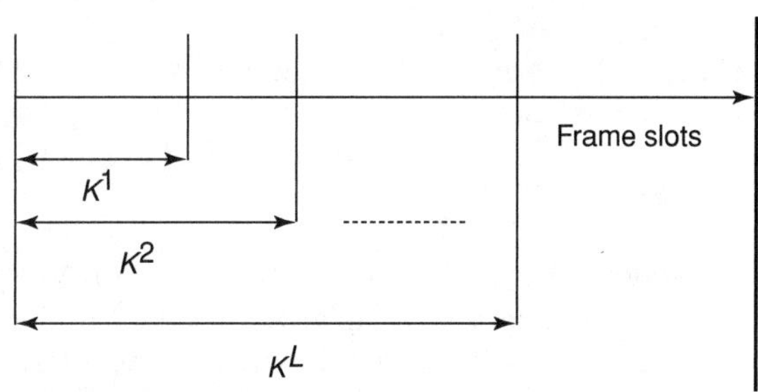

FIGURE 5.32 Relationship between K^p.

$$\sum_{t=1}^{k} x_{i,t}^{p} - \sum_{t=1}^{k-1}\sum_{l=1}^{|N_i|} x_{N_i^l,t}^{p} \leq d_i^p, k = 2, \ldots, K^p, i = 1, \ldots, N, p = 1, \ldots, L. \tag{5.30}$$

$$\sum_{t=1}^{k}\sum_{l=1}^{|N_i|} y_{i,t}^{l,p} \leq \sum_{t=1}^{k-1} y_{F_{i,t}}^{l,p} \quad k = 2, \ldots, K^p, i = 1, \ldots, N, i = N_{F_i}^l, p = 1, \ldots, L. \tag{5.31}$$

$$y_{i,k}^{l,p} = 0, k \leq h(i) \quad i = 1, \ldots, N. \quad l = 1, \ldots, |N_i|, p = 1, \ldots, L. \tag{5.32}$$

$$x_{i,k}^{p} = 0, k \geq K^p + 2 - h(i), i = 1, \ldots, N, \\ p = 1, \ldots, L. \tag{5.33}$$

$$x_{0,k}^{p} = 0, k = 1, \ldots, K^p, p = 1, \ldots, L \tag{5.34}$$

$$K^p \leq K^{p+1}, p = 1, \ldots, L - 1, \tag{5.35}$$

Constraint Eq. (5.27) represents that the uplink traffic of priority p via node i to root node 0 must be transmitted before the $(K^p + 1 - h(i))$th frame in order to reach the root node after $h(i)$ hops.

Equation (5.28) means that the downlink traffic of priority p from root node to node i should reach node i inside the schedule validity K^p.

Constraint Eq. (5.29) requires that the total traffic to and from node i cannot exceed the total capacity P of node i.

Constraint Eq. (5.30) means that the uplink traffic of priority p sent by node i during $k, k = 1, \ldots, K^p$ frame slots should not exceed that received from its neighboring children and its own originated traffic during the k frames. And constraint Eq. (5.31) requires that the downlink traffic of priority p sent by node i during $k, k = 1, \ldots, K^p$ frames should not exceed that received from its neighboring parent node.

For root node 0, it has no uplink traffic of priority p and also has no parent node. So, $x_{0,k}^{p} = 0, k = 1, \ldots, K^p, p = 1, \ldots, L$. For any $i \neq 0$, there is no downlink traffic of priority p unless it has received some traffic from its parent nodes, and all uplink traffic of priority p should be sent prior to the $K^p + 2 - h(i)$ frame to ensure that traffic reaches the root node before frame K.

Constraint Eq. (0.35) determines that traffic with higher priority should be transmitted earlier than that of lower priority.

When $L = 1$, this model is reduced to the case where network traffic has no priority difference (Jianfeng et al., 2005).

Scheduling Algorithm This optimization model with priority support is denoted as M. Solution to M does not specify how to arrange the $x_{i,k}^{p}$ uplink and $\sum_{l=1}^{|Ni|} y_{i,k}^{l,p}$ downlink traffic in the P mini-slots of each frame slot k such that no interference occurs. A scheduling algorithm is proposed in the following to allocate the traffic of each node to its mini-slots. In the scheduling algorithm, $z_{i,k}^{t}$ represents the t-th mini-slot of node i at frame slot

k. $z^t_{i,k} = NULL$ means that node i is free in min-slot t of k frame slot. $z^t_{i,k} = s^p(j)$ represents that node i sends traffic to node j in mini-slot k, and $z^t_{i,k} = r^p(j)$ represents that node i receives traffic from node j in mini-slot k. For $p = 1, \ldots, L$, the smaller number represents higher priority. If $L = 1$, there is no priority difference between the network traffic and the algorithm can be simplified further (Jianfeng et al., 2005).

Algorithm: Time Slot Allocation Algorithm

Input: $G = (V, E), x^p_{i,k}, y_{i,k}, P, i \in V, k = 1, \ldots, K$.

Output: Mini slots assignment $z^t_{i,k}, t = 1, \ldots, P$ for each i, k.
begin

1. For each $k, k = 1, \ldots, K^L$, get $x^p_{i,k}, y^{l,p}_{i,k}, i \in V, l = 1, \ldots, |N_i|. z^t_{i,k} = NULL, \forall t \leq P. T_i = \{t|z^t_{i,k} = NULL, t = 1, \ldots, P\}, p = 1, \ldots, L$.
2. $i := 0, V_0 := V, l = 1, p = 1$/* Begin from BS node.*/
3. Select $t \in T_i$, if $y^{l,p}_{i,k} > 0, z^t_{i,k} = s^p(N^l_i), y^{l,p}_{i,k} = y^{l,p}_{i,k} - 1, T_i := T_i - \{t\}, j = N^l_i$, $z^t_{j,k} = r^p(i), T_j = T_j - \{t\}$. If $y^{l,p}_{i,k} > 0$, go to step 3, otherwise l++, if $l \leq |N_i|$, go to step 3.
4. For $j = N^l_i, l = 1, \ldots, |N_i|, t \in T_i \bigcap T_j$, if $x^p_{j,k} > 0, z^t_{j,k} = s^p(i), z^t_{i,k} = r^p(j)$, $x^p_{j,k} = x^p_{j,k} - 1, T_j = T_j - \{t\}, T_i = T_i - \{t\}$, go to step 4, otherwise $p++$, if $p \leq L$ go to step 3.
5. $V_0 := V - \{i\}$. If $V_0 \neq NULL$, select $i \in V_0$, and $F_i \notin V_0$, set $l = 1$, go to step 3.

end

Let us use Fig. 5.33 as an example to illustrate the scheduling algorithm when $L = 2$. In this example, one unit is the bandwidth assumed for each frame whereas one unit is the uplink and downlink traffic assumed for each network node. Out of one unit of traffic, 0.5 is with high priority and 0.5 is with low priority. Figure 5.34 shows the bandwidth allocation result from M. Figure 5.34 shows that the total number of scheduling frames in this schedule validity is 18, the traffic with higher priority completes its transmission within 9 frame slots whereas the transmission of traffic with lower priority is deferred.

Figure 5.36 shows the bandwidth allocation method, which is calculated from the optimization model when $L = 1$. In this result, BS has the traffic volume in downlink

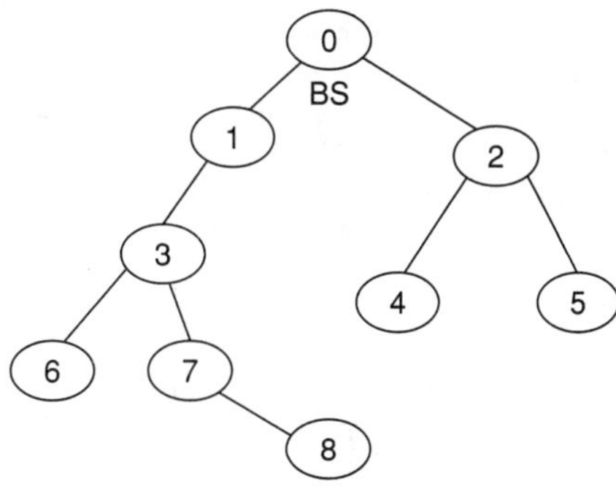

FIGURE 5.33 Simulation topology.

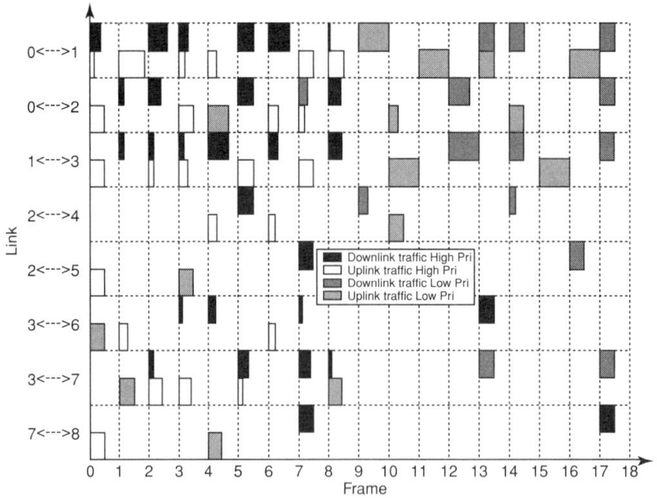

FIGURE 5.34 Bandwidth allocation ($L = 2$).

as in uplink generated by SSs. As shown in Fig. 5.33, the total traffic is 8 units in each direction.

White blocks in Fig. 5.35 represent uplink traffic through the link, and black blocks represent downlink traffic. The total number of scheduling frame slots is 18, which means that all data traffic can be transmitted to its destination inside 18 frame slots. Generally speaking, the delay upper bound in each direction is 18 frame slots. Figure 5.37 shows that many links are active simultaneously in the same frame slot.

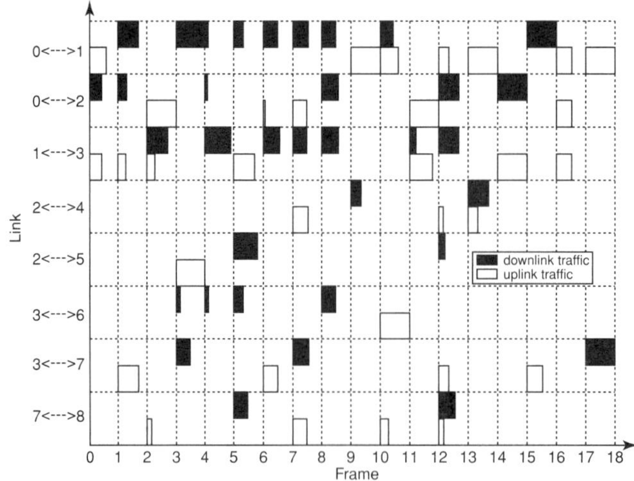

FIGURE 5.35 Bandwidth allocation ($L = 1$).

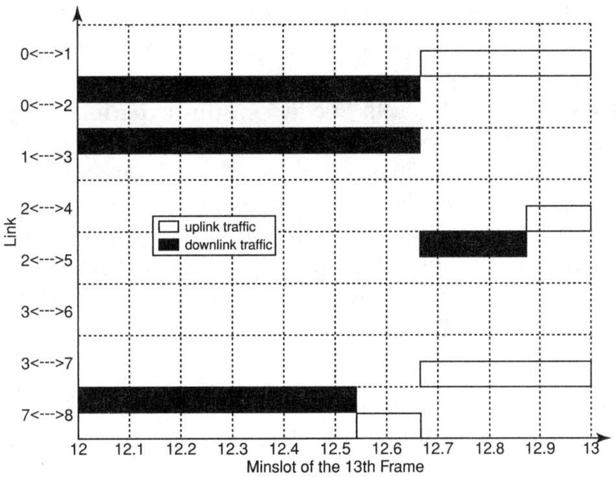

FIGURE 5.36 Mini-slot allocation ($L = 1$).

5.9.1.2 Experiment Result

In this section, experiment results based on the network shown in Fig. 5.33 are reported. The total bandwidth and the duration for each frame are assumed to be 10 Mbps and 10 ms, respectively. For simplicity, the total bandwidth is allocated to the data subframe (divided into 256 minislots). Also, the bandwidth consumed from the control subframe is not considered. The system is designed similar to a gated system and only the packets that arrive prior to the start of schedule validity can be transmitted. All packets arriving during the time of one schedule validity are buffered for next round scheduling. Each node sends *MSH_CSCH* message according to the traffic grooming result, and after receiving the *MSH_CSCH:Grant* message from BS, packets are transmitted following the time slot allocation scheme. Matlab is the software used for simulation modeling purposes. The

Node	Downlink Traffic Average Bandwidth(Kbit)		Uplink Traffic Average Bandwidth(Kbit)	
	High Pri	Low Pri	High Pri	Low Pri
	CID {Peiority =111}	CID {Peiority =000}	CID {Peiority =111}	CID {Peiority =000}
1	60	40	50	50
2	40	60	70	30
3	40	60	40	60
4	60	40	60	40
5	50	50	70	30
6	30	70	50	50
7	70	30	30	70
8	50	50	30	70

FIGURE 5.37 Simulation environment configuration.

simulation environment is the same as that mentioned in the work of Jianfeng et al. (2006).

Experiment Results for Traffic with Priority In this experiment, traffic follows a Poisson distribution with mean as shown in Fig. 5.37. Traffic of two priorities is assumed. The priority/class parameter embedded in the Mesh CID specifies the priority class of the MAC SDU, and represents the two priorities shown in Fig. 5.37.

The average network delay for the traffic based on scheduling algorithm is evaluated via simulations. Based on the scheduling algorithm, each node knows the mini-slot where it is free to send its traffic. Throughout the simulation experiments, the following local policy is applied:

(1) When a node is assigned a mini-slot to transmit its downlink traffic, it sends the traffic with maximum hop first.

(2) For the traffic with the same number of hops, they are served from the left to right.

Figures 5.38 and 5.39 show that the delay of the traffic with higher priority is about half of the traffic with lower priority. For example, the mean value of the average latency for downlink high priority service is 7 frame slots, while the latency for downlink low priority is 15 frame slots. Figure 5.38 also shows that the traffic with higher priority receives access for transmission earlier than the traffic with lower priority. Therefore, model M together with the scheduling algorithm can guarantee that the high priority traffic always takes precedence over low priority traffic during the scheduling process. As indicated in Fig. 5.39, all nodes can complete their transmission within their schedule validity, with the delay upper bound being the schedule validity of different priority.

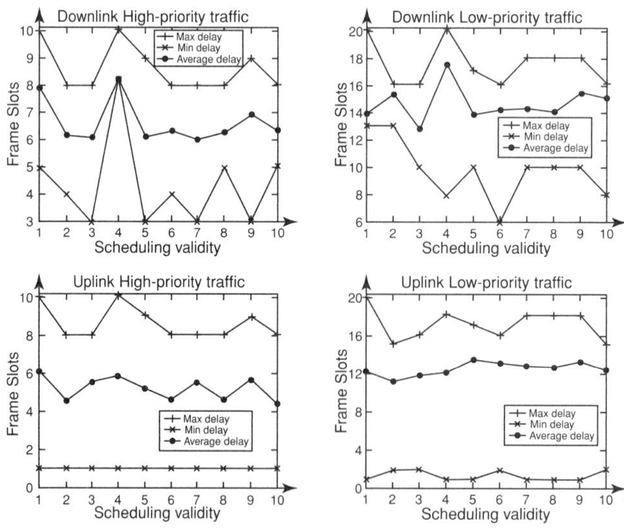

FIGURE 5.38 Delay analysis of algorithm.

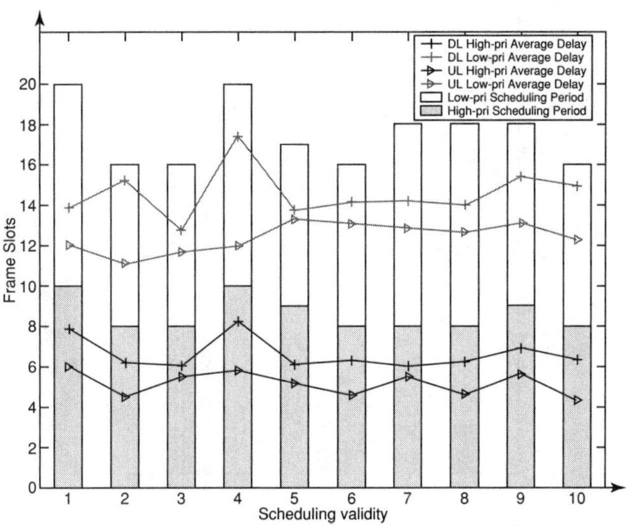

FIGURE 5.39 Calculated period and average delay.

The effect of schedule validity between the proposed scheduling scheme and FIFO queue is compared in the following. Bandwidth for data subframe in each frame slot is taken as 1 unit (summation of all traffic under different priority). FIFO queue serves the request according to the sequence of each node defined in *MSH_CSCH* message. The following optimization is added in FIFO mode: if an active node's Uplink/Downlink traffic is less than 1 unit, the remaining bandwidth can be allocated to its neighboring father node.

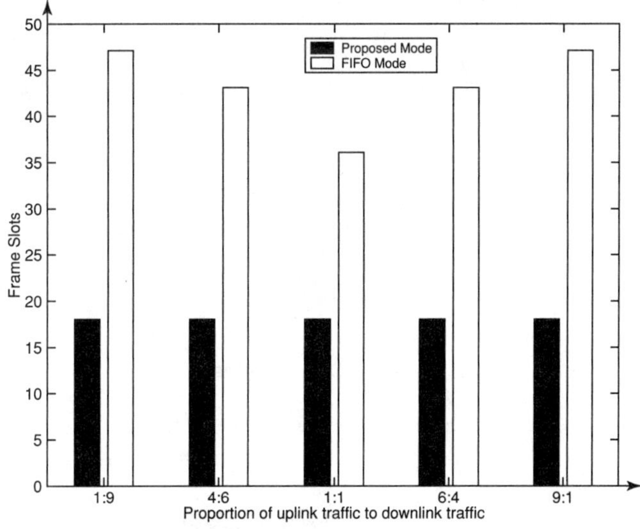

FIGURE 5.40 Frame slots vs. Traffic ratios.

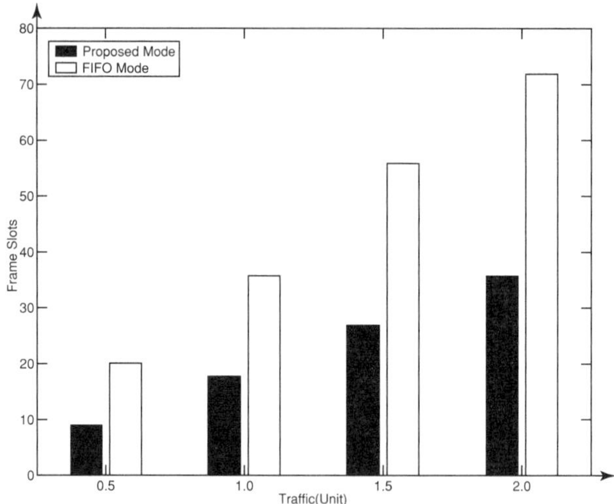

FIGURE 5.41 Frame slots vs. Traffic load.

The schedule validity in various traffic ratio scenarios is also compared and five different traffic ratios of uplink and downlink are considered: 1:1, 4:6, 6:4, 1:9 and 9:1. In each scenario, all nodes have the same amount of traffic. Figure 5.40 shows that the proposed scheduling method needs only half of the centralized scheduling time needed in the FIFO mode whereas the number of scheduling time slots does not change with varying traffic ratio. However, in FIFO mode, when ratio changes from 1:1 to 4:6 or 6:4, 7 more time slots are needed. When the ratio is 1:9 or 9:1, 11 additional time slots are needed. Another comparison is that of scheduling time variety when traffic grows from 0.5 unit per node to 1.5 unit per node. Figure 5.41 shows both modes need more scheduling time when traffic increases.

CHAPTER 6

Capacity Principles in WM²Nets

6.1 Introduction[1]

In a wireless mobile mesh network (WM^{2}Net) context, the issue of capacity is considered in the light of two intrinsically different types of traffic: (1) the extra-mesh traffic and (2) the intra-mesh traffic. When extra-mesh traffic becomes the overwhelming traffic in a WM^{2}Net, the network performance is very akin to that of a traditional cellular network where the available capacity of each cell is fixed and largely defined from the base station. In a single-cell network, for instance, the mean traffic throughput per subscriber is proportional to $k_1 B/N$, where N is the number of subscribers in the cell, B is the capacity of the base station covering that area and k_1 is a function of the system overheads for multiple access, channel coding, etc. Capacity scales up with higher AP populations and/or with cell sectorization techniques, such that the mean capacity per subscriber is of the order of $k_1 \cdot k_2 \cdot M \cdot B/N$, where M is the number of sectors/cell and k_2 is a function of the inter-sector/cell interference. Thus, in terms of scalability, the per-user throughput in the context of a cellular network decreases nominally as $1/N$ until additional base stations are deployed.

When intra-mesh traffic becomes the overwhelming traffic in a WM^{2}Net, it is argued that the underlying network capacity increases as the nodal population increases. This argument is ratified by the fact that intra-mesh traffic is basically the forwarding (relaying) data volume. The network capacity needed to cope with this traffic theoretically increases with the nodal population. This observation is factored, however, by the traffic demands of the newly coming users. This further translates to a net reduction in the available per-node throughput as the node population increases. Indeed, as shown in Section 6.3, there appears to be very little prospect of avoiding the asymptotic reduction in per-user throughput with increasing subscriber base.

Besides, in pure mesh configurations with no central entity being in place to instrument the allocation of resources, a number of unwanted conditions may occur. The so-called "tragedy of the commons," for instance, is a reality when resources are shared among multiple users. Such a tragedy relates to the days when common land was used

[1] The author warmly thanks Dr. Ahmad Atefi from Plextek Ltd (http://www.plextek.co.uk) for his spirited discussions on the topic of wireless mesh networking and especially on WM^{2}Net issues related to capacity ("Study of efficient mobile mesh networks," Ofcom Contract Ref. C31400/001).

for the grazing of livestock with free access for all. The danger is that the free access to a finite resource can result in that resource being fully consumed or compromised further such that it loses its usefulness to all. What then, if users could somehow add grazing capacity as they joined the common?

The situation is even worse in hybrid WM^{2}Nets, where a much higher volume of traffic is aggregated around the APs and the last-hop nodes to APs. In the absence of any sectoring of the AP coverage, this traffic concentration forces all neighbor hops into reuse contention. Hence, the maximum combined throughput of all these hops does not exceed the maximum relaying capacity of each individual MH.

The work by Gupta and Kumar (2000) provides much of the pivotal theoretical analysis of the *average* throughput capacity in multihop wireless networks. Their main result indicates that as the number of nodes per unit area (n) increases, the throughput capacity decreases approximately as $1/\sqrt{n}$. The general conclusions drawn are:

1. For a network of n identical randomly located nodes, each capable of relaying data at W bps over a fixed range, the upper bound for the throughput capacity is of order of $c_1 W\sqrt{n}$ bps.[2] Thus, the network's capacity increases in proportional to the square root of the node population.
2. This capacity is shared among the nodes such that the upper bound for the average throughput $\lambda(n)$ achieved by each node for a randomly chosen destination is of order of $c_2 W/\sqrt{(n \cdot \log n)}$ bps for the Random Network. Thus, the per-user throughput decreases with increasing node population.
3. The parameters c_1 and c_2 are functions of the signal-to-noise ratio (SNR) threshold, β, required of the carrier modulation scheme and the rate of decay of RF signal power (the propagation law), γ, such that for a high SNR threshold the capacity limits are reduced, whilst for a high propagation law the capacity limits are increased.

Even though these results are frequently quoted by researchers, these are, however, idealized theoretical on performance *upper bounds*. Numerous other works have attempted to analyze more practicable scenarios. An example is the theoretical work by Arpacioglu and Zygmunt (2004). This work concludes that the average per-user throughput $\lambda(n)$ has a faster rate of decay approximately proportional to $1/n$, rather than $1/\sqrt{n}$ or $1/\sqrt{(n \cdot \log n)}$ predicted by Gupta and Kumar (2000). In essence, the primary reason for the more pessimistic performance figures is that in the study of Gupta and Kumar (2000), the path loss is modeled as d^{γ}. However, as the density of nodes is increased, d can be reduced to zero; hence, there appears an anomalous decrease in path loss for $r < 1$. Arpacioglu and Zygmunt (2004) eliminate this by setting the path loss as $(1 + d)^{\gamma}$. This leads to the result that the network's capacity does not increase monotonically with n, but there is rather an upper limit of simultaneous transmissions that can be supported in a given area. In fact, *this upper limit is independent of n. Intuitively, this is correct when one sets practical limits on propagation attenuation law, and required signal-to-interference margin.*

[2] In this context, the phrase "of order of" means that to a first-order approximation the value tends towards this value as n approaches infinity.

But, regardless of these disagreements over the order of proportionality with n, this family of models all agree that the average per-user throughput diminishes towards zero as the number of nodes increases. We conclude then that

A mesh network does not scale indefinitely.

With these in mind, it is useful to identify the primary reasons why per-user throughput decreases with increasing population in a mesh that supports *intra-mesh traffic*, and what parameters, if any, might be altered to avoid this demise.

Consider the activity around a single node. The use of an element of time/bandwidth resource to communicate across a hop will impose an interference boundary around the transmitting node within which that resource cannot be reused. Other nodes that intend to communicate within this interference zone must use other elements of time/bandwidth resource. If the transmission rate of nodes is W bps then the maximum total throughput through this interference zone is of order W. Other traffic paths can pass through this zone, but the total throughput is limited to W bps.

If node density is such that there are m other nodes in this zone, then the zone's throughput can be shared—providing a mean of W/m to each. If this zone is of area a, then one can consider having a maximum throughput of order W/a (bps/unit area). Clearly, the area a must be retained as small as possible. This confirms the conclusion that short hop lengths and high propagation attenuation factor are conducive to high throughput network capacity (but subject to other limitations such as routing overheads, route volatility, susceptance to mobility, etc.). Without loss of generality, we can conclude then that throughput is primarily limited by the throughput capacity of the interference zones and not by other factors such as the node relay-throughput as illustrated in http://www.plextek.co.uk.

The fact that node relay-throughput is not a limiting factor is illustrated in a system simulation by Hekmat and Mieghem (2004). This work addresses mesh throughput from the standpoint of signal-to-interference levels within a network and, based on this, determines values for hop-count, capacity, and per-node throughput. *Note*: Its weakness is that it uses a regular lattice of nodes rather than random distribution, but it does attempt to model some practical values for data rate and bandwidth—based on 802.11b (code division multiple access (CDMA)).

Performance plots usually show some key trends. Figure 6.1 illustrates the "saturation point" of a system where the achievable per-node throughput equals the throughput required to support a given service level. This achievable capacity is approximately half of the "stand-alone" throughput capacity of a node, W ($W = 2$ Mbps in Hekmat and Mieghem's (2004) simulation); that is, network saturation is caused by mutual interference and not by node saturation.

Returning to the concept of the interference zone around a node, one can extend this concept to a multihop route passing through the mesh. Figure 6.2 illustrates such a route as comprising a sequence of transmit/receive boundaries. The different colors and letters (A, B, and C) illustrate the usage of different bandwidth/time resources along the route. The smaller circles indicate the omnidirectional boundary of the wanted signal on each link and the larger circles indicate the interference zones corresponding to each of these (for clarity only the interference boundaries for resource A are shown).

In this example, the interference zone around a transmitter is assumed to extend to nominally twice the communication distance. As a consequence, three resource elements

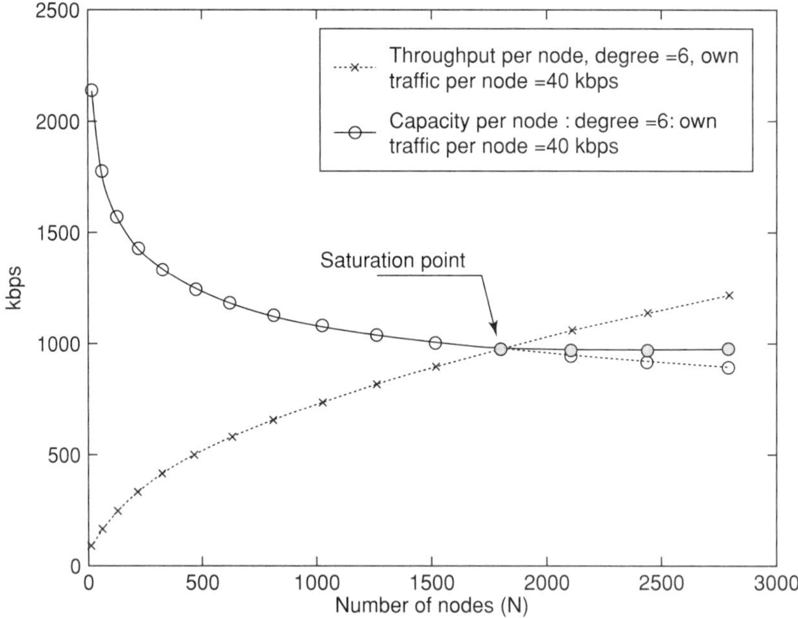

FIGURE 6.1 Comparing available and required relay throughput per node (802.11 system with 22 MHz channel width, before CDMA de-spreading, and 2 Mbps relay rate).

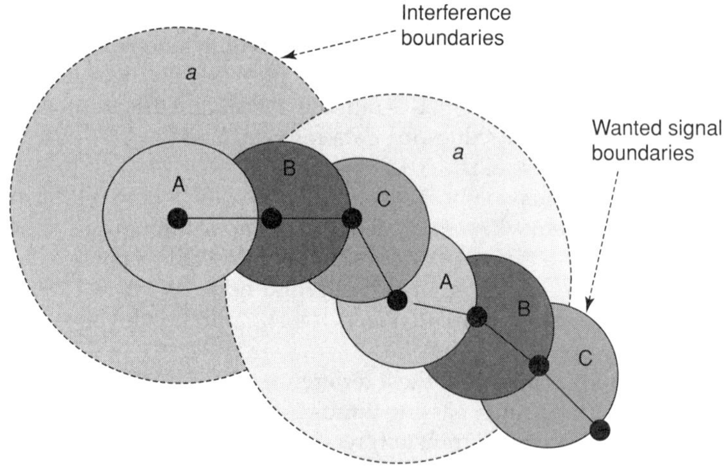

FIGURE 6.2 Spectrum resource reuse along a traffic route and associated interference zones around transmitters.

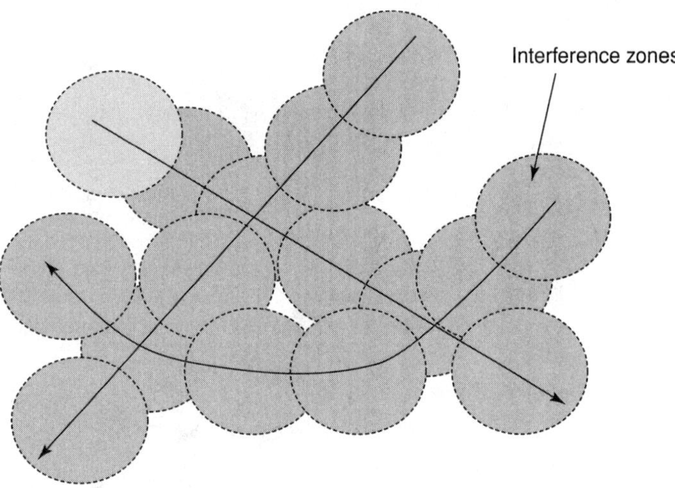

Figure 6.3 Routing three paths to avoid a three-route occupancy of one interference zone.

are needed whereas a minimum of three hop lengths separation is required between coresource users[3]—this represent a near-optimal situation for a "string of pearls" route in which hops are nominally aligned and are of the same length. In practice, the resource utilization is likely to be substantially higher: Li et al. (Grossglauser and Tse, 2001) suggest a theoretical lower limit of four hop lengths. Simulation results using a modified 802.11 showed, however, that a lower limit of seven hops is required.

If the number of different resource elements needed per route is represented as b ($b = 3$–7, as illustrated above), then for a traffic rate of T bps a resource capacity of bT is needed for each route. One can then conclude that the total number of such traffic routes, m, is limited by:

$$\sum_{i=1}^{m} b\,T_i \leq W$$

Let us assume a system that employs terminals with 20 Mbps relay throughput, supports traffic rates up to 1 Mbps, and uses an average $b = 5$ spectrum resource elements per traffic route. This system could support only four such traffic routes passing through each interference zone around active nodes. Furthermore, one can easily observe from Fig. 6.3 that a single traffic route lays down a footprint of adjoining interference zones along its path. This route hence extends this problem throughout its length. The crossover "bottleneck" caused by a specific interference zone could be reduced to some extent by diverse routing around it, using a suitable load-balancing routing protocol. Since, however, the interference zone is relatively large (e.g., circa 2-hop radius in this example),

[3] This factor 3 is derived on the following basis. Let the required SNR be β and the propagation attenuation factor be γ. The theoretical ratio of interference range versus preferred link length, Δ, is then given by $10\gamma\log\Delta = \log\beta$. For example, if $\beta = 13$ dB and $\gamma = 4$ then the range-ratio is approximately 2. This analysis neglects log-normal fading.

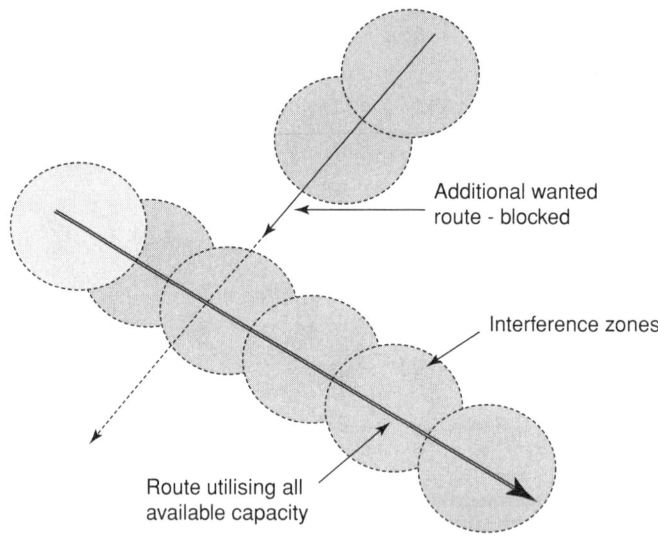

Figure 6.4 Single high throughput route causing partitioning of the network.

routing around it is likely to impose a considerable increase on the route length (via hop-count) and a relative degradation of performance is then expected. This issue is illustrated for just three crossing routes.

From this analysis, one can also remark yet another issue: as the offered traffic rate, T, approaches a high proportion of the node relay rate, W, such that bT approaches W, then all spectrum resource is consumed along this route. Thus, a single traffic route imposes an uncrossable boundary through its length and the mesh becomes disconnected as illustrated in Fig. 6.4.

6.2 Add Fixed Nodes to Enhance Capacity

One of the primary factors that cause the reduction in capacity with increasing number of nodes is the fact that the network performance of a pure WM^{2}Net is a function of the mobility of nodes and their traffic carried. In this context, it seems strictly impossible for an operator to guarantee a certain service level, unless the dependence on users is somehow mitigated. Adding to this, the length of the communication path in a WM^{2}Net can extend from nearest neighbor (one-hop) to the full diameter of the area covered (many hops); hence, as the network size increases geographically and/or in terms of node density the number of hops per path increases too. It becomes clear that if traffic flows were more localized among neighboring nodes, regardless of the network size, the number of hops per path would not increase pro rata with size and hence the network would scale much finer. The last argument is indeed true as the more localized the traffic flows are, the more capacity can be supported in a localized scope and, as a result, the less the available capacity is affected by population growth.

For the case, however, where local scope is not feasible and the end-to-end path length does increase as the network scales up, an alternative technique is to route this longer-range traffic over a hierarchical fixed network. A means to address this conflict is

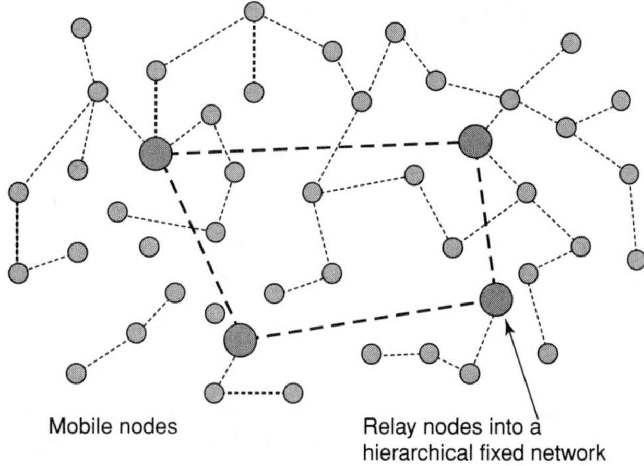

Figure 6.5 Hybrid Network: Intra-mesh traffic with Infra-structure support.

to add a fixed infrastructure in the form of an overlaid fixed mesh network. This results in a hybrid network, as illustrated in Fig. 6.5.

The use of fixed nodes can help the system to:

- enhance connectivity or coverage when user nodes are sparsely distributed. This may be the case during early roll out of the service when there is inadequate customer base to provide sufficient connectivity of the mesh or geographical coverage (in this context, they are often referred to as "seed nodes") (http://www.radiantnetworks.com/meshworks/how.asp).
- ensure a minimum degree of coverage and connectivity, independent of customer density. This may be required, for example, to address the lack of subscriber nodes that arises when users commute in and out of city/residential/recreational areas.
- enhance throughput capability in regions with high customer density.
- enhance coverage by aiding routing around obstacles, such as in the urban environment.

A question that remains to be answered at this stage is, "How many fixed nodes to add?" This is a tricky question and certainly difficult to answer in absolute terms as this is a function of the projected user behavior as well as of the traffic volume. In practical situations, it is very unlikely to achieve accurate prediction of user behavior; hence, a precautionary design margin concept is often used to cope with this situation.

A useful foundation theoretical framework that relates the number of relay nodes and capacity is presented by Liu et al. (2003). The followings are concluded:

1. If the number of relay nodes m increases at a rate less than $\sqrt{n}$ then, although there is a substantial increase in the capacity of the network, the rate of decay of per-user throughput with increasing nodes, n does not improve substantially; that is, capacity is improved but scalability is not. Specifically:

The network capacity is nominally proportional to

$$\sqrt{\frac{n}{\log\left(n/m^2\right)}} \tag{6.1}$$

The capacity gain factor over that in the study of Gupta and Kumar (2000) is of the order

$$\sqrt{\frac{\log n}{\log\left(n/m^2\right)}} \tag{6.2}$$

2. If the population of relay nodes m increases at a rate greater than $\sqrt{n}$ then there is substantial improvement in scalability and capacity. Specifically,
   Network capacity increases polynomially with m (i.e., proportional to $\sqrt[\alpha]{m}$ where α is less than 1)
   The capacity gain factor is of the order

$$\sqrt[m]{\frac{\log n}{n}} \tag{6.3}$$

3. If the population of relay nodes m increases at the same rate as n, the per-user throughput remains constant—that is, the network is fully scalable. This characteristic is self-evident from the fact that in this case each relay node serves a constant number of nodes; therefore, each node retains a constant share of the total transmission bandwidth available. At this point one has the equivalent of a cellular network where the network capacity scales with the number of base stations, where:
   Network capacity is proportional to m (i.e., proportional to n in the limit).

6.3 Using Smart Antenna Technology and Beamforming Techniques to Increase Capacity

WM^{2}Net technologies need be able to cope with the intricacies of the wireless medium; namely hostile propagation, interference conditions, and limited spectrum. In this regard, antenna systems are an effective and promising solution. More specifically, antenna systems can help mitigate the impairments caused by propagation channel, mainly fast fading and cochannel interference inherent in a multiple access environment.

Traditionally, wireless portable terminals are equipped with single omnidirectional antennas (Fig. 6.6). A commonly used type of omnidirectional antenna is the rubber duck. The rubber duck antenna provides 360° coverage and comes with a range of antenna gains varying from 2–3 dBi to 9–10 dBi for special applications.

Using omni-transmissions, only a single interference-free communication link can be established in a given channel contention area, resulting to limited network throughput efficiency. An improvement over omnidirectional antennae is *antenna diversity* (Murch

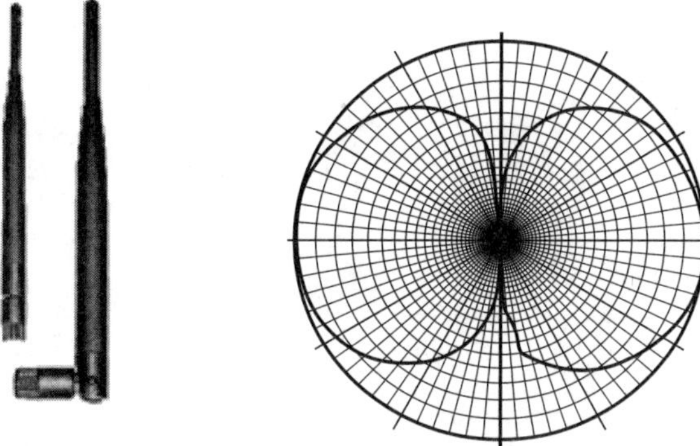

Figure 6.6 Omnidirectional antennas and their radiation pattern.

and Letaief, 2002). In the past years, antenna diversity has been widely used in commercial base station deployments. It is mainly used in the form of receive diversity (i.e., in the uplink direction). Transmit diversity can also be deployed in the downlink direction.

Current technology has made it possible to build cheap and compact smart (or adaptive) antenna technologies (see Fig. 5.4). Smart antennas have the ability to couple energy along a desired direction and at the same time suppress interference induced from unintended users. A smart antenna operates as an adaptive antenna array that modifies its radiation pattern, frequency response, and/or other parameters, aiming to mitigate cochannel interference.

Recently antenna systems equipped with multiple antennas (MAs) in both transmitter and receiver sides are introduced. This type of antenna system is widely known as multiple input multiple output (MIMO) system. In practice, there have been generally two trends to exploit MIMO capability: diversity increase and rate increase. Diversity increase can be realized by systems with MAs in only one side: single input multiple output (SIMO) and multiple input single output (MISO). However, rate increase is usually targeted for MIMO systems with MA at both sides. Obviously, there could be a tradeoff between the two approaches that is well treated by Zheng and Tse (2003). Examples of diversity increasing techniques are space–time trellis codes (STTC) (Tarokh et al., 1998; Jamali and Le-Ngoc, 1994; Naguib, 2000; Hammons and Gamal, 2000; Chen, 2001a) and space–time block codes (STBC) (Alamouti, 1998a; Tarokh and Jafarkhani, 1999). Rate increase approach is usually realized by establishing parallel data streams between transmitter and receiver and is generally known as spatial multiplexing. The V-BLAST algorithm is a prominent example of spatial multiplexing (Hochwald and Marzetta, 2000; Foschini et al., 1999).

6.3.1 Background on Smart Antenna Technologies

Traditionally, wireless communications in a multiuser environment are separated by frequency, as in frequency division multiple access (FDMA); by time, as in time division multiple access (TDMA); or by code, as in CDMA. Smart antennas, on the other hand, add a new way of separating users, namely by space through space division multiple access (SDMA). In SDMA, users of the same cell can use the same physical communication

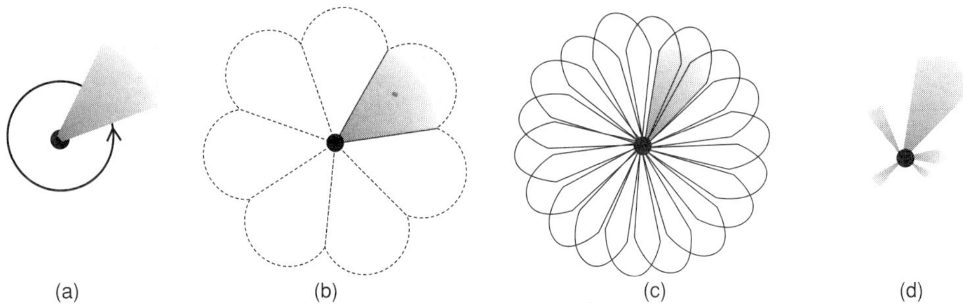

FIGURE 6.7 Example beampatterns for (a) steered, (b) sectorized, and (c) switched fixed-beam directional antennas, and (d) digital beamforming arrays. (The latter two are often referred to as smart antennas.)

channel.[4] This is achieved from using MAs to provide more accurate directional targeting and fine-tune antenna coverage patterns that match the traffic conditions and complex radio environments. This capability of smart antennas to adjust their patterns to the changing network traffic, or RF conditions, provides network operators maximum flexibility in controlling and customizing sector antenna pattern beamwidth and azimuthal orientation over that of standard sector antennas. Thereby, adaptive directional reception and transmission are achieved on the uplink and adaptive directional transmission on the downlink. At the same time, less interference is received from other directions on the uplink, or transmitted towards the other directions on the downlink. Hence, more users can be accommodated in the system; this is a direct result of the spectral efficiency achieved.

Smart antennas can be broadly classified into two groups—*fixed-beam directional antennas* and *digital beamforming arrays* (Fig. 6.7c and 6.7d, respectively).

6.3.1.1 Fixed Beam Directional Antennas

This group consists of passive antennas with a fixed beampattern designed to concentrate the energy in a particular direction. Azimuthal 360° coverage is obtained by either steering a single directional antenna, or using multiple sectorized directional antennas, or a switched-beam array (Fig. 6.7). Fixed-beam directional antennas work well when there is a single dominant line-of-sight (LoS) path between the transmitter and the receiver, for example, in flat and rural terrain or terrestrial to satellite communication. Examples of directional antennas include patch (or panel), yagi, and parabolic antennas. Patch antennas (Fig. 6.8) have high gain and are suitable for long halls, walkways, and corridors. Their radiation pattern is semi-directional. Yagi and parabolic antennas have highly directional propagation pattern with high gain. The parabolic shape antennas (Fig. 6.9) have a very narrow radiation pattern and high gain. For these reasons, they are mostly used for point-to-point communication over long distances.

While the main advantages of a fixed-beam architecture are its simplicity and low cost, its spatial flexibility is restricted due to hardware mechanics. This limits their use in

[4] The physical communication channel here is defined as a combination of carrier frequency, time slot, and spreading code.

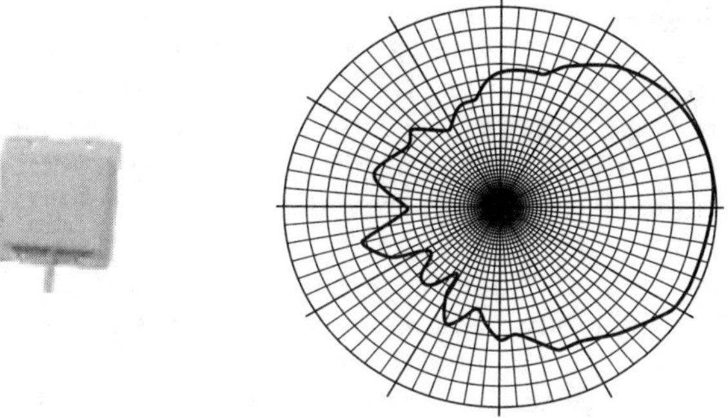

Figure 6.8 Patch antenna and its radiation pattern.

rich scattering environments where multiple paths between the transmitter and receiver lead to a large angular spread of the signal.

6.3.1.2 Digital Beamforming Arrays

This group consists of omnidirectional antenna elements that are spaced a fraction of a wavelength apart. This spatial sampling, along with the phasing of the current on each element (typically adjusted by multiplying the signal with an antenna weight), causes constructive and destructive interference, generating the desired radiation pattern. More specifically, digital arrays can adapt their beam pattern by forming a weighted combination of signals from the antenna elements and place precise nulls towards interferers. Very precise and adaptive control of the beam and the nulls can be obtained by adjusting the antenna weights using signal-processing techniques (Monzingo and Miller, 1980). Adaptive arrays are capable of providing *array gain*, *adaptive interference cancellation*, and also *diversity* or *spatial multiplexing* gain in multipath environments, as discussed next.

Array Gain: If the same signal is sent through each antenna, the independent copies can be coherently combined given that the channel is known at the receiver. This leads to an increase in mean SNR, referred to as *array gain*. If the channel is also known at the transmitter (due to symmetry or using feedback mechanism), maximum array gain can be extracted by setting transmit and receive weights as

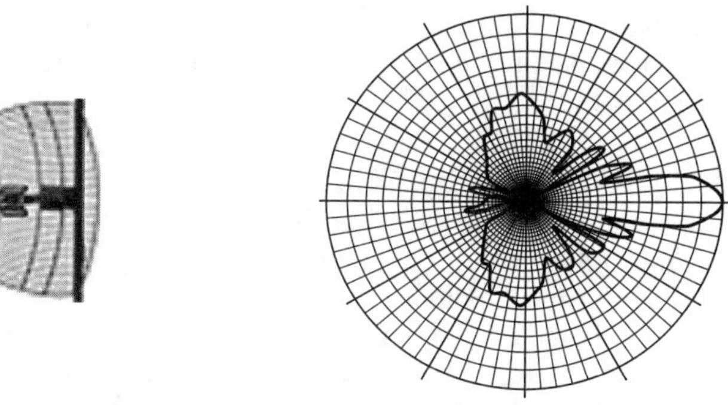

Figure 6.9 Parabolic antenna and its radiation pattern.

the singular vectors corresponding to the dominant singular value of the channel matrix, thus coupling energy into the best spatial subchannel.

Interference Cancellation: The receiver can also use its N degrees of freedom (N receive weight parameters corresponding to N antennas) to provide a gain of 1 to the desired signal and a gain of zero up to $N-1$ interferers. This is known as zero forcing (ZF) (Biglieri, 2007). More generally, the receive weights can be adaptively tuned in real-time using a signal processing algorithm to maximize the receive signal-to-interference plus noise ratio (SINR) (Monzingo and Miller, 1980), and possibly suppress more interferers depending on their strength and correlation of channels to the receiver.

Diversity Gain: Adaptive arrays exploit multiple, independently faded copies of the signal and are thus capable of providing spatial diversity in a multipath environment. Diversity gain can be extracted by efficient signal processing techniques like selection diversity combining (selecting the highest SNR component) or maximal ratio combining (forming a weighted combination of the received signals in proportion to their SNR) at the receiver, and space–time coding (STC) at the transmitter (Biglieri et al., 2007). This reduces the variance of the SNR (*diversity gain*), as the probability that the signal is faded at all antenna elements is much lower than in a single antenna system. Diversity gain is additional.

Spatial Multiplexing: By selecting the N transmit and N receive antenna weights based on the channel information, an adaptive array can create N noninterfering spatial subchannels that can be used to send multiple signals between the same transmit–receiver pair concurrently. This provides an N-fold capacity increase known as *spatial multiplexing gain*. As each subchannel has a different SNR, a power allocation scheme such as water filling (scaling power by inverse singular value of the channel) must be used (Biglieri et al., 2007).

Some of the benefits of adaptive arrays are mutually exclusive. This is so, as the degrees of freedom (antenna weights) at the transmitter and receiver that are used to provide one form of gain cannot simultaneously be used to provide another form of gain. For example, antenna weights designed for spatial multiplexing cannot provide diversity gain or suppression of interference from other users, and vice versa.

Comparing digital beamforming arrays with fixed beam arrays, the latter require a LoS path between the transmitter and receiver whereas digital arrays work well under rich scattering conditions like dense urban or indoor environments where the signal arrives at the receiver from multiple paths. Further, in a multipath environment digital arrays provide spatial diversity from exploiting the linearly independent channels created by multipath.

6.3.2 Background on Multiple-Antenna Systems

Let us consider the communications setting in Fig. 6.10. Node A is assumed to have M antennas for transmission and N antennas for reception, while node B deploys K antennas for transmission and L antennas for reception. If MAs are deployed in the receiver but single channel in the transmitter (i.e., $K=1$, $M=1$ and either $L>1$ or $N>1$), techniques such as antenna diversity and adaptive/smart antennas are commonly used for a multiantenna system. *Antenna diversity* is based on the observation that signals received from uncorrelated antennas have independent fading. Thus, at least one good signal can

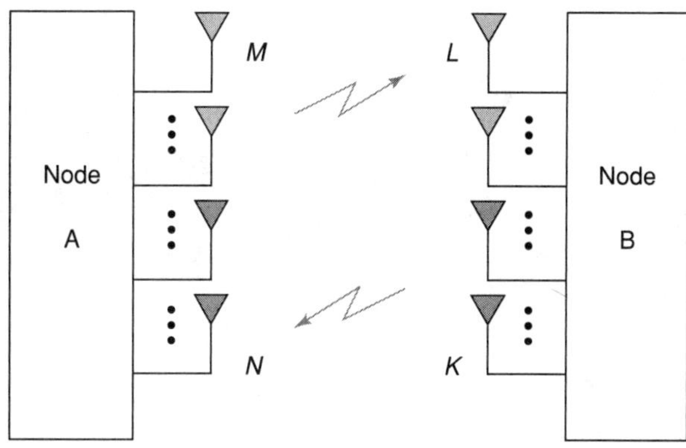

Figure 6.10 Multiple-antenna systems (Murch and Letaief, 2002; Blostein and Leib, 2003).

be received from the receiver with high probability. Antenna uncorrelation is usually achieved through space, polarization, or pattern diversity, and the processing technologies for diversity include switch diversity, equal gain, and maximum ratio combining (Murch and Letaief, 2002). When strong interference is present, diversity processing alone is not sufficient to receive high quality signals.

To resolve this issue, *adaptive antenna array processing* is used to shape the antenna beamform so as to enhance the desired signals while nullifying the interfering signals. The technique for adaptive antenna processing is called *optimum combining*. It assumes that some part of the desired signal can be acquired through a training sequence. Antenna diversity and smart antenna techniques are also applicable to WM²Nets. Due to complexity and cost, a fully adaptive smart antenna system is used only in base stations (access points (APs)). To implement fully adaptive smart antenna systems on a mobile terminal, research and development efforts are still on process. See Section 5.2.4 for further discussions on this issue.

As a consequence, with low cost being a challenging issue in WM²Nets, directional antennas have been actively researched in the area of ad hoc networks. A mechanically or electronically steerable or switched *directional antenna system* can be tuned to a certain direction. By using directional transmission, interference between network nodes can be mitigated and thus, network capacity can be improved (Spyropoulos and Raghavendra, 2003; Ramanathan et al., 2004). However, directional antennas challenge medium access control (MAC) protocol design (Yum and Hung, 1992; Nasipuri et al., 2000; Ko et al., 2000; Choudhury et al., 2002).

If MAs are deployed in the transmitter whereas single antenna in the receiver, that is, $N = 1$, $L = 1$, and either $K > 1$ or $M > 1$, antenna diversity or smart antenna cannot be applied unless the channel state information (CSI) is available. However, usually partial information of channel state is available at the transmitter. To achieve diversity under this situation, a commonly used technique is STC (Alamouti, 1998), where signals are transmitted at different antennas in different symbol periods and processed with a certain coding technique. The received signals are then combined at the receiver through an appropriate algorithm such as the maximum likelihood detection (MLD). STC is a promising technique that achieves second order diversity with no bandwidth expansion (Murch and Ben Letaief, 2002).

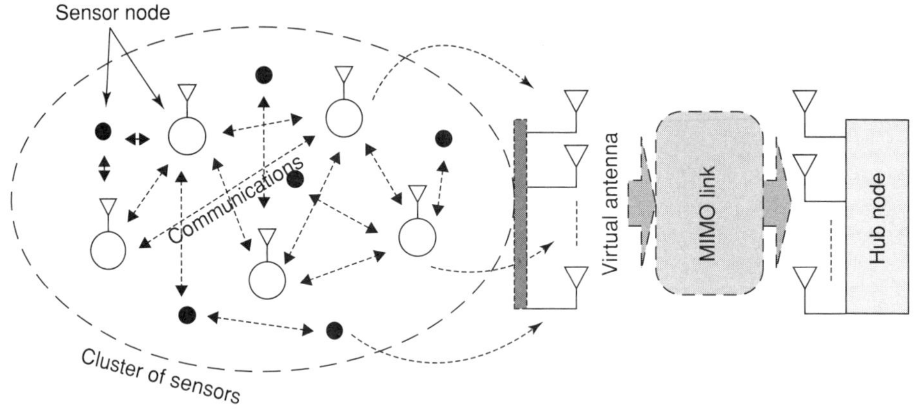

FIGURE 6.11 Cooperative MIMO transmission in a multihop WM²SNet.

If MAs are deployed in both the transmitter and the receiver, that is, $M > 1, L > 1$ or $K > 1, N > 1$, the multiple-antenna system is known as a MIMO system, where both diversity and simultaneous transmissions are in place. These systems promise linear increase in information theoretic capacity in terms of the number of antenna elements (Telatar, 1995; Foschini, 1996). MA facility promises high capacity and spectrum efficiency and provide extra degree of freedom to combat and control the hostile conditions of the radio channel, and also to control and manage the self-interference caused by wireless system operation. This extra degree of freedom is known as space domain and establishes the third dimension along the time and frequency for system radio resource.

The MIMO concept can be extended to cover cooperative communication scenarios with a number of transmitting and receiving nodes involved. These nodes may belong to more than one user or device, and create a cooperative environment to establish all the required data transfer among related users or devices.

Notwithstanding the above concerns, owing to the limited physical size of a mesh node, it is a very difficult task to apply multiantenna techniques in WM²Nets. If individual single-antenna mesh nodes are arranged to cooperate on signal transmission and reception, a virtually cooperative MIMO system as illustrated in Fig. 6.11 can be constructed such that the energy-efficient objective of MIMO communications can be realized in a multihop wireless mesh network.

The traditional expectation of MIMO systems for higher energy efficiency than single input single output (SISO) systems in a Rayleigh fading environment may be overthrown when both the transmission energy and the circuit energy consumption are considered. This is because the circuit complexity of a MIMO transceiver is much higher than that of a SISO transceiver and sometimes it is not clear whether or not MIMO systems are more energy-efficient than SISO systems due to the high circuit complexity associated with the MIMO structure. In short-range transmission, especially when the data rate and the modulation scheme are fixed, SISO systems may outperform MIMO systems as far as the energy efficiency is concerned. However, by optimizing the constellation size, the superiority of MIMO systems in energy efficiency can be extended to the scenario of short-range transmission (Cui et al., 2004).

Jayaweera (2006) investigated a variation of virtually cooperative MIMO transmission for wireless sensor networks proposed by Cui et al. (2004) that is suitable for a commonly encountered sensor network model. This work first investigates the dependence of energy efficiency of a virtual MIMO scheme on system and propagation parameters such as transmission distance, constellation size (transmission rate), and link path loss exponent. Then, accounting for the training overhead required in a MIMO-based system the results are finely tuned (Cui et al., 2004). It is also illustrated that a rigorous energy optimization model needs to account for the energy consumption during the training period since the knowledge of CSI is crucial for the proper operation of MIMO-based techniques. Its analysis and numerical results reveal that the proposed cooperative MIMO communication in wireless sensor networks can provide a significant improvement in energy efficiency with judicious choice of the system-level design parameters.

6.3.3 Spatial Diversity Coding

For those communication systems whose user terminals have physical limitation in having MA elements, reception diversity (i.e., the signal transmitted by a single transmitting element is captured by a number of receiving elements) can be used only in uplink the direction. In these cases, transmission diversity will be an option to improve downlink performance. In this scheme, the signals of the transmitting elements will interfere at the receiver and an appropriate kind of space–time processing should be used at the transmitter in order to simplify receiver detection. This processing will be in form of introducing appropriate correlation between data streams of the transmit elements through a space–time encoder. STC can be viewed as a two dimensional code where its coded symbols are transmitted in both space and time domains. The key development of this concept was originally revealed by Tarokh et al. (1998). They adopted multiple trellis-coded modulation (MTCM), where each state transition of the trellis code generates a multiple of coded symbols to be transmitted by transmission antennas. The decoding of the STTC is accomplished by the same kind of Viterbi algorithm used for MTCM decoding (Tarokh et al., 1998). These codes were shown to provide a diversity gain, which is linearly related to the number of transmission antennas. Since the original STTC were introduced by Tarokh et al. (1998), there has been extensive research to improve the performance of the original STTC designs. These original STTC designs were hand crafted (according to the proposed design criteria) and, therefore, are not optimum designs. In recent years, a large number of research findings have been published that propose new code constructions or perform systematic searches for different convolutional STTC or some variant of the original design criteria proposed by Tarokh et al. (1998). Examples of such work can be found in the work by Naguib et al. (2000), Hammons and Gamal (2000), and Chen et al. (2001a).

The popularity of STTC significantly diminished with the discovery of the so-called space–time block codes (STBCs) (Alamouti, 1998a). This is because in contrast to the vector Viterbi required for STTC, STBC can be decoded using simple linear processing at the receiver. Although STBC codes produce the same diversity gain as the STTC, their coding gain is less than STTC's.

Performance of all the spatial diversity coding techniques relies upon the propagation channel; their promised diversity is achievable only if the fading channels assigned to different transmission (Tx)–receiving (Rx) antenna pairs are largely uncorrelated. While most of the spatial diversity coding techniques require channel knowledge at the receiver, some variants of these codes do not. Examples include the differential techniques proposed by Tarokh and Jafarkhani (2000), Hughes (2000), Yuan and Shao, (2003), Chiavaccini and Vitetta (2003), and Schlegel and Grant (2003) or the nondifferential approaches introduced by Hochwald and Marzetta (2000) and Hochwald et al. (2000a).

Similar to ordinary SISO, transmission of space–time coded signals over frequency selective channels causes intersymbol interference (ISI) and requires an appropriate kind of equalization. Due to the special structure of some space–time codes, the use of classical equalization methods is not a straightforward practice. Initial attempts to address the problem for STTC were made by Fragouli (2002), Naguib (2000a), and Bauch and Naguib (1999), in which the structure of the code was used to convert the problem into one that can be solved using known equalization schemes. For orthogonal STBCs, the ISI channel will destroy the orthogonal property of codes, which is their most attractive property in their linear detection process. Notably, for multicarrier transmission, each subchannel is frequency-flat and the Alamouti scheme can be applied over two consecutive subchannels or two consecutive orthogonal frequency division multiplexing (OFDM) blocks (Liu et al., 1999). However, fast channel variation along time or frequency domains will again degrade the orthogonality of the Alamouti scheme.

6.3.4 Spatial Multiplexing

The MA elements can help to establish parallel data streams between transmitter and receiver, and as a result to increase data rate; this kind of technique is known as spatial multiplexing (Golden et al., 1999; Foschini et al., 1999; Foschini, 1996). For a MIMO system, each transmission antenna sends an independent data stream. The vector signal captured by reception antennas will be the mixture of all transmitted data streams. Each transmitted data stream will be received via a spatial signature thanks to MA reception. Therefore, separation of these data streams at receiver can be recast as a multisource detection (similar to multiuser detection in a CDMA system) problem. In a richly scattered propagation environment with appropriate antenna spacing at both transmission and reception sides, and with the number of receive antennas at least equal to the number of data streams, the system will be full rank, and all data streams can be separated. In this regard, all the available multiuser detection techniques such as decorrelator, MMSE, parallel interference canceller (PIC), serial interference canceller (SIC), sphere decoder (SD), and advanced turbo multiuser detection techniques can be used for spatial multiplexing. Actually, the V-BLAST algorithm initially used for spatial multiplexing was using a kind of serial interference cancellation. It is noteworthy that a turbo detection approach can even work in a low rank condition as it is able to benefit from the redundancy of the error correction coding used prior to spatial multiplexing.

In the absence of channel knowledge at receiver (as well as transmitter) still spatial multiplexing is applicable. The blind source separation concept (Comon and Chevalier, 2000) can be applied in this case. In *blind array processing* techniques, the input sources are mixed linearly by a mixing matrix (here corresponding to the MIMO channel) and separated by exploiting higher order statistics of the receive array signals (Cardoso and Souloumiac, 1993; Papadias, 2000) or covariance subspace estimation (Loubaton et al., 2001) and/or some alphabet (modulation format related) information (van der Veen and Paulraj, 1996) to cite just a few of the many contributions there. The price paid for avoiding channel training in blind approaches is a slight degradation of BER performance and more often in the increased computational complexity.

6.3.5 Beamforming

The simplest example of beamforming is antenna sectorization, which increases antenna gain over the area (sector) that the desired user is present, and nulls/reduces antenna gain over other sectors. A smart antenna is more sophisticated than sectorization; it adaptively switches/steers a narrow antenna beam over the intended user or places some nulls over some strongest interferers. Beamforming requires some channel knowledge about the intended user and also in some cases about interferers (when interference nulling is used). Beamforming is usually implemented using an antenna array at the base station that transmits/receives in downlink/uplink, respectively. This technique is mostly efficient in LoS condition with enough angular separation (larger than beamwidth) between the intended user and interferers. More intelligence is required in multipath propagation conditions to receive desired user's signals from different directions.

Transmission and reception beamforming methods aim to suppress multiple user interference (MUI) by using proper linear processing techniques aiming to maximize the SINR of all the involved users. These approaches set space domain as a new dimension for system multiple access, which is the SDMA, as they enable multiuser transmission/reception at the same time, frequency, and with same spreading code (in CDMA systems). Significant performance improvement could be achieved if all signals are jointly processed with appropriate power allocation/control and linear spatial signal processing (Chang et al., 2002). Due to the extended degrees of freedom, that is, multitude of users with different channel conditions, besides power control and spatial processing of signals, adjusting data rates will allow the system to drastically improve the aggregate data rate of all users. This approach is known as sum-capacity maximization. Appropriate spatial precoding approaches like ZF, block digitalization (when receiving users have antenna array), and dirty paper coding are necessary to maximize the sum-capacity of the multiuser system (Caire and Shamai, 2003). Multiuser diversity can be considered as a special case of multiuser rate adjustment. In this approach, a radio resource is granted only to a user with the best channel condition. This approach does not require an antenna array, however, its generalized form for MA transmission can be realized through sum-capacity maximization approach.

6.3.5.1 Eigenbeamforming

In the case of the availability of instantaneous channel knowledge at transmission side (ICIT), appropriate power and phase adjustment on transmission antennas will increase the information theoretic capacity of the MIMO link; this kind of technique is known as water filling and eigenbeamforming (Paulraj et al., 2003). Rather than using equal power transmission over the transmission antennas, the corresponding eigenmodes of the MIMO channel can be excited with appropriate amount of power that is determined by water-filling theorem (Telatar, 1995) using appropriate linear processing at transmission. In practice, for a specific coding and modulation scheme other approaches rather than water filling may be assumed. In contrast to beam steering and interference nulling, eigenbeamforming is more effective in richly scattered environment forming a full/high rank MIMO channel with strong eigenmodes.

Channel knowledge is very crucial for all beamforming techniques. Under erroneous channel knowledge conditions, the system has to resort to other kinds of processing techniques, such as equal power transmission rather than single-user eigenbeamforming and blind source separation in multiuser Reception beamforming. Hybrid approaches are also possible when only partial transmission channel knowledge is available, for example, combining space–time diversity coding with single-user beamforming (Jongren et al., 2002). In multipath propagation conditions, more sophisticated processing is necessary to control and equalize channels in domains of both space and time or space and frequency. For multicarrier transmission, every subchannel is frequency flat and original techniques used for flat channels can be used.

6.3.6 Capacity Enhancements Using Directional Antenna Techniques

Two potential incentives for the use of directional antennas in relevance to *fixed wireless mesh networks* are the reduced mutual interference, which is the primary mechanism to increase their spectral efficiency (Radiocommunications Agency, 2000). Antenna gain is a significant issue in WM^{2}Nets for reducing transmitted power consumption, and far more because the gain benefit is realized at both ends of the wireless link. Hence, the net effect on link budget could be equivalent to twice the antenna gain.

When the use of directional antennas is considered in a mobile wireless mesh network context, these potential benefits are diluted by the following limitations:

(1) Achievable gain and directionality of practical antennas for hand-held products—particularly below frequencies in the region of 5 GHz.

(2) Selective "deafness" over some radio paths, resulting in confusion and ambiguities in the MAC layer for route exploration and carrier sense multiple access with collision avoidance (CSMA/CA) schemes.

(3) Overheads for antenna control.

As a consequence, whilst antenna directionality can increase network spectral efficiency in fixed wireless mesh configurations, the benefits may not be substantial in a practical deployment of mobile mesh devices at the carrier frequencies of interest. Nevertheless, the following paragraphs present an analysis that considers directional antennas as a means for reducing mutual interference. It is assumed that the effective radiated power (ERP) in the main beam and the receiver sensitivity referred to the antenna ("radiated sensitivity") are the same as the omnidirectional case.

A starting point in the analysis is to consider an idealized antenna having negligible side-lobe responses. This can be represented using a "flat top" model—where the antenna beam in the azimuth (horizontal) plane is represented as an arc of a circle subtending an angle equal to the 3 dB beamwidth of a polar response. The leads to a simplistic interfering/noninterfering alignment of beams as illustrated in Fig. 6.12.

Figure 6.12 Interference model for directional antennas.

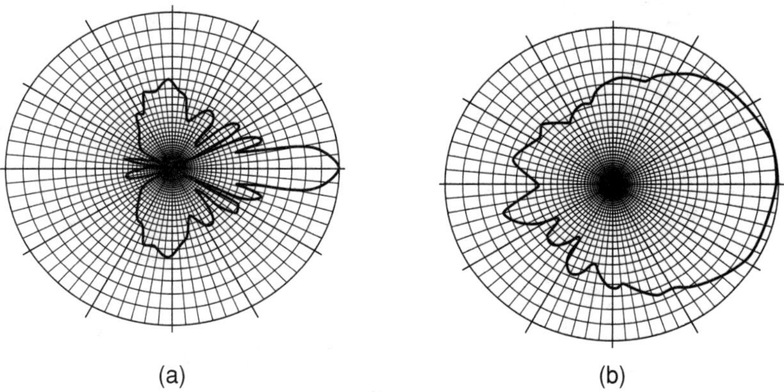

FIGURE 6.13 Radiation pattern for composite antenna model.

For a network of randomly deployed nodes equipped with such antennas, the theoretical upper limit on the improvement of throughput capacity is as large as $4\pi^2/\alpha\beta$ (where α and β are the beamwidths of the transmit and receive antennas, respectively). This is derived by Yi et al. (2003), and can be deduced directly by noting that the probability of nodes falling within beams is reduced by a factor $\alpha/2\pi$ and $\beta/2\pi$ compared to the omnidirectional case, and thus the reduction in mutual interference is proportional to the product of these two.

To illustrate the magnitude of improvement, consider that for mobile mesh products in the bands of interest the minimum achievable antenna beamwidth is likely to be in the region of 90° ($\pi/2$). The above upper bound on capacity improvement from the idealized antenna is then ×16. However, for any practical antenna, and more so for mobile mesh products in the bands of interest there will be finite side-lobe responses, which will seriously erode the above potential gains. As a first step towards analyzing side-lobe effects one can model the antenna beam as one having a uniform side-lobe response outside the main beam, as illustrated in Fig. 6.13.

The key manifestation of this finite side-lobe response in the network is to extend the interference boundary around nodes (Yi et al., 2003). If an antenna has a mean side-lobe level which is κ dB below the main beam then given a propagation environment with attenuation rate γ (i.e., path loss proportional to (range)$^\gamma$) the differential coverage range, Δ_R, between main beam and side lobe is given by:

$$\kappa = 10 \cdot \gamma \cdot \log(1/\Delta_g), \text{ that is, } 10 \cdot \log(\Delta_g) = -\kappa/\gamma \qquad (6.4)$$

A practical work at http://www.plextek.comand data from the antenna-supply industry postulates that for mobile/hand-held products operating below approximately 6 GHz the side-lobe response is unlikely to be more than about 10–15 dB below the main beam. Hence, assuming a likely figure for side-lobe level of $\kappa = 13$ dB, in a fourth-law propagation environment R is only 0.5. Thus, the interference boundary for the side lobes is only, nominally, half that of the main beam.

Some pivotal theoretical analysis of a mesh network with directional antennas has been carried out by Yi et al. (2003). As shown the theoretical maximum capacity gain

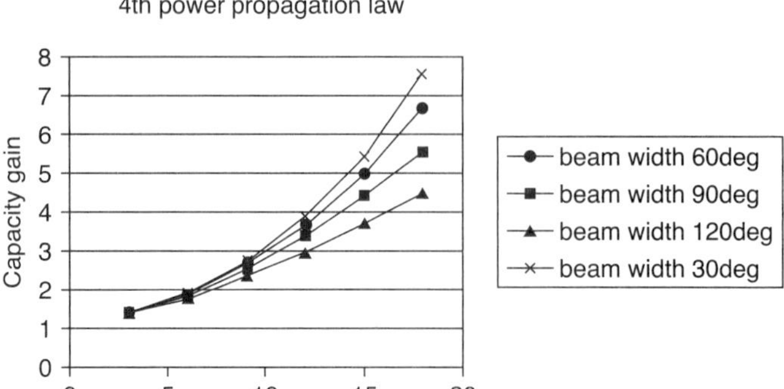

FIGURE 6.14 Theoretical capacity gain vs. Antenna performance.

factor for an idealized random network is of the order of:

$$1/[(\Delta_g)^2 + \{(1 - \Delta_g)^2\}\, \alpha\beta/4\pi^2] \tag{6.5}$$

Again, considering the case of 90° beamwidths with −13 dB side lobes this implies a capacity gain in the region of ×3.3 (compared to a theoretical gain of ×16 for the zero side-lobes case). This illustrates the detrimental effect of finite side-lobe levels. Theoretical capacity-gain vs. antenna beamwidth and side-lobe level is illustrated in Fig. 6.14.

As shown, capacity gain is more sensitive to the side-lobe level than it is to the beamwidth. Furthermore, as beamwidth is reduced, the side-lobe level dominates performance. Without loss of generality, this indicates that there is little benefit in decreasing beamwidth without equal attention to reducing side-lobe levels.

However, this capacity-gain performance is also a function of the propagation environment. The range difference, R, between main beam and side lobes decreases with increasing propagation attenuation law and so the benefit of side-lobe attenuation diminishes. This follows from the premise that for a given density of nodes the ratio of the number of nodes residing inside the main beam coverage area to the number residing in the side-lobe coverage area diminishes with higher propagation law. This is illustrated in Fig. 6.15, for a beamwidth of 90° when Eq. (6.5) is applied. From this, one can see that the benefit from antenna directionality decreases with increasing propagation attenuation factor. (*Note*: all of the curves are asymptotic to the theoretical maximum gain of ×16 noted above.)

It must be noted that for each propagation law the capacity-gain curves in Fig. 6.15 are normalized to the omnidirectional antenna case. The curves thus imply that propagation law impacts the network capacity for omnidirectional antennas, which then implies that there are scale factors to be applied to the vertical axis for each curve. In fact, the corollary to this is that the high attenuation environments enable greater spatial reuse and hence higher spectral efficiency than a low attenuation environment (Gupta and Kumar, 2000; Arpacioglu and Zygmunt, 2004); however, the low attenuation environment will reap more benefit from the use of directional antennas, because of reduced interference induced over the longer propagation ranges.

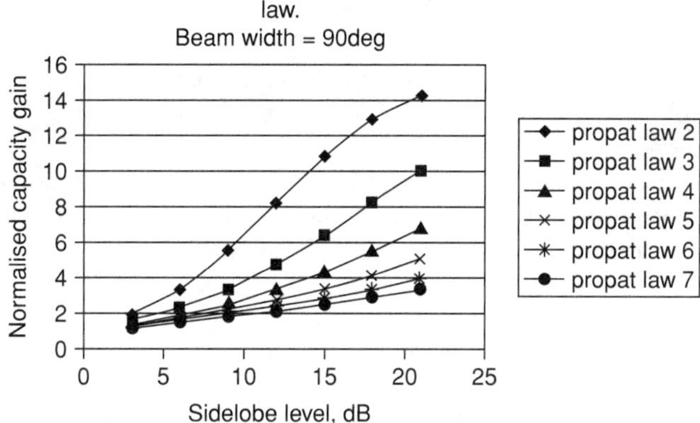

Figure 6.15 Capacity gain vs. propagation environment.

6.4 Spatiotemporal Correlation Properties and Data Fusion

In WM^{2}Net, a number of different devices (e.g., micro/nano sensors) distributed over an area may be involved in a multiparty communication (e.g., for the observation and distribution of a single phenomenon). By means of *fusion*, different sources of information or data are combined to improve the performance of a system. Multimesh data fusion is about combining and relating data from several meshes to achieve more accurate inferences than could be achieved from a single mesh. The different inputs may originate from a single mesh at different moments (fusion in time) or from several meshes at the same time (fusion in space). In addition, dense deployment of mesh nodes makes the mesh observations highly correlated in time and space.

For example, a number of meshes can only detect the proximity of an observed object. Higher-level information, such as speed, size, or shape of an object can then be obtained only by correlating data from multiple mesh nodes whose locations are known. The velocity of a mobile object, for example, can be estimated by the ratio of the spatial and temporal distances between two consecutive object sightings by different mesh nodes. As another example, the union of the coverage areas of mesh nodes that concurrently detect the object can approximate its size and shape.

The existence of spatial and temporal correlations brings significant potential advantages for the development of efficient communication protocols well suited for the WM^{2}Net paradigm. For example, intuitively, due to the spatial correlation, data from spatially separated sensors is more useful to the AP than highly correlated data from nodes in proximity. Therefore, it may not be necessary for all meshes to communicate their data to the AP. Instead, a smaller number of mesh readings may suffice to communicate the information on the sensed phenomenon to the AP within a certain reliability/fidelity level. Similarly, for a certain target tracking application, the measurement reporting frequency, at which sensors communicate their observations, can be adjusted such that temporally correlated phenomenon signal is captured at the AP within a certain distortion level and with minimum energy expenditure.

Clearly, understanding the spatiotemporal correlation characteristics of WM^{2}Net brings potential advantages to be exploited in the design of efficient communication protocols, which may help overcome the severe energy and processing limitations of WM²Nets.

Furthermore, spatiotemporal coordination among mesh nodes may also be necessary to ensure correctness and consistency of distributed measurements (Ganesan et al., 2004). For example, if the sampling rate of sensors is low compared to the temporal frequency of an observed phenomenon, it may be necessary to ensure that sensor readout occurs concurrently at all sensor nodes in order to avoid false observation results. Likewise, the spatial distribution of sensors has an impact on the correctness of observation results. For example, in order to estimate the average of a certain physical quantity over a certain physical area (e.g., average room temperature), it is typically not sufficient to simply calculate the average over all sensor nodes covering the area, because then areas with higher node density would be overrepresented in the resulting average value. As another example, one way of extending network lifetime is to periodically switch off radio transceivers of mesh nodes and place them into power-saving sleep modes. Operating mesh nodes in power-saving sleep mode temporal coordination among meshes may then be required in order to ensure seamless and continuous operation of the WM²Net. Temporal coordination is required to ensure that activity periods of meshes overlap in time in order to enable communication.

6.4.1 On Maximizing Capacity in Fixed Mesh Networks with MIMO Links[5]

6.4.1.1 Preliminaries

The study investigates the improvement of the normalized capacity offered by MIMO's beamforming capacities. As shown, in a high SNR environment, the behavior of MIMO is similar to the behavior of single antenna systems, and hence the techniques and insights used to maximize network capacity with SISO links can be easily extended to networks with MIMO links. On the other hand, in low SNR settings, the full capabilities of MIMO are required and hence the computational complexity of maximizing the capacity of networks with MIMO links is considerably greater than the computational complexity of capacity maximization of networks with SISO links.

The key findings are that beamforming provides performance improvement only when the SNR and the signal-to-interference ratio (SIR) are relatively low with $SNR < 10$ dB and SIR $< 2.5 - \text{SNR}/2$ dB. Otherwise, the beamforming capacities of MIMO can be neglected, the MIMO link can be treated as a single link, and the traditional techniques for network capacity can be applied.

6.4.1.2 Optimization Techniques for MIMO Links

Throughout this study it is assumed that all channel gains are known. Also, it is assumed that the transmitter and receiver both have N antennas. Due to space limitations, this study only examines the two-link case. However, the results and insights can be applied to the multilink setting.

Let us denote with $\vec{x}_l$ the vector of signals transmitted from the transmitters of link l. The correlation matrix of this vector is $Q_l = E(\vec{x}_l \vec{x}_l^\dagger)$, where $Q_l \in H^{+,0}$ and $H^{+,0}$ is

[5] Excerpt from the invited article "On maximizing capacity in fixed mesh networks with MIMO links," Stephan Bohacek, Department of Electrical and Computer Engineering, University of Delaware, Newark, DE, USA, E-mail: bohacek@udel.edu

the space of nonnegative definite Hermitian matrices. The correlation matrix, Q_l, is a design parameter. The total transmission power over link l is $trace(Q_l)$. It is assumed that the channel gains are normalized so that the maximum allowable transmission power corresponds to $trace(Q_l) = 1$. Given a set of correlation matrices Q_1, Q_2, it can be shown (Cover and Thomas, 1991) that the maximum data rate over link l is

$$B_l(Q_1, Q_2) = \log_2\left(\det\left(R_l + H_{l,l} Q_l H_{l,l}^{\dagger}\right)\right) - \log_2(\det(R_l)),$$

where $R_l = I + \sum_{k \neq l} H_{k,l} Q_k H_{k,l}^{\dagger}$ and $H_{k,l}$ is the channel gain matrix from the transmit antennas of link k to the receive antennas of link l. Thus, $H_{l,l}$ is the matrix of channel gains across link l and $H_{k,l}$ are the channel gains from the transmit antennas of link k to the receive antennas of link l. Hence, R_l represents the noise plus interference. Note that the channel gains are normalized so that the noise power is one.

MIMO can be used to both maximize the data rate and reduce interference. However, in general, the highest data rate across one link will increase the interference floor to other links. Hence, it is not possible to simultaneously maximize the data rate across each link. Instead, a trade-off between data rate and beamforming must be achieved. For example, given weights λ_1 and λ_2, the following weighted capacity problem can be solved:

$$\begin{aligned} &\max_{Q_1, Q_2 \in \mathrm{H}^{+,0}} \quad \lambda_1 B_1(Q_1, Q_2) + \lambda_2 B_2(Q_1, Q_2) \\ &\text{subject to} : \mathrm{trace}(Q_1) \leq 1 \\ &\qquad\qquad\quad \mathrm{trace}(Q_2) \leq 1. \end{aligned}$$

Ye and Blum (2003) investigated this optimization problem. It was found that for high SNR, the optimization is convex and can be solved with the projected gradient approach (Bertsekas, 1999). It was also found that even in low SNR this technique still works well.

Since optimizing the data rate over $Q_l \in H^{+,0}$ with $trace(Q_l) \leq 1$ is able to take advantage of all the capabilities of MIMO communication, we refer to such optimization as *full optimization*. Full optimization can achieve high data rates and/or employ beamforming to reduce interference. However, since the space $\{Q_l \in H^{+,0} : trace(Q_l) \leq 1\}$ is quite large, full optimization is computationally complex, which increases with the number of links. Due to these reasons, we then consider a scheme based on *eigenchannels*.

If N denotes the number of antennas, the MIMO link can be divided into N eigenchannels as follows.[6] Let x' denote the vectors of signal transmitted across N eigenchannels. Given x', the vector of signals transmitted by the antennas is $x = V_l x'$, where V_l is from the singular value decomposition of $H_{l,l}$, that is, $U_l \Lambda_l V_l^{\dagger} = H_{l,l}$. Let y be the vector of signals received by the receive antennas, hence $y = H_{l,l} x$. The received signal across the eigenchannel is denoted by y' and is given by $y' = \mathrm{U}_l^{\dagger} y = U_l^{\dagger} H_{l,l} V_l x' = \Lambda_l x'$, where the Λ_l is diagonal matrix of the singular values of $H_{l,l}$, which are denoted as $h_{l,l,i,i}$. Hence, $y_i' = h_{l,l,i,i}\, x_i'$. The total data rate achievable across this MIMO link is $\sum_{i=1}^{N} \log_2(1 + h_{l,l,i,i}^2 P_{l,i})$, where $P_{l,i}$ is the transmission power allocated to the i-th eigenchannel of the l-th link.

Based on this eigenchannel representation, a MIMO link is a collection of parallel, noninterfering channels. However, eigenchannels from other links will cause interference.

[6] See pages 291–293 in the study of Tse and Viswanath (2005) for in-depth background on eigenchannels.

Specifically, the gain across the interfering channel from the transmitter of the i-th eigenchannel of the k-th link to the receiver of the j-th eigenchannel of the l-th link is $h_{k,l,I,j} = u_{l,j}^{\dagger} H_{k,l}^{\dagger} v_{k,i}$, where $u_{l,j}$ and $v_{k,i}$ are the j-th column of U_l and the ith column of V_k respectively, and U_l and V_k are from the singular value decomposition of the channel gain matrices $H_{l,l}$ and $H_{k,k}$ respectively. Given P, a vector of transmit power assigned to different eigenchannels, the data rate across the l-th link is

$$E_l(P) := \sum_{i=1}^{N} \log_2 \left(1 + \frac{h_{l,l,i,i} P_{l,i}}{1 + \sum_{k \neq l} h_{k,l,j,i} P_{k,j}}\right).$$

If the eigenchannel approach is used, then the weighted capacity problem is replaced with

$$\max_{\vec{p}} \lambda_1 E_1(P) + \lambda_2 E_2(P)$$
$$\text{subject to: } \sum_{i=1}^{N} P_{l,i} \leq 1 \text{ for all } l.$$

This optimization problem is similar to those in networks with SISO links, except that instead of a limit on the power transmitted over each channel, here there is a constraint on the *total* power allocated to eigenchannels of a link.

When either $\lambda_1 = 0$ or $\lambda_2 = 0$, then the optimal solution to the above weighted capacity problems is given by water filling. Specifically, the powers allocated to the eigenchannels are $P_{l,i}^{*} = \frac{1}{\mu_l} - \frac{1}{h_{l,l,i,i}}$, where μ_l is the largest number such that $\sum_{i=1}^{N} P_{l,i}^{*} = 1$ (see page 293 of the work by Tse and Viswanath, 2005 for details).

To further decrease the computational complexity, we consider treating the MIMO link as a single channel with the total power allocated to the link denoted by p_l, where p_l scales the power allocated to each eigenchannel. In other words, the power allocated to the i-th eigenchannel of the l-th link is allocated $P_{l,i}^{*} p_l$, where $P_{l,i}^{*}$ is the water-filling capacity for the i-th eigenchannel of the l-th link. In this case, the weighted capacity problem becomes

$$\begin{aligned} &\max \lambda_1 E_1(P) + \lambda_2 E_2(P) \\ &\text{subject to}: P_{l,i} = P_{l,i}^{*} p_l \text{ for all } l \text{ and } i \\ &\qquad\qquad 0 \leq p_l \leq 1. \end{aligned}$$

Since this scheme models the MIMO link as a single channel, we refer to this approach as the *single channel* approach.

The following points should be emphasized.

- When only a single transmitting link is considered, these three approaches achieve the same data rate as given by water filling. We denote the water-filling data rates as $\overline{BR1}$ and $\overline{BR2}$, respectively. This study emphasizes data rates relative to these maximum data rates. Hence, when water filling is applied, the link achieves a normalized data rate of one.
- When multiple transmitting (and thus interfering) links are considered, the single channel approach has fewer degrees of freedom than the eigenchannel

approach, and the eigenchannel approach has fewer degrees of freedom than full optimization. Hence, we expect that the single channel approach is the least computationally complex, but yields the worst performance. On the other hand, full optimization is computationally complex, but provides optimal results. The next sections examine the performance of these schemes.

6.4.1.3 Capacity of Simultaneously Transmitting Links

When sending data across two links, transmissions can occur across both links simultaneously, or time division multiplexing (TDM) can be used and links transmit at different times. Which of these approaches yields the highest capacity depends on the channels. The *simultaneously transmitting normalized capacity region* can be defined as $nBR2(BR1)$, which denotes the maximum normalized data rate across link 2 given the normalized data rate across link 1. In the case of full optimization, we consider

$$\begin{aligned} nBR2_{Full}(BR1) := \max \frac{B_2(Q_1, Q_2)}{\overline{BR2}} \\ \text{subject to}: \frac{B_1(Q_1, Q_2)}{\overline{BR1}} \geq BR_1 \\ Q_i \in H_{l,l}^{\dagger} \\ \text{trace}(Q_i) = 1. \end{aligned}$$

In the eigenchannel approach, we consider

$$\begin{aligned} nBR2_{\text{Eigenchannel}}(BR1) := \max \frac{E_2(P)}{\overline{BR2}} \\ \text{subject to}: \frac{E_1(P)}{\overline{BR1}} \geq BR_1 \\ \sum_{i=1}^{N} P_{1,i} \leq 1 \text{ for } i = 1, 2. \end{aligned}$$

Examples of these capacity regions are shown in Fig. 6.16, for several different values of SNR and for the SIR fixed at 10 dB. The upper left frame shows that when the SNR is considerably lower than the SIR, the interference can be ignored, and hence both links can transmit at their maximum data rate (i.e., a normalized data rate of one). This is true for both full optimization and the eigenchannel approach. However, as the SNR becomes comparable with the SIR, the system becomes interference limited and hence interference reduces the data rate. In the eigenchannel approach, when the SNR is high, if both links are transmitting, the data rate for each link is relatively small. However, if full optimization is used, then relatively high data rate across both links is still possible at high SNR. This is due to the ability of MIMO to mitigate interference as well as transmit at relatively high data rates.

6.4.1.4 Multiplexing Versus Simultaneous Transmissions

While the lower right frame of Fig. 6.16 seems to indicate that the eigenchannel approach performs considerably worse than the full optimization, the difference is subtler than

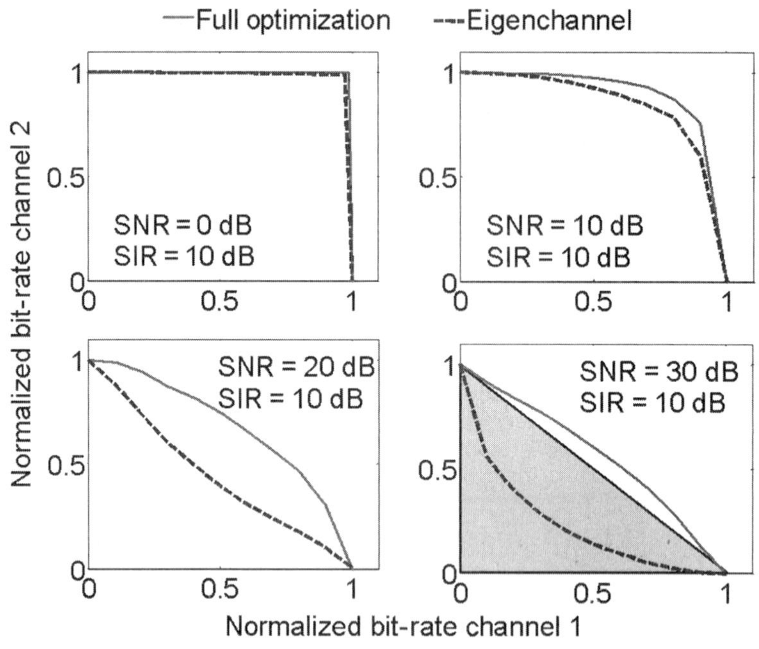

FIGURE 6.16 Capacity regions for several different values of SNR and SIR.

Fig. 6.16 indicates. Specifically, Fig. 6.16 shows the data rate when both links transmit simultaneously. It is also possible to use TDM and have each link transmit individually. The capacity region when TDM is allowed is the convex hull of the capacity region when links transmit simultaneously. The lower right-hand frame of Fig. 6.16 shows the normalized capacity region for the eigenchannel approach when TDM is allowed. Consequently, when the SNR is high or low, the eigenchannel approach with TDM achieves nearly the same capacity as full optimization. For moderate values of SNR, the full optimization provides higher capacity than the eigenchannel approach. This section investigates the performance behavior when multiplexing is adopted.

In order to determine whether multiplexing or simultaneous transmissions provide higher capacity, we consider the *normalized capacities*

$$\begin{aligned} \max \frac{B_1(Q_1, Q_2)}{\overline{BR1}} &+ \frac{B_2(Q_1, Q_2)}{\overline{BR2}} \\ \text{subject to} : Q_1 \in H_{l,l}^{\dagger},\ & Q_2 \in H_{l,l}^{\dagger} \\ \operatorname{trace}(Q_1) \leq 1,\ & \operatorname{trace}(Q_2) \leq 1, \end{aligned}$$

$$\max \frac{E_1(P)}{\overline{BR1}} + \frac{E_2(P)}{\overline{BR2}}$$
$$\text{subject to} : \sum_{i=1}^{N} P_{1,i} \leq 1 \text{ for } l = 1, 2, \text{and } i = 1, \ldots, N,$$

and

$$\max \frac{E_1(P)}{\overline{BR1}} + \frac{E_2(P)}{\overline{BR2}}$$
$$\text{subject to}: P_{l,i} = p_l P^*_{l,i} \text{ for } l = 1,\ 2 \text{ and } i = 1, \dots, N$$
$$0 \leq p_l \leq 1 \text{ for } l = 1,\ 2.$$

Note that these maximums can be never less than one and also never greater than two. For example, if only one link transmits at its maximum rate, $\overline{BRl}$, then the weighted capacity is one. Furthermore, this same capacity is achieved by multiplexing. Hence, when the solutions to these problems are equal to one, then multiplexing achieves a weighted capacity that is no worse than simultaneous transmissions. The lower right frame of Fig. 6.16 is an example for this ratio being one for the eigenchannel approach and slightly larger than one for full optimization. On the other hand, when the maximum is two, then both links can simultaneously achieve their maximum data rates. The upper left frame of Fig. 6.16 is an example for the solution to these problems being two. For intermediate values between one and two, the maximum weighted capacity region is given by both links transmitting simultaneously. However, since interference cannot be neglected, optimization is required in order to achieve the maximum capacity.

Figure 6.17 shows the average values of the above normalized capacities for several values of SNR and as a function of SIR. Here the channel gains are

$$H_{l,l} = \left(R_{l,l} + \sqrt{-1} I_{l,l}\right) \times 10^{\text{SNR}/10}$$
$$H_{l,k} = \left(R_{l,k} + \sqrt{-1} I_{l,k}\right) \times 10^{(\text{SNR}-\text{SIR})/10},$$

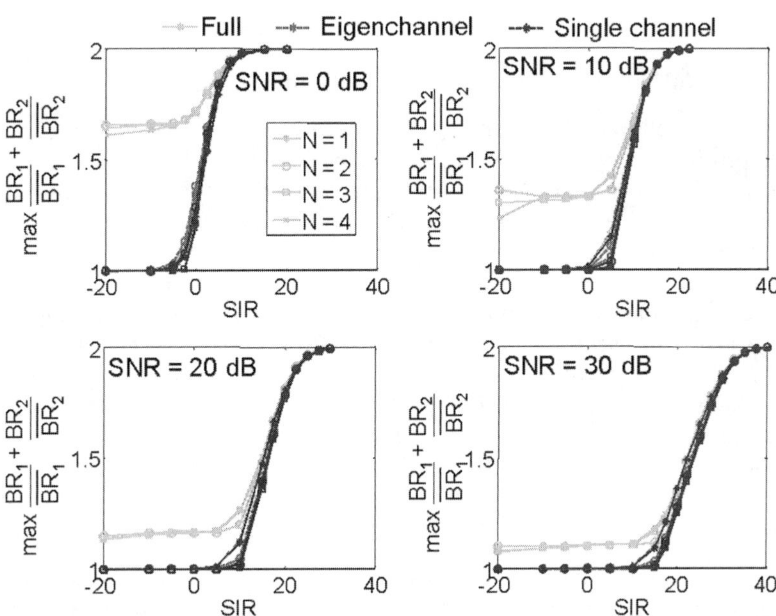

Figure 6.17 Average values of normalized capacities for several values of SNR and SIR.

where $R_{l,l}$, $I_{l,l}$, $R_{l,k}$, and $I_{l,k}$ are $N \times N$ normally distributed random matrices with zero mean and unit variance. In Fig. 6. 17, averaging of 100 trials for each combination of SNR and SIR is pursued.

Several observations can be drawn from Fig. 6.17. First, the eigenchannel approach has a *normalized* capacity region similar to the single antenna case and the eigenchannel approach is the same as the single channel approach. Hence, while the eigenchannel approach has more degrees of freedom than the single channel approach, it has no ability to reduce interference. Therefore, we conclude that the eigenchannel approach should not be used. On the other hand, in some cases, due to MIMO's interference mitigating abilities, full optimization is able to reduce interference and hence provides a larger normalized capacity region than the single channel case. For example, if the SNR is low (e.g., 0 dB) and the SIR is even lower (e.g., less than −10 dB), then full optimization is able to achieve a considerably larger normalized capacity region than SISO links. However, in the case of a planned mesh network, it is unlikely that links would have such low SNRs. In the case of high SNR (e.g., greater than 20 dB), the impact of MIMO's ability to reduce interference is limited.

These observations have important implications: in high SNR environments, full optimization is not needed but the single channel approach is sufficient. Furthermore, full optimization is needed only when SIR is considerably lower than the SNR. Based on these results, we find that full optimization provides significant improvement in capacity only when SNR < 10 dB and IR < 2.5 − SNR/2 dB. Otherwise, the single channel approach is sufficient, and, of course, considerably less computationally complex. Since mesh networks often have high SNR, we conclude that network capacity optimization with MIMO links can be easily accomplished. However, if the links have low SNR, then capacity maximization requires solving computationally intensive optimization problems.

6.4.2 WM^{2}Snet Deployment: An Experimental Approach[7]

6.4.2.1 Deployment: A Challenge to Overcome

The physical placement of WM^{2}Snet sensor nodes in a real environment is one of the most critical phases in WM^{2}Snet-based application deployment. This study addresses some of the parameters that affect the deployment phase.

The *accessibility* of the monitoring scenario has a significant impact on the deployment strategies available. Environments such as an active volcano crater are hardly accessible to humans. Optimum deployment of nodes is an almost impossible practice then. Strategies such as dropping the devices from an aircraft could be a viable solution.

Additionally, WM^{2}Snet *sensor node characteristics* like memory, batteries, processor, weight, and antenna may also impact network behavior, alongside the various sensors attached to the node, which can vary in type (e.g., temperature and radioactivity), range, or accuracy. Sensors can be classified into four classes (Beutel et al., 2004): Spec—tiny

[7] Excerpt from the invited article "Wireless mesh sensor network deployment: An experimental approach," T. Camilo, A. Rodrigues, J. Sá Silva, *F. Boavida, P. Melo, L. Pedrosa, R. Neves, and †R. Rocha. (*Department of Informatics Engineering, University of Coimbra, Polo II, Pinhal de Marrocos, 3030-290 Coimbra-Portugal, E-mail: {tandre, arod, sasilva, boavida}@dei.uc.pt; †Instituto de Telecomunicações, Instituto Superior Técnico, Technical University of Lisbon, Campus IST-Taguspark, 2744-016 Porto Salvo, Portugal, E-mail: {pedro.melo, luis.pedrosa, rui.neves, rui.rocha}@tagus.ist.utl.pt).

nodes only capable of monitoring actions; Mote—monitoring nodes with forwarding capabilities; Imote—motes with enhanced monitoring features, which normally perform video and audio sensing; and Stargate—nodes acting as a gateway between the WM^{2}Snet and the wireline network.

Node density reflects the coverage requirement for a given application. Sensor over-provisioning will certainly increase fault tolerance and gathered data, thus increasing the overall measurement accuracy of the sensing process at the expense of increased costs and energy consumption. In theory, the minimum node density to monitor a specific area (A) can be calculated through Eq. (6.6) (Slijepcevic and Potkonjak, 2001), where r represents the sensor range of each node. However, this estimation does not consider the use of more than one kind of sensor and it ignores the environment characteristics (e.g., trees and mountains).

$$NS = \frac{2A \cdot \pi}{r^2\sqrt{27}} \tag{6.6}$$

Generally, a WM^{2}Snet solution should focus on the definition of the application for which it is intended, the specific application monitoring scenario, and the desired application accuracy level. Measurements of real sensor node communications by Fanimokun and Frolik (2003) led to the conclusion that some obstacles can indeed enhance node's communications, basically due to radio reflections that reinforce the signal strength at the receiver. Furthermore, some WM^{2}Snet solutions take advantage of the environment elements, for example, an application built to monitor city traffic, where city cabs are used to retrieve monitoring data (Shah et al., 2003).

One important aspect in efficiently deploying WM^{2}Snets is to find an optimal node placement strategy. Nodes can either be deployed manually by an operator at predetermined locations, or be spread by a random strategy (e.g., being dropped from an aircraft).

6.4.2.2 WM^{2}Snets at Work

The study presented in this section provides results from real deployments of WM^{2}Snet, covering several aspects such as sensor node antenna orientation and characteristics, node height, communication range, and diverse monitoring environments and deployment strategies. To cover all these issues several experiments were carried out simultaneously with both platforms. The first group of experiments was enacted in a controlled environment to properly assess the mote's radio capabilities without outside interference. Afterwards, the effect of the environment was evaluated for several different test scenarios. Finally, six different representative placement strategies were exercised in order to compare performance metrics between them. Two types of sensor nodes were used in these studies: MicaZ (http://www.xbow.com/Products/Product) (Fig. 6.18)

Figure 6.18 MicaZ sensor node.

Figure 6.19 ESB sensor node.

based on a ZigBee (ZigBee Alliance, 2004) radio; and Embedded Sensor Boards (ESB) http://www.scatterweb.net (Fig. 6.19) using the TR-1001 transceiver. Table 6.1 presents the main characteristics of the two studied sensor nodes.

The WM²Snet Nodes Antenna The type of antenna employed in a WM²Snet node can greatly affect its performance. MicaZ is equipped with a semi-hard $\lambda/4$ monopole antenna, connected with a 90° MMCX to the radio module, while ESB has a similar yet simpler antenna, with about twice the size of the former, soldered to the mote. The latter antenna can easily take the most awkward forms and orientations, rendering useless any detailed study of its characteristics. On the other hand, the MicaZ antenna can rotate only around the MMCX connector axis, maintaining its shape. Thus, it is relevant to study its radiation properties. For this purpose, we used an anechoic chamber, which allowed us to measure the antenna's radiating properties in a controlled environment. With these measurements, we gathered the necessary data to trace the main radiation diagrams for the MicaZ antenna, shown in Figs. 6.20 and 6.21.

In these diagrams, we can see the radiation properties for both vertical and horizontal planes and polarizations of the receiver. These confirm what is expected from such radiating elements and are in line with the results published by Scott (2004), as well as the theoretical dipole's radiation diagrams. One may note the impact of the connector upon the radiation pattern, which appears to be similar to that originating from the sensor's motherboard.

Next, while still in the anechoic chamber,[8] we evaluated the radio receiving power when the receiver antenna was in different positions. Figure 6.22 presents the various angles considered, either bending along the plane that include the sender (S90°, S45°, S-45°, S-90°) or along the plane that is perpendicular to the sender (P90°, P45°, P-45°, P-90°). Both sensor nodes were 1.5 m high and at a 5 m distance; the values presented in Fig. 6.23 represent the average received power over 60 packets.

Both transmitters were configured to send traffic at maximum power. In general, the worst results were achieved when the antenna was at a P90° or P-90° angle (perpendicular to the sender).

Monitoring Scenario Characterization The myriad applications for WM²Snets imply that the monitoring environment, where the sensor node deployment takes place, can vary within a broad range of scenarios. The following study illustrates the impact of the

[8] The authors would like to thank Instituto Superior Técnico for making this facility available and especially to Profs. Carlos Fernandes and Luís Correia for their assistance in pursuing the antenna characterization test.

Microcontroller				Communication			
Model	Flash RAM EEPROM	Active Power (mA)	Sleep Power (μA)	Radio	Frequency/ Modulation	Max. Data (Rate kbps)	Transmit/Receive Current
ATMega 128L	128 KB 4 KB (4 + 512) KB	8	<15	CC2420	2.4 GHz/DSSS	250	17.4 mA (0 dbm)/19.7 mA
MSP430 F149	60 KB 2 KB 256 B + 64 KB	8	8	TR1001	868.35 MHz/ASK	115.2	12 mA (0 dbm)/4.8 mA

Table 6.1 Main Characteristics of MicaZ and ESB Platforms (http://www.xbow.com/Products/Product; http://www.scatterweb.net; Polastre et al., 2005).

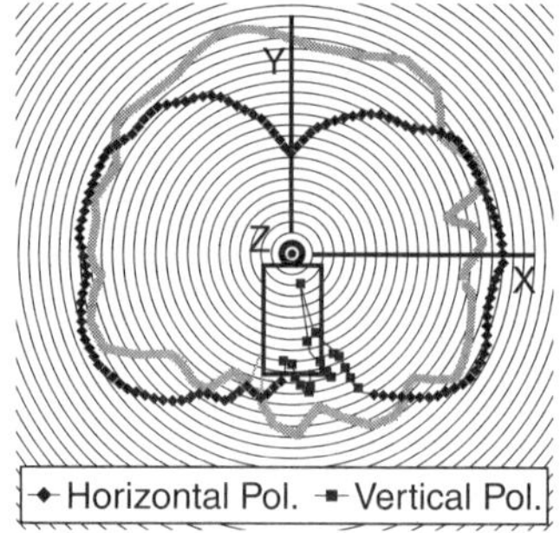

FIGURE 6.20 Radiation diagram along the horizontal plane.

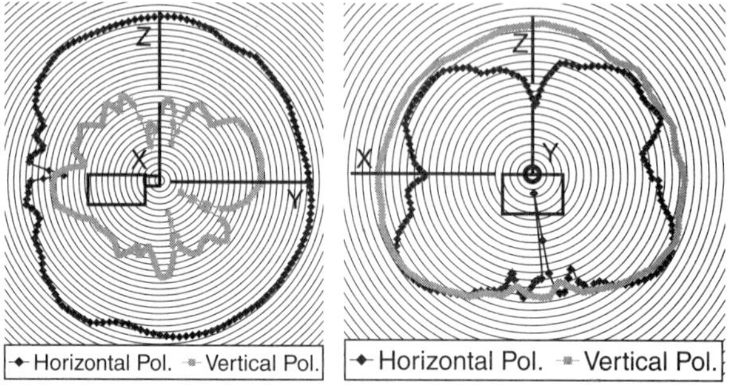

FIGURE 6.21 Radiation diagrams along the vertical plane.

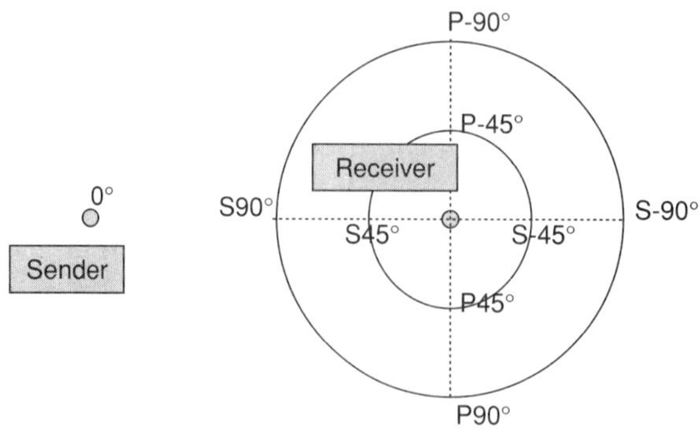

FIGURE 6.22 Receiving antenna positions.

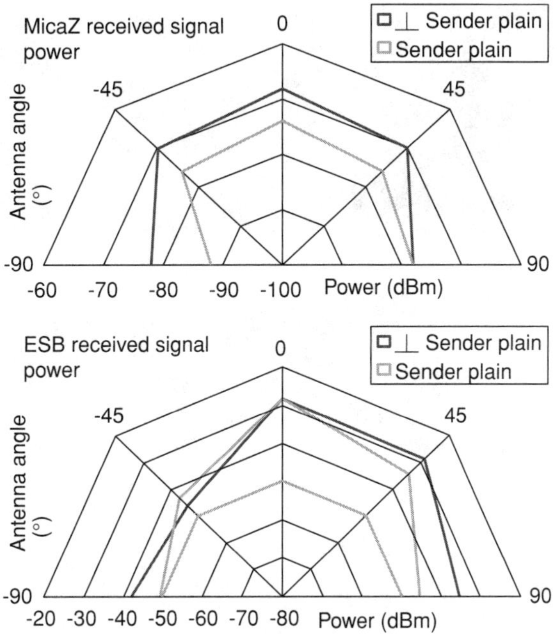

FIGURE 6.23 Received power at different antenna positions.

monitoring scenario on the radio range. Various operating conditions like the temperature (T) and air relative humidity (RH) were accounted for throughout the experiments. The scenarios considered are illustrated in Fig. 6.24: (a) **indoor**: a wall corridor (T 20°C, RH 45%); (b) **grass**: a moist land with short grass (T 25°C, RH 50%); (c) **forest**: a Mediterranean forest with vegetation up to 50 cm high (T 18°C, RH 62%); (d) **urban**: a common street (T 25°C, RH 50%); (e) **plain**: a moist flat stretch of land (T 17°C, RH 57%).

We present here the results obtained from several field trials. For each of the above environments we transferred 60 packets between two nodes, while varying the distance between them (from 1 to 15 m) and their relative heights (the receiver changed from 0 m up to 2 m). For both platforms, the average received power was obtained and is illustrated in Fig. 6.25.

The first conclusion that we can draw when comparing the results shown in Figs. 6.23 and 6.25 is that the environment greatly impacts the performance of the WM²Snet communications. The most irregular radio measurements were achieved with the indoor scenario (Fig. 6.25a and 6.25b). The multipath fading fluctuations observed are likely due to reflections on the walls. Some reinforcement of received signals may be observed, which can be explained by the well-known guide effect of hallways (Fanimokun and Frolik, 2003).[9] When it comes to outdoor scenarios a similar effect is observed for the urban case where objects (parked cars), located close by, can interfere with the radio transmission (Fig. 6.25g). In contrast, the forest scenario shows the opposite effect, where trees represent an obstacle to radio propagation (Fig. 6.25e).

[9] Thanks to its spread spectrum transmission technique MicaZ is less sensitive to this multipath effect.

Figure 6.24 Monitoring scenarios. (a) Indoor. (b) Grass. (c) Forest. (d) Urban. (e) Plain.

A clear conclusion that can be drawn from Fig. 6.25 is that it is not advisable to deploy sensor nodes at ground level, since their performance is highly hindered by environmental factors. (Yet, this may not always be possible to avoid.) In fact, as highlighted in Figs. 6.26 and 6.27, it can severely compromise the packet delivery ratio, the MicaZ being the most affected platform.

To further assess the impact of the mote's height on radio performance, a study of the effect of this factor on the received signal strength was carried out for each scenario. The results are presented in Fig. 6.28. For both platforms, the gain of raising the motes above ground level varied from 10% to 50%, depending on the scenario.

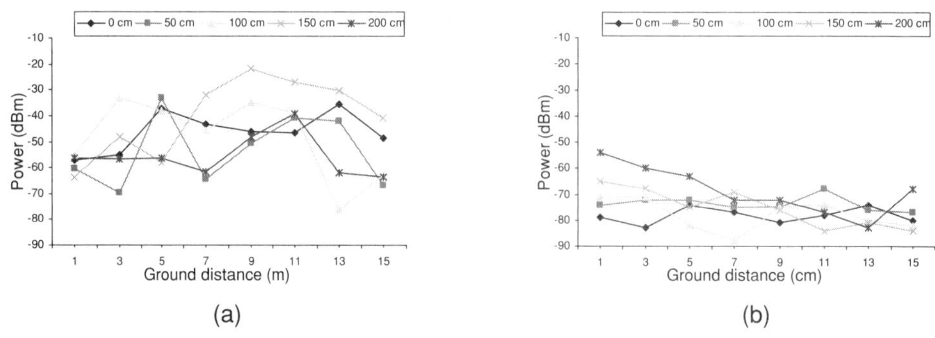

Figure 6.25 Receiving power in different environments. (a) Indoor ESB. (b) Indoor MicaZ. (c) Grass ESB. (d) Grass MicaZ. (e) Forest ESB. (f) Forest MicaZ. (g) Urban ESB. (h) Urban MicaZ. (i) Plain ESB. (j) Plain MicaZ. (*continued*)

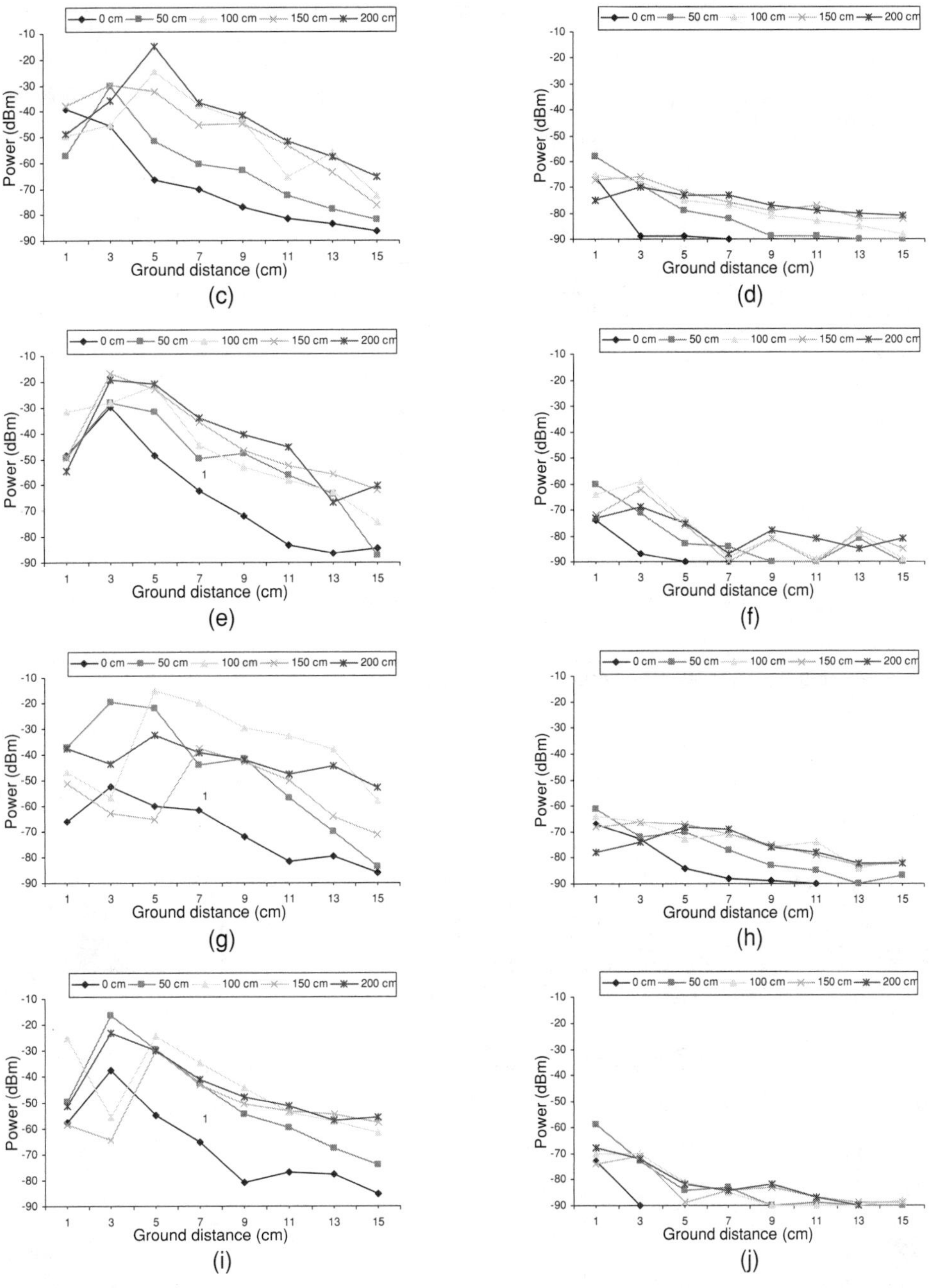

Figure 6.25 *(Continued)*

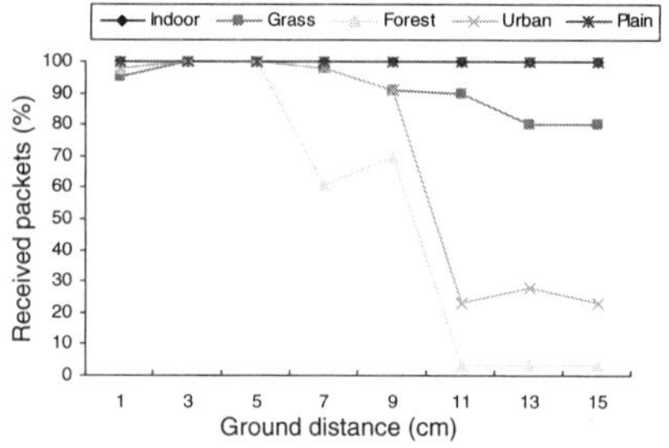

FIGURE 6.26 Received packets for each environment (ESB).

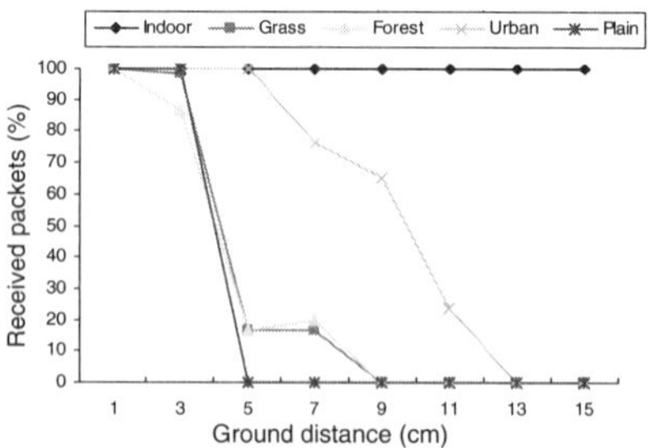

FIGURE 6.27 Received packets for each environment (MicaZ).

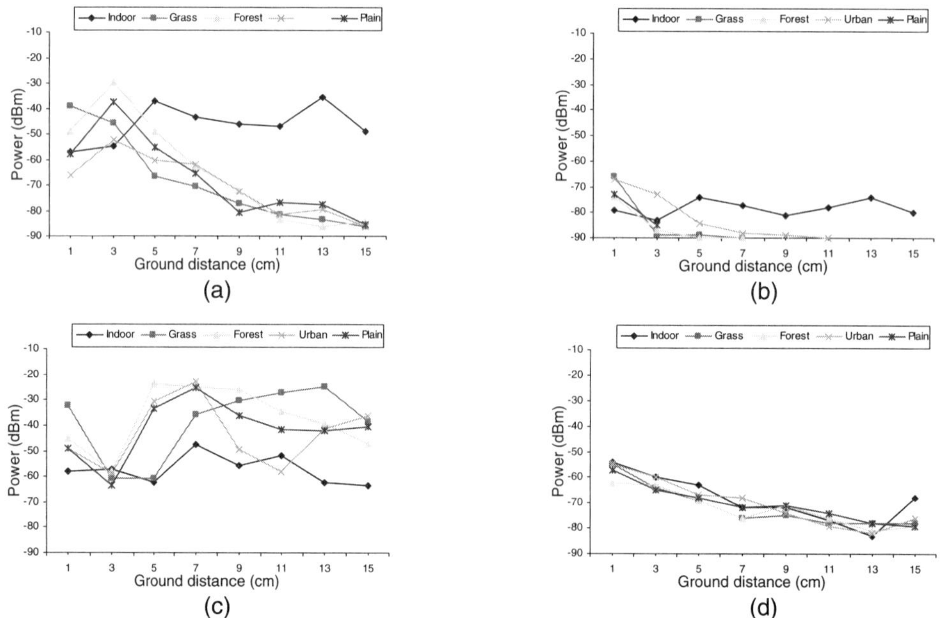

FIGURE 6.28 Receiving power at different heights for each environment. (a) Nodes at ground level (ESB). (b) Nodes at ground level (MicaZ). (c) Nodes at 2 m high (ESB). (d) Nodes at 2 m high (MicaZ).

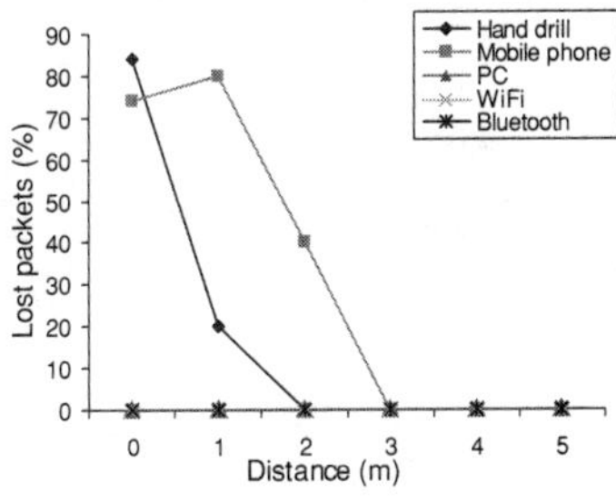

FIGURE 6.29 Interferences in ESB.

Radio Noise Communications on wireless channels are greatly affected from environmental factors such as interference from other coexisting radio networks. The following study compares the performance of both platforms under the impact of five different sources of radio noise: electric hand drill; mobile phone (900 MHz); personal computer; WiFi AP (2.4 GHz); Bluetooth handset. In this experiment, two sensor nodes were placed 5 m apart and 0.5 m high. A total of 60 packets were sent from one node to the other (at maximum transmission power, as before). The average packet loss ratio in the presence of each noise source is illustrated in Fig. 6.29 for ESB sensor node and in Fig. 6.30 for the MicaZ platform. The noise source was placed at several distances from the receiver, ranging from 0 m to 5 m.

The results clearly show that MicaZ motes suffer less loss in the presence of all noise sources tested, except the WiFi AP. The ESB sensor node, on the other hand, appears to be significantly affected by the electric hand drill (84% packet loss) and the mobile phone (76% packet loss) when they are operating close to the receiver. Again, the broadband spread spectrum transmission system provides an advantage when it comes to noise and interference.

Deployment Strategies When implementing real-world sensor networks the deployment strategy adopted is of utmost importance and must take into consideration several factors: accessibility, reliability, and available resources. The following study mimics node placement in a battlefield; the sensor network should detect enemy troops and warn the user application. When deploying nodes in such a rough environment, it is necessary to perform it as quickly as possible, avoiding enemy contact. Suitable strategies were therefore considered, each one requiring different deployment times: **grid**: the soldier

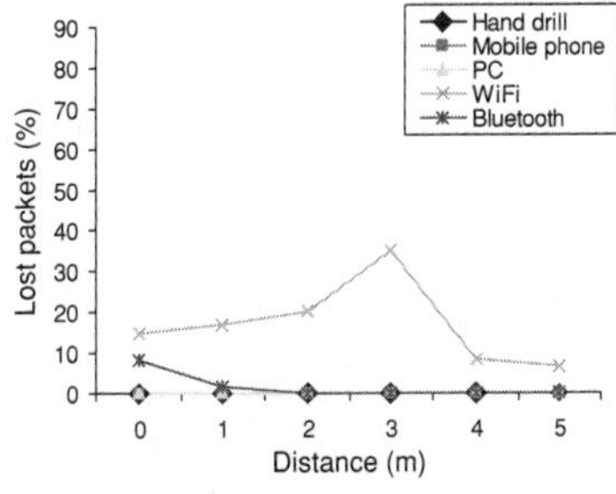

FIGURE 6.30 Interferences in MicaZ.

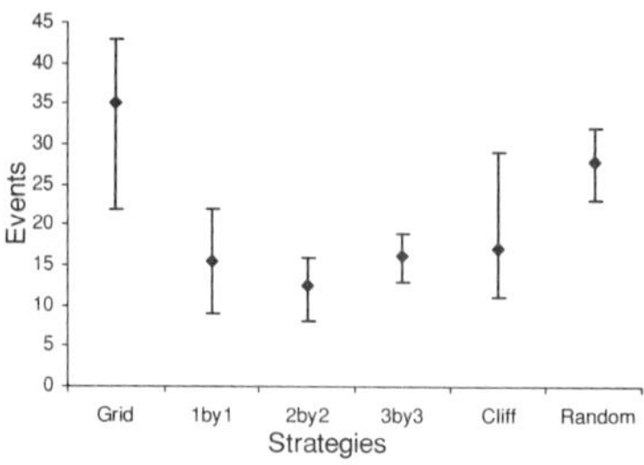

Figure 6.31 Deployment strategy with ESB.

has time to perform the node placement; consequently the network is geographically placed as a grid. A measuring tool was used to insure correct placement[10]; **one-by-one**: the soldier throws the sensor nodes one by one not caring about accurate node position while still trying to cover as much area as possible; **two-by-two** and **three-by-three**: the same as in one-by-one but in pairs or trios, respectively; **cliff**: in this case, the sensors (represented by test dummies) were all dropped at the same time from a cliff. In order to study the differences between an authentic randomly positioned scenario and the random distributions used in simulations an additional type of strategy was considered, **random:** the sensor locations were randomly calculated with a computer program and all nodes were placed in the chosen locations. The monitored area considered was 5 m wide × 11 m long, and a total of eight nodes were used, plus the sink node, which was able to directly communicate with all the ground nodes. The phenomenon used to stimulate sensor nodes, corresponding to the enemy troops crossing the monitored area, was repeated seven times for each strategy. Figures 6.31 and 6.32 present the number of events that were monitored by the sensor network and reported to the sink node for each strategy.

In both platforms, the **random** strategy presents good results, when compared to the more realistic approaches. As expected, the **grid** strategy tends to be one of the best solutions. Both achieved a good coverage of the sensing area; however, the **grid** strategy provides an optimal coverage, whereas the **random** is slightly less effective.

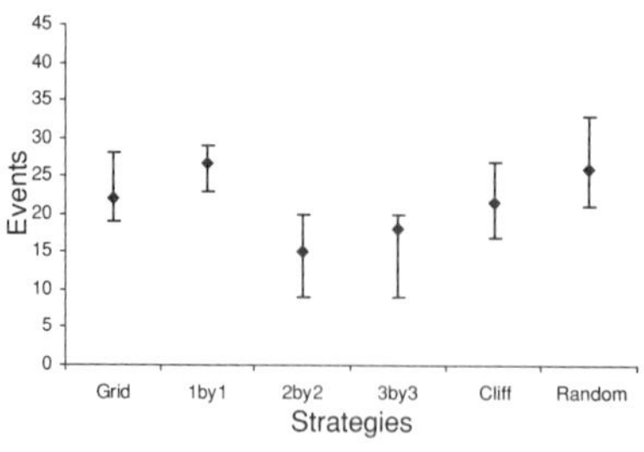

Figure 6.32 Deployment strategy with MicaZ.

[10] Distance between sensors corresponds to exactly twice the sensing range.

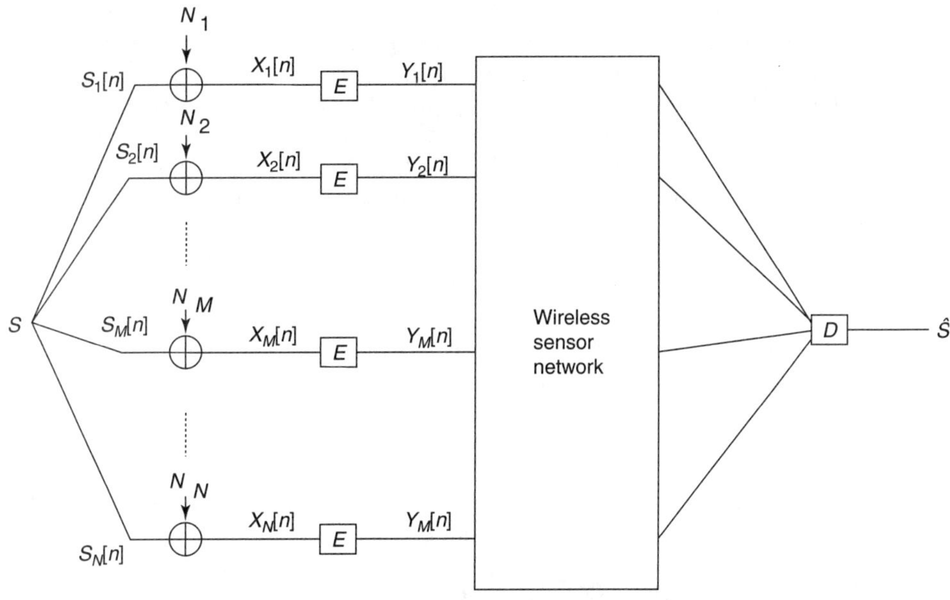

FIGURE 6.33 Correlation model and architecture.

6.4.3 Spatiotemporal Correlation Theory in WM²Snets[11]

6.4.3.1 Communication Architecture and Correlation Model

Throughout this study, a WM²Snet is assumed. In a sensor field, each sensor observes the noisy version of a physical phenomenon. The sink is interested in observing the physical phenomenon using the observations from sensor nodes with the highest accuracy. The physical phenomenon of interest can be modeled as a spatiotemporal process $s(t, x, y)$ as a function of time t and spatial coordinates (x, y).

The model for the information gathered by N sensors in an event area is illustrated in Fig. 6.33. The sink is interested in estimating the event source, S, according to the observations of the sensor nodes, n_i, in the event area. Each sensor node n_i observes $X_i[n]$, the noisy version of the event information, $S_i[n]$, which is spatially correlated to the event source, S. In order to communicate this observation to the sink, each node has to encode its observation. The encoded information, $Y_i[n]$, is then sent to the sink through the sensor network. The sink, at the other end, decodes this information to get the estimate, $\hat{S}$, of the event source S. The encoders and the decoders are labeled as E and D in Fig. 6.33, respectively. Using this model, we will exploit various aspects of correlation among sensor readings both in terms of time and space.

Each observed sample, $X_i[n]$, of sensor n_i at time n is represented as $X_i[n] = S_i[n] + N_i[n]$ where the subscript i denotes the spatial location of node n_i, that is, (x_i, y_i), $S_i[n]$

[11] Excerpt from the invited article "Spatiotemporal correlation theory in wireless mesh sensor networks," Özgür B. Akan, Next Generation Wireless Communications Laboratory, Department of Electrical and Electronics Engineering, Middle East Technical University, Ankara, Turkey, 06531, E-mail: akan@eee.metu.edu.tr

is realization of space–time process $s(t, x, y)$ at time $t = t_n$[12] and $(x, y) = (x_i, y_i)$, and $N_i[n]$ is observation noise. $\{N_i[n]\}_n$ is a sequence of i.i.d Gaussian random variables of zero mean and variance σ_N^2. We further assume that the noise each sensor node encounters is independent of each other, that is, $N_i[n]$ and $N_j[n]$ are independent for $i \neq j$ and $\forall n$.

As it is shown in Fig. 6.33, each observation $X_i[n]$ is then encoded into $Y_i[n]$ by the source coding at the sensor node as $Y_i[n] = f_i(X_i[n])$ and then sent through the network to the sink. The sink decodes the received data to reconstruct an estimation $\hat{S}$ of the source S

$$\hat{S} = g(Y_1[n_1], \ldots, Y_1[n_\tau]; \ldots; Y_N[n_1], \ldots, Y_N[n_\tau]) \tag{6.7}$$

based on the data received from N nodes in the event area over a time period $\tau = t_{n_\tau} - t_{n_1}$. The sink is interested in reconstructing the source S according to a distortion constraint

$$D = E[d(S, \hat{S})] \tag{6.8}$$

Next, the general distortion function in Eq. (6.8) is used to independently obtain the distortion functions for physical phenomena modeled by both point and field sources.

6.4.3.2 Spatiotemporal Correlation in WM^{2}Snets

Spatiotemporal Characteristics of Point Sources In many WM^{2}Snet applications such as target detection and fire detection, the goal is to estimate the properties of an event generated by a single point source, through collective observations of sensor nodes. Here, we first introduce a model for the point source and formulate its spatiotemporal characteristics. Next, we derive the distortion function for the estimation of the point source.

Here, the point source is assumed to generate a continuous signal, which is modeled as a random process $f_S(s, t)$, where s denotes the outcome and t denotes time. For ease of illustration, we use $f_S(t)$ hereafter. We model a point source, $f_S(t)$, as a Gaussian random process such that $f_S(t)$ is first-order stationary, that is, $\mu_S(t) = \mu_S$ and has a variance σ_S^2. Without loss of generality, we assume $\mu_S = 0$.

For ease of illustration, we assume the coordinate axis is centered at the point source. As a result, the received signal, $f(x, y, t)$, at time t at a location (x, y) can be modeled as

$$f(x, y, t) = f_S\left(t - \frac{\sqrt{x^2 + y^2}}{v}\right) e^{-\frac{\sqrt{x^2+y^2}}{\theta_S}} \tag{6.9}$$

which is the delayed and attenuated version of the signal $f_S(t)$. In this model, we assume that the event signal travels with the speed v, and is attenuated based on an exponential law, where θ_S is the attenuation constant. Note that the function $f(x, y, t)$ is also a Gaussian random process and the samples taken by the sensors are jointly Gaussian random variables (JGRVs). Since $\mu_S = 0$, the mean of the received signal $\mu_E = 0$.[13] The variance

---

[12] A discrete-time model is used since each node is assumed to sample the physical phenomenon synchronously after the initial wake-up.

[13] The subscripts S and E denote the *source* and *event*, respectively.

of the received signal is also given as follows:

$$\sigma_E^2(x, y) = E[f^2(x, y, t)] = \left(\sigma_S e^{-\sqrt{x^2+y^2}/\theta_s}\right)^2 \tag{6.10}$$

An interesting result from Eq. (6.10) is that the variance of the signal observed at location (x, y) depends on the distance between the observation location and the point source. As in Fig. 6.33, the received signal at time t_k by a sensor n_i at location (x_i, y_i) is given by

$$S_i[k] = f(x_i, y_i, t_k). \tag{6.11}$$

Assuming wide-sense stationarity, the *spatiotemporal correlation function* for two samples of a point source taken at locations (x_i, y_i) and (x_j, y_j), and at times t_k and t_l, respectively, is given by

$$\rho_p(i, j, k, l) = \frac{E\lfloor S_i[k] S_j[l] \rfloor}{\sigma_E(x_i, y_i)\sigma_E(x_j, y_j)} = \rho_S(\Delta_t) \tag{6.12}$$

where $\Delta_t = \left| t_k - t_l - (d_i - d_j)/v \right|$, $d_i = \sqrt{x_i^2 + y_i^2}$ is the distance of the sensor n_i to the point source, and $\rho_S(\Delta_t) = E[f_S(t) f_S(t + \Delta_t)]/\sigma_S^2$ is the correlation function of the point source, which is given by $\rho_S(\Delta_t) = e^{-\Delta_t/\theta_t}$, where θ_t is a constant governing the degree of correlation. Note that the spatiotemporal correlation between two samples, $\rho_p(i, j, k, l)$, depends mainly on the difference between sample times t_k and t_l since generally $v \gg (d_i - d_j)$.

In WM^{2}SNet, we are interested in estimating the signal generated by the point source using the samples collected by the sensor nodes. The expectation of generated signal, $f_S(t)$, over an interval τ is given by

$$S(\tau) = \frac{1}{\tau} \int_0^\tau f_S(t)\, dt. \tag{6.13}$$

Each sensor node, n_i, receives the attenuated and delayed version of the generated signal $f_S(t)$, that is, $S_i[k]$. Due to the impurities in the sensor circuitries, the sampled signal is the noisy version of this received signal, which is given by

$$X_i[k] = S_i[k] + N_i[k], \tag{6.14}$$

where the subscript i denotes the location of the node n_i, that is, (x_i, y_i), k denotes the sample index, which corresponds to time $t = t_k$, $Xi[k]$ is the noisy version of the actual sample $Si[k]$, and $Ni[k]$ is the observation noise, that is, $N_i[k] \sim \mathrm{N}(0, \sigma_N^2) \cdot Si[k]$ is given by Eqs. (6.9) and (6.11).

The observed information, $X_i[k]$, is then encoded and sent to the sink through the WM^{2}SNet. It has been shown that joint source-channel coding outperforms separate coding. Moreover, for WM^{2}SNet with finite number of nodes, uncoded transmission outperforms any approach based on the separation paradigm leading to the optimal solution for infinite number of nodes (Vuran et al., 2004). Therefore, we assume that uncoded transmission is deployed in each node. Hence, transmitted observation, $Y_i[k]$, is given by

$$Y_i[k] = \sqrt{\frac{P_E}{\sigma_S^2 + \sigma_N^2}} X_i[k], \qquad i = 1, \ldots, N \tag{6.15}$$

where σ_S^2 and σ_N^2 are the variances of event information $S_i[k]$ and observation noise $N_i[k]$, respectively.

Transmitted information is decoded at the sink. Since uncoded transmission is used, it is well known that minimum mean square error (MMSE) estimation is the optimum decoding technique (Vuran et al., 2004). Hence, the estimation, $Z_i[k]$, of event information $S_i[k]$ is simply MMSE estimation of $Y_i[k]$, given by

$$Z_i[k] = \frac{\sigma_E^2(x_i, y_i)}{\sigma_E^2(x_i, y_i) + \sigma_N^2}(S_i[k] + N_i[k]) \tag{6.16}$$

The sink is interested in estimating the expected value of the event during a decision interval τ, which is given by Eq. (6.13). Assuming each sensor node sends information at a rate of f samples/s, this estimation can simply be found by

$$\hat{S}(\tau, f, M) = \frac{1}{\tau f M} \sum_{i=1}^{M} \sum_{k=1}^{\tau f} Z_i[k] \tag{6.17}$$

where M is the number of sensor nodes that send samples of the observed point source. M nodes are chosen among the nodes in the network to represent the point source, and hence are referred to as *representative nodes*. Consequently, the distortion achieved by this estimation is given by (Vuran and Akan, 2006a)

$$D_p(\tau, f, M) = E[(S(\tau) - \hat{S}(\tau, f, M))^2] \tag{6.18}$$

where subscript p denotes the point source. Using Eqs. (6.6), (6.7), (6.10), (6.16), and (6.17), (6.18) can be expressed as

$$D_p(\tau, f, M) = \sigma_S^2 - \frac{2}{\tau^2 f M} \sum_{i=1}^{M} \sum_{k=1}^{\tau f} \frac{\sigma_S^4 e^{-3d_i/\theta_s}}{\sigma_S^2 e^{-2d_i/\theta_s} + \sigma_N^2} \theta_t [2 - e^{-(t_k + d_i/c)} - e^{-(\tau - t_k - d_i/c)/\theta_t}]$$
$$+ \frac{\sigma_N^2}{\tau f M^2} \sum_{i=1}^{M} \frac{\sigma_S^4 e^{-2d_i/\theta_s}}{\left(\sigma_S^2 e^{-d_i/\theta_s} + \sigma_N^2\right)^2} + \frac{1}{\tau^2 f^2 M^2} \sum_{i=1}^{M} \sum_{j=1}^{M} \sum_{k=1}^{\tau f} \sum_{l=1}^{\tau f} \alpha \rho(i, j, k, l), \tag{6.19}$$

where

$$\alpha = \frac{\sigma_S^8 e^{-2(d_i + d_j)/\theta_s}}{\left(\sigma_S^2 e^{-d_i/\theta_s} + \sigma_N^2\right)\left(\sigma_S^2 e^{-2d_j/\theta_s} + \sigma_N^2\right)},$$

$d_i = \sqrt{(x_i + y_i)}$, and $\rho(i, j, k, l)$ is the spatiotemporal correlation function given in Eq. (6.12).

Spatiotemporal Characteristics of Field Sources In some WM²SNet applications such as temperature monitoring and seismic monitoring, the physical phenomenon is dispersed over the sensor field, and hence can be modeled as a field source. Thus, here, we explore the spatiotemporal characteristics of observing such a phenomenon in WM²SNets.

As in Section 6.4.3.2.A, the event signal $f(x, y, t)$ is assumed to be a Gaussian random process with $N(0, \sigma_S^2)$. The sink is interested in estimating the signal $f(x_0, y_0, t)$ over the decision interval τ at location (x_0, y_0). Assuming the observed signal $f(x, y, t)$ is wide-sense stationary (WSS), expectation of the signal over decision interval τ that is $S(\tau)$ can be calculated by the time average of observed signal as

$$S(\tau)\frac{1}{\tau}\int_0^{\tau} f(x_0, y_0, t)\, dt \tag{6.20}$$

where (x_0, x_0) is the event location. Signal $S_i[k]$ received at time t_k by a node at location (x_i, y_i) is defined as in Eq. (6.11) and $Si[k]$'s are JGRV with $N(0, \sigma_S^2)$. Covariance of two samples, $S_i[k]$ and $S_j[l]$, is given by

$$\text{cov}\{S_i[k], S_j[l]\} = \sigma_S^2\, \rho_S(i, j)\rho_t(\tau) \tag{6.21}$$

where $\rho_S(i, j) = e^{-d_{i,j}/\theta_s}$ and $\rho_t(\delta) = e^{-|\delta|/\theta_t}$ are spatial and temporal correlation functions, respectively, $\delta = (k - l)/f$, f is the sampling rate, $d_{i,j} = \sqrt{(x_i - x_j)^2 + (y_i - y_j)^2}$ is the distance between two nodes n_i and n_j, and θ_s and θ_t are spatial and temporal correlation coefficients, respectively.

Following the discussion and derivations in Section 6.4.3.2.A, the noisy version of the signal, $X_i[k]$, and transmitted signal, $x_0[k]$, are given by Eqs. (6.14) and (6.15), respectively. The estimation $Z_i[k]$ can be found as

$$Z_i[k] = \frac{\sigma_S^2}{\sigma_S^2 + \sigma_N^2}(S_i[k] + N_i[k]) \tag{6.22}$$

After collecting the samples of the signal in the decision interval τ from M nodes, the sink estimates the expectation of the signal over the last decision interval as given in Eq. (6.17). As a result, the distortion achieved by this estimation is given as in Eq. (6.18). Using the definitions above and substituting Eqs. (6.20), (6.21), and (6.17) into Eq. (6.18), the distortion function can be derived as (Vuran and Akan, 2006a)

$$\begin{aligned} D_f(\tau, f, M) = \sigma_S^2 &- \frac{2\sigma_S^2}{\tau^2 f M(\sigma_S^2 + \sigma_N^2)} \sum_{i=1}^{M} \rho_s(i, s) \sum_{k=1}^{\tau f} \theta_t [2 - e^{-k/(f\theta_t)} - e^{-\left(t - \frac{k}{f}\right)/\theta_t}] \\ &+ \frac{\sigma_S^4 \sigma_N^2}{\tau f M(\sigma_S^2 + \sigma_N^2)^2} + \frac{\sigma_S^6}{(\tau f M(\sigma_S^2 + \sigma_N^2))^2} \sum_{i=1}^{M}\sum_{j=1}^{M}\sum_{k=1}^{\tau f}\sum_{l=1}^{\tau f} \rho_s(i, j)\rho_t(|k - l|/f) \end{aligned} \tag{6.23}$$

6.4.3.3 Results and Exploiting Correlation in WM²SNet

To provide further insight into the spatiotemporal correlation characteristics and distortion analysis derived here, we present some analytical results and discuss possible

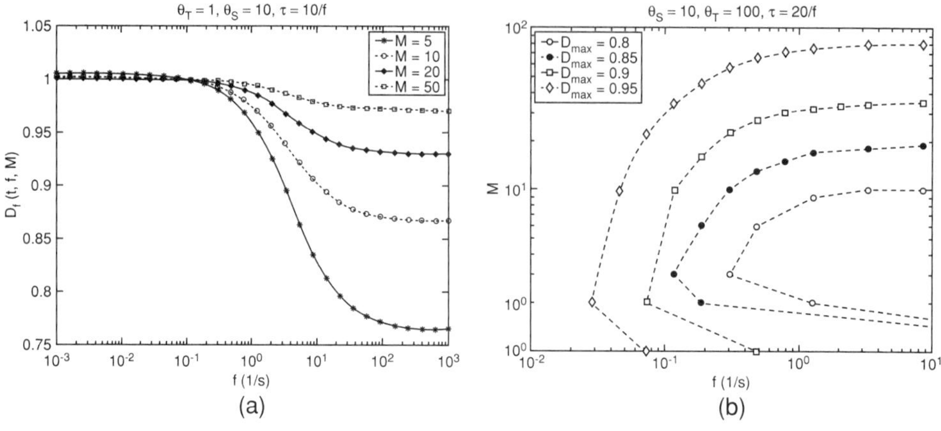

FIGURE 6.34 (a) (Point source) distortion vs. Sampling rate for different values of M. (b) (Field source) Number of nodes vs. Sampling rate (M, f) tuples meeting various D_{max} constraints.

approaches for the design of efficient communication techniques exploiting the spatiotemporal correlation in WM^{2}SNet.

Analysis Results Here, we present some numerical simulation results for spatiotemporal correlation characteristics of point and field sources using the distortion functions given by Eqs. (6.19) and (6.23). A WM^{2}Snet of a grid topology of 50 m $\times$ 50 m with 120 nodes is used for the evaluations. For each evaluation, the closest M nodes to the source are chosen to send information. For a point source, the behavior of the distortion function in Eq. (6.19) for various values for sampling rate f and the number of representative nodes M, are shown in Fig. 6.34a. It is clearly seen that, as the sampling rate increases, distortion decreases, which shows the effect of temporal resolution on event estimation. Moreover, above a specific range of f values, the distortion remains relatively constant. This observation reveals that there is an optimal value f_{opt} for temporal resolution such that further increase in sampling rate f does not affect the distortion.

The effect of the number of representative nodes M on distortion is also shown in Fig. 6.34a. It is clear that increasing M increases distortion for high values of sampling rate f due to decrease in spatial correlation. On the other hand, for lower sampling rates, for example, $f < 0.1\,\text{s}^{-1}$, an increase in M improves the distortion since the temporal resolution is not sufficient in this case. As a result of increased M, the spatial correlation helps build a more accurate estimation of the signal. However, increasing M above a specific value, for example, $M = 10$, has no impact on distortion. This result reveals that there is an optimal value M_{opt}, to be determined for energy-efficient communication in WM^{2}SNet.

In Fig. 6.34b, the tradeoff between spatial and temporal resolution is shown for field sources. Each point represents the boundary of the feasible region for (M, f) values that meet a certain distortion constraint D_{max}. The figure can be read as follows: For each allowed distortion D_{max}, the tuples represent the boundary of the feasible region inside which the distortion constraint is guaranteed.

An important result is that, for each D_{max} value, there is an optimum operating point, where a minimum number of nodes can be used with low sampling rate. Increasing M above this value also requires increase in temporal resolution. Consequently, aggressively collecting information from each sensor node in the field does not necessarily correspond to more accurate estimation. This figure serves as an important guideline for communication protocols design, and deployment for a particular distortion requirement.

Exploiting Spatiotemporal Correlation in WM^{2}SNet In WM^{2}SNet, due to the spatial correlation among the individual sensor readings, it may not be necessary for every sensor node to transmit its data to the sink. Instead, a smaller number of sensor measurements might be adequate to communicate the event features to the sink within a certain reliability constraint. Clearly, a smaller number of nodes transmitting information reduces contention in the wireless medium resulting in decreased energy consumption. Consequently, energy consumed from both transmission of packets and collision penalties can be reduced drastically if the spatial correlation is exploited.

Vuran and Akyildiz (2006) proposed a correlation-based MAC protocol that exploits the spatial correlation in WM^{2}SNet. Based on the spatial correlation among sensor nodes, the developed MAC protocol collaboratively regulates medium access so that redundant transmissions from correlation neighbors are suppressed. Experimental results by Vuran and Akyildiz (2006) reveal that significant performance gains and energy savings are obtained by exploiting spatial correlation in WM^{2}Snets.

As observed in the previous section, due to the spatiotemporal correlation in WM^{2}SNet, the event estimation distortion at the sink decreases with increasing f conveying more information to the sink from the event area. Note that after a certain reporting frequency f, the observed event distortion cannot be further reduced. Therefore, a significant energy saving can be achieved by selecting small enough f that achieves the desired level of distortion and does not lead to an overutilization of the scarce sensor resources. On the other hand, any arbitrarily small f determined using Eq. (6.19) or Eq. (6.23) to achieve a certain distortion bound may not necessarily suffice since sensor samples may be lost in the network due to link errors and network disconnection. Moreover, very high values of f may endanger the event transport reliability by leading to congestion in the WM^{2}Snet.

In fact, to achieve a desired distortion level in the estimation with minimum energy expenditure, an event-to-sink reliable transport (ESRT) protocol is developed by Akan and Akyildiz (2005) based on the spatiotemporal correlation and *event-to-sink reliability* notion for WM^{2}SNet. The objective of this scheme is to achieve reliable event transport with minimum energy expenditure and congestion control by exploiting the correlation and the collaborative nature of the WM^{2}SNet. As observed from the performance evaluation results provided by Akan and Akyildiz (2005), spatiotemporal correlation conveyed in the physical characteristics of the phenomenon and deployment of the WM^{2}Snet can be exploited in addressing the energy-efficient reliable event communication problem in WM^{2}Snets.

6.4.4 Order-Optimal Data Aggregation in WM^{2}Nets[14]

6.4.4.1 Background

The design and analysis of wireless mesh networks differs from that of more general data communication networks, such as the Internet or wireless MANets, in that the

[14] Excerpt from the invited article "Order-optimal data aggregation in wireless mesh networks," *Richard J. Barton and †Rong Zheng (*ERC, Inc., ESC Group, NASA Johnson Space Center, Houston, TX 77058, USA, E-mail: richard.j.barton@nasa.gov; †Department of Computer Science, University of Houston, Houston, TX 77204, USA, E-mail: rzheng@cs.uh.edu).

predominant traffic patterns in a WM^{2}Net are many-to-one and one-to-many communication. The performance of WM^{2}Nets is thus characterized by the rate at which data can be disseminated from or aggregated to the AP (or IGW). The problem of the maximum sustainable rate at which each mesh can transmit data to the AP under a power constraint is called the *data aggregation problem*.

Two common ways for data traffic reduction are *data aggregation* and *data fusion*. Whereas in data aggregation mesh nodes aggregate data collected from other nodes in order to reduce the data traffic in the network (Krishnamachari et al., 2002), in data fusion, instead of reducing redundant information, nodes process the data locally before relaying it further.

Capacity bounds for the data aggregation problem have been established by Barriac et al. (2004) and Barton et al. (2005, 2005a). Barton et al. (2005a) investigate the capability of large-scale networks to measure and transport a two-dimensional stationary random field using nodes equipped with fixed scalar quantizers. As shown, as the density of nodes increases to infinity, the total number of bits transmitted to the sink also increases to infinity. At the same time, the single-receiver transport capacity of the network remains constant as the density increases. Barton et al. (2005) considers a WSNet and investigate the more general problem of computing and communicating a symmetric function of sensor measurements. As shown, for a certain class of functions, called divisible functions, the maximum rate at which the function f can be computed and communicated to the sink satisfies $\Theta\left(1/\log\left(|\mathcal{R}(f, n)|\right)\right)$, where n is the number of sensors in the network and $\mathcal{R}(f, n)$ is the range of the function f. Since computation of the identity function is equivalent to transporting all raw data, and the identity function is a divisible function with $\mathcal{R}(f, n) = |\mathcal{X}|^n$ for some $|\mathcal{X}| < \infty$, $\Theta(1/n)$ is a tight bound on the achievable throughput for each sensor.

Both of the studies discussed above assume a simplified protocol model for the wireless channel. In particular, each link has a fixed capacity of at most W, and transmissions between linked nodes are successful as long as other transmitters are sufficiently distant. This model clearly does not consider the time-varying or nondeterministic nature of channels. For example, a Raleigh fading model is often more appropriate for nodes dispersed over a large region in an environment subject to multipath propagation, and link capacity is more accurately modeled as a function of SINR. Nonetheless, the $\Theta(1/n)$ upper bound is not at all surprising, and it reflects the basic observation that for data aggregation in a multihop environment, the traffic load increases for nodes closer to the sink (AP), and the total achievable rate is limited by the maximum rate at which the sink can receive information from its neighbors.

Barriac et al. (2004) studied the transport capacity of many-to-one dense wireless networks subject to a total average power constraint. As shown, for nodes randomly placed on a sphere of unit radius, the transport capacity of $\Theta(\log(n))$ can be achieved as the number of nodes n grows to infinity. This result is used to derive necessary and sufficient conditions that characterize the set of observable random fields by dense networks.

Alternatively, one approach to improve the data aggregation rate is to employ cooperative communication techniques, in which multiple nodes in the network cooperate to deliver the same piece of information to the sink. Cooperative relay strategies (Barton and Zheng, 2006; Barton and Zheng, 2006a; Barton and Zheng, 2006b; EdelMann et al., 2001) and cooperative transmission strategies (El Gamal, 2005; Giridhar and Kumar, 2005; Gupta and Kumar, 2000) are two common cooperative communication techniques. The study that follows considers a new cooperative transmission strategy based on a technique called *time reversal communication* (TRC). As demonstrated, a rate of $\Theta(\log(n/n))$ is achievable using cooperative TRC whereas this rate is in fact order optimal for the data aggregation problem with a single-antenna sink.

6.4.4.2 Physical-Layer and Network-Layer Models

The study considers a WM²Net G_n consisting of a group of n nodes, $N = 1, 2, \ldots, n\}$ located on the plane at constant density ρ. Without loss of generality, let us set $\rho = 1$, and assume that nodes are placed on a regular grid. It is also assumed that all wireless links are baseband channels impaired by circularly symmetric, complex-valued additive white Gaussian noise (AWGN) with power spectral density N_0 as well as additive interference, which is also assumed to be Gaussian. Depending on the nature of the communication link being considered, two different channel propagation models are assumed thereafter:

Non-Cooperative Multihop Relay: In a non-cooperative multihop relay, the power is assumed to decay with distance at an exponential rate with path-loss exponent $\alpha > 2$. Hence, the maximum achievable rate (in bps) for communication from node i to node j in a non-cooperative multihop relay is given by $r_{ij} = B \log\left(1 + SINR_{ij}\right)$, where B denotes the shared bandwidth for all links on the network, P_i is the power transmitted by node i, ρ_{ij} is the distance between nodes i and j, I is the set of interfering users, and $SINR_{ij} = \frac{P_i \rho_{ij}^{-\alpha}}{BN_0 + \sum_{k \in I} P_k \rho_{ik}^{-\alpha}}$. Similarly, when common information is broadcast from node i in a □ to a set of nodes non-cooperative multihop relay, the maximum achievable rate for the broadcast is given by $r_i = \min_{j \in R} B \log(1 + SINR_{ij})\}$.

Cooperative TRC: An example of cooperative TRC is illustrated in Fig. 6.35. As shown, the cluster of network nodes identified as Group A cooperates to transmit a common data stream to the sink node identified as the receiver in figure. During a training phase, the sink node transmits a short sequence of training pulses, which are received from all the nodes in the cluster. After the transmission of the training sequence, the receiving nodes in the cluster independently perform pulse estimation in order to estimate the exact arrival time, duration, and shape of the received pulse. After completion of the training phase, information is transmitted from an arbitrary node in the cluster to the sink using the following cooperative TRC scheme. Whenever a node in the cluster has data to transmit to the sink node, the source node first disseminates the information to all of the other nodes in the cluster. The ensemble of nodes in the cluster then cooperate to transmit the information to the sink node by synchronously transmitting a stream of identical data symbols modulated onto the time reverse of their respective estimated received waveforms.

In modeling cooperative TRC links, all channels are characterized from a deterministic path-loss component with exponent $a > 2$ cascaded with a Raleigh fading channel. The fading channels are modeled as i.i.d. wide-sense-stationary-uncorrelated-scattering Rayleigh fading channels that are random but fixed for a very large number of packet intervals on the network. Based on the work of Marco et al. (2003), the following abstraction

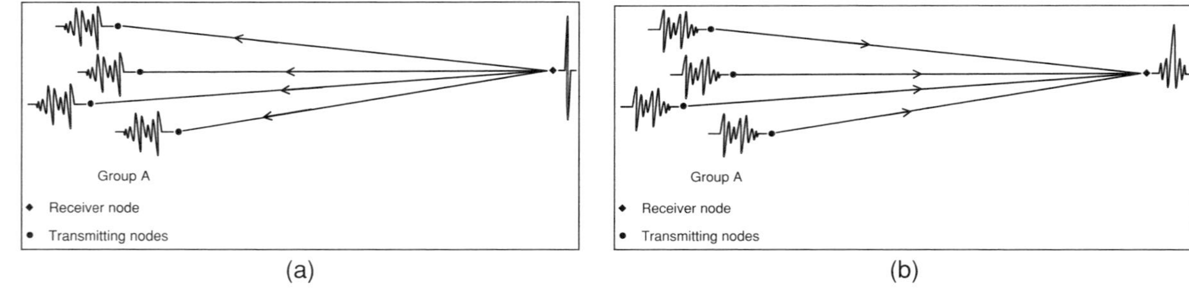

Figure 6.35 (a) Training for transmission from the cooperating cluster (Group A) to the sink (Receiver Node), (b) Transmission from the cooperating cluster to the sink.

to model TRC is introduced:

$$r_{Tj} = B \log\left(1 + \frac{\left(\sum_{i \in T} A_i \rho_{ij}^{-\alpha/2}\right)^2}{BN_0 + \sum_{k \in I} P_k \rho_{kj}^{-\alpha}}\right), \tag{6.24}$$

where $A_i = \sqrt{P_i}$ is the average amplitude of transmitted signal, and T the set of cooperative transmitters. The numerator term reflects coherent aggregation of peak signal strength at the receiver node from multiple transmitters whereas the sum term in the denominator corresponds to non-coherent aggregation of interference signals.

6.4.4.3 Asymptotic Data Aggregation Rate

To prove the main result, the following technical lemmas are considered.

> *Lemma 1:* Consider a grid with the data aggregation point O at the center, as illustrated in Fig. 6.36 below. Nodes are labeled in a 2-D coordinate system with O at the origin. To route from a node u located at (x, y) to the aggregation point O, the following rules apply.

A Voronoi diagram is constructed on the grid. Given a set of n nodes, a Voronoi tessellation is the partitioning of a plane into convex polygons such that each polygon contains exactly one generating point, and every point in a given polygon is closer to its generating point than to any other. For nodes on a grid, the Voronoi diagram is also a grid (shown in dotted lines in Fig. 6.36).

Let D_{uO} be a straight line from u to point O. Let $\{u = u_0, u_1, u_2, \ldots, u_k = O\}$ be the set of nodes whose Voronoi cells intersect with D_{uO}.

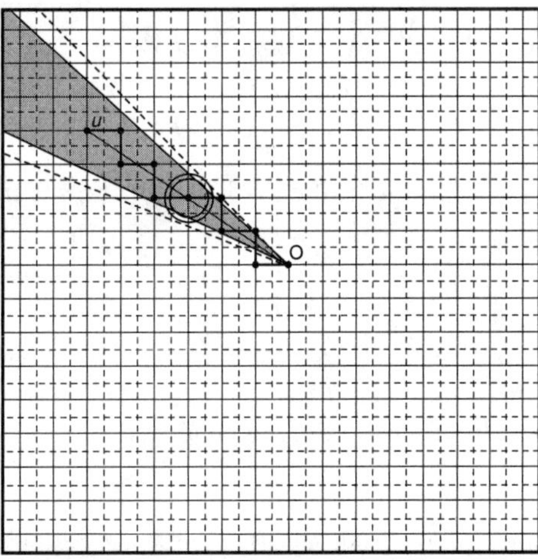

FIGURE 6.36 Routing on a grid.

The nodes $\{u_1, u_2, \ldots, u_{k-1}\}$ form the sequence of relays[15] from u to O.

Since data are relayed through Voronoi cells with common edges, communication takes place between neighboring nodes[16] on the grid. Since each node has a unique path to O, data aggregation follows a tree rooted at O.

Let λ be the rate of data generated by each WM^{2}Net node. The total amount of traffic $T(x, y)$ that a node u at (x, y) must relay (including its own data) is bounded by:

$$\left\lfloor \frac{n}{\sqrt{x^2+y^2}} \cdot \frac{\sqrt{2}}{4} \right\rfloor < T(x, y) < \left\lceil \frac{n}{\sqrt{x^2+y^2}} \right\rceil \tag{6.25}$$

for $1 \leq x^2 + y^2 \leq n$.

Proof: Consider two concentric disks centered at u with radius $\sqrt{2}/2$ and 1, respectively. Clearly, the Voronoi cell containing u is fully contained in the outer disk and completely contains the inner disk. Therefore:

1. The sufficient condition for a node v to route through u is that v is further from O than u and falls in the cone formed by O, the two tangent lines of the inner disk, and the boundary of the $\sqrt{n} \times \sqrt{n}$ grid (indicated by the shaded area in Fig. 6.36), denoted by A.
2. The necessary condition set for a node v to route through u is that v is further from O than u and falls in the cone formed by O, the two tangent lines of the outer disk, and the boundary of the $\sqrt{n} \times \sqrt{n}$ grid, which we denote by B.

Hence, to determine the traffic load at node u, it is equivalent to compute the number of nodes falling into the regions A and B defined above. For ease of computation, we simplify the boundary of $\sqrt{n} \times \sqrt{n}$ grid using the circumscribed and inscribed circles. We have,

$$\text{Area}(A) > \left(\frac{n}{x^2+y^2} - 1\right)\sqrt{x^2+y^2-\tfrac{1}{2}} \cdot \frac{\sqrt{2}}{4}, \text{ and}$$

$$\text{Area}(B) < \left(\frac{2n}{x^2+y^2} - 1\right)\sqrt{x^2+y^2-1} \cdot \frac{1}{2}.$$

When $1 \leq x^2 + y^2 \leq n$, the traffic load $T(x, y)$ at u satisfies:

$$\left\lfloor \frac{n}{\sqrt{x^2+y^2}} \cdot \frac{\sqrt{2}}{4} \right\rfloor \lambda < T(x, y) < \left\lceil \frac{n}{\sqrt{x^2+y^2}} \right\rceil \lambda.$$

This result demonstrates that for nodes placed at constant distance from the sink, the traffic loads differ *at most* by a constant factor (i.e., $\sqrt{2}/4$) under the proposed routing strategy. In other words, the proposed routing is load balanced. ■

[15] In cases where node u lies on the diagonals of the square area, we tilt the line D_{uO} slightly to avoid intersections with the Voronoi cells at the grid points.

[16] Neighboring nodes are those that are separated by unit Manhattan distance.

The following lemma describes the transmission achievable in non-cooperative unicast transmission.

Lemma 2: (Lemma 4 provided by Zheng, 2006): Under the physical-layer model discussed above for non-cooperative multihop relay, for any integer $k \geq 0$, there exists a TDMA scheduling scheme such that one node per square of edge length l can broadcast concurrently to all nodes located within a radius of k squares (in Manhattan distance) with fixed rate $R(l, k)$ satisfying

$$R(l, k) \geq \frac{B}{4(k+1)^2} \log\left(1 + \frac{P_{\max}}{BN_0 [l(k+1)]^{\alpha} + K_1 P_{\max}}\right) \tag{6.26}$$

where K_1 is a constant independent of k and l.

Using these lemmas, the main result on the achievable data aggregation rate is then established. To maximize the rate of data fusion at the sink, the $\sqrt{n} \times \sqrt{n}$ grid is divided into three areas as shown in Fig. 6.37.

In Areas I and III, data are aggregated using non-cooperative multihop relay on sink trees. In Area II, nodes are organized into $R \times R$ square clusters. Data are first broadcast among all nodes inside the cluster (termed *intra-cluster communication*), then all nodes in a cluster cooperatively perform TRC towards the sink (termed *inter-cluster communication*). Communication in each distinct area is carried out in *nonoverlapping time slots* of equal length. This allows different communication strategies to be used without interfering with one another. The rate sustainable for each node in the network is determined by the minimum of the achievable rates in each area.

In the following, let λ be the rate achievable from each node. Let us consider the collection of clusters comprising Area II with centers at distance no greater than $d' = d + R/\sqrt{2}$ from the sink. Each cluster is a square with sides of length R. The number

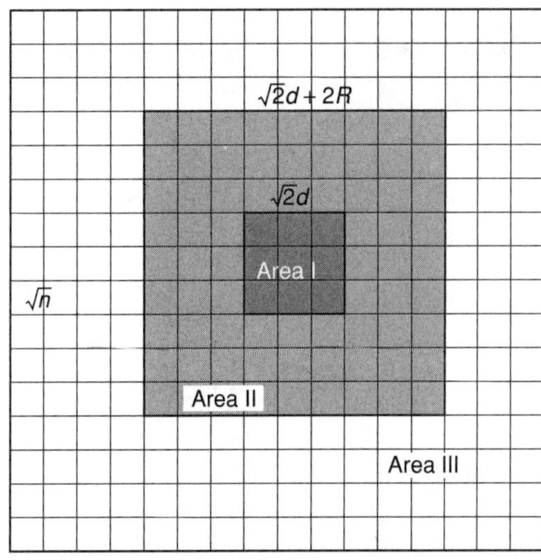

FIGURE 6.37 Partition of the network into three distinct areas.

of clusters in Area II is $M = 4(\sqrt{2}d + R)/R$ and the number of nodes in each cluster is $m = R^2$, whereas the number of nodes in Area I is $n_1 = 2d^2$, and the number of nodes in Area III is $n_3 = n - 2(\sqrt{2}d + 2R)^2$. We are primarily concerned with the rate of growth in achievable rate as $m, n \to \infty$ as well as with the noise-limited rather than interference-limited regime of network operation. Hence, Eq. (6.24) can be rewritten in more compact notation as

$$r_{\mathrm{TRC}} \approx B \log\left(1 + \frac{m^2 d'^{-\alpha} P_{\max}}{B N_0 K'}\right) = B \log\left(1 + \frac{R^4 d'^{-\alpha} P_{\max}}{B N_0 K'}\right),$$

where K' is appropriately defined. Similarly, the result of Lemma 2 can be rewritten as

$$R(l, k) = R(k) \geq \frac{B}{4(k+1)^2} \log\left(1 + \frac{P_{\max}}{B N_0 K''}\right),$$

where K'' is appropriately defined. We can now state the following theorem.

Theorem 1: If $2 < \alpha < 4$, the achievable data aggregation rate for the network $\mathcal{G}_n$, with $n \to \infty$, is $\Theta(\log n/n)$.

Proof: To show that the data aggregation rate is also $\Omega(\log n/n)$ for $2 < \alpha < 4$, we first derive the achievable rates in each of the three areas of the network. ■

Achievable Rate in Areas I and III In Area III, nodes forward their data towards the sink according to the above routing strategy until a node in Area II (a *root node*) is reached. Letting $\sqrt{x^2 + y^2} = (\sqrt{2}d + 2R)/2$ in Eq. (6.24), it follows from Lemma 4 that the traffic load for a node close to the border of Area II is upper bounded by $\lceil 2n/(\sqrt{2}d + 2R) \rceil \lambda$.

By Lemma 3 with $d' \approx d$, each root node can receive information from its closest neighbor at rate $R(1) = \frac{B}{16} \log\left(1 + \frac{P}{B N_0 K''}\right)$. Hence, since communication in each area is carried out at independent time slots, the sustainable rate in Area III satisfies

$$\left\lceil \frac{2n}{\sqrt{2}d + 2R} \right\rceil \lambda \geq \frac{B}{48} \log\left(1 + \frac{P}{B N_0 K''}\right). \tag{6.27}$$

Similarly, in Area I, following the same argument, we have

$$(2d - R)^2 \lambda \geq \frac{B}{48} \log\left(1 + \frac{P}{B N_0 K''}\right). \tag{6.28}$$

Achievable Rate in Area II From Corollary 1, we know that each cluster of size R^2 can transmit at rate $\frac{\Delta B}{\Delta + B} \log\left(1 + \frac{R^4 (d')^{-\alpha} P}{\frac{\Delta B}{\Delta + B} N_0 K'}\right)$ to a node at distance d'. Using TDMA to separate transmissions within each cluster as well as within each of the three areas, the *effective rate* of inter-cluster communication for each cluster is thus

$r_{\text{inter}} = \frac{\Delta B}{3M(\Delta+B)} \log\left(1 + \frac{R^4(d')^{-\alpha} P}{\frac{\Delta B}{\Delta+B} N_0 K'}\right)$[17]. On the other hand, assuming that intra-cluster and inter-cluster communication are placed in different time slots, allowing thus concurrent intra-cluster communications, the intra-cluster broadcast rate is given by $r_{\text{intra}} = \frac{B}{48} \log\left(1 + \frac{P}{BN_0K''}\right)$, as in Area I. If $M \to \infty$, it follows that the total effective achievable rate for each cluster is given by

$$\frac{r_{\text{intra}}\, r_{\text{inter}}}{r_{\text{intra}} + r_{\text{inter}}} \approx r_{\text{inter}} = \frac{\Delta B}{3M(\Delta + B)} \log\left(1 + \frac{R^4\,(d')^{-\alpha}\, P}{\frac{\Delta B}{\Delta+B} N_0 K'}\right)$$

Now, since nodes in Area II must transport all of the traffic from Area III to the sink, the amount of traffic that each cluster must carry is no greater than

$$\lambda R^2 + \left\lceil \frac{2n}{\sqrt{2}d + 2R} \right\rceil 2R\lambda \approx \lambda R^2 + \frac{16n\lambda}{M+4},$$

where the first term corresponds to the amount of traffic generated from the cluster itself, and the second term corresponds to the aggregated load from Area III. If we let $d = n^\beta$, $R = n^\gamma$, with $0 < \gamma < \beta < \frac{1}{2}$, then as $n \to \infty$, we have $M \approx 4\sqrt{2}n^{\beta-\gamma}$. It follows that the sustainable rate in Area II satisfies

$$\lambda R^2 + \frac{16n\lambda}{M+4} \approx \frac{16n}{M}\lambda \geq \frac{\Delta B}{3M(\Delta + B)} \log\left(1 + \frac{R^4(d')^{-\alpha}P}{\frac{\Delta B}{\Delta+B} N_0 K'}\right),$$

or equivalently

$$\lambda \geq \frac{\Delta B}{48n\,(\Delta + B)} \log\left(1 + \frac{R^4(d')^{-\alpha}P}{\frac{\Delta B}{\Delta+B} N_0 K'}\right) \qquad (6.29)$$

as $n \to \infty$.

Achievable Rate for the Network Comparing Eq. (6.27) to Eq. (6.29) and noting that $d' \approx d$ as $n \to \infty$, we see that the entire network can sustain a rate of

$$\Omega\left(\frac{\Delta B}{48n\,(\Delta + B)} \log\left(\frac{1 + R^4(d')^{-\alpha}\, P}{\frac{\Delta B}{\Delta+B} N_0 K'}\right)\right).$$

Furthermore, if $\alpha < 4$ and we choose $0 < \gamma < \beta < \frac{4}{\alpha}\gamma$ with $\beta < \frac{1}{2}$, then we can achieve a sustainable rate of $\Omega(\log n/n)$ bit/s per node, as claimed.

---

[17] In fact, since we are using TDMA to separate the cluster transmissions, there will be no interference at the sink, and the constant K' in this expression actually satisfies $K' = 1$. Since this does not affect the result, we continue to use the more general expression.

6.4.5 Transmit-Diversity Techniques for MIMO–OFDM Mesh Networks[18]

6.4.5.1 Introduction

MIMO and OFDM are two technologies that help wireless communications networks to achieve the high date-rate and extended coverage requirements of future wireless applications. MIMO–OFDM systems can provide diversity gains in the space and frequency domains, which help system designers overcome the challenges of fading in wireless channels. In this study, we focus on transmit-diversity design with partial channel feedback at the transmitter.

The following notations are used: I_N denotes the $N \times N$ identity matrix, $\mathbf{1}_{\mathbf{n}\times\mathbf{m}}$ denotes an $n \times m$ all one matrix, $\mathbf{0}_{\mathbf{n}\times\mathbf{m}}$ denotes a $n \times m$ all zero matrix, the superscripts T, H, and $*$ represent the transpose, conjugate transpose and element-wise conjugation respectively, and $\otimes$ represents the tensor product. Finally, vec($\mathbf{C}$) transforms a matrix $vec(C) = [C_1^{\mathrm{T}} \ldots C_M^{\mathrm{T}}]^{\mathrm{T}}$, where $\boldsymbol{c}_i$ is the i-th column.

6.4.5.2 System Model

In this section, we introduce the MIMO–OFDM system model used throughout the study. We consider a MIMO frequency selective fading channel model with M_{t} transmit antennas and M_{r} receive antennas. OFDM is utilized, as it provides an attractive means to lower the complexity of equalization and decoding in a frequency selective environment (Cimini, 1985), and it has N subcarriers. The multipath channel has L significant delay paths between each transmit–receive antenna pair. The path gains for different delays are assumed to be independent. The channel impulse response from transmit antenna i to receive antenna j can be modeled as

$$h_{ij}(\tau) = \sum_{l=0}^{L-1} \alpha_{ij}(l)\delta(\tau - \tau_l) \tag{6.30}$$

where τ_l is the delay of the l-th path and $\alpha_{ij}(l)$ is the complex path gain between transmit antenna i and receive antenna j. The $\alpha_{ij}(l) \sim CN(0, \beta_l^2)$ are modeled as zero mean, circularly symmetric complex Gaussian random variables with variance β_l^2. The channel gains are assumed jointly Gaussian. The time delay τ_l and the variance β_l^2 are the same for each transmit receive link (Li and Rus, 2002). The power of the L paths are normalized such that $\sum_{l=0}^{L-1} \beta_l^2 = 1$. From Eq. (6.30), the frequency response of the channel is given by

$$\widetilde{H}_{ij}(f) = \sum_{l=0}^{L-1} \alpha_{ij}(l)e^{-j2\pi f \tau_l} \tag{6.31}$$

where $j = \sqrt{-1}$. We consider MIMO–OFDM systems with spatial correlation at the transmitter side. Receive antennas are assumed independent and with the same fading

[18] Excerpt from the invited article "Transmit-diversity techniques for MIMO–OFDM mesh networks," Ahmed K. Sadek, Senior Engineer, Corporate Research and Development, Qualcomm Incorporated, San Diego, CA 92121, USA. E-mail: asadek@qualcomm.com

statistics, that is,

$$E[a_{ij}(l)a_{ij}^*(l)] = \beta_l^2 r(i-p)\delta(j-q) \tag{6.32}$$

where $r(i-p)$ is the spatial correlation factor between transmit antennas i and p, and $\delta(\cdot)$ is the delta function. This model generally arises when the transmitter is unobstructed with a lot of scatterers, while the receiver is surrounded by a rich scattering environment.

At the transmitter, the input bits to the SF-beamformer coder are divided into b-bits long, which are then mapped into a SF-beamformer symbol. Each SF-beamformer symbol can be expressed as an $M_t \times N$ matrix

$$\mathbf{B} = [\mathbf{b}(0)\,\mathbf{b}(1)\ldots\mathbf{b}(N-1)], \tag{6.33}$$

where $b(n) = [b_1(n)\ \ b_2(n)\ldots\ b_M(n)]^T$ is an $M_t \times 1$ column vector. The SF-beamformer symbol $\mathbf{B}$ is assumed to satisfy the energy constraint $E[||B||_F^2] = NM_t$, where $E[\cdot]$ denotes expectation, and $||B||_F$ is the Frobenius norm of B. The OFDM transmitter applies IFFT to each row of the matrix B. By appending a cyclic prefix, it transmits the OFDM symbol corresponding to the i-th row of B at the i-th antenna. We stress that till this point we are assuming a joint design of the SF-beamformer symbol B and no specific relation between the SF code and the beamformer is assumed.

At the receiver, after matched filtering, removing the cyclic prefix, and applying FFT, the received signal at the n-th subcarrier at receive antenna j is given by

$$y_j(n) = \sqrt{\frac{\rho}{M_t}}h_j^T(n)b(n) + u_j^{(n)} \tag{6.34}$$

where $h_j(n) = [H_{1j}(n)\,H_{2j}(n)\ldots H_{M_{tj}}(n)]^T$; in which $H_{ij}(n) = \sum_{l=0}^{L-1}\alpha_{ij}e^{-j2\pi n\Delta f\tau_l}$ represents the channel frequency response at the n-th subcarrier between transmit antenna i and receive antenna j, where $\Delta f = 1/T$ is the subcarrier frequency separation, and T is the OFDM symbol period. In Eq. (6.34), $b(n)$ is the n-th column of the matrix B and represents the channel symbol vector transmitted on the n-th subcarrier. The term $u_j(n) \sim CN(0,1)$ in Eq. (6.34) denotes the additive white circularly symmetric complex Gaussian noise, with zero mean and unit variance, at the n-th subcarrier at receive antenna j. Thus, the average SNR at each receive antenna is just ρ.

Performance Analysis and General Beamformer Design In this section, we analyze the performance of the MIMO–OFDM system with arbitrary channel correlation conditions at the transmitter as specified in the previous section. We derive the average pairwise error probability, which will give insights into the factors that affect the design of the beamformer. Then, we formulate a general optimization problem to design a joint SF-beamformer symbol. The goal is to design a matrix B that minimizes the pairwise error probability of the system under the energy constraint.

System Performance Analysis The pairwise error probability between two channel symbols B and $\widetilde{B}$ for a given channel realization can be upper bounded by (Sadek, forthcoming)

$$P_r(B \to \widetilde{B}|H) \le \exp\left(-\frac{\rho}{4M_t}||\Phi||^2\right) \tag{6.35}$$

where Φ is an $NM_r \times 1$ vector given by

$$\Phi = H[\text{vec}(b) - \text{vec}(\widetilde{B})] \tag{6.36}$$

Since the channel coefficients are jointly Gaussian, the vector Φ, for fixed channel symbols, has a Gaussian distribution with zero mean and covariance matrix $\mathbf{R}_\Phi = E[\Phi\Phi^H]$, which is of size $NM_r \times NM_r$. Since N is usually greater than LM_t, the matrix $\mathbf{R}_\Phi$ can be shown to be rank deficient. Averaging the pairwise error probability over all channel realizations we get

$$P_r(B \to \widetilde{B}) \le \left(\frac{\rho}{4M_t}\right)^{-r(\mathbf{R}_\Phi)} \left(\prod_{i=0}^{r(\mathbf{R}_\Phi)-1} \mu_i(\mathbf{R}_\Phi)^{-1}\right) \tag{6.37}$$

where $r(\mathbf{R}_\Phi)$ and $\mu_i(\mathbf{R}_\Phi)$ are the rank and the i-th eigenvalue of the covariance matrix $\mathbf{R}_\Phi$, respectively. The diversity gain of the system is given by $r(\mathbf{R}_\Phi)$, and the coding gain is given by the product term $\prod_{i=0}^{r(\mathbf{R}_\Phi)-1} \mu_i(\mathbf{R}_\Phi)$. According to Eq. (6.37), the performance of the system is mainly determined by the matrix $\mathbf{R}_\Phi$, which contains information about the SF-beamformer symbol and the spatial correlation structure of the channel. In this study, we assume spatial correlation at the transmitter side, and uncorrelated fading with identical statistics at the receiver (Eq. 6.32).

Theorem 1 (*Sadek, forthcoming*): The covariance matrix $\mathbf{R}_\Phi$ can be decomposed as

$$\mathbf{R}_\Phi = I_{M_r} \otimes F\, diag\,[R_{a,0}, R_{a,1}, \ldots, R_{a,L-1}]\, F^H, \tag{6.38}$$

where $R_{a,l}$ and F are defined as follows:

$$F = \left[D^{\tau_0}(B - \widetilde{B})^T \ldots D^{\tau_{L-1}}(B - \widetilde{B})^T\right] \tag{6.39}$$

$$R_{a,l} = E\left[a_j(l)\, a_j^H(l)\right] \tag{6.40}$$

Theorem 1 determines the covariance matrix $\mathbf{R}_\Phi$ as a function of the spatial correlation structure of the channel and the SF-beamformer symbol B. This theorem will serve as a basis for the design of the beamformer. In the following, we formulate the general optimization problem to jointly design a SF-beamformer symbol that minimizes the average pair-wise error probability under the energy constraint.

We try to jointly design a general SF-beamformer matrix $\mathbf{B}$ that minimizes the system pairwise error probability Eq. (6.37) with the energy constraint $E[||B||_F^2] = NM_t$. More specifically, the optimization problem can be stated as follows

$$\text{Min}\left(\frac{\rho}{4M_t}\right)^{-r(\mathbf{R}_\Phi)} \left(\prod_{i=0}^{r(R_\Phi)-1} \mu_i(\mathbf{R}_\Phi)^{-1}\right), \qquad s.t.\quad E\left[||B||_F^R\right] = NM_t$$

where the minimization is over all possible SF-beamformer symbol pairs (minimize the worst case codewords pair) and the matrix $\mathbf{R}_\Phi$ is specified in Eq. (6.8). We emphasize at this point that the above design criteria for SF-beamformer are general in the sense that we do not impose any structure on the SF-beamformer B. We try to design a SF-beamformer matrix that matches to the channel statistical information available at the transmitter. If there is no spatial correlation at the transmitter side, that is, the spatial correlation matrices $\mathbf{R}_{\alpha,l}$ are identity, the above conditions reduce to the design criteria of SF codes (Su et al., 2003).

The general optimization problem is difficult to tackle analytically. To overcome this problem, we will adopt another transmitting scheme in which a SF code is already designed to achieve full diversity for a spatial correlation-free channel, and then we try to design a beamformer to match to the channel correlation matrix. The beamformer can be thought of as a linear transformation applied to the SF code in order to improve its performance under the CSI available at the transmitter. In particular, for any fixed SF code, we try to design a beamformer **W** that matches to the channel covariance structure. Thus, the SF-beamformer matrix **B** is obtained via applying a linear transformation on the SF code. This approach of splitting the transmitter design problem into a predesigned SF code and a beamformer is similar to the one suggested by Jöngren et al. (2002) for STC over flat fading channels.

In the sequel, we denote the SF code by a $M_t \times N$ matrix **C**. The linear transformation, or beamformer **W**, can take various forms, for example: (1) $\text{vec}(\mathbf{B}) = \mathbf{W}\text{vec}(\mathbf{C})$ in which the beamformer matrix **W** is of size $N M_t \times N M_t$; (2) $\mathbf{B} = \mathbf{WC}$ in which the beamformer **W** is of size $M_t \times M_t$. In general, we represent the relation between the SF-beamformer symbol B and the SF-code **C** as follows

$$f(\mathbf{B}) = Wf(\mathbf{C}), \tag{6.41}$$

where $\mathbf{f}(\cdot)$ is a function that can, for example, take the form $f(\mathbf{B}) = \text{vec}(\mathbf{B})$ or $f(\mathbf{B}) = \mathbf{B}$. Note that Eq. (6.41) is a general representation of all possible linear transformations between **C** and **B**. Since we can think of $\mathbf{f}(\cdot)$ as a rearrangement of **B**, we can write **B** as a product of a function of **W** and a function of **C** as follows

$$\mathbf{B} = \text{g}(\mathbf{W})q(\mathbf{C}), \tag{6.42}$$

where $\mathbf{g}(\cdot)$ is an $M_t \times K$ matrix, $\text{q}(\cdot)$ is a $K \times N$ matrix, and K depends on the function $\text{f}(\cdot)$.

Substituting Eq. (6.42) into Eq. (6.38), we get

$$R_\Phi = I_{M_r} \otimes \hat{F}_{diag}[g^{\text{T}}(W)R_{\alpha,0}g^*(W),\ g^{\text{T}}(W)R_{\alpha,1}g^*(W), \ldots,\ g^{\text{T}}(W)R_{\alpha,L-1}g^*(W)]\,\hat{F}^{\text{H}} \tag{6.43}$$

where

$$\hat{F} = [D^{\tau_0}(q(C) - q(\tilde{C}))^{\text{T}} \ldots D^{\tau_{L-1}}(q(C) - q(\tilde{C}))^{\text{T}}] \tag{6.44}$$

In order to apply the design criteria we need to find the rank and eigenvalues of the matrix $\mathbf{R}_\Phi$ in terms of the beamformer matrix **W**. In order to simplify the notations, let the $LK \times LK$ matrix $\hat{R}$ denote the block diagonal matrix in Eq. (6.43) as

$$\tilde{R} = diag[g^{T}(W)R_{\alpha,0}g^{*}(W), g^{T}(W)R_{\alpha,1}g^{*}(W), \ldots, g^{T}(W)R_{\alpha,L-1}g^{*}(W)] \tag{6.45}$$

Then, the rank of $\mathbf{R}_{\Phi}$ can be given by $r(\mathbf{R}_{\Phi}) = M_{r}\, r(\hat{F}\tilde{R}\hat{F}^{H})$. The matrix $\hat{F}$ is of size $N\times LK$, and we assume that $LK \leq N$, which is typically true in OFDM systems as N is usually designed much larger than LM_{t}. Assuming that the SF code is designed to achieve full diversity in the case of no spatial correlation, we rewrite $\hat{F}\tilde{R}\hat{F}^{H}$ after row and column reordering in the form

$$J = \left[\frac{\hat{F}_1}{\hat{F}_2}\right]\tilde{R}\left[\hat{F}_1^{H}\ \hat{F}_2^{H}\right] = \begin{bmatrix} \hat{F}_1\tilde{R}\hat{F}_1^{H} & \hat{F}_1\tilde{R}\hat{F}_2^{H} \\ \hat{F}_2\tilde{R}\hat{F}_1^{H} & \hat{F}_2\tilde{R}\hat{F}_2^{H} \end{bmatrix}, \tag{6.46}$$

where J is the reordered matrix, $\hat{F}_1$ is of size $LK \times LK$ and is full rank, and the matrix $\hat{F}_2$ takes the rest of the matrix. Since the ordered singular values of a matrix are not smaller than the corresponding singular values of any square submatrix obtained by deleting equal number of rows and columns of the original matrix (Horn and Johnson, 1991), we get

$$\mu_i(\hat{F}\tilde{R}\hat{F}^{H}) \geq \mu_i(\hat{F}_1\tilde{R}\hat{F}_1^{H}), \tag{6.47}$$

where $\mu_i()$ denotes the i-th eigenvalue of a matrix and are ordered in nonincreasing order.

The eigenvalues of $\hat{F}_1\tilde{R}\hat{F}_1^{H}$ are given by

$$\mu_i(\hat{F}_1\tilde{R}\hat{F}_1^{H} = \theta_i\mu_i(\tilde{R}) \tag{6.48}$$

where μ_i is a nonnegative real number such that $\mu_{min}(\hat{F}_1\hat{F}_1^{H}) \leq \theta_i \leq \mu_{max}(\hat{F}_1\hat{F}_1^{H})$, which follows by Ostrowski (Horn and Johnson, 1985), and μ_{min} and μ_{max} denote the smallest and largest eigenvalues, respectively. Applying Ostrowski's theorem along with Eq. (6.47), we can find the rank of the matrix $\hat{F}\tilde{R}\hat{F}^{H}$ as follows:

$$r(\hat{F}\tilde{R}\hat{F}^{H}) = r(\tilde{R}) = \sum_{i=0}^{L-1} r(g^{T}(W)\,R_{\alpha,l}g^{*}(W)), \tag{6.49}$$

where the second equality comes from the block diagonal structure of $\tilde{R}$. Similarly, the eigenvalues of the matrix $\hat{F}\tilde{R}\hat{F}^{H}$ can be lower bounded as follows:

$$\mu_i(\hat{F}\tilde{R}\hat{F}^{H}) \geq \mu_{min}(\hat{F}_1\hat{F}_1^{H})\mu_i(\tilde{R}) \tag{6.50}$$

Note that maximizing the coding gain of the system corresponds to maximizing the product of the nonzero eigenvalues of the matrix $\mathbf{R}_{\Phi}$, which is equivalent to maximizing the product of the nonzero eigenvalues of $\hat{F}\widetilde{R}\hat{F}^{\mathrm{H}}$. From Eq. (6.50), this product can be lower bounded as follows:

$$\prod_{i=1}^{r(\widetilde{R})} \mu_i(\hat{F}\widetilde{R}\hat{F}^{\mathrm{H}}) \geq \gamma \prod_{i=1}^{r(\widetilde{R})} \mu_i \widetilde{R} \tag{6.51}$$

where γ is a constant that depends on $\mu_{\min}(\hat{F}_1 \hat{F}_1^{\mathrm{H}})$. If the matrix $\widetilde{R}$ is full rank, the product of its eigenvalues corresponds to its determinant. Thus, the goal now is to maximize the determinant of the matrix $\widetilde{R}$ under the energy constraint on the SF-beamformer symbol. According to Hadamard's inequality (Horn and Johnson, 1985), the determinant of the matrix $\widetilde{R}$ is upper bounded by the product of its diagonal elements, and the upper-bound is achieved if and only if $\widetilde{R}$ is diagonal. More specifically

$$\det(\widetilde{R}) \leq \prod_{i=1}^{LK} \widetilde{R}_{ii} \tag{6.52}$$

where $\widetilde{R}_{ii}$ is the i-th diagonal element of the matrix $\widetilde{R}$. Hence, the equality holds when the matrix $\widetilde{R}$ is diagonalized and this can be achieved only if the L block diagonal entries of the matrix $\widetilde{R}$ are diagonalized (Eq. 6.45). This corresponds to choosing **W** to diagonalize $\mathrm{g}^{\mathrm{T}}(\mathrm{W})\mathrm{R}_{\alpha,l}\mathrm{g}^*(\mathrm{W})$, for all $0 \leq l \leq L-1$.

A beamformer that achieves the upper bound in Eq. (6.52) is considered optimal. However, according to Eq. (6.43), irrespective of the form of the function $\mathrm{g}(\mathbf{W})$ the same beamformer should match to the covariance matrices of all the L delay paths simultaneously in order to achieve the upper bound. This can not be achieved, in general, except for the special cases when all of the L delay paths have the same spatial correlation matrix, or when $L = 1$ which corresponds to the flat fading case. As a result, it is very difficult, if not impossible, to find a closed form solution for the optimal beamformer.

Therefore, in order to solve the optimization problem we must, in general, employ numerical search techniques for the beamformer matrix, which will be exhaustive. In order to provide some insights, we will render to suboptimal solutions for the problem. For simplicity of exposition, we will adopt the conventional definition of the beamformer, which corresponds to $\mathbf{B} = \mathbf{WC}$ in the rest of this study.

6.4.5.3 Suboptimal Designs of Beamformers

In this section, two approaches for designing the beamformer are proposed. The proposed approaches, although suboptimal, are well motivated by the derived performance criteria and the understanding of the underlying physics of the problem. Intuitively, the beamforming is done in the eigenspace of the channel to whiten the effect of the spatial correlation by beamforming in the directions corresponding to the channel eigenvectors, this corresponds to diagonalizing the channel spatial correlation matrices. The power loading along the eigenbeams is proportional to the channel eigenvalues associated with the eigenbeams, that is, more power is sent along directions with better channel conditions.

Eigenvalue Selection Scheme In this subsection, we design the beamformer jointly for all subcarriers, and propose the eigenvalue selection scheme.

The optimal beamformer **W** should satisfy the upper bound in Eq. (6.52). Two main components constitute any beamformer: the directions along which the information is being sent, and the power loading along each of these directions. We represent the beamformer **W** in the following way:

$$\mathbf{W} = \mathbf{U}\Gamma, \tag{6.53}$$

where the i-th column in **U** corresponds to the i-th beamforming direction, and Γ is a diagonal matrix with the i-th diagonal element representing the power loading along this direction. According to the beamformer definition in Eq. (6.53), the beamformer directions U should be designed to simultaneously whiten the effects of the spatial correlation matrices $\{R_{\alpha,l}\}_{l=0}^{L-1}$ in order to satisfy the upper bound in Eq. (6.52). However, as discussed before, this can be achieved only for very special cases. To solve the general optimization problem we need to employ exhaustive search techniques.

One intuitive, but suboptimal approach to overcome this problem is to select from among the LM_t-dimensional space an M_t space and design the beamformer to match this smaller space. One can choose this smaller space according to different criteria. In the eigenvalue selection approach, we choose the largest M_t eigenvalues λ from the LM_t eigenvalues available from the eigendecomposition of the L covariance matrices, and the corresponding M_t eigenvectors $\mathbf{V}_i$. The beamformer is then designed to transmit in the directions of these eigenvectors and the power loading is distributed proportional to the eigenvalues along these directions. The rationale behind doing the power loading in this way is that, in general, the available power should be distributed according to the channel conditions, that is, more power should be allocated to channels with better quality.

The algorithm can be summarized in the following steps: (Alamouti, 1998) Let the eigendecomposition of the spatial correlation matrix at the l-th path be given by $R_{\alpha,l} = V_l \Lambda_l V_l^{\mathrm{H}}$, where $0 \leq l \leq L-1$ (Palomar et al., 2003). Choose the largest M_t eigenvalues and the corresponding eigenvectors from the LM_t available eigenvalues and eigenvectors in Λ_l and V_l, $l = 0, 1, \ldots, L-1$ (Narula et al., 1998). Arrange the M_t selected pairs in matrix format as follows:

$\Lambda = dia\log(\lambda_1, \ldots, \lambda_{M_t})$, $V = dia\log(v_1, \ldots, v_{M_t})$ (Visotsky and Madhow, 2001). The beamformer **W** is determined as

$\mathbf{W} = \mathbf{U}\Gamma$, in which

$$\mathbf{U} = \mathbf{V}^*, \Gamma = dia\log(\sigma_1, \ldots, \sigma_{M_t})$$

and $\sigma_\iota^2 = \frac{\lambda_\iota}{\sum_{l=0}^{L-1} \lambda_\iota}$.

The SF-beamformer symbol B is required to satisfy the energy constraint $E[||B||_F^2] = NM_t$ as stated before, then we can normalize the resultant SF-beamformer symbol to satisfy $B = \frac{WC}{||wc||_F}\sqrt{NM_t}$, which guarantees that the energy of the SF-beamformer symbol **B** does not exceed NM_t.

It can be expected that choosing the directions with the largest eigenvalues, that is, with the most reliable channel conditions, enhances the coding gain. However, since the directions associated with the beamformer belong to different eigenspaces (they belong to different eigendecompositions), they are no more orthogonal and the matrix **V** is not, in general, full rank. Accordingly, full diversity is not guaranteed in the eigenvalue selection approach. We will explore this more in another approach described next.

Eigenspace Selection Scheme In this scheme, we try to jointly select the eigenvalues and eigenvectors, not only based on coding gain, but also based on the diversity order of the system. The criterion that we suggest is maximizing the volume occupied by the beamformer matrix, which is given by the absolute value of the determinant of the beamformer matrix

$$W = \arg\max_{\substack{\lambda_i, v_i \\ i=1,\ldots,L_t M}} \mid \det(W), \tag{6.54}$$

To understand the intuition behind using this cost function, let us investigate the $M_t = 2$ case, in which the beamformer can be written as follows:

$$\mathbf{W} = [u_1 u_2] diag[\sigma_1, \sigma_2], \tag{6.55}$$

In this case, the criterion is proportional to the area spanned by the matrix **W**. This area is given by $\sigma_1, \sigma_2 \sin(<u_1;u_2>)$, where u_i and σ_ι are the i-th eigenbeam and associated allocated power respectively, and $< \cdot, \cdot >$ denotes the angle between the two vectors. Clearly, the coding gain is controlled by the part $\sigma_1\sigma_2$, which corresponds to the power loading and the magnitude of the channel eigenvalues. The diversity gain is controlled by $\sin(<u_1;u_2>)$. Note that full diversity corresponds to the case when the two eigenvectors u_1 and u_2 are orthogonal, while diversity order one results when these two vectors are parallel.

In a higher dimensional space, the volume occupied by the beamformer, given by $\mid \det(\mathbf{W}) \mid$, is merely the volume spanned by a parallelepiped in an M_t-dimensional space. Hence maximizing det(**W**) provides a tradeoff between the coding gain and the diversity order achieved by the system, and it would be expected that such a scheme would provide a performance tradeoff between pure SF coding that achieves full diversity and the eigenvalue selection scheme that maximizes the coding gain of the system. We summarize the algorithm for the eigenspace selection scheme in the following steps Alamouti (1998): Let the eigendecomposition of the spatial correlation matrix at the l-th path be given by $R_{\alpha,l} = V_l \Lambda_l V_l^H$, where $0 \leq 1 \leq L-1$ (Palomar et al., 2003). Choose every possible combination of M_t eigenvalue and eigenvector pairs from the LM_t pairs available from the eigendecomposition in the previous step (Narula et al., 1998). Arrange the M_t selected pairs in matrix format as in the eigenvalue selection algorithm (Visotsky and Madhow, 2001). The beamformer **W** is determined as $\mathbf{W} = \mathbf{U}\Gamma$ [5_AHMED]. Calculate $\mid \det(\mathbf{W}) \mid$ (Jöngren et al., 2002). From among all possible combinations, choose **W** with the largest determinant.

Similar to the eigenvalue selection scheme, the columns of the matrix **U** in the eigenspace selection algorithm are not, in general, orthogonal as the directions of the beamformer belong to different eigenspaces.

6.4.5.4 Simulation Results

To demonstrate the performance improvement due to applying the proposed algorithms compared to that of SF coding without beamforming, we performed some computer simulations. The channel model used is a two-ray, equal-power delay profile, with a delay of 20 μs between the two rays. The MIMO–OFDM system has $N = 128$ subcarriers, and QPSK modulation is used. The total bandwidth of the system is 1 MHz. Two antennas

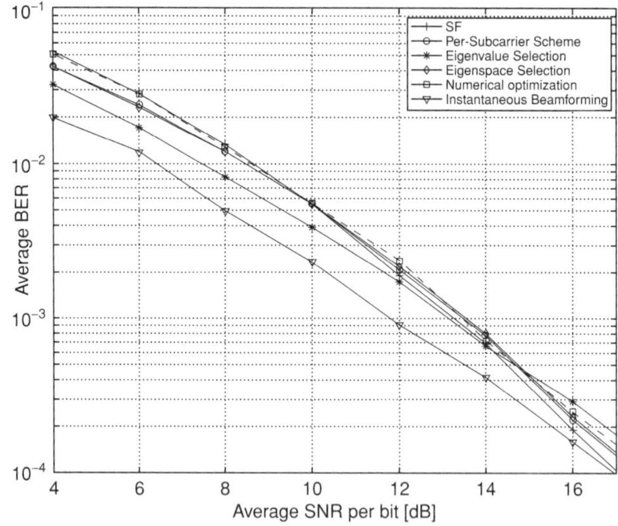

FIGURE 6.38 BER performance comparison: $M_t = 2$ transmit and $M_r = 1$ receive antennas. The dashed line represents the performance results when using numerical optimization to design the beamformer. Performance of instantaneous beamforming is also depicted.

are used at the transmitter side, and the receiver can be equipped with either one or two antennas.[19]

We choose the full-diversity SF code from (Su et al., 2003) to conduct the simulations. The 2×2 Alamouti's code (Alamouti, 1998) with repetition two times is used throughout the simulation experiments. To generate the spatial correlation channel coefficients, we use the following model $a_l = A_l \widetilde{a}_l$ where $l \in \{0, \ldots, L-1\}$, the vector a_l is defined as $a_l = [a_1^{\mathrm{T}}(l), \ldots, a_{M_r}^{\mathrm{T}}(l)]^{\tau}$, $\widetilde{a}_l$ is an $M_r M_t \times 1$ vector with i.i.d entries chosen from a complex Gaussian distribution with zero mean and variance β_2^l, and the matrix A_l contains the correlation coefficients, as follows $A_l A_l^H = I_{M_r} \otimes R_{\alpha,l}$. The eigenvalues for $\mathbf{R}_{\alpha,0}$ are 0,13 and 0,8, and for $\mathbf{R}_{\alpha,1}$ are 0,7 and 0,2.

In Fig. 6.38, we have considered two more scenarios in the comparisons. In particular, we applied numerical optimization techniques to find the optimal beamforming design that satisfies Eq. (6.52). Since the problem is highly nonconvex, the numerical algorithm is not guaranteed to converge to a global optimal, and only local optima are reached. The local optima depend heavily on the initial conditions selected and can lead to performance inferior to the proposed algorithms.

Since comparison of the proposed heuristic to the optimal solution is not feasible, we compared the performance of the proposed algorithms to the best scenario possible when perfect instantaneous channel information is available at the transmitter. In this case, transmitting along the eigenvector with the largest eigenvalue minimizes the error

[19] In the simulations, exhaustive search for the optimal global solution of the beamformer is not considered due to the overhead induced from computations. For example, for the case of $M_t = 2$ antennas we have 8 independent variables.

probability. We follow the same approach by designing a beamformer separately at each subcarrier. The results are depicted in Fig. 6.38. It is clearly shown that the gap between this best scenario and the considered heuristics is not large.

6.4.6 UWB Mesh Networks in Hostile Environment: Interference Analysis and Performance Study[20]

6.4.6.1 System Model

We assume a system with N_F nonoverlapping FH bands, each with bandwidth B_h, where B_h is the bandwidth required to transmit a time hopping-pulse position modulated (TH-PPM) signal in the absence of frequency hopping (FH). Let N_u denote the user population and $s^k(t)$ the k-th user's signal at time t in this FH/TH-PPM ultra wideband (UWB) system. $s^k(t)$ takes the form

$$s^k(t) = \sqrt{\frac{E_b}{N_s}} \sum_{j=-\infty}^{+\infty} c_j^{fh}(k) p\left[t - j T_f - c_j^{th}(k) T_c - d_j(k)\delta\right] \tag{6.56}$$

where $p(t)$ is a chip waveform, which can take arbitrary time-limited pulse shapes specifically tailored for use in UWB communication systems. $p(t)$ is normalized to satisfy $\int_{-\infty}^{+\infty} p^2(t)\, dt = 1$.

The notations and parameters are:

- N_s denotes the number of pulses used to transmit a single information bit. T_f is the frame duration. In general case, $N_s \geq 1$ pulses carry one bit of information. The bit duration T_b should satisfy $T_b \geq T_f N_s$.
- E_b is the energy per information bit. $\sqrt{\frac{E_b}{N_s}}$ is the normalized energy in each symbol.
- $c_j^{th}(k)\, T_c$ is the time shift introduced by the TH code. T_c is the chip duration. $c_j^{th}(k)$ is the j-th coefficient of the TH sequence used by user k; it is pseudo-random with each element taking on an integer in the range $[0; N_h - 1]$, where N_h is the number of hops. $T_c \leq T_f / N_h$ should be satisfied.
- The $d_j(k)\delta$ term represents the time shift introduced by PPM modulation. Here, only 2PPM is considered. Therefore, $d_j(k)$ represents the j-th binary data bit (0 or 1) transmitted by the k-th user; δ is the PPM shift.
- $c_j^{fh}(k) = \sqrt{2} \cos(2\pi f_k j)$ is the k-th user's spreading code during j-th frame.

Notice that each symbol chooses one of the N_F subbands to transmit the signal; however, in each subband, the transmitted signal is TH-2PPM.

---

[20] Excerpt from the invited article "UWB mesh networks in hostile environment: Interference analysis and performance study," Lingming Wang and Qilian Liang. Department of Electrical Engineering, University of Texas at Arlington, Arlington, TX 76019-0016, USA. E-mail: wang@wcn.uta.edu; liang@uta.edu (This work was supported by the Office of Naval Research (ONR) Young Investigator Award under Grant N00014-03-1-0466).

In wireless communications, wireless nodes toggle between active (communication) and idle status. In order to save energy, wireless nodes choose to be idle for most of the time. The number of nodes that are actually in the status of active communication is unknown. However, the total number of mesh nodes in the network and the access rate λ, that is, the rate that a node in the communication status, for each node are assumed to be known. Clearly, the event of status of node is a Bernoulli distribution with mean as λ and variance as $\lambda(1-\lambda)$. The number of active nodes, $N_{\mathrm{u}}^{\mathrm{T}}$, can be encountered as the sum of N_{U} independent, identically distributed Bernoulli random variables, which is a Binomial distribution. As the WM^{2}Net population is often very large, we can approximate the Binomial distribution with a Gaussian random variable with the mean $N_{\mathrm{U}}\lambda$ and variance $N_{\mathrm{U}}\lambda(1-\lambda)$ as

$$f_{N_u^T}\left(n_u^t\right) = \frac{1}{\sqrt{2\pi\, N_{\mathrm{U}}\lambda(1-\lambda)}} e^{-(n_u^t - N_{\mathrm{U}}\lambda)^2/2N_{\mathrm{U}}\lambda(1-\lambda)} \tag{6.57}$$

For the $N_{\mathrm{u}}^{\mathrm{T}}$ users, they randomly choose one of the subbands to transmit the signal according to $c_j^{\mathrm{fh}}(k)$ symbol by symbol. It is also a Binomial random variable with the coefficient $1/N_{\mathrm{F}}$. To simplify the problem, we assume that the users are optimally distributed, so the number of users sharing the same channel P_J should be expressed as

$$N_u = N_u^T / N_F \tag{6.58}$$

6.4.6.2 Multiuser Interference Analysis

In this section, we will first focus on the analysis of MUI in the absence of hostile jammer interference. We assume there is no inter-channel interference. Therefore, the received signal of first user's j-th symbol can be expressed as:

$$r_j(t) = r_j^{(1)}(t) + r_{j,\mathrm{mui}}(t) + n(t) \tag{6.59}$$

where $r_{j,\mathrm{mui}}(t)$ is the MUI contribution at the receiver input. For a large number of users with comparable powers, we can approximate the MUI with a white Gaussian process using the central limit theorem (Win and Scholtz, 2000) and, as such, it can be lumped into the additive Gaussian noise,

$$w_{\mathrm{tot}}(t) = r_{j,\mathrm{mui}}(t) + n(t) \tag{6.60}$$

and $w_{\mathrm{tot}}(t)$ still remains a white Gaussian process. Since the system is asynchronous, we need to consider all cases where a pulse originated by all transmitters but (say) T $\times$ 1 is detected by the receiver. First of all, we need to analyze the noise generated from an interfering pulse at the output of the receiver by using the method provided by Benedetto and Giancola (2004),

$$\mathrm{mui}^{(k)}(\tau^{(k)}) = \sqrt{E_{\mathrm{RX}}^{(k)}} \int_0^{T_c} p(t - \tau^{(k)}) v_t \, dt \tag{6.61}$$

where $E_{\mathrm{RX}}^{(k)} = \alpha^{(k)}(E_{\mathrm{b}}/N_{\mathrm{s}})$. We assume here that $\alpha^{(k)} = 1\ \forall k$.

Since $\tau^{(k)}$ is uniformly distributed over $[0, T_f)$, and identically distributed for different τ [1], under the hypothesis of perfect power control, for example, $E_{RX}^{(k)} = E_{RX}|\forall k$, the total MUI energy is

$$\sigma_{mui}^2 = \frac{E_{RX}}{T_f}\sum_{k=2}^{N_u}\left(\int_0^{T_f}\left(\int_0^{T_c} p(t-\tau)v(t)\,dt\right)^2 d\tau\right)\Bigg| \tag{6.62}$$

Defining $\sigma_M^2|$ as

$$\sigma_M^2 = \int_0^{T_f}\left(\int_0^{T_c} p(t-\tau)v(t)\,dt\right)^2 d\tau \tag{6.63}$$

we get

$$\sigma_{mui}^2 = \frac{E_{RX}}{T_f}(N_u - 1)\sigma_M^2\Big| \tag{6.64}$$

Let SNR_{ref} denote the equivalent signal to noise and MUI ratio over one symbol. It is then

$$SNR_{ref} = \left(SNR_n^{-1} + SIR_{mui}^{-1}\right)^{-1}\Big| \tag{6.65}$$

where SNR_n is the SNR over one symbol.

Hence,

$$\begin{aligned} SNR_{ref} &= \left(\left(\frac{E_{RX}}{N_0}\right)^{-1} + \left(\frac{T_f}{(N_u - 1)\sigma_M^2}\right)^{-1}\right)^{-1} \\ &= \frac{T_f}{N_s T_f + (N_u - 1)\sigma_M^2\left(\frac{E_b}{N_0}\right)}\left(\frac{E_b}{N_0}\right) \end{aligned} \tag{6.66}$$

where E_b/N_0 is the system SNR.

6.4.6.3 Performance Analysis with Multitone/Pulse Interference

In this section, the SIR for the hostile interference part is calculated. The following assumptions are made (Table 6.2): The multitone/pulse interference has a total power P_J, which is transmitted in a total of q equal power interfering tones spread randomly over the spread spectrum bandwidth. The time duration for the interference pulse is the same as the time duration of the transmitted signal pulse $p(t)$, which is denoted as T_p. To simplify the problem, we suppose $T_c = 2T_p$ and $\delta = T_p$. The hop period of the interference is also T_p, and each hop is independent. The multitone/pulse interference can catch the signal pulse with perfect timing. We consider the scenario where there is at most one interferer per FH subband. Hence, in a single hop, the probability that a FH band includes an interference tone/pulse is q/N_F. Observing the transmitted signal as in

Parameter	Notation	Second-order monocycle
Shaping factor for the pulse	ϵ	0.25 ns
Time shift introduced by PPM	δ	0.5 ns
Pulse duration	T_p	0.5 ns
Frame duration	T_f	8 ns
Chip duration	T_c	1 ns
Number of hops	N_h	6

TABLE 6.2 Parameters of the Example FH/TH-PPM UWB System

Eq. (6.56), the signal hops both in the frequency domain and in the time domain symbol by symbol. Therefore, the analysis below will first focus on one symbol.

We partition the symbol duration as two time slots, hence, for the multitone/pulse interference, there are two hops. In each hop, the interference is independently distributed. Overall, there are four cases with regard to the jammer interference for each symbol: There is no jammer interference in either of two slots, with the probability as $P\{case\,1\} = (1 - \frac{q}{N_F}) \cdot (1 - \frac{q}{N_F})$; there is jammer interference in each slot, with the probability as $P\{case\,2\} = \frac{q}{N_F} \cdot \frac{q}{N_F}$; there is one and only one jammer interference pulse, and it is at the same slot as the signal pulse.

The probability of case 3 is $P\{case\,3\} = (1 - \frac{q}{N_F}) \cdot \frac{q}{N_F}$; there is one and only one jammer interference pulse, and it is not at the same slot as the signal pulse. The probability of case 4 is $P\{case\,4\} = (1 - \frac{q}{N_F} \cdot \frac{q}{N_F})$.

The received signal of the j-th symbol of first user can be expressed as

$$r_j'(t) = r_j^1(t) + I_{\text{jammer}}(t) + w_{\text{tot}}(t)|$$

where $r_j^k(t)$ and $I_{\text{jammer}}(t)$ are the jammer interference contributions at the receiver input, and $w_{\text{tot}}(t)$ accounts for both the thermal and MUI noise contributions, and is still a white Gaussian process as proved above. Hence, a maximum a posteriori (MAP) approach can be adopted here to get the minimum error probability. For different cases of the jammer interference, the detection boundaries are shown in Fig. 6.39.

Hence, we can get the $SINR_{\text{jammer}}$ straightforwardly:

$$SINR_{\text{jammer|case 1,2}} = \frac{E_{RX}}{N_0'} \tag{6.67}$$

$$SINR_{\text{jammer|case 3}} = \frac{\left(\sqrt{E_{RX}} + \sqrt{\frac{p_j}{q}}\right)^2}{N_0'} \tag{6.68}$$

$$SINR_{\text{jammer|case 4}} = \frac{\left(\sqrt{E_{RX}} - \sqrt{\frac{p_j}{q}}\right)^2}{N_0'} \tag{6.69}$$

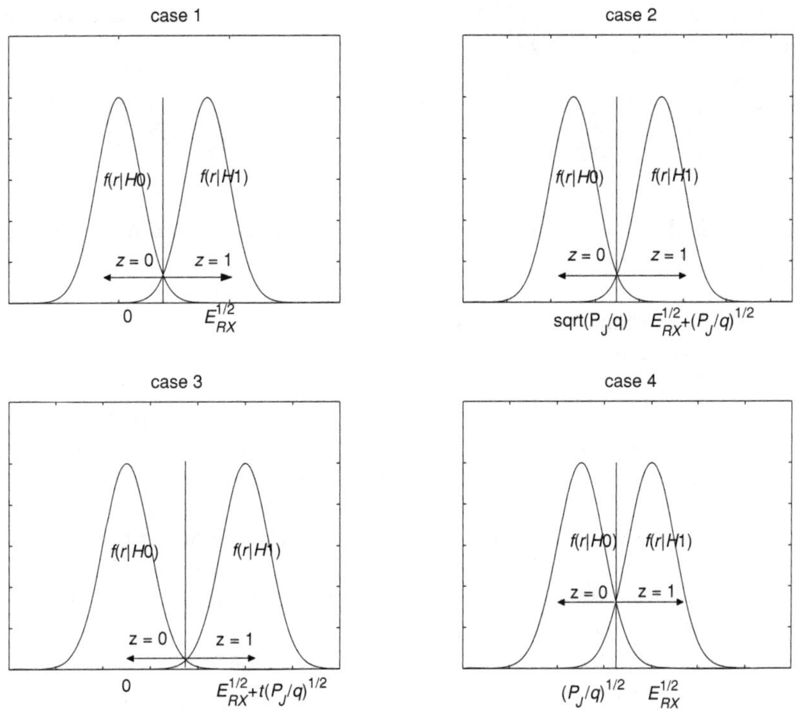

FIGURE 6.39 The MAP detection rule for all the cases.

Since the error probability of a 2-PPM signal is (Benedetto and Giancola, 2004) $Pr = Q(\sqrt{SNR_{\text{spec}}})\big|$, and after removing the conditioning on cases, we get

$$
\begin{aligned}
Pr'_s = {} & \left(\left(\frac{N_F - q}{N_F}\right)^2 + \left(\frac{q}{N_F}\right)^2\right) Q\left(\sqrt{\left(\frac{\frac{E_b}{N_s}}{N'_0}\right)}\right) \\
& + \left(\frac{N_F - q}{N_F}\right)\left(\frac{q}{N_F}\right) Q\left(\sqrt{\left(\frac{\left(\sqrt{\frac{E_b}{N_s}} + \sqrt{\frac{P_j}{q}}\right)^2}{N'_0}\right)}\right) \\
& + Q\left(\sqrt{\left(\frac{\left(\sqrt{\frac{E_b}{N_s}} + \sqrt{\frac{P_j}{q}}\right)^2}{N'_0}\right)}\right).
\end{aligned}
\tag{6.70}
$$

Considering only N_u is a random variable, we should substitute Eqs. (6.57) and (6.58) into Eq. (6.70), obtaining

$$
Pr_s = \frac{1}{N_F\sqrt{2\pi N_U \lambda(1-\lambda)}} \int_1^{N_U} Pr'_s e^{-(n_u - N_U\lambda)^2/2N_U\lambda(1-\lambda)} dn_u \tag{6.71}
$$

After we got the symbol error rate (SER) Pr_s, it is easy for us to obtain the bit error rate (BER) Pr_b by majority law.

$$P_{rb} = \sum_{k=\lceil \frac{N_s}{2} \rceil}^{N_s} C_{N_s}^k Pr_s^k (1 - Pr_s)^{N^s} - k \tag{6.72}$$

where $\lceil \rceil$ is the ceiling operation, and $C_{N_s}^k$ is an N_s-choose-k Binomial coefficient, that is, $C_{N_s}^k = \frac{k!(N_s-k)!}{N_s!}$.

6.4.6.4 Numerical Results and Comparisons

- The discussion on N_s:
  We fix $N_F = 20, q = 8, N_u = 10$ and the energy of the signal and jammer interference ratio $E_b/P_J = 5$ dB, and compare the SER and BER among $N_s = 1,3,5,7$. The results are shown in Figs. 6.40 and 6.41, respectively. As illustrated, SER is increasing as N_s increases; this is attributed to the fact that the higher the number of symbols used to transmit one bit, the less the per symbol energy. However, BER is more meaningful here. The higher the number of symbols used to transmit one information bit, the better the performance achieved. The curves drop quickly from $SNR = 0$ dB to 15 dB. However, beyond 15 dB, performance becomes flat, owing to the jammer interference.
- The discussion on $E_b = P_J$:
  We set $N^F = 20$, $N_s = 3$, $q = 8$, and $N_u = 10$, and compare the SER and BER among $E_b/P_J = 0, 5, 10$ dB. Figures 6.42 and 6.43 show the results. For both SER and BER, larger E_b/P_J can guarantee better performance. From $E_b/P_J = 5$ dB to $E_b/P_J = 10$ dB, the performance gain is very limited, the reason of which is

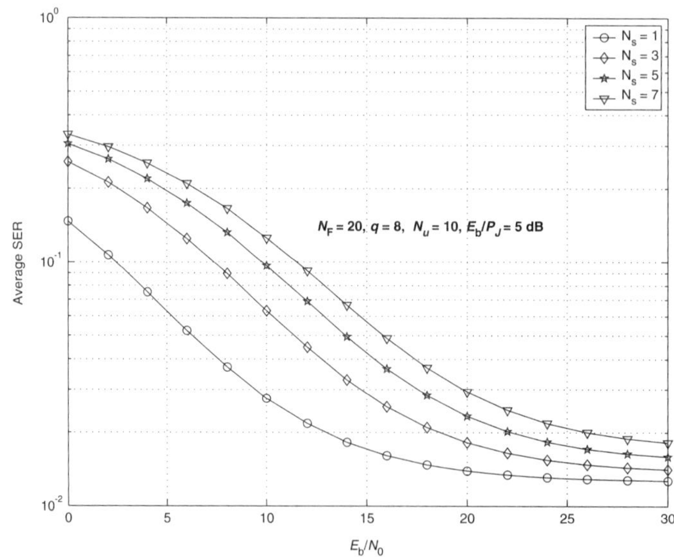

FIGURE 6.40 The average SER for different N_s.

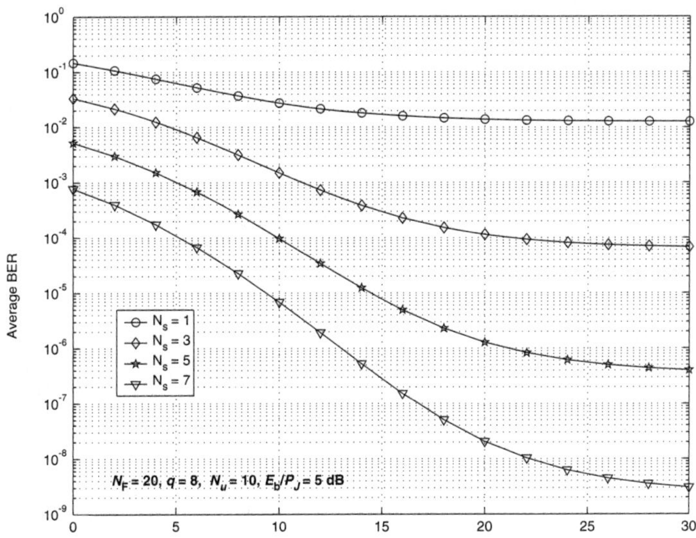

FIGURE 6.41 The average BER for different N_s.

when E_b/P_J is higher than some better threshold, the jammer interference is too weak to harm the system.

- The discussion on q:
  N_F, N_s, N_u and E_b/P_J are fixed at 20, 5, 10, and 5 dB,respectively. We attempt to evaluate the performance for $q = 2, 8, 18$. At the first glance, larger q may be thought to yield worse performance, because large q means the probability

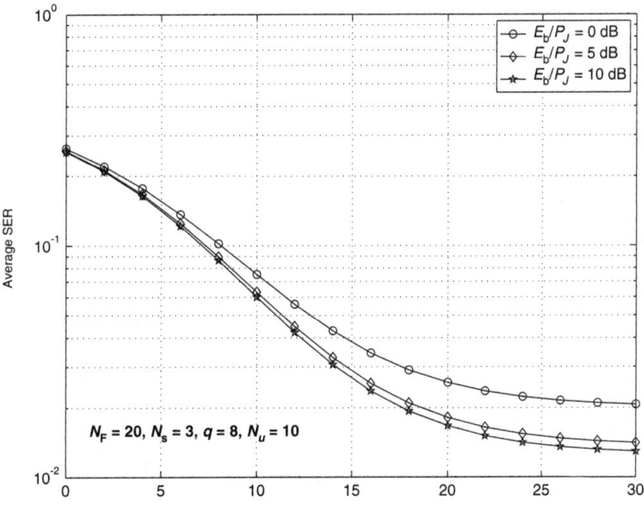

FIGURE 6.42 The average SER for different E_b/P_J.

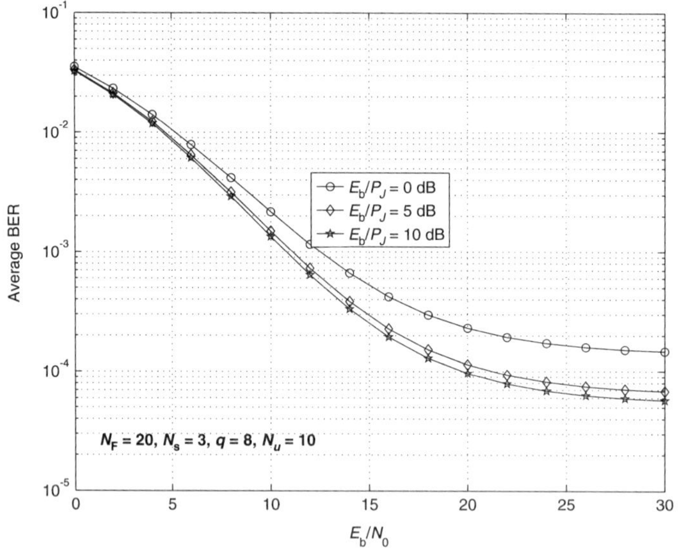

FIGURE 6.43 The average SER for different E_b/P_J.

that a jammer interference bumps the information signal is higher. However, we notice that high q also means that the energy of the jammer interference for each subband is reduced, which is because the total jammer interference power is fixed. We can get the same conclusion in Figs. 6.44 and 6.45. The worst and best performances obtained are for $q = 2$ and $q = 8$, respectively.

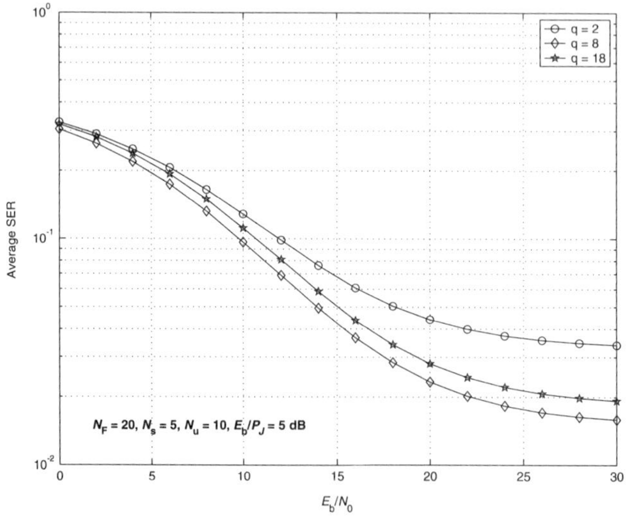

FIGURE 6.44 The average SER for different q.

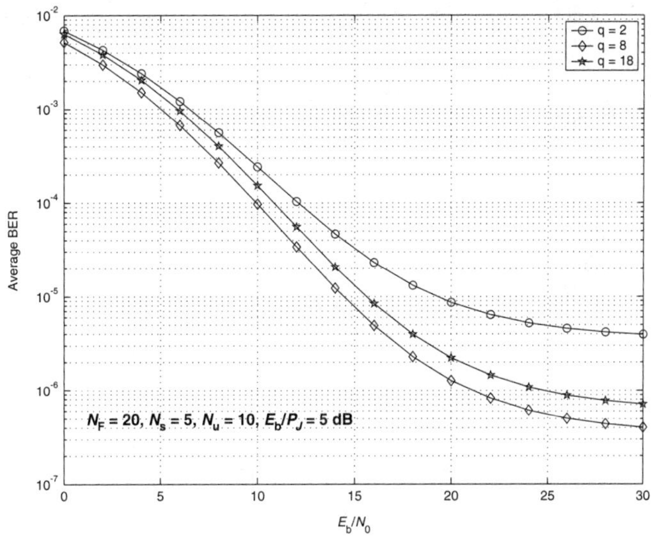

FIGURE 6.45 The average BER for different q.

- The discussion on N_F:
  We evaluate the performance for $N_F = 1, 5, 10, 20$ when $N_s = 1$, and the number of communicating users is 100. We need to evaluate how partitioning N_F can decrease the MUI. For this purpose, we set it as a jammer interference free channel. As Fig. 6.46 shows, performance is improved when N_F is higher.

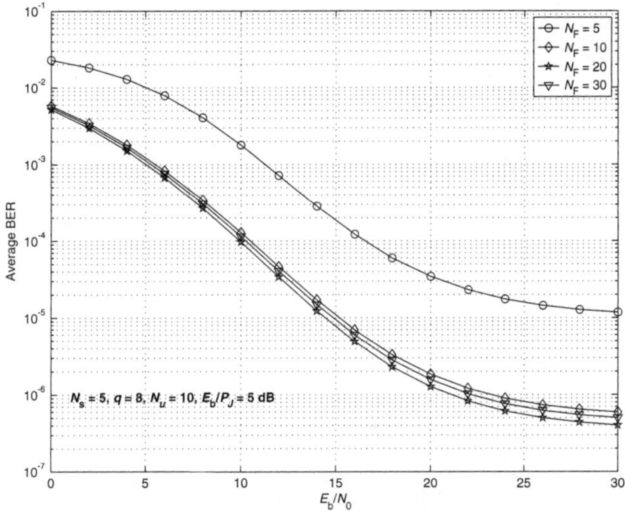

FIGURE 6.46 The average BER for different N_F.

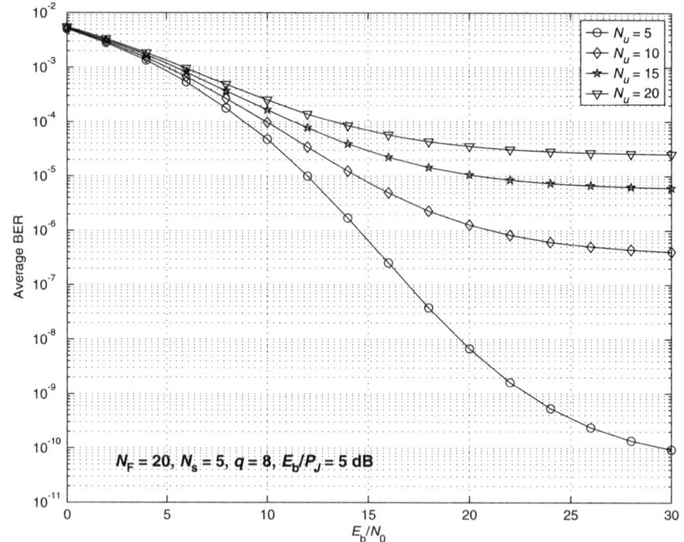

FIGURE 6.47 The average SER for different N_u.

- The discussion on N_u:
  We set $N_F = 20$, $q = 8$, $N_s = 5$ and $E_b/P_J = 5$ dB. N_u is equal to 5, 10, 15, 20, respectively. As shown in Figs. 6.47 and 6.48, at high SNR, the performance is degraded quickly as N_u increases. This is because MUI is related to E_b, assuming that each user has comparable power.

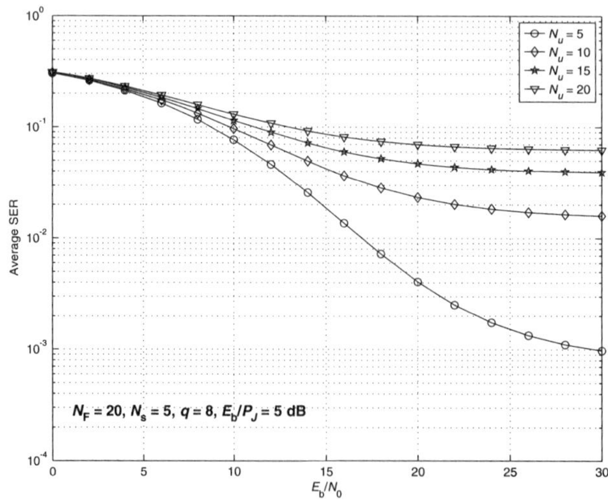

FIGURE 6.48 The average BER for different N_u.

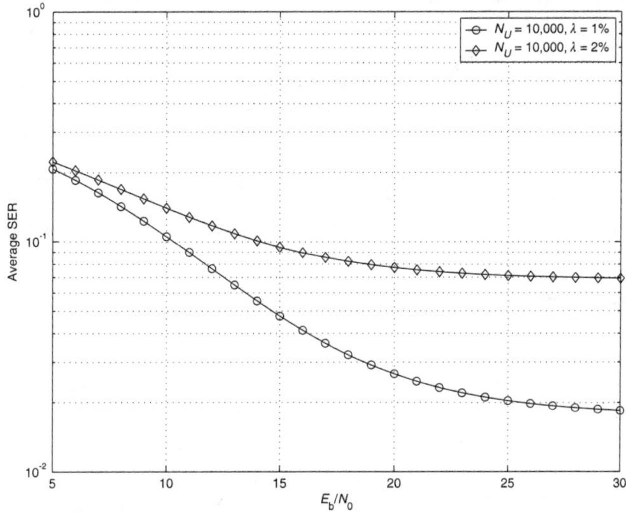

FIGURE 6.49 The average SER for different λ.

- The discussion on λ:

  We proved that N_u approximates a Gaussian RV. With N_U known, the number of WM²Net nodes, but N_u unknown, with N_u being the number of users that would share the same subband, we need to calculate the SER as in Eq. (6.71). We set $N_u = 10{,}000$, N_F is set to 20 and assume that users are optimally distributed. For different access rate $\lambda = 0.01$ and $\lambda = 0.02$, the performances are shown in Figs. 6.49 and 6.50. We can see that although N_u are the same, the difference in λ would yield totally different performance figures.

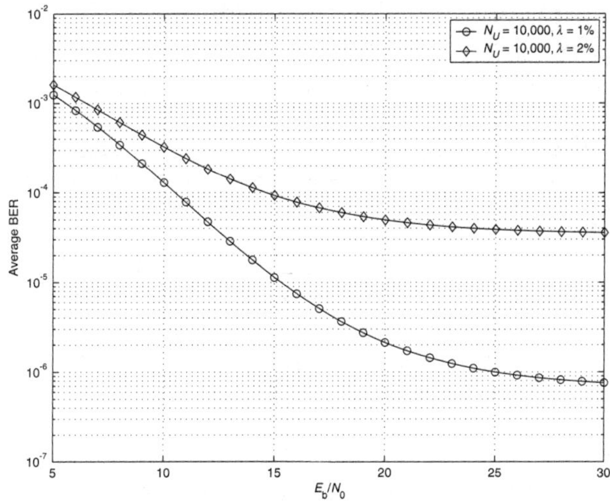

FIGURE 6.50 The average BER for different λ.

Number of bytes: 4	1	1	Max: 127
Preamble	SFD	Frame length	PSDU
Synchronization header		PHY header	PHY payload

Figure 6.51 ZigBee packet format.

6.5 Principles of Communications in WM^{2}Nets—The Physical Layer[21]

6.5.1 PHY Layer Specifications

6.5.1.1 ZigBee

The IEEE-802.15.4 standard (IEEE, 2003) uses a single MAC layer above two possible physical layers, one for the 868/915 MHz bands and one for the 2.4 GHz band. The standard has been amended by the 802.15.4b version to add new PHY layers in the 868/915 MHz bands enabling higher data rates. The worldwide availability and larger bit rates explain the predominance of the 2.4 GHz PHY use in the wireless industry. The structure of a packet (Fig. 6.51) is common to all PHY layers. The synchronization header is composed of a preamble (32 binary zeros) and a start-of-frame delimiter (SFD, one byte). The PHY header specifies the number of bytes in the PHY payload, the maximum being 127. The PHY service data unit (PSDU) carries the payload data.

The 2.4 GHz PHY air interface uses an orthogonal-quadrature phase shift keying (O-QPSK) modulation with a direct-sequence spread spectrum. Each 4-bit symbol corresponds to a 32-chip sequence with a 2 Mchips/s chip rate, implying a bit rate of 250 kbps. The 16 sequences are pseudo-orthogonal. Chips are modulated using O-QPSK with half-sine pulse shaping (Fig. 6.52), spectrally similar to constant envelope MSK, thereby making possible the use of nonlinear amplification.

The required receiver sensitivity is −85 dBm with a maximum acceptable input level better than −20 dBm. The sensitivity is defined as the minimum received power for a packet error rate (PER) below 1%, considering a 20-byte PSDU. This corresponds to a BER of $1.2*10^{-4}$.

The 2.4 GHz band is divided into 16 channels of 5 MHz each. The carrier and symbol frequency offset must be less than 40 ppm. The rejection of channels $n \pm 1$ and $n \pm 2$, n being the desired channel, is respectively 0 dB and 30 dB relative to channel n. Clearly, low power consumption is targeted by the relaxed channel spacing and blocking requirements as well as by the low and adjustable output power (−3 dBm).

6.5.1.2 Ultra Wide Band—Low Data Rate

The IEEE-802.15.4a standard defines an alternative to IEEE-802.15.4 with the same objectives of low power, low cost, and low complexity, but improved performance. Two PHY are proposed, one using impulse radio with a precise ranging service and one using a robust chirp-based spread spectrum modulation. The packet format is similar to that of IEEE-802.15.4, but the header fields are matched to the new features. The preamble is a

[21] Excerpt from the invited article "Principles of communications in wireless sensorNets—The Physical Layer," Eric Mercier and François Dehmas, CEA-Léti - Minatec, 17 av. des Martyrs, 38054 Grenoble Cedex 9, France. E-mail: eric.mercier@cea.fr; francois.dehmas@cea.fr

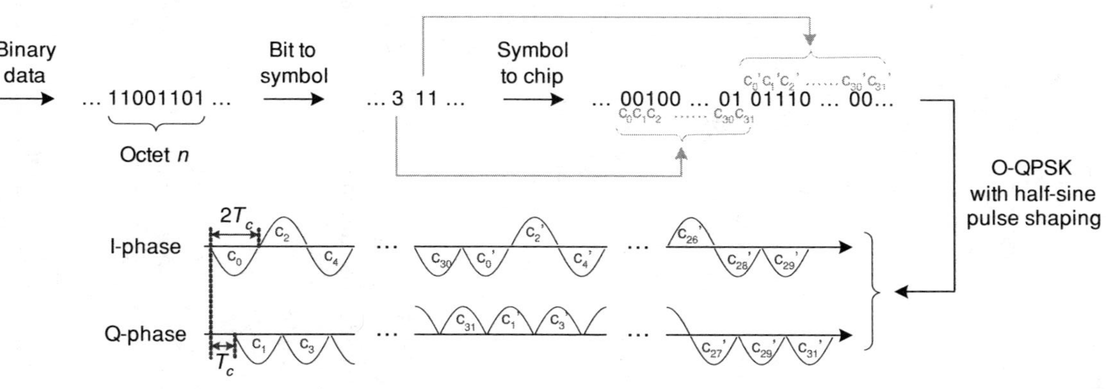

FIGURE 6.52 Modulation and spreading.

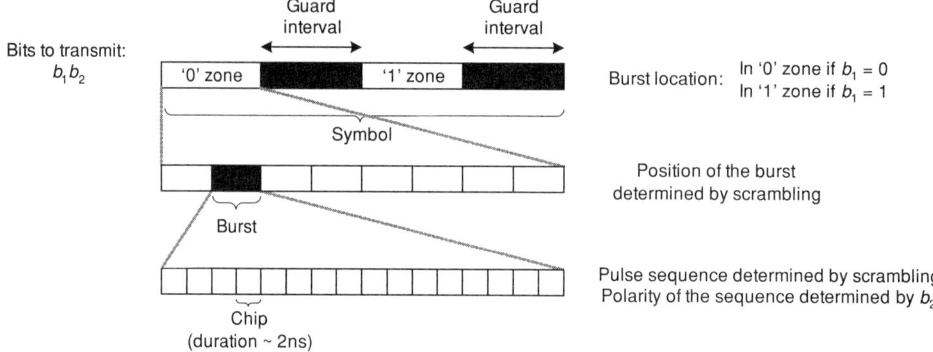

FIGURE 6.53 Symbol structure.

ternary sequence repeated 16, 64, 1024, or 4096 times and one byte in the PHY header indicates the preamble length and bit rate.

The IR-UWB PHY waveform is composed of pulses transmitted in short bursts. The mean pulse repetition frequency (PRF) is the ratio of the number of pulses per symbol to the symbol duration and is 3.90 MHz, 15.6 MHz, or 62.4 MHz, the latter being optional. The pulse is a root-raised cosine with a roll-off factor of 0.6. One symbol is made of two bits (Fig. 6.53). The first bit determines the position of the burst in the symbol and the second bit the polarity of the burst. The sequence of chips in a burst and the position of a burst in the half-symbol are given by a scrambling code. Time and polarity scrambling are added for better interference robustness and spectrum smoothing.

A forward error correction is added to improve the link budget. The selected Reed-Solomon code RS(63,55), working on 6-bit symbols, and the optional systematic convolutional code with a short constraint length of 3 and a rate of ½ keep the computational complexity at a reasonable level. The systematic bits determine the pulse position and the other bits the polarity. A noncoherent receiver can thus be used by simply ignoring the inner convolutional code. The minimum receiver sensitivity specifications are the same as in IEEE-802.15.4, −85 dBm.

Among the proposed bit rates (Table 6.3), only the 0.85 Mbps rate is mandatory. For a given mean PRF, the bit rate is proportional to the number of pulses per burst and the symbol length.

The band plan was designed for maximum compliance with international regulations and bearing in mind low-cost implementation constraints and coexistence with other incumbents. For each band group (sub GHz, low-band and high-band), only one

Mean PRF (MHz)	**15.60**				**3.90**			
Convolutionnal coding	Yes, rate = 1/2			No	Yes, rate = 1/2			No
Pulses per burst	128	**16**	2	1	32	**4**	2	1
Symbol rate (MHz)	0.12	**0.98**	7.80	15.60	0.12	**0.98**	1.95	3.90
Bit rate (Mbps)	0.11	**0.85**	6.81	27.24	0.11	**0.85**	1.70	6.81

TABLE 6.3 Bit Rate Options (Mandatory Rates in Bold)

	Channel number	Center Frequency (MHz)	Bandwidth (MHz)
Sub GHz	**0**	**499.2**	**499.2**
Low-band	1; 2	3494.4; 3993.6	499.2
	3	**4492.8**	**499.2**
	4	3993.6	1331.2
High-band	5; 6; 8; 10; 12; 13; 14	6489.6; 6988.8; 7488.0; 8486.4; 8985.6; 9484.8; 9984.0	499.2
	9	**7987.2**	**499.2**
	7; 11; 15	6489.6; 7987.2; 9484.8	1081.6; 1331.2; 1354.97

Table 6.4 Band Plan for IEEE-802.15.4a PHY (Mandatory Bands in Bold).

single channel is mandatory in order to maintain a compromise between complexity and flexibility (Table 6.4).

6.5.1.3 Expected Performance

For IEEE-802.15.4, simulations can be performed over an AWGN channel. Indeed, the signal has a relatively small bandwidth with a 3 MHz main lobe such that the spectrum of the channel can be considered to be flat in this band. Optimal performance for orthogonal codes with noncoherent detection is given by (Proakis, 1995):

$$BER = \frac{8}{15}\frac{1}{16}\sum_{n=2}^{16}(-1)^n\binom{16}{n}\cdot\exp\left[4\frac{E_b}{N_0}\cdot\left(\frac{1}{n}-1\right)\right]$$

A shift of 0.5 dB is observed between the theoretical and the simulated curves (Fig. 6.54) due to the *quasi*-orthogonality of the codes. In addition, simulation results are given for a nonoptimal receiver for which correlation values are obtained by making a noncoherent sum of 8 partial correlations. Although a loss of 3 dB is measured at the receiver, it is shown to be more resistant to carrier frequency offset. In the UWB case, the performance is given for a mean PRF of 15.6 MHz and a bit rate of 850 kbps. The theoretical performance is given over an AWGN channel whereas the simulated results consider a channel model of a LoS indoor residential environment (Ouvry, 2005) (Fig. 6.54).

The coherent receiver has been simulated assuming a perfect RAKE and a known channel impulse response. The noncoherent receiver integrates the energy during the 32 ns burst. The target raw BER is about 10^{-3} to achieve a PER of 10^{-2} after error corrections.

6.5.2 Implementation Issues

6.5.2.1 ZigBee

A review of current Zigbee compliant devices shows that full CMOS implementations have been preferred for low-cost and high-volume reasons. The number of external components is minimized and the bill-of-material count is limited to a crystal and a few passives. The antenna input/output port is a simple L/C based matching network and no SAW filter is usually necessary. Typical receiver chains (Fig. 6.55) are based on a low-IF architecture (typically 2 MHz), linear analogue amplification with AGC, and 2 MHz

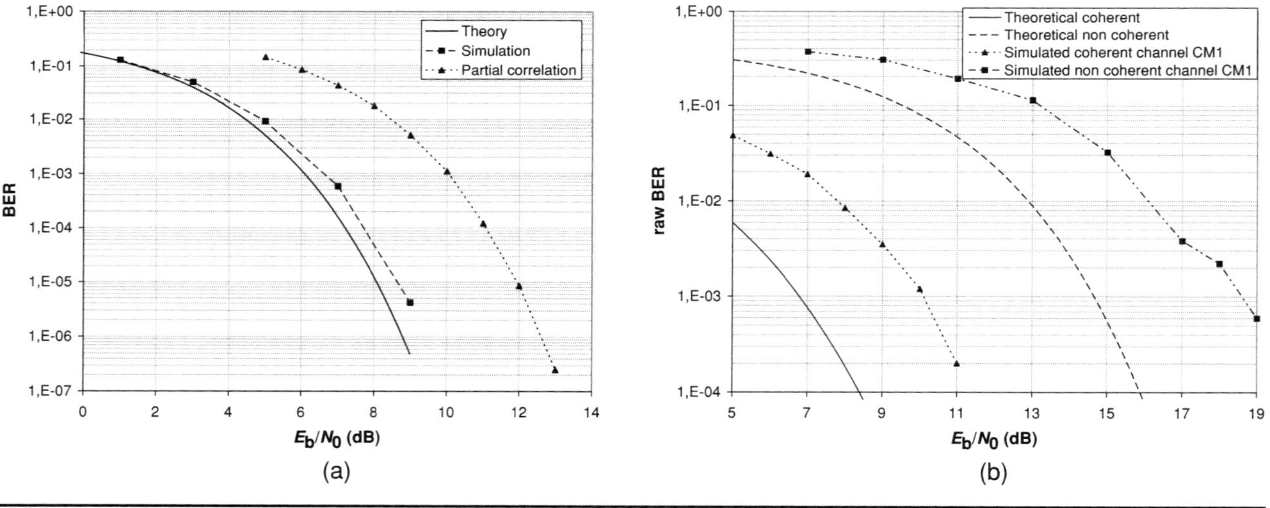

FIGURE 6.54 BER vs. E_b/N_0.

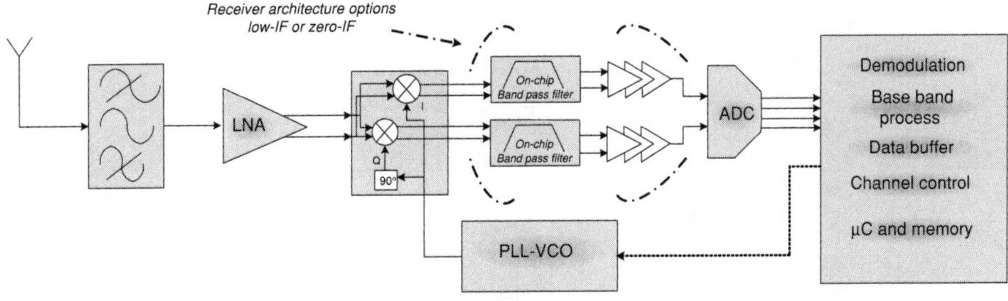

FIGURE 6.55 Receiver architecture for IEEE-802.15.4.

bandwidth filtering. The ADC resolution goes from 1 bit up to 6 bits with sampling rates from 4 to 16 MHz. The demodulation and de-spreading are operated after the digital conversion and the received bytes are stored on chip.

The IEEE-802.15.4 PHY layer specification was specifically defined in order to make these low complexity, low power, highly digital architectures possible. Indeed, the adjacent channel and image-rejection specifications reduce the phase noise specification, in the range of −85 dBc/Hz at 1 MHz, and the use of an O-QPSK modulation alleviates the analogue processing thanks to digital demodulation and low required A/D conversion resolution. On the transmitter side, the modulation is applied directly on the PLL, either on the frac-N divider or on the VCO, or the baseband signal issued from a DAC is directly up-converted to the carrier frequency.

Whereas most implementations reviewed aim for much better performance than the minimum addressed from the IEEE-802.15.4 standard (IEEE, 2003) aiming to extend the range of the communication link, the Letibee (Bernier, 2004) approach targets the minimum standard requirements but with the lowest power consumption possible. To this end, a nonlinear Zero-IF receiver architecture with 3-bit ADC resolution was chosen.

As illustrated in Table 6.5, the best consumption is $I_{Rx} = 16$ mA at $V_{CC} = 1.8$ V for −101 dBm sensitivity whereas less than $I_{Tx} = 3$ mA for the Letibee device for −85 dBm

Manufacturer ***Part n°***	**Sensitivity (dBm)** ***Std min −85 dBm***	**I_{Rx} (mA)**	**Output power (dBm)** ***Std min −3 dBm***	**I_{Tx} (mA)**	**V_{min} (V)**
Atmel *AT86RF230*	−101	16	+3	17*(+3 dBm)*	1.8
TI-Chipcon *CC2430*	−94	27	0	24.7 *(0 dBm)*	2
ST/Ember *ST260/ EM250*	−98	29	+5	24.3 *(0 dBm)*	1.7
Oki-Integration *ML7065*	−90	57	+3	56 *(0 dBm)*	2.3
Freescale *MC1321x*	−92	37	+4	30 *(0 dBm)*	2
Jennic *JN5121*	−93	40	+1	35 *(0 dBm)*	2.2
ZMD *44101*	−100	28	+3	32 *(0 dBm)*	2.2
CEA–Leti *Letibee*	−85	3	−3	5 *(−3 dBm)*	1

TABLE 6.5 Examples of Existing Devices for IEEE-802.15.4

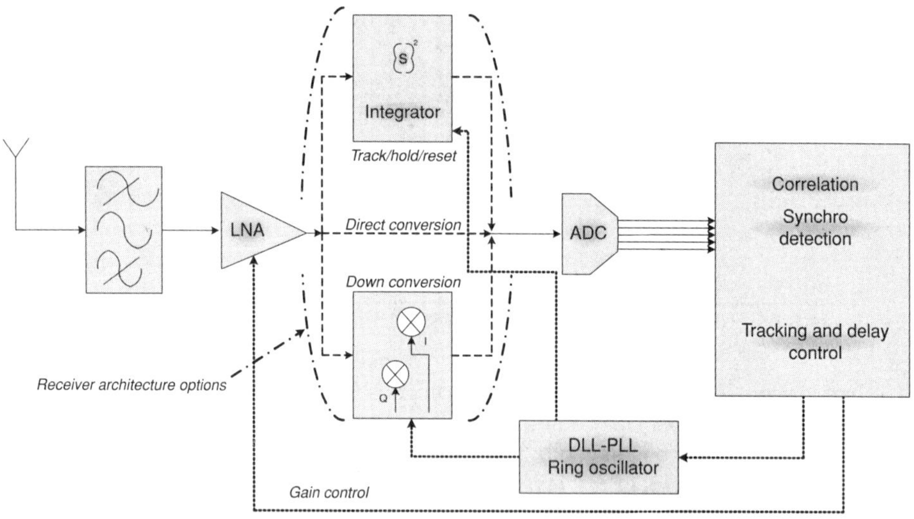

FIGURE 6.56 Receiver architecture for IEEE-802.15.4a.

sensitivity. These figures prove that the Zigbee standard is well suited for low-power, low-cost implementations, especially considering that these digital oriented architectures will benefit from future reductions of CMOS node sizes.

6.5.2.2 Ultra Wide Band—Low Data Rate (LDR)

LDR UWB, covering several hundred of kbps up to 10 Mbps, is a new technology and, as of date, few solutions have emerged. Impulse Response–UWB, where one bit is coded by a set of pulses, is the outcome from a joint study pursued by IMEC (Ryckaert, 2006), CWC-University of Oulu (Stoica et al., 2005), Berkeley Wireless Research Center, and CEA-Leti. An original approach based on FM-UWB, where a double FM modulation spreads the signal spectrum, has been proposed by CSEM (Gerrits et al., 2005). The UWB system architectures are still under study, but all aim low-power dissipation.

The trend of the forthcoming IEEE-802.15.4a (Fig. 6.56) is to use 2-Pulse Position Modulation, making the use of energy detection thus possible. The typical front end has a delay or phase-locked loop (DLL/PLL) or multiphase clocks for the time base, a pulse generator, and a band-pass filter on the transmitter side, and a LNA, quadratic detector or I/Q down-converter, ADC, and correlation process—based on either auto-correlation or correlator banks—on the receiver side. The BWRC (Otis et al., 2005) has also published an immediate-conversion front end (O'Donnell and Brodersen, 2005). Whatever the option, a complex digital core is used. The key point is the target accuracy of the time base, which typically is in the order of tens of picoseconds or, if relaxed, the rather heavy digital processing needed to compensate for drifts. Based on time-of-arrival or time-of-flight, the accuracy of detection also impacts localization information, with a desired ranging performance threshold being better than 1 m. The power consumption could theoretically be lower than Zigbee, since few pulses represent one bit compared to tens of carrier periods for Zigbee. Nevertheless, higher frequencies in UWB as well as the complexity of the base-band processing in Rx mode require nearly $I = 30$ mA.

CHAPTER 7

Security Issues in WM^{2}Nets

7.1 Introduction

The wireless multihop nature of wireless mobile mesh networks (WM^{2}Nets) inherits all the *security issues* of multihop radio systems. What makes security uniquely challenging for WM^{2}Nets, however, is that the computational, memory, and bandwidth costs of the security mechanisms must be carefully balanced against the limited resources of the individual nodes. If all meshes had sufficient memory and processing power, commercial off-the-shelf (COTS) security approaches could be a viable option.

The fundamental security primitives of authentication, integrity, and confidentiality are very much essential in a WM^{2}Net (Salem and Hubaux, 2005a). Authentication refers to the verification of identities of the communicating entities; integrity refers to the validity of the original message such that it is not tampered by an adversary; and confidentiality refers to the establishment of a secure channel to transmit encrypted text such that it seems a garble to an eavesdropper.

Due to the multihop nature of WM^{2}Nets, data are transiting third party equipment not belonging to either the user or the operator. Data capture, delay, and manipulation will therefore be more vulnerable to various malicious attacks. Among others, these include: (1) *Eavesdropping on data from unauthorized third parties*—sending unencrypted data over the air means that it is easy for an attacker with a properly tuned receiver to snoop on (or overhear) data exchange transmissions, (2) *Modification or injection of data from a third party without the knowledge of the communicating parties.* This is more harmful than simple eavesdropping, as data are actually changed. The attacker, for instance, may intercept data on patient vital signs and randomly change it, rendering it meaningless, (3) *Replay of previous queries/data.* Even if encryption is used, the attacker may pick up an encrypted message in its entirety and replay it at a later time. The attacker might repeatedly flood the network with a query message, causing thus nodes to waste valuable energy in (re)transmitting data continuously, (4) *Spoofing an access point (AP) to obtain non-legitimate access to data.* The attacker pretends to be an AP and ask meshes to send him their data, and (5) *Spoofing a mesh node to report forged data.* The attacker pretends to be a mesh node, fabricating thus meaningless data.

With these in mind, when using a wireless mobile mesh sensor network (WM^{2}Snet) in critical domains such as detection of chemical or biological agents or tracking of enemy vehicles, incorrect or maliciously corrupted data can have disastrous consequences. For example, an adversary can easily sniff into the data exchange communicated among mesh nodes and gain access to mission critical information to either monitor the data exchange or, worse, inject false messages into the network. Security services in this context are essential to ensure the authenticity, confidentiality, freshness, and integrity of the critical information collected and processed by such networks.

In addition, as in every wireless communication system, in a WM^{2}Net context a number of malicious attacks may occur. These are broadly classified as impersonation attacks, anti-integrity attacks, and anti-confidentiality attacks (Xie et al., 2005). An adversary can conduct an impersonation attack by introducing a rogue mobile router (MR) that sends forged/replayed registration messages to entice mesh clients. The mesh clients can accept the advertised false connection and assume that it is connected to the Internet. The rogue MR can also send a fake registration message by masquerading as another node (man-in-the-middle attack) so that all the packets are tunneled to it. Thus, it can gain unauthorized access to the network information by a simple MAC address spoofing. A rogue MR can conduct an anti-integrity attack by poisoning the route tables. It can cause other MRs to redirect their traffic towards itself by advertising a higher rate link/less congested link to the AP (or Internet gateway (IGW)). An attacker can conduct an anti-confidentiality attack and reveal critical information (like keys) by eavesdropping, brute-force attack, or cryptanalysis. Thus, it is very critical to establish an unbreakable trust relationship between the MRs, the mesh clients, and the AP/IGW.

In this context, to provide a secure WM^{2}Net, a multitude of security techniques must then be integrated. Encryption is the foundation for ensuring the confidentiality of data. In modern cryptographic systems, data encryption relies on the exchange of keys such that the legitimate users (those with the correct key) can decrypt the corresponding data items.

7.2 Security Overview of ZigBee[1]

7.2.1 Security Architecture of ZigBee

ZigBee is a set of communication protocols that operate on the application (APL) and network (NWK) layers. It works on top of the low-power MAC and PHY layers, which are standardized in the IEEE 802.15.4 standard (IEEE, 2003) for digital radios of low-rate wireless personal area network (WPAN) devices (The wireless personal area network working group, IEEE 802.15: http://www.ieee802.org/15/)

ZigBee provides two levels of authentication: *network level authentication* (NLA) and *application level authentication* (ALA). The first level secures the network from outsider attacks; this is achieved by using a common *network key*. The ALA is achieved by sharing a unique *link key* between pair of devices, preventing insider and outsider attacks but at a price of a higher memory cost. Link keys are shared among two devices and may be used only by the APS sublayer; however, network keys are shared across the entire network for securing broadcast messages and may be used by the MAC, NWK, and APL layers. In both cases, the key size is 128 bit. A device shall acquire the link key via key-transport, key-establishment, or preinstallation, while the network key can be acquired via key-transport or preinstallation. The key-establishment service uses the symmetric-key key establishment (SKKE) (ZigBee Alliance, 2004) protocol to share link keys between two

[1] Excerpt from the invited article "Security overview of ZigBee," *Dave Singelée and Bart Preneel (*SCD—COSIC, Department of Electrical Engineering, Katholieke Universiteit Leuven Kasteelpark Arenberg 10, 3001 Heverlee-Leuven, Belgium. E-mail: Dave.Singelee@esat.kuleuven.be).

devices; in this protocol, a trust relationship is established using a shared, secret, and symmetric key, referred to as a *master key*, typically preinstalled during manufacturing, or derived using some cryptographic method.

The ZigBee device may function in any of the following three modes of operation:

- **ZigBee Coordinator**: The most capable device, the coordinator, forms the root of the network tree and might bridge to other networks. There is exactly one ZigBee coordinator in each network. It is able to store information about the network, including acting as the repository for keys. It configures the security level of the network and the address of the *trust center* device. Each network has exactly one ZigBee trust center. This device is trusted by all other devices within the ZigBee network and is responsible for distributing and establishing keys in the network. By default, the ZigBee coordinator is the ZigBee trust center. The coordinator can always designate an alternate trust center.
- A **ZigBee Router** can act as an intermediate router, passing data from other devices.
- A **ZigBee End Device** contains just enough functionality to talk to its parent node (either the coordinator or a router). It cannot relay data from other devices.

In ZigBee, the layer that originates a frame is responsible for securing it. So, if a NWK command frame needs protection, NWK layer security shall be employed. Figure 7.1 shows an example of the security fields that may be included in a NWK frame. The auxiliary header contains security information (security control, frame counter, ...), the payload can be encrypted or not, and the message integrity code (MIC) is used to protect the integrity of both header fields and the payload (the *security control field* in the auxiliary header specifies the level of security that is applied to the frame). Both encryption and message integrity are provided by one building block: the *CCM** algorithm. Security information is stored in *access control lists* (ACLs). Each ACL entry contains the following security information: destination address, security control field, key, nonce, and the key and frame counter. The frame counter is incremented by one for every outgoing frame. The maximum value is $2^{32}-1$. When a new key is used, the frame counter is reset to 0. There is always a default ACL entry that is used if there is no specific ACL entry for the destination. There can be maximally 255 ACL entries. The exact number of ACL entries is vendor specific.

ZigBee uses the *open trust model* (http://www.zigbee.org/). This implies that all different layers of the communication stack and all applications running on a single device trust each other. Keys can be reused in each layer. To simplify interoperability, the security level used by all devices in a given network and by all layers of a device shall be the same. If protection from theft of service is required, NWK layer security shall be used for all frames. The NWK key is a broadcast key that is used by all devices in the same

NWK Header	Auxiliary Header	(Encrypted) NWK Payload	MIC

FIGURE 7.1 (Part of) ZigBee frame with security at the NWK level.

network. As a consequence, using a NWK key does not prevent insider attacks. If APL layer security is applied, a link key is used to protect outgoing frames. Link keys are employed to enable end-to-end security (between source and destination device).

7.2.1.1 CCM* Algorithm

*CCM** is a generic combined encryption and authentication block cipher mode. *CCM** is only defined for use with block ciphers with a 128-bit block size. The block cipher that is used in the ZigBee specification is the AES-128. The *CCM** mode is a minor modification of the *CCM* mode specified in the IEEE 802.15.4 MAC layer specification (IEEE, 2003). *CCM** includes all of the features of *CCM* and additionally offers encryption-only and integrity-only capabilities. In total, there are 8 possible security levels: the payload of a frame can be encrypted or not, and the length of the MIC, which protects the integrity of the header fields and the payload of a frame, can be 0, 32, 64, or 128 bits. The security control field in the header specifies which security level is used to secure the frame. As the *CCM* mode, the *CCM** mode requires only one 128-bit key. Together with this key, a unique 104-bit nonce *N* is used. This nonce is a function of the security control field, the frame counter, and the address of the sender. Within the scope of a key, the nonce value should be unique. The frame counter prevents reusing a nonce under the same key.

Figure 7.2 shows the computation of an authentication tag T. E is the block cipher AES-128, $B_1 || \ldots || B_t$ are the t data blocks that have to be integrity protected (each block has a length 128 bits), and B_0 is a data block that contains the nonce N and some constants. The authentication tag T holds the M leftmost bits of the output X_{t+1}. The value M specifies the length (in bytes) of the MIC. Encryption of the data blocks M_i is demonstrated in Fig. 7.3. The 128-bit blocks A_i contain the constant value *flags* (8-bit representation of the value 1), the nonce N, and a 16-bit counter i. The results of the encryption are the t cipher text blocks C_i.

7.2.1.2 Key Hierarchy

Several types of keys are used in ZigBee, forming a key hierarchy. Typically, the security manager of a device (situated in the APL layer) will perform the following steps:

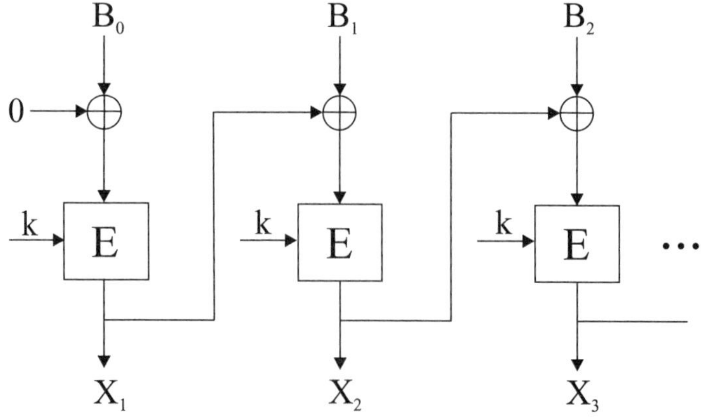

FIGURE 7.2 CCM* authentication block cipher mode.

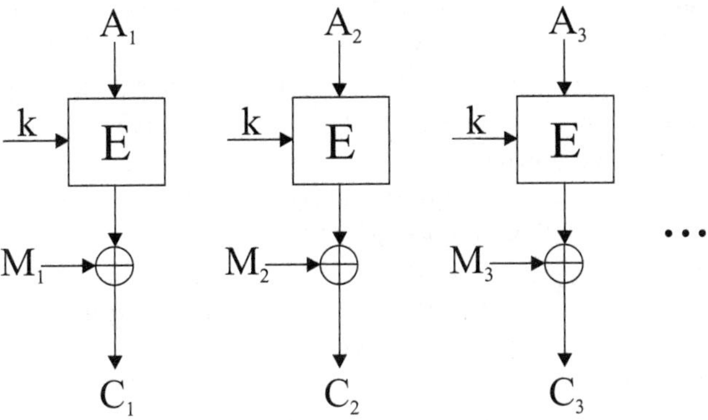

FIGURE 7.3 CCM* encryption block cipher mode.

(1) **Obtain the trust center master key**: Initially, each device shares a trust center master key with the trust center. The device can obtain this trust center master key (together with the address of the trust center) in two ways. Or the device acquires the trust center master key via insecure key-transport (e.g., it is sent in clear from the trust center to the device at low power), or it acquires this key via preinstallation (e.g., factory installation or based upon data entered by a user). It is very important that no other device can obtain this trust center master key, as the security of all other keys used in ZigBee depends on the confidentiality of the trust center master key.

(2) **Establish link key with trust center**: The trust center and the device, which shares a trust center master key, will execute the symmetric-key authenticated key agreement (SKKE) protocol to establish a link key with each other. First, both devices generate a random 128-bit challenge (*QEU* and *QEV*, respectively) and send it to the other device. These challenges are fed, together with the trust center master key, to a key derivation function. The results are two 128-bit keys: the *MacKey* and the *KeyData*. The former is the key of a MIC, used to mutually authenticate the challenges *QEU* and *QEV*. After a successful authentication, both devices will use the*KeyData* key as shared link key. This link key will be employed to secure the communication between the trust center and the device.

(3) **Compute key-load key**: The key-load key is derived from the link key as follows:

$$key - load\,key = HMAC_{link\,key}\,(0x02)$$

Here, *HMAC* is a keyed message authentication code (Menezes et al., 1996). This type of message authentication function uses a cryptographic hash function in combination with a secret key. The trust center uses the key-load key to transport an application master key securely to a device.

(4) **Compute key-transport key**: The key-transport key is derived from the link key as follows:

$$key - transport\;\;key = HMAC_{link\;\;key}(0x00)$$

The trust center uses the key-transport key to transport an application link key or a NWK key securely to a device.

(5) **Obtain the NWK key**: The trust center puts the NWK key (that is currently being used in the network) in a specially constructed command frame, secures it with the key-transport key and transmits it to the device. The NWK key is used to encrypt broadcast communication in the network. Note that command frame are always encrypted and integrity protected (with a 128-bit MIC).

(6) **Obtain the application link key**: When two devices in a network want to communicate securely (end-to-end), they need an application link key. One way to obtain such an application key is as follows: the trust center generates the application link key and puts it in a specially constructed command frame. This frame is sent securely to each device. The security of the frame is protected by employing the key-transport key. The advantage of the trust center sending out the application link keys directly is that key-escrow can be implemented.

(7) **Obtain the application master key**: Instead of directly transmitting the application link key to both devices, the trust center can also generate an application master key. It puts this key in a specially constructed command frame, and sends this securely to both devices. The security of this frame is protected by employing the key-load key.

(8) **Establish application link key with other devices**: After the devices obtained the application master key, they execute the SKKE protocol. This is done exactly as described above. The only difference is that the application master key is used to derive the link key, instead of the trust center master key. The output of the SKKE protocol is the application link key, which is used for secure end-to-end communication between both devices.

The above is valid only if the trust center is working in commercial mode. When the trust center works in residential mode, the device will not establish a link key with other devices. A more detailed discussion on the modes of operation of the ZigBee trust center is now presented.

7.2.1.3 ZigBee Trust Center

There is one trust center deployed in the ZigBee network. This device is often the ZigBee coordinator and is trusted by all devices in the network. It is responsible for the distribution of keys (link keys and NWK keys) among the ZigBee devices. The ZigBee trust center also enforces the policies in the network. These policies state how a device can join or leave the network (securely or insecurely), if and when keys have to be updated, etc. The trust center can be configured to operate in either *commercial* or *residential mode*:

- **The commercial mode** of the trust center is designed for high-security commercial applications. In this mode, the trust center maintains a list of devices, master keys, application link keys, and NWK keys that it needs to control. It also enforces the policies of NWK key updates and network admittance. In this mode, the memory required for the trust center grows with the number of devices in the network. When the trust center works in commercial mode, it shall follow the steps of the key hierarchy described above.

- **The residential mode** of the trust center is designed for low-security residential applications. In this mode, the trust center maintains a list of the NWK key and controls the policies of network admittance. It does not have to maintain a list of devices, master keys, or application link keys. When operating in residential mode, the NWK key is never updated and therefore, the memory required for the trust center does not grow with the number of devices in the network. This limits the implementation complexity, but also reduces the security. When the trust center works in residential mode, it shall not follow the steps of the key hierarchy described above. Instead, it will just send the NWK key to a device joining the network via insecure key transport. This key is used to secure communication. Master keys and link keys are not employed when the trust center works in residential mode.

In commercial (Fig. 7.4) and residential (Fig. 7.5) mode authentication, a device joins the network in following steps:

1. After a network discovery, the device sends a NLME-ASSOCIATE request to the nearest router.
2. The router responds with an NMLE-ASSOCIATE response. The device now joins the network but is not authenticated yet.
3. With an update-device command, the router informs the trust center that a new device needs be authenticated.
4. The trust center decides whether to accept the new device or not.

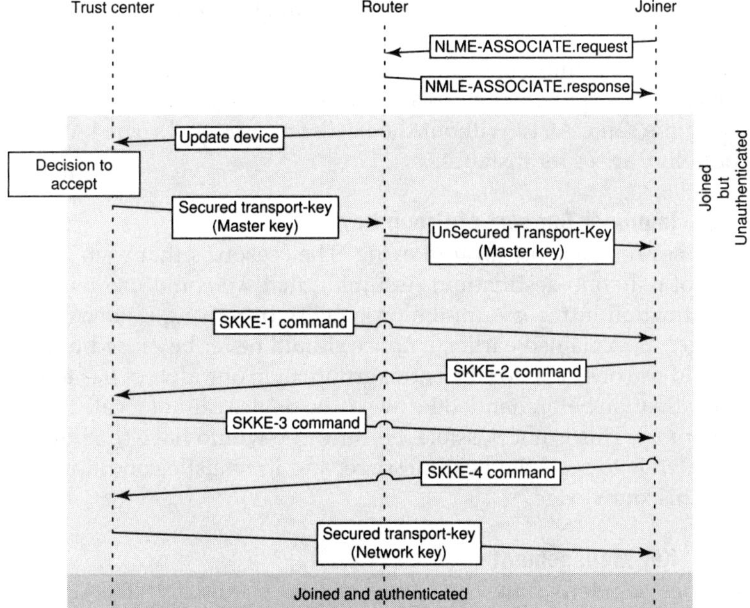

FIGURE 7.4 Commercial mode authentication.

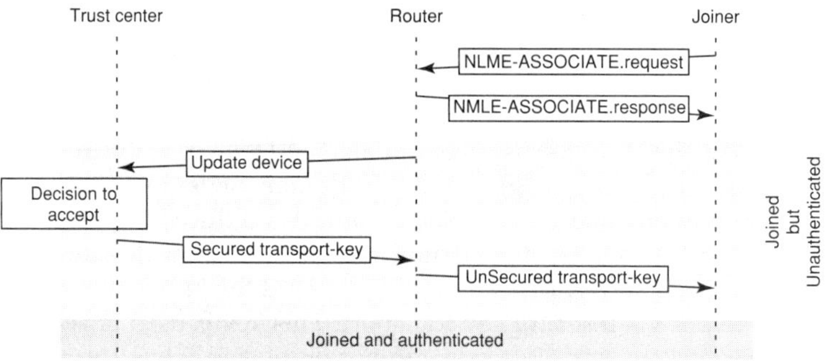

Figure 7.5 Residential mode authentication.

7.2.2 Weaknesses in the ZigBee Security Architecture

Improper use of the security mechanisms in ZigBee can raise several security concerns (Perez, 2006; Sastry and Wagner, 2004). Although some that were present in the IEEE 802.15.4 standard (http://www.zigbee.org) are tackled in newer versions of ZigBee, several security limitations still remain. An overview of these is given below.

7.2.2.1 (Nonce) Management Problems

As already stressed in the previous section, security information is stored in ACLs. Each ACL entry contains the following security information: destination address, security control field, key, nonce, and the key and frame counters. The nonce is a function of the security control field, the frame counter, and the address of the sender. In fact, only the frame counter is variable; hence, the nonce is derived directly from the frame counter. Encrypting two messages with the same key and the same nonce is very dangerous from a cryptographic point of view and should certainly be avoided. A problem arises if a key is used in two different ACLs (because in this case, the frame counter in each ACL is updated independently and this could result in the reuse of a nonce) or if a nonce is reused in the same ACL (without the key being updated). The latter can occur when a power failure arises for instance.

7.2.2.2 Improper Support of Group Keying

ZigBee does not support group keying. The reason is that each ACL can contain the address of only one destination. Assuming that we could use multiple ACLs, one for each destination in the group, the probability of reusing a nonce would then become very large. As explained earlier, a nonce should never be reused under the same key. If one would use one ACL for the entire group, then one always has to update the address of the destination beforehand; otherwise, the device cannot locate the correct ACL entry in its memory. This is not possible, because one would have to know in advance which device intends to send the next message, and in realistic conditions, a device does not possess this knowledge.

7.2.2.3 Key Management

The ZigBee standard states that there can be maximally 255 ACL entries. The exact number of ACL entries is vendor specific, and often much lower than 255. As an example, the Chipcon CC2420 supports only two ACL entries (Sastry and Wagner, 2004). The number of application link keys a device can maximally share with other devices is equal to the number of ACL entries. So in the best case, it can share a key with only 255 other ZigBee devices, which is considerably less than the maximum number of 65,536 devices in a ZigBee network. A better support for secure end-to-end communication is certainly required.

7.2.2.4 Replay Attacks

On a single message transmission, the frame counter is incremented by one. This prevents replay attacks, as frames with a lower frame counter than that stored in the ACL will be discarded. This can, however, cause a security problem in broadcast communication. In a ZigBee network, broadcast communication is secured using the NWK key, which is stored in the default ACL. Every time a message is broadcast, all devices in the network should increment the frame counter in its default ACL. If a device goes to *sleep mode*, and does not receive broadcast messages for a certain time, the device refrains from sending broadcast messages in the future. The frame counter in its default ACL will have a lower value than the one in the default ACL of the other devices. The other devices will discard a message with a lower frame counter, as they wrongfully detect this event as a replay attack. As a consequence, a device can never go to sleep mode, and this can have an immediate impact on its battery lifetime.

7.2.2.5 Initialization Procedure

The secure initialization and installation of the master key determines the security of the other keys. When an attacker obtains the trust center master key, this would compromise the security of the other keys used in ZigBee, as they are all derived from the trust center master key.

A device can obtain the trust center master key (and the address of the trust center) in two ways: via insecure key-transport or via preinstallation. The former is the easiest method, but also the most insecure one. Transmitting a key at low power, as suggested in the ZigBee standard, does not provide sufficient protection. The attacker can build a ZigBee device with a strong directional antenna and intercept communication from a long distance. Assuming that there is no attacker present during the insecure key-transport is a very dangerous assumption. Theoretically, insecure key-transport is secure only when it is conducted in a Faraday cage. This is, however, not very practical. That is why it is recommended to obtain the trust center master key via preinstallation. This is more awkward, but provides more security. For example, one could install the trust center address and master key during the fabrication of the ZigBee device. There are, however, some practical problems. One does not always know in advance in which network the ZigBee device will be employed. Deriving the trust center master key from data entered by a user (a password) can be dangerous, because users tend to use low-entropy passwords, and an attacker can try all passwords or perform a dictionary attack. Since the SKKE protocol, used to establish a link key, contains a key confirmation step, an attacker can easily verify every guess of the password.

7.2.2.6 Location Privacy

The header of a ZigBee frame, which is never encrypted, contains the address of the source and destination device. This address is either the 64-bit IEEE address, or a 16-bit short address (used once the network is set up). An attacker who eavesdrops on the transmitted data knows the addresses of the devices that are communicating. It is possible for an attacker to construct a stronger antenna to intercept ZigBee communication from a greater distance. As a consequence, an eavesdropper does not have to be physically close to the communicating devices. This way, the attacker can keep track of the place and time that ZigBee devices are communicating. This is a violation of the privacy.

7.2.2.7 Insufficient Integrity Protection

In total, there are 8 security levels that can be employed to secure a frame. The payload can be encrypted or not, and the frame can contain a MIC of 0, 32, 64, or 128 bits. As a consequence, it is possible to apply encryption and no integrity protection on a frame. This is a dangerous mode of security, and should never be used. Encryption in itself does not provide integrity protection. As shown in Fig. 7.3, the cipher text C_i is the XOR of the plaintext message M_i and an encryption of a block A_i. This means that if the attacker changes the j-th bit of C_i, the same bit will change in the message M_i. This can have disastrous consequences. Nevertheless, the ZigBee standard states that all ZigBee command frames should be encrypted and integrity protected with a 128-bit MIC.

7.3 Coordinated Packet Traceback in WM^{2}Nets[2]

7.3.1 Related Work on Traceback

Network attacks can be either persistent or sporadic (Vetter et al., 1997). In persistent attacks, the offenders must frequently launch attack packets to bombard the victims, which can be relatively easy to trace by mechanisms such as probabilistic packet marking, traffic logging, data mining, etc.; whereas in sporadic attacks, a single packet can render havoc at the potential victim, such as the WinNuke, Ping of Death, and Teardrop attacks.

Intrusion detection systems (IDSs) use attack signature or pattern to help distinguish malicious packets from normal traffic. At the very least, an attack signature is defined by the IP address or address range of the entity that is being attacked. A variety of methods for IDSs were discussed in (Templeton and Levitt, 2003) that can help detect if received packets have spoofed source addresses.

After malicious attacks are detected, the subsequent traceback to the attack origins requires the network participate in the preattack tracking and postattack tracing operations, in which packet tracking refers to the recording of packets or packet flows when the packets are forwarded from their sources to their destinations, and packet tracing refers to the operations to find out the source route of a packet, which is the sequence of nodes that have forwarded the packet or the flow. This term is slightly different from what is commonly known by the source routing in IP forwarding.

The source route identification problem is commonly referred to as network traceback, or IP packet traceback in the Internet arena. Two network tracing problems are currently being studied: "single-domain IP traceback" and "traceback across stepping-stones" (or a "connection chain"). Traceback across steppingstones is to identify the origin of an anonymous attacker through a chain of connections before the attacker interacts with a victim host. The decentralized source identification system (DECIDUOUS) uses IPsec

[2] Excerpt from the invited article "Coordinated packet traceback in wireless mesh networks," Denh Sy and Lichun Bao, Computer Science Department, University of California, Irvine, CA 92697. E-mail: lbao@ics.uci.edu

security associations (SAs) and authentication headers to dynamically deploy secure authentication tunnels, and traces back to the attack origins (Chang and Chen, 2000a; Chang et al., 1999). The premise of DECIDUOUS is that if an attack packet has been correctly authenticated by a certain router, the attack packet must have been transmitted through that router. It utilizes IPsec security associations to dynamically deploy secure authentication tunnels in order to further trace down the possible attackers' locations. Other approaches such as correlating the inter-packet delay were also proposed (Wang and Reeves, 2003).

A number of approaches have been proposed to trace single-domain IP traceback. The IP marking approaches enable routers to probabilistically mark packets with partial path information, and try to reconstruct the complete path from the packets with the markings (Goodrich, 2002; Park and Lee, 2002; Savage et al., 2000; Song and Perrig, 2001; Waldvogel, 2002). A more explicit approach using IP marking is to add the IP address of the router to the IP header by turning on the Record Route option of the IP header (Doeppner et al., 2000). An algebraic approach is proposed to transform the IP traceback problem into a polynomial reconstruction problem, and uses techniques from algebraic coding theory to recover the true origin of spoofed IP packets by having routers embed information randomly into packets (Dean et al., 2001). This is similar to the technique used in (Savage et al., 2000), which uses algebraic techniques to encode the path information as points on polynomials, and then reconstruct these polynomials at the victim. Li et al. (2004) proposed a hybrid approach to selective mark packets for tracking in the intermediate routers using Bloom filters.

ICMP traceback (iTrace) proposes to introduce a new ICMP message (or an iTrace Traceback message) so that routers can, with a low probability, generate iTrace messages to help the victim or its upstream ISP to identify the source of spoofed IP packets (Bellovin et al., 2001). With enough ICMP Traceback messages from enough routers along the path, the traffic source and path can be determined. An intention-driven iTrace is also introduced to reduce unnecessary iTrace messages to improve the performance of iTrace systems (Mankin et al., 2001). An IP overlay network-based traceback system, named CenterTrack, selectively reroutes interesting IP packets directly from edge routers to special tracing routers, and the hop-by-hop input debugging is then used (Stone, 2000). Input debugging refers to the diagnostic features required to recursively determine from which adjacency a packet arrived that matches an attack signature on an individual router, until the edge of the network is reached and the edge ingress adjacency is identified.

Snoeren et al. (2001) proposed an architecture, source path isolation engine (SPIE), that integrates the IDS and single-packet traceback engines to identify the attach graph. According to SPIE, once the IDS detects an abnormal attack event, the attack packet is fed into the traceback manager to generate a traceback request, which is sent to multiple network agents for constructing the regional attack graph based on the attack packet. Afterward, the regional attach graphs are assembled into a complete attack graph at the traceback manager, and fed back the IDS. SPIE is based on a Bloom filter (Bloom, 1970) for packet logging purposes (Sanchez et al., 2001).

7.3.2 Multidimensional Hash Table

7.3.2.1 Bloom Filter and Its Variants

A Bloom filter is a space-efficient probabilistic data structure that is used to test whether or not an element is a member of a set (Bloom, 1970). Bloom filters are used in myriad of applications. Wherever a list or set is used, and space is a consideration, a Bloom filter is commonly considered (Broder and Mitzenmacher, 2002).

Typically, elements are only added to the set, but not removed. Given a finite set, false-positive judgment of the membership is possible, but false-negative judgment is not in the traditional Bloom filters without refreshing. When the more elements are added to the set, the probability of false positive becomes greater. Several variants of Bloom filters have been proposed. Attenuated Bloom filters (Rhea and Kubiatowicz, 2002) use arrays of Bloom filters to store shortest path distance information. Spectral Bloom filters (Matias and Cohen, 2003) extend the data structure to support estimates of frequencies. In counting Bloom filters (Fan et al., 1998) each entry in the filter need not be a single bit but rather a small counter. Insertions and deletions to the filter increment or decrement the counters respectively. When the filter is intended to be passed as a message, compressed Bloom filters (Mitzenmacher, 2002) may be used, where parameters can be adjusted to the desired trade-off between size and false-positive rate.

Space-code Bloom filter (SCBF) provides frequency estimation of an element by probabilistically filling up multiple normal Bloom filters, from which to statistically infer the frequency of the element (Kumar et al., 2003). In SCBF, a randomly chosen Bloom filter of the SCBF module takes the flow ID (source and destination IP addresses and port numbers) as the inputs, and records the membership in the table. When the Bloom filters in the SCBF module saturate, the contents of the Bloom filters are paged out into a log file. Later, according to the number of positive answers to a flow query into the log files offline, the flow volume can be probabilistically inferred.

7.3.2.2 Multidimensional Hash Table

It is easy to see that a Bloom filter can be reduced to a hash table based on a single hash function, where $k = 1$. Alternatively, a Bloom filter is the augmentation of a single hash table by the multiplicity of hashing functions. We generalize the construction of Bloom filters by introducing the concept of dimensions in hash algorithms, in which a dimension can expand by the number of either hash functions, hash tables, or both. We count the instances of expansions, and denote the number of dimensions, accordingly.

For instance, a hash table based on a single hashing operation is one-dimensional. The Bloom filter is a 2-dimensional hash table because a Bloom filter is implemented using k hashing functions in an array. In addition, the Bloom filter table can be split into k separate hash tables, respectively, as shown in Fig. 7.6a, which is another 2-dimensional hash table. In addition, the individual hash functions in Fig. 7.6a can again be augmented by k different hash functions, thus making it a 3-dimensional hash table, as shown in Fig. 7.6b.

In particular, we refer to the 3-dimensional hash table in Fig. 7.6b as a space–time Bloom filter (STBF) in CAPTRA (CoordinAted Packet TRAceback protocol) for WM^2Net packet traceback because the 2-dimensional Bloom filter data structure is replicated and scattered among multiple WM^2Net nodes, and updated asynchronously for packet tracking in different parts of the WM^2Net. It is straightforward that the aforementioned SCBF is another 3-dimensional Bloom filter in Fig. 7.6b for a different purpose.

By defining the dimensions in hash tables, we have generalized the concept of Bloom filter to a simpler but more comprehensive form that captures, however, all the hash algorithms exposed throughout this study. Basically, the dimension of hashing operation is unlimited. Depending on how each Bloom filter in the dimensions is utilized, we can

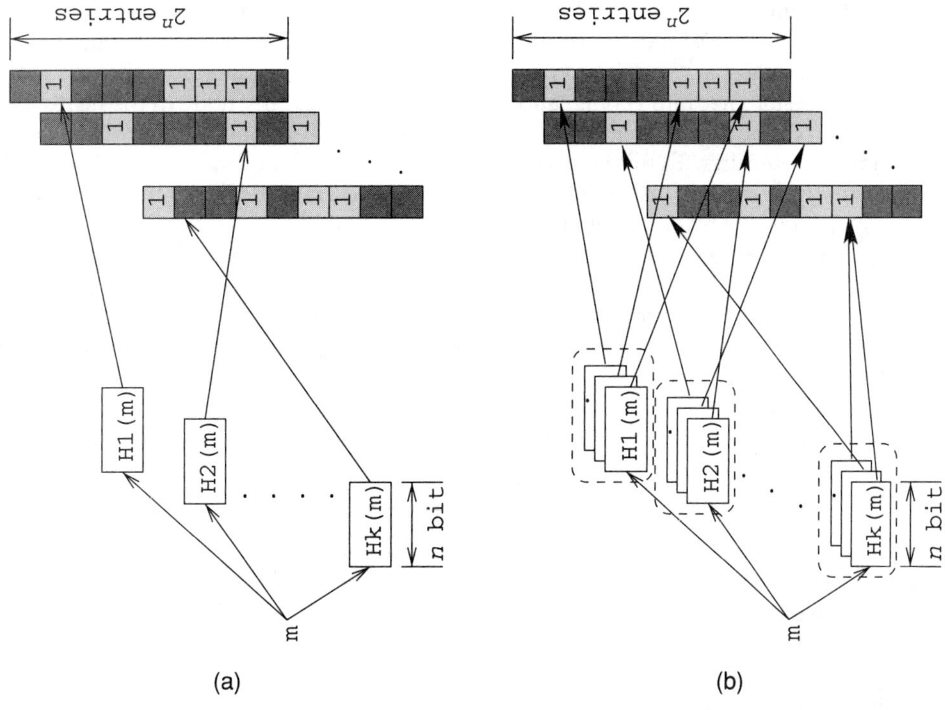

FIGURE 7.6 Dimensions of hash functions in filters. (a) Two-dimensional hash table. (b) Three-dimensional hash table.

apply the appropriate dimensions of hashing operations for membership query, such as packet tracing, routing, and accounting purposes (Kumar et al., 2003).

7.3.2.3 Coordinated Packet Traceback (CAPTRA)

We assume that a WM²Net consists of low-power devices that are interconnected via a shared wireless channel using the same channel access control protocol, such as the simplified IEEE 802.11 (IEEE, 1999; IEEE, 1999a) without collision avoidance mechanism using RTS/CTS. In a number of WM²Net deployments, the use of APs as an aggregation point and the exit to the Internet is a common practice. APs possess sufficient computing power to implement the intrusion detection mechanisms and initiate traceback operations. In addition, we assume that WM²Net node memories are limited so that the Bloom filters cannot be arbitrarily large, nor can there be permanent storage for logging purposes.

The choice of the Bloom filter algorithm is a local choice dependent on the memory resources and computational tasks of nodes, and its parameters on each mesh node are configurable, such as the table sizes. The reporting of a hit in the Bloom filter at a wireless node is also dependent on the capability of the node. Nodes in relatively static and dense wireless environments can report the hits less frequently, because the voting quorum can easily be satisfied, than nodes in highly mobile and sparse networks for the same level of traceback accuracy.

For simplicity, we assume that nodes are homogeneous in WM^{2}Net in terms of memory allocation, hashing algorithm, and hit reporting frequency. The essential difference of packet traceback in wireless networks from wireline networks operations is that the transmission medium is open and can be overheard by any node in the nearby vicinity of the packet transmitter. Therefore, we can take advantage of this fact in the space–time Bloom filter (STBF) constructions. That is, the Bloom filters of a single STBF are virtually dispersed among adjacent wireless nodes, and the Bloom filter lookup is now a majority-vote by all the Bloom filters of the potential nodes.

The distribution of the virtual STBF is especially useful in WM^{2}Nets because (1) the aggregated STBF provides larger memory capacity and parallel computing power than a single node, (2) the traffic intensity is uneven in different locations of WM^{2}Nets, thus distributing the Bloom filters off-loads the memory requirements of the nodes with high traffic in its neighborhood, and (3) nodal mobility often invalidates the source routes constructed from the packet traceback. By distributing the STBFs to multiple nodes in vicinity, the traceback operations can tolerate certain node losses along the source route because the predecessor of the missing node can still be detected if there are sufficient witnesses to the packet being sent from the predecessor.

Hash Functions Because the main purpose of packet traceback is to capture the path traversed by a specific packet through the network, we extract the certain pieces of information to identify the packet. For instance, the identification of an IP packet may include IP version, header length, source and destination addresses, fragment information, and portions of the payload (Snoeren et al., 2001), which is then concatenated and packed into a bit string as the input to hash functions.

In order to use the same hash algorithm for different hash functions in the STBF, we feed additional information to the hash functions for packet tracing purpose:

1. An identifier for each hash function, which includes two pieces of information: (1) the host ID and (2) a unique ID for the hash function.
2. The predecessor from which the packet is received.

By including the hash function ID, we can use the same hashing algorithm, such as MD5 (Rivest, 1992), to compose different hash functions in the STBF. By including the predecessor ID, a Bloom filter hit for a packet also reveals the previous hop of the packet. In the actual computations, all neighbor IDs have to be checked in the Bloom filter in order to find out the predecessor.

Query in the STBF As STBF continuously monitors network traffic, the hash tables can gradually saturate, and the probability of false positive increases. To guarantee the best performance trade-off between the memory utilization and false-positive rate, a "50% Golden Rule" is derived and applied so that the Bloom filter can hold the maximum number of elements (Bloom, 1970; Kumar et al., 2003) with lowest false-positive rate.

Previous applications of Bloom filters for accounting or tracing purposes reset the whole Bloom filter memory when the filter saturation crosses the threshold. Therefore, the false-positive rate is a variable, while the false-negative rate is either 0 before flushing or 1 after flushing for certain elements. In the STBF, individual Bloom filter elements are dispersed among multiple nodes. When the saturation ratio of an individual hash table crosses the 50% threshold, we wipe out only that single hash table in the STBF.

Therefore, we define a majority-vote principle for STBF that allows STBF to yield a "hit" response when the number of "hit" responses from different Bloom filter elements satisfies a quorum. This is drastically different from the membership query operation in the traditional Bloom filter applications, where a hit requires unanimous hits on all the hash functions.

Due to the reset function on the STBF, the majority-vote quorum is a trade-off between the false-negative and false-positive rates in STBFs. If the quorum is low, the false-positive rate increases, and if high, the false-negative rate increases.

Intuitively, the majority-vote mechanism in STBF maintains gradually fading memory of the packet events and allocates more acute memory to recent traffic flows. We see increasing probability of false negatives because the individual hash tables are asynchronously being reset. In contrast, previous Bloom filter resets abruptly change the false-positive and false-negative rates of the Bloom filter to 0 or 1 in previous applications.

In order to avoid possible synchronized hash table resets, we add certain randomness to the reset timing when the saturation rate reaches the threshold.

Traceback Messages In support of packet traceback operations, three traceback messages are used, which jointly provide the majority-vote mechanism to corroborate the conviction of a node as one on the packet source route.

Message $TRAC_{REQ}$ initiates a traceback query to discover the predecessor of a packet, used for pinging a network host. It is initialized by the AP and recursively carried out by all nodes along the source route. In the payload field of the traceback request message, the packet identification is carried.

Upon each message $TRAC_{REQ}$ arrival, the packet traceback protocol extracts the packet identification, and runs the hash functions on all the current one-hop neighbors to determine the potential predecessor of the packet. If such predecessor exists, the traceback operation can continue by sending the traceback request to the predecessor. Message $TRAC_{VERD}$ is for a node to indicate that it has witnessed the packet, and to issue a verdict to the sending node of the packet.

Message $TRAC_{CONF}$ is sent to the WM²Net AP by a node c to tell that node c has forwarded the packet before if it has collected enough verdicts from its neighbors. In addition, node c propagates the traceback request farther to its candidate neighbors for farther traceback.

Packet Tracking In CAPTRA, each node maintains a Bloom filter using k hashing functions. As mentioned before, upon overhearing a packet from a sender, a node hashes a concatenated bit stream containing the packet identification information, the sender's ID, and the node's ID to a table index, and sets the corresponding bits to 1 in the Bloom filter. For instance, if node i hears a packet *pkt* with identification information *pkt.id*, the packet is tracked by a set of k bits in node i's Bloom filter, indexed by using Eq. (7.1).

$$\bigcup_{j=1}^{k} Hash_j(pkt.id \,||\, pkt.tx \,||i||\, j) \tag{7.1}$$

in which $Hash_j$ is a hash function with index number j, *pkt.id* is the packet identification information, *pkt.tx* is the transmitter ID of the packet, the symbol "k" represents the concatenation of the operants, and k is the number of hash functions in the Bloom filter.

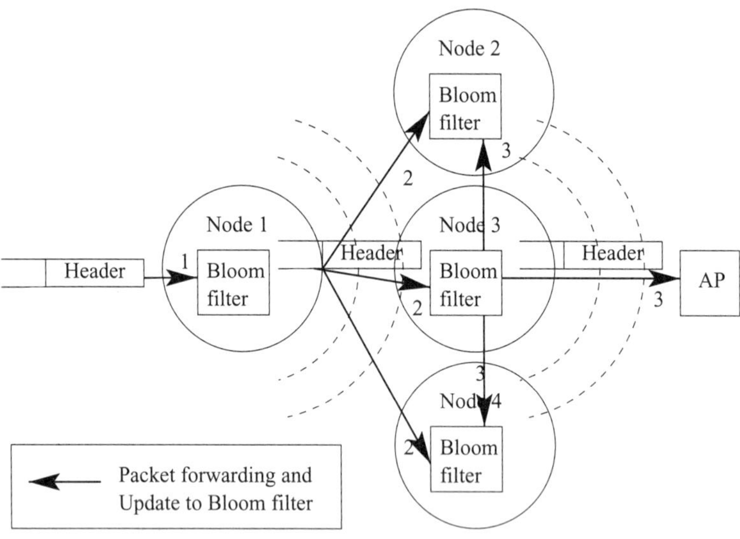

Figure 7.7 STBF for packet tracking.

A packet is tracked when a node either forwards the packet, or overhears and successfully receives the packet. Because the overhearing node's ID is involved in the hashing algorithm, each node overhearing the packet maps the packet to a different Bloom filter entry. Therefore, the Bloom filter fill-up rates are different at different nodes.

Figure 7.7 illustrates a partial mesh network with three nodes and an AP for data collection purposes. The arrows indicate the packet transmissions and receptions. The senders and the receivers of the packet record the packet in their Bloom filters.

According to the "50% Golden Rule," the Bloom filters are refreshed once the saturation ratio reaches 50% of the Bloom filter capacity, and every bit of the Bloom filter is reset to 0.

Packet Tracing In order to describe the packet traceback operations in CAPTRA, we again use the partial network in Fig. 7.7 as an example.

Suppose that the AP in Fig. 7.7 initiates a traceback query on packet *pkt* due to a security breach detected by an IDS. The AP sends out the traceback request with the packet identification *pkt.id* in the newly defined message TRAC_{REQ} with *pkt.id* as payload. Because the AP can look up its own Bloom filter for a hit, and find out the sender of the packet identified by *pkt.id*, the traceback request is sent directly to the packet predecessor—mesh node 3, and is overheard from all nodes around the AP.

Due to the broadcast nature of the wireless medium, nodes 2 and 4 overhear the traceback request. This trend is illustrated in Fig. 7.8 where packet *pkt* is overheard from nodes 2 and 4 on its transition to node 2. Nodes 2 and 4 record then the *pkt.id* along with the sender ID in their Bloom filters. If there is a positive hit, node 2 or 4 sends the $\text{TRAC}_{\text{VERD}}$ packet to 3 to reinforce the tracing request.

When enough verdicts are collected after a period of time, including the Bloom filter hits at nodes 2, 3, and 4 and the AP, node 3 sends back a $\text{TRAC}_{\text{CONF}}$ packet to the AP

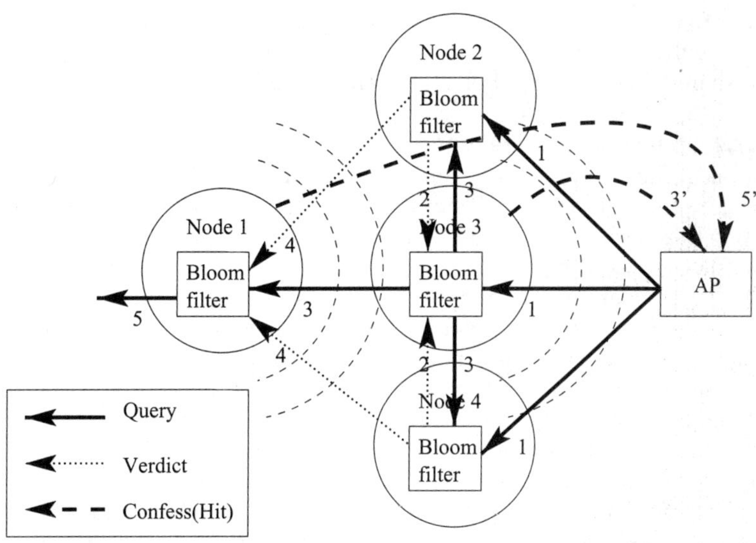

FIGURE 7.8 Packet tracing process using STBF.

for attack source route construction. The threshold of the number of verdicts for further traceback is a tunable parameter, and is at least 2 because the traceback request sender and receiver also generate verdicts. Note that each traceback request is a verdict to the receiver of the traceback request.

In addition, node 3 generates further traceback request TRAC_{REQ}(*pkt.id*) based on its current neighbors, and in Fig. 7.8, a new request is generated to node 1, after which the verdict generation process follows the same as above. If in rare cases, multiple one-hop neighbors of node 3 appear to be the predecessor of the packet, each of them receives the traceback request. If a node is wrongly convicted of the packet transfer, it is expected the wrong traceback request gradually die out because of lack of evidence further away from the convicted node.

Loop Avoidance in Packet Traceback With no additional mechanisms, the above traceback protocol can cause loops while forwarding the traceback request, one instance in Fig. 7.8 being that node 1 sends the traceback request again to node 3 because node 1 has heard about node 3's transmission of the packet, which causes unnecessary traceback activities.

To avoid the formation of potential loops, the traceback request is appended with distance information, indicating the distance from the current sender of traceback request to the origin of the traceback request. If a node receiving the traceback request has a shorter distance to the request origin than the sender of the request, the traceback request is discarded. Note that the criteria based on the reverse path to the origin may be incorrect owing to network mobility.

Termination of Tracing Query When a node has no other one-hop suspect but the one who transferred the packet, the tracing ends. Some optimizations are possible in CAPTRA. In case the suspect that receives the traceback request is an uncooperative node, and refuses to respond to the request for packet tracing, the request generator may be able to conclude that the suspect has actually transfer the packet according to the verdicts from their shared one-hop neighbors, for example, nodes 1 and 3 in Fig. 7.8.

7.3.2.4 Simulation Evaluations

We used J-Sim (http://www.j-sim.org/) to simulate the packet traceback protocol in WM^2Nets. J-Sim is a network simulator constructed entirely in Java. The MAC layer of the wireless mesh nodes was modified to implement CAPTRA, RTS/CTS was turned off to simulate a WM^2Net.

In the simulations, twenty wireless nodes, numbered from 0 to 19, are deployed in a linear fashion, spaced 8 meters apart. Each node has a 25-meter transmission range, so that each node has up to 7 neighbors. The propagation and path-loss model use the free-space model. AODV is used as the underlying routing protocol. Each node contains a Bloom filter of size 4,096 bits, which is refreshed using the "50% Golden Rule." The number of hash functions for each packet, k, is a system-wide variable, and varies in two simulation scenarios from 2 to 3 for comparison purposes.

The node traffic source is attached to node 0, and destined to node 19. A CBR traffic generation model with 512 bytes per packet at a rate of 10 packets per second was used. We change the location node 0 in the network so as to change the hop distance to node 19 in order evaluate the performance packet tracing with regard to the hop distance.

In each simulation round, the packet tracking capacity of the network is tested by collecting the false-positive and false-negative rates when different amounts of traffic are forwarded between the times when the packet is generated at the source and when the packet is being traced back from node 19. As discussed earlier, the CAPTRA voting system is used to convict a node, and have also a node confess.

Simulation Results and Analysis Figure 7.9 presents the false-positive and false-negative rates under the various simulation scenarios. Two types of Bloom filters are implemented and compared side-by-side, in which one type of Bloom filter uses 2 hash functions in all the simulation scenarios, and the other uses 3 hash functions in each. The false-positive rates are simply the number of false positives divided by the hop count in the WM^2Net. Similarly, false-negative rates are the number of false negatives divided by the hop count.

In Figs. 7.9a, 7.9c, and 7.9e, the traceback performance is measured for 2-, 4- and 8-hop source routes of a particular packet. As we can see, when the number of hops increases, the network has shorter and shorter memory of the traced packet. This is indicated from the number of packets interleaved from the time when the traced packet is generated to the time when the packet is traced back. In addition, the duration of the network memory about the packet depends on the network density, and the size of the Bloom filters, which are 4,096 bits.

On the other hand, Figs. 7.9b, 7.9d, and 7.9f show the false-positive and false-negative rates of similar simulations when the number of hash functions is three. Comparing with the other three corresponding sub-diagrams, these three settings show a more stable traceback performance, but shorter memory of the traced packet due to the faster fill-up rate of the Bloom filters. Therefore, the choice of Bloom filters depends on the application trade-offs between accuracy of traceback and the duration of the traceback validity.

The unstable false rates that appear on the curves of the false-positive and false-negative rates are due to the Bloom filter collisions.

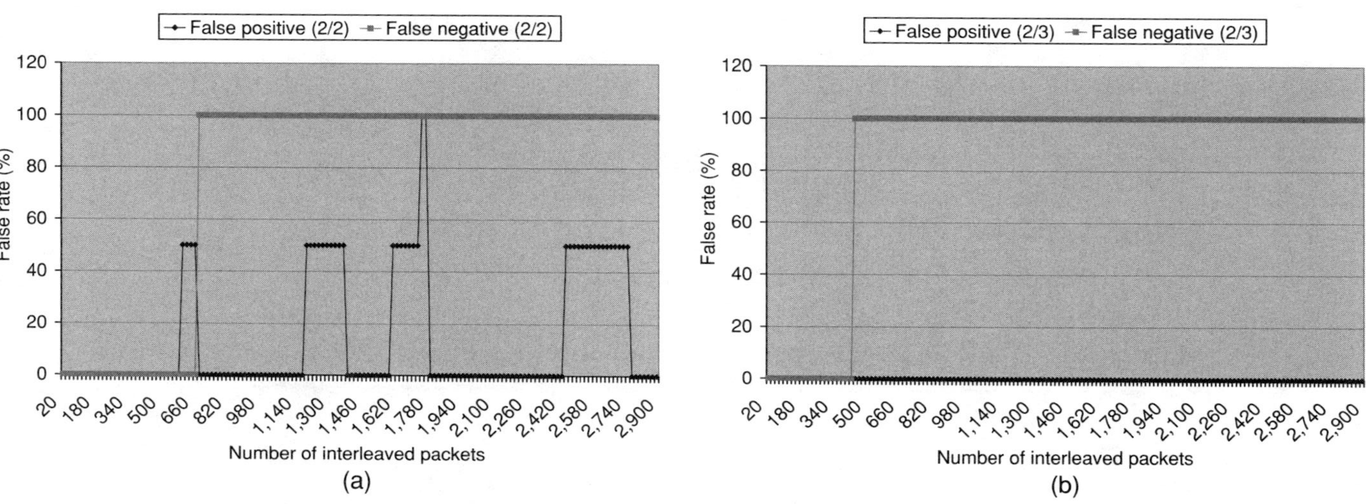

Figure 7.9 The false-positive and false-negative rates in multihop packet traceback. (a) 2-hop traceback with two hash keys. (b) 2-hop traceback with three hash keys. (c) 4-hop traceback with two hash keys. (d) 4-hop traceback with three hash keys. (e) 8-hop traceback with two hash keys. (f) 8-hop traceback with three hash keys.

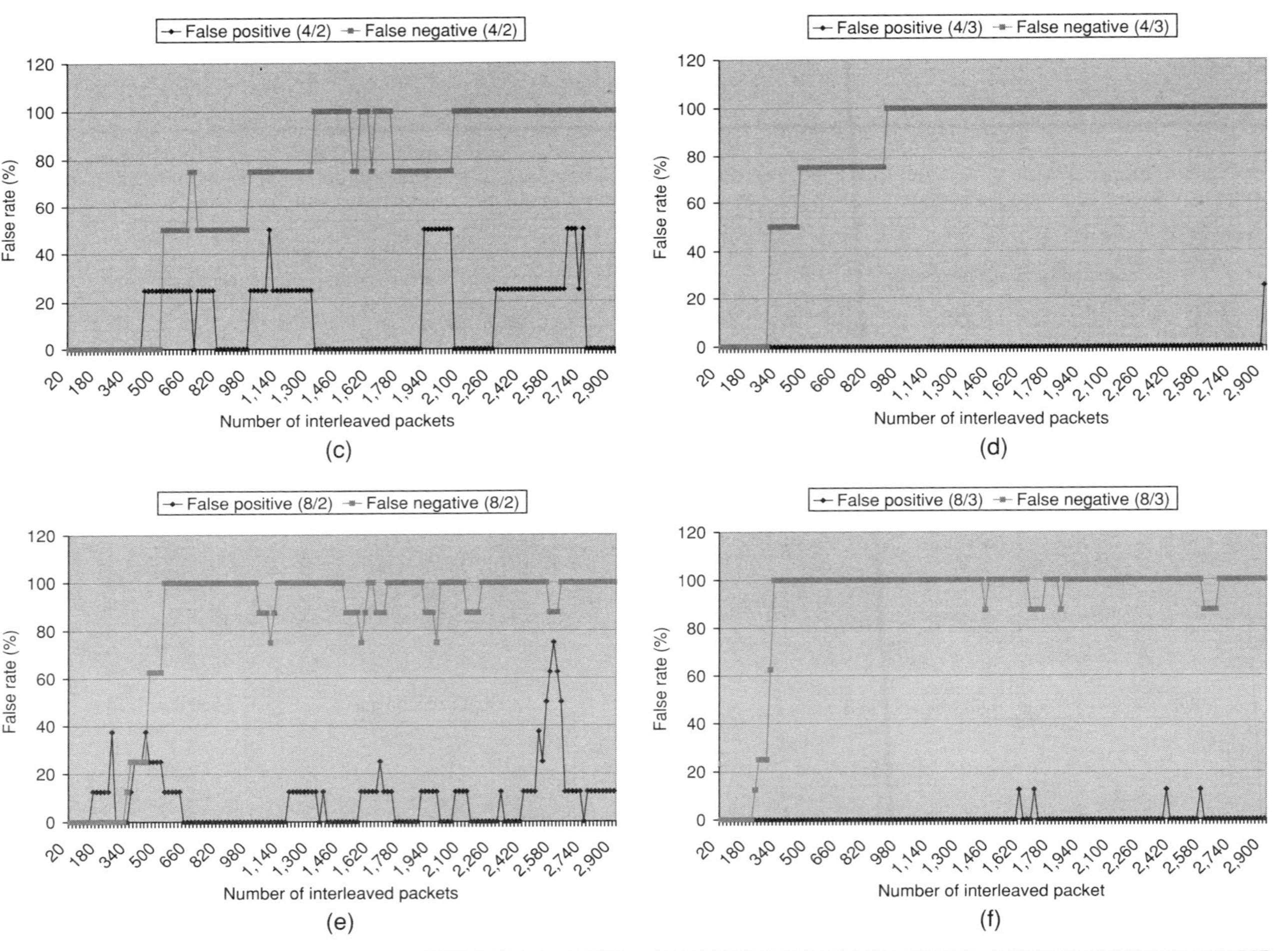

FIGURE 7.9 *(Continued)*

7.4 On the Identity-Based Encryption for WM²Nets[3]

7.4.1 Related Work on Key Management

A considerable number of works (Eschenauer and Gligor, 2002; Perrig et al., 2002; Zhu et al., 2003; Liu et al., 2003; Di Pietro et al., 2003; Oliveira et al., 2007; Oliveira et al., 2006) have focused on efficient key management of symmetric cryptosystems. Perrig et al. (2002) proposed SPINS, a suite of efficient symmetric key-based security building blocks. Eschenauer et al. (2002) looked at random key predistribution schemes, whereas (Zhu et al., 2003) proposed LEAP, a rather efficient scheme based on local distribution of secret keys among neighboring nodes.

The studies specifically targeted to PKC have tried either to use adequate conventional algorithms (e.g., RSA) to WSNets, or to employ more efficient techniques (e.g., ECC). Watro et al. (2004) proposed TinyPK. To perform key distribution, TinyPK assigns RSA efficient public operations to nodes and RSA expensive private operations to better suited external parties. To perform key distribution, TinyPK assigns RSA efficient public operations to nodes and expensive private operations to better suited external parties. Gura et al. (2004) reported results for ECC and RSA on the ATmega128 and demonstrated that the first outperforms the latter. Their ECC implementation uses prime fields. Malan et al. (2004) implemented ECC using binary fields and polynomial basis and presented results for the Diffie-Hellman protocol based on the ECDLP.

The above works have shown that nodes are able to compute PKC operations, but public key authentication has not been their focus of research. Motivated by that, proposals by Du et al. (2005), Zhang et al. (2005), and Doyle et al. (2006) have been made to address this issue. Du et al. (2005) proposed a scheme based in Merkle trees; this is able to authenticate public keys using only symmetric operations. Zhang et al. (2005) have made use of IBE (Identity-based Encryption) for key distribution in WSNets. They hoped that pairings would be soon feasible in resource-constrained nodes and were not concerned with implementation issues. The work of Doyle et al. (2006) also focused on IBE and presented some simulation results on pairings. The work, however, has considered a class of nodes more powerful than those found in resource-constrained nodes.

7.4.2 Pairings: Concepts

Bilinear pairings—or pairings for short—were first used in the context of cryptanalysis (Menezes et al., 1993), but their pioneering use in cryptosystems is due the works of Sakai et al. (2000) and Joux (2000). In this section, we first present some paring concepts and then define the Tate pairing. (For more on these definitions, see for instance, Galbraith, 2005.) In what follows, let E/F_q be an elliptic curve over a finite field F_q, $E(\mathrm{F}_q)$ be the group of points of this curve, and $\#E(\mathrm{F}_q)$ be the group order.

---

[3] Excerpt from the invited article "On the identity-based encryption for wireless mesh networks," *Leonardo B. Oliveira, Diego F. Aranha, Eduardo Morais, Felipe Daguano, Julio López, and Ricardo Dahab (*Institute of Computing, UNICAMP, Brazil, E-mail: leob@ic.unicamp.br). The work was supported by FAPESP under grant 2005/00557-9.

7.4.2.1 Bilinear Pairing

Let ℓ be a positive integer. Let G_1 and G_2 be additively-written groups of order ℓ with identity 0, and let G_T be a multiplicatively-written group of order ℓ with identity 1.

A *bilinear pairing* is a computable, non-degenerate function $e : G_1 \times G_2 \to G_T$. The most important property of pairings in cryptographic constructions is the bi-linearity, namely:

$$\forall P \in G_1, \forall Q \in G_2 \text{ and } \forall a, b \in Z^*, \text{ we have } e([a]P, [b]Q) = e(P, Q)^{ab}.$$

7.4.2.2 Embedding Degree

A subgroup G of $E(F_q)$ is said to have an *embedding degree* k with respect to ℓ if k is the smallest integer such that $\ell | q^k - 1$.

7.4.2.3 Bilinear Diffie-Hellman Problem

Most of the PBC applications rely on the hardness of the following problem for their security (Galbraith, 2005): Given $P, [a]P, [b]P$, and $[c]P$ for some $a, b \in Z^*$, compute $e(P, P)^{abc}$.

This problem is known as the *bilinear Diffie-Hellman problem*. The hardness of the bilinear Diffie-Hellman problem depends on the hardness of the Diffie-Hellman problems both on $E(F_q)$ and in F_{q^k}. So, for most PBC applications, the parameters q, ℓ, and k must satisfy the following security requirements:

1. ℓ must be large enough so that solving the elliptic curve discrete logarithm problem (ECDLP) in an order-ℓ subgroup of $E(F_q)$ is infeasible (e.g., using Pollard's rho algorithm);
2. k must be large enough so that solving the discrete logarithm problem (DLP) in F_{q^k} is infeasible (e.g., using the index-calculus method).

7.4.2.4 The Tate Pairing

Let $E(F_q)$ contain a subgroup of prime order ℓ coprime with q and with embedding degree k. (In most applications, ℓ also is a large prime divisor of $\#E(F_q)$.) The *Tate pairing* is the bilinear pairing:

$$\hat{e}: E(F_{q^k})[\ell] \times E(F_{q^k})/[\ell]E(F_{q^k}) \to F^*_{q^k}/(F^*_{q^k})^\ell.$$

7.4.3 Applying IBE to WM^{2}Nets

Today, IBE (Boneh and Franklin, 2003; Cocks, 2001) seems to be the only truly practical mean of providing public key encryption in WM^{2}Snets. IBE would employ nodes' identification (e.g., node IDs) as public keys and PKI's expensive operations would be thus unnecessary.

We go further and argue that IBE is not only ideal for WM^{2}Nets, but the converse is also true. For example, IBE schemes have strong requirements such as the existence of an unconditionally trusted entity, which is responsible for issuing users' private keys. WM^{2}Nets, however, possess intrinsically such an entity, namely the AP. Another requirement is that the keys must be delivered over confidential and authentic channels to users. In most of the WM^{2}Nets applications, however, nodes' private keys can be distributed *off-line*, that is, they can be generated and preloaded directly into nodes prior to deployment.

The various symbols denote:

id_X :	Node X's ID	$\text{mac}_k()$:	MAC computed using key k
$\wp_X$:	Group of nodes in node X's neighborhood	$\text{enc}_k()$:	Encryption computed using key k
k_{XY} :	Secret key shared between nodes X and Y	m :	Message information
P_X :	Node X's public key	$\Rightarrow, \rightarrow$:	Broadcast and unicast, respectively

FIGURE 7.10 Key distribution protocol.

In spite of all its advantages, IBE still is a public key cryptosystem and thus it is orders of magnitude more complex than symmetric cryptosystems. Because of this, as usual, IBE would be used only for setting up pairwise secret keys among nodes.

In Fig. 7.10, we show how IBE can be used to establish secret keys among communicating nodes. (In WM²Nets, where the communication is in general multihop from nodes to the BS, communicating nodes are often the neighboring nodes.) The protocol works as follows.

Prior to deployment, each node X is assigned the following information: the node's ID id_X, the node's IBE private key S_X, and a function ϕ that takes an ID (e.g., id_Y) as input and outputs the corresponding IBE public key to the ID (e.g.,P_Y).

After deployment, each node broadcasts its ID and a nonce (Step 1). Neighboring nodes thus use the function ϕ together with the received ID to derive the corresponding public key. After that, neighboring nodes generate a secret key and respond to the original node by including this key in the message (Step 2). The transmission of the message is protected from using IBE's public and private keys. To prevent replay attacks, the nonce from the original node's broadcast in Step 1 is also included in the message. Finally, subsequent communications among nodes are protected with MACs[4] computed using the secret keys (Step 3). A value computed from the nonce (nonce') is also included as input to the MAC to prevent from message replays; in fact, the value of the "freshness token" nonce' needs to be updated during each interaction (Step 3).

7.4.4 Implementation and Evaluation

The time consuming part while evaluating IBE is the pairing computation. In this section, we describe implementation issues and present results on computing pairings over MICAz, the new generation of MICA mote node (Hill, 2002). MICAz is powered with the ATmega128 microcontroller (8-bit/7.38 MHz processor, 4 KB SRAM, 128 KB flash memory).

7.4.4.1 Implementing Issues

Recall that E/F_q is an elliptic curve defined over F_q, ℓ is a large prime divisor of $\#E(\mathrm{F}_q)$ coprime to q, and is the embedding degree.

7.4.4.2 The Pairing

The two most important pairings in ECC are the Tate and the Weil pairings. According to Galbraith (2005), the Tate pairing seems to be more efficient than the Weil pairing. Therefore, the Tate pairing appears to be more adequate to WM²Nets than the Weil pairing.

[4] MAC = message authentication code.

7.4.4.3 The Field

Given a cryptosystem, the hardness of its underlying problem dictates the size of the security parameters. Notably, the harder the problem, the smaller the parameter size becomes. The parameter size, in turn, dictates the efficiency, that is, the smaller the parameter size, the faster the computation time. The DLP in prime fields is considered to be harder than the DLP in binary fields and thus it seems that prime fields are more adequate to WM^2Nets.

7.4.4.4 Curve Selection

Supersingular curves have been shown empirically to be faster (Scott, 2005) than non-supersingular curves. Authors, however, tend to choose nonsupersingular curves rather than supersingular curves because they feel that the latter have security advantages compared to the formers. Since until now no concrete evidence for that has appeared (Scott, 2005), supersingular curves seem to be more adequate to WM^2Nets.

7.4.4.5 Parameters q and ℓ

The choice of the parameters q and ℓ is a key factor in the efficiency of pairing computation, as curve operations are performed using arithmetic of the underlying field. In prime fields, by choosing q a Mersenne prime (i.e., a number of the form $2^p - 1$) helps in computing modular reduction operations efficiently. However, it has been shown recently that such technique also decreases the hardness of the DLP in F_q (http://eprint.iacr.org/) and is potentially unsafe in the context of PBC. For ℓ, on the other hand, it is possible to choose a Solinas prime, which decreases the number of point additions and makes the pairing computation faster.

7.4.4.6 Embedding Degree k

We have chosen $k = 2$ since it provides a number of benefits while computing pairings (Scott, 2005). For example, $k = 2$ allows the denominator elimination optimization and makes F_{q^k} arithmetic easier to implement.

7.4.4.7 Parameter Sizes

Parameter sizes often pose a tradeoff between security level and efficiency. For most PBC schemes (including IBE), the security requirements highlighted above can be satisfied by choosing $\ell > 2^{160}$ and $q^k > 2^{1024}$. However, security requirements in WM^2Nets are often relaxed (Perrig et al., 2002) to meet their needs for efficiency. This is possible because of their short lifetimes and because the goal is not to protect each node individually, but the network operation as a whole. Until now, the largest parameter sizes for which the ECDLP and the DLP in prime fields are known to be solved are 2^{109} and 2^{448}, respectively. Therefore, it seems that $\ell \geq 2^{128}$ and $q^k \geq 2^{512}$ are able to meet the current security requirements of WM^2Nets.

7.4.4.8 Point Coordinates

The two most common coordinate systems are the *projective* system (x, y, z) and the *affine* (x, y) system. The affine system requires inversions while performing point addition or doubling operations. The inverse operation, in turn, is commonly expensive. The projective system, on the other hand, reduces the need for inverse and thus seems to be more adequate to the target processor.

Tate Pairing		
Time (s)	**RAM (bytes)**	**ROM (bytes)**
(30,21)	(1,831)	(18,384)

Table 7.1 Costs to Evaluate the Tate Pairing on MICAz

7.4.4.9 Twists

Let d be a quadratic non-residue in F_q. The twist of an elliptic curve $E/F_q : y^2 = x^3 + ax + b$ is given by $E^t/F_q : y^2 = x^3 + d^2ax + d^3b$. For $k = 2$, there exists an isomorphism $\lambda : E(F_{q^2}) \to E^t(F_q)$ such that $\lambda[(a, 0), (0, d)] \to (-a, d)$, and arithmetic in $E(F_{q^2})$ can be thus carried out faster in the group $E^t(F_q)$.

7.4.5 Result

In this section, we describe the results of TinyTate, an implementation of the Tate pairing for resource constrained nodes. The implementation is based on Barreto et al.'s (2002) work and uses the Miller's algorithm (Miller, 1986) for pairing computation.

We use the following parameters: (1) the Tate Pairing on elliptic curves defined over fields with a large prime characteristic; (2) the embedding degree $k = 2$, q is a 256-bit prime, and ℓ a 128-bit Solinas prime; and (3) group field arithmetic uses projective coordinates. To be concrete, we use the curve $E/F_q : y^2 = x^3 + x$ with the parameters[5]:

$$q = 37781606889598235856745576472658394721481625071533302983957476142038207746163;$$

$$\ell = 170141188531071632644604909702696927233;$$

$$h = 222060320700642449943812747791145685108;$$

where h stands for the cofactor of the curve order $\#E(F_q)$.

Results in Table 7.1 were measured on a MICAz node running TinyOS. The average execution time to compute a pairing is 30.21s. The costs concerning RAM and ROM (flash) memories are 1,831 and 18,384 bytes, respectively.

Since we use IBE only to distribute secret keys among neighboring nodes, the costs above are not a heavy burden to the whole system.

7.5 Key Management for WM²Snets[6]

7.5.1 Taxonomy of Key Management Schemes

The key management process involves three phases, *key setup phase*, *key discovery phase*, and *key update phase*. In the key setup phase, secret keys or secret information are generated and carefully preloaded into mesh nodes. In the key discovery phase, the deployed mesh nodes communicate with each other and exchange information to discover or generate

[5] Note that in this particular case there is a twist with same equation of the original curve.

[6] Excerpt from the invited article "Key management for wireless mesh sensor networks," Zhen Yu, Yawen Wei, and Yong Guan. Department of Electrical and Computer Engineering, Iowa State University, Ames, IA 50011, USA, E-mail: {yuzhen@iastate.edu, weiyawen@iastate.edu, yguan@iastate.edu}.

secret keys. Moreover, two neighbors may establish secret keys through other discovered secure path(s) if they cannot find secret keys directly. In the key update phase, compromised or obsolete keys are revoked and new keys are distributed or generated. We notice that key updating (or rekeying) is a hard problem and has not been fully solved so far.

Key management schemes for WSNets can be classified in various categories (see Table 7.2 for a taxonomy of key management schemes for WSNets). Depending on the type of keys used, they are classified in three classes: pairwise key schemes, group key schemes, and global key schemes. These keys are used for encryption or authentication of communications between a pair of nodes, among a group (or cluster) of nodes, and over the whole networks. We further classify the schemes into probabilistic, deterministic, and hybrid, depending on whether the keys can be established with some probability or deterministically. Hybrid schemes apply some deterministic approach over a probabilistic one.

7.5.2 Pairwise Key Management Schemes

7.5.2.1 Probabilistic Schemes with no Deployment Knowledge

Eschenauer and Gligor (2002) proposed the first probabilistic key predistribution scheme, called a *basic scheme*. In the key setup phase, each node is randomly preloaded with m keys from a global key pool of size M. Hence, a pair of nodes can establish a secure link with probability $p = 1 - \frac{\binom{M-m}{m}}{\binom{M}{m}}$. From random graph theory, the desired probability P_c that the entire network is connected can be achieved by choosing a proper p. The drawback of this scheme is that a pairwise key may be used by multiple pair of nodes. Given x nodes compromised, the adversaries can compromise an additional link with probability $1 - (1 - \frac{m}{M})^x$.

To improve resilience, Chan et al. (2003) extended the basic scheme to a *q-composite scheme*, which requires a pair to share at least $q > 1$ keys to establish a secure link. However, this scheme sacrifices the achievable connectivity. Hence, the authors proposed a *random pairwise-key scheme*, which keeps the achievable connectivity when improving resilience. The idea is that for each node, a set of m nodes is randomly chosen, and a unique pairwise key is assigned for this node and every node in the set. Since all pairwise keys are distinct, the scheme has a perfect resilience against node capture, but the size of network is limited by $\frac{m}{p}$.

In the basic scheme, each node is preloaded with m keys and has to broadcast all m key IDs to discover the shared key(s). Hwang et al. (2004) proposed a *cluster-grouping scheme* to reduce the communication overhead of sensor nodes. In this scheme, the keys stored by each node are divided into c ($c < m$) equal-length clusters where each cluster has a start key ID. The remaining key IDs within the cluster are implicitly known from the start key ID. Hence, each node needs to broadcast only c start key IDs, instead of m key IDs. Zhu et al. (2003) presented a pairwise key scheme that can reduce not only the communication overhead and but also the required storage. In this scheme, each sensor is assigned with a unique ID and a pseudorandom function f, where its key IDs can be determined by executing f (*ID*). Thus, each node only needs to broadcast its IDs to discover the shared key(s). Similarly, Ren et al. (2006) also proposed to use hash chains to construct the global key pool. Hence, for each chain assigned to a node, the node

	Pairwise Key				Group Key	Global Key
	Deterministic					
	Probabilistic	Predeployment Security	Postdeployment Security	Hybrid	Deterministic	
No deployment knowledge	Eschenauer and Gligor (2002); Chan et al. (2003); Zhu et al. (2003); Ren et al. (2006); Hwang et al. (2004); Hwang and Kim (2004a)	Blom (1985); Blundo (1992); Chan, and Perrig, (2005)	Dutertre et al. (2004); Zhu et al. (2003)	Du et al. (2003); Liu and Ning (2003)	Blundo (1992); Zhu et al. (2003)	Perrig et al. (2001)
Deployment knowledge	Du et al. (2004); Liu and Ning (2003a); Liu and Zhao (2005); Huang et al. (2004)			Liu and Ning (2003a); Yu and Guan (2005) Zhou et al. (2005); Zhou et al. (2005a)		

Table 7.2 Taxonomy of Key Management Schemes For WSNet

needs to store only the corresponding generation key and the hash function. In addition, Hwang and Kim (2004a) revisited the basic scheme and its derivatives, and proposed to reduce the number of keys stored by each node while still keeping a certain probability of sharing a key between two nodes. The basic idea is that probability is sufficient to guarantee the largest component, instead of the whole network, to be almost connected.

7.5.2.2 Deterministic Schemes

Blom (1985) proposed a deterministic scheme to set up pairwise keys for a group of N nodes.

First, a key distribution center constructs a $(t+1) \times (t+1)$ symmetric matrix D and a $(t+1) \times n$ public matrix G over a finite field, where $(DG)^T$ is called the secret matrix. All pairwise keys of these N nodes are stored in a symmetric matrix $K = (DG)^T G$. Then, each node i is preloaded with the i-th row of the secret matrix and the i-th column of the public matrix. After deployment, any two nodes i and j can individually derive their pairwise key $k_{ij} = k_{ji}$ by only exchanging their columns. This scheme is t-secure, that is, no additional key is revealed given up to t nodes compromised. Otherwise, the whole matrix will be broken. Although we can improve the resilience of the scheme by increasing t, each node has to store a larger amount of secret information, which is $O(t)$.

Blundo (1992) presented a *polynomial-based key management scheme*, which is a special case of Blom's scheme when matrix D is a Vandermonde matrix. The basic component of this scheme is a t-degree bivariate polynomial $f(x, y) = \sum_{i=0}^{t} \sum_{j=0}^{t} a_{ij} x^i y^j$ over a finite field, where we have $f(x, y) = f(y, x)$ by choosing $a_{ij} = a_{ji}$. Each node i is preloaded with a polynomial share $f(i, y)$. After deployment, any two nodes i and j can individually derive their pairwise key $f(i, j) = f(j, i)$ by evaluating their own share at the point of the peer's IDs. Similar to Blom's scheme, Blundo's scheme is also t-secure.

Chan and Perrig (2005) introduced a deterministic scheme called *peer intermediaries for key establishment* (PIKE), in which all N nodes are organized into a two-dimensional space and each node has a unique coordinate (x, y), where $x, y \in [0, \sqrt{N} - 1]$. Each node shares unique pairwise keys with $2(\sqrt{N} - 1)$ nodes that have the same x or y coordinate. If two nodes with no common x or y coordinate, they need to choose an intermediate node that has common x or y coordinate with both of them to help them establish a pairwise key securely. The problem of PIKE is its high communication overhead, because a node can establish only pairwise keys with $2(\sqrt{N} - 1)$ nodes directly and needs to find multilink path for any of other neighbors.

All deterministic schemes discussed above have a common assumption; that is, the adversaries can compromise nodes as long as they are deployed. However, this assumption might be too strong. Although the nodes are not tamper-resist, the adversaries need at least some short period of time to find, break, and control a node. We call that period the *postdeployment security window*. Based on this weak threat model, key management schemes can be designed more efficiently and effectively. One solution is the *localized encryption and authentication protocol* (LEAP) proposed by Zhu et al. (2003). Each node i is first preloaded with an initial key K_I and a pseudorandom function f, which can determine the node's master key $k_i = f_{KI}(i)$. Within the security window after deployment, two nodes i and j exchange their IDs and can calculate the pairwise key $K_{ij} = f_{kj}(i)$. At the end of the security window, all nodes remove K_I from their memory, so the adversaries cannot calculate the pairwise keys of other links, even when they compromise some nodes. With this scheme, node addition is also simple. Each new node maintains K_I within the period of security window after deployment and hence it can calculate the pairwise keys with its neighbors. Moreover, Dutertre et al. (2004) proposed a similar solution with the same postdeployment security assumption.

7.5.2.3 Hybrid Schemes

Deterministic schemes such as Blom's and Blundo's provide perfect resilience as long as the number of compromised nodes is below the threshold value. They can be applied over probabilistic key predistribution schemes to further improve the resilience of these schemes. Du et al. (2003) combined the basic scheme and *Blom*'s scheme together and designed a *multispace key predistribution scheme*. First, one public matrix G and ω symmetric matrices D_i (where $i = 1, \ldots, \omega$) are constructed. These matrices form ω spaces (D_i, G). Then, each node randomly selects τ spaces and stores the corresponding rows of spaces. After deployment, any two nodes can establish a pairwise key if they happen to select the same space. This scheme has a similar threshold property as that of Blom's original scheme, which is that no additional links will be compromised given the number of compromised nodes less than a threshold value. However, after that threshold value, the whole network will be quickly broken. Similarly, Liu and Ning (2003) proposed to use Blundo's scheme to improve the security of the basic scheme. In one approach, each node randomly selects a subset of polynomials from a pool and stores the corresponding polynomial shares of the selected polynomials. In another approach, N nodes are organized into $m \times m$ grids, where each node is assigned with a coordinate (i, j) and each row i or column j of grids is associated with a polynomial $f_i^c(x, y)$ or $f_j^r(x, y)$. Each node then is preloaded with the polynomial shares corresponding to its row and column polynomials. Both approaches improve the resilience.

7.5.2.4 Probabilistic and Hybrid Schemes with Deployment Knowledge

Deployment knowledge is *a priori* information about the expected locations of WM²Snet nodes or the distribution of nodes' location. It can tell us which mesh nodes are more likely to become neighbors and in which local area a mesh node is more likely to reside. With the aid of deployment knowledge, significant improvements of performance can be achieved. This is attributed to the fact that nodes do not need to establish pairwise keys with far-away nodes.

Liu and Ning (2003a) proposed a *closest pairwise key scheme* in which each node shares pairwise keys with its c closest neighbors whose expected locations are closest to the node. In the extension version of this scheme, each node A has a unique key K_A. For node A and its c closest neighbors $B_1, B_2, \ldots, B_c$, the pairwise keys are $f(K_{Bi}|ID_A)$, where f is a pseudorandom function. Node A stores all c pairwise keys, while node B_i stores only its key K_{Bi} and f. This scheme has good resilience and connectivity and it also reduces the memory requirement of meshes. But its performance will be degraded with the growth of location error and it is hard to estimate nodes' expected locations accurately.

Du et al. (2004) first proposed the concept of deployment knowledge. They described a *group-based deployment model* in which the target field is divided into square grids and nodes are categorized into groups. Each group of nodes picks their keys from a corresponding sub key pool and they are supposed to be deployed into a fixed grid. Sub key pools should be carefully designed. Specifically, the sub key pool of one group overlaps with the pools of the group's two horizontally and vertically neighboring groups with ratio α and with the group's four diagonally neighboring groups with ratio β, where $0 < \beta < \alpha < 1$. Since neighboring key pools overlap, the nodes from neighboring groups have higher probability to establish pairwise keys than those nodes whose groups are far from each other. This scheme outperforms the basic one in term of resilience. However, if smarter adversaries selectively compromise nodes within the same group, the scheme cannot keep the same good performance in resilience. That is, it cannot deal with selective node captures very well. In (Liu and Zhao, 2005), the authors proposed a general *group-based key predistribution framework*. In the framework, the nodes from the same group establish pairwise keys using some existing approaches, while every group has an exact node to form cross-groups, which are used to bridge neighboring groups. Whenever two nodes from different groups want to establish a pairwise key, they always need to exploit the bridging nodes. Hence, one problem of the framework is that those bridging nodes may consume up their energy earlier than other nodes. Huang et al. (2004) introduced a location-aware key management scheme, which is similar to the framework proposed by Liu and Ning (2003a).

We have already discussed some schemes that use deterministic approaches to improve resilience of probabilistic schemes, and further introduced some schemes that exploit deployment knowledge to improve the performance of schemes. Intuitively, both deterministic approaches and deployment knowledge can be combined together to facilitate the design of key management schemes with better performance. Liu and Ning (2003a) integrated Blundo's polynomial-based technique with a group-based deployment model and designed *location-based key predistribution using bivariate polynomials*. First, a bivariate polynomial is assigned to each grid of the target field. Then, for every node to be deployed into some grid, it is preloaded with the polynomial shares of polynomials from its own grid and four directly neighboring grids (i.e., two horizontally and two vertically neighboring grids). This scheme can easily achieve high connectivity with good resilience.

Similarly, Yu and Guan (2005) depicted an approach to combine Blom's scheme with group-based deployment model. There are two significant differences between Yu's approach and other schemes. First, the authors divided the target field into hexagon grids, instead of square grids. Since hexagon grids are symmetric and have fewer neighboring grids, this scheme can achieve high connectivity with more efficient use of memory. Second, the authors proposed to use a *geometric random graph* (Penrose, 2003), instead of the *(Bernoulli) random graph*, to model WM^{2}Snets. The reason is that the geometric random graph takes into consideration the limited communication range of WM^{2}Snet nodes; however, the (Bernoulli) random graph does not. In addition, Zhou et al. (2005, 2005a) discussed how to incorporate Blundo's polynomial-based technique and group-based deployment model with hexagon and triangle grids.

7.5.3 Group Key Schemes

Based on established pairwise keys, establishing a group key becomes straightforward. As depicted by Zhu et al. (2003), one node can directly send a group key to its neighbors through the links secured with pairwise keys.

Another approach is to exploit the predistributed polynomial shares of mesh nodes to generate a common group key. Blundo (1992) proposed two models. The first model is non-interactive, where users compute a common key without any interaction. A random symmetric polynomial $f(x_1, \ldots, x_t)$ with t variables of degree λ is selected initially, where the coefficients come from a finite field $GF(q)$. Each user i receives share $f(i, x_2, \ldots, x_t)$. Users $j_1, \ldots, j_t$ can generate the conference key $K_{j1,\ldots,jt}$ by evaluating their polynomial shares. That is, each user j_i can obtain the conference key $K_{j1,\ldots,jt}$ independently by evaluating $f(j_i, j_1, \ldots, j_{i-1}, j_{i+1}, \ldots, j_t)$. In the second model, interactions are allowed in key computation. A symmetric polynomial $f(x, y)$ of degree $(\lambda + t - 2)$ is selected initially. Each user i receives share $f(i, y)$. Users $j_1, \ldots, j_t$ can establish the conference key K as follows: (1) the user with the largest identity, that is, user j_t, selects a random key K, (2) j_t calculates $K_{jt,jl} = f(j_t, j_l)$ for each $l = 1, \ldots, t-1$, (3) j_t sends $x_l = K_{j_t,j_l} \oplus K$ to each j_l, and (4) each j_l generates $K_{jl,jt} = f(j_l, j_t) = K_{jt,jl}$, and derives the secret $K = x_l \oplus K_{j_l,j_t}$. In this model, user j_t performs $(t-1)$ polynomial evaluations, and sends $(t-1)$ messages, which carry a single x value to establish the group key.

7.5.4 A Global Key Management Scheme

Perrig et al. (2001) proposed μ TESLA for authenticated broadcast and global key update. It requires the base station and mesh nodes to be loosely time synchronized. First, the base station picks the last key K_n of a chain and generates the remaining keys $K_0, K_1, \ldots, K_{n-1}$ of the chain using a one-way hash function H such that $K_i = K_{i+1}$. Given K_i, each node can generate the sequence $K_0, K_1, \ldots, K_{i-1}$, but not $K_{i+1}, \ldots, K_n$. At the i-th time slot, the base station broadcasts message M along with its MAC, MAC_{Ki} (M). Meshe nodes need to store this message until the base station discloses the authentication key K_i at the $(i+1)$-th time slot. This is called delayed disclosure. Receiving the $(i+1)$-th message, WM²Snet nodes can verify the disclosed authentication key K_i by using the previous disclosed key K_{i-1} as $K_{i-1} = K_i$. This scheme requires mesh nodes to store a message until the authentication key is disclosed. This introduces communication delays, causes storage problems, and may well be used by the adversaries to launch denial-of-service (DoS) attacks. For example, the adversaries may jam key disclosure messages aiming to saturate the WM²Snet nodes' storage.

7.6 Lightweight Key Management in WM²Nets by Leveraging Initial Trust[7]

7.6.1 Notation and Cryptographic Primitives

The following table summarizes the notation used throughout this study.

A, B, C ...	Node Identities
N_a, N_b	Random numbers (nonces) generated by A or B
R_a	Random number stored in node A before deployment
$G_k(m)$	Keyed one-way hash function applied to string m using key k
$MAC_k(m)$	MAC for message m, generated using key k
bk_1	Group authentication key used for bootstrapping
bk_2	Key generation key used for bootstrapping
gk_i	Key shared by all mesh nodes of generation i and used for authentication with previous generations
K_{ab}	Pairwise key established by neighbors A and B

[7] Excerpt from the invited article "Lightweight key management in wireless mesh networks by leveraging initial trust," Bruno Dutertre, Steven Cheung, and Joshua Levy, Computer Science Laboratory, SRI International, 333 Ravenswood Aveue, Menlo Park, CA 94025, USA.

G is a keyed one-way hash function. It has the property that, given a random quantity r and a data string m, it is computationally infeasible to find the key k such that $r = G_k(m)$. Moreover, given m and k, one can compute $G_k(m)$ efficiently. More formally, **G** is assumed to form a pseudorandom function family. That is, a polynomial time adversary cannot distinguish between the function G_k for a randomly chosen key k, and a true random function f of same domain and range as G_k. For example, the notion of undistinguishability is defined rigorously by Bellare et al. (1997).

MAC is an algorithm for constructing secure message-authentication codes using k. Given both k and a message m, $MAC_k(m)$ can be efficiently computed. We also assume that the MAC is collision resistant. Knowing m and $MAC_k(m)$, it is computationally intractable to construct a message m' such that $MAC_k(m') = MAC_k(m)$. Similar to G, MAC can be constructed from a pseudorandom function family.

7.6.2 Bootstrapping Service

Authentication and key management require initial trust between some of the parties involved. For example, a public-key certificate is accepted as valid if signed by a trusted authority. If only symmetric-key cryptography is used, the parties that trust each other must somehow acquire a common shared secret that will enable them to communicate securely. In traditional networks, the initial secrets that are necessary to bootstrap the authentication services are typically set up by hand. For example, if a central authentication server is used, an administrator distributes an initial shared key when the client is registered with the server. This initial key is typically communicated offline to ensure secrecy.

In the case of large networks of embedded devices, the manual setting of a large number of keys is not practical. In many scenarios, access to the devices for administration is impossible once the devices are deployed. Mesh nodes, such as nanosensors, could be dropped from a plane over an inaccessible region or deployed in a toxic environment (Estrin et al., 1999). In such cases, device configuration is possible only prior to deployment, and there are no secure offline channels. Once deployed, the network must be autonomous and self-organizing. Without manual intervention, the devices themselves should then set up the initial keys securely.

The typical scenario is for a set S of wireless mesh nodes. At this point, the devices must discover their neighbors and self-organize in an ad hoc manner. During this initial phase, the main security concerns are external attacks and possibly malicious devices already present in the environment. The nodes from S themselves may be assumed initially trustworthy, as it takes time for an adversary to compromise them. As the risk of device compromise increases with time, it is crucial to very quickly establish the initial secure links. This calls for an efficient localized algorithm with minimal communication overhead. The bootstrapping protocol considered is a localized algorithm that builds initial trusted links between WM^{2}Net nodes that are within direct communication range of each other. It is executed in a short time window after the nodes have been deployed.

7.6.2.1 Protocol Description

Since all node members of S are assumed initially trustworthy, two neighbors A and B can trust each other and establish a secure link if and only if they can determine that they both belong to S. Hence, a lightweight form of authentication is sufficient, namely, the ability for a node to prove that it belongs to S. This is implemented cheaply by loading a secret group authentication key bk_1 into all the members of S. Another secret key, the key generation key denoted by bk_2, is also stored in all node members of S. This is used by the neighbors A and B to generate a pairwise key K_{ab}. Loading these two keys in all devices can be done easily when mesh nodes are programmed, whereas it shows a very minimal administration overhead.

The protocol is straightforward. A node, say A, initiates the protocol by generating a random nonce N_a and broadcasting a *hello* message. The *hello* is of the following form:

$$<\text{hello, A}, N_a, \text{MAC}_{bk1}(\text{hello, A}, N_a)>$$

The message contains A's identity, the nonce N_a, and a MAC generated using bk_1. On reception of such a message, a node member of S can check whether the MAC is valid and hence assures that the sender possesses the secret key bk_1. Let B be such a node. Once B has verified the MAC, it generates a random nonce N_b and sends the following reply to A:

$$<\text{Ack, A, B}, N_b, \text{MAC}_{bk1}(\text{Ack, A, B}, N_b, N_a)>$$

This *acknowledgment* communicates the nonceN_b to A and assures that B knows bk_1 and also that B has receivedN_a. When A receives the message can check whether the MAC is valid, and if so, extracts the nonce N_b.

After this exchange, A and B have proven to each other that they know the group authentication key, and they are also both in possession of nonces N_a and N_b. They construct a pairwise symmetric key as follows:

$$K_{ab} = G_{bk2}(N_a, N_b)$$

where G is a keyed one-way hash function. This pairwise key enables them to communicate securely in the future. The key K_{ab} is actually split into two subkeys,K_{ab}^1 and K_{ab}^2, used for encryption and authentication of future messages, respectively.

7.6.2.2 Security

This bootstrapping protocol is a variant of the implicit key exchange protocol AKEP2 by Bellare and Rogaway (1994). Using the models and techniques introduced by Bellare and Rogaway (1994), it can be proven that it is secure against an adversary who does not know the keys bk_1 and bk_2. The proof relies on the assumption that MAC and G are pseudorandom function families. Under this assumption, one can show that the following properties are satisfied for any adversary E, initiator A, and responder B.

- The probability that B accepts a *hello* message that appears to be from A but was not sent by A is negligible.
- The probability that A accepts an *acknowledgment* message that appears to be from B but was not sent by B is negligible.
- E cannot distinguish between the key K_{ab} and a random bit string of the same length.

These properties can be stated precisely and proven rigorously as shown in (Bellare and Rogaway, 1994). This proof shows that the protocol is secure against an adversary E who can listen to traffic and inject messages, as long as E does not know the bootstrapping keys bk_1 and bk_2.

Since mesh nodes are typically not tamperproof, an adversary could potentially obtain the keys by physically compromising a node. Clearly, if an adversary obtains bk_1 and bk_2 during the bootstrapping time window, then it can compute the pairwise keys K_{ab} from the messages it intercepts, or interfere with bootstrapping by forging *hello* and *acknowledgment* messages. This risk is small if the bootstrapping window is kept short. However, an additional risk exists if the adversary can record the messages exchanged between A and B during bootstrapping and later discover the key bk_2. Since all nodes use the same key-generation key, compromise of bk_2 can lead to the compromise of a large number of pairwise keys. The counter-measure to this attack is to erase both bk_1 and bk_2 as soon as possible after the bootstrapping window has elapsed.

7.6.2.3 Robustness and Cost

The unreliability of the communication link is a major issue in designing protocols for WM^{2}Nets. We use several mechanisms to make the bootstrapping protocol robust to message loss.

First, all nodes that are deployed together will function both as initiator and responder. All these will initiate the bootstrapping protocol at least once by broadcasting a *hello* message. Two neighbors A and B have then at least two chances to establish a secure link: once with B and once with A as the initiator. Optionally, mesh nodes can be programmed to send more than one *hello* message, thus executing the bootstrapping protocol more than once. This increases the probability that bootstrapping succeeds between two neighbors even if some messages are lost.

In addition, several timing mechanisms are employed to reduce the probability of message collisions. Randomization is used to prevent all nodes from sending a *hello* at the same time. When first started, a node will wait for a random period of time after deployment before sending its *hello* message. A similar technique is used to reduce the risk of collisions between several *acknowledgments* to a *hello*. When a node A broadcasts a *hello* message, neighbors of A that already share a pairwise key with A do not respond. Such nodes either already responded to a previous *hello* from A, or they have sent a *hello* to which A responded. Except for these nodes, all neighbors of A that received the *hello* is expected to respond. To reduce the probability of collisions between *acknowledgments* from different responders, replies to A are sent after a randomized wait time.

A final mechanism reduces the risk of collisions between *hello* messages and *acknowledgments*. Once a *hello* message is transmitted at time t, a time interval $[t, t + \Delta]$ is then reserved for *acknowledgments* to this *hello*. A timer triggers the transmissions of *hello* messages. If a node B receives a *hello* at time t, it will not broadcast its own *hello* until after $t + \Delta$. If B's timer expires in the interval, then B will not send its *hello* but restart the timer, with a randomized delay, to retry later.

All these mechanisms improve the robustness of the protocol in a network composed of unreliable radio links. Besides, the protocol generates low communication overhead as messages are exchanged only with neighbors. For a node A, the cost is one broadcast message per *hello*, and at most one reply from A to each of its neighbors. A more economical approach could be envisaged that requires only one *hello* message per node. A protocol that relies on this approach to exchange session keys is discussed in (Lai et al.,

2002). In such protocols, the key K_{ab} must be constructed from nonces attached to *hello* messages from A and B. This is very cheap in terms of communication, but also very unreliable if *hello* messages are lost (e.g., due to collisions or radio noise).

7.6.3 Multiphase Deployment

We assume that mesh nodes are deployed in successive generations. The bootstrapping protocol applies to mesh nodes of a single generation. This section presents an extension of bootstrapping that enables a node A of generation i to establish a secure link with a node B of a later generation $j > i$.

The basic idea is for A to store a random quantity, R_a, and a secret $S_{a,j}$ derived from R_a. The secret, $S_{a,j}$, has the property that no other node of generation i, or earlier generation, can efficiently compute $S_{a,j}$ from R_a. On the other hand, a node of generation j can efficiently compute $S_{a,j}$ from R_a. The secret is used to establish a secure link between A and nodes of generation j. The construction of $S_{a,j}$ relies on a keyed one-way hash function such as the function G used previously. For authentication across multiple generations, we add an extra key gk_j that is shared by all nodes of generation j, and the secret $S_{a,j}$ is constructed by

$$S_{a,j} = G_{gkj}(R_a)$$

Thus, under the assumption that G is a secure one-way function, only nodes of generation j can construct $S_{a,j}$ from R_a. Node A itself knows $S_{a,j}$ and R_a, but it does not possess gk_j. Several secrets such as $S_{a,j}$ must be stored in A before deployment; each corresponds to one generation between i+1 and $i + n$, where $n > 0$ is the number of future generations with which A can establish secure links.

Node A of generation i and B of generation j use the following protocol, called cross-generation bootstrapping (XGB). When B is first deployed advertises the event by broadcasting a *hello* message:

$$<hello, \mathrm{B}, j, N_b>$$

The *hello* message is constructed of B's identity and generation, and a randomly generated nonce N_b. Upon receiving the message, A extracts the generation number j and obtains the corresponding secret $S_{a,j}$. Then A sends the following *acknowledgment* to B:

$$<\mathrm{Ack, A, B}, R_a, MAC_{Sa,j}(\mathrm{Ack, A, B}, R_a, N_b)>$$

When B receives this message it computes $S_{a,j}$ using gk_j and R_a. Then B verifies whether the MAC is valid; if so, B assures that the sender possesses the secret $S_{a,j}$. If the MAC is determined to be valid, B completes the protocol by sending a second *acknowledgment* that B can authenticate using the secret:

$$<\mathrm{Ack2, B, A}, MAC_{Sa,j}(\mathrm{Ack2, B, A})>$$

After XGB, A and B will derive a new session key based on $S_{a,j}$, R_a, and N_b for securing their communication in a way similar to that of the bootstrapping protocol.

Because of the one-way property of function G, A cannot obtain gk_j. Thus, A may not tamper with the communication between a node of generation j and another node other than itself. Also, A cannot masquerade as another node Z of generation i, of an earlier generation, or of a later generation when communicating with a node of generation j because A cannot efficiently compute $G_{gkj}(R_z)$.

As before, the security of the XBG protocol relies on the assumption that nodes of generation j are trustworthy when deployed and remain trustworthy for as long as it takes for the protocol to complete. It is also crucial for all nodes of generation j to erase the key gk_j as soon as the cross-generation protocol terminates.

Using the secure local links established by the XGB protocol, one can securely transmit a group key, K_g, from generation i and pregeneration i nodes to generation $(i+1)$ nodes. In other words, we have a set of old nodes of generations smaller than $(i+1)$ that share a secret group key K_g. This set may be strictly smaller than the set of all generation i and pregeneration i nodes. For example, some nodes may be excluded because they are detected to be compromised or misbehaving. When generation $i+1$ is deployed, we want them to obtain the group key K_g so that all nodes can participate in a common application.

Again, we assume that mesh nodes are not compromised shortly after they are deployed. After generation $(i+1)$ nodes are deployed, there exists a time window during which all generation $(i+1)$ nodes are trusted and no adversary can obtain the secrets stored in these nodes. During this time window, old nodes can establish secure local links with new nodes of generation $i+1$ using XBG, and they can transmit K_g to them using the secure local links. To prevent misbehaving old nodes from forcing generation $(i+1)$ nodes to use an incorrect group key, generation $(i+1)$ nodes can exchange the group keys they receive among themselves to filter out incorrect group key(s). We assume that the majority of the group keys obtained from distinct (based on the R_a values) pregeneration-$(i+1)$ nodes are correct. Moreover, thanks to its inability to obtain $S_{g,i+1}$ for another nodes Z, a misbehaving pregeneration-$(i+1)$ node cannot masquerade as Z in this process. Thus the misbehaving nodes cannot perform a Sybil attack (Douceur et al., 2002) to outnumber the correct nodes by presenting themselves as multiple pregeneration-$(i+1)$ nodes.

7.6.4 Using Secure Local Links

Once neighbors can communicate via secure local links, other security services can be built inexpensively. As a simple example, chaining can be used to secure communication between distant nodes. We present a group-key distribution protocol built on top of the secure local links.

A relatively easy way of adding security to a WM^{2}Net is to rely on a common group key that is known by all nodes. For example, this approach is supported by TinySec (Karlof and Wagner, 2003), a link-layer encryption service for TinyOS. Using a global key, messages between nodes can be encrypted for confidentiality, or protected against corruption by using a MAC. An alluring property of this approach is that secure multicast is very efficient. The sender of a multicast message encrypts the message and computes the MAC once using the group key. Every recipient decrypts the message and checks the MAC only once.

A limitation of using a shared group key is that it compromises of a single node is sufficient to obtain the key, which gives an adversary access to all network traffic. To recover from such an attack, one needs the means to distribute a new group key to all group members except those that are considered compromised. This can be easily implemented by exploiting the secure local links.

Our key refresh protocol provides this service. It can be initiated from any group member, although this is typically done from a base station. The initiator generates a new, random group key and optionally constructs a list of nodes to be excluded from the group. The new key together with the exclusion list, a sequence number, and the initiator's identity is distributed via the secure local links to all nodes, except those on the exclusion list. First, the initiator securely sends a copy of the key and exclusion list to its neighbors that are not on the list, using the pairwise key it shares with each of these neighbors.

A key-refresh message sent by A to B is of the following form:

$$<\text{KeyRefresh, B, A, O}, N, \{K_g\}_{K_{ab}^1}, \text{L}, MAC_{K_{ab}^2}(\ldots)>$$

In this message, O is the originator of the new key, N is the sequence number of the group key, K_g the new group key, and L the exclusion list. The message is protected by using the pairwise key K_{ab} that A and B set up during bootstrapping. More precisely, the subkey K_{ab}^1 is used to guarantee confidentiality of K_g, while K_{ab}^2 is used for authentication and integrity.

When B receives such a key-refresh message it checks the message integrity using K_{ab}^2 as whether the message is fresh, using the sequence number N and the originator identity O. If both checks succeed, B accepts the new group key carried in the message and forwards it to all its neighbors except A and any node on the exclusion list. This requires a re-encryption and MAC computation for each of B's good neighbors.

This protocol distributes the new group key securely and robustly. As long as the good group members are connected, the flooding-like procedure distributes the new key to all good members in a robust manner. However, this procedure is expensive in terms of communication and computation overhead. The key-refresh message is decrypted once but encrypted multiple times from each node, and sent in separate messages to all neighbors. This may not be a significant issue if the group key is not changed very often, but more efficient solutions are certainly desirable.

The identity of the originator along with a sequence number provides the means to arbitrate between conflicting key-refresh messages. This can occur if multiple nodes initiate the protocol almost at the same time. Key-refresh messages are ordered using the lexicographic order on the pair (N, O). When a key-refresh message is received from B it is accepted and forwarded only if it is higher in the lexicographic order than all key-refresh messages seen by B in the past.

7.6.5 Implementation

We have implemented and experimented with the bootstrapping and key-refresh protocols using Mica devices (Hill and Culler, 2002). The Mica platform is based on an Atmel ATmega 103L or Atmega 128 microcontroller and the RF Monolithics TR100 radio transceiver. The microcontroller is an 8-bit processor that runs at 4 MHz, and includes 4 KB of RAM and 128 KB of flash program memory. Mica supports a variety of sensor boards with photo-diode, thermistor, microphone and sounder, and magnetic and acceleration sensors. The radio has a fixed frequency of 916.5 MHz and a range that can be varied from inches to hundreds of feets, depending on power.

The Mica platform runs UC Berkeley's TinyOS operating system (Hill et al. 2000). TinyOS is a modular operating system designed for small sensor platforms. In the TinyOS model, an application consists of a set of software components that interact using event passing and a simple tasking mechanism. The TinyOS infrastructure provides a collection of low-level components for interaction with sensor hardware, which can be flexibly assembled and integrated with application components. Since version 1.0, TinyOS and application components can be written in NesC, an extension of the C programming language that supports the TinyOS component and composition model. The implementation was based on TinyOS 1.0.

7.6.5.1 Radio Stack

Implementing the proposed security protocols using TinyOS required significant extensions to the TinyOS radio stack. In version 1.0, TinyOS provides two different radio stacks for the Mica platform. The first is the standard radio implementation with no security handlers. In this implementation, radio messages consist of a header, a payload, and a cyclic redundancy check (CRC) that is used to detect message corruption. The header includes fields such as destination address, message type, and length. This version of the radio stack was not suitable for the proposed protocols, as the message formats they require do not match TinyOS messages. For example, all the protocols use cryptographic MAC for authentication and integrity, which means that a CRC is unnecessary. Also, the protocols do not use some of the header fields provided from TinyOS.

The second radio stack available with TinyOS is TinySec (Karlof and Wagner, 2003). It provides link-layer security based on a fixed network-wide key. In TinySec, MAC replaces the CRC whereas the payload is encrypted. This use of cryptography for securing radio communication could address some of the requirements accounted for in this study but is not sufficiently flexible for the proposed protocols. TinySec relies on a fixed key that is used for all messages and provides no interface for changing the key. In particular, we consider several keys maintained per neighbor. Some messages require different keys depending on the destination. Conversely, checking a received message requires identifying the sender to find the correct pairwise keys to use. Furthermore, some MAC computations require information that is not included in the messages sent (e.g., the *acknowledgments* to a *hello* message during bootstrapping). For these reasons, we need a radio stack that provides flexible per-message formatting and encryption.

We have developed a new radio stack for TinyOS that provides these services. This stack is an extension and combination of the standard TinyOS and TinySec stacks. It provides four communication services that use the following four types of messages:

- *Plain messages* in a format similar to that used from the standard TinyOS stack. Messages are sent in clear. A CRC is added for error detection.
- *Encrypted messages*, similar to the format used by TinySec. The message payload is encrypted, and a MAC is added for integrity and authentication.
- *Authenticated messages*, a variant of the TinySec format in which the payload is sent in clear and a MAC is added.
- *Raw messages* intended to be formatted by the application. A raw message consists of a single header byte that specifies the message length and a payload.

Thus, two of the communication services provided from the radio stack adopted are the same with these provided from the TinyOS and TinySec stacks. Authenticated messages are a variant of TinySec messages. The raw-message interface gives the application full responsibility for formatting and error checking. All four types of communication

Stack	Bytes in ROM	Bytes in RAM
TinyOS	9,440	356
TinySec	14,630	1,078
Our stack	11,818	914

Table 7.3 Code Size with Different Stacks

services are available within the same radio stack, and can be accessed via different interfaces. By default, the encrypted and authenticated message services use group keys that are fixed at compilation time; the radio stack used, however, provides an interface for changing these keys at runtime.[8]

An application that sends a message via the raw-message interface is free to format the payload in any way. Conversely, when a raw-message is received, the radio stack forwards it to the application performing no further checking. This interface allows maximum flexibility and is the one we use for the bootstrapping and key refresh protocols.

Our radio stack reuses a number of components from TinyOS and TinySec, aiming to remain compatible with them. In this sense, we use, for example, the same MAC algorithm as TinySec and encrypt the payload in CBC mode using the cipher stealing technique also employed in TinySec. The block ciphers we use for computing MACs and for encryption are also inherited from TinySec.

The size and performance of the adopted radio stack are similar to the TinySec stack. Table 7.3 shows the code size and RAM usage of the same example application compiled with the TinyOS stack, the TinySec stack, and the newly developed stack. The data were obtained with the TinyOS 1.0 distribution. In this example, both TinySec and the stack used the SkipJack block cipher. The application is one of the demo applications distributed with TinyOS; it periodically increments a counter and sends its value on the radio. As expected, compared with the nonsecure TinyOS stack, using cryptography increases the code size and RAM usage of both the proposed stack and TinySec. However, the code size fits easily within the Mica program memory. On the other hand, the RAM consumption is close to 25% of the total Mica RAM, which may be a lot for certain applications. Several optimizations are possible to reduce the memory used by the block cipher. For example, the SkipJack implementation stores a constant table of 256 bytes in RAM. It is possible to move this table into ROM, at the cost of a slight reduction in performance. With the table stored in ROM, SkipJack is about 7% slower than it would if the table was stored in RAM.

7.6.5.2 Protocol Implementation

Our bootstrapping protocol is intended for authentication and key distribution between neighbor nodes in a network. We have implemented this protocol on the Mica platform using the radio stack described above. The *hello* and *acknowledgment* messages are sent and received via the raw-message interfaces since they require special formatting and MAC construction.

The bootstrapping protocol is implemented using a NesC component called SecureLinkManager. The main role of the SecureLinkManager module is to build a table of

[8] This implementation uses version 1.0 of TinyOS. A more recent version of TinySec (Karlof et al., 2003) includes some of the same extensions as our radio stack.

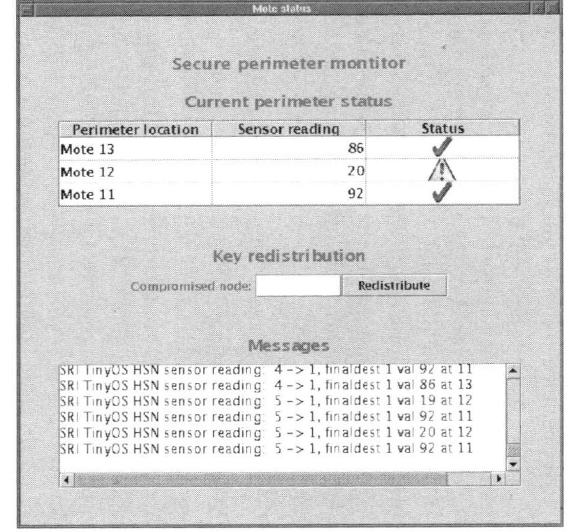

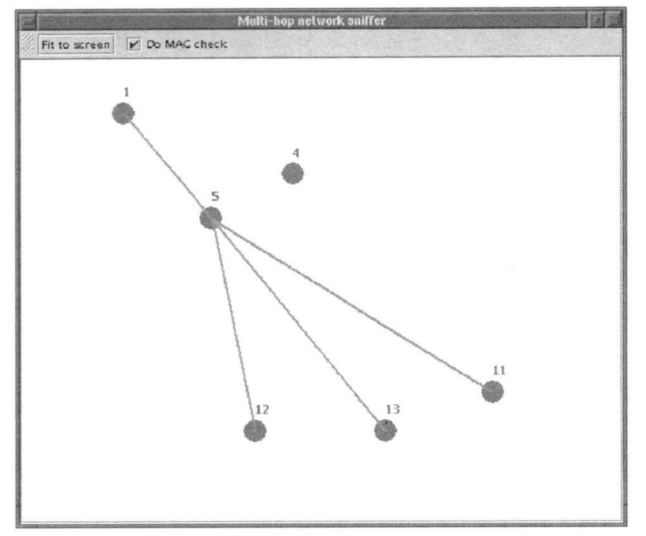

FIGURE 7.11 Perimeter monitoring demonstration.

authenticated neighbors. At the end of bootstrapping, the table contains the identity of each authenticated neighbor and the two pairwise keys (i.e., K_{ab}^1 and K_{ab}^2) established with this neighbor, and other bookkeeping data.

The bootstrapping protocol uses a different block cipher than those available with the TinySec distribution, namely, AES. The main reason for developing a new cipher implementation was to reduce the memory space needed to store the pairwise keys. TinySec provides implementation of two block ciphers—RC5 and SkipJack—but these implementations are optimized for speed. They use buffers to store intermediate data derived from the cryptographic keys to speed up encryption and decryption. Storing this data requires 128 bytes of memory for SkipJack and 104 bytes for RC5. This is too much if one needs to store cryptographic material equivalent to two keys per neighbor. We have developed an AES implementation that requires less RAM. This implementation uses 128-bit keys, has a block size of 128 bits, and is optimized for space. Using this implementation, the neighbor table requires only 48 bytes per neighbor for storing cryptographic material.

All the cryptographic operations performed by the SecureLinkManager module rely on this AES implementation. This includes MAC computation and generation of the pairwise keys as discussed earlier. In addition, we use the AES cipher for implementing a secure pseudorandom generator for generating nonces. This generator is initialized using a random AES key, which must be different for each node, and that is constructed when the Mica nodes are programmed.

We have also developed a prototype implementation of the key refresh protocol. This protocol is used to change the group keys used by the TinySec-like services of our radio stack. The implementation of this key-refresh protocol relies on the neighbor table constructed by bootstrapping to flood key-refresh messages. These key-refresh messages are formatted at the application level and are transmitted via the raw-message interface. The encryption and MAC applied to these messages use the pairwise key stored in the neighbor table. The aforementioned implementation of AES is thus employed.

7.6.5.3 An Example Application

We have tested the proposed bootstrapping and key refresh implementations in a demonstration application: a perimeter monitoring scenario in which sensors along a perimeter communicate sensor readings (in this study, light levels) via an ad hoc network of other nodes. The routing layer is an implementation of destination-sequenced distance-vector (DSDV) routing (Perkins and Bhagwat, 1994) written for TinyOS by Intel Research's heterogeneous sensor networks project (Intel Research, "Heterogenous sensor networks": http://www.intel.com/ research/exploratory/heterogeneous.htm).

During normal operation, sensor readings are sent along dynamically updated multihop paths to a base station. However, the routing protocol is vulnerable to malicious route update messages. For instance, a compromised "black hole" node can falsely advertise that it is close to the base station. As a result, the node will not forward sensor readings. Even in the case where the messages are signed with a group key (as in TinySec), all sensor measurements can be thwarted by a single malicious node that knows the group key. However, with the fallback of pairwise keys obtained via bootstrapping, we can—if we are aware of the identity of the malicious node—refresh the group key to trusted nodes only. Upon assembly of the TinyOS components for bootstrapping, key refresh, and routing, the implementation demonstrated this capability with success. A screenshot of this application is shown in Fig. 7.11.

CHAPTER 8

Autonomic Selfware WM^{2}Net Communications

8.1 Introduction

The recent advances in communication and networking technologies and the way in which these are being integrated in the human, working, and social framework have made it evident that there are a number of related technical and socioeconomic areas whose understanding is still less than satisfactory. In particular, it can be observed that the increasingly higher density mesh of communications systems and the resulting growing complexity of control require more and more distributed and self-organizing structures that rely on simple, dependable, evolvable, and collaborative behaviors.

The main feature of emerging communication paradigms will be the ability to adapt to evolving situations, where new resources can become available, administrative domains can change, and economic models can vary accordingly. One of the major trends in the emerging communication landscape is related to the arising of pervasive communication/computing environments, characterized by an extremely large number of embedded devices (Weiser, 1999; Kahn et al., 1999). Such devices will possess sensing/identifying capabilities, making it possible for user-situated services to interface directly with the surrounding environment.

The vision is of a world pervaded by ubiquitous communication facilities, offering their services to the users and capable of self-organizing and self-preserving their functionalities without any direct human intervention. The network elements should have the capability to self-organize and self-configure themselves, to observe and to react to context changes without explicit user interaction. Such a vision entails fundamental advances both in the architecture and functionality of the network, and in the characterization and understanding of the common communication medium.

The new discipline proposed to define such service-driven, situated, autonomously controlled, self-organized, distributed, technology independent, and scalable communication architectures is widely known as *autonomic communication* (AC) (http://cordis.europa.eu/ist/fet/comms.htm). AC is an emerging paradigm in which the applications and the services are not ported onto a pre-existing network, but where the network itself grows out of the applications and the services that end users want. AC has a broad scope, addressing all facets of communication—human-to-human, human-to-cyber, business-to-business, cyber-to-cyber, etc.—by empowering network elements to best fit communication intentions, to observe and to react by self-organization to context changes without explicit user interaction.

AC is centered around networking ***selfware***—a novel approach to perform network control, as well as management, middle box communication, service creation and composition of network functionalities, etc. based on universal and fine-grained multiplexing of numerous policies, rules, and events that is done autonomously but facilitates desired behavior of groups of network elements.

The vision of self-ware ACs identifies a number of new research challenges:

1. ***AC service architecture paradigms***—Allows new services, bottom up services, composed services, self-adaptable and self-configurable growing infrastructure, context management and data mining, synergies between peer-to-peer and context awareness, and awareness of services situations enabling semantic mediation for transport peering and collaboration.
2. ***Zero-effort deployment*** ("spray deployment")—Facilitates at any scale (smart dust, PAN, Internet nodes) self-assembly of communication–computational particles; programing paradigms should focus on local, autonomic cooperation and propagation fields.
3. ***Programming of self-organization***—Includes both architectural programmability enabled by extended van Neumann paradigm—with a virtual layer above and *quantic* layer below to reduce the visible complexity and to add stochastic aspect and include reflection of communication knowledge, context networks, and communication applications. The main challenge is that knowledge of communication is reflected inside the network and implemented by autonomic network elements.
4. ***Self-management***—Addresses consistency of coordinated distributed decisions, security boundaries, boundaries of controllability, emergent reliability, conflicts, conflict resolution through negotiation, need to analyze issues with mobile code, dynamic composition of protocols and services, export of functions to different layers, and the need not only to deploy but also to discard obsolete or undesirable functions.
5. ***AC contribution to network information theory***—Addresses fundamental problems of multiple media and multiple resources (not only bandwidth, but also storage, processing, power, mobility, context, and their possible trade-offs) in a cross-layer optimized setting.
6. ***Security and Protection in AC***—comes from the embedding of security and trust requirements into the communication system's functionality (*model driven security*) and by investigating how simple security components are developing complex behaviors. These behaviors in turn result from the process of negotiation of protection level agreements. Further, since autonomic nodes could self-program their networking behavior out of the cooperative standard to gain maximum advantage against others, this calls for novel techniques for trusted software and trusted flow on untrusted hosts that cannot be based on existing security paradigms such as public key infrastructures or controversial hardware trusted computing platforms.
7. ***Coordination and intelligence in service provisioning for AC***—Enables AC systems to support strategic and business service and trust requirements and to adapt flexibly to evolutions at the business level. This also facilitates the

so-called *cooperative networking* (CoNet), where network operators with different and not easily compatible business models can jointly offer networking services. Specifically for wireless networks, AC along with cross layer optimization needs to consider tensions and conflicts in scenarios of mass device coordination; this can be based on genetic approaches for example.

8. ***Behavior knowledge and knowledge execution in AC***—Enable AC systems to capture and follow the *Concept drift* over time. Requires knowledge of ensemble management, introspection, mediation, ontology acquisition and use, optimization, contextualization, including identification of technologies suitable for standardization.
9. ***Programing of self-organization***—Includes reflection of communication knowledge, context networks, and communication applications. The main challenge is that knowledge of communication is reflected inside the network and implemented by autonomic network elements. This effort considers the structure and dynamics of metaphysical networks (e.g., trust networks) in their context as well.

8.2 Related Standardization Efforts

Defense Advanced Research Projects Agency (**DARPA**) next generation (XG) (http://www.darpa.mil/ato/programs/XG) communication program is developing the technology to allow multiple users to share use of the spectrum through adaptive mechanisms that deconflict users in terms of time, frequency, code, and other signal characteristics. DARPA's goals are to enable an increase of a factor of ten in the usage of typical spectrum. The key technologies are centered on an *autonomous dynamic spectrum utilization* function that is surrounded by four support functions, namely *sensing* (real-time low power wideband monitoring), *characterization* (rapid waveform determination), *reaction* (formulate best course of action), and *adaptation* (transition network to new emission plan). Opportunistic frequency sharing is planned to achieve by imposing *policy* (e.g., rules of frequency, time to vacate, maximum power, maximum transmit time, etc.) on all users. The policies define a set of abstract behaviors currently deployed manually by spectrum managers.

The Internet Engineering Task Force (**IETF**) (http://www.ietf.org) is engineering solutions for the operational Internet to ensure its further growth and usage. The guiding principle Internet architects are preserving is known as the end-to-end (E2E) principle; it attempts to facilitate innovation by rejecting functionality placement within the network. However, all-purpose deployment of the Internet that started in early 1990s has greatly challenged the dominance of the E2E principle. A significant amount of IETF effort is currently devoted to non-E2E engineering, such as firewalls, network address translators (NAT), and protocol translators (PT), and their traversal, signaling, transport and content caching proxies, etc. The largest ever attempt to fix the Internet model was undertaken by the IPng working group in standardizing Internet Protocol version 6 (IPv6) that, however, itself currently needs multiple fixes to meet the reality challenge (multihoming, mobility, QoS, NAT, etc.). The emerging understanding within the Internet users and designers is that IPv6 is just another addressing realm; the future reality is the coexistence of multiple addressing realms. More than 20 years of IETF engineering for the quality-of-service (QoS) resulted in more state (intelligence) within the Net. This intelligence needs management, often by complex multilayer architectures. The IETF approach to simplify management is based on policy, which, following the Distributed Management Task Force, Inc. (DMTF) (http://www.dmtf.org) Common Information Model (CIM), has to be based on the overall top-down description of the network and by definition is network operator driven.

Internet Research Task Force **(IRTF)** (http://www.irtf.org) is a research branch of the IETF. In a number of its closed working groups it attempts to find long-term solutions. The one closely related to multiple addressing realms is the recently created Searchable Internet Resource Names (SIREN) group. SIREN seeks to find a tractable option of enhancing existing domain name system (DNS) with layers above to allow "directory-like search (using qualified natural language strings rather than names)."

DARPA's **NewArch project** (http://www.isi.edu/newarch/) has defined a number of novel paradigms and constructs to meet the challenges of the current Internet; these include FARA—innovative addressing architecture, new Internet routing architecture (NIRA), and role-based architecture (RBA).

World Wide Web Consortium **(W3C)** (http://www.w3c.org) is a global organization aiming to develop the "Semantic Web," where, based on metadata definitions, websites will be able to perceive the meaning of communication and to self-organize to perform complex and composite functionalities. The W3C is making a set of specifications that are being already used not only for pure communication but, for example, for workflow management and system integration.

Wireless World Research Forum (**WWRF**) (http://www.wireless-world-research.org) is a global forum coordinating research for next generation wireless communication. Within WG3', "cooperative and ad hoc networks" a number of white papers have been produced addressing research challenges and approaches of CoNet, that is, interworking of networks with different communication technologies and business models. In December 2003, a special interest group (SIG3) "Self-organization in wireless-world systems" has been created within WWRF.

8.3 Related Industrial Initiatives

8.3.1 IBM's Autonomic Computing

IBM's autonomic computing is an emerging approach to self-managed computing systems with a minimum of human interference. The term derives from the body's autonomic nervous system, which controls key functions without conscious awareness or involvement.

The IBM's Autonomic Computing (http://www.research.ibm.com/autonomic) initiative suggests the following defining characteristics of an autonomic system:

- An autonomic computing system needs to "know itself"—its components must also possess a system identity. Since a "system" can exist at many levels, an autonomic system will need detailed knowledge of its components, current status, ultimate capacity, and all connections to other systems to govern itself. It will need to know the extent of its "owned" resources, those it can borrow or lend, and those that can be shared or should be isolated.

- An autonomic computing system must configure and reconfigure itself under varying (and in the future, even unpredictable) conditions. System configuration or "setup" must occur automatically, as well as dynamic adjustments to that configuration to best handle changing environments.
- An autonomic computing system never settles for the status quo—it always looks for ways to optimize its workings. It will monitor its constituent parts and fine-tune workflow to achieve predetermined system goals.
- An autonomic computing system must perform something akin to healing—it must be able to recover from routine and extraordinary events that might cause some of its parts to malfunction. It must be able to discover problems or potential problems, then find an alternate way of using resources or reconfiguring the system to keep functioning smoothly.
- A virtual world is no less dangerous than the physical one, so an autonomic computing system must be an expert in self-protection. It must detect, identify, and protect itself against various types of attacks to maintain overall system security and integrity.
- An autonomic computing system must know its environment and the context surrounding its activity, and act accordingly. It will find and generate rules for how best to interact with neighboring systems. It will tap available resources, even negotiate the use by other systems of its underutilized elements, changing both itself and its environment in the process—in a word, adapting.
- An autonomic computing system cannot exist in a hermetic environment. While independent in its ability to manage itself, it must function in a heterogeneous world and implement open standards—in other words, an autonomic computing system cannot, by definition, be a proprietary solution.
- An autonomic computing system will anticipate the optimized resources needed while keeping its complexity hidden. It must marshal I/T resources to shrink the gap between the business or personal goals of the user, and the I/T implementation necessary to achieve those goals—without involving the user in that implementation.

This new paradigm shifts the fundamental definition of the technology age from one of computing to one defined by data. Access to data from multiple, distributed sources, in addition to traditional centralized storage devices will allow users to transparently access information when and where they need it. At the same time, this new view of computing will necessitate changing the industry's focus on processing speed and storage to one of developing distributed networks that are largely self-managing, self-diagnostic, and transparent to the user.

This new computer paradigm means the design and implementation of computer systems, software, storage, and support must exhibit these basic fundamentals from a user perspective:

- *Flexible.* The system will be able to sift data via a platform- and device-agnostic approach.
- *Accessible.* The nature of the autonomic system is that it is always on.
- *Transparent.* The system will perform its tasks and adapt to a user's needs without dragging the user into the intricacies of its workings.

8.3.2 Hitachi's Harmonious Computing

Hitachi's harmonious computing is the vision that combines progression, collaboration, and trust (http://www.hitachi.co.jp/Prod/comp/soft1/global/vision/harmonious. html): *Progression* is optimization of system expansion and operational cost; *Collaboration* is acceleration of business speed and concentration on core business; and *Trust* is nonstop and secure business/public infrastructures. This vision seems to be a further expansion of Hitachi-developed technology called autonomous decentralized systems (ADS) that was and is highly recognized by the international community.

8.3.3 NTT's Resonant Communication Network Architecture

NTT's Resonant Communication Network Architecture (RENA)[1] is essentially a plan to standardize within the ITU-T a multitechnology network (optical, IP, multiprotocol label switching (MPLS), modem, sensor, etc.) glued together by ubiquitous broadband connectivity and richness of service portfolio that is controlled by a common *service/network control platform.*

The initiative aims to make a resonant communication network environment with exceptional usability and create diverse feature-rich new services and business opportunities based on it. The network must satisfy several advanced requirements to create such a public infrastructure. The most important ones are the following:

- Support for tens of millions of broadband subscribers: The number of IP network users will continue to increase, and there will be a vast increase in the number of terminals (including information appliances, radio-frequency ID tags, and sensors) connected to the network.
- Ability to accommodate changing patterns of network use: It must support access by a diverse range of different types and modes of communication including E2E and multipoint communications.
- QoS assurance: It must offer a range of reliability conditions and QoS options from best effort to assured-quality services.
- Assured network security and reliability: Users must have total confidence that their privacy will be fully protected when they use the network.
- Good usability: The network must support safe, secure communications without requiring any technical knowledge or difficult setup procedures on the part of users.

---

[1] Takashi Hanazawa, Nippon Telegraph and Telephone Corporation, Resonant Communication Network Architecture; *NTT's Plans for NGN and Proposals for the Areas of Standardization,* presentation at Workshop on Next Generation Networks: What, When & How? Geneva, July 9–10, 2003.

8.4 Related R&D Projects

- **Ambient Networks** (http://www.ambient-networks.org) is the WWI integrated project driven by the Ericsson-lead consortium which addresses the issues of seamless composition of heterogeneous networks with strong emphasis on wireless and mobile communication. The project plans to perform intensive work on ambient network architecture and ambient network context management.
- **Security Expert INITiative (SENIT) (**http://www.ist-world.org/ProjectDetails.aspx?ProjectId=c4658c78a0eb489095f09638a8291832**)** introduces a security approach closer to real life, easy to use and flexible enough to adapt the complexity of the security mechanisms to the level of the risk faced. To information systems and network designers, SEINIT delivers new security models to design innovative security architectures, the security policies addressing the existing and coming threats, and the components to build networks infrastructure.
- The **A**utonomic **C**ommunication: **C**oordination **A**ction (ACCA) project (http://www.autonomic-communication.org/projects) coordinates and integrates within a harmonized R&D program targeting new proactive initiatives within EU's IST Future and Emerging Technologies (FET)[2] major studies in the area of self-organization (self-management, self-healing, self-awareness, etc.) in application to a network element's autonomic behavior exposed by innovative (cross-layer optimized, context-aware, and securely programmable) protocol stack in its interaction with numerous often-dynamic network communities.

  The ACCA (Fig. 8.1) evaluates a critical mass of relevant known and emerging paradigms and builds an AC R&D community prepared to undertake practical steps in realizing the R&D program. Recognizing that ACCA aims to solve the problem of communication infrastructure evolvability through self-organization and that this research requires a broad interdisciplinary approach, the Action explores concurrently multiple paradigm spaces addressing the problem from the viewpoints of software and hardware developments, radio technology advances, design methodology, control theory, formal methods, distributed systems research, etc.
- The **A**utonomic **N**etwork **A**rchitecture **(ANA)** project (http://www.ana-project.org) aims at exploring novel ways of organizing and using networks beyond legacy Internet technology. The scientific objective of this proposal is to identify fundamental autonomic network principles. The ultimate goal is to design and develop a novel autonomic network architecture that enables flexible, dynamic, and fully autonomous formation of network nodes as well as whole networks. The key attribute is that such a network scales in a functional way—that is, the network can extend both horizontally (more functionality) as well as vertically (different ways of integrating abundant functionality). The resulting autonomic network architecture will allow dynamic adaptation and reorganization of the network according to the working, economical, and social needs of the users. This is expected to be especially challenging in a mobile context where

[2] http://cordis.europa.eu/ist/fet/

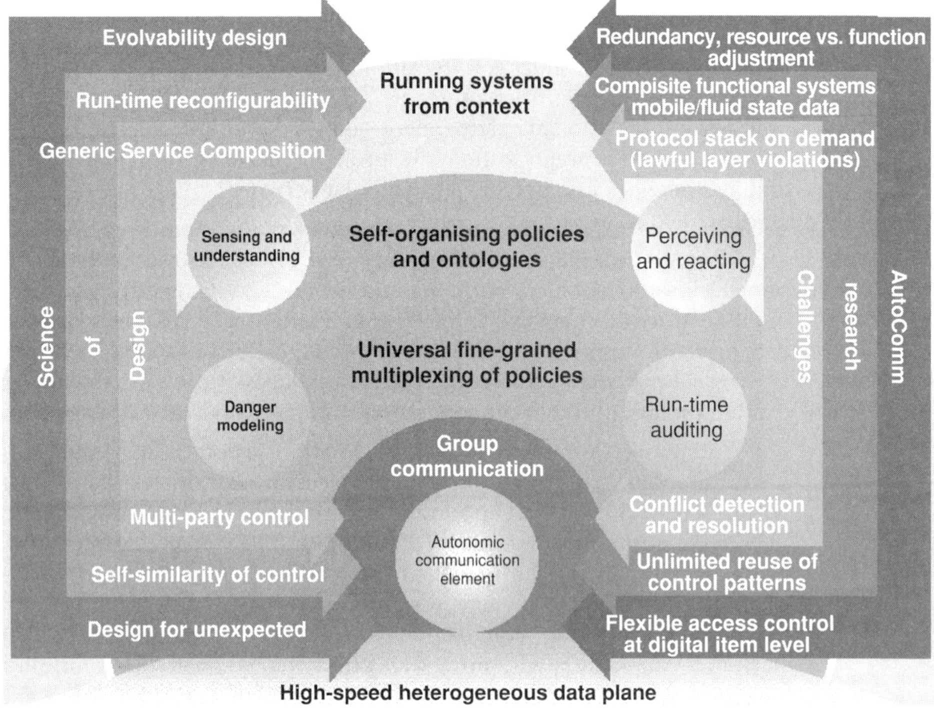

FIGURE 8.1 ACCA Framework (http://www.autonomic-communication.org/projects).

new resources become available dynamically, administrative domains change frequently, and the economic models may vary.

- **HAGGLE** (http://www.haggleproject.org/index.php/Main_Page) is a new autonomic networking architecture designed to enable communication in the presence of intermittent network connectivity, which exploits autonomic opportunistic communications (i.e., in the absence of E2E communication infrastructures).

  The project introduces a radical departure from the existing Transmission Control Protocol/Internet Protocol (TCP/IP) suite, completely eliminating layering above the data-link, and exploiting and application-driven message forwarding, instead of delegating this responsibility to the network layer. To this end, the proposed architecture goes beyond already innovative cross-layer approaches, defining a system that uses real best effort, context-aware message forwarding between ubiquitous mobile devices, in order to provide services when connectivity is local and intermittent. The project uses only functions that are absolutely necessary and common to all services, but that are sufficient to support a large range of current and future applications, more oriented to the human way of communicating (and, more generally, the way communities of any type of entities communicate), rather than related other technological aspect of the communication.

- **E-NEXT**—Network of Excellence in **E**merging **N**etworks **EX**periments and **T**echnologies (http://www.ist-e-next.net/) lists 41 research labs with the potential of up to 400 researchers involved in various programs and projects coordinated by a joint research program including self-aware and service-aware networking pilot research.

8.5 Potential Impact of AC on Future Communication Paradigms

The complexity of communication systems continues to grow thus creating a major obstacle to the introduction of new services and capabilities. The way out is seen within the emerging self-ware AC communications and computing paradigms. This is reflected by the motto: *Technology has to manage itself.* What does self-management mean with respect to communication? What should we do in order to allow communications messages enhanced with a kind of metadata carrying the intention, and meaning? What is self-similarity? Is it possible in principle to design a self-similar control plane, where virtually unlimited complexity of control will be implemented by simple concatenation, recursion, concurrency, etc., of the same control modules? What these modules are to be?

The following paragraphs summarize and comment upon the main impacts of AC on future communication paradigms and global computing.

8.5.1 Removing isolation and patchwork from network control plane

AC envisages programmable and multipurpose group communication to act as glue between separated control, signaling, and middle box communication protocols running within the Internet. Today they are designed to serve partial purposes, they often replicate state data pertaining to the same connectivity session; interaction between different entities within a control plane is an issue being addressed by midcom (IETF Working Group on Middle Box Communication) (common approach to controlling IP layer functionality) and by ccamp (IETF Working Group on Common Control and Management Plane) (common approach to controlling a lower layer functionality). Group communication realized by native IP multicast has never been widely deployed because of the lack of proper business model—providers have no incentives to save bandwidth for transit flows. At the same time, group communication at the control plane is always limited to a single administrative domain and thus offers a possibility of controlled and securely programmable infrastructure. The impact of group communication is hard to underestimate for it brings new values to all participating entities (i.e., to network functionalities), such as sharing and multipurpose use of state data, and coordinated adaptation of a group behavior patterns.

8.5.2 Facilitating Design for Evolvability

We want systems to scale and to evolve, however, traditional system design concentrates on meeting *multiple* requirements (cost/efficiency, robustness, fairness, versatility), and the result that is always constrained by multiple design trade-offs is inevitably limiting system's time to live and to evolve. We need to understand how to assess during the design the need of yet unknown requirements, what kind of redundancy a system might need to evolve, and what are the relations of the evolvability requirement to other traditional ones at different steps of system evolution. We need to understand successful evolution patterns that utilize resource adjustment, functionality adjustment, or both;

those are to be formulated as behavior definitions and rules of their applicability for ranges of systems.

8.5.3 Reconfigurability on the Fly

Communication needs to be robust; communication systems must be resilient; these and similar requirements clearly demand that *alternatives* to perform each and every function are to be *readily* available. However, even if an alternative is available it has to be configured to take over a task that otherwise will fail or degrade. Often, data needed to configure an alternative are either already lost or unavailable (invocation parameters), or are distributed among multiple entities that are not directly coordinating their behavior. Indirectly they might participate in common service provisioning; however, there will be typically no entity that would be able to make use of this fact. On the other hand, parameterization of all possible alternatives will be not only extremely expensive but also often just impossible because of resource limitations. Thus, there is a problem of keeping *task state data readily available for multiple possible implementations*. We need to understand how to define state data invariance and implementation specifics at the same time.

8.5.4 Distributed and Autonomous But Globally Optimal Control

Policy-based Management is recognized as the mainstream R&D area in network and systems management. Yet, it was also recognized fairly recently that network systems and devices were not designed with this management approach in mind. To apply PBM in its current form requires multiple adaptation steps. Furthermore, PBM introduces new problems. Namely, the more devices and systems are being controlled by policies the more fragile networks and systems are becoming.

Policy is a rule defining a choice in the behavior of a system; hence, a system is composed of a rulebase (policy part) and a functionality (with externalized behavior choices). If we allow multiple sources of policies, we will need policy multiplexing with resolution of inevitable conflicts and adaptation for functionality implementation (hence, embedding). Multiple policy sources might use different ontologies to produce policies for the same functionality; from experience we know that only the leaves of the ontology tree are actually used, hence the challenge is to develop self-organization at the level of ontologies so that (actually same) leaves of different trees could be understood as similar.

Current PBM requires that all policies are designed within common semantics, that is, they require similar views at the managed system and its environment. CIM designed by DMTF and adopted by the IETF is a useful approach, however, every practical use case requires its extensions for particular management task. We need to understand, to what extent we really need a *common* information model, describing the world in a top-down fashion, while we use practically only leaves of the ontology tree. Is it possible to design *partial ontologies* in such a way that they self-organize when and if a particular use case becomes obvious?

8.5.5 Design for Unexpected

The lack of dynamic trust and security became the most visible phenomenon of the Internet these days. Over decades, a multitude of proprietary access control systems has been developed, and now we face the issue of their interworking. There is a strong practical need to allow such interworking. At different levels, we see this need on daily basis. In emergency situations when parts of the communication infrastructure might be broken, authorities need to use all of the available channels that are protected from unauthorized use, where authorization often does not consider emergency situations. Even without emergency, there is a strong case for unexpected access, simply because legacy access is static and any ad hoc usage is by default outlawed. Consider, for example the case of a virtual organization: two companies decide to run a joint project with project-specific access rights to some of the resources. It will be very cost inefficient to modify basic access control to account for these short-lived project access policies. Yet, at the level of a single organization there are plenty of access control issues if only the organization runs its network on a heterogeneous operating systems platform.

8.5.6 How to Allow Every Party to Express Her Interests and Be Heard

Very few policies are expressed explicitly; the majority of policies (preferences, desires, intentions, etc.) are expressed in implicit form. Things are being *tried out* by sending requests. If a system is designed to serve these requests and the request is well formed then it eventually is served. What should we do in order to allow discovery requests, requests enhanced with a kind of metadata carrying the intention, and meaning? How should we design systems that can handle *every* request as a discovery request? Where should we draw boundaries of respected discovery requests?

8.5.7 Generic Service Composition

Searching for "clear and clean layer implementation" there is a need to investigate generic service composition models in which layering becomes a special case. It is not a matter of standardizing only packet format and message exchange protocols but also protocol/service interfaces, such that they can export data to other components and import data from other components in a clear way. Traditional layering is de facto violated by many Internet protocols (e.g., mobility support in IPv6 keeps host route table at layer three; wireless hosts with multiple radio interfaces are injecting layer three to layer two commands from layer four, etc.), as well as of sublayers (e.g., MPLS is actually layer 2.5; IPSec is layer 3.5, and transport layer security is 4.5, etc.). Originally, layering was introduced to cope with potential multiplicity of implementations, however, practically reports on successful layer *implementation substitutions* are virtually unknown. Over decades, layered architecture (originally aiming at flexibility) became the major obstacle of flexibility because of layer secrets—the knowledge of how particular layer functions were implemented in this particular layer implementation—hence layer violations. We need to understand how to break this routine of patching the layered protocol stack, how to come back to clear and clean layer implementation, maybe using different concepts and paradigms, derived from today's understanding of the communication needs and requirements.

8.5.8 Running Systems from Context

Probably the most wanted key on contemporary keyboards is the one that will launch a command "DO WHAT I MEAN." Seriously speaking, this key will mean a multitude of things depending on what a user wants/means, but also depending on what is the current user session aiming at. DWIM can mean "apply formatting to a text (when a user is editing the text)," or "find a site with this content," or "encrypt my communication session," etc. To enable this kind of customization we would need to understand how to combine generic user preferences, common communication conditions, and media characteristics pertaining to a current session. The above composition of profiles, preferences, settings, assumptions, and parameters will form a context that eventually will define what a user means. When we will know how to define and use this context, we'll need to understand how we can operate on this context, namely, how context hierarchies, recursions, inference, etc. should be deployed in communication.

8.5.9 How to Reuse Successful Designs

Self-similarity (e.g., for traffic model meaning high variability at all time scales) is known to be bad because it does not allow any efficient modeling and prediction; however, *self-similarity of the network control plane* would be extremely good, for it could allow unlimited reuse of control blocks, models, formats, behavior patterns, etc. We need to understand, whether it is possible in principle to design a self-similar control plane, where virtually unlimited complexity of control will be implemented by simple concatenation, recursion, concurrency, etc. of the very same control modules. What are these modules to be? How should they relate to controlled functionality, and to controlling agents? We need to refine the major concerns in this prospective design, and how can we successfully separate these concerns in order to obtain really powerful mechanisms for control and management pattern reuse.

8.5.10 Immunity and Model-Driven Security

Taking all of the above as features we envisage a future communication systems as being extremely intelligent: they will do what a user means, react on context, combine preferences that are not even explicitly exposed, be self-aware, and carefully sense the environment to reconfigure themselves. All these features are possible only within an *open communication paradigm.* Given the openness, it is important to address the issue of immunity as well. Immunity is understood here as a property of a system to resist autonomously both unsafe and insecure, and privacy-violating operation attempts.

8.6 Principles of AC Network Architectures

To form a multiservice self-ware network, we need to define a functional architecture for the interconnection and interoperability of the different autonomous elements and functions (i.e., *self-associate* in new network contexts), for individual network nodes to be able to automatically bootstrap and *self-organize,* for the overall network to autonomously organize itself. In this view, the network architecture has to take into account the following constrains for autonomy:

- **Self-configuration:** The autonomic network elements must be able to configure themselves once into the home network domain. Auto-configuration includes such aspects as IP address, security, QoS. Auto-configuration should also deal with the technology handover (e.g., going from WiFi to universal mobile telecommunications system (UMTS)) and with the parameterization of each technology to obtain the optimal resource usage interaction.
- **Self-management:** The autonomic home network must be able to self-manage in order to ensure a stable operation state. Whenever a new service needs be

deployed or a new terminal comes into the network, the self-management functions must drive the network to a stable operation state. This state has to be optimal with respect to the current operational conditions and the requirements of all services within the available resources.

- **Self-diagnostics:** The network as a whole must be able to identify its operational state and take action to drive itself to a desired stable state. The network must be able to identify the users accessing the service domain and recognize of heterogeneous home appliances must be able to manage their interoperation (e.g., interference from one appliance to the others) as well as precedence and priorities.
- **Self-protection:** An autonomic home network must be able to identify security threats to the content, such as intrusion or denial of service attacks. The network must take appropriate action to protect itself against such threats and must ensure a transparent experience for the user.
- **Self-organization and self-management:** The creation and maintenance of customized communication structures should not necessitate manual intervention (apart from high level expression of policies). The configuration process itself must be flexible, such that it can adapt to the existing topology requirements, topology changes (i.e., topology aware), available resources (i.e., resource aware), and particular needs and goals of the application (i.e., context awareness).
- **On-demand creation and removal:** Customized communication structures are a natural vehicle for the implementation of distributed network services, such as distributed collaborative monitoring. In AC, such custom communication structures should be possible to establish on-demand, upon request and without any further manual preparation. In case a particular communication structure is no longer needed, it should resolve itself and release all resources it occupies.
- **Security**: Security is a fundamental aspect in such customized structures that should be strongly self-protecting as opposed to add-on patches commonly used in current VPNs and overlays.

Specific to networking principles, selfware AC/computing environments present three main challenges that conventional networking approaches are not able to effectively face, namely device and technology heterogeneity, cooperation and misbehavior, and complexity.

8.6.1 Device and Technology Heterogeneity

One of the main features of selfware AC devices is the huge heterogeneity arising at the device level. These all-embracing networked environments comprise devices with very different technical features, ranging from simple passive radio-frequency identification (RFIDs)[3] tags to standard sensor nodes up to smartphones, laptops, and PDAs.

[3] RFID is an automatic identification method, relying on storing and remotely retrieving data using devices called RFID tags or transponders. An RFID tag is an object that can be applied to or incorporated into a product, animal, or person for the purpose of identification using radiowaves. Some tags can be read from several meters away and beyond the line of sight of the reader.

All this heterogeneity represents a harmful feature for conventional IP-based networking paradigms, where the hourglass model imposes severe constraints on the resources needed from the various networked entities. The other possible approach is to actually exploit this heterogeneity, by means of a multitiered architecture, to envision different logical roles in the network that can then be mapped onto the different device classes. Also, heterogeneity could be exploited in many fashions for achieving such goals, the most successful one being the inclusion of such information into the routing metric to be used to perform multihop communications.

8.6.2 Cooperation and Misbehavior

Operations and services in an AC framework are completely organized in a distributed and decentralized way. Cooperation among entities represents one of the main requirements to make the system working properly. However, non-cooperative behaviors may be exhibited by individual users. An entity that does not cooperate is called *misbehaving*. Cooperation misbehavior can be caused by entities that are malicious or self-interested. The main aim of self-interested or selfish entities is not to *damage the overall functioning of the system; rather, a selfish entity does not cooperate in the sense that it is unwilling to spend its resources on behalf of others.*

Node selfishness represents a class of intentional but not malicious misbehavior that can affect the cooperative environment. Specifically, a node may act selfishly whenever it is required to forward packets on behalf of others, without being directly rewarded. The presence of selfish nodes may significantly degrade the performance of the entire system. Network services may not be available and cooperative nodes may become overloaded, possibly leading to new cooperation misbehavior. If a mesh user behaves selfishly by removing itself from the mesh grid, then the whole mesh may potentially suffer from this loss of connectivity. In the worst-case scenario, the network becomes partitioned.

A malicious entity breaks the cooperative paradigm to intentionally damage other nodes. As a consequence, several security problems, mainly denial-of-service (DoS) attacks, can affect the network. For example, an attacker can send forged routing packets to create a routing loop or to partition the network (routing disruption attack) or can inject extra data packets into the network consuming bandwidth resources (resource-consumption attack) (Hu et al., 2002).

Selfish behavior may be encountered at different levels.

- **MAC-layer misbehavior**—All nodes within a WM^{2}Net use the same frequency band as a shared medium for receiving and transmitting data. A selfish node may fail to adhere to the contention resolution protocol aiming to gain a bigger than its fair share of the channel bandwidth (Kyasanur and Vaidya, 2003) (this may be realized in selecting smaller backoff values, or using a different retransmission strategy that is not doubling the Contention Window value after a collision occurence). As a result, the node is unwilling to wait for a fair time before transmitting.
- **Network-layer misbehavior**—In WM^{2}Nets, basic network functions like routing and forwarding are distributed over all the participating nodes (Royer and Toh, 1999). Every node must act as a router, and far-off nodes communicate using intermediate nodes as relays. Selfishness at the network layer mainly regards:

1. *Forwarding*: the node does not perform the packet forwarding function. Packets with a source or destination address different from the current node are discarded.
2. *Routing*: The node does not perform the routing function. It discards every routing packet that is not of its interest. There will be no route including that node, and hence it will never be asked to forward a packet on behalf of others.

- **Transport-layer misbehavior**—At the transport layer, cooperation misbehaviors are mainly identified with the TCP congestion control mechanism inherited from the wireline context (Savage et al., 1999). Specifically, if the sending node does not apply the appropriate congestion control algorithm, it may send data more quickly than the other nodes, forcing competing traffic to be delayed or discarded. As a result, the presence of a node running differently creates unfairness between active traffic connections. In addition, research on cooperation issues has mainly focused on the improvement of TCP performance over wireless links since events such as route failures and route changes due to nodes' mobility can cause serious problems. Route failures can cause packet drops at intermediate nodes and are wrongly interpreted as congestion problems; route changes can introduce frequent out-of-order delivery and confuse the TCP control mechanism. To improve performance, current solutions are based on cooperation of intermediate nodes using some feedback to notify the sender about route failures (Chandran et al., 2001; Holland and Vaidya, 1999; Liu and Singh, 2001). Nevertheless, these solutions may introduce a new kind of misbehavior. Since intermediate nodes do not get any advantage from this cooperation, they may act selfishly and decide not to send route failure notifications to the sender.

8.6.2.1 Enforcing Cooperation Solutions at Network Layer

The following, cooperative enforcing mechanisms are commonly proposed in the literature for resolving the selfishness problems at the network layer. These are classified in two categories: *pricing-based schemes* and *reputation-based schemes*.

Pricing-based schemes stimulate packet forwarding by means of virtual currency (credit). The key idea is that nodes providing a service should be remunerated, while nodes receiving a service should be charged. Based on this concept, Buttyan and Hubaux (2003) propose a tamper-resistant security module that maintains a nuglet counter. The proposed protocol requires the node to pass each packet (generated as well as received for forwarding) to the nuglet counter which is decreased when the node wants to send a packet as originator, and increased when the node forwards a packet. Hence, if a node wants to send its own packets, it must forward packets for the benefit of others. To overcome the requirement of a tamper-proof hardware, in SPRITE (Zhong et al., 2003) a credit clearance service (CCS) is inserted to handle credits. When a node receives a message, the node keeps a receipt of the message. Later, the node reports to the CCS the messages received/forwarded by uploading its receipts. The CCS then determines the charge and credit to each node involved in the transmission of a message, depending on the reported receipts of a message.

Reputation-based schemes discourage misbehavior by estimating reputation of nodes and punishing nodes with bad behavior. Marti et al. (2000) present a solution aimed at detecting and avoiding misbehaving nodes through a mechanism based on a watchdog and a reputation system. The watchdog identifies misbehaving nodes by performing a neighborhood monitoring: it observes the behavior of neighbors by promiscuously listening to communications of nodes in the same transmission range. According to collected information, the reputation system maintains a value for each observed node that represents a reputation of its behavior. The reputation mechanism allows avoiding sending packets through misbehaving nodes. In this way, malicious nodes are rewarded and strengthened, while cooperation enforcing is thoroughly absent. CONFIDANT (Buchegger and Boudec, 2002) and CORE (Michiardi and Molva, 2002) are two extensions extending the previous scheme with a punishment mechanism that isolates misbehaving nodes by not serving their requests. When a neighbor's reputation falls down a predefined threshold, service provision to the misbehaving node is interrupted. In such a way, there is no advantage for a node to misbehave because any resource utilization will be forbidden.

8.6.2.2 Self-Organization

The starting point for the self-organization paradigm is based on the assumption that the system takes birth, evolves, and dies dynamically in ad hoc fashion, depending on the particular operating scenarios. In addition, the underlying composition of nodes and legacy static infrastructures may interact with each other to accomplish the intuitive behavior required by legacy applications. This dynamic life cycle is assumed to require inherent capability for self-organization. The aim of this section is to analyze the degree at which the self-organization functions are required and, if self-organization functions are needed, what kind of self-organization functions are the most essential requirements.

The number of wireless devices is increasing (Weiser, 1993; Kephart and Chess, 2003), and they are not only mobile phones and legacy computers but also a very many different kinds of small devices, sensors, actuators, or other mesh devices. The increasing number of devices creates possibilities for new applications, but also adds the complexity and dynamics of the networks. The key question is how this complexity and dynamics can be handled in a sensible way without requiring users to become technical experts in the field, and the network owners to spend time and resources in managing dynamic networks (Bettstetter, 2005). One potential solution approach is to increase the degree of self-organization in the system. A number of examples exist in nature where the self-organization solves very challenging problems, for example, in the human body, and in society, etc.

The aim of the self-organization functions is to minimize human intervention by building capabilities into the system to automatically organize itself as far as possible. In the literature, self-organization has been defined to be a process, where the organization of the system spontaneously increases, without this increase being controlled by the environment or an encompassing or otherwise external system. A self-organizing system does not only regulate or adapt its behavior, but it also creates its own organization (Heylighen and Gersheson, 2003). The self-organization refers to negotiating and reaching the organizational structure in a distributed manner, and enabling it to survive even if changes are continuously happening either in the environment or in the nodes themselves. The individual entities exchange information locally and there is no need for any external control entities. In addition, self-organization applies high-level behavioral rules in the individual entities that lead to sophisticated functionality in the system.

The self-organization function is divided into several subfunctions. These are clarified in the Table 8.1. Note that these subfunctions are not necessarily independent from each other; however, they are all part of the self-organization. The problem that each function aims to solve is defined, and the related technology requirements are briefly described.

8.6.3 Building Blocks for Self-Evolving Communication Systems

Recent trends in theoretical biology suggest that the "survival of the fittest" paradigm (based on the mutation, crossover, and selection operators) cannot be understood as the sole driving force of evolution towards more and more complex forms of life. Other phenomena and mechanisms (e.g., cooperation and coalitions, autopoiesis, gene expression, etc.) may play a key role in shaping the order of self-organizing systems out of the potential chaos generated by the high system complexity matched to the maximum entropy principle. Nothing but environmental pressure guides evolution, so that, without loss of generality, it can be seen as a form of self-organization. Nevertheless, self-organization is a phenomenon observed in a wide variety of contexts in both natural and physical sciences, but there is no apparent consensus either on its general meaning nor on its relation to evolution. Indeed, whilst many of the sophisticated (self-organizing) mechanisms that characterize biological systems evolved over long periods of time, they do not rely today on evolutionary processes in order to operate. *Evolution* could, in principle, be regarded as the means through which self-organization is achieved in complex systems.

In a pervasive environment, distributed on-line evolution can be viewed from a *macroscopic* or *microscopic* perspective. At a microscopic level we see individual programs implementing a given service or protocol. A macroscopic perspective offers a system-wide view, in which different services coexist and interact (cooperating or competing) as in a digital ecosystem ("Digital Business Ecosystem (DBE) Project": http://www.digital-ecosystem.org). The two scales have different dynamics and require different tools to be modeled and analyzed. At the *microscale*, we deal with program transformations, from self-tuning of parameters up to self-generation of code. At the *macroscale*, we observe how successfully services replicate and propagate throughout the network, how competing services share resources, whether and how cooperation is built in some regions, etc.

Both perspectives are needed to engineer a system that can be then left to evolve on its own. Since a WM^{2}Net is a large-scale complex system, where a large number of mutually interacting dynamical entities coexist, the study of the microlevel is not sufficient for understanding/predicting the behavior on a system-wide scale. On the other hand, the study of the macroscopic behavior alone is not sufficient to derive the necessary program transformations that will lead to optimum performance at both macro- and microlevels. Therefore, both perspectives must be integrated in the evolution framework.

8.6.3.1 Microevolution Building Blocks

Given that we are dealing with *on-line* or *runtime* evolution, it is essential that the system is able to produce new generations of programs that are superior or at least as good as their predecessors. At the microlevel, we need to provide an *open-ended* evolutionary framework for a continuous evolution of the system in response to changes in the environment or in user requirements. In this view, new techniques are sought to manipulate and transform the implementation of protocols and services at runtime.

Self-configuration	**Problem:** Integration of multiple vendors' products and platforms is time consuming and error prone especially when required to be done dynamically at run time. **Requirement:** Configuration of components and systems shall follow their high-level policies, and rest of system should adjust automatically and seamlessly into the situation.
Self-management	**Problem:** The distributed nature of the system implies that a single node is limited to its own role and capabilities, and it cannot carry out all the required tasks. **Requirement:** The roles of each node need to be discovered, allocated dynamically, and environment need to be continuously monitored to keep the system in the living state.
Self-optimization	**Problem:** Systems have hundreds of manually set nonlinear tuning parameters, and their number continuously increases. **Requirement:** Components and systems modeling shall seek opportunities to improve their own performance and efficiency.
Self-healing	**Problem:** Problem determination in large, complex systems can take weeks from a team of programers. **Requirement:** System shall automatically detect, diagnose, and repair localized software and hardware problems.
Self-protection	**Problem:** Detection and recovery from attacks and cascading fading is manual. **Requirement:** The system shall automatically defend against malicious attacks or cascading failures. It uses early warning to anticipate and prevent system-wide failures.
Self-adaptation	**Problem:** The environment of the system is continuously under changing situations, which require changes in the system internal and external behaviors. **Requirement:** The system shall automatically detect the meaningful changes happening in the environment, and adapt the system internal and external behavior accordingly.
Self-awareness	**Problem:** Each node cannot rely only on the information received from the other nodes, because they are not necessarily available in the next moment of time. **Requirement:** System shall be automatically situation aware and detect the environment conditions to be able to act in a stand-alone situation in a proactive and self-aware way.
Self-localization	**Problem:** The localization of the node itself, the other nodes and the information stored therein is not always directed forward, because of continuous changes happening in to the system. **Requirement:** System shall automatically localize the node itself and the other nodes and discover the information stored in them.

Table 8.1 The Analyzed Self-Organization Functions

Genetic algorithms (GAs) and genetic programing (GP) are two common bio-inspired algorithms for evolving programs and their parameters. However, we see classical linear or tree-based GP as not suitable for on-line evolution of programs, since the programs generated by crossover or mutation may have a low rate of success (Langdon and Poli, 2006). We foresee that a chemical language lends itself more easily to transformations by genetic operators, and the resulting programs can be more robust to potentially defective execution paths (Tschudin and Yamamoto, 2004). We have therefore paid considerable attention to chemical computing as presented in Section 8.6.3.1. Some chemical computing approaches such as artificial chemistries (Dittrich et al., 2001) have the additional advantage of lending themselves to macrolevel studies by employing techniques similar to those used in systems biology. This provides useful tools for understanding the macrobehaviors stemming from microscale evolution algorithms.

Evolutionary Computing *Evolutionary computing* (Eiben and Smith, 2003) encompasses a range of problem-solving techniques collectively called *Evolutionary algorithms* (EAs). EAs are inspired by biological evolution principles such as genetics and natural selection. These are algorithms for searching optimum solutions within a given search space of all possible solutions. EAs employ ideas from genetics and natural selection to perform a *beam search* that restricts the search space to a "beam" area of retained promising solutions. They are typically applied when the search space is so vast that conventional optimization techniques cannot be efficiently applied.

The main EAs are *GAs*, *GP*, *Evolutionary programing* (EP), and *Evolution strategy* (ES). They all model candidate solutions as a population of individuals with a genotype that is transformed and evaluated against a given fitness criterion until an optimum solution is found. The difference among them lies in the way candidate solutions are represented, and on the search operators applied to obtain new solutions. In a GA (Holland, 1975) candidate solutions are represented as a population of individuals whose genotype is a fixed-length binary string. The goal of a GA is to find the optimum value of such binary string that optimizes a given fitness criterion. An initial population of candidate strings is generated and evaluated against the fitness criterion. The best strings (highest fitness) are then mutated and recombined (crossover) to produce new strings that are then reevaluated, and so on, until an optimum solution is found, or until some stop criterion is satisfied (for example, distance to the optimum closer than a threshold).

GP (Banzhaf et al., 1998; Koza, 1992; Langdon and Poli, 2002) applies the GA idea to evolve computer programs automatically. GP can be considered as a subfield of *machine learning* in the sense that it seeks to obtain populations of programs that improve automatically (Banzhaf et al., 1998), so they can somehow "learn" good behaviors. A GP algorithm is essentially the same as a GA, but instead of simple fixed-length binary strings, the candidate solutions are variable-length computer programs, and their genotype is the representation of the program implementation itself. GP typically evolves programs encoded in a linear (similar to assembly language) or tree language (similar to functional languages such as Lisp (http://www.lisp.org/HyperSpec/Body/sec_1-1-2.html), but other representations are also possible, such as general graphs (Poli, 1999), finite state machines (Araujo et al., 2003; Sharples, 2001; Sharples and Wakeman, 2000), neural networks (Nolfi and Floreano, 2000), and more recently, chemical programs (Matsumaru et al., 2006a; Yamamoto and Tschudin, 2005). Programs synthesized by GP are known to have even generated patented or patentable inventions (Koza et al., 2003).

However, GP has been mostly applied to obtain solutions in an off-line manner: once deployed, the programs evolved by GP in general do not continue to evolve during their operation. Comparatively little has been achieved in the evolution of running systems, but some significant results can be found in domains such as evolvable hardware (Sipper et al., 1997) and robotics (Andersson et al., 2000; Nolfi and Floreano, 2000; Steels, 1994).

EP (Fogel, 1966) is similar to GA but focuses on the evolution of finite state machines and does not employ crossover to generate new solutions. Instead, it concentrates on mutation as the main search operator. *Evolutionary strategy* (ES) (Beyer and Schwefel, 2002a; Rechenberg, 1973) is similar to EP in the sense that mutation is the main operator, but focuses on evolving solutions represented as real vectors. We consider EP and ES as specific variations of the more generic GA and GP forms, so EP and ES will not be further discussed in this chapter.

Let us now look at GAs and GP in more detail. The difference between GA and GP lies in the representation of the evolved solutions, and sometimes this distinction can be blurred: GA bit strings can represent parameters of a protocol or service; or the identification of links and nodes in a graph or state machine (Araujo et al., 2003; Poli, 1999; Sharples 2001; Sharples and Wakeman, 2000); or the encoding of which building block is present or absent in a solution, in order to evolve combinations of building blocks ("Digital Business Ecosystem (DBE) Project": http://www.digital-ecosystem.org). So, the approach of evolving protocols as finite state machines as in (Sharples 2001; Sharples and Wakeman, 2000) can be seen as a form of GP, but solution candidates were represented as bit strings, so it can be regarded as a GA too. Similarly, the approach in (Yamamoto and Tschudin, 2005) for evolving protocols by combination of their building blocks can be regarded as GP since the representation used was a variable-length chemical program. However, recombining previously existing modules can also be achieved by encoding each possible module as a bit in a string and then applying GA. Similarly in robotics it is very common to produce robot controllers by means of evolving the weights of neural networks (Nolfi and Floreano, 2000). Given that the evolution of these weights produces new behavior in a robot, this can be regarded as a form of GP. The representation of the candidate solutions can, however, be achieved with a simple GA bit string. Instead of classifying an approach into GP or GA, it seems wiser to look at the complexity and flexibility of the representations and their suitability for the problem at hand. A continuous space of evolutionary computing techniques is more accurate, in which solutions have a degree of adaptability given by the portion of the program that is represented as an evolvable genotype.

GAs and GP follow essentially the same steps:

1. Representing candidate solutions. Candidate solutions to a proposed problem are modeled as a population of individuals. Each individual (candidate solution) is encoded using a genetic (genome-like) representation or genotype. The genotype usually takes the form of a bit stream for GAs and a program tree for GP.
2. Fitness evaluation. Each generated candidate solution is rated by means of a *fitness* function that evaluates the suitability of the solution to solve the proposed problem.
3. Selection. A *selection* mechanism operates on the result of the fitness evaluation to select a subset of solutions to "survive" to the next generation.

4. Generating new candidate solutions. One or more *genetic operators* (mutation and crossover being the most common) modify the genome to generate new candidate solutions. A *mutation* is a modification of a small portion of the genotype chosen at random. *Crossover* consists in randomly selecting genotype segments in two individuals and swapping these segments between them. The result is two new individuals, which are inserted in the population.
5. Iterative search and optimization. The generation of solutions, evaluation, and selection process continues in an iterative way until a solution with maximum fitness is found, or until the fitness of the generated solutions remains stable or shows little improvement over generations. The solution with maximum fitness is then taken as the output of the algorithm. GAs and GP are usually performed as off-line tasks in a centralized manner.

Extensions of GAs and GP to a distributed context include distributed genetic algorithms (DGAs) (Belding, 1995) and parallel distributed genetic programing (PDGP) (Poli, 1999). DGA is essentially a synchronous extension of GA to support parallel computing, and PDGP is a type of DGA that operates on graphs. These systems are not meant for on-line operation.

An example of a hybrid on-line/off-line GAs is the environment identifying genetic algorithm (EIGA) (Mori and Matsumoto, 2003), in which an on-line module takes care of adaptation to the real-world environment using a neural network, while the off-line part performs more intensive parallel search on a population of individuals, based on feedback from the on-line module. Superior individuals found at the off-line module are sent to the on-line module at regular intervals. This approach is promising since only the best solutions found off-line are executed in the real world, protecting the system from most unsuitable variants. However, in a pervasive environment, it might be difficult to find a sufficiently powerful node able to execute the intensive off-line simulations required to evaluate the fitness of a whole population of candidate solutions. A grid-like DGA would be needed, so that candidates could be dispatched to different machines for off-line (simulated) evaluation. The resulting fitness could then be recollected for evaluation and selection of best candidates. Although in principle feasible, we feel that such a solution could make the system unnecessarily heavy by requiring a grid middleware and a lot of coordination, and as a result, it could discourage users from adopting it.

The position paper (Eiben, 2004) argues that evolutionary computing is a promising technique for achieving the much-wanted self-managing properties of autonomic systems, and presents an evolutionary approach to runtime self-optimization in a distributed system. The position is illustrated with a web service example, in which several web servers offer the same service to a large number of visitors. Each web server behaves according to a parameter vector that determines a session handling strategy. The objective is to maximize the QoS experienced by the visitors, which is a function of the delay to obtain the service and the degree of request satisfaction. An evolutionary approach is presented in which the population to be evolved is the set of parameter vectors, and servers using a given parameter vector are evaluated on-line during the web session. The paper points out the asynchronous nature of such evolutionary approach, which resembles natural evolution more closely than traditional EAs: at a given point in time in a runtime EA, each individual may be in a different state (some being mutated, some reproducing, some being evaluated), while in traditional EAs, all operations are usually performed in a predefined sequence. However, there seems to be no indication so far of whether the approach has been already implemented or not.

Few results are available that shown on-line evolution of distributed software. The Bio-Networking Architecture Project (Bio-Net) ("Bio-Net: Bio-Networking Architecture Project": http://netresearch.ics.uci.edu/bionet/) is an example. It is building a biologically-inspired middleware platform to support adaptive agent-based distributed services (Suzuki and Suda, 2005). It applies GAs in a decentralized way to evolve the behavior of agents that provide network services (Nakano and Suda, 2004; Nakano and Suda, 2005). In the Bio-Net platform (Nakano and Suda, 2004; Nakano and Suda, 2005), an agent invokes behavior x (such as replication, reproduction, migration, and death) based on factors that include the agent's internal state (such as age, energy level, user demand for its services, distance from the user) and environmental conditions (available resources, agent population size on the platform, etc.). Each factor v_i has a weight w_{xi} associated to behavior x. A behavior is invoked when the weighted sum of factors exceeds a threshold θ_x, that is, when

$$\sum u_i \cdot w_{xi} > \theta_x$$

Note that this behavior representation is similar to neurons in artificial neural networks: a neuron "fires" when the weighted sum of input signals exceeds a threshold. The weights w_{xi} represent the genome of an agent. During reproduction, the genome of two parent agents are recombined by crossover, and then mutated to produce new behavioral weights in their offspring. Simulation results show that on-line evolution can improve agent performance, and that agents are able to adapt to their environment.

Chemical Computing The term *chemical computing* refers to two different areas (Calude and Paun, 2001; Dittrich, 2005): The first one is to derive computation models inspired by chemical reactions, which nevertheless run on traditional computers. The second one is the use of real (organic) molecules and (bio)chemistry knowledge to build computational devices, for example, in molecular/DNA computing.

Chemical computing models have been applied to diverse fields such as image processing applications (Banâtre and M'etayer, 1996), operating systems, compilers, dynamic software reconfiguration (Wermelinger, 1999), multiagent systems (Stamatopoulou and Gheorghe, 2004), distributed computing (Syropoulos, 2004), and, more recently, robotics (Ziegler and Banzhaf, 2001), grid computing (Banâtre et al., 2006), and autonomic computing (Banâtre et al., 2004a).

Three main chemical computing models for their relevance to self-ware AC are further analyzed: *gamma*, *membrane computing*, and *artificial chemistries*.

The Gamma Formalism *Gamma* (Banâtre et al., 2005a; Banâtre and M'etayer, 1986) was proposed in 1986 as a programing formalism based on a chemical reaction metaphor. More recently, γ-calculus has been introduced as a formalism that extends the original Gamma model to a higher-order calculus (Banâtre et al., 2005). Banâtre et al. (2004a) applied this new calculus to specify Autonomic Computing systems, including a mailing system as an example.

In Gamma (Banâtre et al., 2005, 2005a, 2004; Banâtre and M'etayer, 1996), computations are modeled as interactions among atomic values that "float" freely in a chemical solution. These values are represented as elements in a *multiset*, an unordered set within which elements may occur more than one. The number of occurrences of a given element within the multiset is called the *multiplicity* of the element. The multiset contains the data to be processed as well as the reaction rules. Reaction rules specify a reaction condition and an action. Computations replace elements satisfying the condition by those specified in the action. A computation terminates when no more chemical reactions can take place. As an example, this is a reaction rule to compute the maximum of a set of numbers:

$$\textbf{replace } x, y \textbf{ by } x \textbf{ if } x \geq y$$

When inserted in a multiset containing several numbers, this rule will replace any two numbers x and y that satisfy the condition $x \geq y$ by x. This computation will proceed until only the greatest of all numbers is left. Note that the rule imposes no order in which numbers should be compared. Moreover, if more numbers are injected into the multiset after the computation is finished, the reaction rule will immediately restart "consuming" the smaller numbers until, again, only the maximum one is left. The rule could also be applied in parallel, by comparing disjoint pairs of numbers. This model therefore enables highly parallel programs to be expressed in a way that is very close to their specification, that is, without the artificial sequentiality constraints imposed by traditional programing languages.

Membrane Computing *Membrane computing* (Calude and Paun, 2001; Paun, 2000) is another chemical computing model in which computations (chemical reactions among *objects*) occur inside a cell-like *membrane structure*. Membranes can be recursively nested, and the outside-most membrane is called the *skin membrane*. As in Gamma, objects are represented as elements of a multiset. They can be transformed into other objects and can cross membranes. A membrane can dissolve under the effect of one of its objects; when this happens the content of the membrane is release to its parent membrane. The resulting computing device is called a *P system*. Similar to Gamma, computations inside a *P system* finish when no more reactions can take place inside its constituent membranes.

Membranes are an important abstraction to enable chemical computing models to scale to large and complex computations. They not only encapsulate complexity from the point of view of rules and name space, but also from the point of view of the (virtual) reaction vessel: if no encapsulation mechanism is available, the implementation of a reaction vessel in a classical von Neumann architecture becomes extremely inefficient: every time step, the execution engine has to check which reaction rules are applicable to which elements of the multiset. For large multisets and complex reaction rules, the execution engine would simply not scale.

Moreover, membranes provide a natural mechanism of communication between different execution engines. Communication is modeled as the exchange of chemicals between membranes. If we allow membranes to be situated at different nodes, we can then implement a communication channel between these remote membranes, and ship objects between membranes.

Artificial Chemistries In *artificial chemistries* (Dittrich et al., 2001), computations are modeled as chemical reaction networks that are represented in the same way as real chemical reaction networks in systems biology: as bipartite graphs with two types of nodes: substrates and reaction rules. Contrary to gamma and membrane computing, which focus on microevolution at the level of programing language theory and execution metaphors, artificial chemistries aim at offering methods at the macrolevel, in order to understand large-scale reaction systems, either natural or artificial. Chemical organization theory (Matsumaru et al., 2006) is used to reduce the complexity of the system from a large network to smaller organizations.

In Hjelmfelt et al. (1991) neural networks are implemented using a reversible reaction mechanism in an artificial chemistry that closely emulates real chemistry. Each neuron consists of eight chemical species coupled by four reversible reactions. The resulting concentrations of chemicals determine the behavior of the neuron (firing or not), and these are connected to other neurons forming a network. Logical gates are implemented using this system, and a possible extension towards a universal Turing machine is discussed.

Evolving chemical programs have been shown to be possible in artificial chemistries: Ziegler and Banzhaf (2001) implemented a robotic control system using an artificial chemistry. The structure of the reaction graph evolves by GP. The genetic operations on the graph are a form of mutation (adding or removing nodes and links) and crossover (exchanging portions of two graphs) that obey mass conservation laws and reaction balance, so that the graphs are kept consistent across transformations. Matsumaru et al. (2006a) investigate how chemical evolution appears in a chemical computing system, using organization theory. This is a step towards understanding global (macro) system behavior as in systems biology.

Fraglets The *Fraglets* chemical programing language (Tschudin, 2003) is based on a chemical model where "molecules" interact with each other or undergo some internal transformation. An implementation of a fraglet interpreter in C is available for download at the Fraglets website (http://www.fraglets.net/).

Fraglets stand for "computation fragments" (Tschudin, 2003) and are used for representing both data carriers and executable rules (code) in a unified way. A fraglet is a string of symbols $(s_1 : s_2 : \ldots : s_n]$ representing data and/or protocol logic. It is a fragment of a distributed computation that may be carried in packets or stored inside a network node. Like molecules, fraglets operate on each other and undergo transformations like splitting up or transmitting themselves to a remote node. The result can be regarded as a distributed chemical reaction network, where the execution of one fraglet rule leads to the creation of the next fraglet to be executed. Code and data are then constantly being transformed and regenerated to perform a given task, typically related to communication protocols.

When fraglets are carried in packets that traverse a network, the successive fraglet symbols can be interpreted as successive header fields in today's regular data packets. Upon arrival, a fraglet packet is injected into the local fraglet store or context, where an execution environment processes the fraglet "headers" in a similar way as a header processing treats packets in traditional networks. The fraglet processing engine continuously executes tag-matching operations on the fraglets in the store, in order to determine the actions that should be applied to them. Fraglet operations, except for the transmission, have the property that they can be carried out in constant time. The fraglet store is a multiset, as in Gamma (see Section 8.6.3.1.B.1): several instances of the same fraglet may be simultaneously present.

8.6.3.2 Macroevolution Building Blocks

Macroevolution considers techniques that look at the system as a whole, in order to understand its global properties. Three main building blocks are identified:

- *Evolutionary Games*: In evolutionary game theory, the evolution of entire populations can be studied by modeling the interactions among individuals that receive a certain amount of "reward" related to their reproduction capability (survival of the fittest). Populations that receive a higher reward will naturally grow, while others might disappear. Populations may also be affected by mutations, or invasions by other populations. The evolutionary games formalism is one of the basic mathematical tools designed for predicting population dynamics in this context. The subsection that follows provides a brief overview of evolutionary games.
- *Stochastic processes*: The behavior of large-scale long-term distributed evolutionary processes can be studied using stochastic methods.
- *Systems biology*: The field of systems biology (Kitano, 2001) makes heavy use of mathematical models to understand real biological systems and the complexity of the interactions between different biological cycles and processes. It builds heavily upon graph theory, control theory, and probability, and includes experimental, theoretical, and modeling techniques. We can learn from the techniques used in systems biology to study our artificial evolving systems in order to understand their global behavior. Indeed, as discussed in Section 8.6.3.1.B.3, artificial chemistries (Dittrich et al., 2001) use graph representations, which are very similar to those used in systems biology, therefore similar tools could be used to analyze them. Matsumaru et al. (2006) presented a chemical organization theory as a theoretical base for studying large-scale artificial chemistry systems, such that full, complex systems can be divided into smaller, manageable organizations. This theory has been applied to study how evolution might emerge spontaneously in chemical organizations (Matsumaru et al., 2006a).

Basic Features of Evolutionary Games The evolutionary games formalism is one of the basic mathematical tools designed for predicting population dynamics in the context of *interactions between populations*. This formalism identifies and studies two concepts. The first is called *evolutionary stable strategy* (ESS), and the second is the *replicator dynamics*.

The equilibrium is characterized by a property of robustness against invaders, which, in the biological context are called *mutations*. The equilibrium has the following properties. First, if equilibrium is reached, then the proportions of each population do not change in time. Secondly, at equilibrium, the populations are immune from being invaded by other small populations. The replicator dynamics describes the evolution of the interacting populations. This dynamics can lead to convergence to the ESS, but in general, even if an ESS exists, there is no guarantee for convergence to it. Even in the absence of convergence to ESS, one may be interested in identifying weaker forms of convergence: convergence to a limit trajectory (possibly to a periodic one) that may occur from any initial sizes of populations within some subset of states (this is known as an asymptotically stable system). If the subset includes all initial states then the system is said to be globally asymptotic stable. For the definitions of these and other notions of convergence and stability, see the work by Khalil (2002). Trajectories may of course diverge, so that any two initial states give rise to distinct trajectories whose distance from each other need not converge to zero. Yet, even then, some form of stability can be defined: one in which all trajectories remain in some bounded set.

Definition of Evolutionary Games Consider a large population of players. Let us assume that each individual occasionally needs to take some action (such as power control decisions, or forwarding decision). Occasionally, the action of some (random number of) other individuals interacts with the action of that individual (e.g., other neighboring nodes transmit at the same time). For simplicity, assume that each individual has only two available actions: 1 and 2. We say that the whole population uses a mixed strategy q^*, if a fraction q^* of the population plays one strategy and the remainder $\overline{q}^*$ plays the others. (This can be realized, for example, if each individual toggles randomly between the strategies.) Let $J(p, q)$ define the expected payoff for our tagged individual if it uses a mixed strategy p while the rest of the population (with which it interacts) uses the mixed strategy q^*.

Suppose that the population uses a mixed strategy q^* and that a small fraction (called "mutations") adopts another distribution p over the two strategies. If for all $p \neq q^*$,

$$J(q^*, q^*) > J(p, q^*), \tag{8.1}$$

then the mutations fraction in the population will tend to decrease (as it has a lower reward, meaning a lower growth rate). q^* is then immune to mutations. If there are n pure strategies ($n = 2$ in our case) denoted by $s_1, \ldots, s_n$, then a sufficient condition for Eq. (8.1) is that

$$J(q^*, q^*) > J(s_i, q^*), s = 1, \ldots, n \tag{8.2}$$

In the special case that the following holds,

$$J(q^*, q^*) = J(p, q^*) \text{ and } J(q^*, p) > J(p, p) \quad \forall p \neq q^*, \tag{8.3}$$

a population using q^* are "weakly" immune against a mutation using p since if the mutant's population grows, then we shall frequently have individuals with strategy q^* competing with mutants; in such cases, the condition $J(q^*,p) > J(p, p)$ ensures that the growth rate of the original population exceed that of the mutants. q^* that satisfies Eq. (8.1) or Eq. (8.3) is called an evolutionary stable equilibrium (ESS).

ESS has been defined in 1972 by M. Smith strategy in (Smith, 1972). In 1982, Maynard Smith's seminal text "Evolution and the Theory of Games" (Smith, 1982) appeared, followed shortly thereafter by Axelrod's (1984) famous work. Although ESS have been defined in the context of biological systems, it is highly relevant to engineering as well (see Vincent and Vincent, 2000). In particular, in the context of competition in the access to a common medium, we can expect that a technology that provides better performance will gain more market shares on the expense of less-performant technologies.

In addition to identifying ESS, the evolutionary game theory community is often interested in the actual evolution dynamics, that is, of the actual convergence to an ESS (when it exists). Various models, called "replicator dynamics," have been proposed (see for example: Cabrales, 2000 and Hofbauer and Sigmund, 2003). We can learn and adopt notions from biology not only through the concept of evolutionary games, but also in applications related to energy issues that have a central role both in biology as well as in mobile networking. The long-term animal survival is directly related to its energy strategies (competition over food, etc.), and a population of animals that have good strategies for avoiding starvation is more fit and is expected to survive (Houston and McNamara, 1991; McNamara, 1990; McNamara et al., 1991). By analogy, we may expect sensor networks whose components have efficient energy strategies to live longer and to have more chances to survive (Mhatre and Rosenberg, 2004).

8.6.3.3 Self-Organized Criticality and Evolution

This section addresses some models of real biological systems that play an important role in understanding and designing self-evolving services. The first link, which is apparent, concerns the problem of describing interactions by means of graph-based models. This concept has quite a long successful history in systems biology, being driven by the need to understand the complex machinery underpinning the gene expression mechanism and interactions among molecules. A good survey of the application of graph-based models to such problems is reported by Barabasi and Oltvai (2004). It is worth remarking that such models are extensively used in bio-informatics, and in particular, in application of (stochastic) ¼-calculus, where they are used to model the interactions among molecules/genes.

An interesting property that biological systems share with (some) computer networks is related to the degree distribution of such graphs of interactions. In particular, it has been extensively shown that a great number of such systems show a node degree that is distributed according to a power law (Albert and Barab'asi, 2002). Such models apply, in particular, to a rather wide class of evolving (or growing) networks, in which the attachment of new entities to existing ones follows a preferential model, that is, newcomers attach more easily to well-connected nodes. Preferential attachment is, however, only one of the mechanisms that give rise to networks with power-law node degree distributions. The group of A. L. Barab'asi at the University of Notre Dame in Indiana coined the term *scale-free* to describe such graphs.

The characterization in terms of node degree distribution alone is, nonetheless, not sufficient to explain some of the key properties of many naturally arising networks that possess such a scale-free property. Indeed, such graphs can be characterized as *robust yet fragile*. This is due to the fact that such systems appear to be extremely robust in terms of fault-tolerance against random failures, but extremely fragile to targeted attacks. This comes from the fact that, in scale-free networks, there are "hubs" that tend to constitute a kind of backbone of the system, holding the system together. And these hubs represent the Achilles heel of such systems.

Another related property of most graphs arising in the study of biological and socioeconomic processes is self-similarity (also referred to as scale-invariance). This refers to the fact that the system, observed at different scales, shows the same properties (i.e., shows a fractal structure) (Song et al., 2005). This involves the definition of a coarse-graining operation, which can be defined in terms of nodes pruning and collapsing. As an example, we could think of a graph of social contacts, where nodes correspond to individuals and links correspond to some form of relationship. We could then, according to some well-defined rule, group such individuals in communities. We would therefore get another graph, where the nodes set consists of communities and links exist if individuals from two different communities are related in the original graph. The process can then be repeated to get the picture, at a different scale (or level of granularity) of the social network under consideration. The fact that the nodes degree follows a power law is—in general—not sufficient to ensure that the network show a self-similar behaviors (and the term itself "scale-free" suggests inherently this self-similarity). A more detailed model can be given as follows (Li et al., 2005). Let us consider a network with associated graph G, and edge set ε. We denote by d_i the degree of node i. Then, we introduce

$$s(G) = \sum_{(i,j)\in\varepsilon} d_i\, d_j$$

Such a quantity takes a large value if nodes with high degree are connected to nodes with high degree. We define by $s_{\max}$ the maximum value of $s(\cdot)$, taken over the set of all graphs with an identical node degree distribution to G. We then define:

$$S(G) = \frac{s(G)}{s_{\max}}$$

Graphs with low $S(G)$ are said to be scale-rich, whereas a graph is said to be scale-free if $S(G)$ is close to 1. In this way, the definition of scale-free incorporates fractal-like behavior. In general, the term "self-similar" refers to the fact that some correlation functions show nontrivial power law behavior. This is clearly an interesting property, from the point of view of a system engineer, in that it ensures that the resulting system is fully scalable. Interestingly, most biological systems are able to reach a steady state presenting self-similarity properties without the need of any external control. Such property is broadly referred to as self-organized criticality (SOC), after the seminal work of Bak et al. (1987). While the suitability of SOC as a target goal for system engineers is currently debated in the complex systems community (Li et al., 2005), it is understood that SOC is one of the mechanisms by which complexity may arise in nature (Frigg, 2002).

Systems exhibiting SOC are, in general, not in equilibrium. They do present a steady state (in the broad sense that some average characteristics keep unchanged over time), but they do present variation in time. They are open dissipative systems, which require energy from the outside to offset the dissipation (Dhar, 1999). It is also particularly appealing that some simple models of evolution in an ecosystem lead to a nonequilibrium steady state exhibiting SOC. The Bak-Sneppen model considers N cospecies, represented by points on a circle. Initially, each species is assigned a fitness value uniformly taken in the set [0,1]. The system dynamic works in rounds. At every round, the species with the minimum fitness is eliminated from the system, together with its two adjacent species on the circle (this accounts for interactions among different species). These are replaced by new species, with fitness taken independently and uniformly in [0,1]. This model can be shown to present SOC features (Meester and Quant, 2005).

8.6.4 Self-Management and Resilience

The traditional network management model is orthogonal to the layered protocol stack structure. As such, the correct operation of the management system and the accuracy of the managed information rely on a mirror-image database of network state, which is defined and standardized separately from the actual protocols and services being managed. This database is complex to maintain, and it is not straightforward to extract results from it that can be used as feedback to steer the network in an autonomic way.

The management of services, infrastructure, storage, configurations, traffic, etc., still relies on humans to extract relevant information from the database and take the appropriate management and control decisions. This empirical management style is one of the reasons why network management has achieved poor results in tasks such as automated troubleshooting and self-configuration.

Network security suffers from a similar problem: security is added to systems as an afterthought, and as such is not an integrated part of the service structure. This may expose the system to security holes, since once the security add-on is compromised; the underlying nonsecure system is easily exposed to attacks. Today to support the first generation of adaptive infrastructures, intelligent management and security tools use deductive reasoning to predict the effect of discrete changes (what-if modeling). Deductive modeling tools are only an intermediate step. They are not sophisticated enough to make decisions across a large number of elements and layers.

An autonomic network requires a completely new approach to network management and security, in which each functional building block inside each node comes with built-in self-management and self-resilience capabilities. In an autonomic network, malicious or erroneous entities could try to disturb an autonomic element in any possible way, and this element should recover and heal itself to continue providing the required service. In case of failures, alternative service blocks should replace the non-functioning ones in a reactive and unsupervised way. These localized selfware features must extend themselves to the granularity of a node, and then to the whole network in a fully decentralized way.

Environment monitoring is a first important building block for autonomic network components. The challenging aspect of monitoring is related to the fact that a component can monitor only its local environment, when a lot of tasks need a more global view of the environment. The challenge here consists of merging information coming from different sources (local or distant) and to end in a global view of network. This problem is strategic in the context of global monitoring of the network as well as for self-protecting nodes, as anomalies (or changes) observed by an autonomic nodes might be coming from a problem not directly observable by the node.

To complement monitoring, inductive (i.e., predictive) modeling is a more elegant solution than deductive modeling. It requires a good understanding of the problem and a lot of monitored data. The big difference is that inductive modeling starts with the desired outcome, and the model returns the optimal configurations for all the elements of the infrastructure. Distributed artificial intelligence agents can monitor critical elements, or even learn which information is more relevant to monitor to achieve a given goal, and train themselves to make the right decisions in complex situations, such as failures or attacks. The inductive approach is "self-aware" in the sense that humans ultimately give up control to the network.

All compartments that make an autonomic network element must also incorporate self-healing capabilities. This involves self-monitoring to detect and recover from internal failures. In addition to internal self-healing, compartments need mechanisms for recovery from malfunctioning of their environment: tolerance to external failures. Upon malfunctioning of a given compartments that make use of it, detect the problem via their internal monitoring mechanism and trigger the self-healing mechanism of the malfunctioning compartment. If this strategy is not successful, a replacement of this compartment either in the local network node or its neighborhood is used. Once found, the alternative compartment is deployed and executed in place of the failed one.

Nodes, compartments, and the network as a whole should also be resilient. *Resilience* is defined as the ability for an entity to tolerate (resist and autonomically recover from) three types of severe impacts on the network and its applications:

1. *Challenging network conditions* that consist of (1) weakly and episodically connected paths primarily due to wireless links, (2) dynamic topologies and relationships due to the mobility of nodes, subnets, and compartments, and (3) high delay due to long paths (e.g., satellite links) or store-and-forward across episodically connected mobile nodes.
2. *Coordinated attack* including wide-scale destruction of infrastructure (attack or natural disaster), attacks against the network control protocols and software, and traffic attacks including distributed denial of service attacks.
3. *Traffic anomalies* that are legitimate but unpredictable, such as flash crowds and extremely bursty high-bandwidth applications such as distributed storage and Grid.

The first two of these impacts drive the need for *survivability* (Sterbenz et al., 2002) (or disruption tolerance); the additional need to tolerate legitimate traffic anomalies drives the need for *resilience*. It is important to note that it is impossible to distinguish a sufficiently sophisticated distributed DoS[4] (DDoS) attack from legitimate traffic, and thus the architecture and mechanisms to tolerate both are identical.

[4] A DoS attack or DDoS attack is an attempt to make a computer resource unavailable to its intended users. Although the means to, motives for, and targets of a DoS attack may vary, it generally consists of the concerted, malevolent efforts of a person or persons to prevent an Internet site or service from functioning efficiently or at all, temporarily or indefinitely.

References

Abdesslem, F.B., L. Iannone, M.D. de Amorim, K. Kabassanov, and S. Fdida, *On the Feasibility of Power Control in Current IEEE 802.11 Devices*, In: Proceedings of IEEE PERCOMW, 2006.

Adler, M., and C. Scheideler, "Efficient Communication Strategies for Ad Hoc Wireless Networks," *Theory of Computing Systems*, vol. V33, no. N5-6, pp. 337–391, 2000.

Aggélou, G. (ed.), *Mobile Ad-Hoc Networks: Design and Integration*, McGraw-Hill, New York, NY, USA, 2004.

Aggélou, G., "Forecasting Network Disconnections in Mobile Wireless Mesh Networks," Submitted for publication, *IEEE Journal on Selected Areas in Communications*, February 2008.

Aggélou, G., and R. Tafazolli, "On the Relaying Capability of Next Generation GSM Cellular Network," *IEEE Personal Communications—Special Issue on Advances in Mobile Ad Hoc Networking*, vol. 8, no. 1, pp. 6–13, 2001.

Aguayo, D., J. Bicket, S. Biswas, G. Judd and R. Morris, *Link-Level Measurements from an 802.11b Mesh Network*, In: Proceedings of ACM SIGCOMM, 2004.

Akan, O.B., and I.F. Akyildiz, "Event-to-Sink Reliable Transport in Wireless Sensor Networks," *IEEE-ACM Transactions on Networking*, vol. 13, no. 5, pp. 1003–1016, 2005.

Alamouti, S., "Space Block Coding: A Simple Transmitter Diversity Technique for Wireless Communications," *IEEE Journal on Selected Areas in Communications*, vol. 16, pp. 1451–1458, 1998.

Alamouti, S.M., "A Simple Transmit Diversity Technique for Wireless Communications," *IEEE Journal on Selected Areas in Communications*, vol. 16, no. 8, pp. 1451–1458, 1998.

Albert, R., and A.L. Barab'asi, "Statistical Mechanics of Complex Networks," *Reviews of Modern Physics*, vol. 74, pp. 47–97, 2002.

Albert, R., H. Jeong, and A.L. Barab'asi, "Error and Attack Tolerance of Complex Networks," *Nature*, vol. 406, 2000.

http://www.ana-project.org

Anderson, B.D.O., P.N. Belhumeur, T. Eren, D.K. Goldenberg, A.S. Morse, W. Whitley, and Y.R. Yang, "Graphical Properties of Easily Localizable Sensor Networks," Wireless Networks, Springer Netherlands, ISSN 1572–8196 (Online), April 06, 2007.

Andersson, B., P. Svensson, P. Nordin, and M. Nordahl, *On-line Evolution of Control for a Four-Legged Robot Using Genetic Programming*, In: Real-World Applications of Evolutionary Computing, EvoWorkshops 2000, Edinburgh, Scotland, Springer LNCS 1803, pp. 319–326, 2000.

Andreopoulos, Y., N. Mastronarde, and Mihaela van der Schaar, "Cross-Layer Optimized Video Streaming Over Wireless Multihop Mesh Networks," *IEEE Journal on Selected Areas in Communications*, vol. 24, no. 11, 2006.

Araujo, S.G., A.C.P. Pedroza, and A.C. Mesquita, *Evolutionary Synthesis of Communication Protocols*, The Tenth International Conference on Telecommunications (ICT 2003), February-March 2003, pp. 986–993.

Arkin, E., E. Demaine, and J. Mitchell, *The Puddle-Jumper Problem*, Personal Communication, 2005.

Arpacioglu, O., and J.H. Zygmunt: *On the Scalability and Capacity of Wireless Networks with Omnidirectional Antennas*, The Third International Symposium on Information Processing in Sensor Networks (IPSN 2004), Berkeley, California, CA, USA, April 27–28, 2004.

Aslam, J., Q. Li, and D. Rus, "A Lifetime-Optimizing Approach to Routing Messages in Ad-Hoc Networks," In: X. Cheng, X. Huang, and D.-Z. Du (eds.), *Ad Hoc Wireless Networking*, Kluwer Academic Publishers, pp. 1–43, 2003.

Aspnes, J., T. Eren, D.K. Goldenberg, A.S. Morse, W. Whiteley, Y.R. Yang, B.D.O., Anderson, and P.N. Belhumeur, "A Theory of Network Localization," *IEEE Transactions on Mobile Computing*, 12(5):1663–1678, December 2006.

Auletta, V., Y. Dinitz, Z. Nutov, and D. Parente, "A 2-Approximation Algorithm for Finding an Optimum 3-Vertex Connected Spanning Subgraph," *Journal of Algorithms*, vol. 32, pp. 21–30, 1999.

Aurenhammer, F., "Voronoi Diagrams—A Survey of a Fundamental Geometric Data Structure," *ACM Computing Surveys*, vol. 23, no. 3, pp. 345–405, 1991.

Axelrod, R, *The Evolution of Cooperation*, Basic Books, New York, NY, USA, 1984.

Baccarelli, E, M. Biagi, N. Cordeschi, and C. Pelizzoni, *Optimal Download for Energy-Limited Wireless Proxy-servers*, INFO-COM Tech. Report no.01TR07infocom, available at http://infocom.uniroma1.it/□biagi/OptimDown.pdf.

Bahl, P., and V.N. Padmanabhan, *RADAR: An In-Building RF-Based User Location and Tracking System*, In: Proceedings of the IEEE INFOCOM 2000, March 2000.

Bahl, P., R. Wattenhofer, L. Li, and Y. Wang, *Distributed Topology Control for Power Efficient Operation in Multihop Wireless Ad Hoc Networks*, In: Proceedings 20th Annual Joint Conference of IEEE Computer and Communications Societies (INFOCOM 2001), 2001.

Bak, P., C. Tang, and K. Wiesenfeld, "Self-Organized Criticality: An Explanation of the 1/f Noise," *Physical Review Letters*, vol. 59, no. 381–384, 1987.

Ballardie, T., Paul Francis, and Jon Crowcroft, *Core Based Trees (CBT): An Architecture for Scalable Inter-Domain Multicast Routing*, In: Proceedings of ACM Special Interest Group on Computer Communication (SIGCOMM '93), San Francisco, CA, USA, pp. 85–95, October 1993.

Banâtre, J.-P., P. Fradet, and Y. Radenac, *Principles of Chemical Programming*, The Fifth International Workshop on Rule-Based Programming (RULE'04), Aachen, Germany, June 2004.

Banâtre, J.-P., P. Fradet, and Y. Radenac, *A Generalized Higher-Order Chemical Computation Model with Infinite and Hybrid Multisets*, In: The 1st International Workshop on New Developments in Computational Models (DCM'05), pp. 5–14, 2005.

Banâtre, J.-P., P. Fradet, and Y. Radenac, *Generalized Multisets for Chemical Programming*, Research Report RR-5743, INRIA, November 2005.

Banâtre, J.-P., P. Fradet, and Y. Radenac, *Towards Grid Chemical Coordination*, In: Proceedings of Symposium on Applied Computing (SAC), 2006, short paper.

Banâtre, J.-P., and D.L. M'etayer, *A New Computational Model and its Discipline of Programming*, September 1986. Technical Report RR0566, INRIA.

Banâtre, J.-P., and D.L. M'etayer, *Gamma and the Chemical Reaction Model*, Internal Publication PI-984, INRIA, February 1996.

Banâtre, J.-P., Y. Radenac, and P. Fradet, *Chemical Specification of Autonomic Systems,* In: Proceedings of the 13th International Conference on Intelligent and Adaptive Systems and Software Engineering (IASSE'04), pp. 72–79, July 2004.

Banerjee, S., and A. Misra, *Minimum Energy Paths for Reliable Communication in Multi-Hop Wireless Networks,* In: Proceedings of Mobihoc, June 2002.

Banerjee, S., A. Misra, J. Yeo, and A. Agrawala, *Energy-Efficient Broadcast and Multicast Trees for Reliable Wireless Communication,* In: Proceedings of the IEEE Wireless Communications and Networking Conference (WCNC '03), vol. 1, pp. 660–667, March 2003.

Banzhaf, W., P. Nordin, R.E. Keller, and F.D. Francone, *Genetic Programming, An Introduction,* Morgan Kaufmann Publishers, Inc., 1998.

Bao Lichun and J. J. Garcia-Luna-Aceves, *Transmission Scheduling in Ad Hoc Networks with Directional Antennas,* In: Proceedings of the ACM MobiCom, pp. 48–58, New York, NY, USA, September 2002.

Barabasi, A.-L., and Z. N. Oltvai, "Network Biology: Understanding the Cell's Functional Organization," *Nature Reviews,* pp. 101–113, 2004.

Barret, C.L., Drozda, M., and Marathe, *M.V.: A Comparative Experimental Study of Media Access Protocols for Wireless Radio Networks,* Research Report LA-UR-01-2879, Los Alamos National Laboratory, Los Alamos, NM, USA, 2001.

Barreto, P.S.L.M., H.Y. Kim, B. Lynn, and M. Scott, "Efficient Algorithms for Pairing-Based Cryptosystems," In: M. Yung (ed.), *Advances in Cryptology—CRYPTO 2002,* no. 2442 in *Lecture Notes in Computer Science,* Springer, pp. 354–368, 2002.

Barriac, G., R. Mudumbai, and U. Madhow, *Distributed Beamforming for Information Transfer in Sensor Networks,* In: Proceedings of the Third International Symposium on Information Processing in Sensor Networks, pp. 81–88, April, 2004.

Barton, R.J., J. Chen, and K. Huang, *Cooperative Time Reversal for Communication in Power-Constrained Wireless Sensor Networks,* Proceedings of the Forty-Third Annual Allerton Conference on Communication, Control, and Computing, Monticello, IL, September 28–30, 2005.

Barton, R.J., J. Chen, K. Huang, and D. Wu, *Cooperative Time-Reversal Communication in Wireless Sensor Networks,* Proceedings of the IEEE Workshop on Statistical Signal Processing, Bordeaux, France, July 17–20, 2005.

Barton, R.J., and R. Zheng, "Order-Optimal Data Aggregation in Wireless Sensor Networks—Part I: Regular Networks," *IEEE Transactions on Information Theory,* submitted April 2006.

Barton, R.J., and R. Zheng, *Order-Optimal Data Aggregation in Wireless Sensor Networks Using Cooperative Time-Reversal Communication,* In: Proceedings of the 40th Annual Conference on Information Sciences and Systems (CISS), Princeton, NJ, pp. 1050–1055, March 22–24, 2006.

Barton, R.J., and R. Zheng, *The Impact of Time-Reversal Modulation on the Performance of Cooperative Relaying Strategies in Wireless Networks,* In: Proceedings of the Information Theory and Applications Workshop, San Diego, CA, USA, 2006.

Basagni, S., I. Chlamatac, v. Syrotiuk, and B. Woodward, *A Distance Routing Effect Algorithm for Mobility (DREAM),* In: Proceedings of the 4th Annual ACM/IEEE International Conference on Mobile Computing and Networking (MobiCom) '98, Dallas, TX, USA, October 25–30, 1998, pp. 76–84.

Basagni, S., I. Chlamtac, and V.R. Syrotiuk, *Geographic Messaging in Wireless Ad Hoc Networks*, In: Proceedings of the 4th Annual ACM/IEEE International Conference on Mobile Computing and Networking (MobiCom '98), Dallas, TX, USA, October 25–30, 1998, pp. 76–84.

Batayneh, F.A., *Location Management in Wireless Data Networks*, Available at: www.cs.wustl.edu/~jain/cse574-06/wireless_location.htm, 2006.

Bauch, G., and A.F. Naguib, *Map Equalization of Space–Time Coded Signals Over Frequency Selective Channels*, In: Proceedings of the IEEE Wireless Communication and Networking Conference (WCNC'99), vol. 1, New Orleans, LA, September 1999, pp. 261–265.

Behzad, A., and I. Rubin, *Multiple Access Protocol for Power Controlled Wireless Access Nets*, In: IEEE Transactions on Mobile Computing, vol. 3, no. 4, pp. 307–316, 2004.

Belding, T.C, *The Distributed Genetic Algorithm Revisited*, In: Proceedings of the Sixth International Conference on Genetic Algorithms, Morgan Kaufmann, San Francisco, CA, USA, pp. 114–121, 1995.

Bellare, M., A. Desai, E. Jokipii, and P. Rogaway, *A Concrete Security Treatment of Symmetric Encryption*, In: Proceeedings of the 38th Symposium on Foundations of Computer Science (IEEE Computer Society), Miami Beach, FL, October 20–22, 1997.

Bellare, M., and Rogaway, P., *Entity Authentication and Key Distribution*, In: *Advances in Cryptology—Crypto'93, Lecture Notes in Computer Science*, vol. 773, Springer-Verlag, Santa Barbara, CA, USA, pp. 232–249, 1994.

Bellovin, S.M., M. Leech, and T. Taylor, *ICMP Traceback Messages*, Technical Report, Internet Draft, IETF, March 2001.

Benedetto, M.D., and G. Giancola, *Understanding Ultra Wide Band Radio Fundamentals*, Prentice Hall Technical Reference, NJ, USA, 2004.

Berg, M., M. van Kreveld, M. Overmars, and O. Schwarzkopf, *Computational Geometry: Algorithms and Applications*, 2nd ed., Springer-Verlag, 2000.

Berners-Lee, T., J. Hendler, and O. Lassila, "The Semantic Web: A New Form of Web Content That Is Meaningful to Computers Will Unleash a Revolution of New Possibilities," *Scientific American*, vol. 284, no. 5, pp. 34–43, pp. 34–43, 2001.

Bertsekas, D.P., *Nonlinear Programming*, Athena Scientific, Belmont, MA, 1999.

Bernier, C., *Ultra-Low Power Radio Links Using MEMS Technology*, In 6th Annual Review, 10–11 June 2004, Leti, France.

Bettstetter, C., *Self-Organization in Communication Networks. Overview and State of the Art*, 2005, WWRF White Paper.

Beutel, J., M. Dyer, L. Meier, M. Ringwald, and L. Thiele, *Next-Generation Deployment Support for Sensor Networks*, TIK Report No. 207, ETH Zurich, November 2004.

Beyer, D., *Wireless Mesh Networks for Residential Broadband*, National Wireless Engineering Conference, San Diego, November 2002.

Beyer, H.G., and H.P. Schwefel, "Evolution Strategies: A Comprehensive Introduction," *Journal of Natural Computing*, vol. 1, no. 1, pp. 3–52, 2002.

Bharghavan, V., S. Demers, S. Shenker, and L. Zhang. *MACAW: A Media Access Protocol for Wireless LANs*. In: Proceedings of the ACM SIGCOMM, 1994.

Bianchi, G., "Performance Analysis of the IEEE 802.11 Distributed Coordination Function," *IEEE Journal on Selected Areas in Communications— Wireless Series*, vol. 18, no. 3, 2000.

Biglieri, E., R. Calderbank, A. Constantinides, A. Goldsmith, A. Paulraj, and H.V. Poor, *MIMO Wireless Communication*, Cambridge University Press, 2007.

Biswas, R.M.S, *Exor: Opportunistic Multi-Hop Routing for Wireless Networks*, In: Proceedings of the 2005 Conference on Applications, Technologies, Architectures, and Protocols for Computer Communication, pp. 133–143, 2005.

Blom, R., *An Optimal Class of Symmetric Key Generation Systems*, in Eurocrypt'84, LNCS 209, 1985.

Bloom, B.H, "Space/Time Trade-offs in Hash Coding with Allowable Errors," *Communications of ACM*, 13(7):422–426, Jul. 1970.

Blostein, S.D., and H. Leib, "Multiple Antenna Systems: Their Role and Impact in Future Wireless Access," *IEEE Communications Magazine*, pp. 94–101, 2003.

Blum, B.M., T. He, S. Son, and J. A. Stankovic, *IGF: A Robust State-Free Communication Protocol for Sensor Networks*, Technical Report CS-2003-11, CS Department, University of Virginia, Charlottesville, VA, 2003.

Blundo, C., A. Santis, A. Herzberg, S. Kutten, U. Vaccaro, and M. Yung, *Perfectly-Secure Key Distribution for Dynamic Conferences*, in CRYPTO'92.

Boneh, D., and M., Franklin, "Identity-Based Encryption from the Weil Pairing," *SIAM Journal on Computing*, vol. 32, no. 3, pp. 586–615, 2003. Also appeared in CRYPTO '01.

Bose, P., P. Morin, I. Stojmenovic, and J. Urrutia, *Routing with Guaranteed Delivery in Ad Hoc Wireless Networks*, In: Proceedings of 3rd ACM International Workshop on Discrete Algorithms and Methods for Mobile Computing and Communications (DIAL M '99), Seattle, WA, USA, August 20, 1999, pp. 48–55.

Botts, M., *Sensor Model Language (SensorML) for In-situ and Remote Sensors*, OGC 04-019, OGC Inc., 2004.

Boyd, S., and L. Vandenberghe, *Convex Optimization*, Cambridge University Press, Cambridge, UK, 2004.

Bray, J., and F.C. Sturman, *Bluetooth: Connect Without Cables*, Prentice-Hall, 2000.

Bredin, J.L. E. D. Demaine, M. Hajiaghayi, and D. Rus, *Deploying Sensor Networks with Guaranteed Capacity and Fault Tolerance*, ACM MobiHoc, pp. 309–319, 2005.

Broch, J., D.A. Maltz, D.B. Johnson, Y.C. Hu, and J. Jetcheva, *A Performance Comparison of Multi-Hop Wireless Ad-Hoc Network Routing Protocols*, In: Proceedings of the Fourth Annual ACM/IEEE International Conference on Mobile Computing and Networking (MOBICOM '98), Dallas, TX, USA, October 25–30, 1998, pp. 85–97.

Broder, A., and M. Mitzenmacher, *Network Applications of Bloom Filters: A Survey*, In: Proceedings of the 40th Annual Allerton Conference on Communication, Control, and Computing, 2002.

Bruno, R., M. Conti, and E. Gregori, "Mesh Networks: Commodity Multi-hop Ad Hoc Networks," *IEEE Communications Magazine*, vol. 43, no. 5, pp. 123–131, 2005.

Buchegger, S., and J.Y.L. Boudec, *Performance Analysis of the CONFIDANT Protocol*, In: Proceedings of the ACM MobiHoc 2002, Lausanne, June 2002.

Buttyan, L., and J. Hubaux, *Stimulating Cooperation in Self-Organizing Mobile Ad Hoc Networks*, ACM/Kluwer Mobile Networks and Applications (MONET), vol. 8, no. 5, October 2003.

Cabrales, A., *Stochastic Replicator Dynamics*, International Economic Review, Department of Economics, University of Pennsylvania and Osaka University Institute of Social and Economic Research Association, vol. 41, no. 2, pp. 451–481, 2000.

Cabri, G., L. Leonardi, and F. Zambonelli, "Mobile-Agent Coordination Models for Internet Applications," *IEEE Computer*, vol. 33, no. 2, pp. 82–89, 2000.

Caire, G., and S. Shamai, "On the Achievable Throughput of a Multiantenna Gaussian Broadcast Channel," *IEEE Transactions on Information Theory*, vol. 49, no. 7, pp. 1691–1706, 2003.

Calude, C.S., and G. Paun, *Computing with Cells and Atoms: An Introduction to Quantum, DNA and Membrane Computing*, Taylor & Francis, 2001.

Cao, Q., T. He, L. Fang, T. Abdelzaher, J. Stankovic, and S. Son, "*Efficiency Centric Communication Model for Wireless Sensor Networks*, In: IEEE INFOCOM, 2006.

Capkun, S. M. Hamdi, and J.P. Hubaux, *GPS-free Positioning in Mobile Ad-Hoc Networks*, In: Proceedings of the 34th Annual Hawaii International Confernce on System Sciences, January 2001.

Cardoso, J., and A. Souloumiac, "Blind Beamforming for Non-Gaussian Signals," *IEEE Proceedings*, pt. F, vol. 140, pp. 362–370, 1993.

Carman, D., P. Kruus, and B. Matt, *Constraints and Approaches for Distributed Sensor Network Security*, Technical Report No. 00-010, NAI Labs, September 2000.

Cesana, M., D. Maniezzo, P. Bergamo, and M. Gerla, *Interference Aware (IA) MAC: An Enhancement to IEEE802.11b DCF*, In: Proceedings of the IEEE VTC 2003-Fall, 2003.

Chambers, B.A. *The Grid Roofnet: A Rooftop Ad Hoc Wireless Network*, Master's Thesis, June 2002, Available at: http://www.pdos.lcs.mit.edu/grid/pubs.html.

Chan, H., and A. Perrig, *PIKE: Peer Intermediaries for Key Establishment in Sensor Networks*, In: IEEE INFOCOM, 2005.

Chan, H., A. Perrig, and D. Song, *Random Key Predistribution Schemes for Sensor Networks*, In: IEEE Symposium on Research in Security and Privacy, 2003.

Chandran, K., S. Raghunathan, S. Venkatesan, and R. Prakash, "A Feedback Based Scheme for Improving TCP Performance in Ad Hoc Networks," *IEEE Personal Communication Systems Magazine, Special Issue on Ad Hoc Networks*, vol. 8, no. 1, pp. 34–39, 2001.

Chandrasekaran, B., J. Josephson, and V. Benjamins, "What are Ontologies and Why do We Need Them?" *IEEE Intelligent Systems*, vol. 14, no. 1, pp. 20–26, 1999.

Chang, H.Y., P. Chen, A. Hayatnagarkar, R. Narayan, P. Sheth, N. Vo, C. L. Wu, S.F. Wu, L. Zhang, X. Zhang, F. Gong, F. Jou, C. Sargor, and X. Wu, "*Design and Implementation of a Real-Time Decentralized Source Identification System for Untrusted IP Packets*, In: Proceedings of the DARPA Information Survivability Conference and Exposition, January, 2000.

Chang, H.Y., R. Narayan, C. Sargor, F. Jou, S.F. Wu, B.M. Vetter, F. Gong, X. Wang, M. Brown, and J.J. Yuill, *DECIDUOUS: Decentralized Source Identification for Network-Based Intrusions*, In: Proceeding of 6th IFIP/IEEE International Symposium on Integrated Network Management, pp. 702–714, 1999.

Chang, J., and L. Tassiulas, *Energy Conserving Routing Wireless Ad Hoc Networks*, In: Proceedings of the 19th Annual Joint Conference of IEEE Computer and Communications Societies (INFOCOM 2000), vol. 1, 2000, pp. 22–31.

Chang, J.-H., L. Tassiulas, and F. Rashid-Farrokhi, "Joint Transmitter Receiver Diversity for Efficient Space Division Multi-access," *IEEE Transactions on Wireless Communications*, vol. 1, no. 1, pp. 16–27, 2002.

Chang, J.-H., and L. Tassiulas, Maximum Lifetime Routing in Wireless Sensor Networks, *IEEE/ACM Transactions on Networking*, vol. 12, no. 4, pp. 609–619, 2004.

Chatterjee, M., S.K. Das, and D. Turgut, "WCA: A Weighted Clustering Algorithm for Mobile Ad Hoc Networks," *Journal of Cluster Computing (Special Issue on Mobile Ad hoc Networks)*, vol. 5, pp. 193–204, April 2002.

Chatzigiannakis, I., S. Nikoletseas, and P. Spirakis, *An Efficient Communication Strategy for AdHoc Mobile Networks*, In: Proceedings of ACM Symposium on Principles of Distributed Computing, pp. 320–322, August 2001.

Chen, B.O.P., and E. Callaway, *Energy Efficient System Design with Optimum Transmission Range for Wireless Ad Hoc Networks*, In: Proceedings of IEEE International Conference on Communications, vol. 2, 2002, pp. 945–952.

Chen, C., E. Seo, H. Kim, and H. Luo, *Self-Learning Collision Avoidance for Wireless Networks*, In: Proceedings of the IEEE INFOCOM, 2006

Chen, D., and P. K. Varshney, "A Survey of Void Handling Techniques for Geographic Routing in Wireless Networks," *IEEE Communications Surveys and Tutorials*, First Quarter, 2007.

Chen, D., and P. K. Varshney, *On Demand Geographic Forwarding for Data Delivery in Wireless Sensor Networks*, Elsevier Computer Communications, Special Issue on Network Coverage and Routing Schemes for Wireless Sensor Networks, December 2006.

Chen, D., D-Z. Du, X.-D. Hu, G-H. Lin, L. Wang, and G. Xue, "Approximations for Steiner Trees with Minimum Number of Steiner Points," *Theoretical Computer Science*, vol. 262, pp. 83–99, 2001.

Chen, D., J. Deng, and P. K. Varshney, *A State-Free Data Delivery Protocol for Wireless Sensor Networks*, In: Proceedings of the IEEE WCNC 2005, New Orleans, LA, March 2005.

Chen, D., J. Deng, and P. K. Varshney, *On the Forwarding Area of Contention-Based Geographic Forwarding for Ad Hoc and Sensor Networks*, In: Proceedings of the IEEE SECON 2005, Santa Clara, California, September, 2005.

Chen, G., J.W. Branch, and B. K. Szymanski, *Local Leader Election, Signal Strength Aware Flooding, and Routeless Routing*, In: Proceedings of the IEEE IPDPS 2005, Denver, Colorado, April 2005.

Chen, Z., J. Yuan, and B. Vucetic, *Improved Space–Time Trellis Coded Modulation Scheme on Slow Fading Channels*, In: Proceedings on ISIT, 2001.

Cheriyan, J., and R. Thurimella, "Algorithms for Parallel k-Vertex Connectivity and Sparse Certificates," *SIAM Journal of Computing*, vol. 22, no. 1, pp. 157–174, 1993.

Cheung, M., and W. Whiteley, *Transfer of Global Rigidity Results Among Dimensions, Graph Powers and Coning*, York University, Toronto, ON, Canada, Technical Report 2005.

Chiang, C.C., M.Gerla, and L. Zhang, "Forwarding Group Multicast Protocol (FGMP) for Multihop, Mobile Wireless Networks," *Cluster Computing*, vol. 1, no. 2, pp. 187–196, 1998.

Chiasserini, C.F., and R.R. Rao, *Routing Protocols to Maximize Battery Efficiency*, IEEE MILCOM, Los Angeles, USA, October 2000.

Chiasserini, C.F., and R.R. Rao, "Improving Battery Performance by Using Traffic Shaping Techniques," *IEEE Journal on Selected Areas in Communications—Wireless Series*, vol. 19, no. 7, pp. 1385–1394, 2001.

Chiasserini, C.F., and R.R. Rao, "Energy Efficient Battery Management," *IEEE Journal on Selected Areas in Communications—Wireless Series*, vol. 19, no. 7, pp. 1235–1245, 2001.

Chiavaccini, E., and G.M. Vitetta, "Further Results on Differential Space–Time Modulations," *IEEE Transactions on Communications*, vol. 51, no. 7, pp. 1093–1101, 2003.

Cho, S., and J.P. Hayes, *Impact of Mobility on Connection Stability in Ad Hoc Networks*, In: Proceedings of the IEEE WCNC, March 2005.

Chou, C.C., and Archan Misra, *Low Latency Multimedia Broadcast in Multi-Rate Wireless Meshes*, In: The First IEEE Workshop on Wireless Mesh Networks, Held in conjunction with SECON-2005, Santa Clara, CA, USA, September 26, 2005.

Chou, C.T., A. Misra, and J. Qadir, "Low Latency Broadcast in Multi-Rate Wireless Mesh Networks," *IEEE Journal on Selected Areas in Communications—Special Issue on Wireless Mesh Networks*, 2006.

Choudhury, R.R., X. Yang, R. Ramanathan, and N.H. Vaidya, *Using Directional Antennas for Medium Access Control in Ad Hoc Networks*, In: The ACM Annual International Conference on Mobile Computing and Networking (MOBICOM), 2002, pp. 59–70.

Chiang, C.-C., and M. Gerla, *Routing and Multicast in Multihop, Mobile Wireless Networks*, In: The IEEE International Conference on Universal Personal Communications (ICUPC.97), pp. 28–33, 1997.

Chiang, C.-C., M. Gerla, and L. Zhang, *Shared Tree Wireless Network Multicast*, In: Proceedings of the IEEE 6th International Conference on Computer Communincations and Networks (ICCCN.97), Las Vegas, Nevada. pp. 28–33, September 1997.

Chiang, C.-C., M. Gerla, and L. Zhang, *Adaptive Shared Tree Multicast in Mobile Wireless Networks*, In: Proceedings of the IEEE Global Communications Conference, Sydney, Australia, pp. 1817–1822, November 1998.

Chiang, C.-C., M. Gerla, and Lixia Zhang, "Tree Multicast Stategies in Mobile, Multihop Wireless Networks," *ACM/Baltzer Mobile Networks and Applications*, vol. 4, no. 3, pp. 193–207, 1999.

Chung, F.R.K., *Spectral Graph Theory*, CBMS Regional Conference Series in Mathematics, no. 92, AMS, 1997.

Cimini, L.J. "Analysis and Simulation of a Digital Mobile Channel Using Orthogonal Frequency Division Multiplexing," *IEEE Transactions on Communications*, vol. 33, pp. 665–675, 1985.

Clarricoats, P.J.B., Y. Rahmatt-Samii, and J.R. Wait, *Handbook of Microstrip Antenna, vol. 1, IEEE Electromagnetic Waves Series 28*, 1989.

Clausen, T., P. Jacquet, C. Adjih, A. Laouiti, P. Minet, P. Muhlethaler, A. Qayyum, and L.Viennot, "Optimized Link," In: S.C. Rhea and J. Kubiatowicz (eds.), *Probabilistic Location and Routing*, INFOCOM, 2002.

Cocks, C., *An Identity Based Encryption Scheme Based on Quadratic Residues*, In: Proceedings of the 8th IMA International Conference on Cryptography and Coding, Springer-Verlag, London, UK, pp. 360–363, 2001.

Comon, P., and P. Chevalier, "Source Separation: Models, Concepts, Algorithms and Performance," In: S. Haykin, (ed.), *Unsupervised Adaptive Filtering. Adaptive and Learning Systems for Communications Signal Processing and Control Series, vol. I, Blind Source Separation*, Wiley, New York, NY, USA, 2000, pp. 191–236.

Conti, M., E. Gregori, and L. Lenzini, *Metropolitan Area Networks, Springer Limited Series on Telecommunication Networks and Computer Systems*, November 1997.

Corson, M.S., and A. Ephremides, *A Distributed Routing Algorithm for Mobile Wireless Networks*, ACM/Baltzer Wireless Networks Journal, Vol. 1, No. 1, pp. 61–81, February 1995.

Costa, M. "Writing on Dirty Paper, " *IEEE Transactions on Information Theory*, vol. 29, no. 3, pp. 439–441, May 1983.

Cover, T.M., and J.A. Thomas, *Elements of Information Theory*, Wiley, New York, NY, USA, 1991.

Cui, S., A.J. Goldsmith, and A. Bahai, "Energy-Efficiency of MIMO and Cooperative MIMO Techniques in Sensor Networks," *IEEE Journal on Selected Areas in Communication*, vol. 22, no. 6, pp. 1089–1098, August 2004.

Cui, S., R. Madan, A.J. Goldsmith, and S. Lall, *Joint Routing, MAC, and Link Layer Optimization in Sensor Networks with Energy Constraints*, In: Proceedings of the IEEE International Conference on Communications, May 2005, pp. 725–729.

Cui, S., A. J. Goldsmith, and A. Bahai, "Energy-Constrained Modulation Optimization," *IEEE Transactions on Wireless Communications*, vol. 4, no. 5, pp. 2349–2360, September 2005.

Das, S.R., R. Castaneda, and J. Yan, "Simulation Based Performance Evaluation of Mobile," Ad Hoc Network Routing Protocols, *ACM/Baltzer Mobile Networks and Applications Journal*, July 2000, pp. 179–189.

Das, S.R., R. Castaneda, J. Yan, and R. Sengupta, *Comparative Performance Evaluation of Routing Protocols for Mobile Ad Hoc Networks*, In: Proceedings of the IEEE 7th International Conference on Computer Communication and Networks (I3CN), Lafayette, LA, October 1998, pp. 153–161.

Das, S.R., C.E. Perkins, and E. Royer, *Performance Comparison of Two On-demand Routing Protocols for Ad Hoc Networks*, In: Proceedings of the 19th Annual Joint Conference of IEEE Computer and Communications Societies (INFOCOM 2000), Tel Aviv, March 2000, pp. 3–12.

Davis, J.A., A.H. Fagg, and B.N. Levine, *Wearable Computers as Packet Transport Mechanisms in Highly-Partitioned Ad-Hoc Networks*, In: Proceedings of International Symposium on Wearable Computers, pp. 141–148, October 2001.

De Couto, D.S.J., D. Aguayo, J.C. Bicket, and R. Morris, *A High-Throughput Path Metric for Multi-Hop Wireless Routing*, In: Proceedings of the ACM MobiCom, 2003.

De Couto, D.S.J., and R. Morris, *Location Proxies and Intermediate Node Forwarding for Practical Geographic Forwarding*, In: Proceedings of the 4th Annual ACM/IEEE International Conference on Mobile Computing and Networking (MobiCom '98), Dallas, TX, USA, October 25–30, 1998.

Dean, D., M. Franklin, and A. Stubblefield, *An Algebraic Approach to IP Traceback*, In: Proceedings of Network and Distributed System Security Symposium, February 2001.

DeBeasi, R., "Military Research Aims to Develop Self-Configuring, Secure Wireless Nets," *Network World*, vol. 16, 2006.

Deering, S.E., D. Estrin, D. Farinacci, V. Jacobson, C.-G. Liu, and L.Wei, "The PIM Architecture for Wide-Area Multicast Routing," *IEEE/ACM Transactions on Networking*, vol. 4, no. 2, pp. 153–162, April 1996.

Deng, J., Y.S. Han, W.B. Heinzelman, and P.K. Varshney, *Balanced-Energy Sleep Scheduling Scheme for High Density Cluster-Based Sensor Networks*, Elsevier Computer Communications Journal, Special Issue on ASWN '04, 2004.

Deng, J., Y.S. Han, W.B. Heinzelman, and P.K. Varshney, *Scheduling Sleeping Nodes in High Density Cluster Based Sensor Networks*, ACM/Kluwer Mobile Networks and Applications (MONET), Special Issue on Energy Constraints and Lifetime Performance in Wireless Sensor Networks, 2004.

de Silva, C.W., *Control Sensors and Actuators*, Prentice-Hall, New Jersey, USA, 1989.

Dhar, D., *Studying Self-Organized Criticality with Exactly Solved Models*, arxiv: cond-mat/9909009, 1999.

Diao, Y., J. L. Hellerstein, S. Parekh, R. Griffith, G. E. Kaiser, and D. Phung, "A Control Theory Foundation for Self-Managing Computing Systems," *IEEE Journal on Selected Areas in Communications*, vol. 23, no. 12, pp. 2213–2222, December 2005.

Dinitz, Y., and Z. Nutov, "A 3-Approximation Algorithm for Finding Optimum 4,5-Vertex Connected Spanning Subgraphs," *Journal of Algorithms*, vol. 32, pp. 31–40, 1999.

Dittrich, P., *Chemical Computing*, In: Unconventional Programming Paradigms (UPP 2004), Springer LNCS 3566, pp. 19–32, 2005.

Dittrich, P., J. Ziegler, and W. Banzhaf, "Artificial Chemistries—A Review," *Artificial Life*, vol. 7, no. 3, pp. 225–275, 2001.

Doeppner, T.W., P.N. Klein, and A. Koyfman, *Using Router Stamping to Identify the Source of IP Packets*, In: Seventh ACM Conference on Computer and Communications Security, Athens, Greece, November 2000, pp. 184–189.

Douceur, J.R., *The Sybil Attack*, In: Proceedings of the 1st International Workshop on Peer-to-Peer Systems (IPTPS '02), Cambridge, MA, March 2002.

Doyle, B., S. Bell, F.S. Alan, K. McCusker, and N. O'Connor, "Security Considerations and Key Negotiation Techniques for Power Constrained Sensor Networks," *The Computer Journal* (Oxford University Press), vol. 49, no. 4, pp. 443–453, 2006.

Doyle, M., T.F. Fuller, and J. Newman, "Modeling of Galvanostatic Charge and Discharge of the Lithium/Polymer/Insertion Cell," *Journal of Electrochemical Society*, vol. 140, no. 6, pp. 1526–1533, 1993.

Draves, R., J. Padhye, and B. Zill, *Routing in Multi-Radio, Multi-Hop Wireless Mesh Networks*, In: Proceedings of the ACM Mobicom, 2004.

Driggers, R., P. Cox, and T. Edwards, *Introduction to Infrared and Electro-Optical Systems*, Artech House, Boston, MA, 1999.

Drossel, B., and F. Schwabl, "Self-Organized Critical Forest-Fire Model," *Physical Review Letters*, vol. 69, no. 11, pp. 1629–1632, 1992.

Du, W., J. Deng, Y.S. Han, S. Chen, and P.K. Varshney, *A Key Management Scheme for Wireless Sensor Networks Using Deployment Knowledge*, In: IEEE INFOCOM, 2004.

Du, W., J. Deng, Y.S. Han, and P.K. Varshney, *A Pairwise Key Pre-Distribution Scheme for Wireless Sensor Networks*, In: ACM CCS, 2003.

Du, W., R. Wang, and P. Ning, *An Efficient Scheme for Authenticating Public Keys in Sensor Networks,"* In: The 6th ACM International Symposium on Mobile Ad Hoc Networking and Computing (MobiHoc'05), pp. 58–67, New York, NY, USA, 2005.

Duda, R., P.E. Hart, and D.G. Stork, *Pattern Classification*, 2nd ed., Wiley Interscience, November 2000.

Dutertre, B., S. Cheung, and J. Levy, *Lightweight Key Management in Wireless Sensor Networks by Leveraging Initial Trust*, SRI International, Technical Report, SRI-SDL-04-02, 2004.

EdelMann, G.F., W.S. Hodgkiss, S. Kim, W.A. Kupeman, and H.C. Song, "*Underwater Acoustic Communication Using Time Reversal*, In: Proceedings of the MST/IEEE OCEANS Conference and Exhibition, pp. 2231–2235, November 2001.

Eiben, A., *Evolutionary Computing and Autonomic Computing: Shared Problems, Shared Solutions?* In: International Workshop on Self-Star Properties in Complex Information Systems, LNCS 3460 (2005), pp. 36–48, Bertinoro (Forli), Italy, May–June 2004.

Eiben, A., and J. Smith, *Introduction to Evolutionary Computing*, Springer, 2003.

Eren, T., D. Goldenberg, W. Whiteley, R.Y. Yang, A.S. Morse, B.D.O. Anderson, and P.N. Belhumeur, *Rigidity, Computation, and Randomization in Network Localization*, In: IEEE INFOCOM, 2004, pp. 2673–2684.

Erin, C., and H.H. Asada, *Energy Optimal Codes for Wireless Communications*, In: Proceedings of the 38th IEEE Conference of Decision and Control, 1999, pp. 4446–4453.

Eschenauer, L., and V.D. Gligor, *A Key-Management Scheme for Distributed Sensor Networks*, In: ACM CCS, 2002.

Eschenauer, L., and V.D. Gligor, *A Key Management Scheme for Distributed Sensor Networks*, In: The 9th ACM conference on Computer and Communications Security (CCS'02), pp. 41–47, 2002.

Esseling, N., E. Weiss, A. Kramling, and W. Zirwas, *A Multi Hop Concept for HiperLAN/2: Capacity and Interference*, In: Proceedings of the European Wireless 2002, vol. 1, pp. 1–7, Florence, Italy, February, 2002.

Estrin, D., R. Govindan, J. Heidemann, and S. Kumar, *Next Century Challenges: Scalable Coordination in Sensor Networks*, In: ACM/IEEE International Conference on Mobile Computing and Networking (MobiCom), pp. 263–270, Seattle, WA, USA, August 1999.

Fan, L., P. Cao, J. Almeida, and A. Broder, *Summary Cache: A Scalable Wide-Area Web Cache Sharing Protocol*, In: Proceeding of the SIGCOMM, 1998.

Fang, J., M. Cao, A.S. Morse, and B.D.O. Anderson, *Sequential Localization of Networks*, In: Proceedings of the MTNS2006, 2006.

Fang, J., M. Cao, A.S. Morse, and B.D.O. Anderson, *Localization of Sensor Networks Using Sweeps*, In: Proceedings of the 2006 CDC, 2006.

Fang, L., P.J. Antsaklis, L. Montestruque, et al., "Design of a Wireless Dead Reckoning Pedestrian Navigation System—The Navmote Experience," *IEEE Transactions on Instrumentation and Measurement*, vol. 54, pp. 2342–2358, 2005.

Fang, Q., J. Gao, and L.J. Guibas, *Locating and Bypassing Routing Holes in Sensor Networks*, In: Proceedings of the IEEE Infocom, Hong Kong, March 2004.

Fanimokun, A., and J. Frolik, *Effects of Natural Propagation Environments on Wireless Sensor Network Coverage Area*, In: Proceedings of the 2003 Southeastern Symposium on System Theory (SSST03), Morgantown, WV, March 16–18, 2003.

Faruque, J., K. Psounis, and A. Helmy, *Analysis of Gradient-Based Routing Protocols in Sensor Networks*, In: Proceedings of the IEEE/ACM International Conference on Distributed Computing in Sensor Systems, June 2005.

Feeney, L.M., and M. Nilsson, *Investigating the Energy Consumption of a Wireless Network Interface in an Ad Hoc Networking Environment*, In: Proceedings of the IEEE INFOCOM, pp. 1548–1557, 22–26 April 2001.

Fensel, D., F. van Harmelen, and I. Horrocks (2003) "OIL and DAML+OIL: Ontology Languages for the Semantic Web," In: *Towards the Semantic Web: Ontology-Driven Knowledge Management*, John Wiley & Sons, West Sussex, England, pp. 11–31.

Ferrara, D., L. Galluccio, A. Leonardi, G. Morabito, and S. Palazzo, *MACRO: An Integrated MAC/Routing Protocol for Geographic Forwarding in Wireless Sensor Networks*, In: Proceedings of IEEE Infocom 2005, Miami, FL, March 2005.

Fiedler, M, "Algebraic Connectivity of Graphs," *Czechoslovak Mathematical Journal*, vol. 23, pp. 298–305, 1973.

Fili, S., *Fixed, Nomadic, Portable and Mobile Applications for 802.16-2004 and 802.16e WiMAX Networks*, WiMAX Forum, November 2005.

Fogel, L. J., *Artificial Intelligence Through Simulated Evolution*, John Wiley & Sons Inc., New York, NY, USA, 1966.

Fok, C.-L., G.-C. Roman, and C. Lu, *Rapid Development and Flexible Deployment of Adaptive Wireless Sensor Network Applications*, In: Proceedings of International Conference on Distributed Computing Systems, June 2005.

Foschini, G.J, "Layered Space–Time Architecture for Wireless Communication in Fading Environment When Using Multi Element Antennas," *Bell Labs Technical Journal*, vol. 1, no. 2, pp. 41–59, 1996.

Foschini, G. J., G.D. Golden, P.W. Wolniansky, and R.A. Valenzuela, "Simplified Processing for Wireless Communication at High Spectral Efficiency," *IEEE Journal on Selected Areas in Communications, Wireless Communications Series*, vol. 17, pp. 1841–1852, 1999.

www.fractus.com/

***"Fraglets Home Page,"* http:/www.fraglets.net/.**

Fragouli, C., N. Al-Dhahir, and S. Diggavi, "Pre-Filtered Space–Time M-BCJR Equalizer for Frequency Selective Channels," *IEEE Transactions on Communications*, vol. 50, pp. 742–753, May 2002.

Fragouli, C., J.Y. Boudec, and J. Widmer, "Network Coding: An Instant Primer," *ACM SIGCOMM Computer Communication Review*, vol. 36, no. 1, pp. 63–68, 2006.

Frank, R., *Understanding Smart Sensors*, 2nd ed., Artech House, Norwood, MA, 2000.

Friedman, M., and A. Kandel, *Introduction to Pattern Recognition: Statistical, Structural, Neural and Fuzzy Logic approaches*, chapter 3, pp. 55–98, Imperial College Press, London, UK, 1999.

Frigg, R., *Self-Organized Criticality—What It Is and What It Isn't*, Technical Report CPNSS-19/02, London School of Economics, London, UK, 2002.

Fu, A., *Energy Allocation and Transmission Scheduling in Satellite and Wireless Networks*, Ph.D. Thesis, Massachusetts Institute of Technology, January 2003.

Füßler, H., J. Widmer, M. Käsemann, M. Mauve, and H. Hartenstein, "Contention-Based Forwarding for Mobile Ad Hoc Networks," *Elsevier Ad Hoc Networks*, vol. 1, no. 4, pp. 351–369, 2003.

Gage, D.W., *Command Control for Many-Robot Systems*, In: Proceedings of the 19th Annual AUVS Technical Symposium; Reprinted in: *Unmanned Systems Magazine*, vol. 10, no. 4, p. 2834, 1992.

Galbraith, S., "Pairings," In: I. Blake, G. Seroussi, and N. Smart, ed., *Advances in Elliptic Curve Cryptography, London Mathematical Society Lecture Notes*, Chapter IX, pp. 183–213, Cambridge University Press, New York, NY, USA, 2005.

Gamal, H.E., "On the Scaling Laws of Dense Wireless Sensor Networks: The Data Gathering Channel," *IEEE Transactions on Information Theory*, vol. 51, no. 3, pp. 1229–1234, March, 2005.

Ganesan, D., S. Ratnasamy, H. Wang, and D. Estrin, "Coping with Irregular Spatio-Temporal Sampling in Sensor Networks," *SIGCOMM Computer Communication Review*, vol. 34, no. 1, pp. 125–130, 2004.

Garcia-Luna-Aceves, J.J., and E.L. Madruga, *A Multicast Routing Protocol for Ad-Hoc Networks,"* In: Proceedings of the IEEE Conference on Computer Communications, INFOCOM, 99:784–792, 1999.

Genesereth, M., and N. Nilsson, *Logical Foundations of Artificial Intelligence*, Morgan Kaufmann, Palo Alto, CA:, 987.

Gerla, M., and J.T.-C. Tsai, "Multicluster, Mobile, Multimedia Radio Network," *ACM-Baltzer Journal of Wireless Networks*, vo l. 1, no. 3, pp. 255–265, 1995.

Gerla, M., and K. Xu, "Multimedia Streaming in Large-Scale Sensor Networks with Mobile Swarms," *ACM SIGMOD*, vol. 32, no. 4, pp 72–76, 2003.

Gerrits, J.F.M., M.H.L. Kouwenhoven, P.R. van der Meer, J.R. Farserotu, and J.R. Long, "Principles and Limitations of Ultra-Wideband FM Communications Systems," *EURASIP Journal on Applied Signal Processing*, vol. 2005, no. 3, pp. 382–396.

Gerrits, J., M. Kouwenhoven, P.R. van der Meer, J. Farserotu, J. Long "Principles and Limitations of Ultra-Wideband FM Communications Systems," *EURASIP Journal on Applied Signal Processing*, vol. XXXX, no. 3, pp. 382–396, 2005.

Gibson, J.D., *The Mobile Communications Handbook*, CRC Press, 1996.

Gilhousen, K., I.M. Jacobs, et al., "On the Capacity of a Cellular CDMA System," *IEEE Transactions on Vehicular Technology*, vol. 40, no. 303–312, 1991.

Giordano S., "Mobile Ad-Hoc Networks," In: I. Stojmenovic, ed., *Handbook of Wireless Networks and Mobile Computing*, Wiley, Inc., 2000.

Giridhar, A., and P.R. Kumar, "Computing and Communicating Functions over Sensor Networks," *IEEE Journal on Selected Areas in Communications*, vol. 23, no. 4, pp. 755–764, April 2005.

Girod, L., M. Lukac, V. Trifa, and D. Estrin. *The Design and Implementation of a Self-Calibrating Distributed Acoustic Sensing Platform*, In: Proceedings of the SenSys, 2006.

Gnawali, O., M. Yarvis, J. Heidemann, and R. Govindan, *Interaction of Retransmission, Blacklisting, and Routing Metrics for Reliability in Sensor Network Routing*, In: Proceedings of the First IEEE Conference on Sensor and Adhoc Communication and Networks, pp. 34–43, Santa Clara, CA, USA, October 2004.

Goel, A., S. Rai, and B. Krishnamachari, *Sharp Thresholds for Monotone Properties in Random Geometric Graphs*, In: Proceedings of the 36th Annual ACM Symposium on Theory of Computing, Chicago, IL, USA, 2004, pp. 580–586.

Goldberg, L., "Wireless LANs: Mobile Computing's Second Wave," *Electronic Design*, vol. 43, pp. 55–72, 1995.

Golden, G.D. G.J. Foschini, R.A. Valenzuela, and P.W. Wolniansky, "Detection Algorithm and Initial Laboratory Results Using the V-BLAST Space–Time Communication Architecture," *Electronics Letters*, vol. 35, no. 1, pp. 14–15, 1999.

Goldenberg, D.K., *Fine-Grained Localization in Sensor and Ad-Hoc Networks*, Faculty of the Graduate School, Yale University, 2006.

Goldenberg, D.K. P. Bihler, M. Cao, J. Fang, B.D.O. Anderson, A.S. Morse, and Y.R. Yang, *Localization in Sparse Networks Using Sweeps*, In: Proceedings of the ACM MOBICOM, 2006.

Goldenberg, D.K., A. Krishnamurthy, W.C. MAness, R.Y. Yang, A. Young, A.S. Morse, A. Savvides, and B.D.O. Anderson, *Network Localization in Partially Localizable Networks*, In: IEEE INFOCOM, 2005, pp. 313–326.

Gomez, J., A.T. Campbell, M. Naghshineh, and C. Bisdikian, *Conserving Transmission Power in Wireless Ad Hoc Networks*, In: Proceedings of the *ICNP'01*, 2001.

Gong, L., "Increasing Availability and Security of an Authentication Service," *IEEE Journal on Selected Areas in Communications*, vol. 11, no. 5, pp. 657–662, 1993.

Goodrich, M.T., *Efficient Packet Marking for Large-Scale IP Traceback*, In: The 9th ACM Conference on Computer and Communications Security (CCS), pp. 117–126, 2002.

Gordon, D., *Ants at Work: How an Insect Society is Organized*," The Free Press, New York, NY, USA, 1999.

Gosh, A., S.K. Das, "Coverage and Connectivity Issues in Wireless Sensor Networks," In: R. Shorey, A.L. Ananda, M.C. Chan, and W.T. Ooi (eds), *Mobile, Wireless, and Sensor Networks: Technology, Applications, and Future Directions*, John Wiley & Sons, 2006.

Govil, K., E. Chan, H. and Wasserman, *Comparing Algorithms for Dynamic Speed-Setting of a Low-Power CPU*, Proceedings of ACM MobiCom'95, Berkeley, CA, USA, November 1995, pp. 13–25.

Goyal, D., and J. Caffery Jr., *Partitioning Avoidance in Mobile Ad Hoc Networks Using Network Survivability Concepts*, In: Proceedings of International Symposium on Computers and Communications, pp. 553–448, July 2002.

Groenevelt, R., *Stochastic Models for Mobile Ad Hoc Networks*, Ph.D. Thesis, INRIA, Sophia Antipolis, France, April 2005.

Grossglauser M., and D. Tse, *Mobility Increases the Capacity of Ad Hoc Networks*, In: Proceedings of the IEEE INFOCOM '01, Anchorage, Alaska, April 2001.

Gupta, H., S. Das, and Q. Gu, *Connected Sensor Cover: Self-Organization of Sensor Networks for Efficient Query Execution*, In: Proceedings of International Symposium on Mobile Ad Hoc Networking and Computing (ACM MOBIHOC), June 2003.

Gupta, P., and P. Kumar, "Critical Power for Asymptotic Connectivity in Wireless Networks," In: W.M. McEneaney, G. Yin, and Q. Zhang (eds.), *Stochastic Analysis, Control, Optimization and Applications: A Volume in Honor of W.H. Fleming*, Birkhauser, Boston, MA, USA, 1998.

Gupta, P., and P.R. Kumar, "The Capacity of Wireless Networks," *IEEE Transactions on Information Theory*, vol. IT-46, no. 2, pp. 388–404, 2000.

Gura, N., A. Patel, A. Wander, H. Eberle, and S.C. Shantz, *Comparing Elliptic Curve Cryptography and RSA on 8-bit CPUs*, In: Workshop on Cryptographic Hardware and Embedded Systems (CHES'04), pp. 119–132, 2004.

Haas, Z.J. *The Zone Routing Protocol (ZRP) for Ad Hoc Networks*, IETF Internet Draft, draft-zone-routing-protocol-00.txt, November 1997.

Haas, Z.J., and B. Liang, "Ad-Hoc Mobility Management with Uniform Quorum Systems," *IEEE/ACM Transactions on Networking*, vol. 7, no. 228–240, 1999.

Haas, Z.J., and S. Tabrizi, *On Some Challenges and Design Choices in Ad-Hoc Communications*, In: IEEE MILCOM '98, Bedford, MA, USA, October 1998.

Hajek, B., A. Krishna, and R.O. LaMaire, "On the Capture Probability for a Large Number of Stations," *IEEE Transactions on Communications*, vol. 45, no. 2, pp. 254–260, 1997.

Hamdi, M., N. Boudriga, and M.S. Obaidat, *Designing a Wireless Sensor Network for Mobile Target Localization and Tracking*, IEEE Global Communications Conference, SatComm Symposium, San Francisco, CA, USA, November–December 2006.

Hammons, A.R., and H.E. Gamal, "On the Theory of Space–Time Codes for PSK modulation," *IEEE Transactions on Information Theory*, vol. 46, pp. 524–542, 2000.

Han, X., X. Cao, E.L. Lloyd, and C.-C. Shen, *Fault-Tolerant Relay Node Placement in Heterogeneous Wireless Sensor Networks*, Personal Communication, 2006.

Han, Y., R.J. La, and A.M. Makowski, *Distribution of Path Durations in Mobile Ad-Hoc Networks—PALM's Theorem at Work*, In: Proceedings of the 16th ITC Specialist Seminar, August 2004.

He, T., J.A. Stankovic, C. Lu, and T. Abdelzaher, *SPEED: A Stateless Protocol for Real-time Communication in Sensor Networks*, In: Proceedings of ICDCS, Providence, RI, USA, 2003.

Heissenbüttel, M., T. Braun, T. Bernoulli, and M. Wälchli, "BLR: Beacon-Less Routing Algorithm for Mobile Ad-Hoc Networks," *Elsevier Computer Communications*, vol. 27, no. 11, 2004.

Hekmat, R., and P. Mieghem, "Interference in Wireless Multi-hop Ad-hoc Networks and its Effect on Network Capacity," *Wireless Networks*, vol. 10, no. 4, pp. 389–399, 2004.

Hekmat, R., and P.V. Mieghem, *Interference in Wireless Multi-Hop Ad-Hoc Networks and its Effect on Network Capacity*, Med-hoc-Net, 2002.

Hendrickson, B., "Conditions for Unique Graph Realizations," *SIAM Journal on Computing*, vol. 21, pp. 65–84, 1992.

Heylighen, F., and C. Gersheson, "The Meaning of Self-Organization in Computing," *IEEE Intelligent Systems*, May–June 2003.

Hightower, J., G. Boriello, and R. Want, *SpotON: An Indoor 3D Location Sensing Technology Based on RF Signal Strength*, University of Washington CSE Report No. 2000-02-02, February 2000.

Hill, J., R. Szewczyk, A. Woo, S. Hollar, D. Culler, and K. Pister, *System Architecture Directions for Networked Sensors*, In: Proceedings of the International Conference on Architectural Support for Programming Languages and Operating Systems (ASPLOS 2000), Cambridge, Massachusetts, November, 2000, pp. 93–104.

Hjelmfelt, A., E. Weinberger, and J. Ross, *Chemical Implementation of Neural Networks and Turing Machines*, In: Proceedings of the National Academy of Sciences of the United States of America, vol. 88, pp.10983–10987, 1991.

Hochwald, B.M., and T.L. Marzetta, "Unitary Space–Time Modulation for Multiple Antenna Communications in Rayleigh Flat Fading," *IEEE Transactions on Information Theory*, vol. 46, pp. 543–564, 2000.

Hochwald, B.M., and T.L. Marzetta, T.J. Richardson, W. Sweldons, and R. Urbanke, "Systematic Design of Unitary Spacetime Constellation," *IEEE Transactions on Information Theory*, vol. 46, pp. 1962–1973, 2000.

Hofbauer, J., and K. Sigmund, "Evolutionary Game Dynamics," *American Mathematical Society*, vol. 40, no. 4, pp. 479–519, 2003.

Holland, G., and N. Vaidya, *Analysis of TCP Performance Over Mobile Ad Hoc Networks*, In: Proceedings of the IEEE/ACM MOBICOM'99, Seattle, WA, USA, August 1999.

Holland, G., N. Vaidya, and P. Bahl., *A Rate-Adaptive MAC Protocol for Multi-Hop Wireless Networks*, In: Proceedings of the ACM MobiCom, 2001.

Holland, J., *Adaptation in Natural and Artificial Systems*, MIT Press, 1992 (1st ed., 1975).

Hong, X., M. Gerla, G. Pei, and C. Chiang, *A Group Mobility Model for Ad-Hoc Wireless Networks*, In: Proceedings of the 2nd ACM International Workshop on Modeling and Simulation of Wireless and Mobile Systems, 1999.

Horan, B., *The Use of Capability Descriptions in a Wireless Transducer Network*, SML Technical Report Series: SMLI TR-2005-131, Sun Microsystems Laboratories, pp. 1–3, 2005.

Horn, R.A., and C.R. Johnson, *Matrix Analysis*, Cambridge University Press, New York, NY, USA, 1985.

Horn, R.A., and C.R. Johnson, *Topics in Matrix Analysis*, Cambridge University Press, New York, NY, USA, 1991.

Hou, T.-C., and V.O.K. Li, "Transmission Range Control in Multihop Packet Radio Networks," *IEEE Transactions Communications*, vol. 34, no. 1, pp. 38–44, 1986.

Houston, A.I., and J.M. McNamara, "Evolutionarily Stable Strategies in the Repeated Hawkdove Game," *Behavioral Ecology*, pp. 219–227, 1991.

Hu, Y., D. Johnson, and A. Perrig, *Ariadne: A Secure On-Demand Routing Protocol for Ad Hoc Networks*, In: Proceedings of the MobiCom 2002, Atlanta, GA, USA, September 2002.

Huang, D., M. Mehta, D. Medhi, and L. Harn, *Location-Aware Key Management Scheme for Wireless Microsensor Networks*, In: ACM SASN, 2004.

Huang, C., and Y. Tseng, *The Coverage Problem in a Wireless Sensor Network*, In: Proceedings of the ACM WSNA, September 2003.

Hughes, B.L., "Differential Space–Time Modulation," *IEEE Transactions on Information Theory*, vol. 46, pp. 145–149, 2000.

Hwang, D., B. Lai, and I. Verbauwhede, *Energy-memory-security Tradeoffs in Distributed Sensor Networks*, In: ADHOCNOW, LNCS 3158, 2004.

Hwang, J., and Y. Kim, *Revisiting Random Key Pre-distribution for Sensor Networks*, In: ACM SASN, 2004.

Iannone, L., and S. Fdida, *MRS: A Simple Cross-Layer Heuristic to Improve Throughput Capacity in Wireless Mesh Networks*, In: Proceedings of CoNEXT 2005, October 2005.

Iannone, L., and S. Fdida, *Can Multi-Rate Radios Reduce End-to-End Delay in Mesh Networks? A simulation Case Study*, In: Mesh Networking: Realizing the Wireless Internet (Meshnets '05), July 2005.

Iannone, L., K. Kabassanov, and S. Fdida, "The Real gain of Cross-Layer Routing in Wireless Mesh Networks," In: Proceedings of Second International Workshop on Multihop Ad Hoc Networks: From Theory to Reality (ACM/SIGMOBILE RealMan'06), May 2006.

IEEE 802.15.4-2003 Standard, *Wireless Medium Access Control and Physical Layer Specifications for Low-Rate Wireless Personal Area Networks*, 2003.

IEEE 802.16-2004 Standard, *IEEE Std 802.16-2004. Part 16: Air Interface for Fixed Broadband Wireless Access Systems*, 2004.

IEEE 802 LAN/MAN Standards Committee, *Wireless LAN Medium Access Control (MAC) and Physical Layer (PHY) Specifications*, IEEE Standard 802.11, 1999.

IEEE Standard for Local and Metropolitan Area Networks Part 16: *Air Interface for Fixed Broadband Wireless Access Systems*, October 2004.

Institute of Electrical and Electronics Engineers (IEEE), *Supplement to 802.11-1999, Wireless LAN MAC and PHY Specifications: Higher Speed Physical Layer (PHY) Extension in the 2.4 GHz Band*, IEEE Standard 802.11, 1999.

Intanagonwiwat, C., R. Govindan, and D. Estrin, *Directed Diffusion: A Scalable and Robust Communication Paradigm for Sensor Networks*, In: Proceedings of the Sixth ACM/IEEE International Conference on Mobile Computing and Networking, August 2000.

dast.nlanr.net/Projects/Iperf/

Iwata, A., C-C. Chiang, G. Pei, M. Gerla, and T-W. Chen, "Scalable Routing Strategies for Ad Hoc Wireless Networks," *IEEE Journal on Selected Areas in Communications*, vol. 17, no. 8, pp. 1369–1379, 1999.

Jackson, B., and T. Jordan, "Connected Rigidity Matroids and Unique Realizations of Graphs," *Journal of Combinatorial Theory Series B*, vol. 94, pp. 1–29, 2005.

Jacobs, D., and B. Hendrickson, "An Algorithm for Two Dimensional Rigidity Percolation: The Pebble Game," *Journal of Combinatorial Physics*, vol. 137, pp. 346–365, 1997.

Jacobsmeyer, J.M., "Congestion Relief on Power-controlled CDMA networks," *IEEE Journal on Selected Areas in Communications*, vol. 18, no. 9, pp. 1758–1761, 1996.

Jacquet, P., and L. Viennot, *Overhead in Mobile Ad-hoc Network Protocols*, INRIA Res. Rept. RR3965, INRIA, Rocquencourt, France, 2000.

Jafar, S.A., and A. Goldsmith, *On Optimality of Beamforming for Multiple Antenna Systems with Imperfect Feedback*, In: Proceedings of the International Symposium on Information Theory, June 2001, p. 321.

Jaggi, N., *Robust Threshold Based Sensor Activation Policies Under Spatial Correlation*, In: Proceedings of the WIOPT, April 2006.

Jaggi, N., A. Krishnamurthy, and K. Kar, *Utility Maximizing Node Activation Policies in Networks of Partially Rechargeable Sensors*, In: The 39th Annual Conference on Information Sciences and Systems (CISS), March 2005.

Jain, R., D. Chiu, and W. Hawe, *A Quantitative Measure of Fairness and Discrimination for Resource Allocation in Shared Computer Systems*, DEC Research Report TR-301, September 1984.

Jain, R., A. Puri, and R. Sengupta, "Geographical Routing Using Partial Information for Wireless Ad Hoc Networks," *IEEE Personal Communications*, vol. 8, no. 1, pp. 48–57, 2001.

Jain, S., and S.R. Das. *Exploting Path Diversity in the Link Layer in Wireless Ad Hoc Networks*, In: Proceedings of the IEEE WoWMoM Symposium, 2005.

Jakllari, G., T. Korakis, and Leandros Tassiulas, *Amac Protocol for Full Exploitation of Directional Antennas in Ad-Hoc Wireless Networks*, In: ACM Mobihoc'03, June 1–3, 2003.

Jamali, S.H., and T. Le-Ngoc, *Coded-Modulation Techniques for Fading Channels*, Kluwer Academic Publishers, 1994.

Jason L.H., and D.E. Culler, "Mica: A Wireless Platform for Deeply Embedded Networks," *IEEE Micro*, vol. 22, no. 6, pp. 12–24, 2002.

Jayaweera, S.K. "Virtual MIMO-Based Cooperative Communication for Energy-Constrained Wireless Sensor Networks," *IEEE Transactions on Wireless Communications*, vol. 5, no. 5, pp. 984–989, 2006.

Jeffrey E., Wieselthier, Gam D. Nguyen, and Anthony Ephremides, "Energy-Efficient Broadcast and Multicast Trees in Wireless Networks," *Mobile Networks and Applications*, vol. 7, no. 6, pp. 481–492, 2002.

Ji, L., and M.S. Corson, *A Lightweight Adaptive Multicast Algorithm*, Global Telecommunications Conference, 1998. GLOBECOM 98. The Bridge to Global Integration. IEEE, 2, 1998.

Jianfeng, C., C. Chi, and Q. Guo, *A Bandwidth Allocation Model with High Concurrence Rate in IEEE802.16 Mesh Mode*, In: Proceedings of the APCC, October 2005.

Jianfeng, C., C. Chi, and Q. Guo, *Modelling and Performance Analysis of the Centralized Scheduling in IEEE 802.16 Mesh Mode*, In: Proceedings of International Network Conference (INC), July 2006.

Johansson, P., T. Larsson, N. Hedman, B. Mielczarek, and M. Degermark, *Scenario-Based Performance Analysis of Routing Protocols for Mobile Ad-hoc Networks*, In: Proceedings of the 5th Annual ACM/IEEE International Conference on Mobile Computing and Networking (MobiCom '99), Seattle, WA, USA, August 15–19, 1999, pp. 195–206.

Johnson, D.B., *Routing in Ad Hoc Networks of Mobile Hosts*, In: Proceedings of the IEEE Workshop on Mobile Computing Systems and Applications (WMCSA), IEEE Computer Society, Santa Cruz, CA, USA, December 1994, pp. 158–163.

Johnson, D.B., and D.A. Maltz, "Dynamic Source Routing in Ad-Hoc Wireless Net*works*, In: T. Imielinski and H. Korth (ed.), *Mobile Computing*, Kluwer Academic Publishers, pp. 153–181, 1996.

Johnston, D., and Walker, J., "Overview of IEEE 802.16 Security," *IEEE Security and Privacy Magazine*, vol. 2, no. 3, May-June 2004, pp. 40–48.

Jöngren, G., M. Skoglund, and B. Ottersten, "Combining Beamforming and Orthogonal Space–Time Block Coding," *IEEE Transctions on Information Theory*, vol. 48, no. 3, pp. 611–627, 2002.

Joux, A., "A One Round Protocol for Tripartite Diffie-Hellman," *Journal of Cryptology*, vol. 17, no. 4, pp. 63–276, 2004; Proceedings of ANTS-IV, 2000.

Jung, E., and N. Vaidya. *A Power Control MAC Protocol for Ad Hoc Networks*. In: Proceedings of the ACM MobiCom, 2002.

Kahn, J.M., R.H. Katz, and K.S.J. Pister, *Next Century Challenges: Mobile Networking for 'Smart Dust'*, In: Proceedings of the ACM MobiCom, Seattle, WA, USA, 1999, pp. 271–278.

Karlof, C., and D. Wagner, "Secure Routing in Wireless Sensor Networks: Attacks and Countermeasures," *Elsevier's AdHoc Networks Journal, Special Issue on Sensor Network Applications and Protocols*, 1(2–3):293–315, 2003; Also appeared in IEEE WSNA'03.

Karumanchi, G., S. Muralidharan, and R. Prakash, *Information Dissemination in Partitionable Mobile Ad Hoc Networks*, In: Proceedings of IEEE Symposium on Reliable Distributed Systems, pp. 4–13, October 1999.

Kyasanur, P., and N. Vaidya, *Detection and Handling of MAC Layer Misbehavior in Wireless Networks*, In : Proceedings of the Dependable Computing and Communications Symposium (DCC) at the International Conference on Dependable Systems and Networks (DSN), San Francisco, CA, USA, June 2003.

Kar, K., A. Krishnamurthy, and N. Jaggi, "Dynamic Node Activation in Networks of Rechargeable Sensors," In: *IEEE/ACM Transactions on Networking*, vol. 14, no. 1, pp. 15–26, 2006.

Karl, H., and Andreas Willig, *Protocols and Architectures for Wireless Sensor Networks*, John Wiley & Sons, Chichester, West Sussex, UK, 2005.

Karlof, C., N. Sastry, and D. Wagner, *TinySec 0.91: User Manual*, Manuscript, February 11, 2003.

Karp, B., and H. T. Kung, *GPSR: Greedy Perimeter Stateless Routing for Wireless Networks*, In: Proceedings of the 6th Annual ACM/IEEE International Conference on Mobile Computing and Networking (MobiCom '00) New York, NY, USA: ACM Press, 2000, pp. 243–254.

Karp, B.N., *Geographic Routing for Wireless Networks*, Thesis, Harvard University, Cambridge, MA, 2001.

Kashyap, A., *Robust Design of Wireless Networks*, Ph.D. dissertation, University of Maryland, College Park, 2006.

Kashyap, A., S. Khuller, and M. Shayman, *Relay Placement for Higher Order Connectivity in Wireless Sensor Networks*, IEEE INFOCOM, 2006.

Kawadia, V., and P.R. Kumar, *Power Control and Clustering in Ad Hoc Networks*, INFOCOM 2003, San Francisco, CA, USA, March 30–April 3, 2003.

Kephart, J.O., and D.M. Chess, "The Vision of Autonomic Computing," *IEEE Computer Magazine*, vol. 36, no. 1, pp. 41–50, January 2003.

Khalil, H.K., *Nonlinear Systems*, Prentice-Hall, Upper Saddle River, NJ, USA, 2002.

Khuller, S., and B. Raghavachari, "Improved Approximation Algorithms for Uniform Connectivity Problems," *Journal of Algorithms*, vol. 21, no. 2, pp. 434–450, 1996.

Khuller, S., and U. Vishkin, "Biconnectivity Approximations and Graph Carvings," *Journal of the ACM*, vol. 41, no. 2, pp. 214–235, 1994.

Kim, T.-S., H. Lim, and J.C. Hou, *Improving Spatial Reuse Through Tuning Transmit Power, Carrier Sense Threshold, and Data Rate in Multihop Wireless Networks*, In: Proceedings of ACM MobiCom,2006.

Kim, T.-S., H. Lim, and J.C. Hou, *A Coordinate-Based Approach for Exploiting Temporal-Spatial Diversity in Wireless Mesh Networks*, In: Proceedings of the ACM MobiCom, 2006.

Kitano, H. (ed.), *Foundations of Systems Biology*, MIT Press, 2001.

Klein, A., R. Pirhonen, J. Sköld, and R. Suoranta, *FRAMES Multiple Access Mode 1—Wideband TDMA with and without Spreading*, In: Proceedings of PIMRC97, pp 37–41, Helsinki, Finland, September 1997.

Kleinrock, L., and F.A. Tobagi, "Packet Switching in Radio Channels: Part I—Carrier Sense Multiple-Access Modes and Their Throughput-Delay Characteristics," *IEEE Transactions of Commun*ications, vol. COM-23, no. 12, pp. 1417–1433, 1975.

Klir, George J., Yuan Bo, *Fuzzy Sets and Fuzzy Logic: Theory and Applications*, Prentice-Hall, 1995.

Ko, Y.B., V. Shankarkumar, N.H. Vaidya, *Medium Access Control Protocols Using Directional Antennas in Ad Hoc Networks*, In: IEEE Annual Conference on Computer Communications (INFOCOM), 2000, pp. 13–21.

Ko, Y., and N., Vaidya, *Location-aided Routing (LAR) in Mobile Ad Hoc Networks*, In: Proceedings on 4th ACM/ IEEE Mobile Computing and Networking, Dallas, TX, USA, 1998, p. 66.

Ko, Y.-B., and N.H. Vaidya, *Using Location Information in Wireless Ad Hoc Networks*, IEEE Vehicular Technology Conference (VTC '99), May 1999.

Kohl, J., and C.Neuman, *The Kerberos Network Authentication Service (V5)*, IETF RFC 1510, September 1993.

Kortsarz and Z. Nutov, "Approximating Node Connectivity Problems Via Set Covers," *Algorithmica*, vol. 37, pp. 75–92, 2003.

Koza, J., *Genetic Programming: On the Programming of Computers by Means of Natural Selection,"* MIT Press, 1992.

Koza, J., M. Keane, M. Streeter, W. Mydlowec, J. Yu, and G. Lanza, *Genetic Programming IV: Routine Human-Competitive Machine Intelligence*, Springer, July 2003.

Krishna, P., N.H. Vaidya, M. Chatterjee, D. K. Pradhan, *A Cluster-Based Approach for Routing in Dynamic Networks*, ACM SIGCOMM Computer Communication Review, vol. 27, no. 2, pp. 49–64, April 1997.

Krishnamachari, B., D. Estrin, and S. B. Wicker, *The Impact of Data Aggregation in Wireless Sensor Networks*, In: Proceedings of the 22nd International Conference on Distributed Computing Systems, ICDCSW, pp. 575–578, Washington, DC, USA, 2002, IEEE Computer Society.

Kumar, S., T.H. Lai, and J. Balogh, *On k-Coverage in a Mostly Sleeping Sensor Network*, In: Proceedings of the 10th Annual International Conference on Mobile Computing and Networking (Mobicom'04), 2004, pp. 144–158.

Kumar, A., D. Majunath, and J. Kuri, *Communication Networking—An Analytical Approach*, Elsevier, M. K. Publishers, 2004.

Kumar, A., J. Xu, E.L. Li, and J. Wang, *Space-Code Bloom Filter for Efficient Traffic Flow Measurement*, In: Proceedings of the 3rd ACM SIGCOMM conference on Internet measurement, pp. 167–172, Miami Beach, FL, USA, 2003.

Lai, B., S. Kim, and I. Verbauwhede, *Scalable Session Key Construction Protocol for Wireless Sensor Networks,"* In: IEEE Workshop on Large Scale Real-Time and Embedded Systems (LARTES), Austin, TX, USA, December 2002.

Lam, S., "A Carrier Sense Multiple Access Protocol for Local Networks," *Computer Networks*, vol. 4, pp. 21–32, 1980.

Laman, G., "On Graphs and Rigidity for Plane Skeletal Structures," *Journal of Engineering Mathematics*, vol. 4, pp. 331–340, 1970.

Langdon, W.B., and R. Poli, "*Foundations of Genetic Programming*, Springer, 2002.

Langdon, W.B., and R. Poli, *The Halting Probability in von Neumann Architectures*, In: P. Collet, M. Tomassini, M. Ebner, S. Gustafson, and A. Ek'art (eds), Proceedings of the 9th European Conference on Genetic Programming, Springer LNCS 3905, pp. 225–237, Budapest, Hungary, April 2006.

Larsson, P., N. Johansson, and K.-E. Sunell, *Coded Bi-Directional Relaying*, In: The 5th Scandinavian Workshop on Ad Hoc Networks (ADHOC'05), Stockholm, Sweden, May 2005.

Lax, M.C., and Annelie Dammander, *WiMAX—A Study of Mobility and a MAC-layer Implementation in GloMoSim*, Master's Thesis in Computing Science, April 6, 2006.

Lee, S.J., W. Su, and M. Gerla, *On-Demand Multicast Routing Protocol in Multihop Wireless Mobile Networks*, Mobile Networks and Applications, vol. 7, no. 6, pp. 441–453, 2002.

Leguay, J., T. Friedman, and V. Conan, *Evaluating Mobility Pattern Space Routing for DTNs*, In: Proceedings of the IEEE INFOCOM, April 2006

Leiner, B.M., D.L. Nielson, and F.A. Tobagi, "Issues in Packet Radio Network Design," *Proceedings of the IEEE*, vol. 75, no. 1, pp. 6–20, January 1987.

Lenat, D., and R.V. Guha, *Building Large Knowledge-Based Systems: Representation and Inference in the CYC Project*, Addison-Wesley, Reading, MA, USA, 1989.

Leonardo B., H., Oliveira, C. Wong, M. Bern, R. Dahab, and A.A.F. Loureiro, "*SecLEACH—A Random Key Distribution Solution for Securing Clustered Sensor Networks*, In: The 5th IEEE International Symposium on Network Computing and Applications (NCA'06), pp. 145–154, 2006.

Li, J. J.D. Jannotti, S.J. De Couto, D.R. Karger, and R. Morris, *A Scalable Location Service for Geographic Ad Hoc Routing*, In: Proceedings of the 6th Annual ACM/IEEE International Conference on Mobile Computing and Networking (MobiCom '00), Boston, August 2000, pp. 120–130.

Li, J., M. Sung, J. Xu, and L. Li, *Large-Scale IP Traceback in High-Speed Internet: Practical Techniques and Theoretical Foundation*, In: IEEE Symposium on Security and Privacy, Berkeley, CA, USA, May 2004.

Li, L., D. Alderson, R. Tanaka, J. C. Doyle, and W. Willinger, *Towards a Theory of Scale-Free Graphs: Definition, Properties, and Implications*, Journal of Internet Mathematics, Vol. 2, No. 4, 2005.

Li, Q., Javed Aslam, and D. Rus, *Online Power-Aware Routing in Wireless Ad-Hoc Networks*, In: Proceedings of the 7th Annual International Conference on Mobile Computing and Networking, Rome, Italy, pp. 97–107, July 2001.

Li, Q., and D. Rus, *Sending Messages to Mobile Users in Disconnected Ad-Hoc Wireless Networks*, In: Proceedings of ACM/IEEE International Conference on Mobile Computing and Networking (MOBICOM), pp. 44–55, August 2000.

Li, Q., and D. Rus, *Message Relay in Disconnected Ad-Hoc Networks*, Proceedings of International Mobility and Wireless Access Workshop, pp. 14–21, October 2002.

Li, Y., J.H. Winters, and N.R. Sollenberger, "MIMO-OFDM for Wireless Communications: Signal Detection With Enhanced Channel Estimation," *IEEE Transactions on Communications*, vol. 50, pp. 1471–1477, 2002.

Liao, Y., K. Tan, Z. Zhang, and L. Gao, *Estimation Based Erasure-coding Routing in Delay Tolerant Networks*," Microsoft Technical Report, 2006.

Lin, G.-H., and L. Wang, "Steiner Tree Problem with Minimum Number of Steiner Points and Bounded Edge-Length," *Information Processing Letters*, vol. 69, pp. 53–57, 1999.

Lin, R., and M. Gerla, "Adaptive Clustering for Mobile Wireless Networks," *IEEE Journal on Selected Areas in Communications*, vol. 15, no. 7, 1997.

Lin, X., and I. Stojmenovic, *GEDIR: Loop-Free Location Based Routing in Wireless Networks*, In: Proco IASTED International Conference on Parallel and Distributed Computing and Systems, 1999, pp. 1025–1028.

Lindgreny, A., A. Doria, and O. Scheleny, *Probabilistic Routing In Intermittently Connected Networks*, In: The Fourth ACM Symposium on Mobile Ad Hoc Networking and Computing (MobiHoc 2003), 2003.

Littman, M.L., and J.A. Boyan, "A Distributed Reinforcement Learning Scheme for Network Routing," *Advances in Neural Infromation Processing Systems*, vol. 6, pp. 670–678, 1993.

Liu, A., G. B. Giannakis, A. Scaglione, and S. Barbarossa, *Decoding and Equalization of Unknown Multipath Channels Based on Block Precoding and Transmit Diversity*, In: Proceedings of the Asilomar Conerence on Signals, Systems, and Computers, 1999, pp. 1557–1561.

Liu B., Z. Liu, and D. Towsley, *On the Capacity of Hybrid Wireless Networks*, In: IEEE InfoCom 2003.

Liu, C.-H., and H.H. Asada, *A Source Coding and Modulation Method for Power Saving and Interference Reduction in DS-CDMA Sensor Network Systems*, In: Proceedings of the American Control Conference, May 2002, vol. 4, pp. 3003–3008.

Liu, D., P. Ning, and R. Li, "Establishing Pairwise Keys in Distributed Sensor Networks," *ACM Transactions on Information and System Security (TISSEC)*, vol. 8, no. 1, pp. 41–77, 2005; Also appeared in ACM CCS'03.

Liu, D., and P. Ning, *Group-based Key Pre-distribution in Wireless Sensor Networks*, In: ACM WiSe, 2005.

Liu, D., and P. Ning, *Location-based Pairwise Key Establishment for Static Sensor Networks*, In: ACM SASN, 2003.

Liu, J., and F. Zhao, *Towards Semantic Services for Sensor-Rich Information Systems*, In: The 2nd IEEE/CreateNet International Workshop on Broadband Advanced Sensor Networks, Boston, MA., USA, 2005.

Liu, J., and S. Singh, "ATCP: TCP for Mobile Ad Hoc Networks," *IEEE Journal on Selected Areas in Communications*, vol. 19, no. 7, pp. 1300–1315, 2001.

Liu, J., E. Cheong, and F. Zhao *Semantics-Based Optimization Across Uncoordinated Tasks in Networked Embedded Systems*, In: Proceedings of the 5th ACM Conference on Embedded Software, Jersey City, NJ, USA, 2005.

Liu, W., W.L. Zhang, and Y. Fang, *Securing Sensor Networks with Location-Based Keys*, In: IEEE Wireless Communications and Networking Conference (WCNC'05), 2005.

Lorch, J., and A. Smith, *Reducing Processor Power Consumption by Improving Processor Time Management in a Single-User Operating System*, In: Proceedings of the ACM MobiCom'96, 1996.

Loubaton, P., E. Moulines, and P. Regalia, "Subspace Methods for Blind Identification and Deconvolution," In: G. Giannakis, J. Hua, P. Stoica, and L. Tong (eds), *Signal Processing Advances in Wireless and Mobile Communications*, Prentice-Hall, Englewood Cliffs, NJ, USA, 2001.

Lun, D.S., M. M'edard, and R. Koetter., *Efficient Operation of Wireless Packet Networks Using Network Coding*, In: Proceedings of the International Workshop on Convergent Technologies (IWCT) 2005, June 2005, Invited paper.

Luo, H., S. Lu, and V. Bharghavan, *A New Model for Packet Scheduling in Multihop Wireless Networks*, In: Proceedings of the ACM MobiCom, 2000.

Luo, J., B. Montrose, and M. Kang (2005) *Adding Semantic Support to Existing UDDI Infrastructure*, Report No. NRL/MR/5540–05-8918, Naval Research Laboratory, Code 5542.

Ma, C., M. Ma, and Y. Yang, *Data-Centric Energy-Efficient Scheduling in Densely Deployed Sensor Networks*, IEEE ICC 2004, pp. 3652–3656, June 2004.

Ma, C., and Y. Yang, *Battery Aware Routing in Wireless Ad Hoc Networks—Part II: Battery-Aware Routing*, The 19th International Teletraffic Congress (ITC-19), September 2005.

Malan, D.J., M. Welsh, and M.D. Smith, *A Public-Key Infrastructure for Key Distribution in TinyOS Based on Elliptic Curve Cryptography*, In: The 1st IEEE International Conference on Sensor and Ad Hoc Communications and Networks (SECON'04), Santa Clara, CA, USA, October 2004.

Malkin, G., RIP version 2, RFC 2453. IETF, November 1998.

Maltz, D., J. Broch, J. Jetcheva, and D.B. Johnson, "The Effects of On-Demand Behavior In Routing Protocols for Multi-Hop Wireless Ad Hoc Networks," *IEEE Journal on Selected Areas in Communications, Special Issue on Mobile And Wireless Networks*, August 1999.

Măndoui, I.I., and A. Zelikovsky, "A Note on the MST Heuristic for Bounded Edge-Length Steiner Trees with Minimum Number of Steiner Points," *Information Processing Letters*, vol. 75, no. 4, pp.165–167, 2000.

Mankin, A., D. Massey, C. Wu, S. F. Wu, and L. Zhang.*On Design and Evaluation of Intention-Driven ICMP Traceback*," In: Proceedings of IEEE International Conference on Computer Communications and Networks (IC3N), 2001.

Mao, G., B. Fidan, and B.D.O. Anderson, "Wireless Sensor Network Localization Techniques," *Computers Networks*, Vol. 51, Issue 10, pp. 2467–2483, 11 July 2007.

Marco, D., E.J. Duarte-Melo, M. Liu, and D.L. Neuhoff, *On the Many-to-One Transport Capacity of a Dense Wireless Sensor Network and the Compressibility of Its Data*, In: Proceedings of the Second International Workshop on Information Processing in Sensor Networks (IPSN), 2003.

Matias, Y., and S. Cohen, *Spectral Bloom Filters*, In: SIGMOD Conference on Management of Data, pp. 241–252, 2003.

Matsumaru, N., F. Centler, P.S. di Fenizio, and P. Dittrich, "Chemical Organization Theory as a Theoretical Base for Chemical Computing," *International Journal of Unconventional Computing*, 3(4), 285–309, 2006; Earlier version in: Workshop on Unconventional Computing, pp. 71–82, Luniver Press, Beckington, Somerset, UK.

Matsumaru, N., P.S. di Fenizio, F. Centler, and P. Dittrich, *On the Evolution of Chemical Organizations*, In: Proceedings of the 7th German Workshop on Artificial Life, pp. 135–146, 2006.

Matula, D.W., *Determining Edge Connectivity in O(nm),*" IEEE Symposium on Foundations of Computer Science, pp. 249–251, 1987.

Mauve, M., J. Widmer, and H. Hartenstein, "A Survey on Position-based Routing in Mobile Ad Hoc Networks," *IEEE Network Magazine*, vol. 15, no. 6, pp. 30–39, November 2001.

McDonald, A.B., and T., Znati, *Predicting Node Proximity in Ad-Hoc Networks: A Least Overhead Adaptive Model for Electing Stable Routes*, MobiHoc 2000, Boston, MA, USA, August 4, 2000.

McDonald A.B., and Taieb Znati "A Mobility-Based Framework for Adaptive Clustering in Wireless Ad-Hoc Networks," *IEEE Journal on Selected Areas in Communication*, vol. 17, no. 8, 1999.

McGuinness, D., "Ontologies Come of Age," In: D. Fensel, J.A. Hendler, H. Lieberman, and W. Wahlster (eds), *Spinning the Semantic Web*, MIT Press, Cambridge, MA, USA, 2003, pp. 171–195.

McNamara, J.M., "The Policy Which Maximizes Long-Term Survival of an Animal Faced with the Risks of Starvation and Predation," *Advances of Applied Probability*, vol. 22, pp. 295–308, 1990.

McNamara, J.M., S. Merad, and E.J. Collins, "The Hawk-Dove Game as an Average Cost Problem," *Advances of Applied Probability*, vol. 23, pp. 667–682, 1991.

Medepalli, K., *Design, Analysis and Optimization of CSMA/CA based Wireless Networks*, Ph.D. Thesis, Department of Electrical Engineering, Stanford University, Stanford, CA, USA, September 2006.

Medepalli, K., and F.A. Tobagi, *Towards Performance Modeling of IEEE 802.11 based Wireless Networks: A Unified Framework and Its Applications*, In: Proceedings of IEEE INFOCOM 2006.

Medepalli, K., and F.A. Tobagi, *On Optimization of CSMA/CA Based Wireless LANs: Part I—Impact of Exponential Backoff*, In: Proceedings of the IEEE ICC 2006, Istanbul, Turkey, 2006.

Medepalli, K., F.A. Tobagi, D. Famolari, and T. Kodama, *On Optimization of CSMA/CA based Wireless LANs: Part-II —Mitigating Efficiency Loss*, In: Proceedings of ICC 2006, Istanbul, Turkey, 2006.

Meester, R., and C. Quant, "Connections Between 'Self-Organized' and 'Classical' Criticality," *Markov Processes Related Fields*, vol. 11, pp. 355–370, 2005.

Melodia, T., D. Pompili, and I. Akyildiz, *Optimal Local Topology Knowledge for Energy Efficiency Geographical Routing in Sensor Networks*, In: IEEE Infocom, March 2004.

Menezes, A., T. Okamoto, and St Vanstone, "Reducing Elliptic Curve Logarithms to Logarithms in a Finite Field," *IEEE Transactions on Information Theory*, vol. 39, no. 5, pp. 1639–1646, 1993.

Menezes, A., P.V. Oorschot, and S. Vanstone, *Handbook of Applied Cryptography*, CRC Press, Boca Raton, FL, USA, 1996.

Menger, K., "Zur allgemeinen Kurventheorie," *Fundamenta Mathematicae*, vol. 10, pp. 96–115, 1927.

Metcalfe, R., and D. Boggs, "Ethernet: Distributed Packet Switching for Local Computer Networks," *Communications of the ACM*, vol. 19, no. 7, 1976, pp. 395–404.

Mhatre, V., and C. Rosenberg, *Energy and Cost Optimizations in Wireless Sensor Networks: A Survey*, In: The 25th Anniversary of GERAD, Kluwer Academic Publishers, January 2004.

Michiardi, P., and R. Molva, *CORE: A Collaborative Reputation Mechanism to Enforce Node Cooperation in Mobile Ad Hoc Networks*, In: Proceedings of Communication and Multimedia Security 2002 Conference, September 2002.

http:/www.microstrain.com

Miller, V., *Short Program for Functions on Curves*, 1986. Unpublished manuscript.

Min, R., M. Bhardwaj, N. Ickes, A. Wang, and A. Chandrakasan, *The Hardware and the Network: Total System Strategies for Power Aware Wireless Microsensors*, In: Proceedings of the IEEE CAS Workshop on Wireless Communications and Networking, September 2002.

Min, R., T. Furrer, and A. Chandrakasan, *Dynamic Voltage Scaling Techniques for Distributed Microsensor Networks*, In: Proceedings of ACM MobiCom'95, August 1995.

Minar, N., K.H. Kramer, and P. Maes, *Cooperating Mobile Agents for Dynamic Network Routing*," Software Agents for Future Communications Systems, Springer, 1999.

Mitzenmacher, M., "Compressed Bloom Filters," *IEEE/ACM Transactions on Networks*, vol. 10, no. 3, pp. 613–620, October 2002.

Monks, J.P., V. Bharghavan, W. Mei, and W. Hwu, *A Power Controlled Multiple Access Protocol for Wireless Packet Networks*, In: Proceedings of the IEEE INFOCOM, 2001.

Monma, C., and S. Suri, "Transitions in Geometric Minimum Spanning Tree," *Discrete Computational Geometry*, vol. 8, pp. 265–293, 1992.

Montgomery, D.C., and G.C. Runger, *Applied Statistics and Probability for Engineers*, 2nd ed., John Wiley, 1999.

Monzingo, R.A., and T.W. Miller, *Introduction to Adaptive Arrays*, Wiley, 1980.

Mori, N., and K. Matsumoto, *Adaptation to a Dynamical Environment by Means of the Environment Identifying Genetic Algorithm*, In: Congress on Evolutionary Computation (CEC '03), vol. 3, pp. 1626–1631, December 2003.

Mouly, M., and M.B. Pautet, *The GSM System for Mobile Communications*, M. Mouly, 49 rue Louise Bruneau, Palaiseu, France, 1992.

Moy, J., *OSPF version 2*, RFC 2328. IETF, April 1998.

Muqattash, A., and M. Krunz, *A Single-Channel Solution for Transmission Power Control in Wireless Ad Hoc Networks*, In: Proceedings of the ACM MobiHoc, 2004.

Murch, R.D., and K.B. Letaief, "Antenna Systems for Broadband Wireless Access," *IEEE Communications Magazine*, vol. 40, no. 4, pp. 76–83, 2002.

Naguib, A., *Equalization of Transmit Diversity Space–Time Coded Signals*, In: Proceedings of the IEEE Global Telecommunications Conference, 2000 (GLOBECOM '00), vol. 2, 2000, pp. 1077–1082.

Naguib, A., N. Seshadri, and R. Calderbank, "Increasing Data Rate over Wireless Channels," *IEEE Signal Processing Magazine*, vol. 17, pp. 76–92, May 2000.

Nakano, T., and T. Suda, *Adaptive and Evolvable Network Services*, In: Proceedings of the Genetic and Evolutionary Computation Conference (GECCO-2004), Springer LNCS 3102, pp. 151–162, 2004.

Nakano, T., and T. Suda, "Self-Organizing Network Services with Evolutionary Adaptation," *IEEE Transactions on Neural Networks*, vol. 16, no. 5, pp. 1269–1278, 2005.

Nandagopal, T., T-E. Kim, X. Gao, and V. Bhargavan, *Achieving MAC Layer Fairness in Wireless Packet Networks*, In: Proceedings of the ACM Mobicom 2000.

Narula, A., M.J. Lopez, M.D. Trott, and G.W. Wornell, "Efficient Use of Side Information in Multiple-Antenna Data Transmission Over Fading Channels," *IEEE Journal in Selected Areas Communications*, vol. 16, pp. 1423–1436, 1998.

Nasipuri, A., S. Ye, and R.E. Hiromoto, *A MAC Protocol for Mobile Ad Hoc Networks Using Directional Antennas*, In: IEEE Wireless Communications and Networking Conference (WCNC), 2000, pp. 1214–1219.

National Semiconductor Corporation, LMX3162 Single Chip Radio Transceiver, Evaluation Notes and Datasheet, March 2000.

Navidi, W., and T. Camp, "Stationary Distributions for the Random Waypoint Mobility Model," *IEEE Transactions on Mobile Computing*, vol. 3, no. 1, 2004.

Niculescu, D., and B. Nath, *Ad Hoc Positioning System (APS)*, In: IEEE GLOBECOM, vol. 1, pp. 2926–2931, 2001.

Niculescu, D., S. Ganguly, K. Kim, and R. Izmailov, *Performance of VoIP in an 802.11-Based Wireless Mesh Network*, In: Proceedings of the IEEE INFOCOM, 2006.

Niles, I., and A. Pease, *Towards a Standard Upper Ontology*, In: The 2nd International Conference on Formal Ontology in Information Systems (FOIS-2001), Ogunquit, ME, USA, 2001.

Nolfi, N.S., and D. Floreano, *Evolutionary Robotics: The Biology, Intelligence, and Technology of Self-Organizing Machines*, MIT Press, 2000.

Obraczka, K., G. Tsudik, and K. Viswanath, *Pushing the Limits of Multicast in Ad Hoc Networks*, International Conference on Distributed Computing Systems, April 1, 2001.

O'Donnell, I.D., and R. W. Brodersen, "An Ultra-Wideband Transceiver Architecture for Low Power, Low Rate, Wireless Systems," In: *IEEE Transactions on Vehicular Technology*, vol. 54, no. 5, 2005.

Olexa, R., *Implementing 802.11, 802.16, and 802.20 Wireless Networks, Planning, Troubleshooting and Operations*, Elsevier, 2004.

Oliveira, L.B., Hao Chi Wong, Antonio A. F. Loureiro, and Ricardo Dahab, "On the design of secure protocols for hierarchical sensor networks," *International Journal of Security and Networks (IJSN), Special Issue on Cryptography in Networks*, vol. 2, no. 3-4, pp. 216–227, 2007.

Otis, B., Y.H. Chee, and J. Rabaey: *A 400 μW-RX, 1.6mW-TX Super- Regenerative Transceiver for Wireless Sensor Networks*, In: "Solid-State Circuits Conference, 2005. Digest of Technical Papers," ISSCC 2005, vol. 1, February 6–10, 2005, pp. 396–397.

Ouvry, L., S. Dubouloz, B. Denis, and S. de Rivaz, *Performance Analysis of LDR UWB Non-Coherent Receivers in Multipath Environments*, In: Proceedings of the IEEE International Conference on Ultra-wideband (IEEE ICU'05), Zurich, September 2005, pp. 491–496.

Ottersten, B., M. Viberg, P. Stoica, and A. Nehorai, "Exact and Large Sample ML Techniques for Parameter Estimation and Detection in Array Processing," In: S. S. Haykin, J. Litva, and T. Shepherd (eds), *Radar Array Processing*, Springer-Verlag, 1993, pp. 99–151.

Padhye, J., S. Agarwal, V. Padmanabhan, L. Qiu, A. Rao, and B. Zill, "*Estimation of Link Interference in Static Multi-hop Wireless Networks*, In: Proceedings of the IMC, 2005.

Palomar, D.P., J.M. Cioffi, and M.A. Lagunas, "Joint Tx-Rx Beamforming Design for Multicarrier MIMO Channels: A Unified Framework for Convex Optimiztion," *IEEE Transactions on Signal Processing*, vol. 51, pp. 2381–2401, September 2003.

Pandana, C., and K.J.R. Liu, "Near-Optimal Reinforcement Learning Framework for Energy-Aware Sensor Communications," *IEEE Journal on Selected Areas in Communications*, vol. 23, no. 4, pp. 788–797, April 2005.

Panigrahi, D., et al., *Battery Life Time Estimation of Mobile Embedded Systems*, In: The 14th International Conference on VLSI Design, pp. 57–63, 2001.

Papadias, C., *A multiuser Kurtosis Algorithm for Blind Source Separation*, In: Proceedings of the International Conference on Acoustics, Speech, and Signal Processing (ICASSP), 2000, pp. 3144–3147.

Park, K., and H. Lee, *On the Effectiveness of Probabilistic Packet Marking for IP Traceback*, In: Proceedings of the SIGCOMM, pp. 15–26, 2001.

Park, S.-J., Vedantham, R., Sivakumar, R., and Akyildiz, I., *A Scalable Approach for Reliable Downstream Data Delivery in Wireless Sensor Networks*, In: Proceedings of the international symposium on Mobile Ad Hoc Networking and Computing (ACM MOBIHOC), pp. 78–89, May 2004.

Park, V.D., and M. Scott Corson, *A Highly Adaptive Distributed Routing Algorithm for Mobile Wireless Networks*, In: Proceedings of IEEE INFOCOM '97, Kobe, Japan, April 1997.

Paulraj, A., R. Nabar, and D. Gore, *Introduction to Space Time Wireless Communications*, Cambridge University Press, New York, NY, USA, 2003.

Paun, G., "Computing with Membranes," *Journal of Computer and System Sciences*, vol. 61, no. 1, pp. 108–143, 2000.

Pearlman, M.R., and Z. J. Haas, "Determining the Optimal Configuration for the Zone Routing Protocol," *IEEE Journal on Selected Areas in Communications, Special Issue on Mobile and Wireless Networks*, August 1999.

Penrose, M., *Random Geomertric Graphs*, Oxford University Press, New York, NY, USA, 2003.

Penrose, M.D., "On k-Connectivity for a Geometric Random Graph," *Random Structures and Algorithms*, vol. 15, pp. 145–164, 1999.

Perez, F., *Security in Current Commercial Wireless Networks: A Survey*, 2006, Available at: http://www.hig.no/imt/file.php?id=1098/

Pering, T., T. Burd, and R. Brodersen, *The Simulation and Evaluation of Dynamic Voltage Scaling Algorithms*, In: Proceedings of International Symposium on Low Power Electronics and Design ISLPED'98, August 1998, pp. 76–81.

Perkins, C., and P. Bhagwat, *Highly Dynamic Destination-Sequenced Distance-Vector Routing (DSDV) for Mobile Computers*, In: The ACM SIGCOMM'94 Conference on Communications Architectures, Protocols and Applications, pp. 234–244, 1994.

Perkins, C.E. (ed.), *Ad Hoc Networking*, Addison-Wesley, Reading, MA, USA, 2000.

Perrig, A., R. Szewczyk, V. Wen, D. Culler, and J. Tygar, *SPINS: Security Protocols for Sensor Networks*, In: The ACM MobiCom, 2001.

Perrig, A., R. Szewczyk, V. Wen, D. Culler, and J. D. Tygar. SPINS: "Security Protocols for Sensor Networks," *Wireless Networks*, vol. 8, no. 5, pp. 521–534, September 2002. Also appeared in MobiCom'01.

Pietro, R.D., L.V. Mancini, and A. Mei, *Random Key-Assignment for Secure Wireless Sensor Networks*, In: The 1st ACM workshop on Security of ad hoc and sensor networks (SASN'03), pp. 62–71, 2003.

Polastre, J., R. Szewczyk, and D. Culler, *Telos: Enabling Ultra-Low Power Wireless Research*, In: Proceedings of IPSN/SPOTS, Los Angeles, CA, USA, April 25–27, 2005.

Poli, R., "Parallel Distributed Genetic Programming," In: D. Corne, M. Dorigo, and F. Glover (eds.), *New Ideas in Optimization, Advanced Topics in Computer Science*, pp. 403–431. McGraw-Hill, Maidenhead, Berkshire, UK, 1999.

Popovski, P., and H. Yomo, *Bi-Directional Amplification of Throughput in a Wireless Multi-Hop Network*, In: The IEEE 63rd Vehicular Technology Conference (VTC), Melbourne, Australia, May 2006.

Popovski, P., and H. Yomo, *The Anti-Packets Can Increase the Achievable Throughput of a Wireless Multi-Hop Network*, In: Proceedings of the IEEE International Conference on Communication (ICC 2006), Istanbul, Turkey, June 2006.

Porret, A., T. Melly, C.C. Enz, and E.A. Vittoz, *A Low-Power Low-Voltage Transceiver Architecture Suitable for Wireless Distributed Sensors Network*, IEEE International Symposium on Circuits and Systems'00, Geneva, vol. 1, 2000, pp. 56–59.

Pottie, G.J., and W.J. Kaiser, "Wireless Integrated Network Sensors," *Communications of the ACM*, vol. 43, no. 5, pp. 551– 558, 2000.

Prakash, R., *Unidirectional Links Prove Costly in Wireless Ad-Hoc Networks*, In: Proceedings of the 3rd International Workshop on Discrete Algorithms and Methods for Mobile

Computing and Communication (Dial M '99), Seattle, WA, USA, August 20, 1998, pp. 15–22.

Proakis, J.G. *Digital Communications*, 3rd ed., McGraw-Hill, New York, NY, 1995.

Qadir, J., C.T. Chou, and A. Misra, *Low Latency Broadcast In Multi-Radio Multi-Channel Multi-Rate Wireless Mesh Networks*, Technical Report, Available at: ftp://ftp.cse.unsw.edu.au/pub/doc/papers/UNSW/0608.pdf, 2006.

Qadir, J., C.T. Chou, and A. Misra, *Exploiting Rate Diversity in Wireless Mesh Networks*, In: Proceedings of the 31st IEEE Conference on Local Computer Networks (LCN), Tampa, FL, USA, November 15–16, 2006.

Qadir, J., C.T. Chou, and A. Misra, *Localized Minimum-Latency Broadcasting in Multi-rate Wireless Mesh Networks*," IEEE International Symposium on a World of Wireless, Mobile and Multimedia Networks, June 2007.

Radiocommunications Agency, *Examination of Issues Related to Spectrum Efficiency of Point-to-Multipoint and Mesh Multimedia Wireless System Architectures Proposed for 40.5-43.5 GHz*, Radiocommunications Agency Report Ref. 1205/AE/MWS2/R/3, April 2000.

Rakhmatov, D., and S. Vrudhula, "Energy Management for Battery-Powered Embedded Systems," *ACM Transactions on Embedded Computing Systems*, vol. 2, no. 3, pp. 277–324, August 2003.

Ramachandran, K., E. Belding, K. Almeroth, and M. Buddhikot, "*Interference-Aware Channel Assignment in Multi-Radio Wireless Mesh Networks*, In: Proceedings of the IEEE Infocom, 2006.

Ramachandran, V., "Parallel Open Ear Decomposition with Applications to Graph Biconnectivity and Triconnectivity," In: *Synthesis of Parallel Algorithms*: Morgan-Kaufmann, San Mateo, CA, USA, 1992.

Ramanathan, R., J. Redi, C. Santivanez, D. Wiggins, and S. Polit, *Ad Hoc Networking with Directional Antennas: A Complete System Solutio*n," In: IEEE Wireless Communications and Networking Conference (WCNC), 2004, pp. 375–380.

Ramanathan, R., and M. Steenstrup, "Hierarchically-Organized, Multihop Mobile Wireless Networks for Quality-of-Service Support," *Mobile Networks and Applications*, vol. 3, 1998.

Ramanathan, S., and M. Steenstrup, "A Survey of Routing Techniques for Mobile Communications Networks," *ACM/Baltzer Mobile Networks and Applications*, vol. 1, no. 2, pp. 89–103, 1996.

Raniwala, A., and T. Chiueh, *Architectures and Algorithms for an IEEE 802.11-Based Multi-Channel Wireless Mesh Network*, In: Proceedings of the IEEE Infocom, 2005.

Rankov, B., and A. Wittneben, *Spectral Efficient Protocols for Nonregenerative Half-Duplex Relaying*, In: Proceedings of the Allerton Conference on Communication, Control, and Computing, Sept. 2005.

Rao, R., S. Vrudbula, and D.N. Rakbmatov, "Battery Modeling for Energy-Aware System Design," *IEEE Computer*, vol. 36, pp. 77–87, December 2003.

Rappaport, T.S., *Wireless Communications: Principles and Practice*,Prentice-Hall, Englewood Cliffs, NJ, USA, 1996.

Rechenberg, I., *Evolutionsstrategie: Optimierung technischer Systeme nach Prinzipien der biologischen Evolution*, Ph.D. Thesis, Stuttgart, Fromman-Holzboog, 1973.

Ren, K., K. Zeng, and W. Lou, "A New Approach for Random Key Pre-distribution in Large-scale Wireless Sensor Networks," *Wireless Communications and Mobile Computing*, vol. 6, no. 3, 2006.

Ribeiro, C., *Bringing Wireless Access to the Automobile: A Comparison of Wi-Fi, WiMAX, MBWA, and 3G*, Available at www.rh.edu/~rhb/cs_seminar_2005/SessionB3/ribeiro.pdf

Rivest, R., *RFC 1321—The MD5 Message-Digest Algorithm*, Technical Report, MIT Laboratory for Computer Science and RSA Data Security, Inc., Network Working Group, April 1992.

Robert, C.P., and G. Casella, *Monte Carlo Statistical Methods*," Springer-Verlag, New York, NY, USA, 2004.

Robins, G., and J.S. Salowe, "Low-Degree Minimum Spanning Trees," *Discrete Computational Geometry*, vol. 14, pp. 151–165, 1995.

Rodoplu, V., and T. Meng, "Minimum Energy Mobile Wireless Networks," *IEEE Journal on Selected Areas in Communications*, vol. 17, no. 8, pp. 1333–1344, 1999.

Rom, R., and M. Sidi, *Multiple Access Protocols: Performance and Analysis*, Springer-Verlag, New York, NY, USA, 1990.

Romer, K., *The Lighthouse Location System for Smart Dust*, In: Proceedings of MobiSys, 2003.

Roy, S., D. Koutsonikolas, S.M. Das, and Y.C. Hu, *High-Throughput Multicast Routing Metrics in Wireless Mesh Networks*, In: Proceedings of the ICDCS, 2006.

Royer, E.M., and C.E. Perkins, *Multicast Ad Hoc On-Demand Distance Vector (MAODV) Routing*, draft-ietf-manet-maodv-00, July 2000, IETF Internet Draft.

Royer, E., and C.-K. Toh, "A Review of Current Routing Protocols for Mobile Ad-Hoc Networks," *IEEE Personal Communications*, vol. 6, no. 2, pp. 46-55, April 1999.

Russomanno, D.J., C. Kothari, and O. Thomas (2005) *Building a Sensor Ontology: A Practical Approach Leveraging ISO and OGC Models*, In: The 2005 International Conference on Artificial Intelligence, Las Vegas, NV, USA, pp. 637–643.

Russomanno, D.J., C. Kothari, and O. Thomas, *Sensor Ontologies: From Shallow to Deep Models*, In: Proceedings of the 37th Southeastern Symposium on Systems Theory, Tuskegee, AL, USA, pp. 107–112, 2005.

Ryckaert, J., M. Badaroglu, and V. De Heyn, *A 16mA UWB 3-5GHz 20 Mpulses/s Quadrature Analog Correlation Receiver in 0.18μm CMOS*, In: International Solid State Circuits Conference—ISSCC. IEEE, 2006, February 5–9, 2006, San Francisco, CA, USA.

Sadek, A.K., W. Su, and K.J.R. Liu, *Diversity Analysis for Frequency-Selective MIMO-OFDM Systems with Arbitrary Spatial and Temporal Correlation*, *IEEE Transactions on Communications*, vol. 54, no. 5, pp 878–888, May 2006.

Sakai, R., K. Ohgishi, and M. Kasahara, *Cryptosystems Based on Pairing*, In: Symposium on Cryptography and Information Security (SCIS2000), pp. 26–28, January 2000.

Salem, N.B., and J.-P. Hubaux, *A Fair Scheduling for Wireless Mesh Networks*, In: Proceedings of the WiMesh, 2005.

Salem, N.B., and J.-P. Hubaux, "Securing Wireless Mesh Networks," *IEEE Wireless Communication Magazine*, vol. 13, no. 2, pp. 15–55, April 2005.

Sanchez, L.A., W.C. Milliken, A.C. Snoeren, F. Tchakountio, C.E. Jones, S.T. Kent, C. Partridge, and W.T. Strayer, *Hardware Support for a Hash-Based IP Traceback*, In: Proceedings of DARPA Information Survivability Conference and Exposition, June 2001.

Sanchez, J.A., and P.M. Ruiz, *LEMA: Localized Energy-Efficient Multicast Algorithm Based on Geographic Routing*, In: Proceedings of the 31st IEEE Conference on Local Computer Networks (LCN '06), November 2006.

Sankarasubramaniam, Y., I.F. Akyildiz, and M.S. Mchughlin, *Energy Efficiency Based Packet Size Optimization in Wireless Sensor Networks*, In: Proceedings of the First

IEEE International Workshop on Sensor Network Protocols and Applications, 2003, pp. 1–8.

Santivanez, C., B. McDonald, I. Stavrakakis, and R. Ramanathan, *On the Scalability of Ad Hoc Routing Protocols*, In: Proceedings of the IEEE INFOCOMM 2002, 2002.

Sastry, N., and D. Wagner, *Security Considerations for IEEE 802.15.4 Networks*, ACM Workshop on Wireless Security (WISE 04), 2004, pp. 32–42.

Savage, S., N. Cardwell, D. Wetherall, and T. Anderson, "TCP Congestion Control with a Misbehaving Receiver," *ACM Computer Communications Review*, pp. 71–78, October 1999.

Savage, S., D. Wetherall, A. Karlin, and T. Anderson, *Practical Network Support for IP Traceback*, In: Proceedings of ACM SIGCOMM Conference, August 2000.

Saxe, J.B., *Embeddability of Weighted Graphs in K-Space is Strongly NP-Hard*, In: Proceedings of the 17th Alleron Conference in Communications, Control and Computing, 1979, pp. 480–489.

Scatterweb, *Platform for Self-Configuring Wireless Sensor Networks*, http://www.scatterweb.net

Schirokauer, O., *The Number Field Sieve for Integers of Low Weight*, Cryptology ePrint Archive, Report 2006/107, 2006, http://eprint.iacr.org/

Schlegel, C., and A. Grant, "Differential Space–Time Turbo Codes," *IEEE Transactions on Information Theory*, vol. 49, no. 9, pp. 2298–2306, 2003.

Schurgers, C., and M. Srivastava, *Energy Efficient Routing in Wireless Sensor Networks*, In: Proceedings of the Military Communications Conference, October 2001.

Scott, M., "Computing the Tate Pairing," In: *Topics in Cryptology—CT-RSA*, vol. 3376 of *Lecture Notes in Computer Science*, pp. 293–304, Springer, 2005.

Scott, T., *Mica Mote Antenna Radiation Pattern Analysis*, Technical Report DCS-303-IR, Computer Science Department, University of Victoria, http://www.cs.uvic.ca/~wkui/research/motereport-p.pdf, August 2004.

Seeley, T., *The Wisdom of the Hive*, Harvard University Press, Cambridge, MA, USA, 2005.

Sennot, L.I., *Stochastic Dynamic Programming and Control of Queueing Systems*, Wiley, New York, NY, USA, 1999.

Shah, S.H., K. Chen, and K. Nahrstedt, *Cross-Layer Design for Data Accessibility in Mobile Ad Hoc Networks*, In: Proceedings of the 5th World multiconference on systemics, cybernetics and informatics (SCI 2001), Orlando, FL, USA, July 2001.

Shah, R., S. Roy, S. Jain, and W. Brunette, *Data MULEs: Modeling a Three-Tier Architecture for Sparse Sensor Networks*, IEEE Workshop on Sensor Network Protocols and Applications, May 2003.

Shao, Z., Q. Zhuge, C. Xue, and E. H.-M. Sha, "Efficient Assignment and Scheduling for Heterogeneous DSP Systems," *IEEE Transactions on Parallel and Distributed Systems*, vol. 16, pp. 516–525, 2005.

Sharples, N., *Evolutionary Approaches to Adaptive Protocol Design*, Ph.D. Thesis, University of Sussex, UK, August 2001.

Sharples, N., and I. Wakeman, *Protocol Construction Using Genetic Search Techniques*, In: Real-World Applications of Evolutionary Computing-EvoWorkshops 2000, Springer LNCS 1803, Edinburgh, Scotland, UK, April 2000.

Shih, E., S. Cho, N. Ickes, R. Min, A. Sinha, A. Wang, and A. Chandrakasan, *Physical Layer Driven Protocol and Algorithm Design for Energy-Efficient Wireless Sensor Networks*, In: Proceedings of the ACM MobiCom'01, Rome, Italy, July 2001, pp. 272–286.

Sikka, P., P. Corke, and L. Overs, *Wireless Sensor Devices for Animal Tracking and Control*, In: Proceedings of the 29th Annual IEEE International Conference on Local Computer Networks, pp. 446–454, 2004.

Singh, M., and M. Huhns, *Service-Oriented Computing*, John Wiley & Sons, West Sussex, UK, 2005.

Singh, S., and C.S. Raghavendra, *Power-Efficient MAC Protocol for Multihop Radio Networks*, In: Proceedings of the IEEE PIRMC '98 conference, September 1998, vol. 1, pp. 153–157.

Sinha, A., and A. Chandrakasan, "Dynamic Power Management in Wireless Sensor Networks," *IEEE Design and Test of Computers*, vol. 18, no. 2 pp. 62–74, 2001.

Sipper, M., E. Sanchez, D. Mange, M. Tomassini, A. Perez-Uribe, and A. Stauffer, "A Phylogenetic, Ontogenetic, and Epigenetic View of Bio-Inspired Hardware Systems," *IEEE Transactions on Evolutionary Computation*, vol. 1, no. 1, April 1997.

Skuce, D., and I. Monarch, *Ontological Issues in Knowledge Base Design: Some Problems and Suggestions*, CMU-CMT-90-119, Carnegie Mellon University, Pittsburgh, PA, 1990.

Slijepcevic, S., and M. Potkonjak, *Power Efficient Organization of Wireless Sensor Networks*, In: ICC, Helsinki, Finland, June 2001.

Smith, M., "Game Theory and the Evolution of Fighting," In: J.M. Smith, *Evolution*, Edinburgh University Press, Edinburgh, UK, pp. 8–28, 1972.

Smith, M.,"*Evolution and the Theory of Games*, Cambridge University Press, Cambridge, UK, 1982.

Smith, M., C. Welty, and D.L. McGuinness, *OWL Web Ontology Language Guide: W3C Proposed Recommendation*, Available at: http://www.w3.org/TR/2003/PR-owl-guide-20031215/, 2003.

Small, T., and Z. J. Hass, *Resource and Performance Tradeoffs In Delay-Tolerant Wireless Networks*, In: Proceedings of the WDTN Workshop in association with ACM SIGCOMM, August 2005.

Snoeren, A.C., C. Partridge, L.A. Sanchez, C.E. Jones, F. Tchakountio, S.T. Kent, and W.T. Strayer, *Hash-based IP Traceback*, In: Proceedings of ACM Conference on Applications, Technologies, Architectures and Protocols for Computer Communication (SIGCOMM), pp. 3–14, 2001.

Song, C., S. Havlin, and H.A. Makse, "Self-Similarity of Complex Networks," *Nature*, vol. 7024, pp. 392–395, 2005.

Song, D.X., and A. Perrig, *Advanced and Authenticated Marking Scheme for IP Traceback*, In: Proceedings of the IEEE INFOCOM Conference, 2001.

Spyropoulos, A., and C.S. Raghavendra, *Asymptotic Capacity Bounds for Ad Hoc Networks Revisited: The Directional and Smart Antenna Cases*, In: IEEE Global Telecommunications Conference (GLOBECOM), 2003, pp. 1216–1220.

Spyropoulos, T., K. Psounis, and C.S. Raghavendra, *Single-Copy Routing in Intermittently Connected Mobile Networks*, In: Proceedings of the IEEE SECON, October 2004.

Spyropoulos, T., K. Psounis, and C. S. Raghavendra, *Multi-Copy Routing in Intermittently Connected Mobile Networks*, Technical Report, University of Southern California, 2004.

Stamatopoulou, P.K.I., and M. Gheorghe, *Modelling of Dynamic Configuration of Biology-Inspired Multi-Agent Systems with Communicating X-Machines asnd Population P Systems*, In: Fifth Workshop on Membrane Computing (WMC5), Milan, Italy, 2004.

Stanford Medical Informatics (2004) "*The Protégé Ontology Editor and Knowledge Acquisition System*," Available at: http://protege.stanford.edu/.

Steels, L., *Emergent Functionality in Robotic Agents Through On-Line Evolution*, In: Proceedings of the AlifeIV, MIT Press, Cambridge, 1994.

Sterbenz, J.P.G, R. Krishnan, R.R. Hain, A.W. Jackson, D. Levin, R. Ramanathan, and J. Zao, *Survivable Mobile Wireless Networks: Issues, Challenges, and Research Directions*, In: Proceedings of the ACM Wireless Security Workshop (WiSE) 2002 at MobiCom, Atlanta, GA, USA, September 2002, pp. 31–40.

Stoica, L., S. Tiuraniemi, I. Oppermann, and H. Repo, *An Ultra Wideband Low Complexity Circuit Transceiver Architecture for Sensor Networks*," In: ISCAS 2005, May 23–26, 2005, vol. 1, pp. 364–367.

Stoica, P., and R.L. Moses, *Introduction to Spectral Analysis*, Prentice-Hall, New Jersey, USA, 1997.

Stojmenovic, I., and X. Lin, "Power-Aware Localized Routing in Wireless Networks," *IEEE Transactions on Parallel and Distributed Systems*, vol. 12, no. 10, pp. 1122–1133, 2001.

Stojmenovic, I., and X. Lin, "Loop-Free Hybrid Single-Path/Flooding Routing Algorithms with Guaranteed Delivery for Wireless Networks," *IEEE Transactions on Parallel and Distributed Systems*, vol. 12, no. 10, 2001.

Stoleru, R., T. He, J.A. Stankovic, and D. Luebke *A High-Accuracy Low-Cost Localization System or Wireless Sensor Networks*, In: Proceedings of the SenSys, 2005.

Stone, R., *CenterTrack: An IP Overlay Network for Tracking DoS Floods*, In: Proceedings of the 9th Usenix Security Symposium, August 2000.

Su, W., Z. Safar, M. Olfat, and K.J.R. Liu, "Obtaining Full-Diversity Space-Frequency Codes From Space–Time Code Via Mapping," *IEEE Transactions on Signal Processing*, vol. 51, no. 11, pp. 2905–2916, November 2003.

Subramanian, A.P., M.M. Buddhicot, and S. Miller, *Interference Aware Routing in Multi-Radio Wireless Mesh Networks*, In: Proceedings of the WiMesh, 2006.

Suzuki, J., and T. Suda, "A Middleware Platform for a Biologically Inspired Network Architecture Supporting Autonomous and Adaptive Applications," *IEEE Journal on Selected Areas in Communications*, vol. 23, no. 2, pp. 249–260, 2005.

Syropoulos, A., *On P Systems and Distributed Computing*, In: The Fifth Workshop on Membrane Computing (WMC5), Milan, Italy, 2004.

Takagi, H., and L. Kleinrock, "Optimal Transmission Ranges for Randomly Distributed Packet Radio Terminals," *IEEE Transactions on Communications*, vol. 32, no. 3, pp. 246–57, 1984.

Tarokh, V., and H. Jafarkhani, "A Differential Detection Scheme for Transmit Diversity," *IEEE Journal on Selected Areas in Communications*, vol. 3, pp. 1043–1047, 2000.

Tarokh, V., H. Jafarkhani, and A.R. Calderbank, "Space–Time Block Codes From Orthogonal Designs," *IEEE Transactions on Information Theory*, vol. 45, pp. 1456–1467, 1999.

Tarokh, V., N. Seshadri, and A.R. Calderbank, "Space–Time Codes for High Data Rate Wireless Communication: Performance Criterion and Code Construction," *IEEE Transactions on Information Theory*, vol. 44, pp. 744–765, 1998.

Telatar, E., *Capacity of Multiantenna Gaussian Channels*, AT&T Bell Laboratories, Technical Memorandum, June 1995.

Templeton, S., and K. Levitt, *Detecting Spoofed Packets*, In: Proceedings of the Third DARPA Information Survivability Conference and Exposition (DISCEX), 2003.

Theodoridis, S., and K. Koutroumbas, *Pattern Recognition*, Academic Press, New York, NY, USA, 1998.

Thomson, S., and T. Narten, *IPv6 Stateless Address Autoconfiguration*, RFC 2462, December 1998.

Tobagi, F.A., and L. Kleinrock, "Packet Switching in Radio Channels: Part II—the Hidden Terminal Problem in Carrier Sense Multiple-Access Modes and the Busy-Tone Solution," *IEEE Transactions of Communications*, vol. COM-23, no. 12, pp. 1417–1433, 1975.

Toh, C.-K., "Maximum Battery Life Routing to Support Ubiquitous Mobile Computing in Wireless Ad Hoc Networks," *IEEE Communications Magazine*, vol. 39, no. 6, pp. 138–147, 2001.

Toh, C.-K., *Ad Hoc Mobile Wireless Networks, Protocols and Systems*, Prentice Hall, New York, NY, USA, 2002.

Triantafillou, P., and I. Ackaterinidis, *Proxy Teller: A Proxy Placement Tool For Content Delivery Under Performance Constraints*, In: Proceedings of the Fourth International Conference on Web Information Systems Engineering (WISE'03), 2003.

Tschudin, C., *Fraglets—A Metabolistic Execution Model for Communication Protocols*, In: Proceedings of the 2nd Annual Symposium on Autonomous Intelligent Networks and Systems (AINS), Menlo Park, USA, July 2003.

Tschudin, C., and L. Yamamoto, *A Metabolic Approach to Protocol Resilience*, In: Proceedings of the 1st Workshop on Autonomic Communication (WAC), Springer LNCS 3457, pp. 190–205, Berlin, Germany, October 2004.

Tse, D., and P. Viswanath, *Fundamentals of Wireless Communication*, Cambridge University Press, Cambridge, UK, 2005.

Uschold, M., and M. Gruninger (1996) "Ontologies: Principles, Methods, and Applications," *Knowledge Engineering Review*, vol. 11, no. 2, pp. 93–115.

Vahdat, A., and David Becker, *Epidemic Routing for Partially-Connected Ad Hoc Networks*, Technical Report, Duke University, April 2003.

Vakali, A., G. Pallis, "Content Delivery Networks: Status and Trend," *IEEE Internet Computing*, vol. 7, no. 6, pp. 68–74, 2003.

van der Veen A.J., and A. Paulraj, "An Analytical Constant Modulus Algorithm," *IEEE Transactions on Signal Processing*, vol. 44, pp. 1136–1155, 1996.

Van Veen, B.D., and K.M. Buckley, "Beamforming: A Versatile Approach to Spatial Filtering," *IEEE ASSP Magazine*, vol. 5, no. 2, pp. 4–24, 1988.

Vedantham, R., Z. Zhuang, and R. Sivakumar, "Hazard Avoidance in Wireless Sensor and Actor Networks," *Special Issue in Wireless Sensor Networks, Elsevier Computer Communications*, vol. 29, pp. 2578–2598, Aug. 2006.

Vetter, B., F. Wang, and S.F. Wu, *An Experimental Study of Insider Attacks for the OSPF Routing Protocol*, In: IEEE International Conference on Network Protocols (ICNP), pp. 293–300, October 1997.

Vincent, V.T.L., and T. L. S. Vincent, "Evolution and Control System Design," *IEEE Control Systems Magazine*, 20(5):20–35, October 2000.

Visotsky, E., and U. Madhow, "Space–Time Transmit Precoding with Imperfect Feedback," *IEEE Transactions on Information Theory*, vol. 47, pp. 2632–2639, 2001.

Vuran, M.C., and O.B. Akan, *Spatio-Temporal Characteristics of Point and Field Sources in Wireless Sensor Networks*, In: Proceedings of the IEEE ICC 2006, Istanbul, Turkey, June 2006.

Vuran, M.C., O.B. Akan, and I.F. Akyildiz, "Spatio-Temporal Correlation: Theory and Applications for Wireless Sensor Networks," *Elsevier's Computer Networks Journal*, vol. 45, no. 3, pp. 245–261, 2004.

Vuran, M.C., and I.F. Akyildiz, "Spatial Correlation-Based Collaborative Medium Access Control in Wireless Sensor Networks," *IEEE/ACM Transactions on Networking*, vol. 14, pp. 316–329, 2006.

Wacker, A., T. Heiber, and H. Cermann, *A Key-Distribution Scheme for Wireless Home Automation Networks*, In: Proceedings of the IEEE CCNC 2004, Las Vegas, NV, USA, January 2004.

Wacker, A., T. Heiber, H. Cermann, and P.J. Marrón, *A Fault-Tolerant Key-Distribution Scheme for Securing Wireless Ad-Hoc Networks*, In: Proceedings of Pervasive 2004, Vienna, Austria, April 2004.

Wacker, A., M. Knoll, T. Heiber, and K. Rothermel, *A New Approach for Establishing Pairwise Keys for Securing Wireless Sensor Networks*, In: Proceedings of ACM SenSys'05, San Diego, CA, USA, November 2005.

Waitzman, D., C.Partridge and S. Deering, (eds), *Distance Vector Multicast Routing Protocol*, RFC 1075, BBN STC and Stanford University, November 1998, Available at http://www.ietf.org/rfc/rfc1075.txt

Waldvogel, M.,"*GOSSIB vs. IP Traceback Rumors*, In: Proceedings of the 18th Annual Computer Security Applications Conference (ACSAC), December 2002.

Wang, A.Y., S.H. Cho, C.G. Sodini, and A.P. Chandrakasan, *Energy Efficient Modulation and MAC for Asymmetric RF Microsensor Systems*, In: Proceedings of ICASSP, May 2001.

Wang, F., B. Vetter, and S.F. Wu, *Secure Routing Protocols: Theory and Practice*, Technical Report, University of California at Davis, CA, USA, May 1997, Available at http://shang.csc.ncsu.edu/papers.htm

Wang, K.H., and B. Li, *Group Mobility and Partition Prediction in Wireless Ad-Hoc Networks*, In: Proceedings of IEEE International Conference on Communications (ICC 2002), vol. 2, pp. 1017–1021, New York City, NY, April 2002.

Wang, K.H., and B. Li, *Efficient and Guaranteed Service Coverage in Partitionable Mobile Ad-Hoc Networks*," In: Proceedings of IEEE Joint Conference of the IEEE Computer and Communications Societies (INFOCOM), pp. 1089–1098, June 2002.

Wang, J., H. Zhai, and Y. Fang, *Opportunistic Packet Scheduling and Media Access Control for Wireless LANs and Multi-Hop Ad Hoc Networks*, In: Proceedings of the IEEE WCNC, 2004.

Wang, X., and D.S. Reeves, *Robust Correlation of Encrypted Attack Traffic Through Stepping Stones by Manipulation of Interpacket Delays*, In: ACM Conference on Computer and Communications Security, pp. 20–29, 2003.

Watro., R.J., D. Kong, Cuti, S.-f., C. Gardiner, C. Lynn, and P. Kruus, "*TinyPK: Securing Sensor Networks with Public Key Technology*, In: The 2nd ACM Workshop on Security of Ad Hoc and Sensor Networks (SASN'04), pp. 59–64, Washington, DC, USA, October 2004.

Weiser, M.,"Some Computer Science Issues in Ubiquitous Computing," *Communications of the ACM*, vol. 36, no. 7, pp. 75–84, 1993.

Weiser, M.,"The Computer for the 21st Century, *The ACM Mobile Compututing and Communications Review*, vol. 3, no. 3, pp. 3–11, 1999.

Weiser, M., et al., *Scheduling for Reduced CPU Energy*, In: Proceedings of 1st USENIX Symposium on Operating System Design and Implementation, November 1994, pp. 13–23.

Wermelinger, M.A. *Specification of Software Architecture Reconfiguration*, Ph.D. Thesis, Universidade Nova de Lisboa, Lisbon, Portugal, September 1999.

Wickelgren, I.J., *Local-Area Networks go Wireless*, IEEE Spectrum, 33, 34, 1996.

WiMAX Forum, *Mobile WiMAX—Part I: A Technical Overview and Performance Evaluation*, Prepared on behalf of the WiMAX Forum, February 21, 2006, Available at http://www.wimaxforum.org

Win, M.Z., and R.A. Scholtz, "Ultra-Wide Bandwidth Time-Hopping Spread-Spectrum Impulse Radio for Wireless Multiple Access Communications," *IEEE Transactions on Communications*, vol. 48, no. 4, pp. 679–691, 2000.

Wolnicki, J., *The IEEE 802.16 WiMAX Broadband Wireless Access; Physical Layer (PHY), Medium Access Control Layer (MAC), Radio Resource Management (RRM)*, 2005-01-14.

Woo, A., S. Seth, T. Olson, J. Liu, and F. Zhao (2006) *A Spreadsheet Approach to Programming and Managing Sensor Networks*, In: Proceedings of the 5th International Conference on Information Processing in Sensor Networks, Nashville, TN, USA, pp. 424–431.

Wu, C.W., and Y.C. Tay, *AMRIS: A Multicast Protocol for Ad Hoc Wireless Networks*, Military Communications Conference Proceedings, 1999 (MILCOM 1999), IEEE, vol. 1, 1999.

Wu, J., and H. Li, *On Calculating Connected Dominating Set for Efficient Routing in Ad Hoc Wireless Networks*, Proceedings of the 3rd International Workshop on Discrete Algorithms and Methods for Mobile Computing and Communications, pp. 7–14, 1999.

Wu, Y., P. A. Chou, and S.-Y. Kung, *Information Exchange in Wireless Networks with Network Coding and Physical-Layer Broadcast*, Microsoft, Technical Report, MSR-TR-2004-78, August 2004.

Xie, B., A. Kumar, D.P. Agrawal, and S. Srinivasan, "Securing Macro/Micro Mobility for Multi-Hop Cellular IP," *Elsevier Special Issue of Pervasive and Mobile Computing*, vol. 2, no. 2, pp. 111–136, 2006.

Xie, J., R.R. Talpade, A. McAuley, and M. Liu, "AMRoute: Ad Hoc Multicast Routing Protocol," *Mobile Networks and Applications*, vol. 7, no. 6, pp. 429–439, 2002.

Xu, S., and T. Saadawi, "Does the IEEE 802.11 MAC Protocol Work Well in Multihop Wireless Ad Hoc Networks," *IEEE Communications Magazine*, vol. 39, no. 6, pp. 130–137, 2001.

Xu, Y., J. Heidemann, and D. Estrin, "*Geography-Informed Energy Conservation for Ad-Hoc Routing*, In: Proceedings of MOBICOM, pp. 70–84, July 2001.

Xu, Y., W. Lee, J. Xu, and G. Mitchell, *PSGR: Priority-Based Stateless Geo-Routing in Wireless Sensor Networks*, In: Proceedings of the IEEE MASS 2005, Washington, DC, USA, November 2005.

Xu, Z., S. Dai, and J.J. Garcia-Luna-Aceves, *Hierarchical Routing Using Link Vectors*, IEEE Infocom, March 1998.

Xue, Y., and B. Li, *A Location-Aided Power-Aware Routing Protocol in Mobile Ad Hoc Networks*, In: Proceedings of the IEEE Symposium on Ad Hoc Mobile Wireless Networks/IEEE GLOBECOM 2001, San Antonio, TX, USA, Novembers 25–29, 2001.

Yamamoto, L., and C. Tschudin, *Experiments on the Automatic Evolution of Protocols using Genetic Programming*, In: Proceedings of the 2nd Workshop on Autonomic Communication (WAC), pp. 13–28, Athens, Greece, October 2005.

Yang, X., and N. Vaidya, *On the Physical Carrier Sense in Wireless Ad-Hoc Networks*, In: Proceedings of the IEEE INFOCOM, 2005.

Yang, Y., and C. Ma, *Battery Aware Routing in Wireless Ad Hoc Networks—Part I: Energy Model*, In: The 19th International Teletraffic Congress (ITC-19), September 2005.

Ye, S., and R.S. Blum, "Optimized Signaling for MIMO Interference Systems with Feedback," *Transactions on Signal Processing*, vol. 51, pp. 2839–2848, 2003.

Yi, S., Y. Pei, and S. Kalyanaraman, *On the Capacity Improvement of Ad Hoc Wireless Networks Using Directional Antennas*, In: Proceedings of the 4th ACM International Symposium on Mobile Ad Hoc Networking and Computing, Annapolis, MD, USA, June 01–03, 2003.

Yu, Z., and Y. Guan, *A Key Pre-Distribution Scheme Using Deployment Knowledge for Wireless Sensor Networks*, In: IPSN, 2005.

Yuan, J., and X. Shao, "New Differential Space Time Block Coding Schemes with Two Tthree and Four Transmit Antennas," *IEEE Communication Letters*, vol. 7, no. 9, pp. 437–439, September 2003.

Yum, T.-S., and K.-W. Hung, "Design Algorithms for Multihop Packet Radio Networks with Multiple Directional Antennas Stations," *IEEE Transactions on Communications*, vol. 41, no. 11, pp. 1716–1724, 1992.

Zadeh, L.A. "Fuzzy Sets," *Information and Control*, vol. 8, no. 3, pp. 338–353, 1965.

Zhang, J., Y.P. Chen, and I. Marsic, *Adaptive MAC Scheduling Using Channel State Diversity for Wireless Networks*, In: Proceedings of the IEEE WiCOM, 2006.

Zhao, J., and R. Govindan, *Understanding Packet Delivery Performance in Dense Wireless Sensor Networks*, In: Proceedings of the First International Conference on Embedded Networked Sensor Systems (SenSys '03), ACM Press, New York, NY, USA, 2003, pp. 1–13.

Zhao, Q., and L. Tong, "Energy Efficiency of Large-Scale Wireless Networks: Proactive Versus Reactive Networking," *IEEE Journal on Selected Areas in Communications*, vol. 23, pp. 1100–1112, May 2005.

Zheng, L., and D. Tse, "Diversity and Multiplexing: A Fundamental Tradeoff in Multiple Antenna Channels," *IEEE Transactions on Information Theory*, vol. 49, no. 5, pp. 1073–1095, 2003.

Zheng, R., *Information Dissemination in Power-Constrained Wireless Networks*, In: The 25th Annual Joint Conference of the IEEE Computer and Communications Societies (INFOCOM), 2006.

Zhong, S., J. Chen, and Y. Yang, *Sprite: A Simple, Cheat-Proof, Credit-Based System for Mobile Ad-Hoc Networks*, In: Proceedings of IEEE Infocom 2003, San Francisco, CA, USA, March 2003.

Zhou, L., and Z. J. Haas, "Securing Ad Hoc Networks," *IEEE Network*, vol. 13, no. 6, pp. 24–30, 1999.

Zhou, X., C.T. Chou, J. Guo, and S. Jha, *Protecting Multicast Sessions in Wireless Mesh Network*, In: Proceedings of the 31st IEEE Conference on Local Computer Networks (LCN), Tampa, Florida, USA, November 14–16, 2006.

Zhou, Y., Y. Zhang, and Y. Fang, *Key Establishment in Sensor Networks based on Triangle Grid Deployment Model*, In: IEEE MILCOM, 2005.

Zhou, Y., Y. Zhang, and Y. Fang, *LLK: A Link-layer Key Establishment Scheme in Wireless Sensor Networks*, In: IEEE WCNC, 2005.

Zhu, S., S. Setia, and S. Jajodia, *LEAP: Efficient Security Mechanisms for Large-Scale Distributed Sensor Networks*, In: 10th ACM conference on Computer and communication security (CCS'03), pp. 62–72. ACM Press, 2003.

Zhu, S., S. Xu, S. Setia, and S. Jajodia, *Establishing Pairwise Keys for Secure Communication in Ad Hoc Networks: A Probabilistic Approach*, In: ICNP, 2003.

Ziegler, J., and W. Banzhaf, "Evolving Control Metabolisms for a Robot," *Artificial Life*, vol. 7, no. 2, pp. 171–190, 2001.

ZigBee Alliance, *ZigBee Specification v1.0*, ZigBee Document 053473r00, Version 1.00, December 14, 2004.

Zorzi, M., and R.R. Rao, "Geographic Random Forwarding (GeRaF) for Ad Hoc and Sensor Networks: Multihop Performance," *IEEE Transactions on Mobile Computing*, vol. 2, no. 4, pp. 337–347, 2003.

Zou, L., M. Lu, and Z. Xiong, *PAGER-M: A Novel Location-based Routing Protocol for Mobile Sensor Networks*, In: Proceedings of the Broadwise, San Jose, CA, USA, October 2004.

Index

Note: Page numbers followed by f indicate figure.

A

B

Q

R

S